ASCE STANDARD

ASCE/SEI
7-16
INCLUDES
SUPPLEMENT 1
INCLUDES ERRATA

Minimum Design Loads and Associated Criteria for Buildings and Other Structures

COMMENTARY

PUBLISHED BY THE AMERICAN SOCIETY OF CIVIL ENGINEERS

Library of Congress Cataloging-in-Publication Data

Names: American Society of Civil Engineers.
Title: Minimum design loads and associated criteria for buildings and other structures.
Other titles: Minimum design loads for buildings and other structures. | ASCE standard, ASCE/SEI 7-16, minimum design loads and associated criteria for buildings and other structures
Description: Reston, Virginia : American Society of Civil Engineers, [2017] | Earlier versions of the standard have title: Minimum design loads for buildings and other structures. | "ASCE standard, ASCE/SEI 7-16." | Includes bibliographical references and index.
Identifiers: LCCN 2017018275| ISBN 9780784414248 (softcover : alk. paper) | ISBN 9780784479964 (PDF)
Subjects: LCSH: Structural engineering–Standards–United States. | Buildings–Standards–United States. | Strains and stresses. | Standards, Engineering–United States.
Classification: LCC TH851 .M56 2017 | DDC 624.102/1873–dc23 LC record available at https://lccn.loc.gov/2017018275

Published by American Society of Civil Engineers
1801 Alexander Bell Drive
Reston, Virginia, 20191-4382
www.asce.org/bookstore | ascelibrary.org

This standard was developed by a consensus standards development process that has been accredited by the American National Standards Institute (ANSI). Accreditation by ANSI, a voluntary accreditation body representing public and private sector standards development organizations in the United States and abroad, signifies that the standards development process used by ASCE has met the ANSI requirements for openness, balance, consensus, and due process.

While ASCE's process is designed to promote standards that reflect a fair and reasoned consensus among all interested participants, while preserving the public health, safety, and welfare that is paramount to its mission, it has not made an independent assessment of and does not warrant the accuracy, completeness, suitability, or utility of any information, apparatus, product, or process discussed herein. ASCE does not intend, nor should anyone interpret, ASCE's standards to replace the sound judgment of a competent professional, having knowledge and experience in the appropriate field(s) of practice, nor to substitute for the standard of care required of such professionals in interpreting and applying the contents of this standard.

ASCE has no authority to enforce compliance with its standards and does not undertake to certify products for compliance or to render any professional services to any person or entity.

ASCE, its affiliates, officers, directors, employees, and volunteers disclaim any and all liability for any personal injury, property damage, financial loss, or other damages of any nature whatsoever, including without limitation any direct, indirect, special, exemplary, or consequential damages, resulting from any person's use of, or reliance on, this standard. Any individual who relies on this standard assumes full responsibility for such use.

ASCE and American Society of Civil Engineers—Registered in U.S. Patent and Trademark Office.

Photocopies and permissions. Permission to photocopy or reproduce material from ASCE publications can be requested by sending an e-mail to permissions@asce.org or by locating a title in ASCE's Civil Engineering Database (http://cedb.asce.org) or ASCE Library (http://ascelibrary.org) and using the "Permissions" link.

Errata: Errata, if any, can be found at https://doi.org/10.1061/9780784414248.

Copyright © 2017 by the American Society of Civil Engineers.
All Rights Reserved.
ISBN 978-0-7844-1424-8 (soft cover)
ISBN 978-0-7844-7996-4 (PDF)
Online platform: http://ASCE7.online
Manufactured in the United States of America.

24 23 22 21 20 19   2 3 4 5 6

CONTENTS

COMMENTARY TO STANDARD ASCE/SEI 7-16

Provisions contents appear in first book

C1	GENERAL		405
	C1.1	Scope	405
	C1.3	Basic Requirements	405
		C1.3.1 Strength and Stiffness	405
		C1.3.1.3 Performance-Based Procedures	405
		C1.3.2 Serviceability	409
		C1.3.3 Functionality	409
		C1.3.4 Self-Straining Forces and Effects	410
		C1.3.7 Fire Resistance	410
	C1.4	General Structural Integrity	410
	C1.5	Classification of Buildings and Other Structures	412
		C1.5.1 Risk Categorization	412
		C1.5.3 Toxic, Highly Toxic, and Explosive Substances	414
	C1.7	Load Tests	415
	References		415
	Other References (Not Cited)		416
C2	COMBINATIONS OF LOADS		417
	C2.1	General	417
	C2.2	Symbols	417
	C2.3	Load Combinations for Strength Design	417
		C2.3.1 Basic Combinations	417
		C2.3.2 Load Combinations Including Flood Load	418
		C2.3.3 Load Combinations Including Atmospheric Ice Loads	418
		C2.3.4 Load Combinations Including Self-Straining Forces and Effects	419
		C2.3.5 Load Combinations for Nonspecified Loads	419
		C2.3.6 Basic Combinations with Seismic Load Effects	420
	C2.4	Load Combinations for Allowable Stress Design	420
		C2.4.1 Basic Combinations	420
		C2.4.2 Load Combinations Including Flood Load	421
		C2.4.3 Load Combinations Including Atmospheric Ice Loads	421
		C2.4.4 Load Combinations Including Self-Straining Forces and Effects	421
	C2.5	Load Combinations for Extraordinary Events	421
	References		423
C3	DEAD LOADS, SOIL LOADS, AND HYDROSTATIC PRESSURE		425
	C3.1	Dead Loads	425
		C3.1.2 Weights of Materials and Constructions	425
		C3.1.3 Weight of Fixed Service Equipment	425
		C3.1.4 Vegetative and Landscaped Roofs	425
		C3.1.5 Solar Panels	425
	C3.2	Soil Loads and Hydrostatic Pressure	425
		C3.2.1 Lateral Pressures	425
		C3.2.2 Uplift Loads on Floors and Foundations	431
	Reference		431
C4	LIVE LOADS		433
	C4.3	Uniformly Distributed Live Loads	433
		C4.3.1 Required Live Loads	433
		C4.3.2 Provision for Partitions	435
		C4.3.3 Partial Loading	435

	C4.4	Concentrated Live Loads	435
	C4.5	Loads on Handrail, Guardrail, Grab Bar, and Vehicle Barrier Systems, and on Fixed Ladders.	435
		C4.5.1 Handrail and Guardrail Systems.	435
		C4.5.2 Grab Bar Systems.	435
		C4.5.3 Vehicle Barrier Systems.	435
		C4.5.4 Fixed Ladders.	435
	C4.6	Impact Loads.	435
		C4.6.4 Elements Supporting Hoists for Façade Access and Building Maintenance Equipment.	435
		C4.6.5 Fall Arrest and Lifeline Anchorages.	436
	C4.7	Reduction in Uniform Live Loads.	436
		C4.7.1 General.	436
		C4.7.3 Heavy Live Loads.	437
		C4.7.4 Passenger Vehicle Garages.	437
		C4.7.6 Limitations on One-Way Slabs.	437
	C4.8	Reduction in Roof Live Loads.	437
		C4.8.2 Ordinary Roofs, Awnings, and Canopies.	437
		C4.8.3 Occupiable Roofs.	437
	C4.9	Crane Loads.	437
	C4.11	Helipad Loads.	437
		C4.11.1 General.	437
		C4.11.2 Concentrated Helicopter Loads.	438
	C4.13	Library Stack Rooms.	438
	C4.14	Seating For Assembly Uses.	438
	C4.17	Solar Panel Loads.	438
		C4.17.1 Roof Loads at Solar Panels.	438
		C4.17.3 Open-Grid Roof Structures Supporting Solar Panels.	438
	References		438
C5	FLOOD LOADS.		439
	C5.1	General.	439
	C5.2	Definitions.	439
	C5.3	Design Requirements.	440
		C5.3.1 Design Loads.	440
		C5.3.2 Erosion and Scour.	440
		C5.3.3 Loads on Breakaway Walls.	440
	C5.4	Loads during Flooding.	440
		C5.4.1 Load Basis.	440
		C5.4.2 Hydrostatic Loads.	440
		C5.4.3 Hydrodynamic Loads.	440
		C5.4.4 Wave Loads.	441
		C5.4.4.2 Breaking Wave Loads on Vertical Walls.	441
		C5.4.5 Impact Loads.	441
	References		444
C6	TSUNAMI LOADS AND EFFECTS.		447
	C6.1	General Requirements.	447
		C6.1.1 Scope.	447
	C6.2	Definitions.	456
	C6.3	Symbols and Notation.	456
	C6.4	Tsunami Risk Categories.	457
	C6.5	Analysis of Design Inundation Depth and Flow Velocity.	458
		C6.5.3 Sea Level Change.	460
	C6.6	Inundation Depths and Flow Velocities Based on Runup.	461
		C6.6.1 Maximum Inundation Depth and Flow Velocities Based on Runup.	461
		C6.6.2 Energy Grade Line Analysis of Maximum Inundation Depths and Flow Velocities.	461
		C6.6.3 Terrain Roughness.	462
		C6.6.4 Tsunami Bores.	462
	C6.7	Inundation Depths and Flow Velocities Based on Site-Specific Probabilistic Tsunami Hazard Analysis.	462
		C6.7.1 Tsunami Waveform.	465
		C6.7.2 Tsunamigenic Sources.	466
		C6.7.3 Earthquake Rupture Unit Source Tsunami Functions for Offshore Tsunami Amplitude.	466
		C6.7.4 Treatment of Modeling and Natural Uncertainties.	466

				Page
	C6.7.5	Offshore Tsunami Amplitude.		466
		C6.7.5.1	Offshore Tsunami Amplitude for Distant Seismic Sources.	466
		C6.7.5.2	Direct Computation of Probabilistic Inundation and Runup.	466
	C6.7.6	Procedures for Determining Tsunami Inundation and Runup		466
		C6.7.6.1	Representative Design Inundation Parameters.	466
		C6.7.6.2	Seismic Subsidence before Tsunami Arrival.	466
		C6.7.6.3	Model Macroroughness Parameter.	466
		C6.7.6.4	Nonlinear Modeling of Inundation.	466
		C6.7.6.5	Model Spatial Resolution.	467
		C6.7.6.6	Built Environment.	467
		C6.7.6.7	Inundation Model Validation.	467
		C6.7.6.8	Determining Site-Specific Inundation Flow Parameters.	467
		C6.7.6.9	Tsunami Design Parameters for Flow over Land.	467
C6.8	Structural Design Procedures for Tsunami Effects			467
	C6.8.1	Performance of Tsunami Risk Category II and III Buildings and Other Structures.		468
	C6.8.2	Performance of Tsunami Risk Category III Critical Facilities and Tsunami Risk Category IV Buildings and Other Structures.		468
	C6.8.3	Structural Performance Evaluation.		468
		C6.8.3.1	Load Cases.	468
		C6.8.3.2	Tsunami Importance Factors.	469
		C6.8.3.3	Load Combinations.	469
		C6.8.3.4	Lateral-Force-Resisting System Acceptance Criteria.	469
		C6.8.3.5	Structural Component Acceptance Criteria.	469
	C6.8.4	Minimum Fluid Density for Tsunami Loads.		470
	C6.8.5	Flow Velocity Amplification.		470
		C6.8.5.2	Flow Velocity Amplification by Physical or Numerical Modeling.	471
	C6.8.6	Directionality of Flow		471
		C6.8.6.1	Flow Direction.	471
		C6.8.6.2	Site-Specific Directionality.	471
	C6.8.7	Minimum-Closure-Ratio-for-Load-Determination.		471
	C6.8.8	Minimum-Number-of-Tsunami-Flow-Cycles.		472
	C6.8.9	Seismic Effects on the Foundations Preceding Local Subduction Zone Maximum Considered Tsunami.		472
	C6.8.10	Physical Modeling of Tsunami Flow, Loads, and Effects.		472
C6.9	Hydrostatic Loads			473
	C6.9.1	Buoyancy.		473
	C6.9.2	Unbalanced Lateral Hydrostatic Force.		473
	C6.9.3	Residual Water Surcharge Load on Floors and Walls.		473
	C6.9.4	Hydrostatic-Surcharge-Pressure-on-Foundation.		473
C6.10	Hydrodynamic Loads			473
	C6.10.1	Simplified Equivalent Uniform Lateral Static Pressure.		473
		C6.10.2.1	Overall Drag Force on Buildings and Other Structures.	474
		C6.10.2.2	Drag Force on Components.	474
		C6.10.2.3	Tsunami-Loads-on-Vertical-Structural-Components, F_w.	474
		C6.10.2.4	Hydrodynamic Load on Perforated Walls, F_{pw}.	475
		C6.10.2.5	Walls Angled to the Flow.	475
		C6.10.3.1	Flow Stagnation Pressure.	475
		C6.10.3.2	Hydrodynamic Surge Uplift at Horizontal Slabs	475
		C6.10.3.3	Tsunami Bore Flow Entrapped in Structural Wall-Slab Recesses	475
C6.11	Debris Impact Loads.			477
	C6.11.1	Alternative Simplified Debris Impact Static Load.		477
	C6.11.2	Wood Logs and Poles.		478
	C6.11.3	Impact by Vehicles.		478
	C6.11.4	Impact by Submerged Tumbling Boulder and Concrete Debris.		478
	C6.11.5	Site Hazard Assessment for Shipping Containers, Ships, and Barges.		478
	C6.11.6	Shipping Containers.		479
	C6.11.7	Extraordinary Debris Impacts.		479
	C6.11.8	Alternative Methods of Response Analysis.		479
C6.12	Foundation Design.			480
	C6.12.1	Resistance Factors for Foundation Stability Analyses.		480
	C6.12.2	Load and Effect Characterization.		481
		C6.12.2.1	Uplift and Underseepage Forces.	481
		C6.12.2.2	Loss of Strength.	481
		C6.12.2.3	General Erosion.	482
		C6.12.2.4	Scour.	483

		C6.12.2.6	Displacements.	483
		C6.12.3	Alternative Foundation Performance-Based Design Criteria.	483
		C6.12.4	Foundation Countermeasures	483
			C6.12.4.1 Fill.	483
			C6.12.4.2 Protective Slab on Grade.	483
			C6.12.4.3 Geotextiles and Reinforced Earth Systems.	483
			C6.12.4.4 Facing Systems.	484
			C6.12.4.5 Ground Improvement.	484
	C6.13	Structural Countermeasures for Tsunami Loading		484
		C6.13.2	Tsunami Barriers.	484
			C6.13.2.2 Site Layout.	484
	C6.14	Tsunami Vertical Evacuation Refuge Structures		484
	C6.15	Designated Nonstructural Components and Systems		485
	C6.16	Nonbuilding Tsunami Risk Category III and IV Structures		485
	References			485
	Other References (Not Cited)			488
C7	SNOW LOADS			489
	C7.0	Snow Loads		489
	C7.2	Ground Snow Loads, p_g.		489
	C7.3	Flat Roof Snow Loads, p_f.		493
		C7.3.1	Exposure Factor, C_e.	493
		C7.3.2	Thermal Factor, C_t.	494
		C7.3.3	Importance Factor, I_s.	494
		C7.3.4	Minimum Snow Load for Low-Slope Roofs, p_m.	495
	C7.4	Sloped Roof Snow Loads, p_s		495
		C7.4.3	Roof Slope Factor for Curved Roofs.	495
		C7.4.4	Roof Slope Factor for Multiple Folded Plate, Sawtooth, and Barrel Vault Roofs.	495
		C7.4.5	Ice Dams and Icicles along Eaves.	495
	C7.5	Partial Loading		495
	C7.6	Unbalanced Roof Snow Loads.		496
		C7.6.1	Unbalanced Snow Loads for Hip and Gable Roofs.	496
		C7.6.2	Unbalanced Snow Loads for Curved Roofs.	497
		C7.6.3	Unbalanced Snow Loads for Multiple Folded Plate, Sawtooth, and Barrel Vault Roofs.	497
		C7.6.4	Unbalanced Snow Loads for Dome Roofs.	497
	C7.7	Drifts on Lower Roofs (Aerodynamic Shade).		497
		C7.7.2	Adjacent Structures.	498
		C7.7.3	Intersecting Drifts at Low Roofs.	498
	C7.8	Roof Projections and Parapets.		498
	C7.9	Sliding Snow.		500
	C7.10	Rain-on-Snow Surcharge Load		500
	C7.11	Ponding Instability.		501
	C7.12	Existing Roofs.		501
	C7.13	Snow on Open-Frame Equipment Structures		501
		C.7.13.3	Snow Loads on Pipes and Cable Trays.	501
		C7.13.2	Snow at Levels below the Top Level.	501
		C7.13.4	Snow Loads on Equipment and Equipment Platforms.	501
	C7.14	Other Roofs and Sites.		501
	References			504
	Other References (Not Cited)			505
C8	RAIN LOADS.			507
	C8.1	Definitions and Symbols.		507
	C8.2	Roof Drainage		507
	C8.3	Design Rain Loads.		507
	C8.4	Ponding Instability and Ponding Load		512
	C8.5	Controlled Drainage		512
	References			513
C9	RESERVED FOR FUTURE COMMENTARY.			515
C10	ICE LOADS—ATMOSPHERIC ICING.			517

C10.1	General		517
	C10.1.1	Site-Specific Studies	517
	C10.1.2	Dynamic Loads	518
	C10.1.3	Exclusions	518
C10.2	Definitions		518
C10.4	Ice Loads Caused by Freezing Rain		519
	C10.4.1	Ice Weight	519
	C10.4.2	Nominal Ice Thickness	519
	C10.4.4	Importance Factors	521
	C10.4.6	Design Ice Thickness for Freezing Rain	521
C10.5	Wind on Ice-Covered Structures		521
	C10.5.5	Wind on Ice-Covered Guys and Cables	521
C10.6	Design Temperatures for Freezing Rain		522
C10.7	Partial Loading		522
References			522

C11	SEISMIC DESIGN CRITERIA		525
C11.1	General		525
	C11.1.1	Purpose	526
	C11.1.2	Scope	526
	C11.1.3	Applicability	526
	C11.1.4	Alternate Materials and Methods of Construction	526
	C11.1.5	Quality Assurance	526
C11.2	Definitions		526
C11.3	Symbols		530
C11.4	Seismic Ground Motion Values		530
	C11.4.1	Near-Fault Sites	531
	C11.4.2	Mapped Acceleration Parameters	531
	C11.4.3	Site Class	531
	C11.4.4	Site Coefficients and Risk-Targeted Maximum Considered Earthquake (MCE_R) Spectral Response Acceleration Parameters	531
	C11.4.5	Design Spectral Acceleration Parameters	532
	C11.4.6	Design Response Spectrum	532
	C11.4.8	Site-Specific Ground Motion Procedures	533
C11.5	Importance Factor and Risk Category		535
	C11.5.1	Importance Factor	535
	C11.5.2	Protected Access for Risk Category IV	535
C11.6	Seismic Design Category		535
C11.7	Design Requirements for Seismic Design Category A		537
C11.8	Geologic Hazards and Geotechnical Investigation		537
	C11.8.1	Site Limitation for Seismic Design Categories E and F	537
	C11.8.2	Geotechnical Investigation Report Requirements for Seismic Design Categories C through F	537
	C11.8.3	Additional Geotechnical Investigation Report Requirements for Seismic Design Categories D through F	537
C11.9	Vertical Ground Motions for Seismic Design		539
	C11.9.2	MCE_R Vertical Response Spectrum	539
References			540
Other References (Not Cited)			541

C12	SEISMIC DESIGN REQUIREMENTS FOR BUILDING STRUCTURES		543
C12.1	Structural Design Basis		543
	C12.1.1	Basic Requirements	543
	C12.1.2	Member Design, Connection Design, and Deformation Limit	546
	C12.1.3	Continuous Load Path and Interconnection	546
	C12.1.4	Connection to Supports	546
	C12.1.5	Foundation Design	546
	C12.1.6	Material Design and Detailing Requirements	546
C12.2	Structural System Selection		546
	C12.2.1	Selection and Limitations	546
		C12.2.1.1 Alternative Structural Systems	547
		C12.2.1.2 Elements of Seismic Force-Resisting Systems	547

	C12.2.2	Combinations of Framing Systems in Different Directions.		548
	C12.2.3	Combinations of Framing Systems in the Same Direction.		548
		C12.2.3.1	R, C_d, and Ω_0 Values for Vertical Combinations.	548
		C12.2.3.2	Two-Stage Analysis Procedure.	548
		C12.2.3.3	R, C_d, and Ω_0 Values for Horizontal Combinations.	548
	C12.2.4	Combination Framing Detailing Requirements.		548
	C12.2.5	System-Specific Requirements.		548
		C12.2.5.1	Dual System.	548
		C12.2.5.2	Cantilever Column Systems.	548
		C12.2.5.3	Inverted Pendulum-Type Structures.	548
		C12.2.5.4	Increased Structural Height Limit for Steel Eccentrically Braced Frames, Steel Special Concentrically Braced Frames, Steel Buckling-Restrained Braced Frames, Steel Special Plate Shear Walls, and Special Reinforced Concrete Shear Walls.	549
		C12.2.5.5	Special Moment Frames in Structures Assigned to Seismic Design Categories D through F.	549
		C12.2.5.6	Steel Ordinary Moment Frames.	549
		C12.2.5.7	Steel Intermediate Moment Frames.	549
		C12.2.5.8	Shear Wall–Frame Interactive Systems.	550
C12.3	Diaphragm Flexibility, Configuration Irregularities, and Redundancy			550
	C12.3.1	Diaphragm Flexibility.		550
		C12.3.1.1	Flexible Diaphragm Condition.	551
		C12.3.1.2	Rigid Diaphragm Condition.	551
		C12.3.1.3	Calculated Flexible Diaphragm Condition.	551
	C12.3.2	Irregular and Regular Classification.		551
		C12.3.2.1	Horizontal Irregularity.	551
		C12.3.2.2	Vertical Irregularity.	551
	C12.3.3	Limitations and Additional Requirements for Systems with Structural Irregularities		552
		C12.3.3.1	Prohibited Horizontal and Vertical Irregularities for Seismic Design Categories D through F.	552
		C12.3.3.2	Extreme Weak Stories.	552
		C12.3.3.3	Elements Supporting Discontinuous Walls or Frames.	552
		C12.3.3.4	Increase in Forces Caused by Irregularities for Seismic Design Categories D through F.	554
	C12.3.4	Redundancy.		554
		C12.3.4.1	Conditions Where Value of ρ is 1.0.	554
		C12.3.4.2	Redundancy Factor, ρ, for Seismic Design Categories D through F.	554
C12.4	Seismic Load Effects and Combinations			555
	C12.4.1	Applicability.		555
	C12.4.2	Seismic Load Effect.		555
		C12.4.2.1	Horizontal Seismic Load Effect.	556
		C12.4.2.2	Vertical Seismic Load Effect.	556
	C12.4.3	Seismic Load Effects Including Overstrength.		556
		C12.4.3.1	Horizontal Seismic Load Effect Including Overstrength.	556
		C12.4.3.2	Capacity-Limited Horizontal Seismic Load Effect.	556
	C12.4.4	Minimum Upward Force for Horizontal Cantilevers for Seismic Design Categories D through F.		556
C12.5	Direction of Loading.			557
	C12.5.1	Direction of Loading Criteria.		557
	C12.5.2	Seismic Design Category B.		557
	C12.5.3	Seismic Design Category C.		557
	C12.5.4	Seismic Design Categories D through F.		557
C12.6	Analysis Procedure Selection			557
C12.7	Modeling Criteria			558
	C12.7.1	Foundation Modeling.		558
	C12.7.2	Effective Seismic Weight.		558
	C12.7.3	Structural Modeling.		559
	C12.7.4	Interaction Effects.		559
C12.8	Equivalent Lateral Force Procedure			560
	C12.8.1	Seismic Base Shear.		560
		C12.8.1.1	Calculation of Seismic Response Coefficient.	560
		C12.8.1.2	Soil–Structure Interaction Reduction.	560
		C12.8.1.3	Maximum SDS Value in Determination of Cs and Ev.	560
	C12.8.2	Period Determination.		561
		C12.8.2.1	Approximate Fundamental Period.	561

	C12.8.3	Vertical Distribution of Seismic Forces.	562
	C12.8.4	Horizontal Distribution of Forces.	562
		C12.8.4.1 Inherent Torsion.	562
		C12.8.4.2 Accidental Torsion.	563
		C12.8.4.3 Amplification of Accidental Torsional Moment.	563
	C12.8.5	Overturning.	564
	C12.8.6	Story Drift Determination.	564
		C12.8.6.1 Minimum Base Shear for Computing Drift.	565
		C12.8.6.2 Period for Computing Drift.	565
	C12.8.7	P-Delta Effects.	565
C12.9	Linear Dynamic Analysis		567
	C12.9.1	Modal Response Spectrum Analysis.	567
		C12.9.1.1 Number of Modes.	567
		C12.9.1.2 Modal Response Parameters.	568
		C12.9.1.3 Combined Response Parameters.	568
		C12.9.1.4 Scaling Design Values of Combined Response.	568
		C12.9.1.5 Horizontal Shear Distribution.	568
		C12.9.1.6 P-Delta Effects.	569
		C12.9.1.7 Soil–Structure Interaction Reduction.	569
		C12.9.1.8 Structural Modeling.	569
	C12.9.2	Linear Response History Analysis.	569
		C12.9.2.1 General Requirements.	569
		C12.9.2.2 General Modeling Requirements.	569
		C12.9.2.3 Ground Motion Selection and Modification.	570
		C12.9.2.4 Application of Ground Acceleration Histories.	571
		C12.9.2.5 Modification of Response for Design.	571
		C12.9.2.6 Enveloping of Force Response Quantities.	571
C12.10	Diaphragms, Chords, and Collectors		571
	C12.10.1	Diaphragm Design.	571
		C12.10.1.1 Diaphragm Design Forces.	572
		C12.10.2.1 Collector Elements Requiring Load Combinations Including Overstrength for Seismic Design Categories C through F.	572
	C12.10.3	Alternative Design Provisions for Diaphragms, Including Chords and Collectors.	572
		C12.10.3.1 Design.	572
		C12.10.3.2 Seismic Design Forces for Diaphragms, Including Chords and Collectors.	573
		C12.10.3.3 Transfer Forces in Diaphragms.	574
		C12.10.3.4 Collectors—Seismic Design Categories C through F.	575
		C12.10.3.5 Diaphragm Design Force Reduction Factor.	576
C12.11	Structural Walls and Their Anchorage		580
	C12.11.1	Design for Out-of-Plane Forces.	580
	C12.11.2	Anchorage of Structural Walls and Transfer of Design Forces into Diaphragms or Other Supporting Structural Elements.	580
		C12.11.2.1 Wall Anchorage Forces.	580
		C12.11.2.2 Additional Requirements for Anchorage of Concrete or Masonry Structural Walls to Diaphragms in Structures Assigned to Seismic Design Categories C through F	580
C12.12	Drift and Deformation		581
	C12.12.3	Structural Separation.	582
	C12.12.4	Members Spanning between Structures.	582
	C12.12.5	Deformation Compatibility for Seismic Design Categories D through F.	582
C12.13	Foundation Design		583
	C12.13.1	Design Basis.	583
	C12.13.3	Foundation Load-Deformation Characteristics.	583
	C12.13.4	Reduction of Foundation Overturning.	584
	C12.13.5	Strength Design for Foundation Geotechnical Capacity.	584
		C12.13.5.2 Resistance Factors.	584
		C12.13.5.3 Acceptance Criteria.	584
	C12.13.6	Allowable Stress Design for Foundation Geotechnical Capacity.	584
	C12.13.7	Requirements for Structures Assigned to Seismic Design Category C.	585
		C12.13.7.1 Pole-Type Structures.	585
		C12.13.7.2 Foundation Ties.	585
		C12.13.7.3 Pile Anchorage Requirements.	585
	C12.13.8	Requirements for Structures Assigned to Seismic Design Categories D through F	585
		C12.13.8.1 Pole-Type Structures.	585
		C12.13.8.2 Foundation Ties.	585

		C12.13.8.3 General Pile Design Requirement..	585
		C12.13.8.4 Batter Piles.	585
		C12.13.8.5 Pile Anchorage Requirements..	585
		C12.13.8.6 Splices of Pile Segments.	585
		C12.13.8.7 Pile–Soil Interaction..	585
		C12.13.8.8 Pile Group Effects..	586
	C12.13.9	Requirements for Foundations on Liquefiable Sites.	586
		C12.13.9.1 Foundation Design.	587
		C12.13.9.2 Shallow Foundations.	587
		C12.13.9.3 Deep Foundations.	587
C12.14	Simplified Alternative Structural Design Criteria for Simple Bearing Wall or Building Frame Systems.		588
	C12.14.1	General..	588
		C12.14.1.1 Simplified Design Procedure.	588
	C12.14.3	Seismic Load Effects and Combinations.	589
	C12.14.7	Design and Detailing Requirements.	589
	C12.14.8	Simplified Lateral Force Analysis Procedure	589
		C12.14.8.1 Seismic Base Shear.	589
		C12.14.8.2 Vertical Distribution..	589
		C12.14.8.5 Drift Limits and Building Separation.	589
References			589
Other References (Not Cited)			591

C13	SEISMIC DESIGN REQUIREMENTS FOR NONSTRUCTURAL COMPONENTS.		593
	C13.1	General.	593
		C13.1.1 Scope.	593
		C13.1.2 Seismic Design Category.	595
		C13.1.3 Component Importance Factor.	595
		C13.1.4 Exemptions.	595
		C13.1.5 Premanufactured Modular Mechanical and Electrical Systems.	596
		C13.1.6 Application of Nonstructural Component Requirements to Nonbuilding Structures.	596
		C13.1.7 Reference Documents.	596
		C13.1.8 Reference Documents Using Allowable Stress Design..	597
	C13.2	General Design Requirements	597
		C13.2.1 Applicable Requirements for Architectural, Mechanical, and Electrical Components, Supports, and Attachments.	597
		C13.2.2 Special Certification Requirements for Designated Seismic Systems.	597
		C13.2.3 Consequential Damage.	598
		C13.2.4 Flexibility.	598
		C13.2.5 Testing Alternative for Seismic Capacity Determination..	599
		C13.2.6 Experience Data Alternative for Seismic Capacity Determination..	599
		C13.2.7 Construction Documents.	600
	C13.3	Seismic Demands on Nonstructural Components	600
		C13.3.1 Seismic Design Force.	600
		C13.3.1.4 Dynamic Analysis	601
		C13.3.2 Seismic Relative Displacements.	602
		C13.3.2.1 Displacements within Structures.	602
		C13.3.2.2 Displacements between Structures.	602
		C13.3.3 Component Period..	602
	C13.4	Nonstructural Component Anchorage.	603
		C13.4.1 Design Force in the Attachment.	604
		C13.4.2 Anchors in Concrete or Masonry..	604
		C13.4.3 Installation Conditions..	605
		C13.4.4 Multiple Attachments.	605
		C13.4.5 Power-Actuated Fasteners..	605
		C13.4.6 Friction Clips.	605
	C13.5	Architectural Components	605
		C13.5.1 General..	606
		C13.5.2 Forces and Displacements..	606
		C13.5.3 Exterior Nonstructural Wall Elements and Connections.	606
		C13.5.4 Glass..	607
		C13.5.5 Out-of-Plane Bending.	607
		C13.5.6 Suspended Ceilings.	607
		C13.5.6.1 Seismic Forces..	607

		C13.5.6.2	Industry Standard Construction for Acoustical Tile or Lay-In Panel Ceilings	607
		C13.5.6.3	Integral Construction	610
	C13.5.7	Access Floors		610
		C13.5.7.1	General	610
		C13.5.7.2	Special Access Floors	610
	C13.5.8	Partitions		610
	C13.5.9	Glass in Glazed Curtain Walls, Glazed Storefronts, and Glazed Partitions		610
		C13.5.9.1	General	611
		C13.5.9.2	Seismic Drift Limits for Glass Components	611
	C13.5.10	Egress Stairs and Ramps		611
C13.6	Mechanical and Electrical Components			611
	C13.6.1	General		612
	C13.6.2	Mechanical Components and C13.6.3 Electrical Components		612
	C13.6.4	Component Supports		613
		C13.6.4.1	Design Basis	613
		C13.6.4.2	Design for Relative Displacement	613
		C13.6.4.3	Support Attachment to Component	613
		C13.6.4.5	Additional Requirements	613
	C13.6.5	Distribution Systems: Conduit, Cable Tray, and Raceways		614
	C13.6.6	Distribution Systems: Duct Systems		614
	C13.6.7	Distribution Systems: Piping and Tubing Systems		614
		C13.6.7.1	ASME Pressure Piping Systems	615
		C13.6.7.2	Fire Protection Sprinkler Piping Systems	616
		C13.6.7.3	Exceptions	616
	C13.6.9	Utility and Service Lines		616
	C13.6.10	Boilers and Pressure Vessels		616
	C13.6.11	Elevator and Escalator Design Requirements		616
		C13.6.11.3	Seismic Controls for Elevators	616
		C13.6.11.4	Retainer Plates	616
	C13.6.12	Rooftop Solar Panels		616
	C13.6.13	Other Mechanical and Electrical Components		617
References				617
Other References (Not Cited)				618
C14	**MATERIAL-SPECIFIC SEISMIC DESIGN AND DETAILING REQUIREMENTS**			**619**
C14.0	Scope			619
C14.1	Steel			619
	C14.1.1	Reference Documents		619
	C14.1.2	Structural Steel		619
		C14.1.2.1	General	619
		C14.1.2.2	Seismic Requirements for Structural Steel Structures	619
	C14.1.3	Cold-Formed Steel		619
		C14.1.3.1	General	619
		C14.1.3.2	Seismic Requirements for Cold-Formed Steel Structures	619
	C14.1.4	Cold-Formed Steel Light-Frame Construction		620
		C14.1.4.1	General	620
		C14.1.4.2	Seismic Requirements for Cold-Formed Steel Light-Frame Construction	620
		C14.1.4.3	Prescriptive Cold-Formed Steel Light-Frame Construction	620
	C14.1.5	Cold-Formed Steel Deck Diaphragms		620
	C14.1.7	Steel Cables		620
	C14.1.8	Additional Detailing Requirements for Steel Piles in Seismic Design Categories D through F		620
C14.2	Concrete			620
		C14.2.2.1	Definitions	620
		C14.2.2.2	ACI 318, Section 10.7.6	621
		C14.2.2.3	Scope	621
		C14.2.2.4	Intermediate Precast Structural Walls	621
		C14.2.2.6	Foundations	621
		C14.2.2.7	Detailed Plain Concrete Shear Walls	621
	C14.2.3	Additional Detailing Requirements for Concrete Piles		621
	C14.2.4	Additional Design and Detailing Requirements for Precast Concrete Diaphragms		622
		C14.2.4.1	Diaphragm Seismic Demand Levels	622
		C14.2.4.2	Diaphragm Design Options	623

		C14.2.4.3	Diaphragm Connector or Joint Reinforcement Deformability.	625
		C14.2.4.4	Precast Concrete Diaphragm Connector and Joint Reinforcement Qualification Procedure.	625
	C14.3	Composite Steel and Concrete Structures		627
		C14.3.1	Reference Documents.	627
		C14.3.4	Metal-Cased Concrete Piles.	627
	C14.4	Masonry		628
	C14.5	Wood.		628
		C14.5.1	Reference Documents.	628
References				628
Other References (Not Cited)				628
C15	SEISMIC DESIGN REQUIREMENTS FOR NONBUILDING STRUCTURES			631
	C15.1	General.		631
		C15.1.1	Nonbuilding Structures.	631
		C15.1.2	Design.	631
		C15.1.3	Structural Analysis Procedure Selection.	631
		C15.1.4	Nonbuilding Structures Sensitive to Vertical Ground Motions.	634
	C15.2	This section intentionally left blank; see section C15.8.		634
	C15.3	Nonbuilding Structures Supported by Other Structures.		634
		C15.3.1	Less Than 25% Combined Weight Condition.	635
		C15.3.2	Greater Than or Equal to 25% Combined Weight Condition.	635
	C15.4	Structural Design Requirements.		636
		C15.4.1	Design Basis.	636
		C15.4.2	Rigid Nonbuilding Structures.	637
		C15.4.3	Loads.	637
		C15.4.4	Fundamental Period.	637
		C15.4.7	Drift, Deflection, and Structure Separation.	637
		C15.4.8	Site-Specific Response Spectra.	637
		C15.4.9	Anchors in Concrete or Masonry.	637
		C15.4.10	Requirements for Nonbuilding Structure Foundations on Liquefiable Sites.	638
	C15.5	Nonbuilding Structures Similar to Buildings		638
		C15.5.1	General.	638
		C15.5.2	Pipe Racks.	638
		C15.5.3.1	Steel Storage Racks.	638
		C15.5.3.2	Steel Cantilevered Storage Racks.	638
		C15.5.4	Electrical Power-Generating Facilities.	639
		C15.5.5	Structural Towers for Tanks and Vessels.	639
		C15.5.6	Piers and Wharves.	639
	C15.6	General Requirements for Nonbuilding Structures Not Similar to Buildings		640
		C15.6.1	Earth-Retaining Structures.	640
		C15.6.2	Chimneys and Stacks	640
		C15.6.2.1	General.	640
		C15.6.2.2	Concrete Chimneys and Stacks.	640
		C15.6.2.3	Steel Chimneys and Stacks.	640
		C15.6.4	Special Hydraulic Structures.	640
		C15.6.5	Secondary Containment Systems.	640
		C15.6.5.1	Freeboard.	641
		C15.6.6	Telecommunication Towers.	641
		C15.6.7	Steel Tubular Support Structures for Onshore Wind Turbine Generator Systems.	641
		C15.6.8	Ground-Supported Cantilever Walls or Fences.	641
	C15.7	Tanks and Vessels		642
		C15.7.1	General.	642
		C15.7.2	Design Basis.	642
		C15.7.3	Strength and Ductility.	643
		C15.7.4	Flexibility of Piping Attachments.	644
		C15.7.5	Anchorage.	644
		C15.7.6	Ground-Supported Storage Tanks for Liquids	644
		C15.7.6.1	General.	644
		C15.7.7	Water Storage and Water Treatment Tanks and Vessels.	646
		C15.7.7.1	Welded Steel.	646
		C15.7.7.2	Bolted Steel.	646
		C15.7.7.3	Reinforced and Prestressed Concrete.	646
		C15.7.8	Petrochemical and Industrial Tanks and Vessels Storing Liquids	646

		C15.7.8.1	Welded Steel..	646
		C15.7.8.2	Bolted Steel.	647
	C15.7.9	Ground-Supported Storage Tanks for Granular Materials		647
		C15.7.9.1	General..	647
		C15.7.9.2	Lateral Force Determination..	647
		C15.7.9.3	Force Distribution to Shell and Foundation.	647
	C15.7.10	Elevated Tanks and Vessels for Liquids and Granular Materials.		647
		C15.7.10.1	General..	647
		C15.7.10.4	Transfer of Lateral Forces into Support Tower..	648
		C15.7.10.5	Evaluation of Structures Sensitive to Buckling Failure.	648
		C15.7.10.7	Concrete Pedestal (Composite) Tanks.	648
	C15.7.11	Boilers and Pressure Vessels.		648
	C15.7.12	Liquid and Gas Spheres..		648
	C15.7.13	Refrigerated Gas Liquid Storage Tanks and Vessels..		648
	C15.7.14	Horizontal, Saddle-Supported Vessels for Liquid or Vapor Storage..		655
C15.8	Consensus Standards and Other Referenced Documents			655
References				655
Other References (Not Cited)				656

C16 NONLINEAR RESPONSE HISTORY ANALYSIS 657

C16.1	General Requirements			657
	C16.1.1	Scope.		657
	C16.1.2.	Linear Analysis.		657
	C16.1.3	Vertical Response Analysis.		658
	C16.1.4	Documentation..		658
C16.2	Ground Motions			658
	C16.2.1	Target Response Spectrum.		658
	C16.2.2	Ground Motion Selection.		659
	C16.2.3	Ground Motion Modification.		660
		C16.2.3.1	Period Range for Scaling or Matching.	661
		C16.2.3.2	Amplitude Scaling..	661
		C16.2.3.3	Spectral Matching.	662
	C16.2.4	Application of Ground Motions to the Structural Model.		662
C16.3	Modeling and Analysis			662
	C16.3.1	Modeling..		662
	C16.3.3	P-Delta Effects..		663
	C16.3.4	Torsion..		663
	C16.3.5	Damping..		664
	C16.3.6	Explicit Foundation Modeling.		664
	C16.4.1	Global Acceptance Criteria		664
		C16.4.1.1	Unacceptable Response.	664
		C16.4.1.2	Story Drift.	666
	C16.4.2	Element-Level Acceptance Criteria.		667
		C16.4.2.1	Force-Controlled Actions.	667
		C16.4.2.2	Deformation-Controlled Actions.	670
		C16.4.2.3	Elements of the Gravity Force-Resisting System..	671
References				671
Other References (Not Cited)				671

C17 SEISMIC DESIGN REQUIREMENTS FOR SEISMICALLY ISOLATED STRUCTURES 673

C17.1	General .			673
C17.2	General Design Requirements			674
	C17.2.4	Isolation System		675
		C17.2.4.1	Environmental Conditions..	675
		C17.2.4.2	Wind Forces.	675
		C17.2.4.3	Fire Resistance..	675
		C17.2.4.4	Lateral Restoring Force.	675
		C17.2.4.5	Displacement Restraint.	675
		C17.2.4.6	Vertical-Load Stability.	675
		C17.2.4.7	Overturning.	676
		C17.2.4.8	Inspection and Replacement..	676
		C17.2.4.9	Quality Control.	676
	C17.2.5	Structural System.		676

		C17.2.5.2	Minimum Building Separations.	676
		C17.2.5.4	Steel Ordinary Concentrically Braced Frames.	676
		C17.2.5.5	Isolation System Connections.	676
	C17.2.6	Elements of Structures and Nonstructural Components.		677
	C17.2.8	Isolation System Properties.		678
		C17.2.8.2	Isolator Unit Nominal Properties.	678
		C17.2.8.3	Bounding Properties of Isolation System Components.	678
		C17.2.8.4	Property Modification Factors.	679
		C17.2.8.5	Upper Bound and Lower Bound Force-Deflection Behavior of Isolation System Components.	681
C17.3	Seismic Ground Motion Criteria.			682
	C17.3.1	Site-Specific Seismic Hazard.		682
	C17.3.3	MCE_R Ground Motion Records.		682
C17.4	Analysis Procedure Selection.			682
C17.5	Equivalent Lateral Force Procedure.			682
	C17.5.3	Minimum Lateral Displacements Required for Design.		683
		C17.5.3.1	Maximum Displacement.	683
		C17.5.3.2	Effective Period at the Maximum Displacement.	683
		C17.5.3.3	Total Maximum Displacement.	683
	C17.5.4	Minimum Lateral Forces Required for Design.		683
		C17.5.4.1	Isolation System and Structural Elements below the Base Level.	684
		C17.5.4.2	Structural Elements above the Base Level.	684
		C17.5.4.3	Limits on V_S.	684
	C17.5.5	Vertical Distribution of Force.		684
	C17.5.6	Drift Limits.		686
C17.6	Dynamic Analysis Procedures.			687
	C17.6.2	Modeling.		688
		C17.6.3.4	Response History Analysis Procedure.	688
C17.7	Design Review.			688
C17.8	Testing.			689
	C17.8.2.2	Sequence and Cycles.		689
	C17.8.2.3	Dynamic Testing.		689
	C17.8.2.4	Units Dependent on Bilateral Load.		690
	C17.8.2.5	Maximum and Minimum Vertical Load.		690
	C17.8.2.7	Testing Similar Units.		690
		C17.8.3	Determination of Force-Deflection Characteristics.	690
		C17.8.4	Test Specimen Adequacy.	691
		C17.8.5	Production Tests.	691
References.				692
Other References (Not Cited).				692

C18	SEISMIC DESIGN REQUIREMENTS FOR STRUCTURES WITH DAMPING SYSTEMS			693
	C18.1	General.		693
	C18.2	General Design Requirements.		693
		C18.2.1	System Requirements.	693
			C18.2.1.2 Damping System.	693
		C18.2.2	Seismic Ground Motion Criteria.	693
		C18.2.3	Procedure Selection.	693
			C18.2.4.1 Device Design.	695
			C18.2.4.4 Nominal Design Properties.	696
			C18.2.4.5 Maximum and Minimum Damper Properties.	696
			C18.2.4.6 Damping System Redundancy.	697
	C18.3	Nonlinear Response History Procedure.		697
		C18.3.2	Accidental Mass Eccentricity.	697
	C18.4	Seismic Load Conditions and Acceptance Criteria for Nonlinear Response History Procedure.		698
		C18.4.1	Seismic Force-Resisting System.	698
	C18.5	Design Review.		698
	C18.6	Testing.		698
			C18.6.1.2 Sequence and Cycles of Testing.	698
			C18.6.1.3 Testing Similar Devices.	698
			C18.6.1.4 Determination of Force-Velocity-Displacement Characteristics.	698
		C18.6.2	Production Tests.	698
	C18.7	Alternate Procedures and Corresponding Acceptance Criteria.		699
		C18.7.1	Response-Spectrum Procedure and C18.7.2 Equivalent Lateral Force Procedure.	699

		C18.7.3	Damped Response Modification.	700
			C18.7.3.1 Damping Coefficient.	700
			C18.7.3.2 Effective Damping.	700
		C18.7.4	Seismic Load Conditions and Acceptance Criteria for RSA and ELF Procedures.	700
			C18.7.4.5 Seismic Load Conditions and Combination of Modal Responses	700
	References			701
	Other References (Not Cited)			701

C19 SOIL–STRUCTURE INTERACTION FOR SEISMIC DESIGN. ... 703
- C19.1 General. ... 703
- C19.2 SSI Adjusted Structural Demands. ... 704
- C19.3 Foundation Damping. ... 705
- C19.4 Kinematic SSI Effects ... 707
    - C19.4.1 Base Slab Averaging. ... 707
    - C19.4.2 Embedment. ... 707
- References ... 708

C20 SITE CLASSIFICATION PROCEDURE FOR SEISMIC DESIGN ... 709
- C20.1 Site Classification ... 709
- C20.3 Site Class Definitions ... 709
    - C20.3.1 Site Class F. ... 709
- C20.4 Definitions of Site Class Parameters. ... 710
- References ... 710

C21 SITE-SPECIFIC GROUND MOTION PROCEDURES FOR SEISMIC DESIGN. ... 711
- C21.0 General. ... 711
- C21.1 Site Response Analysis ... 711
    - C21.1.1 Base Ground Motions. ... 711
    - C21.1.2 Site Condition Modeling. ... 711
    - C21.1.3 Site Response Analysis and Computed Results. ... 712
- C21.2 Risk-Targeted Maximum Considered Earthquake (MCE_R) Ground Motion Hazard Analysis. ... 712
    - C21.2.1 Probabilistic (MCE_R) Ground Motions. ... 712
        - C21.2.1.1 Method 1. ... 712
        - C21.2.1.2 Method 2. ... 712
    - C21.2.2 Deterministic (MCE_R) Ground Motions. ... 713
    - C21.2.3 Site-Specific MCE_R. ... 713
- C21.3 Design Response Spectrum ... 713
- C21.4 Design Acceleration Parameters. ... 713
- C21.5 Maximum Considered Earthquake Geometric Mean (MCE_G) Peak Ground Acceleration. ... 714
- References ... 714
- Other References (Not Cited) ... 715

C22 SEISMIC GROUND MOTION, LONG-PERIOD TRANSITION, AND RISK COEFFICIENT MAPS ... 717
- Risk-Targeted Maximum Considered Earthquake (MCE_R) Ground Motion Maps. ... 722
- Long-Period Transition Maps. ... 723
- Maximum Considered Earthquake Geometric Mean (MCE_G) PGA Maps. ... 723
- Ground Motion Web Tool. ... 724
- Risk Coefficient Maps. ... 724
- Uniform Hazard and Deterministic Ground Motion Maps ... 724
- References ... 724
- Other References (Not Cited) ... 724

C23 SEISMIC DESIGN REFERENCE DOCUMENTS (No Commentary). ... 725

C24 RESERVED FOR FUTURE COMMENTARY. ... 727

C25 RESERVED FOR FUTURE COMMENTARY. ... 729

C26	WIND LOADS: GENERAL REQUIREMENTS		731
	C26.1	Procedures	731
		C26.1.1 Scope	731
		C26.1.2 Permitted Procedures	731
	C26.2	Definitions	732
	C26.3	Symbols	733
	C26.4	General	734
		C26.4.3 Wind Pressures Acting on Opposite Faces of Each Building Surface	734
	C26.5	Wind Hazard Map	734
		C26.5.1 Basic Wind Speed	734
		C26.5.2 Special Wind Regions	740
		C26.5.3 Estimation of Basic Wind Speeds from Regional Climatic Data	740
	C26.6	Wind Directionality	741
	C26.7	Exposure	741
		C26.7.4 Exposure Requirements	743
	C26.8	Topographic Effects	744
	C26.9	Ground Elevation Factor	748
	C26.10	Velocity Pressure	748
		C26.10.1 Velocity Pressure Exposure Coefficient	748
		C26.10.2 Velocity Pressure	750
	C26.11	Gust Effects	751
	C26.12	Enclosure Classification	755
	C26.13	Internal Pressure Coefficients	756
	C26.14	Tornado Limitation	757
		C26.14.1 Tornado Wind Speeds and Probabilities	757
		C26.14.2 Wind Pressures Induced by Tornadoes Versus Other Windstorms	758
		C26.14.3 Occupant Protection	759
		C26.14.4 Minimizing Building Damage	759
		C26.14.5 Continuity of Building Operations	764
		C26.14.6 Trussed Communications Towers	764
	References		764
	Other References (Not Cited)		766
C27	WIND LOADS ON BUILDINGS: MAIN WIND FORCE RESISTING SYSTEM (DIRECTIONAL PROCEDURE)		767
	C27.1	Scope	767
		C27.1.5 Minimum Design Wind Loads	767
	Part 1: Enclosed, Partially Enclosed, and Open Buildings of All Heights		767
	C27.3	Wind Loads: Main Wind Force Resisting System	767
		C27.3.1 Enclosed and Partially Enclosed Rigid and Flexible Buildings	767
		C27.3.2 Open Buildings with Monoslope, Pitched, or Troughed Free Roofs	768
		C27.3.5 Design Wind Load Cases	768
	Part 2: Enclosed Simple Diaphragm Buildings with $h \leq 160$ ft ($h \leq 48.8$ m)		769
	C27.5	Wind Loads: Main Wind Force Resisting System	769
		C27.5.1 Wall and Roof Surfaces: Class 1 and 2 Buildings	769
		C27.5.2 Parapets	769
		C27.5.3 Roof Overhangs	770
	References		770
	Other References (Not Cited)		770
C28	WIND LOADS ON BUILDINGS: MAIN WIND FORCE RESISTING SYSTEM (ENVELOPE PROCEDURE)		771
	Part 1: Enclosed and Partially Enclosed Low-Rise Buildings		771
	C28.3	Wind Loads: Main Wind Force Resisting System	771
		C28.3.1 Design Wind Pressure for Low-Rise Buildings	771
		C28.3.2 Parapets	771
		C28.3.4 Minimum Design Wind Loads	773
	Part 2: Enclosed Simple Diaphragm Low-Rise Buildings		773
	References		774
	Other References (Not Cited)		774

C29	WIND LOADS ON BUILDING APPURTENANCES AND OTHER STRUCTURES: MAIN WIND FORCE RESISTING SYSTEM (DIRECTIONAL PROCEDURE)		775
	C29.3	Design Wind Loads: Solid Freestanding Walls and Solid Signs	775
		C29.3.1 Solid Freestanding Walls and Solid Freestanding Signs	775
		C29.3.2 Solid Attached Signs	776
	C29.4	Design Wind Loads: Other Structures	776
		C29.4.1 Rooftop Structures and Equipment for Buildings	776
		C29.4.2 Design Wind Loads: Circular Bins, Silos, and Tanks with $h \leq 120$ ft ($h \leq 36.5$ m), $D \leq 120$ ft ($D \leq 36.5$ m), and $0.25 \leq H/D \leq 4$	776
		C29.4.2.1 External Walls of Isolated Circular Bins, Silos, and Tanks	776
		C29.4.2.2 Roofs of Isolated Circular Bins, Silos, and Tanks	776
		C29.4.2.3 Undersides of Isolated Elevated Circular Bins, Silos, and Tanks	776
		C29.4.2.4 Roofs and Walls of Grouped Circular Bins, Silos, and Tanks	776
		C29.4.3 Rooftop Solar Panels for Buildings of All Heights with Flat Roofs or Gable or Hip Roofs with Slopes Less Than 7°	777
		C29.4.4 Rooftop Solar Panels Parallel to the Roof Surface on Buildings of All Heights and Roof Slopes	778
	C29.5	Parapets	778
	C29.7	Minimum Design Wind Loading	779
	References		779
C30	WIND LOADS: COMPONENTS AND CLADDING		781
	C30.1	Scope	781
		C30.1.1 Building Types	781
		C30.1.5 Air-Permeable Cladding	781
	C30.3	Building Types	783
		C30.3.1 Conditions	783
		C30.3.2 Design Wind Pressures	783
	Part 1: Low-Rise Buildings		786
	Part 3: Buildings with $h > 60$ ft ($h > 18.3$ m)		786
	Part 4: Buildings with 60 ft $< h \leq 160$ ft (18.3 m $< h \leq 48.8$ m) (Simplified)		787
	C30.6	Building Types	787
		C30.6.1 Wind Load: Components and Cladding	788
		C30.6.1.2 Parapets	788
		C30.6.1.3 Roof Overhangs	788
	Part 5: Open Buildings		788
	C30.7	Building Types	788
	Part 7: nonbuilding Structures		788
	C30.12	Circular Bins, Silos, and Tanks with $h \leq 120$ ft ($h \leq 36.5$ m)	788
		C30.12.2 External Walls of Isolated Circular Bins, Silos, and Tanks	788
		C30.12.3 Internal Surface of Exterior Walls of Isolated Open-Topped Circular Bins, Silos, and Tanks	788
		C30.12.4 Roofs of Isolated Circular Bins, Silos, and Tanks	788
		C30.12.6 Roofs and Walls of Grouped Circular Bins, Silos, and Tanks	788
	References		789
	Other References (Not Cited)		790
C31	WIND TUNNEL PROCEDURE		793
	C31.4	Load Effects	794
		C31.4.1 Mean Recurrence Intervals of Load Effects	794
		C31.4.2 Limitations on Wind Speeds	794
		C31.4.3 Wind Directionality	794
	C31.6	Roof-Mounted Solar Collectors for Roof Slopes Less than 7 Degrees	794
		C31.6.1 Wind Tunnel Test Requirements	794
		C31.6.1.1 Limitations on Wind Loads for Rooftop Solar Collectors	794
		C31.6.1.2 Peer Review Requirements for Wind Tunnel Tests of Roof-Mounted Solar Collectors	795
	References		795
APPENDIX C11A QUALITY ASSURANCE PROVISIONS (Deleted)			515

APPENDIX C11B EXISTING BUILDING PROVISIONS (No Commentary)			515
APPENDIX CC SERVICEABILITY CONSIDERATIONS			801
CC.1	Serviceability Considerations		801
CC.2	Deflection, Vibration, and Drift		801
	CC.2.1	Vertical Deflections	801
	CC.2.2	Drift of Walls and Frames	810
	CC.2.3	Vibrations	810
CC.3	Design for Long-Term Deflection		811
CC.4	Camber		811
CC.5	Expansion and Contraction		811
CC.6	Durability		811
References			811
Other References (Not Cited)			811
APPENDIX CD BUILDINGS EXEMPTED FROM TORSIONAL WIND LOAD CASES			813
APPENDIX CE PERFORMANCE-BASED DESIGN PROCEDURES FOR FIRE EFFECTS ON STRUCTURES			815
CE.1	Scope		815
CE.2	Definitions		815
CE.3	General Requirements		816
CE.4	Performance Objectives		816
	CE.4.1	Structural Integrity	817
	CE.4.2	Project-Specific Performance Objectives	817
CE.5	Thermal Analysis of Fire Effects		817
	CE.5.1	Fuel Load	818
	CE.5.2	Structural Design Fires	818
	CE.5.3	Heat Transfer Analysis	818
CE.6	Structural Analysis of Fire Effects		819
	CE.6.1	Temperature History for Structural Members and Connections	820
	CE.6.2	Temperature-Dependent Properties	820
	CE.6.3	Load Combinations	821
References			821
INDEX			Index-1
SUPPLEMENT 1			S1
ERRATA			E1

Supplements, errata, and interpretations may become available in the future. Please check for important new materials at http://dx.doi.org/10.1061/9780784414248. ASCE 7 Online (https://asce7.online/) provides real-time updates of supplements and errata.

COMMENTARY TO STANDARD ASCE/SEI 7-16

This commentary is not a part of the ASCE Standard *Minimum Design Loads and Associated Criteria for Buildings and Other Structures*. It is included for information purposes.

This commentary consists of explanatory and supplementary material designed to assist local building code committees and regulatory authorities in applying the recommended requirements. In some cases it will be necessary to adjust specific values in the standard to local conditions. In others, a considerable amount of detailed information is needed to put the provisions into effect. This commentary provides a place for supplying material that can be used in these situations and is intended to create a better understanding of the recommended requirements through brief explanations of the reasoning employed in arriving at them.

The sections of the commentary are numbered to correspond to the sections of the standard to which they refer. Because it is not necessary to have supplementary material for every section in the standard, there are gaps in the numbering in the commentary.

CHAPTER C1
GENERAL

C1.1 SCOPE

The minimum design loads, hazard levels, associated criteria, and intended performance goals contained in this standard are derived from research and observed performance of buildings, other structures, and their nonstructural components under the effects of loads. These parameters vary depending on the relative importance of the building, other structure, or nonstructural component. The loads provided in this standard include loads from both normal operations and rare hazard events. All loads and associated criteria are prescribed to achieve an intended performance, which is defined by a reliability index or limit state exceedance probability or preservation of function during a specified hazard event.

Loads and load combinations are set forth in this document with the intent that they be used together. If one were to use loads from some other source with the load combinations set forth herein or vice versa, the reliability of the resulting design may be affected.

With the 2016 edition of the standard, the title was modified to include the words "and Associated Criteria" to acknowledge what has been in this standard for many editions. For example, earthquake loads contained herein are developed for structures that possess certain qualities of ductility and postelastic energy dissipation capability. For this reason, provisions for design, detailing, and construction are provided in Chapters 11 through 22. In some cases, these provisions modify or add to provisions contained in design specifications. However, this standard only adds associated criteria when the modification is needed to achieve the intended structural performance when subjected to the loads specified herein.

C1.3 BASIC REQUIREMENTS

C1.3.1 Strength and Stiffness. Buildings and other structures must satisfy strength limit states in which members and components are proportioned to safely carry the design loads specified in this standard to resist buckling, yielding, fracture, and other unacceptable performance. This requirement applies not only to structural components but also to nonstructural elements, the failure of which could pose a substantial safety or other risk. Chapter 30 of this standard specifies wind loads that must be considered in the design of cladding. Chapter 13 of this standard specifies earthquake loads and deformations that must be considered in the design of nonstructural components and systems designated in that chapter.

Although strength is a primary concern of this section, strength cannot be considered independent of stiffness. In addition to considerations of serviceability, for which stiffness is a primary consideration, structures must have adequate stiffness to ensure stability. In addition, the magnitude of load imposed on a structure for some loading conditions, including earthquake, wind, and ponding, is a direct function of the structure's stiffness.

Another important consideration related to stiffness is damage to nonstructural components resulting from structural deformations. Acceptable performance of nonstructural components requires either that the structural stiffness be sufficient to prevent excessive deformations or that the components can accommodate the anticipated deformations.

Standards produced under consensus procedures and intended for use in connection with building code requirements contain recommendations for resistance factors for use with the strength design procedures of Section 1.3.1.1 or allowable stresses (or safety factors) for the allowable stress design procedures of Section 1.3.1.2. The resistances contained in any such standards have been prepared using procedures compatible with those used to form the load combinations contained in Sections 2.3 and 2.4. When used together, these load combinations and the companion resistances are intended to provide reliabilities approximately similar to those indicated in Tables 1.3-1, 1.3-2, and 1.3-3. Some standards known to have been prepared in this manner include the following:

ACI 318, *Building Code Requirements for Structural Concrete*, American Concrete Institute.

AISC 341, *Seismic Provisions for Structural Steel Buildings*, American Institute of Steel Construction.

AISC 358, *Prequalified Connections for Special and Intermediate Steel Moment Frames for Seismic Applications*, American Institute of Steel Construction.

AISC 360, *Specification for Structural Steel Buildings*, American Institute of Steel Construction.

AISI S100-16, *North American Specification for the Design of Cold-Formed Steel Structural Members*, American Iron and Steel Institute.

Aluminium Association. *Specification for Aluminum Structures*, Aluminum Association.

AWC NDS-2015, *National Design Specification for Wood Construction*, American Wood Council.

AWC SDPWS-2015, *Special Design Provisions for Wind and Seismic*, American Wood Council.

SEI/ASCE 8, *Specification for the Design of Cold-Formed Stainless Steel Structural Members*, ASCE.

TMS 402, *Building Code Requirements and Specification for Masonry Structures*, The Masonry Society.

TMS 602, *Specification for Masonry Structures*, The Masonry Society.

C1.3.1.3 Performance-Based Procedures. Section 1.3.1.3 introduces alternative performance-based procedures that may be used in lieu of the procedures of Sections 1.3.1.1 and 1.3.1.2

to demonstrate that a building or other structure, or parts thereof, has sufficient strength. These procedures are intended to parallel the so-called "alternative means and methods" procedures that have been contained in building codes for many years. Such procedures permit the use of materials, design, and construction methods that differ from the prescriptive requirements of the building code, or in this case the standard, that can be demonstrated to provide equivalent performance. Such procedures are useful in that they permit innovation and the development of new approaches before the building codes and standards have an opportunity to provide for these new approaches. In addition, these procedures permit the use of alternative methods for certain special structures, which may not be covered by code but by means of their occupancy, use, or other features, can provide acceptable performance without compliance with the prescriptive requirements.

The reliability of a proposed design does not need to be evaluated when the standard's design procedures in Sections 1.3.1.1 and 1.3.1.2 are applied. However, when performance-based procedures are used, the reliability achieved for the proposed design should be consistent with the target reliabilities stipulated in Section 1.3.1.3.

Alternative design methods have a range of implementation levels. Such methods are addressed in standards or best practice guidelines that address performance-based design goals and methodologies that incorporate the fundamental basis of reliability analysis. A minimum level of alternative design would involve standard design procedures, with analyses based on the requirements of Section 1.3.1.3. For example, a building that exceeds the code limits for building height could be designed with applicable codes and standards, and member demand and capacities would be checked to determine their adequacy for the design loads and conditions. For seismic design, the provisions of the ASCE 41 standard and of the *Tall Buildings Initiative, Guidelines for Performance-Based Seismic Design of Tall Buildings* (PEER 2010) were either calibrated by structural performance level or were demonstrated in comparison with prescriptive design methods to provide reliabilities equal to or better than Table 1.3-2.

Flood load factors in ASCE 7 may not achieve the reliability targets of Table 1.3-1. For structures within the 100-year coastal flood zone, the load factor of 2.0 was based on a beta value of 2.5 (Mehta et al. 1998) rather than 3.0. For structures not in the coastal zone, the load factor of 1.0 reflects the prescriptive minimum 100-year flood elevation for stillwater flooding; thus this flood has a 1% annual chance of being exceeded, which is essentially a beta of 1.3. For Risk Category III and IV structures, no reliability analysis has been performed. For storm events of greater return period, the flood hazard expands both in spatial extent and in depth. For those structures and components of structures that were not subject to the prescriptive flood design requirements, any design resistance depends on that imparted from design for other hazards.

Tsunami and extraordinary events do not have specific reliability tables in this section. Guidance for performance-based design of tsunami can be found in Chapter 6 and its associated commentary. For extraordinary events, the user is directed to Section C2.5 for discussion on reliability consideration for extraordinary events.

The alternative procedures of Section 1.3.1.3 may be used to demonstrate adequacy for one or more design loads, while the standard procedures of Sections 1.3.1.1 and 1.3.1.2 are used to demonstrate adequacy for other design loads. For example, it is relatively common to use the alternative procedures to demonstrate adequate earthquake, fire, or blast resistance, while the standard prescriptive procedures of Sections 1.3.1.1 and 1.3.1.2 are used for all other loading considerations.

The alternative procedures of Section 1.3.1.3 are intended to be used in the design of individual projects, rather than as the basis for broad qualification of new structural systems, products, or components. Procedures for such qualification are beyond the scope of this standard, as the limited number of test data required in Section 1.3.1.3.2 are not appropriate for the application of new materials in a structural system or the development of prefabricated structural assemblies intended for general widespread use in structural systems. A more robust level of testing is needed for new materials or structural assemblies for general use.

Section 1.3.1.3 requires demonstration that a design has adequate strength to provide an equivalent or lower probability of failure under load than that adopted as the basis for the prescriptive requirements of this standard for buildings and structures of comparable Risk Category. Tables 1.3-1, 1.3-2, and 1.3-3 summarize performance goals, expressed in terms of target reliabilities, associated with protection against structural failure that approximate those notionally intended to be achieved using the design procedures of Section 2.3. The target reliability indices are provided for a 50-year reference period, and the probabilities of failure are annual probabilities. Annualized probabilities can be applied to limit states where the loads and member resistance do not vary over the reference period. If a member is subject to degradation, such as corrosion, then degradation effects over the service period should be considered through a time-varying or stochastic process as part of the reliability analysis.

The target reliabilities have been developed and vetted by a number of consensus groups over a period of more than 30 years and have been confirmed through professional practice in AISC 360, ACI 318, and other standards and documents. The target reliabilities for Risk Category II in Table 1.3-1 are based on probabilistic analyses of structural member performance for strength design procedures and are documented in Ellingwood et al. (1980, 1982) and Galambos et al. (1982). The reliabilities are consistent with those adopted by the NBCC (2010), CEN 250 (2002) and ISO (1998). Structural members and connections designed using typical specifications for engineering materials (steel, reinforced concrete, masonry, timber) were analyzed to determine their reliabilities for common limit states, such as yielding in tension members, formation of plastic hinges in compact laterally supported beams, or column buckling and connection fracture for a nominal service period of 50 years. The reliabilities were determined for load combinations involving dead, live, wind, snow, and earthquake loads initially. The target reliabilities listed in Table 1.3-1 are based on strength criteria for structural members. The target reliabilities listed in Table 1.3-2 for strength and deflection limit states are based on strength and deflection criteria for system response to earthquakes where inelastic behavior is assumed. The target reliabilities for Risk Categories I, III, and IV were determined by reviewing the intended performance of structural members and systems, as well as target reliabilities specified by other codes and standards for similar performance criteria.

Seismic design practice has evolved in the past three decades from the original reliability basis mentioned previously. The target reliabilities in Table 1.3-2 for earthquake-resistant structural system design are defined for the response of the structural system as described in NIST (2012), which was prepared by NEHRP Consultants Joint Venture: For Risk Category I and Risk Category II structures, i.e., $I_e = 1.0$, acceptable Life Safety risk is defined by an "absolute" collapse probability of 1% in 50 years and a "conditional" probability of 10% given MCE_R ground motions. The conditional probability of 10% is based on FEMA P-695 methodology. The absolute probability of 1% in 50 years and the conditional probability of 10% given MCE_R ground motions were used by the U.S. Geological Survey to develop the probabilistic MCE_R ground motions of ASCE/SEI 7-10. The conditional probabilities of 5% (Risk Category III structures) and 2.5% (Risk Category IV structures) represent improved reliability anticipated for structures designed with an importance factor, I_e, greater than 1.0. Although not specifically stated by ASCE/SEI 7-10 commentary, it may be presumed that Risk Category III and Risk Category IV structures have absolute collapse probability objectives that are less than 1% in 50 years (i.e., due to design using an importance factor of $I_e > 1.0$).

Engineers may need load criteria for strength design that are consistent with the requirements in this standard for situations that are not covered explicitly within this standard. They may also need to consider load criteria for special situations, as required by the client in performance-based engineering applications, in accordance with Section 1.3.1.3. In addition, groups writing standards and specifications for strength design of structural systems and elements may need to develop resistance factors that, when used with the load requirements in this standard, permit the stipulated reliability to be achieved. Such load criteria should be developed using an accepted procedure, such as that provided in Section 2.3.6, to ensure that the resulting factored design loads and load combinations are consistent with target reliabilities (or levels of performance), the common load criteria in Section 2.3.2, and existing standards and specifications governing strength design for common construction materials. Peer-reviewed statistical data for loads in this standard are provided in Table C1.3-1, adopted from Ellingwood et al. (1980, 1982) and Galambos et al. (1982). The statistics provided are the ratio of the mean, X_m, to nominal, X_n, values of the load and the coefficient of variation, COV, of a cumulative distribution function (CDF) fitted to the 90th percentile and above of the probability distribution of the load. The parameters for S (snow) are based on data for the northeastern quadrant of the United States.

The reliability of structural members can be determined through a reliability analysis, such as a Monte Carlo analysis with random variables assigned probability distributions with mean and COV values based on statistical data. Reliability analyses can also be conducted using a nonparametric hazard curve based on data. Fig. C1.3-1 provides an example of the design equation, limit state equations, and statistical variables for a compact steel flexural member designed for dead plus live load. The statistics used in this example are typical (cf. Table C1.3-1). The user should determine the appropriate probabilistic models for his or her design situation.

Fig. C1.3-1 presents an illustration of how the reliability index, β, is determined for the common case of a compact steel beam with full lateral support, in which the limit state is the formation of the first plastic hinge. The reliability depends on the ratio of nominal live to dead load in the limit state equation. For

Table C1.3-1 Load Distributions and Parameters

Load	X_m/X_n	V_x	CDF
D	1.05	0.10	Normal
L	1.00	0.25	Type I
W	Site-dependent		Type I
S	0.82	0.26	Type II
E	Site-dependent		Type II

<u>Design Equation</u>

$$0.9R_n = 1.2D_n + 1.6L_n$$

<u>Limit State Equation G(X)</u>

$$G(R,D,L) = R - D - L$$

$$G(X) = \left(\left[1.2 + 1.6\left(\frac{L_n}{D_n}\right)\right]/0.9\right)X_1 - X_2 - \left(\frac{L_n}{D_n}\right)X_3$$

where

R = Strength random variable, and R_n = Nominal strength
D = Dead load random variable, and D_n = Nominal dead load
L = Live load random variable, and L_n = Nominal live load

$$X_1 = R/R_n, \quad X_2 = D/D_n, \quad X_3 = L/L_n;$$

Typical range of L/D_n is 0.5 to 4.0.

Statistics

Variable	Mean	Coefficient of Variation (COV)	Probability Density Function (PDF)
X_1	1.08	0.09	Lognormal
X_2	1.05	0.10	Normal
X_3	1.00	0.25	Type I

FIGURE C1.3-1 Equations and Statistics of Load and Resistance Parameters for a Monte Carlo Analysis to Determine Achieved Reliability

$L_n/D_n = 2.0$ (a typical value), the probability of failure (50-year basis) is 0.00298. The corresponding reliability index $\beta = \Phi^{-1}(1 - P_f) = 2.75$ (50-year basis), and $P_f = 6 \times 10^{-5}$ (annual basis). These reliabilities can be compared to the reliability targets in Table 1.3-1. The variation in β with L_n/D_n is very small; unlike ASD, the dead and live load factors of 1.2 and 1.6 were selected so as to properly reflect the differences in variability between dead and live load.

The system reliabilities for earthquake are different than those for other environmental hazards because the design philosophy of the standard is to prevent system collapse in the risk-targeted maximum considered earthquake (MCE_R) shaking. The R, C_d, and Ω_0 coefficients specified in Chapter 12 for seismic loading, together with the systems detailing requirements specified in the referenced standards, are intended to ensure minimum acceptable probabilities of structural collapse, given the occurrence of Maximum Considered Earthquake (MCE_R) ground shaking. As discussed in Section C11.4, for typical structures (Risk Category I and II), the conditional probability of collapse is assumed to be 10%, given the occurrence of the MCE_R. This assumption is based on significant research documented in FEMA P695 (2009). The additional collapse goals of 5% for Risk Category III and 2.5% for Risk Category IV were arrived at by assuming that the seismic fragility (probabilistic model of system strength) is described by a lognormal distribution with a logarithmic standard deviation of 0.6 and adjusting the strength of the structure by the earthquake importance factors of 1.25 and 1.5, respectively. Since collapse is a function of loading (ground shaking) intensity, still lower, but nonnegligible probabilities of collapse also exist at design shaking levels. The collapse risk for design earthquake shaking is approximately 2.5%, 1%, and 0.5% for Risk Category II, III, and IV structures.

This standard also seeks to protect against local failure that does not result in global collapse but could result in injury risk to a few persons. Chapter 16 of the standard defines structural elements according to their criticality as critical, ordinary, and noncritical, where critical elements can lead to global collapse, ordinary elements to endangerment of a limited number of lives, and noncritical elements do not have safety consequences. For ordinary elements in Risk Category II structures, the standard accepts a 25% probability of failure given MCE_R shaking (approximately 10% probability of failure for design earthquake shaking). Failure probabilities for ordinary elements in Risk Category III and IV structures are, respectively, 15% and 9% for MCE_R shaking and 4% and 2% for DE shaking. It is anticipated that the failure probabilities for anchorage of rigid nonstructural components attached at grade to the structure may be in the same range as the probabilities for ordinary elements. However, the uncertainties associated with the reliability for anchorage of rigid nonstructural components that are elevated within a structure are much higher than for structural elements because the methods used to characterize the strength demands on nonstructural components are more approximate than those used for overall building demands, and appropriate reliability levels have not yet been established for them. Furthermore, demands on anchorage of flexible nonstructural components and distributions are significantly more complex, especially when points of attachment of the nonstructural components are elevated within a structure and need to consider both inertial effects and relative displacements. Future study should seek to evaluate nonstructural reliabilities in a rigorous manner, and if consensus can be achieved regarding the appropriate reliability levels for anchorage, to adjust the design procedure for anchorage of these components to achieve these appropriate reliabilities.

It is important to note that provision of adequate strength is not sufficient to ensure proper performance. Considerations of serviceability and structural integrity are also important. Use of the alternative procedures of Section 1.3.1.3 is not intended as an alternative to the requirements of Sections 1.3.2, 1.3.3, 1.3.4, 1.3.5, 1.3.6, or 1.4 of this standard.

The requirements of this standard and its companion referenced standards are intended to provide protection against structural failure. They are also intended to provide property and economic protection for small events, to the extent practical, as well as improve the probability that critical facilities will be functional after severe storms, earthquakes, and similar events. Although these goals are an important part of the requirements of this standard, at the present time there is no documentation of the reliability intended with respect to these goals. Consequently, Tables 1.3-1 and 1.3-2 address safety considerations only.

Compliance with Section 1.3.1.3 may be demonstrated by analysis, testing, or a combination of both methods. It is important to recognize that there is uncertainty as to whether the performance objectives tabulated in Tables 1.3-1 and 1.3-2 can be achieved. There is inherent uncertainty associated with prediction of the intensity of loading that a structure will experience, the actual strength of materials incorporated in construction, the quality of construction, and the condition of the structure at the time of loading. Whether testing, analysis, or a combination of these is used, provision should be made to account for these uncertainties and to ensure that the probability of poor performance is acceptably low. See Ellingwood et al. (1982) and Galambos et al. (1982) for estimates of such uncertainties.

Rigorous methods of reliability analysis can be used to demonstrate that the reliability of a design achieves the targets indicated in Tables 1.3-1 and 1.3-2; a simple illustration of such a method is provided in C1.3-1. While such analyses would certainly constitute an acceptable approach to satisfy the requirements of Section 1.3.1.3, these may not be the only acceptable approaches. Consensus bodies or other standards developing organizations may develop guideline documents that provide alternative performance-based design methods or alternate prescriptive procedures that meet or exceed the reliabilities stated in this section.

Since most building officials and other Authorities Having Jurisdiction do not have the expertise necessary to judge the adequacy of designs justified using the Section 1.3.1.3 procedures, independent peer review is an essential part of this process. Peer review can help to reduce the potential that the design professional of record will overlook or misinterpret one or more potential behaviors that could result in poor performance. Peer review can also help to establish that an appropriate standard of care was adhered to during the design. For peer review to be effective, the reviewers must have the appropriate expertise and understanding of the types of structures, loading, analysis methods, and testing used in the procedures.

The target reliabilities listed in Tables 1.3-1 for members and connections and those in Tables 1.3-2 and 1.3-3 for structural systems are included in this standard specifically for application to performance-based procedures for individual projects that are peer-reviewed by experts in the field. For several reasons, these target reliabilities are not intended to be compared to reliability indices developed by material specification groups for general structural applications. For example, reliability indices for some materials are based on testing of small coupons of the material supplemented by factors to account for scaling up to structural-sized elements, whereas other materials test full-sized structural members as the foundation for this analysis. Additionally, some reliability analyses use default lognormal data distribution assumptions, and others use distributional forms applicable to the material and to each load combination of interest.

C1.3.1.3.2 Testing. Laboratory testing of materials and components constructed from those materials is an essential part of the process of validating the performance of structures and nonstructural components under load. Design resistances specified in the industry standards used with the strength procedures of Section 1.3.1.1 and the allowable stress procedures of Section 1.3.1.2 are based on extensive laboratory testing and many years of experience with the performance of structures designed using these standards in real structures. Similarly, analytical modeling techniques commonly used by engineers to predict the behavior of these systems have been benchmarked and validated against laboratory testing. Similar benchmarking of resistance, component performance, and analytical models is essential when performance-based procedures are used. Where systems and components that are within the scope of the industry standards are used in a design, analytical modeling of these systems and components and their resistances should be conducted in accordance with these standards and industry practice, unless new data and testing suggest that other assumptions are more appropriate. Where new systems, components, or materials are to be used, laboratory testing must be performed to indicate appropriate modeling assumptions and resistances.

No single protocol is appropriate for use in laboratory testing of structural and nonstructural components. The appropriate number and types of tests that should be performed depend on the type of loading the component will be subjected to, the complexity of the component's behavior, the failure modes it may exhibit, the consequences of this failure, and the variability associated with the behavior. Resistances should be selected to provide an acceptably low probability of unacceptable performance. Commentary to Chapter 2 provides guidance on the calculation of load and resistance factors that may be used for this purpose, when LRFD procedures are used.

Regardless of the means used to demonstrate acceptable performance, testing should be sufficient to provide an understanding of the probable mean value and variability of resistance or component performance. For materials or components that exhibit significant variability in behavior, as a result either of workmanship, material variation, or brittle modes of behavior, a very large number of tests may be required to properly characterize both the mean values and dispersion. It is seldom possible to conduct such a large number of tests as part of an individual project. Therefore, for reasons of practicality, this standard permits a small number of tests, with the number based on the observed variability. Users are cautioned to conduct tests on material that is representative of that expected to be used in the specific project and that all significant sources of variability are included in the test samples. When high variability is observed in these test data, the minimum requirement of six tests is not adequate to establish either the true mean or the variability with confidence, and appropriate caution should be used when developing component resistance or performance measures based on this limited testing. This is a primary reason why the procedures of this section are limited to use on the specific projects being analyzed (i.e., they are not "portable" to similar projects) and that data from these tests are not intended as a means of obtaining prequalification of new systems, materials, or components for broad application.

Some industries and industry standards have adopted standard protocols and procedures for qualification testing. For example, AISC 341, Chapter K, specifies the required testing for qualification of connections used in certain steel seismic force-resisting systems. The wood structural panel industry has generally embraced the testing protocols developed by the Consortium of Universities for Research in Earthquake Engineering project (Krawinkler et al. 2002). When a material, component, or system is similar to those for which such an industry standard exists, the industry standard should be used, unless it can be demonstrated to the satisfaction of peer review and the Authority Having Jurisdiction that more appropriate results will be attained by using alternative procedures and protocols.

When data from Section 1.3.1.3.2 testing is used to characterize a variable within the reliability analysis, sample sizes shall be sufficient to define the mean and coefficient of variation of the test results within specific confidence bounds determined by the significance of that variable in the reliability analysis. While most testing being conducted in accordance with Section 1.3.1.3.2 will be used primarily to confirm or supplement engineering analyses, it is possible that test data will also be used to characterize one of the random variables that are part of the reliability analysis. Because each variable in the reliability analysis will influence the final computed reliability index, each test-based variable must be subjected to a reasonable amount of statistical rigor. For very small sample sizes, it is not possible to define the mean or the standard deviation precisely. However, there are well-established methods to compute confidence bounds on those parameters. For example, a conservative estimate of the mean value might be the lower 75% confidence bound.

C1.3.2 Serviceability. In addition to strength limit states, buildings and other structures must also satisfy serviceability limit states that define functional performance and behavior under loads normally experienced during the lifetime of the structure or during a time defined specifically for a project or a particular limit state. Serviceability limit states include such items as deflection and vibration. In the United States, strength limit states have traditionally been specified in building codes because they control the safety of the structure. Serviceability limit states, however, are usually noncatastrophic, define a level of quality of the structure or element, and are a matter of judgment as to their application. Serviceability limit states involve the perceptions and expectations of the owner or user and are a contractual matter between the owner or user and the designer and builder. It is for these reasons, and because the benefits are often subjective and difficult to define or quantify, that serviceability limit states for the most part are not included within the model U.S. building codes, with several notable exceptions, such as member deflection limits. In some cases, material design standards provide serviceability limit states for structural elements composed of their material.

The fact that serviceability limit states are usually not codified should not diminish their importance. Exceeding a serviceability limit state in a building or other structure usually means that its function is disrupted or impaired because of local minor damage or deterioration or because of occupant discomfort or annoyance. Therefore, this section states that serviceability limit states and the service loads associated with those limit states should be defined in the project design criteria, which would often be developed in consultation with the owner of the building or other structure. Appendix C and its commentary provide guidance to the designer on developing serviceability design criteria.

Service loads can vary significantly from the design loads specified in this standard. Often the service loads are dependent on the specific serviceability limit state being investigated. For example, beam deflection for a stiffness serviceability limit state has typically been evaluated using the live load is specified in this standard without a load factor applied to it. While the live load used for evaluating floor vibration caused by footfall has commonly been taken as an estimated average of the actual live load present, often significantly less than the design live load is specified in this standard.

C1.3.3 Functionality. Structures in Risk Category IV are intended to have some measure of protection against damage to the structure and designated nonstructural systems that would preclude the facility from resuming its intended function following the design environmental hazard. For example, in Chapter 13 of the standard, nonstructural systems assigned to Risk Category IV must be tested or verified by analysis to be rugged enough to retain their pre-earthquake function following the design earthquake. There are additional requirements in the seismic chapters, 11 through 23, to limit structural damage and drift to preserve function of the structure. Since the provisions of this standard require that structures and nonstructural components be designed to remain essentially elastic under most other environmental hazards, preservation of function is generally provided.

When a performance-based design is elected over the prescriptive design procedures in the standard, the registered design professional should confirm that the structure has sufficient strength and stiffness to not incur damage during the design environmental hazard event that would prevent the facility from resuming its intended function, or in some cases or hazards, functioning during the design hazard.

Because the nature of function preservation is very broad and encompasses many different structural components and systems subjected to various hazards, there are not specific reliability targets. For that reason, the terms "reasonable probability" and "adequate structural strength" as applied to structural systems are used to indicate that the application of Section 1.1.3 is not an absolute target value. What constitutes reasonable probability depends on many factors, including the recognition that the fragilities of structural systems to ensure function are not well established and should be agreed upon by the user, the client, and if applicable, the Authority Having Jurisdiction and the peer reviewers.

The designated nonstructural systems should be composed of components that have adequate strength, stiffness, and ruggedness and are adequately attached to the structure so that they do not incur damage sufficient to prohibit the function of the system within a specified period of time as the facility is brought back online. This functionality can be demonstrated through analysis or through physical testing of the nonstructural system or components.

Designated nonstructural systems may vary between structures based on the function of the facility. However, systems that are essential to Life Safety are commonly accepted as needed function preservation. Such systems may include fire detection and suppression systems, emergency exit lighting, and systems that contain explosive, toxic, or highly toxic materials. The specific systems and components that should be considered part of the "designated nonstructural system" should be determined by the user and may require approval by the Authority Having Jurisdiction and, if part of the design, the peer reviewers.

The requirements of this section are not to preclude damage to the structural elements or nonstructural components. In fact, there might be considerable damage to nonessential nonstructural systems and some indications of inelastic deformation of the structure. There should be no structural damage that would indicate the structure is unsafe to support the loads with reliability similar to that required to support it during normal operation. There should be no damage to the nonstructural components that prevents function, such as blocked egress routes. Designated nonstructural systems may have cosmetic damage, but the components can function as they did before the hazard.

The statement that the provisions within this standard are deemed to comply with this section means that through following the prescriptive provisions contained herein, the design professional should be able to provide the intended function preservation performance of Risk Category IV structures in a reasonable period of time consistent with the current state of the practice. In some chapters of this standard, there are other terms that refer to essentially the same components as "Designated Nonstructural System." "Designated Seismic System" in Chapter 13 and "Critical Equipment and Systems" in Chapter 6 are two examples. In those cases, the hazard chapter-specific terms are retained to allow for the hazard-specific requirements to be implemented.

C1.3.4 Self-Straining Forces and Effects. Indeterminate structures that experience dimensional changes develop self-straining forces and effects. Examples include moments in rigid frames that undergo differential foundation settlements, pretensioning or post-tensioning forces as well as any relaxation or loss of such forces sufficient to affect structural performance, and shear forces in bearing walls that support concrete slabs that shrink. Unless provisions are made for self-straining forces and effects, stresses in structural elements, either alone or in combination with stresses from external loads, can be high enough to cause structural damage.

In many cases, the magnitude of self-straining forces can be anticipated by analyses of expected shrinkage, temperature fluctuations, foundation movement, and so forth. However, it is not always practical to calculate the magnitude of self-straining forces. Designers often specify relief joints, suitable framing systems, or other details to minimize self-straining forces and effects.

C1.3.7 Fire Resistance. Where there is no applicable building code (the building code under which the structure is designed), the structural fire resistance provisions of the International Building Code should be applied. Appendix E provides performance-based design procedures for alternative means specified in Section 1.3.1.

C1.4 GENERAL STRUCTURAL INTEGRITY

Sections 1.4.1 through 1.4.4 present minimum strength criteria intended to ensure that all structures are provided with minimum interconnectivity of their elements and that a complete lateral force-resisting system is present with sufficient strength to provide for stability under gravity loads and nominal lateral forces that are independent of design wind, seismic, or other anticipated loads. Conformance with these criteria provides structural integrity for normal service and minor unanticipated events that may reasonably be expected to occur throughout their lifetimes. For many structures that house large numbers of persons, or that house functions necessary to protect the public safety or occupancies that may be the subject of intentional sabotage or attack, more rigorous protection should be incorporated into designs than provided by these sections. For such structures, additional precautions can and should be taken in the design of structures to limit the effects of local collapse and to prevent or minimize progressive collapse in accordance with the procedures of Section 2.5, as charged by Section 1.4. Progressive collapse is defined as the spread of an initial local failure from element to element, resulting eventually in the collapse of an entire structure or a disproportionately large part of it.

Some authors have defined resistance to progressive collapse to be the ability of a structure to accommodate, with only local failure, the notional removal of any single structural member. Aside from the possibility of further damage that uncontrolled debris from the failed member may cause, it appears prudent to consider whether the abnormal event will fail only a single member.

Because accidents, misuse, and sabotage are normally unforeseeable events, they cannot be defined precisely. Likewise, general structural integrity is a quality that cannot be stated in simple terms. It is the purpose of Section 1.4 and the commentary to direct attention to the problem of local collapse, present guidelines for handling it that will aid the design engineer, and promote consistency of treatment in all types of structures and in all construction materials. ASCE does not intend, at this time, for this standard to establish specific events to be considered during design or for this standard to provide specific design criteria to minimize the risk of progressive collapse.

Accidents, Misuse, Sabotage, and Their Consequences. In addition to unintentional or willful misuse, some of the incidents that may cause local collapse (Leyendecker et al. 1976) are explosions caused by ignition of gas or industrial liquids, boiler failures, vehicle impact, impact of falling objects, effects of adjacent excavations, gross construction errors, very high winds such as tornadoes, and sabotage. Generally, such abnormal events would not be a part of normal design considerations. The distinction between general collapse and limited local collapse can best be made by example as follows.

General Collapse. The immediate, deliberate demolition of an entire structure by phased explosives is an obvious instance of general collapse. Also, the failure of one column in a one-, two-, three-, or possibly even four-column structure could precipitate general collapse because the local failed column is a significant part of the total structural system at that level. Similarly, the failure of a major bearing element in the bottom story of a two- or three-story structure might cause general collapse of the whole structure. Such collapses are beyond the scope of the provisions discussed herein. There have been numerous instances of general collapse that have occurred as the result of such events as bombing, landslides, and floods.

Limited Local Collapse. An example of limited local collapse would be the containment of damage to adjacent bays and stories following the destruction of one or two neighboring columns in a multibay structure. The restriction of damage to portions of two or three stories of a higher structure following the failure of a section of bearing wall in one story is another example.

Examples of General Collapse

Ronan Point. A prominent case of local collapse that progressed to a disproportionate part of the whole building (and is thus an example of the type of failure of concern here) was the Ronan Point disaster, which brought the attention of the profession to the matter of general structural integrity in buildings. Ronan Point was a 22-story apartment building of large, precast-concrete, load-bearing panels in Canning Town, England. In March 1968, a gas explosion in an 18th-story apartment blew out a living room wall. The loss of the wall led to the collapse of the whole corner of the building. The apartments above the 18th story, suddenly losing support from below and being insufficiently tied and reinforced, collapsed one after the other. The falling debris ruptured successive floors and walls below the 18th story, and the failure progressed to the ground. Better continuity and ductility might have reduced the amount of damage at Ronan Point.

Another example is the failure of a one-story parking garage reported in Granstrom and Carlsson (1974). Collapse of one transverse frame under a concentration of snow led to the later progressive collapse of the whole roof, which was supported by 20 transverse frames of the same type. Similar progressive collapses are mentioned in Seltz-Petrash (1979).

Alfred P. Murrah Federal Building. On April 19, 1995, a truck containing approximately 4,000 lb of fertilizer-based explosive (ammonium nitrate/fuel oil) was parked near the sidewalk next to the nine-story reinforced concrete office building (Weidlinger 1994; *Engrg. News Rec.* 1995; Longinow 1995; Glover 1996). The side facing the blast had corner columns and four other perimeter columns. The blast shock wave disintegrated one of the 20- × 36-in. (508 × 915 mm) perimeter columns and caused brittle failures of two others. The transfer girder at the third level above these columns failed, and the upper-story floors collapsed in a progressive fashion. Approximately 70% of the building experienced dramatic collapse. One hundred sixty-eight people died, many of them as a direct result of progressive collapse. Damage might have been less had this structure not relied on transfer girders for support of upper floors, if there had been better detailing for ductility and greater redundancy, and if there had been better resistance for uplift loads on floor slabs.

There are a number of factors that contribute to the risk of damage propagation in modern structures (Breen 1976). Among them are the following:

1. There is an apparent lack of general awareness among engineers that structural integrity against collapse is important enough to be regularly considered in design.
2. To have more flexibility in floor plans and to keep costs down, interior walls and partitions are often non-load-bearing and hence may be unable to assist in containing damage.
3. In attempting to achieve economy in structure through greater speed of erection and less site labor, systems may be built with minimum continuity, ties between elements, and joint rigidity.
4. Unreinforced or lightly reinforced load-bearing walls in multistory structures may also have inadequate continuity, ties, and joint rigidity.
5. In roof trusses and arches, there may not be sufficient strength to carry the extra loads or sufficient diaphragm action to maintain lateral stability of the adjacent members if one collapses.
6. In eliminating excessively large safety factors, code changes over the past several decades have reduced the large margin of safety inherent in many older structures. The use of higher strength materials permitting more slender sections compounds the problem in that modern structures may be more flexible and sensitive to load variations and, in addition, may be more sensitive to construction errors.

Experience has demonstrated that the principle of taking precautions in design to limit the effects of local collapse is realistic and can be satisfied economically. From a public safety viewpoint, it is reasonable to expect all multistory structures to possess general structural integrity comparable to that of properly designed, conventionally framed structures (Breen 1976; Burnett 1975).

Design Alternatives. There are a number of ways to obtain resistance to progressive collapse. In Ellingwood and Leyendecker (1978), a distinction is made between direct and indirect design, and the following approaches are defined:

Direct Design: Explicit consideration of resistance to progressive collapse during the design process through either
 Alternate Path Method: A method that allows local failure to occur but seeks to provide alternate load paths so that the damage is absorbed and major collapse is averted.
 Specific Local Resistance Method: A method that seeks to provide sufficient strength to resist failure from accidents or misuse.
Indirect Design: Implicit consideration of resistance to progressive collapse during the design process through the provision of minimum levels of strength, continuity, and ductility.

The general structural integrity of a structure may be tested by analysis to ascertain whether alternate paths around hypothetically collapsed regions exist. Alternatively, alternate path studies may be used as guides for developing rules for the minimum levels of continuity and ductility needed to apply the indirect design approach to enhance general structural integrity. Specific local resistance may be provided in regions of high risk because it may be necessary for some element to have sufficient strength to resist abnormal loads for the structure as a whole to develop alternate paths. Specific suggestions for the implementation of each of the defined methods are contained in Ellingwood and Leyendecker (1978).

Guidelines for the Provision of General Structural Integrity. Generally, connections between structural components should be ductile and have a capacity for relatively large deformations and energy absorption under the effect of abnormal conditions. This criterion is met in many different ways, depending on the structural system used. Details that are appropriate for resistance to moderate wind loads and seismic loads often provide sufficient ductility. In 1999, ASCE issued a state of practice report that is a good introduction to the complex field of blast-resistant design (ASCE 1999).

Work with large precast panel structures (Schultz et al. 1977; PCI Committee on Precast Bearing Walls 1976; Fintel and Schultz 1979) provides an example of how to cope with the problem of general structural integrity in a building system that is inherently discontinuous. The provision of ties combined with careful detailing of connections can overcome difficulties associated with such a system. The same kind of methodology and design philosophy can be applied to other systems (Fintel and Annamalai 1979). The ACI *Building Code Requirements for Structural Concrete* (ACI 2014) includes such requirements in Section 4.10.

There are a number of ways of designing for the required integrity to carry loads around severely damaged walls, trusses, beams, columns, and floors. A few examples of design concepts and details are

1. Good plan layout. An important factor in achieving integrity is the proper plan layout of walls and columns. In bearing-wall structures, there should be an arrangement of interior longitudinal walls to support and reduce the span of long sections of crosswall, thus enhancing the stability of individual walls and of the structures as a whole. In the case of local failure, this reduction will also decrease the length of wall likely to be affected.
2. Provide an integrated system of ties among the principal elements of the structural system. These ties may be designed specifically as components of secondary load-carrying systems, which often must sustain very large deformations during catastrophic events.
3. Returns on walls. Returns on interior and exterior walls make them more stable.
4. Changing directions of span of floor slab. Where a one-way floor slab is reinforced to span, with a low safety factor, in its secondary direction if a load-bearing wall is removed, the collapse of the slab will be prevented and the debris loading of other parts of the structure will be minimized. Often, shrinkage and temperature steel will be enough to enable the slab to span in a new direction.
5. Load-bearing interior partitions. The interior walls must be capable of carrying enough load to achieve the change of span direction in the floor slabs.
6. Catenary action of floor slab. Where the slab cannot change span direction, the span will increase if an intermediate supporting wall is removed. In this case, if there is enough reinforcement throughout the slab and enough continuity and restraint, the slab may be capable of carrying the loads by catenary action, though very large deflections will result.
7. Beam action of walls. Walls may be assumed to be capable of spanning an opening if sufficient tying steel at the top and bottom of the walls allows them to act as the web of a beam with the slabs above and below acting as flanges (Schultz et al. 1977).
8. Redundant structural systems. A secondary load path (e.g., an upper-level truss or transfer girder system that allows the lower floors of a multistory building to hang from the upper floors in an emergency) allows framing to survive removal of key support elements.
9. Ductile detailing. Avoid low-ductility detailing in elements that might be subject to dynamic loads or very large distortions during localized failures (e.g., consider the implications of shear failures in beams or supported slabs under the influence of building weights falling from above).
10. Provide additional reinforcement to resist blast and load reversal when blast loads are considered in design (ASCE 1997).
11. Consider the use of compartmentalized construction in combination with special moment-resisting frames (as defined in FEMA 1997) in the design of new buildings when considering blast protection.

Although not directly adding structural integrity for the prevention of progressive collapse, the use of special, nonfrangible glass for fenestration can greatly reduce risk to occupants during exterior blasts (ASCE 1997). To the extent that nonfrangible glass isolates a building's interior from blast shock waves, it can also reduce damage to interior framing elements (e.g., supported floor slabs could be made to be less likely to fail due to uplift forces) for exterior blasts.

C1.5 CLASSIFICATION OF BUILDINGS AND OTHER STRUCTURES

C1.5.1 Risk Categorization. In the 2010 edition of this standard, a new Table 1.5-2 was added that consolidates the various importance factors specified for the several type of loads throughout the standard in one location. This change was made to facilitate the process of finding values of these factors. Simultaneously with this addition, the importance factors for wind loads were deleted as changes to the new wind hazard maps adopted by the standard incorporated consideration of less probable design winds for structures assigned to higher Risk Categories, negating the need for separate importance factors. Further commentary on this issue may be found in the commentary to Chapter 26.

The Risk Categories in Table 1.5-1 are used to relate the criteria for maximum environmental loads or distortions specified in this standard to the consequence of the loads being exceeded for the structure and its occupants. For many years, this standard used the term "Occupancy Category," as have the building codes. However, the term "occupancy" as used by the building codes relates primarily to issues associated with fire and Life Safety protection, as opposed to the risks associated with structural failure. The term Risk Category was adopted in place of the older Occupancy Category in the 2010 edition of the standard to distinguish between these two considerations. The Risk Category numbering is unchanged from that in the previous editions of the standard (ASCE 7-98, -02, and -05), but the criteria for selecting a category have been generalized with regard to structure and occupancy descriptions. The reason for this generalization is that the acceptable risk for a building or structure is an issue of public policy, rather than purely a technical one. Model building codes such as the International Building Code (ICC 2009) and NFPA-5000 (NFPA 2006) contain prescriptive lists of building types by Occupancy Category. Individual communities can alter these lists when they adopt local codes based on the model code, and individual owners or operators can elect to design individual buildings to higher Occupancy Categories based on personal risk management decisions. Classification continues to reflect a progression of the anticipated seriousness of the consequence of failure from lowest risk to human life (Risk Category I) to the highest (Risk Category IV). Elimination of the specific examples of buildings that fall into each category has the benefit that it eliminates the potential for conflict between the standard and locally adopted codes and also provides individual communities and development teams the flexibility to interpret acceptable risk for individual projects.

Historically, the building codes and the standard have used a variety of factors to determine the Occupancy Category of a building. These factors include the total number of persons who would be at risk were failure to occur, the total number of persons present in a single room or occupied area, the mobility of the occupants and their ability to cope with dangerous situations, the potential for release of toxic materials, and the loss of services vital to the welfare of the community.

Risk Category I structures generally encompass buildings and structures that normally are unoccupied and that would result in negligible risk to the public should they fail. Structures typically classified in this category have included barns, storage shelters, gatehouses, and similar small structures. Risk Category II includes the vast majority of structures, including most residential, commercial, and industrial buildings and has historically been designated as containing all those buildings and structures not specifically classified as conforming to another category.

Risk Category III includes buildings and structures that house a large number of persons in one place, such as theaters, lecture halls, and similar assembly uses and buildings with persons having limited mobility or ability to escape to a safe haven in the event of failure, including elementary schools, prisons, and small health-care facilities. This category has also included structures associated with utilities required to protect the health and safety of a community, including power-generating stations and water-treatment and sewage-treatment plants. It has also included structures housing hazardous substances, such as explosives or toxins, which if released in quantity could endanger the surrounding community, such as structures in petrochemical process facilities containing large quantities of H_2S or ammonia.

Failures of power plants that supply electricity on the national grid can cause substantial economic losses and disruption to civilian life when their failures can trigger other plants to go offline in succession. The result can be massive and potentially extended power outage, shortage, or both that lead to huge economic losses because of idled industries and a serious disruption of civilian life because of inoperable subways, road traffic signals, and so forth. One such event occurred in parts of Canada and the northeastern United States in August 2003.

Failures of water- and sewage-treatment facilities can cause disruption to civilian life because these failures can cause large-scale (but mostly non-life-threatening) public health risks caused by the inability to treat sewage and to provide drinking water.

Failures of major telecommunication centers can cause disruption to civilian life by depriving users of access to important emergency information (using radio, television, and phone communication) and by causing substantial economic losses associated with widespread interruption of business.

Risk Category IV has traditionally included structures the failure of which would inhibit the availability of essential community services necessary to cope with an emergency situation. Buildings and structures typically grouped in Risk Category IV include hospitals, police stations, fire stations, emergency communication centers, and similar uses.

Ancillary structures required for the operation of Risk Category IV facilities during an emergency also are included in this Risk Category. When deciding whether an ancillary structure or a structure that supports such functions as fire suppression is in Risk Category IV, the design professional must decide whether failure of the subject structure will adversely affect the essential function of the facility. In addition to essential facilities, buildings and other structures containing extremely hazardous materials have been added to Risk Category IV to recognize the potentially devastating effect a release of extremely hazardous materials may have on a population.

The criteria that have historically been used to assign individual buildings and structures to Occupancy Categories have not

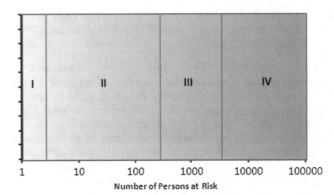

FIGURE C1.5-1 Approximate Relationship between Number of Lives Placed at Risk by a Failure and Occupancy Category

been consistent and sometimes have been based on considerations that are more appropriate to fire and Life Safety than to structural failure. For example, university buildings housing more than a few hundred students have been placed into a higher Risk Category than office buildings housing the same number of persons.

A rational basis should be used to determine the Risk Category for structural design, which is primarily based on the number of persons whose lives would be endangered or whose welfare would be affected in the event of failure. Fig. C1.5-1 illustrates this concept.

"Lives at risk" pertains to the number of people at serious risk of life loss given a structural failure. The Risk Category classification is not the same as the building code occupancy capacity, which is mostly based on risk to life from fire. The lives at risk from a structural failure include persons who may be outside the structure in question who are nonetheless put at serious risk by failure of the structure. From this concept, emergency recovery facilities that serve large populations, even though the structure might shelter relatively few people, are moved into the higher Risk Categories.

When determining the population at risk, consideration should also be given to longer term risks to life than those created during a structural failure. The failure of some buildings and structures, or their inability to function after a severe storm, earthquake, or other disaster, can have far-reaching impact. For example, loss of functionality in one or more fire stations could inhibit the ability of a fire department to extinguish fires, allowing fires to spread and placing many more people at risk. Similarly, the loss of function of a hospital could prevent the treatment of many patients over a period of months.

In Chapters 7, 10, and 11, importance factors are presented for the four Risk Categories identified. The specific importance factors differ according to the statistical characteristics of the environmental loads and the manner in which the structure responds to the loads. The principle of requiring more stringent loading criteria for situations in which the consequence of failure may be severe has been recognized in previous versions of this standard by the specification of mean recurrence interval maps for wind speed and ground snow load.

This section now recognizes that there may be situations when it is acceptable to assign multiple Risk Categories to a structure based on use and the type of load condition being evaluated. For instance, there are circumstances when a structure should appropriately be designed for wind loads with importance factors greater than one, but would be penalized unnecessarily if designed for seismic loads with importance factors greater than one. An example would be a hurricane shelter in a low seismic area. The structure would be classified in Risk Category IV for wind design and in Risk Category II for seismic design.

C1.5.3 Toxic, Highly Toxic, and Explosive Substances. A common method of categorizing structures storing toxic, highly toxic, or explosive substances is by the use of a table of exempt amounts of these materials (EPA 1999b; International Code Council 2000). These references and others are sources of guidance on the identification of materials of these general classifications. A drawback to the use of tables of exempt amounts is the fact that the method cannot handle the interaction of multiple materials. Two materials may be exempt because neither poses a risk to the public by themselves but may form a deadly combination if combined in a release. Therefore, an alternate and superior method of evaluating the risk to the public of a release of a material is by a hazard assessment as part of an overall risk management plan (RMP).

Buildings and other structures containing toxic, highly toxic, or explosive substances may be classified as Risk Category II structures if it can be demonstrated that the risk to the public from a release of these materials is minimal. Companies that operate industrial facilities typically perform hazard and operability (HAZOP) studies, conduct quantitative risk assessments, and develop risk management and emergency response plans. Federal regulations and local laws mandate many of these studies and plans (EPA 1999a). Additionally, many industrial facilities are located in areas remote from the public and have restricted access, which further reduces the risk to the public.

The intent of Section 1.5.2 is for the RMP and the facility's design features that are critical to the effective implementation of the RMP to be maintained for the life of the facility. The RMP and its associated critical design features must be reviewed on a regular basis to ensure that the actual condition of the facility is consistent with the plan. The RMP also should be reviewed whenever consideration is given to the alteration of facility features that are critical to the effective implementation of the RMP.

The RMP generally deals with mitigating the risk to the general public. Risk to individuals outside the facility storing toxic, highly toxic, or explosive substances is emphasized because plant personnel are not placed at as high a risk as the general public because of the plant personnel's training in the handling of toxic, highly toxic, or explosive substances and because of the safety procedures implemented inside the facilities. When these elements (trained personnel and safety procedures) are not present in a facility, then the RMP must mitigate the risk to the plant personnel in the same manner as it mitigates the risk to the general public.

As a result of the prevention program portion of an RMP, buildings and other structures normally falling into Risk Category III may be classified into Risk Category II if means (e.g., secondary containment) are provided to contain the toxic, highly toxic, or explosive substances in the case of a release. To qualify, secondary containment systems must be designed, installed, and operated to prevent migration of harmful quantities of toxic, highly toxic, or explosive substances out of the system to the air, soil, groundwater, or surface water at any time during the use of the structure. This requirement is not to be construed as requiring a secondary containment system to prevent a release of any toxic, highly toxic, or explosive substance into the air. By recognizing that secondary containment shall not allow releases of "harmful" quantities of contaminants, this standard acknowledges that there are substances that might contaminate groundwater but do not produce a sufficient concentration of toxic, highly toxic, or explosive substances during a vapor release to constitute a health or safety risk to the public. Because it represents the "last line of defense," secondary containment does not qualify for the reduced classification.

If the beneficial effect of secondary containment can be negated by external forces, such as the overtopping of dike walls by floodwaters or the loss of liquid containment of an earthen dike because of excessive ground displacement during a seismic event, then the buildings or other structures in question may not be classified into Risk Category II. If the secondary containment is to contain a flammable substance, then implementation of a program of emergency response and preparedness combined with an appropriate fire suppression system would be a prudent action associated with a Risk Category II classification. In many jurisdictions, such actions are required by local fire codes.

Also as the result of the prevention program portion of an RMP, buildings and other structures containing toxic, highly toxic, or explosive substances also could be classified as Risk Category II for hurricane wind loads when mandatory procedures are used to reduce the risk of release of toxic, highly toxic, or explosive substances during and immediately after these predictable extreme loadings. Examples of such procedures include draining hazardous fluids from a tank when a hurricane is predicted or, conversely, filling a tank with fluid to increase its buckling and overturning resistance. As appropriate to minimize the risk of damage to structures containing toxic, highly toxic, or explosive substances, mandatory procedures necessary for the Risk Category II classification should include preventative measures, such as the removal of objects that might become airborne missiles in the vicinity of the structure.

In previous editions of ASCE 7, the definitions of "hazardous" and "extremely hazardous" materials were not provided. Therefore, the determination of the distinction between hazardous and extremely hazardous materials was left to the discretion of the Authority Having Jurisdiction. The change to the use of the terms "toxic" and "highly toxic" based on definitions from Federal law (29 CFR 1910.1200 Appendix A with Amendments as of February 1, 2000) has corrected this problem.

Because of the highly quantitative nature of the definitions for toxic and highly toxic found in 29 CFR 1910.1200 Appendix A, the General Provisions Task Committee felt that the definitions found in federal law should be directly referenced instead of repeated in the body of ASCE 7. The definitions found in 29 CFR 1910.1200 Appendix A are repeated in the following text for reference.

Highly Toxic. A chemical falling within any of the following categories:

1. A chemical that has a median lethal dose [LD(50)] of 50 mg or less per kilogram of body weight when administered orally to albino rats weighing between 200 and 300 g each.
2. A chemical that has a median lethal dose [LD(50)] of 200 mg or less per kilogram of body weight when administered by continuous contact for 24 hr (or less if death occurs within 24 hr) with the bare skin of albino rabbits weighing between 2 and 3 kg each.
3. A chemical that has a median lethal concentration [LC(50)] in air of 200 parts per million by volume or less of gas or vapor, or 2 mg per liter or less of mist, fume, or dust, when administered by continuous inhalation for 1 hr (or less if death occurs within 1 hr) to albino rats weighing between 200 and 300 g each.

Toxic. A chemical falling within any of the following categories:

1. A chemical that has a median lethal dose [LD(50)] of more than 50 mg per kg, but not more than 500 mg per kg of

body weight when administered orally to albino rats weighing between 200 and 300 g each.
2. A chemical that has a median lethal dose [LD(50)] of more than 200 mg per kilogram, but not more than 1,000 mg per kilogram of body weight when administered by continuous contact for 24 hr (or less if death occurs within 24 hr) with the bare skin of albino rabbits weighing between 2 and 3 kg each.
3. A chemical that has a median lethal concentration [LC (50)] in air of more than 200 parts per million but not more than 2,000 parts per million by volume of gas or vapor, or more than 2 mg per liter but not more than 20 mg per liter of mist, fume, or dust, when administered by continuous inhalation for 1 hr (or less if death occurs within 1 hr) to albino rats weighing between 200 and 300 g each.

C1.7 LOAD TESTS

No specific method of test for completed construction has been given in this standard because it may be found advisable to vary the procedure according to conditions. Some codes require the construction to sustain a superimposed load equal to a stated multiple of the design load without evidence of serious damage. Others specify that the superimposed load shall be equal to a stated multiple of the live load plus a portion of the dead load. Limits are set on maximum deflection under load and after removal of the load. Recovery of at least three-quarters of the maximum deflection, within 24 hr after the load is removed, is a common requirement (ACI 2014).

REFERENCES

Aluminum Association. (2015). "Specification for aluminum structures." Arlington, VA.
American Concrete Institute (ACI). (2014). "Building code requirements for structural concrete and commentary." *ACI Standard 318*, Detroit.
American Institute of Steel Construction (AISC). (2016). "Seismic provisions for structural steel buildings." *AISC 341*, Chicago.
AISC. (2016). "Prequalified connections for special and intermediate steel moment frames for seismic applications." *AISC 358*, Chicago.
AISC. (2016). "Specification for structural steel buildings." *AISC 360*, Chicago.
American Iron and Steel Institute (AISI). (2016). "North American specification for the design of cold-formed steel structural members." *AISI S100*, Washington, DC.
ASCE. (1997). *Design of blast resistant buildings in petrochemical facilities*, New York.
ASCE. (1999). *Structural design for physical security: State of the practice*, Reston, VA.
ASCE. (2002). "Specification for the design of cold-formed stainless steel structural members." *SEI/ASCE 8-02*, Reston, VA.
ASCE. (2014). "Seismic evaluation and retrofit of existing buildings." *ASCE 41-13*, Reston, VA.
American Wood Council (AWC). (2015). "National design specification for wood construction." *NDS-2015*, Leesburg, VA.
AWC. (2015). "Special design provisions for wind and seismic." *SDPWS-2015*, Leesburg, VA.
Breen, J. E., ed. (1976). *Progressive collapse of building structures* (summary report of a workshop held at the University of Texas at Austin, Oct. 1975), U.S. Department of Housing and Urban Development. *Report PDR-182*, Washington, DC.
Burnett, E. F. P. (1975). *The avoidance of progressive collapse: Regulatory approaches to the problem*. U.S. Department of Commerce, National Bureau of Standards, Washington, DC, *NBS GCR 75-48* (available from National Technical Information Service, Springfield, VA), October.
Ellingwood, B., Galambos, T. V., MacGregor, J. G., and Cornell, C. A. (1980). "Development of a probability based load criterion for American National Standard A58." National Bureau of Standards, *Special Publication No. 577*, Washington, DC.
Ellingwood, B. R., and Leyendecker, E. V. (1978). "Approaches for design against progressive collapse." *J. Struct. Div.* 104(3), 413–423.
Ellingwood, B., MacGregor, J. G., Galambos, T. V., and Cornell, C. A. (1982). "Probability based load criteria: Load factors and load combinations." *J. Struct. Div.* 108(5), 978–997.
Environmental Protection Agency (EPA). (1999a). "Chemical accident prevention provisions." *40 CFR Part 68*, Washington, DC, July.
EPA. (1999b). "Emergency planning and notification—The list of extremely hazardous substances and their threshold planning quantities." *40 CFR Part 355*, Appendix A, Environmental Protection Agency, Washington, DC, July.
Federal Emergency Management Agency (FEMA). (1997). *NEHRP recommended provisions for seismic regulations for new buildings and other structures*, Washington, DC, *Report No. 302/February 1998*, Part 1–Provisions.
FEMA. (2009). "Quantification of building seismic performance factors." FEMA *P-695*, Washington, DC.
Fintel, M., and Annamalai, G. (1979). "Philosophy of structural integrity of multistory load-bearing concrete masonry structures." *Concrete Int.* 1(5), 27–35.
Fintel, M., and Schultz, D. M. (1979). "Structural integrity of large-panel buildings." *J. Am. Concrete Inst.* 76(5), 583–622.
Galambos, T. V., Ellingwood, B., MacGregor, J. G., and Cornell, C. A. (1982). "Probability based load criteria: Assessment of current design practice." *J. Struct. Div.*, 108(5), 959–977.
Glover, N. J. (1996). *The Oklahoma City bombing: Improving building performance through multi-hazard mitigation*, ASCE/ Federal Emergency Management Agency, *FEMA Report 277*, Washington, DC.
Granstrom, S., and Carlsson, M. (1974). "Byggfurskningen T3: Byggnaders beteende vid overpaverkningar (The behavior of buildings at excessive loadings)." Swedish Institute of Building Research, Stockholm, Sweden.
International Code Council (ICC). (2000). *International building code.* Tables 307.7(1) and 307.7(2), Falls Church, VA.
ICC. (2009). *International building code.* "Table 1604.5 Classification of buildings and other structures for importance factors," Falls Church, VA.
International Standards Organization (ISO). (1998). "General principles on reliability for structures." *ISO 2394*, www.iso.org. Geneva, Switzerland.
Krawinkler, H., Parisi, F., Ibarra, L., Ayoub, A., and Medina, R. (2002). *Development of a testing protocol for woodframe structures*, Consortium of Universities for Research in Earthquake Engineering, Richmond, CA.
Leyendecker, E. V., Breen, J. E., Somes, N. F., and Swatta, M. (1976). *Abnormal loading on buildings and progressive collapse—An annotated bibliography*, U.S. Dept. of Commerce, National Bureau of Standards. *NBS BSS 67*. Washington, DC.
Longinow, A. (1995). "The threat of terrorism: Can buildings be protected?" *Bldg. Operat. Mgmt.*, 46–53, July.
Mehta, K. C., Kriebel, D. L., White, G. J., and Smith, D. A. (1998). An investigation of load factors for flood and combined wind and flood, Report prepared for Federal Emergency Management Agency, Washington, DC.
National Building Code of Canada (NBCC). (2010). National Building Code of Canada, Canadian Commission on Building and Fire Codes, M-23A, National Research Council, Ottawa.
National Fire Protection Association (NFPA). (2006). *Building construction and safety code*, NFPA 5000, Table 35.3.1, "Occupancy category of buildings and other structures for wind, snow and earthquake," Quincy, MA.
National Institute of Standards and Technology (NIST). (2012). Tentative framework for development of advanced seismic design criteria for new buildings." *GCR 12-917-20.* Washington, DC.
Pacific Earthquake Engineering Research Center, University of California (2010). Tall buildings initiative, guidelines for performance-based seismic design of tall buildings, Version 1.0, Report No. 210/05, Berkeley, CA.
PCI Committee on Precast Bearing Walls. (1976). "Considerations for the design of precast bearing-wall buildings to withstand abnormal loads." *J. Prestressed Concrete Inst.*, 21(2), 46–69.
Schultz, D. M., Burnett, E. F. P., and Fintel, M. (1977). *A design approach to general structural integrity, design and construction of large-panel concrete structures*, U.S. Department of Housing and Urban Development, Washington, DC.
Seltz-Petrash, A. E. (1979). "Winter roof collapses: Bad luck or bad design." *Civ. Engrg.* 49(12), 42–45.

Weidlinger, P. (1994). "Civilian structures: Taking the defensive." *Civ. Engrg.* 64(11), 48–50.

OTHER REFERENCES (NOT CITED)

EN 1990. (2002). *Eurocode–Basis of structural design*, CEN 2002. European Committee for Standardization (CEN) http://www.cen.eu.

Engineering News-Record (ENR). (1995). "Moment frames avoid progressive collapse," May 1, 13.

Federal Emergency Management Agency (FEMA). (1993). *Wet floodproofing requirements for structures located in special flood hazard areas in accordance with the national flood insurance program*, Federal Emergency Management Agency, Mitigation Directorate, *Technical Bulletin 7-93*, Washington, DC.

McManamy, R. (1995). "Oklahoma blast forces unsettling design questions." *Engineering News-Record* 234(17), 9.

Occupational Safety and Health Administration (OSHA). (2000). *Standards for general industry*, U.S. Department of Labor, Occupational Safety and Health Administration, *29 CFR* (Code of Federal Regulations) Part 1900 with Amendments as of February 1, 2000, Washington, DC.

Standards Australia. (2005). "General principles on reliability for structures." *AS-5104*, Sydney.

CHAPTER C2
COMBINATIONS OF LOADS

C2.1 GENERAL

Loads in this standard are intended for use with design specifications for conventional structural materials, including steel, concrete, masonry, and timber. Some of these specifications are based on allowable stress design, whereas others use strength (or limit states) design. In the case of allowable stress design, design specifications define allowable stresses that may not be exceeded by load effects caused by unfactored loads, that is, allowable stresses contain a factor of safety. In strength design, design specifications provide load factors and, in some instances, resistance factors. Load factors given herein were developed using a first-order probabilistic analysis and a broad survey of the reliabilities inherent in contemporary design practice (Ellingwood et al. 1982; Galambos et al. 1982). It is intended that these load factors be used by all material-based design specifications that adopt a strength design philosophy in conjunction with nominal resistances and resistance factors developed by individual materials-specification-writing groups. Ellingwood et al. (1982) also provide guidelines for materials-specification-writing groups to aid them in developing resistance factors that are compatible, in terms of inherent reliability, with load factors and statistical information specific to each structural material.

The requirement to use either allowable stress design (ASD) or load and resistance factor design (LRFD) dates back to the introduction of load combinations for strength design (LRFD) in the 1982 edition of the standard. An indiscriminate mix of the LRFD and ASD methods may lead to unpredictable structural system performance because the reliability analyses and code calibrations leading to the LRFD load combinations were based on member rather than system limit states. Registered design professionals often design (or specify) cold-formed steel and open web steel joists using ASD and, at the same time, design the structural steel in the rest of the building or other structure using LRFD. Foundations are also commonly designed using ASD, although strength design is used for the remainder of the structure. Using different design standards for these types of elements has not been shown to be a problem. This requirement is intended to permit current industry practice while, at the same time, not permitting LRFD and ASD to be mixed indiscriminately in the design of a structural frame.

C2.2 SYMBOLS

Self-straining forces and effects can be caused by differential settlement of foundations, creep, shrinkage or expansion in concrete members after placement and similar effects that depend on the material of construction and conditions of constraint, and changes in temperature of members caused by environmental conditions or operational activities during the service life of the structure. See Section C1.3.3 for examples of when self-straining forces and effects may develop.

C2.3 LOAD COMBINATIONS FOR STRENGTH DESIGN

C2.3.1 Basic Combinations. Unfactored loads to be used with these load factors are the nominal loads of this standard. Load factors are from Ellingwood et al. (1982), with the exception of the load factor of 1.0 for E, based on the more recent NEHRP research on seismic-resistant design (FEMA 2004), and for W, based on the wind speed maps at longer return periods for each Risk Category. The basic idea of the load combination analysis is that in addition to dead load, which is considered to be permanent, one of the principal loads (previously referred to as primary variable loads) takes on its maximum lifetime value while the other loads assume "arbitrary point-in-time" values, the latter being loads that would be measured at any instant of time (Turkstra and Madsen 1980). This is consistent with the manner in which loads actually combine in situations in which strength limit states may be approached. However, nominal loads in this standard are substantially in excess of the arbitrary point-in-time values. To avoid having to specify both a maximum and an arbitrary point-in-time value for each load type, some of the specified load factors are less than unity in combinations 2 through 5. Load factors in Section 2.3.1 are based on a survey of reliabilities inherent in existing design practice (Ellingwood et al. 1982; Galambos et al. 1982).

In design where first-order analysis is permitted, superposition of factored loads can be performed either before or after the analysis. However, when a second-order analysis, which considers the effects of structural deformation on member forces, is used to design members and connections, the load factors must be applied before the analysis. The second-order analysis can be accomplished using a computer program with this capability for frame effects and member effects or by amplifying the results of a first-order analysis through the use of coefficients that amplify the first-order moments for the effects of member deformations or joint displacements. Since second-order effects are nonlinear, the second-order analyses must be conducted under factored load combinations (strength design) or load combinations amplified by the factor of safety (allowable stress design). Second-order effects in this context are the effects of loads acting on the deformed configuration of a structure and include P–δ effects and P–Δ effects.

Note that each of the principal loads, including the dead load, is a random variable. The degree of variability in each load is reflected in the associated load factor for the principal load; the other load factors provide the companion values.

The principal loads in the load combinations are identified in Table C2.3-1. The load factor on wind loads in combinations

Table C2.3-1 Principal Loads for Strength Design Load Combinations

Load Combination		Principal Load
1	$1.4D$	D
2	$1.2D + 1.6L + 0.5(L_r$ or S or $R)$	L
3	$1.2D + 1.6(L_r$ or S or $R) + (1.0L$ or $0.5W)$	L_r or S or R
4	$1.2D + 1.0W + 1.0L + 0.5(L_r$ or S or $R)$	W
5	$0.9D + 1.0W$	W
6	$1.2D + E_v + E_h + L + 0.2S$	E
7	$0.9D - E_v + E_h$	E

4 and 5 of Section 2.3.1 and on earthquake loads in combinations 6 and 7 of Section 2.3.6 were changed to 1.0 in previous editions of ASCE 7 when the maps for design wind speed and design seismic acceleration were modified to support a risk-consistent design methodology, as described in Chapters 21 and 26.

Exception 2 permits the companion load S that appears in combinations 2 and 4 to be the balanced snow load defined in Sections 7.3 for flat roofs and 7.4 for sloped roofs. Drifting and unbalanced snow loads, as principal loads, are covered by combination 3.

Load combinations 5 and 7 apply specifically to the case in which the structural actions due to lateral forces and gravity loads counteract one another.

Load combination requirements in Section 2.3 apply only to strength limit states. Serviceability limit states and associated load factors are covered in Appendix C of this standard.

This standard historically has provided specific procedures for determining magnitudes of dead, occupancy live, wind, snow, and earthquake loads. Other loads not traditionally considered by this standard may also require consideration in design. Some of these loads may be important in certain material specifications and are included in the load criteria to enable uniformity to be achieved in the load criteria for different materials. However, statistical data on these loads are limited or nonexistent, and the same procedures used to obtain load factors and load combinations in Section 2.3.1 cannot be applied at the present time. Accordingly, load factors for fluid load (F) and lateral pressure caused by soil and water in soil (H) have been chosen to yield designs that would be similar to those obtained with existing specifications, if appropriate adjustments consistent with the load combinations in Section 2.3.1 were made to the resistance factors. Further research is needed to develop more accurate load factors.

Fluid load, F, defines structural actions in structural supports, framework, or foundations of a storage tank, vessel, or similar container caused by stored liquid products. The product in a storage tank shares characteristics of both dead and live loads. It is similar to a dead load in that its weight has a maximum calculated value, and the magnitude of the actual load may have a relatively small dispersion. However, it is not permanent: Emptying and filling cause fluctuating forces in the structure, the maximum load may be exceeded by overfilling, and densities of stored products in a specific tank may vary.

Uncertainties in lateral forces from bulk materials, included in H, are higher than those in fluids, particularly when dynamic effects are introduced as the bulk material is set in motion by filling or emptying operations. Accordingly, the load factor for such loads is set equal to 1.6.

Where H acts as a resistance, a factor of 0.9 is suggested if the passive resistance is computed with a conservative bias. The intent is that soil resistance be computed for a deformation limit appropriate for the structure being designed, not at the ultimate passive resistance. Thus an at-rest lateral pressure, as defined in the technical literature, would be conservative enough. Higher resistances than at-rest lateral pressure are possible, given appropriate soil conditions. Fully passive resistance would likely not ever be appropriate because the deformations necessary in the soil would likely be so large that the structure would be compromised. Furthermore, there is a great uncertainty in the nominal value of passive resistance, which would also argue for a lower factor on H should a conservative bias not be included.

C2.3.2 Load Combinations Including Flood Load. The nominal flood load, F_a, is based on the 100-year flood (Section 5.1). The recommended flood load factor of 2.0 in V-Zones and Coastal A-Zones is based on a statistical analysis of flood loads associated with hydrostatic pressures, pressures caused by steady overland flow, and hydrodynamic pressures caused by waves, as specified in Section 5.4.

The flood load criteria were derived from an analysis of hurricane-generated storm tides produced along the United States East and Gulf coasts (Mehta et al. 1998), where storm tide is defined as the water level above mean sea level resulting from wind-generated storm surge added to randomly phased astronomical tides. Hurricane wind speeds and storm tides were simulated at 11 coastal sites based on historical storm climatology and on accepted wind speed and storm surge models. The resulting wind speed and storm tide data were then used to define probability distributions of wind loads and flood loads using wind and flood load equations specified in Sections 5.3 and 5.4. Load factors for these loads were then obtained using established reliability methods (Ellingwood et al. 1982; Galambos et al. 1982) and achieve approximately the same level of reliability as do combinations involving wind loads acting without floods. The relatively high flood load factor stems from the high variability in floods relative to other environmental loads. The presence of $2.0F_a$ in both combinations (4) and (6) in V-Zones and Coastal A-Zones is the result of high stochastic dependence between extreme wind and flood in hurricane-prone coastal zones. The $2.0F_a$ also applies in coastal areas subject to northeasters, extratropical storms, or coastal storms other than hurricanes, where a high correlation exists between extreme wind and flood.

Flood loads are unique in that they are initiated only after the water level exceeds the local ground elevation. As a result, the statistical characteristics of flood loads vary with ground elevation. The load factor 2.0 is based on calculations (including hydrostatic, steady flow, and wave forces) with stillwater flood depths ranging from approximately 4 to 9 ft (1.2–2.7 m) (average stillwater flood depth of approximately 6 ft (1.8 m)) and applies to a wide variety of flood conditions. For lesser flood depths, load factors exceed 2.0 because of the wide dispersion in flood loads relative to the nominal flood load. As an example, load factors appropriate to water depths slightly less than 4 ft (1.2 m) equal 2.8 (Mehta et al. 1998). However, in such circumstances, the flood load generally is small. Thus, the load factor 2.0 is based on the recognition that flood loads of most importance to structural design occur in situations where the depth of flooding is greatest.

C2.3.3 Load Combinations Including Atmospheric Ice Loads. Load combinations 2, 4, and 5 in Section 2.3.3 and load combinations 2, 3, and 7 in Section 2.4.3 include the simultaneous effects of snow loads as defined in Chapter 7 and atmospheric ice loads as defined in Chapter 10. Load combinations 2 and 3 in Sections 2.3.3 and 2.4.3 introduce the simultaneous effect of wind on the atmospheric ice. The ice load, D_i, and the wind load on the atmospheric ice, W_i, in combination correspond to an event with approximately a 500-year mean recurrence interval (MRI). Accordingly, the load factors on W_i and D_i are set equal to 1.0 and 0.7 in Sections 2.3.3 and 2.4.3, respectively. The 0.7 load factor on W_i and D_i in Section 2.4.3 aligns allowable stress design to have reliabilities for atmospheric ice loads consistent with the definition of wind and ice loads in Chapter 10 of this standard, which is based on strength principles. The snow loads defined in Chapter 7 are based on measurements of frozen precipitation accumulated on the ground, which includes snow, ice caused by freezing rain, and rain that falls onto snow and later freezes. Thus the effects of freezing rain are included in the snow loads for roofs, catwalks, and other surfaces to which snow loads are normally applied. The atmospheric ice loads defined in Chapter 10 are applied simultaneously to those portions of the structure on which ice caused by freezing rain, in-cloud icing, or snow accrete that are not subject to the snow loads in Chapter 7. A trussed tower installed on the roof of a building is one example. The snow loads from Chapter 7 would be applied to the roof with the atmospheric ice loads from Chapter 10 applied to the trussed tower. Section 2.3.3 load combination 2 ($1.2D + L + D_i + W_i + 0.5S$) or Section 2.4.3 load combination 2 ($D + 0.7D_i + 0.7W_i + S$) are applicable. If a trussed tower has working platforms, the snow loads would be applied to the surface of the platforms, similar to a roof, with the atmospheric ice loads applied to the tower. Section 2.3.3 load combination 2 would reduce to $1.2D + D_i$ for cases where the live load, wind on ice load, and snow load are zero. Section 2.4.3 load combination 2 would similarly reduce to $D + 0.7D_i$. If a sign is mounted on a roof, the snow loads would be applied to the roof and the atmospheric ice loads to the sign.

C2.3.4 Load Combinations Including Self-Straining Forces and Effects. Self-straining forces and effects should be calculated based on a realistic assessment of the most probable values rather than the upper bound values of the variables. The most probable value is the value that can be expected at any arbitrary point in time.

When self-straining forces and effects are combined with dead loads as the principal action, a load factor of 1.2 may be used. However, when more than one variable load is considered and self-straining forces and effects are considered as a companion load, the load factor may be reduced if it is unlikely that the principal and companion loads will attain their maximum values at the same time. The load factor applied to T should not be taken as less than a value of 1.0.

If only limited data are available to define the magnitude and frequency distribution of the self-straining forces and effects, then its value must be estimated conservatively. Estimating the uncertainty in the self-straining forces and effects may be complicated by variation of the material stiffness of the member or structure under consideration.

When checking the capacity of a structure or structural element to withstand the effects of self-straining forces and effects, the following load combinations should be considered.

When using strength design:

$$1.2D + 1.2T + 0.5L$$

$$1.2D + 1.6L + 1.0T$$

These combinations are not all-inclusive, and judgment is necessary in some situations. For example, where roof live loads or snow loads are significant and could conceivably occur simultaneously with self-straining forces and effects, their effect should be included. The design should be based on the load combination causing the most unfavorable effect.

C2.3.5 Load Combinations for Nonspecified Loads. Engineers may wish to develop load criteria for strength design that are consistent with the requirements in this standard in some situations where the standard provides no information on loads or load combinations. They also may wish to consider loading criteria for special situations, as required by the client in performance-based engineering (PBE) applications in accordance with Section 1.3.1.3. Groups responsible for strength design criteria for design of structural systems and elements may wish to develop resistance factors that are consistent with the standard. Such load criteria should be developed using a standardized procedure to ensure that the resulting factored design loads and load combinations will lead to target reliabilities (or levels of performance) that can be benchmarked against the common load criteria in Section 2.3.1. Section 2.3.5 permits load combinations for strength design to be developed through a standardized method that is consistent with the methodology used to develop the basic combinations that appear in Section 2.3.1.

The load combination requirements in Section 2.3.1 and the resistance criteria for structural steel in AISC 360 (2016), for cold-formed steel in AISI S100 (2016), for structural concrete in ACI 318 (2014), for structural aluminum in the *Specification for Aluminum Structures* (Aluminum Association 2015), for engineered wood construction in *AWC NDS-2015 National Design Specification for Wood Construction* (2015), and for masonry in TMS 402-16, *Building Code Requirements and Specifications for Masonry Structures and Companion Commentaries* (2016), are based on modern concepts of structural reliability theory. In probability-based limit states design (PBLSD), the reliability is measured by a reliability index, β, which is related (approximately) to the limit state probability by $P_f = \Phi(-\beta)$. The approach taken in PBLSD was to

1. Determine a set of reliability objectives or benchmarks, expressed in terms of β, for a spectrum of traditional structural member designs involving steel, reinforced concrete, engineered wood, and masonry. Gravity load situations were emphasized in this calibration exercise, but wind and earthquake loads were considered as well. A group of experts from material specifications participated in assessing the results of this calibration and selecting target reliabilities. The reliability benchmarks so identified are *not* the same for all limit states; if the failure mode is relatively ductile and consequences are not serious, β tends to be in the range 2.5 to 3.0, whereas if the failure mode is brittle and consequences are severe, β is 4.0 or more.
2. Determine a set of load and resistance factors that best meets the reliability objectives identified in (1) in an overall sense, considering the scope of structures that might be designed by this standard and the material specifications and codes that reference it.

The load combination requirements appearing in Section 2.3.1 used this approach. They are based on a "principal action–companion action" format, in which one load is taken at its maximum value while other loads are taken at their point-in-time values. Based on the comprehensive reliability analysis performed to support their development, it was found that these load factors are well approximated by

$$\gamma_Q = (\mu_Q/Q_n)(1 + \alpha_Q \beta V_Q) \qquad \text{(C2.3-1)}$$

in which μ_Q is the mean load, Q_n is the nominal load from other chapters in this standard, V_Q is the coefficient of variation in the load, β is the reliability index, and α_Q is a sensitivity coefficient that is approximately equal to 0.8 when Q is a principal action and 0.4 when Q is a companion action. This approximation is valid for a broad range of common probability distributions used to model structural loads. The load factor is an increasing function of the bias in the estimation of the nominal load, the variability in the load, and the target reliability index, as common sense would dictate.

As an example, the load factors in combination 2 of Section 2.3.1 are based on achieving a β of approximately 3.0 for a ductile limit state with moderate consequences (e.g., formation of first plastic hinge in a steel beam). For live load acting as a principal action, $\mu_Q/Q_n = 1.0$ and $V_Q = 0.25$; for live load acting as a companion action, $\mu_Q/Q_n \approx 0.3$ and $V_Q \approx 0.6$. Substituting these statistics into Eq. (C2.3-1), $\gamma_Q = 1.0[1+0.8(3)(0.25)] = 1.6$ (principal action) and $\gamma_Q = 0.3[1+0.4(3)(0.60)] = 0.52$ (companion action). ASCE Standard 7-05 (2005) stipulates 1.60 and 0.50 for these live load factors in combinations 2 and 3. If an engineer wished to design for a limit state probability that is less than the standard case by a factor of approximately 10, β would increase to approximately 3.7, and the principal live load factor would increase to approximately 1.74.

Similarly, resistance factors that are consistent with the aforementioned load factors are well approximated for most materials by

$$\phi = (\mu_R/R_n) \exp[-\alpha_R \beta V_R] \qquad (C2.3-2)$$

in which μ_R = mean strength, R_n = code-specified strength, V_R = coefficient of variation in strength, and α_R = sensitivity coefficient, equal approximately to 0.7. For the limit state of yielding in an ASTM A992 (2011) steel tension member with specified yield strength of 50 ksi (345 MPa), $\mu_R/R_n = 1.06$ (under a static rate of load) and $V_R = 0.09$. Eq. (C2.3-2) then yields $\phi = 1.06 \exp[-(0.7)(3.0)(0.09)] = 0.88$. The resistance factor for yielding in tension in Section D of the *AISC Specification* (2010) is 0.9. If a different performance objective were to require that the target limit state probability be decreased by a factor of 10, then ϕ would decrease to 0.84, a reduction of about 7%. Engineers wishing to compute alternative resistance factors for engineered wood products and other structural components where duration-of-load effects might be significant are advised to review the reference materials provided by their professional associations before using Eq. (C2.3-2).

There are two key issues that must be addressed to use Eqs. (C2.3-1) and (C2.3-2): selection of reliability index, β, and determination of the load and resistance statistics.

The reliability index controls the safety level, and its selection should depend on the mode and consequences of failure. The loads and load factors in this standard do not explicitly account for higher reliability indices normally desired for brittle failure mechanisms or more serious consequences of failure. Common standards for design of structural materials often do account for such differences in their resistance factors (for example, the design of connections under AISC or the design of columns under ACI). Tables 1.3-1 and 1.3-2 provide general guidelines for selecting target reliabilities consistent with the extensive calibration studies performed earlier to develop the load requirements in Section 2.3.1 and the resistance factors in the design standards for structural materials. The reliability indices in those earlier studies were determined for structural members based on a service period of 50 years. System reliabilities are higher to a degree that depends on structural redundancy and ductility. The probabilities represent, *in order of magnitude*, the associated annual member failure rates for those who would find this information useful in selecting a reliability target.

The load requirements in Sections 2.3.1–2.3.3 are supported by extensive peer-reviewed statistical databases, and the values of mean and coefficient of variation, μ_Q/Q_n and V_Q, are well established. This support may not exist for other loads that traditionally have not been covered by this standard. Similarly, the statistics used to determine μ_R/R_n and V_R should be consistent with the underlying material specification. When statistics are based on small-batch test programs, all reasonable sources of end-use variability should be incorporated in the sampling plan. The engineer should document the basis for all statistics selected in the analysis and submit the documentation for review by the Authority Having Jurisdiction. Such documents should be made part of the permanent design record.

The engineer is cautioned that load and resistance criteria necessary to achieve a reliability-based performance objective are coupled through the common term β in Eqs. (C2.3-1) and (C2.3-2). Adjustments to the load factors without corresponding adjustments to the resistance factors will lead to an unpredictable change in structural performance and reliability.

C2.3.6 Basic Combinations with Seismic Load Effects. The seismic load effect, E, is combined with the effects of other loads. For strength design, the load combinations in Section 2.3.6 with E include the horizontal and vertical seismic load effects of Sections 12.4.2.1 and 12.4.2.2, respectively. Similarly, the basic load combinations for allowable stress design in Section 2.4.8 with E include the same seismic load effects.

The seismic load effect including overstrength factor, E_m, is combined with other loads. The purpose for load combinations with overstrength factor is to approximate the maximum seismic load combination for the design of critical elements, including discontinuous systems, transfer beams and columns supporting discontinuous systems, and collectors. The allowable stress increase for load combinations with overstrength is to provide compatibility with past practice.

C2.4 LOAD COMBINATIONS FOR ALLOWABLE STRESS DESIGN

C2.4.1 Basic Combinations. The load combinations listed cover those loads for which specific values are given in other parts of this standard. Design should be based on the load combination causing the most unfavorable effect. In some cases, this may occur when one or more loads are not acting. No safety factors have been applied to these loads because such factors depend on the design philosophy adopted by the particular material specification. The principal load, or maximum variable load, in the load combinations is identified in Table C2.4-1.

Wind and earthquake loads need not be assumed to act simultaneously. However, the most unfavorable effects of each should be considered separately in design, where appropriate. In some instances, forces caused by wind might exceed those caused by earthquake, and ductility requirements might be determined by earthquake loads.

Load combinations (7) in Section 2.4.1 and (10) in Section 2.4.5 address the situation in which the effects of lateral or uplift forces counteract the effect of gravity loads. This action eliminates an inconsistency in the treatment of counteracting

Table C2.4-1. Principal Loads for Allowable Stress Design Load Combinations

Load Combination	Principal Load
1 D	D
2 $D + L$	L
3 $D + (L_r$ or S or $R)$	L_r or S or R
4 $D + 0.75L + 0.75(L_r$ or S or $R)$	L
5 $D + 0.6W$	W
6 $D + 0.75L + 0.75(0.6W) + 0.75(L_r$ or S or $R)$	W
7 $0.6D + 0.6W$	W
8 $D + 0.7E_v + 0.7E_{mh}$	E
9 $D + 0.525E_v + 0.525E_{mh} + 0.75L + 0.75S$	E
10 $0.6D - 0.7E_v + 0.7E_{mh}$	E

loads in allowable stress design and strength design and emphasizes the importance of checking stability. The reliability of structural components and systems in such a situation is determined mainly by the large variability in the destabilizing load (Ellingwood and Li 2009), and the factor 0.6 on dead load is necessary for maintaining comparable reliability between strength design and allowable stress design. The earthquake load effect is multiplied by 0.7 to align allowable stress design for earthquake effects with the definition of E in Section 11.3 of this standard, which is based on strength principles.

Most loads, other than dead loads, vary significantly with time. When these variable loads are combined with dead loads, their combined effect should be sufficient to reduce the risk of unsatisfactory performance to an acceptably low level. However, when more than one variable load is considered, it is extremely unlikely that they will all attain their maximum value at the same time (Turkstra and Madsen 1980). Accordingly, some reduction in the total of the combined load effects is appropriate. This reduction is accomplished through the 0.75 load combination factor. The 0.75 factor applies only to the variable loads because the dead loads (or stresses caused by dead loads) do not vary in time.

Some material design standards that permit a one-third increase in allowable stress for certain load combinations have justified that increase by this same concept. Where that is the case, simultaneous use of both the one-third increase in allowable stress and the 25% reduction in combined loads is unsafe and is not permitted. In contrast, allowable stress increases that are based upon duration of load or loading rate effects, which are independent concepts, may be combined with the reduction factor for combining multiple variable loads. In such cases, the increase is applied to the total stress, that is, the stress resulting from the combination of all loads.

In addition, certain material design standards permit a one-third increase in allowable stress for load combinations with one variable load where that variable is earthquake load. This standard handles allowable stress design for earthquake loads in a fashion to give results comparable to the strength design basis for earthquake loads as defined in Chapter 12 of this standard.

Exception (1) permits the companion load S appearing in combinations (4) and (6) to be the balanced snow load defined in Sections 7.3 for flat roofs and 7.4 for sloped roofs. Drifting and unbalanced snow loads, as principal loads, are covered by combination (3).

When wind forces act on a structure, the structural action causing uplift at the structure–foundation interface is less than would be computed from the peak lateral force because of area averaging. Area averaging of wind forces occurs for all structures. In the method used to determine the wind forces for enclosed structures, the area-averaging effect is already taken into account in the data analysis leading to the pressure coefficients C_p [or (GC_p)]. However, in the design of tanks and other industrial structures, the wind force coefficients, C_f, provided in the standard do not account for area averaging. For this reason, exception 2 permits the wind interface to be reduced by 10% in the design of nonbuilding structure foundations and to self-anchored ground-supported tanks. For different reasons, a similar approach is already provided for seismic actions in Section 2.4.1, Exception 2.

Exception 2 in Section 2.4.5 for special reinforced masonry walls is based upon the combination of three factors that yield a conservative design for overturning resistance under the seismic load combination:

1. The basic allowable stress for reinforcing steel is 40% of the specified yield.
2. The minimum reinforcement required in the vertical direction provides a protection against the circumstance where the dead and seismic loads result in a very small demand for tension reinforcement.
3. The maximum reinforcement limit prevents compression failure under overturning.

Of these, the low allowable stress in the reinforcing steel is the most significant. This exception should be deleted when and if the standard for design of masonry structures substantially increases the allowable stress in tension reinforcement.

C2.4.2 Load Combinations Including Flood Load. See Section C2.3.2. The multiplier on F_a aligns allowable stress design for flood load with strength design.

C2.4.3 Load Combinations Including Atmospheric Ice Loads. See Section C2.3.3.

C2.4.4 Load Combinations Including Self-Straining Forces and Effects. When using allowable stress design, determination of how self-straining forces and effects should be considered together with other loads should be based on the considerations discussed in Section C2.3.4. For typical situations, the following load combinations should be considered for evaluating the effects of self-straining forces and effects together with dead and live loads.

$$1.0D + 1.0T$$

$$1.0D + 0.75(L + T)$$

These combinations are not all-inclusive, and judgment is necessary in some situations. For example, where roof live loads or snow loads are significant and could conceivably occur simultaneously with self-straining forces, their effect should be included. The design should be based on the load combination causing the most unfavorable effect.

C2.5 LOAD COMBINATIONS FOR EXTRAORDINARY EVENTS

Section 2.5 advises the structural engineer that certain circumstances might require structures to be checked for low-probability events such as fire, explosions, and vehicular impact. Since the 1995 edition of ASCE Standard 7, Commentary C2.5 has provided a set of load combinations that were derived using a probabilistic basis similar to that used to develop the load combination requirements for ordinary loads in Section 2.3. In recent years, social and political events have led to an increasing desire on the part of architects, structural engineers, project developers, and regulatory authorities to enhance design and construction practices for certain buildings to provide additional structural robustness and to lessen the likelihood of disproportionate collapse if an abnormal event were to occur. Several federal, state, and local agencies have adopted policies that require new buildings and structures to be constructed with such enhancements of structural robustness (GSA 2003; DOD 2009). Robustness typically is assessed by notional removal of key load-bearing structural elements, followed by a structural analysis to assess the ability of the structure to bridge over the damage (often denoted alternative path analysis). Concurrently, advances in structural engineering for fire conditions (e.g., AISC 2010, Appendix 4) raise the prospect that new structural design requirements for fire safety will supplement the existing deemed-to-satisfy provisions in the next several years. To meet these needs, the load combinations for extraordinary events were moved in ASCE 7-10 (ASCE 2010) to Section 2.5 of this standard from Commentary C2.5, where they appeared in previous editions.

These provisions are not intended to supplant traditional approaches to ensure fire endurance based on standardized time–temperature curves and code-specified endurance times. Current code-specified endurance times are based on the ASTM E119 (2014) time–temperature curve under full allowable design load.

Extraordinary events arise from service or environmental conditions that traditionally are not considered explicitly in design of ordinary buildings and other structures. Such events are characterized by a low probability of occurrence and usually a short duration. Few buildings are ever exposed to such events, and statistical data to describe their magnitude and structural effects are rarely available. Included in the category of extraordinary events would be fire, explosions of volatile liquids or natural gas in building service systems, sabotage, vehicular impact, misuse by building occupants, subsidence (not settlement) of subsoil, and tornadoes. The occurrence of any of these events is likely to lead to structural damage or failure. If the structure is not properly designed and detailed, this local failure may initiate a chain reaction of failures that propagates throughout a major portion of the structure and leads to a potentially catastrophic partial or total collapse. Although all buildings are susceptible to such collapses in varying degrees, construction that lacks inherent continuity and ductility is particularly vulnerable (Taylor 1975; Breen and Siess 1979; Carper and Smilowitz 2006; Nair 2006; NIST 2007).

Good design practice requires that structures be robust and that their safety and performance not be sensitive to uncertainties in loads, environmental influences, and other situations not explicitly considered in design. The structural system should be designed in such a way that if an extraordinary event occurs, the probability of damage disproportionate to the original event is sufficiently small (Carper and Smilowitz 2006; NIST 2007). The philosophy of designing to limit the spread of damage rather than to prevent damage entirely is different from the traditional approach to designing to withstand dead, live, snow, and wind loads but is similar to the philosophy adopted in modern earthquake-resistant design.

In general, structural systems should be designed with sufficient continuity and ductility that alternate load paths can develop after individual member failure so that failure of the structure as a whole does not ensue. At a simple level, continuity can be achieved by requiring development of a minimum tie force, say 20 kN/m (1.37 kip/ft) between structural elements (NIST 2007). Member failures may be controlled by protective measures that ensure that no essential load-bearing member is made ineffective as a result of an accident, although this approach may be more difficult to implement. Where member failure would inevitably result in a disproportionate collapse, the member should be designed for a higher degree of reliability (NIST 2007).

Design limit states include loss of equilibrium as a rigid body, large deformations leading to significant second-order effects, yielding or rupture of members or connections, formation of a mechanism, and instability of members or the structure as a whole. These limit states are the same as those considered for other load events, but the load-resisting mechanisms in a damaged structure may be different, and sources of load-carrying capacity that normally would not be considered in ordinary ultimate limit states design, such as arch, membrane, or catenary action, may be included. The use of elastic analysis underestimates the load-carrying capacity of the structure (Marjanishvili and Agnew 2006). Materially or geometrically nonlinear or plastic analyses may be used, depending on the response of the structure to the actions.

Specific design provisions to control the effect of extraordinary loads and risk of progressive failure are developed with a probabilistic basis (Ellingwood and Leyendecker 1978; Ellingwood and Corotis 1991; Ellingwood and Dusenberry 2005). One can either reduce the likelihood of the extraordinary event or design the structure to withstand or absorb damage from the event if it occurs. Let F be the event of failure (damage or collapse) and A be the event that a structurally damaging event occurs. The probability of failure due to event A is

$$P_f = P(F|A)P(A) \qquad (C2.5\text{-}1)$$

in which $P(F|A)$ is the conditional probability of failure of a damaged structure and $P(A)$ is the probability of occurrence of event A. The separation of $P(F|A)$ and $P(A)$ allows one to focus on strategies for reducing risk. $P(A)$ depends on siting, controlling the use of hazardous substances, limiting access, and other actions that are essentially independent of structural design. In contrast, $P(F|A)$ depends on structural design measures ranging from minimum provisions for continuity to a complete postdamage structural evaluation.

The probability, $P(A)$, depends on the specific hazard. Limited data for severe fires, gas explosions, bomb explosions, and vehicular collisions indicate that the event probability depends on building size, measured in dwelling units or square footage, and ranges from about 0.2×10^{-6}/dwelling unit/year to about 8.0×10^{-6}/dwelling unit/year (NIST 2007). Thus, the probability that a building structure is affected may depend on the number of dwelling units (or square footage) in the building. If one were to set the conditional limit state probability, $P(F|A) = 0.05$–0.10, however, the annual probability of structural failure from Eq. (C2.5-1) would be less than 10^{-6}, placing the risk in the low-magnitude background along with risks from rare accidents (Pate-Cornell 1994).

Design requirements corresponding to this desired $P(F|A)$ can be developed using first-order reliability analysis if the limit state function describing structural behavior is available (Ellingwood and Dusenberry 2005). The structural action (force or constrained deformation) resulting from extraordinary event A used in design is denoted A_k. Only limited data are available to define the frequency distribution of the load (NIST 2007; Ellingwood and Dusenberry 2005). The uncertainty in the load caused by the extraordinary event is encompassed in the selection of a conservative A_k, and thus the load factor on A_k is set equal to 1.0, as is done in the earthquake load combinations in Section 2.3. The dead load is multiplied by the factor 0.9 if it has a stabilizing effect; otherwise, the load factor is 1.2, as it is with the ordinary combinations in Sections 2.3.1 and 2.3.6. Load factors less than 1.0 on the companion actions reflect the small probability of a joint occurrence of the extraordinary load and the design live, snow, or wind load. The companion actions $0.5L$ and $0.2S$ correspond, approximately, to the mean of the yearly maximum live and snow loads (Chalk and Corotis 1980; Ellingwood 1981). The companion action in Eq. (2.5-1) includes only snow load because the probability of a coincidence of A_k with L_r or R, which have short durations in comparison with S, is negligible. A similar set of load combinations for extraordinary events appears in *Eurocode 1* (2006).

The term $0.2W$ that previously appeared in these combinations has been removed and has been replaced by a requirement to check lateral stability. One approach for meeting this requirement, which is based on recommendations of the Structural Stability Research Council (Galambos 1998), is to apply lateral notional forces, $N_i = 0.002 \Sigma P_i$, at level i, in which ΣP_i = gravity force from Eq. (2.5-1) or (2.5-2) act at level i, in combination with the loads stipulated in Eq. (2.5-1) or (2.5-2). Note that Eq. (1.4-1) stipulates that when checking general structural integrity, the lateral forces acting on an intact structure shall equal $0.01W_x$, where W_x is the dead load at level x.

REFERENCES

American Concrete Institute (ACI). (2014). "Building code requirements for structural concrete and commentary." *ACI 318-14 318R-14*, ACI, Farmington Hills, MI.
American Institute of Steel Construction (AISC). (2010). "Specification for structural steel buildings." 14th Ed., Appendix A4, *Structural design for fire conditions*, Chicago.
AISC. (2016). "Specification for structural steel buildings," *ANSI/AISC 360-16*, Chicago.
American Iron and Steel Institute (AISI). (2016). "North American specification for the design of cold formed steel structural members." *AISI S100-16*, American Iron and Steel Institute, Washington, DC.
American Wood Council (AWC). (2015). *National design specification (NDS) for wood construction*, Leesburg, VA.
ASCE. (2005). "Minimum design loads for buildings and other structures." *ASCE/SEI 7-05*, Reston, VA.
ASCE. (2010). "Minimum design loads for buildings and other structures." *ASCE/SEI 7-10*, Reston, VA.
ASTM International. (2011). "Standard specification for structural steel shapes." *ASTM A992/A992M–11*, West Conshohocken, PA.
ASTM. (2014). "Standard test methods for fire tests of building construction and materials." *ASTM E119-14*, West Conshohocken, PA.
Aluminum Association. (2015). "Specification for aluminum structures." *Aluminum design manual*, Arlington, VA.
Breen, J. E., and Siess, C. P. (1979). "Progressive collapse—Symposium summary." *ACI J.*, 76(9), 997–1004.
Carper, K., and Smilowitz, R., eds. (2006). "Mitigating the potential for progressive disproportionate collapse." *J. Perform. Constr. Facil.*, 20(4).
Chalk, P. L., and Corotis, R. B. (1980). "Probability models for design live loads." *J. Struct. Div.*, 106(10), 2017–2033.
Department of Defense. (DOD). (2009). *Design of buildings to resist progressive collapse*, Unified Facilities Criteria (UFC) 4-023-03, July, Washington, DC.
Ellingwood, B. (1981). "Wind and snow load statistics for probabilistic design." *J. Struct. Div.*, 107(7), 1345–1350.
Ellingwood, B., and Corotis, R. B. (1991). "Load combinations for buildings exposed to fires." *Eng. J.*, 28(1), 37–44.
Ellingwood, B. R., and Dusenberry, D. O. (2005). "Building design for abnormal loads and progressive collapse." *Comput.-Aided Civ. Inf. Eng.*, 20(5), 194–205.
Ellingwood, B., and Leyendecker, E. V. (1978). "Approaches for design against progressive collapse." *J. Struct. Div.*, 104(3), 413–423.
Ellingwood, B. R., and Li, Y. (2009). "Counteracting structural loads: Treatment in ASCE Standard 7-05." *J. Struct. Eng.*, 135(1), 94–97.
Ellingwood, B., MacGregor, J. G., Galambos, T. V., and Cornell, C. A. (1982). "Probability-based load criteria: Load factors and load combinations." *J. Struct. Div.*, 108(5), 978–997.
Eurocode 1. (2006). "Actions on structures, Part 1–7: General actions–Accidental actions." *NEN-EN 1991-1-7*.
Federal Emergency Management Agency (FEMA). (2004). "NEHRP recommended provisions for the development of seismic regulations for new buildings and other structures." *FEMA Report 450*, Washington, DC.
Galambos, T. V., ed. (1998). *SSRC guide to stability design criteria for metal structures*, 5th Ed., John Wiley & Sons, New York.
Galambos, T. V., Ellingwood, B., MacGregor, J. G., and Cornell, C. A. (1982). "Probability-based load criteria: Assessment of current design practice." *J. Struct. Div.*, 108(5), 959–977.
General Services Administration. (GSA). (2003). *Progressive collapse analysis and design guidelines for new federal office buildings and major modernization projects*, General Services Administration, Washington, DC.
Kriebel, D. L., White, G. J., Mehta, K. C., and Smith, D. A., (1998). "An investigation of load factors for flood and combined wind and flood." Report prepared for Federal Emergency Management Agency, Washington, DC.
Marjanishvili, S., and Agnew, E. (2006). "Comparison of various procedures for progressive collapse analysis." *J. Perform. Constr. Facil.*, 20(4), 365–374.
Nair, R. S. (2006). "Preventing disproportionate collapse." *J. Perform. Constr. Facil.*, 20(4), 309–314.
National Institute of Standards and Technology. (NIST). (2007). "Best practices for reducing the potential for progressive collapse in buildings." *NISTIR 7396*, Gaithersburg, MD.
Pate-Cornell, E. (1994). "Quantitative safety goals for risk management of industrial facilities." *Struct. Safety*, 13(3), 145–157.
Taylor, D. A. (1975). "Progressive collapse." *Can. J. Civ. Eng.*, 2(4), 517–529.
Turkstra, C. J., and Madsen, H. O. (1980). "Load combinations in codified structural design." *J. Struct. Div.*, 106(12), 2527–2543.
The Masonry Society. (TMS). (2016). "Building code requirements and specifications for masonry structures and companion commentaries," 402–406, Longmont, CO.

CHAPTER C3
DEAD LOADS, SOIL LOADS, AND HYDROSTATIC PRESSURE

C3.1 DEAD LOADS

C3.1.2 Weights of Materials and Constructions. To establish uniform practice among designers, it is desirable to present a list of materials generally used in building construction, together with their proper weights. Many building codes prescribe the minimum weights for only a few building materials, and in other instances no guide whatsoever is furnished on this subject. In some cases, the codes are drawn up so as to leave the question of what weights to use to the discretion of the building official, without providing any authoritative guide. This practice, as well as the use of incomplete lists, has been subjected to much criticism. The solution chosen has been to present, in this commentary, an extended list that will be useful to designer and official alike. However, special cases will unavoidably arise, and authority is therefore granted in the standard for the building official to deal with them.

For ease of computation, most values are given in terms of pounds per square foot, psf (kilonewtons per square meter, kN/m^2) of given thickness in Table C3.1-1a (Table C3.1-1b). Pounds-per-cubic-foot, lb/ft^3 (kN/m^3), values, consistent with the pounds-per-square foot psf (kilonewtons per square meter, kN/m^2) values, are also presented in some cases in Table C3.1-2. Some constructions for which a single figure is given actually have a considerable range in weight. The average figure given is suitable for general use, but when there is reason to suspect a considerable deviation from this, the actual weight should be determined.

Engineers, architects, and building owners are advised to consider factors that result in differences between actual and calculated loads.

Engineers and architects cannot be responsible for circumstances beyond their control. Experience has shown, however, that conditions are encountered that, if not considered in design, may reduce the future utility of a building or reduce its margin of safety. Among them are

1. Dead loads. There have been numerous instances in which the actual weights of members and construction materials have exceeded the values used in design. Care is advised in the use of tabular values. Also, allowances should be made for such factors as the influence of formwork and support deflections on the actual thickness of a concrete slab of prescribed nominal thickness.
2. Future installations. Allowance should be made for the weight of future wearing or protective surfaces where there is a good possibility that such may be applied. Special consideration should be given to the likely types and position of partitions, as insufficient provision for partitioning may reduce the future utility of the building.

Attention is directed also to the possibility of temporary changes in the use of a building, as in the case of clearing a dormitory for a dance or other recreational purpose.

C3.1.3 Weight of Fixed Service Equipment. Fixed service equipment includes but is not limited to plumbing stacks and risers; electrical feeders; heating, ventilating, and air conditioning systems; and process equipment such as vessels, tanks, piping, and cable trays. Both the empty weight of equipment and the maximum weight of contents are treated as dead load.

Section 1.3.6 indicates that when resistance to overturning, sliding, and uplift forces is provided by dead load, the dead load shall be taken as the minimum dead load likely to be in place. Therefore, liquid contents and movable trays shall not be used to resist these forces unless they are the source of the forces. For example, the liquid in a tank contributes to the seismic mass of the tank and therefore can be used to resist the seismic uplift; however, the weight of the liquid cannot be used to resist overturning, sliding, or uplift from wind loads because the liquid may not be present during the wind event.

C3.1.4 Vegetative and Landscaped Roofs. Landscaping elements, such as soil, plants, and drainage layer materials, and hardscaping elements, such as walkways, fences, and walls, are intended to remain in place and are therefore considered dead loads. While the weight of hardscaping materials does not fluctuate, the weight of soil and drainage layer materials used to support vegetative growth is subject to significant variation because of its ability to absorb and retain water. Where the weight is additive to other loads, the dead load should be computed assuming full saturation of soil and drainage layer materials. Where the weight acts to counteract uplift forces, the dead load should be computed assuming dry unit weight of the soil and dry drainage layer materials.

Vegetative and landscaped roof areas may be able to retain more water than the condition where the soil and drainage layer materials are fully saturated. The water may result from precipitation or from irrigation of the vegetation. This additional amount of water should be considered rain load in accordance with Chapter 8 or snow load in accordance with Chapter 7 as applicable.

C3.1.5 Solar Panels. This section clarifies that solar panel-related loads, including ballasted systems that are not permanently attached, shall be considered as dead loads for all load combinations specified in Chapter 2.

C3.2 SOIL LOADS AND HYDROSTATIC PRESSURE

C3.2.1 Lateral Pressures. Table 3.2-1 includes high earth pressures, 85 pcf ($13.36 \ kN/m^2$) or more, to show that certain soils are poor backfill material. In addition, when walls are

Table C3.1-1a Minimum Design Dead Loads (psf)[a]

Component	Load (psf)
CEILINGS	
Acoustical fiberboard	1
Gypsum board (per 1/8-in. thickness)	0.55
Mechanical duct allowance	4
Plaster on tile or concrete	5
Plaster on wood lath	8
Suspended steel channel system	2
Suspended metal lath and cement plaster	15
Suspended metal lath and gypsum plaster	10
Wood furring suspension system	2.5
COVERINGS, ROOF, AND WALL	
Asbestos-cement shingles	4
Asphalt shingles	2
Cement tile	16
Clay tile (for mortar add 10 psf)	
Book tile, 2-in.	12
Book tile, 3-in.	20
Ludowici	10
Roman	12
Spanish	19
Composition:	
Three-ply ready roofing	1
Four-ply felt and gravel	5.5
Five-ply felt and gravel	6
Copper or tin	1
Corrugated asbestos-cement roofing	4
Deck, metal, 20 gauge	2.5
Deck, metal, 18 gauge	3
Decking, 2-in. wood (Douglas fir)	5
Decking, 3-in. wood (Douglas fir)	8
Fiberboard, 1/2-in.	0.75
Gypsum sheathing, 1/2-in.	2
Insulation, roof boards (per inch thickness)	
Cellular glass	0.7
Fibrous glass	1.1
Fiberboard	1.5
Perlite	0.8
Polystyrene foam	0.2
Urethane foam with skin	0.5
Plywood (per 1/8-in. thickness)	0.4
Rigid insulation, 1/2-in.	0.75
Skylight, metal frame, 3/8-in. wire glass	8
Slate, 3/16-in.	7
Slate, 1/4-in.	10
Waterproofing membranes:	
Bituminous, gravel-covered	5.5
Bituminous, smooth surface	1.5
Liquid applied	1
Single-ply, sheet	0.7
Wood sheathing (per inch thickness)	3
Wood shingles	3
FLOOR FILL	
Cinder concrete, per inch	9
Lightweight concrete, per inch	8
Sand, per inch	8
Stone concrete, per inch	12
FLOORS AND FLOOR FINISHES	
Asphalt block (2-in.), 1/2-in. mortar	30
Cement finish (1-in.) on stone–concrete fill	32
Ceramic or quarry tile (3/4-in.) on 1/2-in. mortar bed	16
Ceramic or quarry tile (3/4-in.) on 1-in. mortar bed	23
Concrete fill finish (per inch thickness)	12
Hardwood flooring, 7/8-in.	4
Linoleum or asphalt tile, 1/4-in.	1
Marble and mortar on stone–concrete fill	33

continues

Table C3.1-1a (*Continued*)

Component				Load (psf)
Slate (per mm thickness)				15
Solid flat tile on 1-in. mortar base				23
Subflooring, 3/4-in.				3
Terrazzo (1-1/2-in.) directly on slab				19
Terrazzo (1-in.) on stone–concrete fill				32
Terrazzo (1-in.), 2-in. stone concrete				32
Wood block (3-in.) on mastic, no fill				10
Wood block (3-in.) on 1/2-in. mortar base				16
FLOORS, WOOD-JOIST (NO PLASTER) DOUBLE WOOD FLOOR				
Joint sizes (in.)	12-in. spacing (psf)	16-in. spacing (psf)	24-in. spacing (psf)	
2 × 6	6	5	5	
2 × 8	6	6	5	
2 × 10	7	6	6	
2 × 12	8	7	6	
FRAME PARTITIONS				
Movable steel partitions				4
Wood or steel studs, 1/2-in. gypsum board each side				8
Wood studs, 2 × 4, unplastered				4
Wood studs, 2 × 4, plastered one side				12
Wood studs, 2 × 4, plastered two sides				20
FRAME WALLS				
Exterior stud walls:				
2 × 4 @ 16-in., 5/8-in. gypsum, insulated, 3/8-in. siding				11
2 × 6 @ 16-in., 5/8-in. gypsum, insulated, 3/8-in. siding				12
Exterior stud walls with brick veneer				48
Windows, glass, frame, and sash				8
Clay brick wythes:				
4 in.				39
8 in.				79
12 in.				115
16 in.				155
Hollow concrete masonry unit wythes:				
Wythe thickness (in inches)	4	6	8	10	12
Density of unit (105 pcf) with grout spacing as follows:					
No grout	22	24	31	37	43
48 in. o.c.		29	38	47	55
40 in. o.c.		30	40	49	57
32 in. o.c.		32	42	52	61
24 in. o.c.		34	46	57	67
16 in. o.c.		40	53	66	79
Full grout		55	75	95	115
Density of unit (125 pcf) with grout spacing as follows:					
No grout	26	28	36	44	50
48 in. o.c.		33	44	54	62
40 in. o.c.		34	45	56	65
32 in. o.c.		36	47	58	68
24 in. o.c.		39	51	63	75
16 in. o.c.		44	59	73	87
Full grout		59	81	102	123
Density of unit (135 pcf) with grout spacing as follows:					
No grout	29	30	39	47	54
48 in. o.c.		36	47	57	66
40 in. o.c.		37	48	59	69
32 in. o.c.		38	50	62	72
24 in. o.c.		41	54	67	78
16 in. o.c.		46	61	76	90
Full grout		62	83	105	127
Solid concrete masonry unit wythes (incl. concrete brick):					
Wythe thickness (in mm)	4	6	8	10	12
Density of unit (105 pcf)	32	51	69	87	105
Density of unit (125 pcf)	38	60	81	102	124
Density of unit (135 pcf)	41	64	87	110	133

[a]Weights of masonry include mortar but not plaster. For plaster, add 5 psf for each face plastered. Values given represent averages. In some cases, there is a considerable range of weight for the same construction.

Table C3.1-1b Minimum Design Dead Loads (kN/m²)[a]

Component	Load (kN/m²)
CEILINGS	
Acoustical fiberboard	0.05
Gypsum board (per mm thickness)	0.008
Mechanical duct allowance	0.19
Plaster on tile or concrete	0.24
Plaster on wood lath	0.38
Suspended steel channel system	0.10
Suspended metal lath and cement plaster	0.72
Suspended metal lath and gypsum plaster	0.48
Wood furring suspension system	0.12
COVERINGS, ROOF, AND WALL	
Asbestos-cement shingles	0.19
Asphalt shingles	0.10
Cement tile	0.77
Clay tile (for mortar add 0.48 kN/m²)	
Book tile, 51 mm	0.57
Book tile, 76 mm	0.96
Ludowici	0.48
Roman	0.57
Spanish	0.91
Composition:	
Three-ply ready roofing	0.05
Four-ply felt and gravel	0.26
Five-ply felt and gravel	0.29
Copper or tin	0.05
Corrugated asbestos-cement roofing	0.19
Deck, metal, 20 gauge	0.12
Deck, metal, 18 gauge	0.14
Decking, 51-mm wood (Douglas fir)	0.24
Decking, 76-mm wood (Douglas fir)	0.38
Fiberboard, 13 mm	0.04
Gypsum sheathing, 13 mm	0.10
Insulation, roof boards (per mm thickness)	
Cellular glass	0.0013
Fibrous glass	0.0021
Fiberboard	0.0028
Perlite	0.0015
Polystyrene foam	0.0004
Urethane foam with skin	0.0009
Plywood (per mm thickness)	0.006
Rigid insulation, 13 mm	0.04
Skylight, metal frame, 10-mm wire glass	0.38
Slate, 5 mm	0.34
Slate, 6 mm	0.48
Waterproofing membranes:	
Bituminous, gravel-covered	0.26
Bituminous, smooth surface	0.07
Liquid applied	0.05
Single-ply, sheet	0.03
Wood sheathing (per mm thickness)	
Plywood	0.0057
Oriented strand board	0.0062
Wood shingles	0.14
FLOOR FILL	
Cinder concrete, per mm	0.017
Lightweight concrete, per mm	0.015
Sand, per mm	0.015
Stone concrete, per mm	0.023
FLOORS AND FLOOR FINISHES	
Asphalt block (51 mm), 13-mm mortar	1.44
Cement finish (25 mm) on stone–concrete fill	1.53
Ceramic or quarry tile (19 mm) on 13-mm mortar bed	0.77
Ceramic or quarry tile (19 mm) on 25-mm mortar bed	1.10

continues

Table C3.1-1b (Continued)

Component	Load (kN/m²)
Concrete fill finish (per mm thickness)	0.023
Hardwood flooring, 22 mm	0.19
Linoleum or asphalt tile, 6 mm	0.05
Marble and mortar on stone–concrete fill	1.58
Slate (per mm thickness)	0.028
Solid flat tile on 25-mm mortar base	1.10
Subflooring, 19 mm	0.14
Terrazzo (38 mm) directly on slab	0.91
Terrazzo (25 mm) on stone–concrete fill	1.53
Terrazzo (25 mm), 51-mm stone concrete	1.53
Wood block (76 mm) on mastic, no fill	0.48
Wood block (76 mm) on 13-mm mortar base	0.77

FLOORS, WOOD-JOIST (NO PLASTER)
DOUBLE WOOD FLOOR

Joint sizes (mm):	305-mm spacing (kN/m²)	406-mm spacing (kN/m²)	610-mm spacing (kN/m²)
51 × 152	0.29	0.24	0.24
51 × 203	0.29	0.29	0.24
51 × 254	0.34	0.29	0.29
51 × 305	0.38	0.34	0.29

FRAME PARTITIONS

Component	Load (kN/m²)
Movable steel partitions	0.19
Wood or steel studs, 13-mm gypsum board each side	0.38
Wood studs, 51 × 102, unplastered	0.19
Wood studs, 51 × 102, plastered one side	0.57
Wood studs, 51 × 102, plastered two sides	0.96

FRAME WALLS

Component	Load (kN/m²)
Exterior stud walls:	
51 mm × 102 mm @ 406 mm, 16-mm gypsum, insulated, 10-mm siding	0.53
51 mm × 152 mm @ 406 mm, 16-mm gypsum, insulated, 10-mm siding	0.57
Exterior stud walls with brick veneer	2.30
Windows, glass, frame, and sash	0.38
Clay brick wythes:	
102 mm	1.87
203 mm	3.78
305 mm	5.51
406 mm	7.42

Hollow concrete masonry unit wythes:

Wythe thickness (in mm)	102	152	203	254	305
Density of unit (16.49 kN/m³) with grout spacing as follows:					
No grout	1.05	1.29	1.68	2.01	2.35
1,219 mm		1.48	1.92	2.35	2.78
1,016 mm		1.58	2.06	2.54	3.02
813 mm		1.63	2.15	2.68	3.16
610 mm		1.77	2.35	2.92	3.45
406 mm		2.01	2.68	3.35	4.02
Full grout		2.73	3.69	4.69	5.70
Density of unit (19.64 kN/m³) with grout spacing as follows:					
No grout	1.25	1.34	1.72	2.11	2.39
1,219 mm		1.58	2.11	2.59	2.97
1,016 mm		1.63	2.15	2.68	3.11
813 mm		1.72	2.25	2.78	3.26
610 mm		1.87	2.44	3.02	3.59
406 mm		2.11	2.78	3.50	4.17
Full grout		2.82	3.88	4.88	5.89
Density of unit (21.21 kN/m³) with grout spacing as follows:					
No grout	1.39	1.68	2.15	2.59	3.02
1,219 mm		1.70	2.39	2.92	3.45
1,016 mm		1.72	2.54	3.11	3.69
813 mm		1.82	2.63	3.26	3.83
610 mm		1.96	2.82	3.50	4.12
406 mm		2.25	3.16	3.93	4.69
Full grout		3.06	4.17	5.27	6.37

continues

Table C3.1-1b (Continued)

Component					Load (kN/m²)
Solid concrete masonry unit					
Wythe thickness (in mm)	102	152	203	254	305
Density of unit (16.49 kN/m³)	1.53	2.35	3.21	4.02	4.88
Density of unit (19.64 kN/m³)	1.82	2.82	3.78	4.79	5.79
Density of unit (21.21 kN/m³)	1.96	3.02	4.12	5.17	6.27

[a]Weights of masonry include mortar but not plaster. For plaster, add 0.24 kN/m³ for each face plastered. Values given represent averages. In some cases, there is a considerable range of weight for the same construction.

Table C3.1-2 Minimum Densities for Design Loads from Materials

Material	Density (lb/ft³)	Density (kN/m³)
Aluminum	170	27
Bituminous products		
Asphaltum	81	12.7
Graphite	135	21.2
Paraffin	56	8.8
Petroleum, crude	55	8.6
Petroleum, refined	50	7.9
Petroleum, benzine	46	7.2
Petroleum, gasoline	42	6.6
Pitch	69	10.8
Tar	75	11.8
Brass	526	82.6
Bronze	552	86.7
Cast-stone masonry (cement, stone, sand)	144	22.6
Cement, Portland, loose	90	14.1
Ceramic tile	150	23.6
Charcoal	12	1.9
Cinder fill	57	9.0
Cinders, dry, in bulk	45	7.1
Coal		
Anthracite, piled	52	8.2
Bituminous, piled	47	7.4
Lignite, piled	47	7.4
Peat, dry, piled	23	3.6
Concrete, plain		
Cinder	108	17.0
Expanded-slag aggregate	100	15.7
Haydite (burned-clay aggregate)	90	14.1
Slag	132	20.7
Stone (including gravel)	144	22.6
Vermiculite and perlite aggregate, nonload-bearing	25–50	3.9–7.9
Other light aggregate, load-bearing	70–105	11.0–16.5
Concrete, reinforced		
Cinder	111	17.4
Slag	138	21.7
Stone (including gravel)	150	23.6
Copper	556	87.3
Cork, compressed	14	2.2
Earth (not submerged)		
Clay, dry	63	9.9
Clay, damp	110	17.3
Clay and gravel, dry	100	15.7
Silt, moist, loose	78	12.3

Table C3.1-2 (Continued)

Material	Density (lb/ft³)	Density (kN/m³)
Silt, moist, packed	96	15.1
Silt, flowing	108	17.0
Sand and gravel, dry, loose	100	15.7
Sand and gravel, dry, packed	110	17.3
Sand and gravel, wet	120	18.9
Earth (submerged)		
Clay	80	12.6
Soil	70	11.0
River mud	90	14.1
Sand or gravel	60	9.4
Sand or gravel and clay	65	10.2
Glass	160	25.1
Gravel, dry	104	16.3
Gypsum, loose	70	11.0
Gypsum, wallboard	50	7.9
Ice	57	9.0
Iron		
Cast	450	70.7
Wrought	480	75.4
Lead	710	111.5
Lime		
Hydrated, loose	32	5.0
Hydrated, compacted	45	7.1
Masonry, ashlar stone		
Granite	165	25.9
Limestone, crystalline	165	25.9
Limestone, oolitic	135	21.2
Marble	173	27.2
Sandstone	144	22.6
Masonry, brick		
Hard (low absorption)	130	20.4
Medium (medium absorption)	115	18.1
Soft (high absorption)	100	15.7
Masonry, concrete[a]		
Lightweight units	105	16.5
Medium weight units	125	19.6
Normal weight units	135	21.2
Masonry grout	140	22.0
Masonry, rubble stone		
Granite	153	24.0
Limestone, crystalline	147	23.1
Limestone, oolitic	138	21.7
Marble	156	24.5
Sandstone	137	21.5

continues

Table C3.1-2 (Continued)

Material	Density (lb/ft³)	Density (kN/m³)
Mortar, cement or lime	130	20.4
Particleboard	45	7.1
Plywood	36	5.7
Riprap (not submerged)		
Limestone	83	13.0
Sandstone	90	14.1
Sand		
Clean and dry	90	14.1
River, dry	106	16.7
Slag		
Bank	70	11.0
Bank screenings	108	17.0
Machine	96	15.1
Sand	52	8.2
Slate	172	27.0
Steel, cold-drawn	492	77.3
Stone, quarried, piled		
Basalt, granite, gneiss	96	15.1
Limestone, marble, quartz	95	14.9
Sandstone	82	12.9
Shale	92	14.5
Greenstone, hornblende	107	16.8
Terra cotta, architectural		
Voids filled	120	18.9
Voids unfilled	72	11.3
Tin	459	72.1
Water		
Fresh	62	9.7
Sea	64	10.1
Wood, seasoned		
Ash, commercial white	41	6.4
Cypress, southern	34	5.3
Fir, Douglas, coast region	34	5.3
Hem fir	28	4.4
Oak, commercial reds and whites	47	7.4
Pine, southern yellow	37	5.8
Redwood	28	4.4
Spruce, red, white, and Sitka	29	4.5
Western hemlock	32	5.0
Zinc, rolled sheet	449	70.5

[a]Tabulated values apply to solid masonry and to the solid portion of hollow masonry.

unyielding, the earth pressure is increased from active pressure toward earth pressure at rest, resulting in 60 pcf (9.43 kN/m³) for granular soils and 100 pcf (15.71 kN/m³) for silt and clay type soils (Terzaghi and Peck 1967). Examples of light floor systems supported on shallow basement walls mentioned in Table 3.2-1 are floor systems with wood joists and flooring and cold-formed steel joists without a cast-in-place concrete floor attached.

Expansive soils exist in many regions of the United States and may cause serious damage to basement walls unless special design considerations are provided. Expansive soils should not be used as backfill because they can exert very high pressures against walls. Special soil testing is required to determine the magnitude of these pressures. It is preferable to excavate expansive soil and backfill with nonexpansive, freely draining sands or gravels. The excavated back slope adjacent to the wall should be no steeper than 45° from the horizontal to minimize the transmission of swelling pressure from the expansive soil through the new backfill. Other special details are recommended, such as a cap of nonpervious soil on top of the backfill and provision of foundation drains. Refer to current reference books on geotechnical engineering for guidance.

C3.2.2 Uplift Loads on Floors and Foundations. If expansive soils are present under floors or footings, large pressures can be exerted and must be resisted by special design. Alternatively, the expansive soil can be removed and replaced with nonexpansive material. A geotechnical engineer should make recommendations in these situations.

REFERENCE

Terzaghi, K., and Peck, R. B. (1967). *Soil mechanics in engineering practice*, 2nd Ed., John Wiley & Sons, New York.

CHAPTER C4
LIVE LOADS

C4.3 UNIFORMLY DISTRIBUTED LIVE LOADS

C4.3.1 Required Live Loads. A selected list of loads for occupancies and uses more commonly encountered is given in Section 4.3.1, and the Authority Having Jurisdiction should approve on occupancies not mentioned. Tables C4.3-1 and C4.3-2 are offered as a guide in the exercise of such authority.

In selecting the occupancy and use for the design of a building or a structure, the building owner should consider the possibility of later changes of occupancy involving loads heavier than originally contemplated. The lighter loading appropriate to the first occupancy should not necessarily be selected. The building owner should ensure that a live load greater than that for which a floor or roof is approved by the Authority Having Jurisdiction is not placed, or caused or permitted to be placed, on any floor or roof of a building or other structure.

To solicit specific informed opinion regarding the design loads in Table 4.3-1, a panel of 25 distinguished structural engineers was selected. A Delphi (Corotis et al. 1981) was conducted with this panel in which design values and supporting reasons were requested for each occupancy type. The information was summarized and recirculated back to the panel members for a second round of responses. Those occupancies for which previous design loads were reaffirmed and those for which there was consensus for change were included.

It is well known that the floor loads measured in a live load survey usually are well below present design values (Peir and Cornell 1973; McGuire and Cornell 1974; Sentler 1975; Ellingwood and Culver 1977). However, buildings must be designed to resist the maximum loads they are likely to be subjected to during some reference period T, frequently taken as 50 years. Table C4.3-2 briefly summarizes how load survey data are combined with a theoretical analysis of the load process for some common occupancy types and illustrates how a design load might be selected for an occupancy not specified in Table 4.3-1 (Chalk and Corotis 1980). The floor load normally present for the intended functions of a given occupancy is referred to as the sustained load. This load is modeled as constant until a change in tenant or occupancy type occurs. A live load survey provides the statistics of the sustained load. Table C4.3-2 gives the mean, m_s, and standard deviation, σ_s, for particular reference areas. In addition to the sustained load, a building is likely to be subjected to a number of relatively short-duration, high-intensity, extraordinary, or transient loading events (caused by crowding in special or emergency circumstances, concentrations during remodeling, and the like). Limited survey information and theoretical considerations lead to the means, m_t, and standard deviations, σ_t, of single transient loads shown in Table C4.3-2.

Combination of the sustained load and transient load processes, with due regard for the probabilities of occurrence, leads to statistics of the maximum total load during a specified reference period T. The statistics of the maximum total load depend on the average duration of an individual tenancy, τ, the mean rate of occurrence of the transient load, ν_e, and the reference period, T. Mean values are given in Table C4.3-2. The mean of the maximum load is similar, in most cases, to Table 4.3-1 values of minimum uniformly distributed live loads and, in general, is a suitable design value.

The 150-psf (7.18 kN/m^2) floor loading is also applicable to typical file cabinet installations, provided that the 36-in. (0.92 m) minimum aisle width is maintained. Five-drawer lateral or conventional file cabinets, even with two levels of bookshelves stacked above them, are unlikely to exceed the 150-psf (7.18 kN/m^2) average floor loading unless all drawers and shelves are filled to capacity with maximum density paper. Such a condition is essentially an upper bound for which the normal load factors and safety factors applied to the 150-psf (7.18 kN/m^2) criterion should still provide a safe design.

If a library shelving installation does not fall within the parameter limits that are specified in Section 4.13, then the design should account for the actual conditions. For example, the floor loading for storage of medical X-ray film may easily exceed 200 psf (9.58 kN/m^2), mainly because of the increased depth of the shelves. Mobile library shelving that rolls on rails should also be designed to meet the actual requirements of the specific installation, which may easily exceed 300 psf (14.4 kN/m^2). The rail support locations and deflection limits should be considered in the design, and the engineer should work closely with the system manufacturer to provide a serviceable structure.

For the 2010 version of the standard, the provision in the live load table for "Marquees" with its distributed load requirement of 75 psf (3.59 kN/m^2) was removed, along with "Roofs used for promenade purposes" and its 60-psf (2.87 kN/m^2) loading. Both "marquee" and "promenade" are considered archaic terms that are not used elsewhere in the standard or in building codes, with the exception of the listings in the live load tables. "Promenade purposes" is essentially an assembly use and is more clearly identified as such.

"Marquee" has not been defined in this standard but has been defined in building codes as a roofed structure that projects into a public right of way. However, the relationship between a structure and a right of way does not control loads that are applied to a structure. The marquee should therefore be designed with all of the loads appropriate for a roofed structure. If the arrangement of the structure is such that it invites additional occupant loading (e.g., there is window access that might invite loading for spectators of a parade), balcony loading should be considered for the design.

Balconies and decks are recognized as often having distinctly different loading patterns than most interior rooms. They are often subjected to concentrated live loads from people

Table C4.3-1 Minimum Uniformly Distributed Live Loads

Occupancy or Use	Live Load lb/ft² (kN/m²)	Occupancy or Use	Live Load lb/ft² (kN/m²)
Air conditioning (machine space)	200[a] (9.58)	Laboratories, scientific	100 (4.79)
Amusement park structure	100[a] (4.79)	Laundries	150[a] (7.18)
Attic, nonresidential		Manufacturing, ice	300 (14.36)
Nonstorage	25 (1.20)	Morgue	125 (6.00)
Storage	80[a] (3.83)	Printing plants	
Bakery	150 (7.18)	Composing rooms	100 (4.79)
Boathouse, floors	100[a] (4.79)	Linotype rooms	100 (4.79)
Boiler room, framed	300[a] (14.36)	Paper storage	[e]
Broadcasting studio	100 (4.79)	Press rooms	150[a] (7.18)
Ceiling, accessible furred	10[b] (0.48)	Railroad tracks	[f]
Cold storage		Ramps	
No overhead system	250[c] (11.97)	Seaplane (see Hangars)	
Overhead system		Restrooms	60 (2.87)
Floor	150 (7.18)	Rinks	
Roof	250 (11.97)	Ice skating	250 (11.97)
Computer equipment	150[a] (7.18)	Roller skating	100 (4.79)
Courtrooms	50–100 (2.40–4.79)	Storage, hay or grain	300[a] (14.36)
Dormitories		Theaters	
Nonpartitioned	80 (3.83)	Dressing rooms	40 (1.92)
Partitioned	40 (1.92)	Gridiron floor or fly gallery:	
Elevator machine room	150[a] (7.18)	Grating	60 (2.87)
Fan room	150[a] (7.18)	Well beams	250 lb/ft (3.65 kN/m) per pair
Foundries	600[a] (28.73)	Header beams	1,000 lb/ft (14.60 kN/m)
Fuel rooms, framed	400 (19.15)	Pin rail	250 lb/ft (3.65 kN/m)
Greenhouses	150 (7.18)	Projection room	100 (4.79)
Hangars	150[d] (7.18)	Toilet rooms	60 (2.87)
Incinerator charging floor	100 (4.79)	Transformer rooms	200[a] (9.58)
Kitchens, other than domestic	150[a] (7.18)	Vaults, in offices	250[a] (11.97)

[a]Use weight of actual equipment or stored material when greater. Note that fixed service equipment is treated as a dead load instead of live load.
[b]Accessible ceilings normally are not designed to support persons. The value in this table is intended to account for occasional light storage or suspension of items. If it may be necessary to support the weight of maintenance personnel, this shall be provided for.
[c]Plus 150 lb/ft² (7.18 kN/m²) for trucks.
[d]Use American Association of State Highway and Transportation Officials lane loads. Also subject to not less than 100% maximum axle load.
[e]Paper storage 50 lb/ft² per foot (2.395 kN/m² per meter) of clear story height.
[f]As required by railroad company.

Table C4.3-2 Typical Live Load Statistics

	Survey Load		Transient Load		Temporal Constants			
	m_s	σ_s[a]	m_t[a]	σ_t[a]	τ_s[b]	ν_e[c]	T[d]	
Occupancy or Use	lb/ft² (kN/m²)	lb/ft² (kN/m²)	lb/ft² (kN/m²)	lb/ft² (kN/m²)	(years)	(per year)	(years)	Mean Maximum Load[a] lb/ft² (kN/m²)
Office buildings: offices	10.9 (0.52)	5.9 (0.28)	8.0 (0.38)	8.2 (0.39)	8	1	50	55 (2.63)
Residential								
Renter occupied	6.0 (0.29)	2.6 (0.12)	6.0 (0.29)	6.6 (0.32)	2	1	50	36 (1.72)
Owner occupied	6.0 (0.29)	2.6 (0.12)	6.0 (0.29)	6.6 (0.32)	10	1	50	38 (1.82)
Hotels: guest rooms	4.5 (0.22)	1.2 (0.06)	6.0 (0.29)	5.8 (0.28)	5	20	50	46 (2.2)
Schools: classrooms	12.0 (0.57)	2.7 (0.13)	6.9 (0.33)	3.4 (0.16)	1	1	100	34 (1.63)

[a]For 200 ft² (18.58 m²) area, except 1,000 ft² (92.9 m²) for schools.
[b]Duration of average sustained load occupancy.
[c]Mean rate of occurrence of transient load.
[d]Reference period.

congregating along the edge of the structure (e.g., for viewing vantage points). This loading condition is acknowledged in Table 4.3-1 as an increase of the live load for the area served, up to the point of satisfying the loading requirement for most assembly occupancies. As always, the designer should be aware of potential unusual loading patterns in the structure that are not covered by these minimum standards.

The minimum live loads applicable to roofs with vegetative and landscaped areas are dependent upon the use of the roof area. The 20-psf (0.96 kN/m²) live load for unoccupied areas is the same load as for typical roof areas and is intended to represent the loads caused by maintenance activities and small decorative appurtenances. The 100-psf (4.79 kN/m²) live load for roof assembly areas is the same as prescribed for interior building areas because the potential for a dense grouping of occupants is similar. Other occupancies within green roof areas should have the same minimum live load as specified in Table 4.3-1 for that occupancy. Soil and walkways, fences, walls, and other hardscaping features are considered dead loads in accordance with Section 3.1.4.

C4.3.2 Provision for Partitions. The 2005 version of the standard provided the minimum partition load for the first time, although the requirement for the load has been included for many years. Historically a value of 20 psf (0.96 kN/m^2) has been required by building codes. This load, however, has sometimes been treated as a dead load.

Assuming that a normal partition would be a stud wall with ½-in. (13-mm) gypsum board on each side, 8 psf (0.38 kN/m^2) per Table C3.1-1, 10 ft (3.05 m) high, a wall load on the floor would be 80 lb/ft (1.16 kN/m). If the partitions are spaced throughout the floor area creating rooms on a grid 10 ft (3.05 m) on center, which would be an extremely dense spacing over a whole bay, the average distributed load would be 16 psf (0.77 kN/m^2). A design value of 15 psf (0.72 kN/m^2) is judged to be reasonable in that the partitions are not likely to be spaced this closely over large areas. Designers should consider a larger design load for partitions if a high density of partitions is anticipated.

C4.3.3 Partial Loading. It is intended that the full intensity of the appropriately reduced live load over portions of the structure or member be considered, as well as a live load of the same intensity over the full length of the structure or member.

Partial-length loads on a simple beam or truss produce higher shear on a portion of the span than a full-length load. "Checkerboard" loadings on multistory, multipanel bents produce higher positive moments than full loads, whereas loads on either side of a support produce greater negative moments. Loads on the half span of arches and domes or on the two central quarters can be critical.

For roofs, all probable load patterns should be considered uniform for roof live loads that are reduced to less than 20 lb/ft^2 (0.96 kN/m^2) using Section 4.8. Where the full value of the roof live load (L_r) is used without reduction, it is considered that there is a low probability that the live load created by maintenance workers, equipment, and material could occur in a patterned arrangement. Where a uniform roof live load is caused by occupancy, partial or pattern loading should be considered regardless of the magnitude of the uniform load. Cantilevers must not rely on a possible live load on the anchor span for equilibrium.

C4.4 CONCENTRATED LIVE LOADS

The provision in Table 4.3-1 regarding concentrated loads supported by roof trusses or other primary roof members is intended to provide for a common situation for which specific requirements are generally lacking.

Primary roof members are main structural members such as roof trusses, girders, and frames, which are exposed to a work floor below, where the failure of such a primary member resulting from their use as attachment points for lifting or hoisting loads could lead to the collapse of the roof. Single roof purlins or rafters (where there are multiple such members placed side by side at some reasonably small center-to-center spacing, and where the failure of a single such member would not lead to the collapse of the roof) are not considered to be primary roof members.

C4.5 LOADS ON HANDRAIL, GUARDRAIL, GRAB BAR, AND VEHICLE BARRIER SYSTEMS, AND ON FIXED LADDERS

C4.5.1 Handrail and Guardrail Systems. Loads that can be expected to occur on handrail and guardrail systems are highly dependent on the use and occupancy of the protected area. For cases in which extreme loads can be anticipated, such as long, straight runs of guardrail systems against which crowds can surge, appropriate increases in loading shall be considered.

C4.5.2 Grab Bar Systems. When grab bars are provided for use by persons with physical disabilities, the design is governed by ICC/ANSI A117.1 *Accessible and Usable Buildings and Facilities* (2009).

C4.5.3 Vehicle Barrier Systems. Vehicle barrier systems may be subjected to horizontal loads from moving vehicles. These horizontal loads may be applied normal to the plane of the barrier system, parallel to the plane of the barrier system, or at any intermediate angle. Loads in garages accommodating trucks and buses may be obtained from the provisions contained in AASHTO *LRFD Bridge Design Specifications*, 7th Edition, 2014, with 2015 interim revisions.

C4.5.4 Fixed Ladders. This provision was introduced to the standard in 1998 and is consistent with the provisions for stairs.

Side rail extensions of fixed ladders are often flexible and weak in the lateral direction. OSHA (2014a) requires side rail extensions, only with specific geometric requirements. The load provided was introduced to the standard in 1998 and has been determined on the basis of a 250-lb (1.11 kN) person standing on a rung of the ladder and accounting for reasonable angles of pull on the rail extension.

C4.6 IMPACT LOADS

Grandstands, stadiums, and similar assembly structures may be subjected to loads caused by crowds swaying in unison, jumping to their feet, or stomping. Designers are cautioned that the possibility of such loads should be considered.

Elevator loads are changed in the standard from a direct 100% impact factor to a reference to ASME A17 (2013). The provisions in ASME A17 include the 100% impact factor, along with deflection limits on the applicable elements.

C4.6.4 Elements Supporting Hoists for Façade Access and Building Maintenance Equipment. The Occupational Safety and Health Administration (OSHA) requires that façade access platforms that are used for building maintenance meet the requirements of Standard 1910.66, *Powered Platforms for Building Maintenance* (OSHA 2014b). OSHA requires that building anchors and components be capable of sustaining without failure a load of at least four times the rated load of the hoist (i.e., the maximum anticipated load or total weight of the suspended platform plus occupants and equipment) applied or transmitted to the components and anchors. A design live load of 2.5 times the rated load, when combined with a live load factor of 1.6, results in a total factored load of 4.0 times the rated load, which matches OSHA's requirements. It should also be noted that when using allowable stress design (ASD), 2.5 times the rated load will result in a comparable design when a safety factor of 1.6 is used in determining the allowable stresses. This load requirement is not statistically based but is intended by OSHA to address accidental hang-up-and-fall scenarios as well as starting and stopping forces that the platforms experience on a day-to-day basis. It also provides a small margin of safety relative to situations where a suspended platform gets hung up on a façade while ascending, allowing the hoists to apply large forces on the supporting elements. OSHA permits hoists to generate in-service forces up to three times their rated loads. These loads should be applied in the same direction(s) as they are expected to occur.

OSHA (2014c) provisions (CFR 1926.451) related to "construction" activities also require supporting equipment to be able to carry at least 1.5 times the stall load of the supported hoist. Since OSHA defines "construction" rather broadly (it includes activities such as painting and hanging signs), most equipment is used for "construction" work, which means that it must have the strength required by OSHA construction provisions. The stall load times the live load factor of 1.6 slightly exceeds the OSHA 1.5 times the stall load requirement.

C4.6.5 Fall Arrest and Lifeline Anchorages. The Occupational Safety and Health Administration (OSHA) requires that lifeline anchorages be capable of sustaining without failure an ultimate load of 5,000 lb (22.2 kN) for each attached person. Using a design live load of 3,100 lb (13.8 kN), when combined with a live load factor of 1.6, results in a total factored load of 4,960 lb (22.1 kN), which essentially matches OSHA's requirements for lifeline anchorages. It should also be noted that when using ASD, a design live load of 3,100 lb (13.8 kN) results in a comparable design when a safety factor of 1.6 is used in determining the allowable stresses. This lifeline load is intended by OSHA to address the fall arrest loads that can and do reasonably occur in typical lanyards for body harnesses, which are highly variable.

C4.7 REDUCTION IN UNIFORM LIVE LOADS

C4.7.1 General. The concept of, and methods for, determining member live load reductions as a function of a loaded member's influence area, A_I, was first introduced into this standard in 1982 and was the first such change since the concept of live load reduction was introduced more than 40 years ago. The revised formula is a result of more extensive survey data and theoretical analysis (Harris et al. 1981). The change in format to a reduction multiplier results in a formula that is simple and more convenient to use. The use of influence area, now defined as a function of the tributary area, A_T, in a single equation has been shown to give more consistent reliability for the various structural effects. The influence area is defined as that floor area over which the influence surface for structural effects is significantly different from zero.

The factor K_{LL} is the ratio of the influence area (A_I) of a member to its tributary area (A_T), that is, $K_{LL} = A_I/A_T$, and is used to better define the influence area of a member as a function of its tributary area. Fig. C4.7-1 illustrates typical influence areas and tributary areas for a structure with regular bay spacings. Table 4.7-1 has established K_{LL} values (derived from calculated K_{LL} values) to be used in Eq. (4.7-1) for a variety of structural members and configurations. Calculated K_{LL} values vary for column and beam members that have adjacent cantilever construction, as is shown in Fig. C4.7-1, and the Table 4.7-1 values have been set for these cases to result in live load reductions that

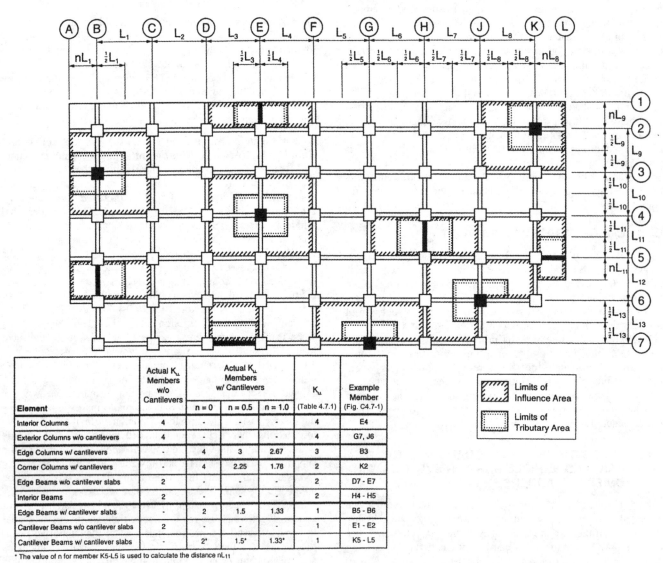

FIGURE C4.7-1 Typical Tributary and Influence Areas

are slightly conservative. For unusual shapes, the concept of significant influence effect should be applied.

An example of a member without provisions for continuous shear transfer normal to its span would be a precast T-beam or double-T beam that may have an expansion joint along one or both flanges or that may have only intermittent weld tabs along the edges of the flanges. Such members do not have the ability to share loads located within their tributary areas with adjacent members, thus resulting in $K_{LL} = 1$ for these types of members. Reductions are permissible for two-way slabs and for beams, but care should be taken in defining the appropriate influence area. For multiple floors, areas for members supporting more than one floor are summed.

The formula provides a continuous transition from unreduced to reduced loads. The smallest allowed value of the reduction multiplier is 0.4 (providing a maximum 60% reduction), but there is a minimum of 0.5 (providing a 50% reduction) for members with a contributory load from just one floor.

C4.7.3 Heavy Live Loads.
In the case of occupancies involving relatively heavy basic live loads, such as storage buildings, several adjacent floor panels may be fully loaded. However, data obtained in actual buildings indicate that rarely is any story loaded with an average actual live load of more than 80% of the average rated live load. It appears that the basic live load should not be reduced for the floor-and-beam design, but that it could be reduced up to 20% for the design of members supporting more than one floor. Accordingly, this principle has been incorporated in the recommended requirement.

C4.7.4 Passenger Vehicle Garages.
Unlike live loads in office and residential buildings, which are generally spatially random, parking garage loads are caused by vehicles parked in regular patterns, and the garages are often full. The rationale behind the reduction according to area for other live loads, therefore, does not apply. A load survey of vehicle weights was conducted at nine commercial parking garages in four cities of different sizes (Wen and Yeo 2001). Statistical analyses of the maximum load effects on beams and columns caused by vehicle loads over the garage's life were carried out using the survey results. Dynamic effects on the deck caused by vehicle motions and on the ramp caused by impact were investigated. The equivalent uniformly distributed loads (EUDL) that would produce the lifetime maximum column axial force and midspan beam bending moment are conservatively estimated at 34.8 psf (1.67 kN/m^2). The EUDL is not sensitive to bay-size variation. In view of the possible impact of very heavy vehicles in the future such as sport utility vehicles, however, a design load of 40 psf (1.95 kN/m^2) is recommended with no allowance for reduction according to bay area.

Compared with the design live load of 50 psf (2.39 kN/m^2) given in previous editions of the standard, the design load contained herein represents a 20% reduction, but it is still 33% higher than the 30 psf (1.44 kN/m^2) one would obtain were an area-based reduction to be applied to the 50 psf (2.39 kN/m^2) value for large bays as allowed in most standards. Also the variability of the maximum parking garage load effect is found to be small, with a coefficient of variation less than 5% in comparison with 20% to 30% for most other live loads. The implication is that when a live load factor of 1.6 is used in design, additional conservatism is built into it such that the recommended value would also be sufficiently conservative for special purpose parking (e.g., valet parking) where vehicles may be more densely parked, causing a higher load effect. Therefore, the 50 psf (2.39 kN/m^2) design value was felt to be overly conservative, and it can be reduced to 40 psf (1.95 kN/m^2) without sacrificing structural integrity.

In view of the large load effect produced by a single heavy vehicle (up to 10,000 lb (44.48 kN)), the current concentrated load of 2,000 lb (8.90 kN) should be increased to 3,000 lb (13.34 kN) acting on an area of 4.5 in. × 4.5 in. (0.11 m × 0.11 m), which represents the load caused by a jack in changing tires.

C4.7.6 Limitations on One-Way Slabs.
One-way slabs behave in a manner similar to two-way slabs but do not benefit from having a higher redundancy that results from two-way action. For this reason, it is appropriate to allow a live load reduction for one-way slabs but restrict the tributary area, A_T, to an area that is the product of the slab span times a width normal to the span not greater than 1.5 times the span (thus resulting in an area with an aspect ratio of 1.5). For one-way slabs with aspect ratios greater than 1.5, the effect is to give a somewhat higher live load (where a reduction has been allowed) than for two-way slabs with the same ratio.

Members, such as hollow-core slabs, that have grouted continuous shear keys along their edges and span in one direction only, are considered as one-way slabs for live load reduction, even though they may have continuous shear transfer normal to their spans.

C4.8 REDUCTION IN ROOF LIVE LOADS

C4.8.2 Ordinary Roofs, Awnings, and Canopies.
The values specified in Eq. (4.8-1) that act vertically upon the projected area have been selected as minimum roof live loads, even in localities where little or no snowfall occurs. This is because it is considered necessary to provide for occasional loading caused by the presence of workers and materials during repair operations.

C4.8.3 Occupiable Roofs.
Designers should consider any additional dead loads that may be imposed by saturated landscaping materials in addition to the live load required in Table 4.3-1. Occupancy-related loads on roofs are live loads (L) normally associated with the design of floors rather than roof live loads (L_r) and may be reduced in accordance with the provisions for live loads in Section 4.7 rather than Section 4.8.

C4.9 CRANE LOADS

All support components of moving bridge cranes and monorail cranes, including runway beams, brackets, bracing, and connections, shall be designed to support the maximum wheel load of the crane and the vertical impact, lateral, and longitudinal forces induced by the moving crane. Also, the runway beams shall be designed for crane stop forces. The methods for determining these loads vary depending on the type of crane system and support. MHI (2009, 2010a,b) and MBMA (2012) describe types of bridge cranes and monorail cranes. Cranes described in these references include top running bridge cranes with top running trolley, underhung bridge cranes, and underhung monorail cranes. AIST (2003) gives more stringent requirements for crane runway designs that are more appropriate for higher capacity or higher speed crane systems.

C4.11 HELIPAD LOADS

C4.11.1 General.
Helipad provisions were added to the standard in 2010. For the standard, the term "helipads" is used to refer specifically to the structural surface. In building codes and other references, different terminology may be used when describing helipads, e.g., heliports, helistops, but the distinctions between these are not relevant to the structural loading issue addressed in ASCE 7.

Although these structures are intended to be specifically kept clear of nonhelicopter occupant loads on the landing and taxi areas, the uniform load requirement is a minimum to ensure a degree of substantial construction and the potential to resist the effects of unusual events.

Additional information on helipad design can be found in Annex 14 to the Convention on International Civil Aviation, Aerodromes, Volume II (ICAO 2013).

C4.11.2 Concentrated Helicopter Loads. Concentrated loads applied separately from the distributed loads are intended to cover the primary helicopter loads. The designer should always consider the geometry of the design basis helicopter for applying the design loads. A factor of 1.5 is used to address impact loads (two single concentrated loads of 0.75 times the maximum takeoff weight) to account for a hard landing with many kinds of landing gear. The designer should be aware that some helicopter configurations, particularly those with rigid landing gear, could result in substantially higher impact factors that should be considered.

The 3,000-lb (13.35-kN) concentrated load is intended to cover maintenance activities, similar to the jack load for a parking garage.

C4.13 LIBRARY STACK ROOMS

Where library shelving installation does not fall within the parameter limits that are specified in Section 4.13 and Table 4.3-1, the design should account for the actual conditions. For example, the floor loading for storage of medical X-ray film may easily exceed 200 psf (9.58 kN/m^2), mainly because of the increased depth of the shelves. Mobile library shelving that rolls on rails should also be designed to meet the actual requirements of the specific installation, which may easily exceed 300 psf (14.4 kN/m^2). The rail support locations and deflection limits should be considered in the design, and the engineer should work closely with the system manufacturer to provide a serviceable structure.

C4.14 SEATING FOR ASSEMBLY USES

The lateral loads apply to "stadiums and arenas" and to "reviewing stands, grandstands, and bleachers." However, it does not apply to "gymnasiums—main floors and balconies." Consideration should be given to treating gymnasium balconies that have stepped floors for seating as arenas, and requiring the appropriate swaying forces.

C4.17 SOLAR PANEL LOADS

C4.17.1 Roof Loads at Solar Panels. These provisions are added to the 2016 edition of the standard to address the installation of rooftop solar panels consistent with current practices (Blaney and LaPlante 2013). These provisions allow the offset of roof live load where the space below the solar panel is considered inaccessible. The dimension of 24 in. (610 mm) was chosen as the clear vertical distance as it is consistent with existing published requirements for solar panel systems and is also a typical minimum height permitted for access into or out of spaces.

C4.17.3 Open-Grid Roof Structures Supporting Solar Panels. This section reduces the uniform roof live load for building structures such as carports and shade structures, which do not include roof deck or sheathing, to the value of the minimum uniform roof live load permitted by Section 4.8.2. The concentrated roof live load requirement in Table 4.3-1 is not modified by this section.

REFERENCES

American Association of State Highway and Transportation Officials (AASHTO). *LRFD bridge design specifications*, 7th Ed., 2014, with 2015 interim revisions, AASHTO, Washington, DC.

American Society of Mechanical Engineers (ASME). (2013). *American national standard safety code for elevators and escalators*. ASME, New York, A17.

Association of Iron and Steel Technology (AIST). (2003). "Guide for the design and construction of mill buildings." *Tech. Report No. 13*, AISE, Warrendale, PA.

Blaney, C., and LaPlante, R. (2013). "Recommended design live loads for rooftop solar arrays." *Proc., SEAOC Convention*, 264–278.

Chalk, P. L., and Corotis, R. B. (1980). "Probability model for design live loads." *J. Struct. Div.*, 106(10), 2017–2033.

Corotis, R. B., Harris, J. C., and Fox, R. R. (1981). "Delphi methods: Theory and design load application." *J. Struct. Div.*, 107(6), 1095–1105.

Ellingwood, B. R., and Culver, C. G. (1977). "Analysis of live loads in office buildings." *J. Struct. Div.*, 103(8), 1551–1560.

Harris, M. E., Bova, C. J., and Corotis, R. B. (1981). "Area-dependent processes for structural live loads." *J. Struct. Div.*, 107(5), 857–872.

International Civil Aviation Organization (ICAO). (2013). *Annex 14 to the convention on international civil aviation, aerodromes, Vol. II: Heliports.*

ICC/ANSI. (2009). "Accessible and Usable Buildings and Facilities." *A117.1*. International Code Council, Washington, D.C.

McGuire, R. K., and Cornell, C. A. (1974). "Live load effects in office buildings." *J. Struct. Div.*, 100(7), 1351–1366.

Metal Building Manufacturers Association (MBMA). (2012). *Metal building systems manual*, MBMA, Cleveland, OH.

MHI. (2010a). "Specifications for top running bridge and gantry type multiple girder electric overhead traveling cranes." *No. 70-2010*, MHI, Charlotte, NC.

MHI (2010b). "Specifications for top running and under running single girder electric overhead traveling cranes utilizing under running trolley hoist." *No. 74-2010*. MHI, Charlotte, NC.

MHI. (2009). "*Specifications for patented track underhung cranes and monorail systems.*" *No. MH 27.1-2009*, MHI, Charlotte, NC.

Occupational Safety and Health Administration (OSHA). (2014a). "Code of federal regulations, Section 1910.27." *OSHA Standards*, Washington, DC.

OSHA. (2014b). "Powered platforms for building maintenance." *Code of federal regulations, Section 1910.66*. OSHA Standards, Washington, DC.

OSHA. (2014c). "Safety standards for scaffolds used in the construction industry." *Code of federal regulations, Section 1926.451*. OSHA Standards, Washington, DC.

Peir, J. C., and Cornell, C. A. (1973). "Spatial and temporal variability of live loads." *J. Struct. Div.*, 99(5), 903–922.

Sentler, L. (1975). "A stochastic model for live loads on floors in buildings." *Report No. 60*, Lund Institute of Technology, Division of Building Technology, Lund, Sweden.

Wen, Y. K., and Yeo, G. L. (2001). "Design live loads for passenger cars parking garages." *J. Struct. Eng.*, 127(3), 280–289.

CHAPTER C5
FLOOD LOADS

C5.1 GENERAL

This section presents information for the design of buildings and other structures in areas prone to flooding. Design professionals should be aware that there are important differences between flood characteristics, flood loads, and flood effects in riverine and coastal areas (e.g., the potential for wave effects is much greater in coastal areas, the depth and duration of flooding can be much greater in riverine areas, the direction of flow in riverine areas tends to be more predictable, and the nature and amount of flood-borne debris varies between riverine and coastal areas).

Much of the impetus for flood-resistant design has come about from the federal government sponsored initiatives of flood-damage mitigation and flood insurance, both through the work of the U.S. Army Corps of Engineers and the National Flood Insurance Program (NFIP). The NFIP is based on an agreement between the federal government and participating communities that have been identified as being flood prone. The Federal Emergency Management Agency (FEMA), through the Federal Insurance and Mitigation Administration (FIMA), makes flood insurance available to the residents of communities provided that the community adopts and enforces adequate floodplain management regulations that meet the minimum requirements. Included in the NFIP requirements, found under Title 44 of the U.S. Code of Federal Regulations (FEMA 1999b), are minimum building design and construction standards for buildings and other structures located in special flood hazard areas (SFHAs).

Special flood hazard areas are those identified by FEMA as being subject to inundation during the 100-year flood. SFHAs are shown on flood insurance rate maps (FIRMs), which are produced for flood-prone communities. SFHAs are identified on FIRMs as zones A, A1-30, AE, AR, AO, and AH, and in coastal high hazard areas as V1-30, V, and VE. The SFHA is the area in which communities must enforce NFIP-compliant, flood damage-resistant design and construction practices.

Prior to designing a structure in a flood-prone area, design professionals should contact the local building official to determine if the site in question is located in an SFHA or other flood-prone area that is regulated under the community's floodplain management regulations. If the proposed structure is located within the regulatory floodplain, local building officials can explain the regulatory requirements.

Answers to specific questions on flood-resistant design and construction practices may be directed to the mitigation division of each of FEMA's regional offices. FEMA has regional offices that are available to assist design professionals.

C5.2 DEFINITIONS

Three new concepts were added with ASCE 7-98. First, the concept of the design flood was introduced. The design flood will, at a minimum, be equivalent to the flood having a 1% chance of being equaled or exceeded in any given year (i.e., the base flood or 100-year flood, which served as the load basis in ASCE 7-95). In some instances, the design flood may exceed the base flood in elevation or spatial extent; this excess will occur where a community has designated a greater flood (lower frequency, higher return period) as the flood to which the community will regulate new construction.

Many communities have elected to regulate to a flood standard higher than the minimum requirements of the NFIP. Those communities may do so in a number of ways. For example, a community may require new construction to be elevated a specific vertical distance above the base flood elevation (this is referred to as "freeboard"); a community may select a lower frequency flood as its regulatory flood; or a community may conduct hydrologic and hydraulic studies, upon which flood hazard maps are based, in a manner different from the Flood Insurance Study prepared by the NFIP (e.g., the community may complete flood hazard studies based upon development conditions at build-out, rather than following the NFIP procedure, which uses conditions in existence at the time the studies are completed; the community may include watersheds smaller than 1 mi^2 (2.6 km^2) in size in its analysis, rather than following the NFIP procedure, which neglects watersheds smaller than 1 mi^2 [2.6 km^2]).

Use of the design flood concept will ensure that the requirements of this standard are not less restrictive than a community's requirements where that community has elected to exceed minimum NFIP requirements. In instances where a community has adopted the NFIP minimum requirements, the design flood described in this standard will default to the base flood.

Second, this standard also uses the terms "flood hazard area" and "flood hazard map" to correspond to and show the areas affected by the design flood. Again, in instances where a community has adopted the minimum requirements of the NFIP, the flood hazard area defaults to the NFIP's SFHA and the flood hazard map defaults to the FIRM.

Third, the concept of a Coastal A-Zone is used to facilitate application of load combinations contained in Chapter 2 of this standard. Coastal A-Zones lie landward of V-Zones, or landward of an open coast shoreline where V-Zones have not been mapped (e.g., the shorelines of the Great Lakes). Coastal A-Zones are subject to the effects of waves, high-velocity flows, and erosion, although not to the extent that V-Zones are. Like V-Zones, flood forces in Coastal A-Zones will be highly correlated with coastal winds or coastal seismic activity.

Coastal A-Zones are not delineated on flood hazard maps prepared by FEMA, but are zones where wave forces and erosion potential should be taken into consideration by designers. The following guidance is offered to designers as help in determining whether or not an A-Zone in a coastal area can be considered a Coastal A-Zone.

For a Coastal A-Zone to be present, two conditions are required: (1) a still-water flood depth greater than or equal to 2.0 ft (0.61 m) and (2) breaking wave heights greater than or equal to 1.5 ft (0.46 m). Note that the still-water depth requirement is necessary, but is not sufficient by itself, to render an area a Coastal A-Zone. Many A-Zones will have still-water flood depths in excess of 2.0 ft (0.61 m), but will not experience breaking wave heights greater than or equal to 1.5 ft (0.46 m), and therefore should not be considered Coastal A-Zones. Wave heights at a given site can be determined using procedures outlined in U.S. Army Corps of Engineers (2002) or similar references.

The 1.5 ft (0.46 m) breaking wave height criterion was developed from post-flood damage inspections, which show that wave damage and erosion often occur in mapped A-Zones in coastal areas, and from laboratory tests on breakaway walls that show that breaking waves 1.5 ft (0.46 m) in height are capable of causing structural failures in wood-frame walls (FEMA 2000).

C5.3 DESIGN REQUIREMENTS

Sections 5.3.4 (dealing with A-Zone design and construction) and 5.3.5 (dealing with V-Zone design and construction) of ASCE 7-98 were deleted in preparation of the 2002 edition of this standard. These sections summarized basic principles of flood-resistant design and construction (building elevation, anchorage, foundation, below design flood elevation [DFE] enclosures, breakaway walls, etc.). Some of the information contained in these deleted sections was included in Section 5.3, beginning with ASCE 7-02, and the design professional is also referred to ASCE/SEI Standard 24 (*Flood Resistant Design and Construction*) for specific guidance.

C5.3.1 Design Loads. Wind loads and flood loads may act simultaneously at coastlines, particularly during hurricanes and coastal storms. This may also be true during severe storms at the shorelines of large lakes and during riverine flooding of long duration.

C5.3.2 Erosion and Scour. The term "erosion" indicates a lowering of the ground surface in response to a flood event, or in response to the gradual recession of a shoreline. The term "scour" indicates a localized lowering of the ground surface during a flood, due to the interaction of currents and/or waves with a structural element. Erosion and scour can affect the stability of foundations and can increase the local flood depth and flood loads acting on buildings and other structures. For these reasons, erosion and scour should be considered during load calculations and the design process. Design professionals often increase the depth of foundation embedment to mitigate the effects of erosion and scour and often site buildings away from receding shorelines (building setbacks).

C5.3.3 Loads on Breakaway Walls. Floodplain management regulations require buildings in coastal high hazard areas to be elevated to or above the design flood elevation by a pile or column foundation. Space below the DFE must be free of obstructions to allow the free passage of waves and high-velocity waters beneath the building (FEMA 1993). Floodplain management regulations typically allow space below the DFE to be enclosed by insect screening, open lattice, or breakaway walls. Local exceptions are made in certain instances for shear walls, firewalls, elevator shafts, and stairwells. Check with the Authority Having Jurisdiction for specific requirements related to obstructions, enclosures, and breakaway walls.

Where breakaway walls are used, they must meet the prescriptive requirements of NFIP regulations or be certified by a registered professional engineer or architect as having been designed to meet the NFIP performance requirements. Meeting the NFIP performance requirements should be understood to mean that the structure to which breakaway walls are attached should withstand both of the following: 1) load combinations, including flood loads acting on the structure and the breakaway walls, up to the point of breakaway wall collapse, and 2) load combinations, including flood loads acting on the structure that remains following breakaway collapse, for flood conditions between those causing breakaway wall collapse and those associated with the design flood.

The prescriptive requirements call for breakaway wall designs that are intended to collapse at loads not less than 10 psf (0.48 kN/m^2) and not more than 20 psf (0.96 kN/m^3). Inasmuch as wind or earthquake loads often exceed 20 psf (0.96 kN/m^2), breakaway walls may be designed for higher loads, provided the designer certifies that the walls have been designed to break away before base flood conditions are reached, without damaging the elevated building or its foundation. FEMA (1999a) provides guidance on how to meet the performance requirements for certification.

C5.4 LOADS DURING FLOODING

C5.4.1 Load Basis. Water loads are the loads or pressures on surfaces of buildings and structures caused and induced by the presence of floodwaters. These loads are of two basic types: hydrostatic and hydrodynamic. Impact loads result from objects transported by floodwaters striking against buildings and structures or parts thereof. Wave loads can be considered a special type of hydrodynamic load.

C5.4.2 Hydrostatic Loads. Hydrostatic loads are those caused by water either above or below the ground surface, free or confined, which is either stagnant or moves at velocities less than 5 ft/s (1.52 m/s). These loads are equal to the product of the water pressure multiplied by the surface area on which the pressure acts.

Hydrostatic pressure at any point is equal in all directions and always acts perpendicular to the surface on which it is applied. Hydrostatic loads can be subdivided into vertical downward loads, lateral loads, and vertical upward loads (uplift or buoyancy). Hydrostatic loads acting on inclined, rounded, or irregular surfaces may be resolved into vertical downward or upward loads and lateral loads based on the geometry of the surfaces and the distribution of hydrostatic pressure.

C5.4.3 Hydrodynamic Loads. Hydrodynamic loads are those loads induced by the flow of water moving at moderate to high velocity above the ground level. They are usually lateral loads caused by the impact of the moving mass of water and the drag forces as the water flows around the obstruction. Hydrodynamic loads are computed by recognized engineering methods. In the coastal high hazard area, the loads from high-velocity currents due to storm surge and overtopping are of particular importance. U.S. Army Corps of Engineers (2002) is one source of design information regarding hydrodynamic loadings.

Note that accurate estimates of flow velocities during flood conditions are very difficult to make, both in riverine and coastal flood events. Potential sources of information regarding velocities of floodwaters include local, state, and federal government agencies and consulting engineers specializing in coastal engineering, stream hydrology, or hydraulics.

As interim guidance for coastal areas, FEMA (2000) gives a likely range of flood velocities as

$$V = d_s/(1 \text{ s}) \quad \text{(C5.4-1)}$$

to

$$V = (gd_s)^{0.5} \quad \text{(C5.4-2)}$$

where

V = average velocity of water in ft/s (m/s)
d_s = local still water depth in ft (m)
g = acceleration due to gravity, 32.2 ft/s/s (9.81 m/s^2)

Selection of the correct value of a in Eq. 5.4-1 will depend upon the shape and roughness of the object exposed to flood flow, as well as the flow condition. As a general rule, the smoother and more streamlined the object, the lower the drag coefficient (shape factor). Drag coefficients for elements common in buildings and structures (round or square piles, columns, and rectangular shapes) will range from approximately 1.0 to 2.0, depending upon flow conditions. However, given the uncertainty surrounding flow conditions at a particular site, ASCE 7-05 recommends a minimum value of 1.25 be used. Fluid mechanics texts should be consulted for more information on when to apply drag coefficients above 1.25.

C5.4.4 Wave Loads. The magnitude of wave forces (lb/ft^2) (kN/m^2) acting against buildings or other structures can be 10 or more times higher than wind forces and other forces under design conditions. Thus, it should be readily apparent that elevating above the wave crest elevation is crucial to the survival of buildings and other structures. Even elevated structures, however, must be designed for large wave forces that can act over a relatively small surface area of the foundation and supporting structure.

Wave load calculation procedures in Section 5.4.4 are taken from U.S. Army Corps of Engineers (2002) and Walton et al. (1989). The analytical procedures described by Eqs. 5.4-2 through 5.4-9 should be used to calculate wave heights and wave loads unless more advanced numerical or laboratory procedures permitted by this standard are used.

Wave load calculations using the analytical procedures described in this standard all depend upon the initial computation of the wave height, which is determined using Eqs. 5.4-2 and 5.4-3. These equations result from the assumptions that the waves are depth limited and that waves propagating into shallow water break when the wave height equals 78% of the local still water depth and that 70% of the wave height lies above the local still water level. These assumptions are identical to those used by FEMA in its mapping of coastal flood hazard areas on FIRMs.

Designers should be aware that wave heights at a particular site can be less than depth-limited values in some cases (e.g., when the wind speed, wind duration, or fetch is insufficient to generate waves large enough to be limited in size by water depth, or when nearby objects dissipate wave energy and reduce wave heights). If conditions during the design flood yield wave heights at a site less than depth-limited heights, Eq. 5.4-2 may overestimate the wave height and Eq. 5.4-3 may underestimate the still-water depth. Also, Eqs. 5.4-4 through 5.4-7 may overstate wave pressures and forces when wave heights are less than depth-limited heights. More advanced numerical or laboratory procedures permitted by this section may be used in such cases, in lieu of Eqs. 5.4-2 through 5.4-7.

It should be pointed out that present NFIP mapping procedures distinguish between A-Zones and V-Zones by the wave heights expected in each zone. Generally speaking, A-Zones are designated where wave heights less than 3 ft (0.91 m) in height are expected. V-Zones are designated where wave heights equal to or greater than 3 ft (0.91 m) are expected. Designers should proceed cautiously, however. Large wave forces can be generated in some A-Zones, and wave force calculations should not be restricted to V-Zones. Present NFIP mapping procedures do not designate V-Zones in all areas where wave heights greater than 3 ft (0.91 m) can occur during base flood conditions. Rather than rely exclusively on flood hazard maps, designers should investigate historical flood damages near a site to determine whether or not wave forces can be significant.

C5.4.4.2 Breaking Wave Loads on Vertical Walls. Equations used to calculate breaking wave loads on vertical walls contain a coefficient, C_p. Walton et al. (1989) provide recommended values of the coefficient as a function of probability of exceedance. The probabilities given by Walton et al. (1989) are not annual probabilities of exceedance, but probabilities associated with a distribution of breaking wave pressures measured during laboratory wave tank tests. Note that the distribution is independent of water depth. Thus, for any water depth, breaking wave pressures can be expected to follow the distribution described by the probabilities of exceedance in Table C5.4-2.

This standard assigns values for C_p according to building category, with the most important buildings having the largest values of C_p. Category II buildings are assigned a value of C_p corresponding to a 1% probability of exceedance, which is consistent with wave analysis procedures used by FEMA in mapping coastal flood hazard areas and in establishing minimum floor elevations. Category I buildings are assigned a value of C_p corresponding to a 50% probability of exceedance, but designers may wish to choose a higher value of C_p. Category III buildings are assigned a value of C_p corresponding to a 0.2% probability of exceedance, while Category IV buildings are assigned a value of C_p corresponding to a 0.1% probability of exceedance.

Breaking wave loads on vertical walls reach a maximum when the waves are normally incident (direction of wave approach is perpendicular to the face of the wall; wave crests are parallel to the face of the wall). As guidance for designers of coastal buildings or other structures on normally dry land (i.e., flooded only during coastal storm or flood events), it can be assumed that the direction of wave approach will be approximately perpendicular to the shoreline. Therefore, the direction of wave approach relative to a vertical wall will depend upon the orientation of the wall relative to the shoreline. Section 5.4.4.4 provides a method for reducing breaking wave loads on vertical walls for waves not normally incident.

C5.4.5 Impact Loads. Impact loads are those that result from logs, ice floes, and other objects striking buildings, structures, or parts thereof. U.S. Army Corps of Engineers (1995) divides impact loads into three categories: (1) normal impact loads, which result from the isolated impacts of normally encountered objects; (2) special impact loads, which result from large objects, such as broken up ice floes and accumulations of debris, either striking or resting against a building, structure, or parts thereof; and (3) extreme impact loads, which result from very large objects, such as boats, barges, or collapsed buildings, striking the building, structure, or component under consideration. Design for extreme impact loads is not practical for most buildings and structures. However, in cases where there is a high probability that a Category III or IV structure (see Table 1.5-1) will be exposed to extreme impact loads during the design flood, and where the resulting damages will be very severe, consideration of extreme impact loads may be justified. Unlike extreme impact loads, design for special and normal impact loads is practical for most buildings and structures.

The recommended method for calculating normal impact loads has been modified beginning with ASCE 7-02. Previous editions of ASCE 7 used a procedure contained in U.S. Army Corps of Engineers (1995) (the procedure, which had been unchanged since at least 1972, relied on an impulse-momentum approach with a 1,000 lb [4.5 kN] object striking the structure at the velocity of the floodwater and coming to rest in 1.0 s). Work (Kriebel et al. 2000 and Haehnel and Daly 2001) has been conducted to evaluate this procedure, through a literature review and laboratory tests. The literature review considered riverine and coastal debris, ice floes and impacts, ship berthing and impact forces, and various methods for calculating debris loads (e.g., impulse-momentum, work-energy). The laboratory tests included log sizes ranging from 380 lb (1.7 kN) to 730 lb (3.3 kN) traveling at up to 4 ft/s (1.2 m/s).

Kriebel et al. (2000) and Haehnel and Daly (2001) conclude that (1) an impulse-momentum approach is appropriate; (2) the 1,000 lb (4.5 kN) object is reasonable, although geographic and local conditions may affect the debris object size and weight; (3) the 1.0-s impact duration is not supported by the literature or by laboratory tests—a duration of impact of 0.03 s should be used instead; (4) a half-sine curve represents the applied load and resulting displacement well; and (5) setting the debris velocity equivalent to the flood velocity is reasonable for all but the largest objects in shallow water or obstructed conditions.

Given the short-duration, impulsive loads generated by flood-borne debris, a dynamic analysis of the affected building or structure may be appropriate. In some cases (e.g., when the natural period of the building is much greater than 0.03 s), design professionals may wish to treat the impact load as a static load applied to the building or structure (this approach is similar to that used by some following the procedure contained in Section C5.3.3.5 of ASCE 7-98).

In either type of analysis—dynamic or static—Eq. C5.4-3 provides a rational approach for calculating the magnitude of the impact load.

$$F = \frac{\pi W V_b C_I C_O C_D C_B R_{max}}{2g\Delta t} \quad \text{(C5.4-3)}$$

where

F = impact force, in lb (N)
W = debris weight in lb (N)
V_b = velocity of object (assume equal to velocity of water, V) in ft/s (m/s)
g = acceleration due to gravity, 32.2 ft/s² (9.81 m/s²)
Δt = impact duration (time to reduce object velocity to zero), in s
C_I = importance coefficient (see Table C5.4-1)
C_O = orientation coefficient, 0.8
C_D = depth coefficient (see Table C5.4-2, Fig. C5.4-1)
C_B = blockage coefficient (see Table C5.4-3, Fig. C5.4-2)
R_{max} = maximum response ratio for impulsive load (see Table C5.4-4)

Table C5.4-1 Values of Importance Coefficient, C_I

Risk Category	C_I
I	0.6
II	1.0
III	1.2
IV	1.3

Table C5.4-2 Values of Depth Coefficient, C_D

Building Location in Flood Hazard Zone and Water Depth	C_D
Floodway or V-Zone	1.0
A-Zone, still-water depth >5 ft	1.0
A-Zone, still-water depth = 4 ft	0.75
A-Zone, still-water depth = 3 ft	0.5
A-Zone, still-water depth = 2 ft	0.25
Any flood zone, still-water depth <1 ft	0.0

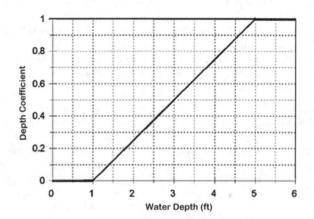

FIGURE C5.4-1 Depth Coefficient, C_D

Table C5.4-3 Values of Blockage Coefficient, C_B

Degree of Screening or Sheltering within 100 ft Upstream	C_B
No upstream screening, flow path wider than 30 ft	1.0
Limited upstream screening, flow path 20 ft wide	0.6
Moderate upstream screening, flow path 10 ft wide	0.2
Dense upstream screening, flow path less than 5 ft wide	0.0

The form of Eq. C5.4-3 and the parameters and coefficients are discussed in the following text:

Basic Equation. The equation is similar to the equation used in ASCE 7-98, except for the $\pi/2$ factor (which results from the half-sine form of the applied impulse load) and the coefficients C_I, C_O, C_D, C_B, and R_{max}. With the coefficients set equal to 1.0, the equation reduces to $F = \pi W V_b / 2g\Delta t$ and calculates the maximum static load from a head-on impact of a debris object. The coefficients have been added to allow design professionals to "calibrate" the resulting force to local flood, debris, and building characteristics. The approach is similar to that employed by ASCE 7 in calculating wind, seismic, and other loads. A scientifically based equation is used to match the physics, and the results are modified by coefficients to calculate realistic load

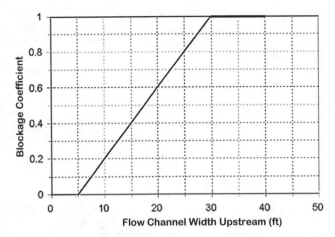

FIGURE C5.4-2 Blockage Coefficient, C_B

Table C5.4-4 Values of Response Ratio for Impulsive Loads, R_{max}

Ratio of Impact Duration to Natural Period of Structure	R_{max} (Response Ratio for Half-Sine Wave Impulsive Load)
0.00	0.0
0.10	0.4
0.20	0.8
0.30	1.1
0.40	1.4
0.50	1.5
0.60	1.7
0.70	1.8
0.80	1.8
0.90	1.8
1.00	1.7
1.10	1.7
1.20	1.6
1.30	1.6
≥1.40	1.5

Source: Adapted from Clough and Penzien (1993).

magnitudes. However, unlike wind, seismic, and other loads, the body of work associated with flood-borne debris impact loads does not yet account for the probability of impact.

Debris Object Weight. A 1,000-lb (4.5-kN) object can be considered a reasonable average for flood-borne debris (no change from ASCE 7-98). This represents a reasonable weight for trees, logs, and other large woody debris that is the most common form of damaging debris nationwide. This weight corresponds to a log approximately 30-ft (9.1-m) long and just under 1-ft (0.3-m) in diameter. The 1,000-lb (4.5-kN) object also represents a reasonable weight for other types of debris ranging from small ice floes, to boulders, to man-made objects.

However, design professionals may wish to consider regional or local conditions before the final debris weight is selected. The following text provides additional guidance. In riverine floodplains, large woody debris (trees and logs) predominates, with weights typically ranging from 1,000 lb (4.5 kN) to 2,000 lb (9.0 kN). In the Pacific Northwest, larger tree and log sizes suggest a typical 4,000 lb (18.0 kN) debris weight. Debris weights in riverine areas subject to floating ice typically range from 1,000 lb (4.5 kN) to 4,000 lb (18.0 kN). In arid or semiarid regions, typical woody debris may be less than 1,000 lb (4.5 kN). In alluvial fan areas, nonwoody debris (stones and boulders) may present a much greater debris hazard. Debris weights in coastal areas generally fall into three classes: in the Pacific Northwest, a 4,000 lb (18.0 kN) debris weight due to large trees and logs can be considered typical; in other coastal areas where piers and large pilings are available locally, debris weights may range from 1,000 lb (4.5 kN) to 2,000 lb (9.0 kN); and in other coastal areas where large logs and pilings are not expected, debris will likely be derived from failed decks, steps, and building components and will likely average less than 500 lb (2.3 kN) in weight.

Debris Velocity. The velocity with which a piece of debris strikes a building or structure will depend upon the nature of the debris and the velocity of the floodwaters. Small pieces of floating debris, which are unlikely to cause damage to buildings or other structures, will typically travel at the velocity of the floodwaters, in both riverine and coastal flood situations. However, large debris, such as trees, logs, pier pilings, and other large debris capable of causing damage, will likely travel at something less than the velocity of the floodwaters. This reduced velocity of large debris objects is due in large part to debris dragging along the bottom and/or being slowed by prior collisions. Large riverine debris traveling along the floodway (the deepest part of the channel that conducts the majority of the flood flow) is most likely to travel at speeds approaching that of the floodwaters. Large riverine debris traveling in the floodplain (the shallower area outside the floodway) is more likely to be traveling at speeds less than that of the floodwaters, for those reasons stated in the preceding text. Large coastal debris is also likely to be traveling at speeds less than that of the floodwaters. Eq. C5.4-3 should be used with the debris velocity equal to the flow velocity because the equation allows for reductions in debris velocities through application of a depth coefficient, C_D, and an upstream blockage coefficient, C_B.

Duration of Impact. A detailed review of the available literature (Kriebel et al. 2000), supplemented by laboratory testing, concluded the previously suggested 1.0-s duration of impact is much too long and is not realistic. Laboratory tests showed that measured impact durations (from initial impact to time of maximum force Δt) varied from 0.01 s to 0.05 s (Kriebel et al. 2000). Results for one test, for example, produced a maximum impact load of 8,300 lb (37,000 N) for a log weighing 730 lb (3,250 N), moving at 4 ft/s (1.2 m/s), and impacting with a duration of 0.016 s. Over all the test conditions, the impact duration averaged about 0.026 s. The recommended value for use in Eq. C5.4-3 is therefore 0.03 s.

Coefficients C_I, C_O, C_D, and C_B. The coefficients are based in part on the results of laboratory testing and in part on engineering judgment. The values of the coefficients should be considered interim, until more experience is gained with them.

The *importance coefficient*, C_I, is generally used to adjust design loads for the structure category and hazard to human life following ASCE 7-98 convention in Table 1.5-1. Recommended values given in Table C5.4-1 are based on a probability distribution of impact loads obtained from laboratory tests in Haehnel and Daly (2001).

The *orientation coefficient*, C_O, is used to reduce the load calculated by Eq. C5.4-3 for impacts that are oblique, not head on. During laboratory tests (Haehnel and Daly 2001) it was observed that while some debris impacts occurred as direct or head-on impacts that produced maximum impact loads, most impacts occurred as eccentric or oblique impacts with reduced values of the impact force. Based on these measurements, an orientation coefficient of $C_O = 0.8$ has been adopted to reflect the general load reduction observed due to oblique impacts.

The *depth coefficient*, C_D, is used to account for reduced debris velocity in shallow water due to debris dragging along the bottom. Recommended values of this coefficient are based on typical diameters of logs and trees, or on the anticipated diameter of the root mass from drifting trees that are likely to be encountered in a flood hazard zone. Kriebel et al. (2000) suggest that trees with typical root mass diameters will drag the bottom in depths of less than 5 ft (1.5 m), while most logs of concern will drag the bottom in depths of less than 1 ft (0.30 m). The recommended values for the depth coefficient are given in Table C5.4-2 and Fig. C5.4-1. No test data are available to fully validate the recommended values of this coefficient. When better data are available, designers should use them in lieu of the values contained in Table C5.4-2 and Fig. C5.4-1.

The *blockage coefficient*, C_B, is used to account for the reductions in debris velocities expected due to screening and sheltering provided by trees or other structures within about 10 log lengths (300 ft, 91.4 m) upstream from the building or structure of interest. Kriebel et al. (2000) quote other studies in which dense trees have been shown to act as a screen to remove debris and shelter downstream structures. The effectiveness of the screening depends primarily on the spacing of the upstream obstructions relative to the design log length of interest. For a 1,000-lb (453.6 kg) log, having a length of about 30 ft (9.1 m), it is therefore assumed that any blockage narrower than 30 ft (9.1 m) would trap some or all of the transported debris. Likewise, typical root mass diameters are on the order of 3 to 5 ft (0.91 to 1.5 m), and it is therefore assumed that blockages of this width would fully trap any trees or long logs. Recommended values for the blockage coefficient are given in Table C5.4-3 and Fig. C5.4-2 based on interpolation between these limits. No test data are available to fully validate the recommended values of this coefficient.

The *maximum response ratio*, R_{max}, is used to increase or decrease the computed load, depending on the degree of compliance of the building or building component being struck by debris. Impact loads are impulsive in nature, with the force rapidly increasing from zero to the maximum value in time Δt, then decreasing to zero as debris rebounds from the structure. The actual load experienced by the structure or component will depend on the ratio of the impact duration Δt relative to the natural period of the structure or component, T_n. Stiff or rigid buildings and structures with natural periods similar to the impact duration will see an amplification of the impact load. More flexible buildings and structures with natural periods greater than approximately four times the impact duration will see a reduction of the impact load. Likewise, stiff or rigid components will see an amplification of the impact load; more flexible components will see a reduction of the impact load. Successful use of Eq. C5.4-3, then, depends on estimation of the natural period of the building or component being struck by flood-borne debris. Calculating the natural period can be carried out using established methods that take building mass, stiffness, and configuration into account. One useful reference is Appendix C of ACI 349 (1985). Design professionals are also referred to Chapter 9 of ASCE 7-10 for additional information.

Natural periods of buildings generally vary from approximately 0.05 s to several seconds (for high-rise, moment frame structures). For flood-borne debris impact loads with a duration of 0.03 s, the critical period (above which loads are reduced) is approximately 0.11 s (see Table C5.4-4). Buildings and structures with natural periods above approximately 0.11 s will see a reduction in the debris impact load, while those with natural periods below approximately 0.11 s will see an increase.

Recent shake table tests of conventional, one- to two-story wood-frame buildings have shown natural periods ranging from approximately 0.14 s (7 Hz) to 0.33 s (3 Hz), averaging approximately 0.20 s (5 Hz). Elevating these types of structures for flood-resistant design purposes will act to increase these natural periods. For the purposes of flood-borne debris impact load calculations, a natural period of 0.5 to 1.0 s is recommended for one- to three-story buildings elevated on timber piles. For one- to three-story buildings elevated on masonry columns, a similar range of natural periods is recommended. For one- to three-story buildings elevated on concrete piles or columns, a natural period of 0.2 to 0.5 s is recommended. Finally, design professionals are referred to Section 12.8.2 of this standard, where an approximate natural period for 1- to 12-story buildings (story height equal to or greater than 10 ft [3 m]), with concrete and steel moment-resisting frames, can be approximated as 0.1 times the number of stories.

Special Impact Loads. U.S. Army Corps of Engineers 1995 states that, absent a detailed analysis, special impact loads can be estimated as a uniform load of 100 lb per ft (1.48 kN/m), acting over a 1 ft (0.31 m) high horizontal strip at the design flood elevation or lower. However, Kriebel et al. (2000) suggest that this load may be too small for some large accumulations of debris and suggests an alternative approach involving application of the standard drag force expression

$$F = (1/2)C_D \rho A V^2 \tag{C5.4-4}$$

where

F = drag force due to debris accumulation, in lb (N)
V = flow velocity upstream of debris accumulation, in ft/s (m/s)
A = projected area of the debris accumulation into the flow, approximated by depth of accumulation times width of accumulation perpendicular to flow, in ft² (m²)
ρ = density of water in slugs/ft³ (kg/m³)
C_D = drag coefficient = 1

This expression produces loads similar to the 100 lb/ft (1.48 kN/m) guidance from U.S. Army Corps of Engineers (1995) when the debris depth is assumed to be 1 ft (0.31 m) and when the velocity of the floodwater is 10 ft/s (3 m/s). Other guidance from Kriebel et al. (2000) and Haehnel and Daly (2001) suggests that the depth of debris accumulation is often much greater than 1 ft (0.31 m), and is only limited by the water depth at the structure. Observations of debris accumulations at bridge piers listed in these references show typical depths of 5 to 10 ft (1.5 to 3 m), with horizontal widths spanning between adjacent bridge piers whenever the spacing of the piers is less than the typical log length. If debris accumulation is of concern, the design professional should specify the projected area of the debris accumulation based on local observations and experience, and apply the preceding equation to predict the debris load on buildings or other structures.

REFERENCES

American Concrete Institute (ACI). (1985). "Code requirements for nuclear safety related concrete structures." *ANSI/ACI 349*. ACI, Farmington Hills, MI.

Clough, R. W., and Penzien, J. (1993). *Dynamics of structures*, 2nd Ed., McGraw-Hill, New York.

Federal Emergency Management Agency (FEMA). (1993). "Free-of-obstruction requirements for buildings located in coastal high hazard areas in accordance with the National Flood Insurance Program." *Technical Bulletin 5-93*. Mitigation Directorate, FEMA, Washington, DC.

FEMA. (1999a). "Design and construction guidance for breakaway walls below elevated coastal buildings in accordance with the National Flood Insurance Program." *Technical Bulletin 9-99*. Mitigation Directorate, FEMA, Washington, DC.

FEMA. (1999b). National Flood Insurance Program, *44 CFR, Ch. 1, Parts 59 and 60*, Washington, DC.

FEMA. (2000). *Coastal construction manual,* 3rd Ed. P-55, FEMA, Washington, DC.

Haehnel, R., and Daly, S. (2001). "Debris impact tests." *Report prepared for the American Society of Civil Engineers by the U.S. Army Cold Regions Research and Engineering Laboratory*, Hanover, NH.

Kriebel, D. L., Buss, L., and Rogers, S. (2000). "Impact loads from flood-borne debris." *Report to the American Society of Civil Engineers*, Reston, VA.

U.S. Army Corps of Engineers (USACE). (1995). "Flood proofing regulations." *EP 1165-2-314*, Office of the Chief of Engineers, U.S. Army Corps of Engineers, Washington, DC.

USACE. (2002). *Coastal engineering manual*, Coastal Hydraulics Laboratory, Waterways Experiment Station, U.S. Army Corps of Engineers, Washington, DC.

Walton, T. L., Jr., Ahrens, J. P., Truitt, C. L., and Dean, R. G. (1989). "Criteria for evaluating coastal flood protection structures." *Tech. Report CERC 89-15*, U.S. Army Corps of Engineers, Waterways Experiment Station, Vicksburg, MS.

CHAPTER C6
TSUNAMI LOADS AND EFFECTS

C6.1 GENERAL REQUIREMENTS

C6.1.1 Scope. The 2016 edition of the ASCE 7 Tsunami Loads and Effects chapter is applicable only to the states of Alaska, Washington, Oregon, California, and Hawaii, which are Tsunami-Prone Regions that have quantifiable probabilistic hazards resulting from tsunamigenic earthquakes with subduction faulting. The Tsunami Design Zone is the area vulnerable to being flooded or inundated by the Maximum Considered Tsunami, which is taken as having a 2% probability of being exceeded in a 50-year period, or 1:2,475 annual probability of exceedance. The Maximum Considered Tsunami constitutes the design event, consisting of the inundation depths and flow velocities taken at the stages of inflow and outflow most critical to the structure (Chock 2015, 2016).

Other causes of local tsunamis include large landslides near the coast or underwater and undersea volcanic eruptions. In Alaska, there is a history of coseismic landslides generated in fjords during major earthquakes; for this reason, the Tsunami Design Zone Maps in Alaska represent the composite hazard of the local tsunami as well as the offshore-generated tsunami. In these cases, the two tsunamis are not simultaneous.

For other states, there is insufficient analysis at present to reliably quantify the probabilistic hazard of landslide-induced local tsunami. As of 2014, Probabilistic Tsunami Hazard Analysis (PTHA) for submarine mass failure slides is ongoing for the Gulf of Mexico. As of 2014, PTHA is lacking for some other regions with historic tsunamis (such as Guam, Commonwealth of the Northern Marianas, American Samoa, Puerto Rico, and U.S. Virgin Islands), so these regions are not covered by ASCE 7-16. However, for Tsunami Risk Category III and IV buildings and other structures, it is recommended that there be consideration of performing probabilistic site-specific tsunami hazard analysis to use as a basis for implementing tsunami-resistant design with these provisions.

Table C6.1-1 provides a general assessment of the exposure of the five western states to tsunami hazard.

Because of the high hydrodynamic forces exerted by tsunamis, one- and two-family dwellings and low-rise buildings of light-frame construction do not survive significant tsunami loading, as indicated by numerous post-tsunami forensic engineering surveys of the major tsunami events of the past decade. Therefore, it is not practical to include tsunami design requirements for these types of low-rise structures. Impractical applications to low-rise light-frame construction should be avoided. As is typical in flood-prone regions, midrise light-frame construction can be built on a pedestal structure to provide sufficient elevation to enable a tsunami-resistant design.

The state or local government should determine a threshold height for where tsunami-resilient design requirements for Risk Category II buildings shall apply in accordance with the state or local statute adopting the building code with tsunami-resilient requirements. The height threshold should be chosen to be appropriate for both reasonable Life Safety and reasonable economic cost to resist tsunami loads. The threshold height would depend on their community's tsunami hazard, tsunami response procedures, and whole community disaster resilience goals. Considerations of evacuation egress time from a local community would also be a consideration. When evacuation travel times exceed the available time to tsunami arrival, there is a greater need for vertical evacuation into an ample number of nearby buildings. The inventory and design vintage of existing taller buildings that may offer some increased safety can also be a consideration for determining the shortfall of such possible refuges. In summary, multiagency stakeholders in coordination with emergency management and design and construction professionals should evaluate the tsunami risks to each local community's public safety and Critical Facilities for response and recovery from the Maximum Considered Tsunami and should use that information to rationally determine the threshold height of application of the tsunami provisions to Risk Category II buildings.

The technique demonstrated by Carden et al. (2015) can be used to evaluate the systemic level of tsunami resistance provided by the ASCE 7 seismic design requirements. Past this point of structural system parity between tsunami and seismic demand, additional investment into the lateral-force-resisting system would become necessary for tsunami resistance. For any structure, it may be necessary to provide enhanced local resistance of structural components subjected to tsunami inundation, in accordance with the provisions of Chapter 6, particularly if the tsunami is close to a shipping port. As documented in Chock et al. (2013a) and Carden et al. (2015), a 65-ft (19.8-m) building was studied for guidance principally based on six factors:

1. 65-ft (19.8-m)-tall buildings of Type I construction subject to the higher Seismic Design Category D requirements of the United States would typically have sufficient systemic strength for overall lateral tsunami forces.
2. Buildings of such height would be expected by the general public to offer safety as a matter of common sense because they would not anticipate U.S. buildings of that height collapsing in a tsunami.
3. Conversely, if a more extreme height were selected, it would imply that buildings of lesser heights are not capable of resisting tsunami effects. Then the public would have an improper justification to prohibit such building heights in the Tsunami Design Zone, resulting in an economic taking not based on technical merits, which would be contradictory to factor No. 1.
4. In the 2011 Tohoku tsunami, the ASCE Tsunami Loads and Effects Subcommittee and its Tohoku Tsunami Reconnaissance Team, in consultation with Japanese

Table C6.1-1 Exposure of the Five Western States to Tsunami Hazard

State	Population at Direct Risk[a,b]	Profile of Economic Assets and Critical Infrastructure
California	275,000 residents plus another 400,000 to 2 million tourists; 840 miles of coastline Total resident population of area at immediate risk of post-tsunami impacts[c]: 1,950,000	> $200 billion plus 3 major airports (SFO, OAK, SAN) and 1 military port, 5 very large ports, 1 large port, 5 medium ports
Oregon	25,000 residents plus another 55,000 tourists; 300 miles of coastline Total resident population of area at immediate risk of post-tsunami impacts[c]: 100,000	$8.5 billion plus essential facilities, 2 medium ports, 1 fuel depot hub
Washington	45,000 residents plus another 20,000 tourists; 160 miles of coastline Total resident population of area at immediate risk of post-tsunami impacts[c]: 900,000	$4.5 billion plus essential facilities, 1 military port, 2 very large ports, 1 large port, 3 medium ports
Hawaii	~200,000[d] residents plus another 175,000 or more tourists and approximately 1,000[d] buildings directly relating to the tourism industry; 750 miles of coastline Total resident population of area at immediate risk of post-tsunami impacts: 400,000[d]	$40 billion, plus 3 international airports, and 1 military port, 1 medium port, 4 other container ports, and 1 fuel refinery intake port, 3 regional power plants; 100 government buildings
Alaska	105,000 residents, plus highly seasonal visitor count; 6,600 miles of coastline Total resident population of area at immediate risk of post-tsunami impacts[c]: 125,000	> $10 billion plus international airport's fuel depot, 3 medium ports, plus 9 other container ports; 55 ports in total

[a]Wood (2007), Wood et al. (2007, 2013), Wood and Soulard (2008), and Wood and Peters (2015).
[b]Lower bound estimates based on present evacuation zones.
[c]National Research Council (2011). The total population at immediate risk includes those in the same census tract whose livelihood or utility and other services would be interrupted by a major tsunami with this inundation.
[d]Updated for exposure to great Aleutian tsunamis (modeling by University of Hawaii per Hawaii Emergency Management Agency).

researchers, does not know of any buildings of this height that systemically collapsed, regardless of whether they were designed as tsunami evacuation refuges (Chock et al. 2013b).
5. Foundations of such buildings would be more substantial in size and mass, thus making them more resistive to localized scour around the building.
6. International Building Code (IBC), International Code Council, Inc. (2015), Table 503 has restrictions on Type of Construction where 65 ft (19.8 m) is commonly used as one of the thresholds over which Type I noncombustible construction is required. (Number of stories is another criterion, which may be even more restrictive.) Since light-frame construction is typically less than 65 ft (19.8 m) tall except for special exceptions to IBC Table 503, selecting this height does not affect most light-frame construction. Although IBC Table 503 has typical fire-resistant height thresholds of 65 ft (19.8 m), the fire-resistant reasons for the IBC height limit do not necessarily relate to the tsunami-resistant threshold.

The baseline study of 65-ft (19.8-m) height was predicated on severe tsunami inundation depths of up to three stories, commensurate to the Pacific Northwest and Alaska. Risk Category II buildings of lesser height may also be designed for tsunamis, with some increase in structural costs. In areas of California south of Cape Mendocino, inundation depths are much lower, and it would be increasingly practical for buildings of lesser height to be tsunami-resilient. Therefore, coastal communities in California may choose a threshold height of less than 65 ft (19.8 m) and achieve tsunami safety with low-rise multistory structures at nominal expense. The state of Hawaii has most of its urban areas in regions of low to moderate seismicity. Accordingly, there could be a height greater than 65 ft (19.8 m) in order for seismic design to provide systemic strength sufficient by parity for overall tsunami loads on the structure. Therefore, the threshold height could be designated to be taller, or it could be designated to be shorter, with the influence on systemic design determined using the methods given in Chock et al. (2013a) and Carden et al. (2015). From the above discussion, note that structural engineering expertise is necessary to evaluate several important technical factors relevant to the jurisdiction's decision to establish a threshold height of applicability for Risk Category II buildings and structures.

Tsunami Risk Category II structures above the designated height threshold, all Tsunami Risk Category III, and all Tsunami Risk Category IV buildings and other structures are subject to these requirements. Risk Category II buildings below the designated height threshold are not required to be designed for tsunami loads. Accordingly, it is important for the community to have Tsunami Evacuation Maps and operational response procedures, recognizing the inventory of Risk Category II buildings not designated to be designed for tsunami effects, which would therefore be vulnerable to tsunami damage.

Mitigation of tsunami risk requires a combination of emergency preparedness for evacuation in addition to providing structural resilience of Critical Facilities, infrastructure, and key resources necessary for immediate response and economic and social recovery. Risk Category I and low-rise Risk Category II buildings are not required to be designed against any tsunami event because they are at a higher risk of being fully inundated and collapsing during a major tsunami. Communities in the Tsunami-Prone Region should be enabled with tsunami warning systems and emergency operations plans for evacuation because Risk Category I and low-rise Risk Category II buildings should not be occupied during a tsunami. A tsunami warning and evacuation procedure consists of a plan and procedure developed and adopted by a community that has a system to act on a tsunami warning from the designated National Oceanic and Atmospheric

Administration (NOAA) tsunami warning center at all hours using two independent means—a 24-hour operational site to receive the warning and established methods of transmitting the warning that will be received by the affected population—and has established and designated evacuation routes for its citizens to either high ground or to a designated Tsunami Vertical Evacuation Refuge Structure. At present, the states of Alaska, Washington, Oregon, California, and Hawaii already have tsunami warning and evacuation procedures. A NOAA TsunamiReady community would include this procedure. In these states, it is recognized by federal, state, and local governments that mitigation of tsunami risk to public safety requires emergency preparedness for evacuation.

The ASCE Board authorized the SEI and COPRI funding necessary for Tsunami Design Zone mapping and Offshore Tsunami Amplitude, within the project for this standard. In 2013, the NOAA National Tsunami Hazard Mitigation Program (NTHMP) Modeling and Mapping Subcommittee convened a Probabilistic Tsunami Hazard Analysis (PTHA) working group and conducted a comparison of prototype probabilistic tsunami inundation maps produced in 2013 by two independent researchers, including a team from the University of Washington as well as the Pacific Marine Environmental Laboratory (PMEL) team performing the work for ASCE, to validate the key procedures of the PTHA methodology. The study was reviewed and advised by a broad panel of experts of the Modeling and Mapping Subcommittee of NTHMP. The USGS team involved in the scientific consensus process for seismic source characterization for the National Earthquake Hazards Reduction Program and ASCE seismic maps were also included in this panel. State geologists and tsunami modelers from all five western states also participated. The results of this 2013–2014 peer review phase (California Geological Survey, 2015) were incorporated into the production process of the 2014–2015 ASCE 7 mapping project. The PTHA methodology used to develop the Tsunami Design Zone Maps is explained in Commentary Section C6.7.

The final step of the PTHA process involves the inundation analysis to determine the runup. Runup geodata define the ground elevation where the tsunami inundation reaches its horizontal limit. The runup data set of ASCE includes these geocoded points defining the locations and elevations of the runup. The inundation limit on land is the smoothed extent line formed by these discrete runup points, and the Tsunami Design Zone essentially consists of the land area between the inundation limit line and the coastline. Review of the ASCE 7-16 Tsunami Design Zone Maps by the Tsunami Loads and Effects Subcommittee also included comparison with the results of two independent modelers conducting region-specific work. *ASCE 7-16 Tsunami Design Zone Maps for Selected Locations* provides 62 Tsunami Design Zone Maps rendered in PDF format and downloadable from https://doi.org/10.1061/9780784480748. The locations of these 62 maps are indicated by circles in Figs. C6.1-1(a) to (i). For the purpose of

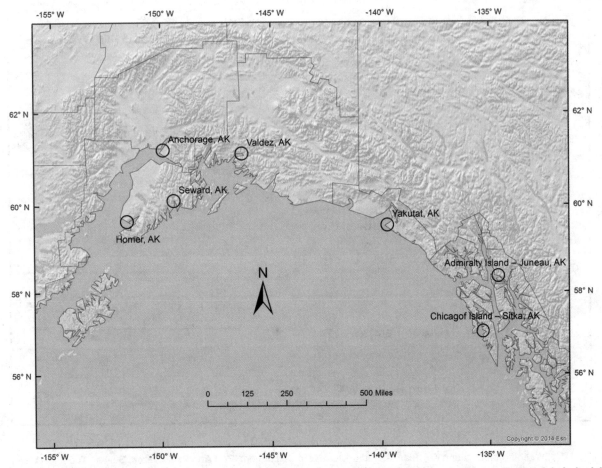

FIGURE C6.1-1(a) Alaska Location Key Plan, Including Seven Areas: Anchorage, Valdez, Seward, Yakutat, Homer, Admiralty Island—Juneau, and Chicagof Island—Sitka

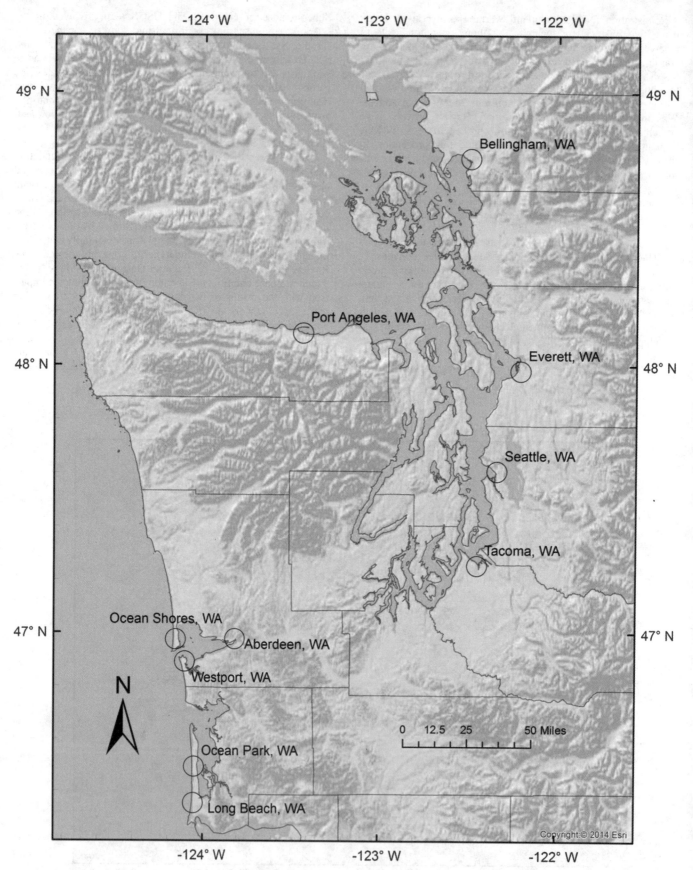

FIGURE C6.1-1(b) Washington Location Key Plan, Including 10 Areas: Bellingham, Port Angeles, Everett, Seattle, Tacoma, Ocean Shores, Aberdeen, Westport, Ocean Park, and Long Beach

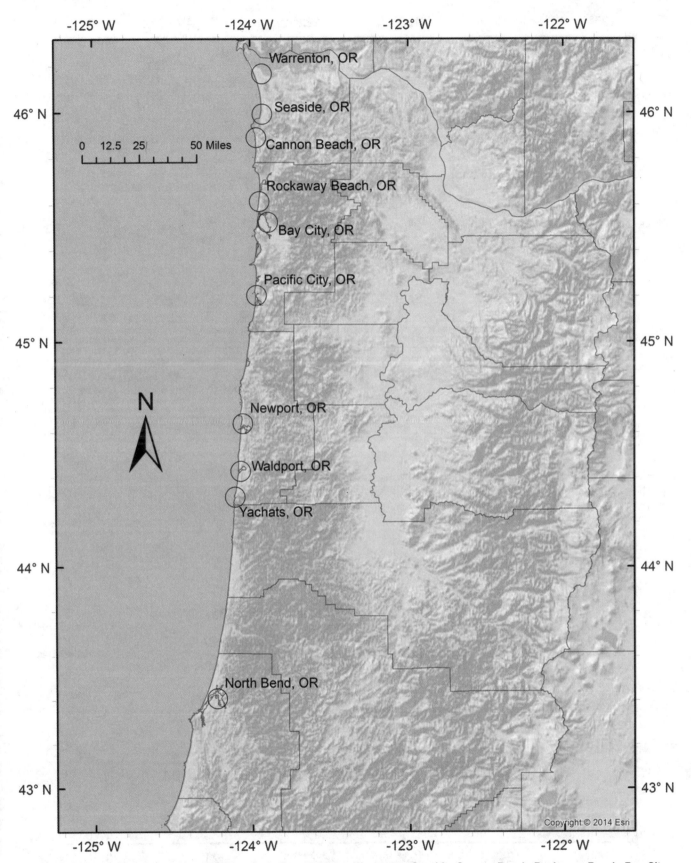

FIGURE C6.1-1(c) Oregon Location Key Plan, Including 10 Areas: Warrenton, Seaside, Cannon Beach, Rockaway Beach, Bay City, Pacific City, Newport, Waldport, Yachats, and North Bend

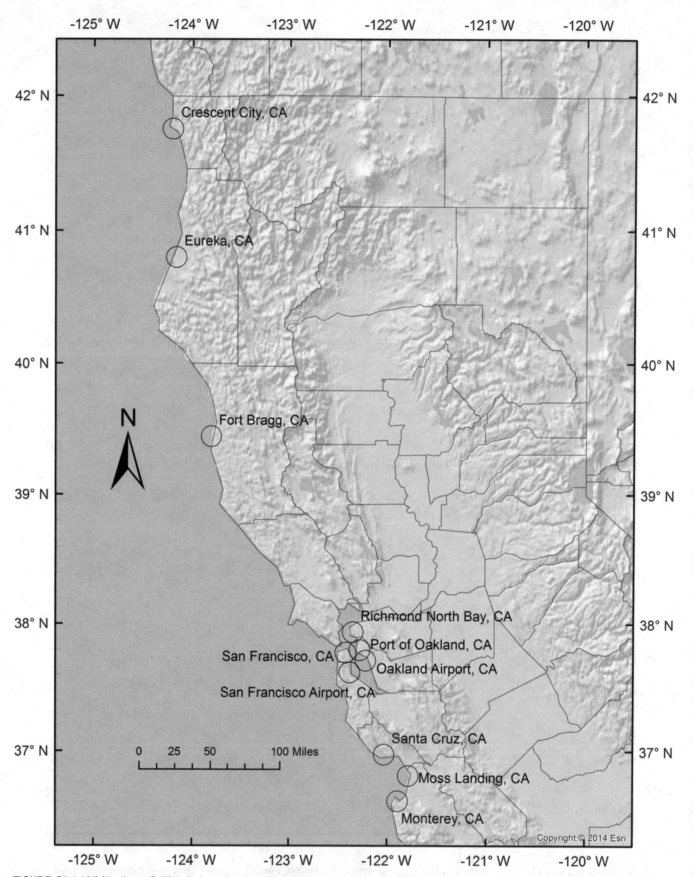

FIGURE C6.1-1(d) Northern California Location Key Plan, Including 11 Areas: Crescent City, Eureka, Fort Bragg, Richmond North Bay, Port of Oakland, San Francisco, Oakland Airport, San Francisco Airport, Santa Cruz, Moss Landing, and Monterey

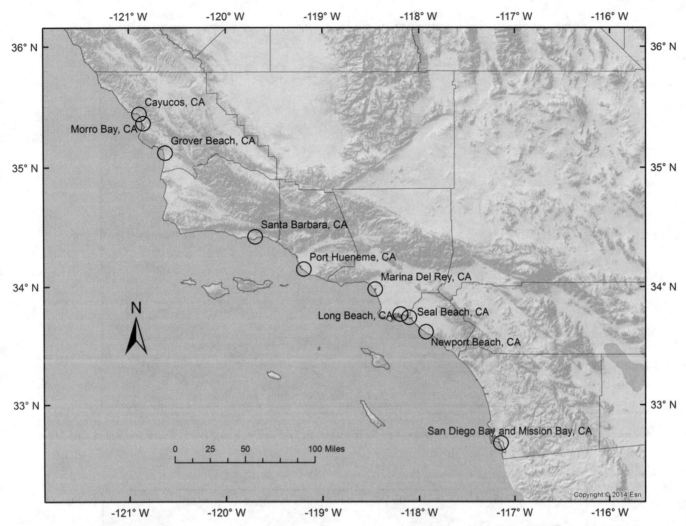

FIGURE C6.1-1(e) Southern California Location Key Plan, Including 10 Areas: Cayucos, Morro Bay, Grover Beach, Santa Barbara, Port Hueneme, Marina del Rey, Long Beach, Seal Beach, Newport Beach, and San Diego Bay and Mission Bay

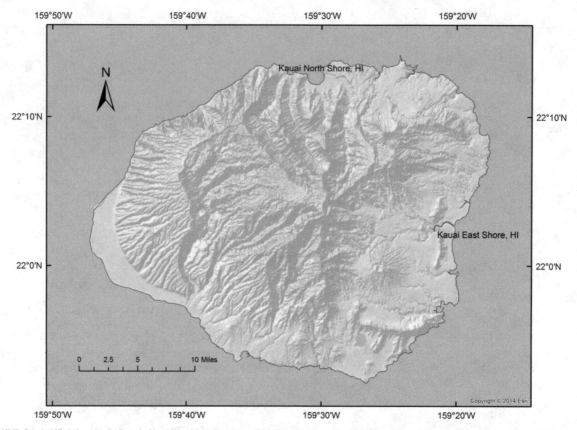

FIGURE C6.1-1(f) Island of Kauai, Hawaii, Location Key Plan, Including Two Areas: Kauai North Shore and Kauai East Shore

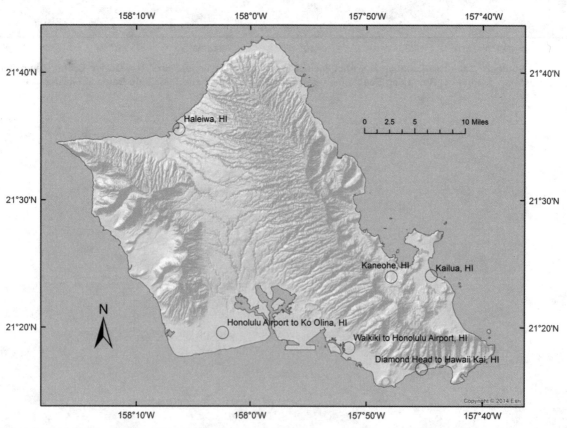

FIGURE C6.1-1(g) Island of Oahu, Hawaii, Location Key Plan, Including Six Areas: Haleiwa, Kaneohe, Kailua, Honolulu Airport to Ko Olina, Waikiki to Honolulu Airport, and Diamond Head to Hawaii Kai

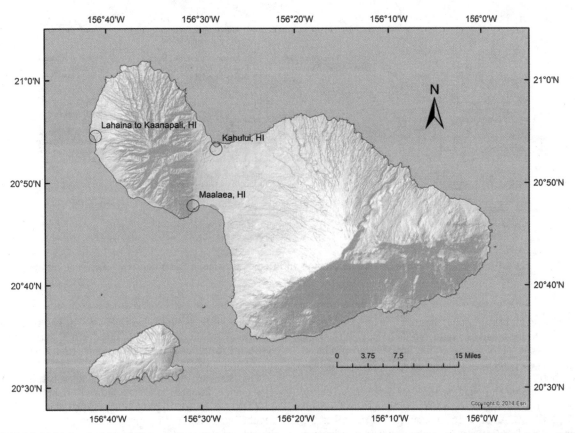

FIGURE C6.1-1(h) Island of Maui, Hawaii, Location Key Plan, Including Three Areas: Lahaina to Kaanapali, Kahului, and Maalaea

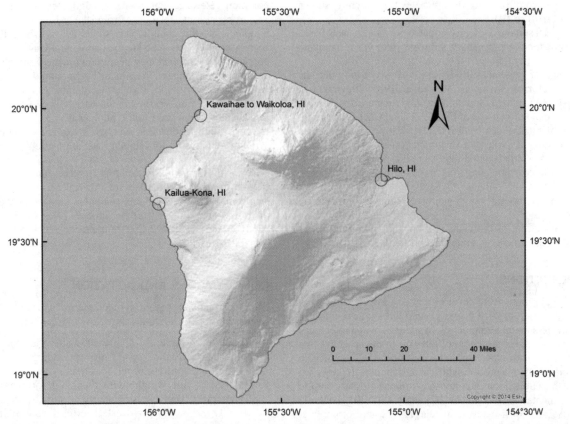

FIGURE C6.1-1(i) Island of Hawaii, Hawaii, Location Key Plan, Including Three Areas: Kawaihae to Waikoloa, Hilo, and Kailua-Kona

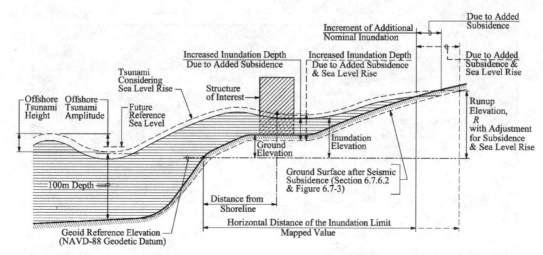

FIGURE C6.1-2 Effects of Relative Sea Level Change and Regional Seismic Subsidence on Tsunami Inundation

defining the Tsunami Design Zone, these PDF maps are considered equivalent to the results served from the ASCE 7 Tsunami Design Geodatabase for the corresponding areas and are produced with the runup, GIS point data, and Tsunami Design Zone data in the Geodatabase.

ASCE 7-16 Tsunami Design Zone Maps for Selected Locations also includes a report prepared for ASCE, "Probabilistic Tsunami Design Maps for the ASCE 7-16 Standard," that describes the development of the 2,500-year probabilistic tsunami design zone maps and is suitable as an accompanying user manual.

The key data sets of the ASCE Tsunami Design Geodatabase are the runup points. These are organized by segments of coastline for each of the five western states (the extent is shown in Fig. C6.1-1(a) to (i)).

The ASCE Tsunami Design Geodatabase data is web-based and does not require proprietary software (http://asce7tsunami.online).

The effect of seismic subsidence, typically following the recognized deformation model of Okada (1985) illustrated in Fig. C6.1-1, is accounted for within the modeling used to develop the mapped inundation limits of Fig. 6.1-1. Fig. C6.1-2 illustrates the effects of relative sea level rise and seismic subsidence on tsunami inundation. Note that in subduction regions, the long-term trend of ground elevation may be uplift, but the engineer is not to consider extrapolating this trend to reduce the design inundation since the subduction earthquake mechanism will result in this temporal uplift being eliminated by seismic subsidence.

A general outline of the main steps in the tsunami analysis and design requirements is given in Fig. C6.1-3.

C6.2 DEFINITIONS

The ASCE definitions were developed after a review of international literature before 2013 that in general did not have uniform consistency of tsunami terminology. In addition, some references may utilize waveform terminology relevant to documenting the local effects of historic tsunami but were not directly applicable to the specification of a probabilistic tsunami for design purposes. However, the publication of a revised tsunami glossary by the Intergovernmental Oceanographic Commission (2013) should lead to greater consistency, and the ASCE terminology is compatible with that document.

In particular, as illustrated in Fig. 6.2-1, key terms for tsunami definition are the Maximum Considered Tsunami, the Offshore Tsunami Amplitude, inundation depth, runup elevation, and the inundation limit.

This standard defines the size of a probabilistic Maximum Considered Tsunami at a standardized offshore depth of 328 ft (100 m), using its amplitude above or below the ambient sea level. This level is different from the peak to trough tsunami height. In site-specific tsunami hazard analysis, a Reference Sea Level of Mean High Water Level is often used in models. However, on land, the North American Vertical Datum 88, also known as NAVD 88 (Zilkoski et al. 1992), is adopted as the common reference datum for ground elevation and tsunami runup elevations resulting from the Maximum Considered Tsunami. By this datum, the "shoreline" is where the NAVD 88 elevation is zero. Outside the continental United States, other adopted reference data may be used in practice, and the results of a site-specific tsunami hazard analysis should be coordinated with the appropriate local reference datum. Another key definition is Froude number, the nondimensionalized flow velocity; in this chapter, when the Froude number is specified, the engineer should recognize that this number is explicitly defining a prescribed relationship of velocity as a function of inundation depth.

As of the 2016 edition of this standard, the Tsunami-Prone Region comprises the states of Alaska, Washington, Oregon, California, and Hawaii; this region is subject to the addition of other areas once further Probabilistic Tsunami Hazard Analyses are performed that demonstrate runup of greater than 3 ft (0.914 m).

C6.3 SYMBOLS AND NOTATION

Particular tsunami design symbology for which the user may desire clarification includes the following:

V_w = general volume of displaced water that the engineer needs to calculate based on the loading condition and structural configuration. The loading equations use the symbol γ for weight density and ρ for mass density. Generally, this notation clarifies and distinguishes the fluid mechanics of hydrostatic effects from hydrodynamic effects.

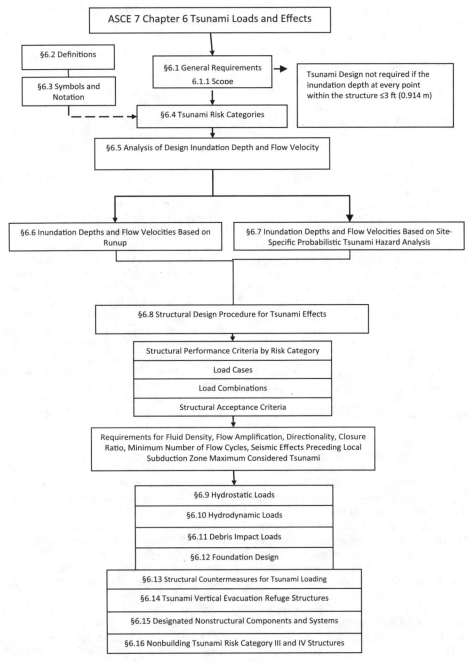

FIGURE C6.1-3 General Organization of Chapter 6

C6.4 TSUNAMI RISK CATEGORIES

This standard classifies facilities in accordance with Risk Categories that recognize the importance or criticality of the facility. In the tsunami chapter, further modified definitions of the Risk Categories for Tsunami Risk Categories III and IV are included with respect to specific occupancy/functional criteria. Critical Facilities designated by local governments are included in Tsunami Risk Category III. Essential facilities are included in Tsunami Risk Category IV.

Critical Facilities are included in Tsunami Risk Category III only if they are so designated by local governments, such as power-generating stations, water-treatment facilities for potable water, wastewater-treatment facilities, and other public utility facilities not included in Risk Category IV. Critical Facilities incorporate facilities needed for post-tsunami mission critical functions or facilities that have more critical roles in community recovery and community services. Essential Facilities are those facilities necessary for emergency response. As approved by the ASCE Board of Direction in 2013, ASCE Policy Statement 518 (2013b) states, "Critical infrastructure includes systems, facilities, and assets so vital that their destruction or incapacitation would have a debilitating impact on national security, the economy or public health, safety, and welfare." FEMA P-543 (2007) states, "Critical facilities comprise all public and private facilities deemed by a community to be essential for the delivery of vital services, protection of special populations, and the provision of other services of importance for that community." In general usage, the term "Critical Facilities" is used to describe all structures or other improvements that, because of their function, size, service area, or uniqueness, have the potential to cause serious bodily harm, extensive property damage, or disruption of vital socioeconomic activities if they are destroyed or damaged, or if their functionality is impaired.

Critical Facilities commonly include all public and private facilities that a community considers essential for the delivery of vital services and for the protection of the community. They usually include emergency response facilities (fire stations, police stations, rescue squads, and emergency operation centers, or EOCs), custodial facilities (jails and other detention centers, long-term care facilities, hospitals, and other health-care facilities), schools, emergency shelters, utilities (water supply, wastewater-treatment facilities, and power), communications facilities, and any other assets determined by the community to be of critical importance for the protection of the health and safety of the population.

The number and nature of Critical Facilities in a community can differ greatly from one jurisdiction to another, and these facilities usually include both public and private facilities. In this sense, each community needs to determine the relative importance of the publicly and privately owned facilities that deliver vital services, provide important functions, and protect special populations.

A number of Essential Facilities do not need to be included in Tsunami Risk Category IV because they should be evacuated before the tsunami arrival. This includes fire stations, ambulance facilities, and emergency vehicle garages. These facilities may be necessarily located within the Tsunami Design Zone because they must serve the public interest on a timely basis, but designing the structures for tsunami loads and effects could be prohibitively expensive with minimal benefit to the resilience of the community. The evacuated emergency response resources would still be available after the tsunami. Also not included in Tsunami Risk Category IV are earthquake and hurricane shelters because these types of shelters for other hazards are not to be used during a tsunami. Earthquake shelters in particular are postdisaster mass care facilities, since earthquakes have no effective pre-event warning. As such, these earthquake shelters do not serve any purpose for tsunami evacuation refuge before the arrival of tsunami. Emergency aircraft hangars are also not included in Tsunami Risk Category IV because these aircraft would need to be deployed outside of the inundation zone to ensure that they are functional after the tsunami. In coordination with the jurisdiction's emergency response and recovery plan, certain facilities could be considered for designation as Tsunami Risk Category II or III, if they are deemed not uniquely required for postdisaster operations or where such functionality can be sufficiently provided from a post-tsunami alternative facility.

Tsunami Vertical Evacuation Refuge Structures are included in Tsunami Risk Category IV because of their function as a safe refuge for evacuees during the tsunami. If health-care facilities with 50 or more resident patients are for some reason located in the Tsunami Design Zone, it is recommended that such facilities should also be designed in accordance with Tsunami Risk Category IV rather than Risk Category III. The height of the structure must afford a sufficient number of elevated floor levels for patient vertical evacuation, similar to a Tsunami Vertical Evacuation Refuge Structure because of the difficulty with evacuating assisted living patients in a timely manner before inundation.

C6.5 ANALYSIS OF DESIGN INUNDATION DEPTH AND FLOW VELOCITY

There are two procedures for determining the inundation depth and velocities at a site: (1) Energy Grade Line Analysis, which takes the runup elevation and inundation limit indicated on the Tsunami Design Zone Map as the target solution point of a hydraulic analysis along the topographic transect from the shoreline to the point of runup; and (2) a site-specific inundation analysis of at least depth-averaged two-dimensional numerical modeling that uses the Offshore Tsunami Amplitude, the wave period that is a conserved property(during shoaling, the timescale of the overall tsunami wave hardly changes), and other waveform parameters as the input to a numerical simulation that includes a high-resolution Digital Elevation Model.

Energy Grade Line Analysis, which has been developed to produce conservative design flow parameters, is always performed for Tsunami Risk Category II, III, and IV structures. The site-specific inundation analysis procedure may or may not be required, depending on the structure Tsunami Risk Category. Site-specific inundation analysis is not required, but may be used, for Tsunami Risk Category II and III structures. The site-specific inundation analysis procedure is performed for Tsunami Risk Category IV structures unless the Energy Grade Line Analysis shows the inundation depth to be less than 12 ft (3.66 m) at the structure. However, Tsunami Vertical Refuge Structures shall always use site-specific inundation analysis, regardless of the inundation depth produced by the Energy Grade Line Analysis.

A precise computer simulation can capture two-dimensional flow and directionality effects that a linear transect analysis cannot, and so it is particularly useful as an additional due diligent investigation of flow characteristics for Risk Category IV structures. However, supposedly exact numerical codes do not include any allowance whatsoever for the uncertainty of the modeling technique, which may lead to underestimation of flow speed. The fundamental responsibility of engineering philosophy requires consideration of the consequences of underestimation to the risk of design failure, as opposed to the academic scientific philosophy that seeks the best mean value rendered with high precision without the burden of other concerns. The less precise Energy Grade Line Analysis developed by the ASCE Tsunami Loads and Effects Subcommitteee purposely includes that allowance for uncertainty, as well as providing a solution in the familiar context of well-established engineering hydraulic fundamentals that can aid an engineer's judgment. Therefore, the Energy Grade Line Analysis provides a measure of engineering statistical reliability that is used for establishing a proportion of the Energy Grade Line Analysis as a "floor value" below which various numerical model simulation techniques should not fall, particularly for flow velocity. To establish the appropriate statistical allowance and conservative bias in the Energy Grade Line Analysis method, 36,000 trial numerical code simulations were performed versus the Energy Grade Line Analysis method. Without this limitation, as is stated in Section 6.7.6.8, there would have been inconsistent reliabilities between the two methods. That is why for Risk Category IV structures of inherently high value, both techniques are valuable to perform, but for quite distinctly different but important reasons (two-dimensional flow characteristics and flow speed).

Table C6.5-1 indicates the necessary inundation depth and flow velocity analysis procedures in accordance with Section 6.5

Table C6.5-1 Inundation Depth and Flow Velocity Analysis Procedures Where Runup Is Given in Fig. 6.1-1

Analysis Procedure Using the Tsunami Design Zone Map of Fig. 6.1-1	Tsunami Risk Category (TRC) Structure Classification			
	TRC II	TRC III	TRC IV (excluding TVERS)	TRC IV—Tsunami Vertical Evacuation Refuge Shelter (TVERS)
Section 6.5.1.1 (R/H_T Analysis)	Not permitted	Not permitted	Not permitted	Not permitted
Section 6.6 (Energy Grade Line Analysis)	✓ If MCT inundation depth ≤3 ft (0.914 m)[a], Chapter 6 does not apply	✓ If MCT inundation depth ≤3 ft (0.914 m)[a], Chapter 6 does not apply	✓	✓
Section 6.7 (Site-Specific Analysis)	Permitted; if MCT inundation depth ≤3 ft (0.914 m)[a], Chapter 6 does not apply	Permitted; if MCT inundation depth ≤3 ft (0.914 m)[a], Chapter 6 does not apply	✓ Required if Energy Grade Line Analysis inundation depth ≥12 ft (3.66 m)[a]	✓

Notes: ✓ indicates a required procedure. MCT means Maximum Considered Tsunami.
[a]MCT inundation depth including sea level rise component, per Section 6.5.3.

Table C6.5-2 Inundation Depth and Flow Velocity Analysis Procedures Where Runup Is Calculated from Fig. 6.7-1

Analysis Procedure	Tsunami Risk Category (TRC) Structure Classification			
	TRC II	TRC III	TRC IV (excluding TVERS)	TRC IV—Tsunami Vertical Evacuation Refuge Shelter (TVERS)
Section 6.5.1.1 (R/H_T) Analysis	✓	✓	Not permitted	Not permitted
Section 6.6 (Energy Grade Line Analysis)	✓ If MCT inundation depth ≤3 ft (0.914 m)[a], Chapter 6 does not apply	✓ If MCT inundation depth ≤3 ft (0.914 m)[a], Chapter 6 does not apply	✓	✓
Section 6.7 (Site-Specific Analysis)	Permitted; if MCT inundation depth ≤3 ft (0.914 m)[a], Chapter 6 does not apply	Permitted; if MCT inundation depth ≤3 ft (0.914 m)[a], Chapter 6 does not apply	✓	✓

Notes: ✓ indicates a required procedure. MCT means Maximum Considered Tsunami.
[a]MCT inundation depth including sea level rise component, per Section 6.5.3.

for each Tsunami Risk Category, for locations shown in the Tsunami Design Zone Map of Fig. 6.1-1.

Table C6.5-2 indicates the necessary inundation depth and flow velocity analysis procedures in accordance with Section 6.5 per Tsunami Risk Category for locations where the inundation limit is not shown in the Tsunami Design Zone Map of Fig. 6.1-1 but where the Offshore Tsunami Amplitude is shown in Fig. 6.7-1.

In certain regions and topographies, for example, in remote areas of Alaska, detailed inundation limit mapping is not available. Where no inundation limit map is available from the Authority Having Jurisdiction, a runup based on the factor given in Eq. (6.5-2), multiplied by the Offshore Tsunami Amplitude, can be used with the Energy Grade Line Analysis for Tsunami Risk Category II and III structures. The guidance of estimating runup elevation by this approximate factor multiplied by the Offshore Tsunami Amplitude is derived from Japanese research reported in Murata et al. (2010), together with runup relations developed by Synolakis (1986), Li (2000), and Li and Raichlen (2001, 2003); numerical simulations; and field observations from the 2011 Tohoku-Oki tsunami.

Since a large volume of literature exists on the prediction of tsunami runup by analytical methods, by laboratory experiments, and by numerical modeling, a method based on peer-reviewed literature may be used to refine the prediction of runup in certain cases. In these cases, the shoreline topography, Nearshore Profile Slope Angle, and offshore bathymetry should correspond to design conditions. Li and Raichlen (2001, 2003) present analytical expressions for calculating runup of solitary waves under breaking and nonbreaking conditions; Madsen and Schäffer (2010) summarize analytical expressions for runup of nonsolitary wave shapes such as N-waves and transient wave trains; and Didenkulova and Pelinovsky (2011) address runup amplification of solitary waves in U-shaped bays.

Fig. C6.5-1 is primarily based on runup relations developed by Li and Raichlen for nonbreaking and breaking waves to predict the ratio of runup, R, to Offshore Tsunami Amplitude, H_T, as a function of Φ, the Nearshore Profile Slope Angle. The experiments were conducted for Nearshore Profile Slope Angles less than 1:50 and show a peak runup ratio near 4 and decreasing runup for shallow slopes. The envelope curve represents nearly an upper bound of predicted runup for the range of wave heights [16.4 ft to 65.6 ft (5 m to 20 m)], $0.05 \leq H_T/h_o \leq 0.2$ in areas where no inundation limit has been mapped. Fig. C6.5-1 shows the runup relations used to determine the envelope curve. The maximum value of $R/H_T = 4.0$ follows the guidance from

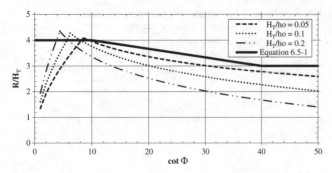

FIGURE C6.5-1 Runup Ratio R/H_T, as a Function of the Mean Slope of the Nearshore Profile, cot Φ for Relative Wave Heights $H_T/h_o = 0.05$, 0.1, and 0.2

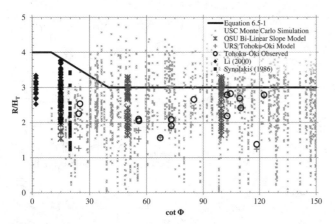

FIGURE C6.5-2 Runup Ratio R/H_T, as a Function of the Mean Slope of the Nearshore Profile, cot Φ, Showing Laboratory, Model, and Field Data

Murata et al. (2010). The minimum value of $R/H_T = 3.0$ is based on field observations from the 2011 Tohoku-Oki tsunami, as well as output from numerical models.

Fig. C6.5-2 presents the envelope equation of Fig. C6.5-1 together with field observations from the 2011 Tohoku tsunami and the results of numerical and physical models of real and idealized runups, together with a large catalog of simulated runups for 36,000 possible combinations of Nearshore Profile Slope Angle and onshore topography. The field and numerical model data suggest that the relation between runup and Offshore Tsunami Amplitude is not greatly reduced as Nearshore Profile Slope Angle increases. In addition, the shallow slopes (high values of cot Φ) observed in the field fall outside of the range possible in physical experiments.

In coastal engineering practice, wave runup is commonly analyzed using a dimensionless parameter called surf similarity, or Iribarren number. The surf similarity parameter characterizes runup using a combination of wave steepness and slope that governs the way the wave evolves as it passes from offshore to onshore. Surf similarity parameter has been applied to tsunamis by Madsen and Schäffer (2010), Hughes (2004), and others. Use of the surf similarity parameter for runup is consistent with coastal engineering design worldwide.

Surf similarity parameter, ξ_{100}, for this application to tsunami engineering is defined as follows:

$$\xi_{100} = \frac{T_{TSU}}{\cot(\Phi)} \sqrt{g/2\pi H_T} \qquad (C6.5-1)$$

where Φ is the mean slope angle of the Nearshore Profile taken from the 328-ft (100-m) water depth to the Mean High Water Level elevation along the axis of the topographic transect for the site. H_T is the offshore tsunami amplitude, and T_{TSU} is the wave period of the tsunami at 328-ft (100-m) water depth. Runup ratios from field observations, as well as from physical and numerical modeling, are shown as a function of ξ_{100} in Fig. C6.5-3 together with an envelope curve that encompasses most of the data. Outliers are in areas where wave focusing is expected to occur and for Monte Carlo simulations with very small crest heights relative to troughs, which alters relative runup values.

Using the surf similarity parameter allows for a reduction in the relative runup height for cases where surf similarity parameter is less than 6 or greater than 20. For the field values presented in Fig. C6.5-3, 30% of the surf similarity values fall into the range that would allow for a reduction in relative runup height below the value of 4.

C6.5.3 Sea Level Change. Sea level rise has not been incorporated into the Tsunami Design Zone Maps, and any additive effect on the inundation at the site should be

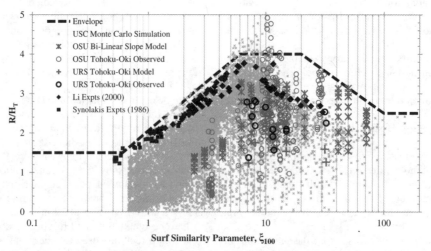

FIGURE C6.5-3 Runup Ratio R/H_T as a Function of Surf Similarity Parameter

explicitly evaluated. The approach taken for consideration of the effects of sea level rise is consistent with that given by the U.S. Army Corps of Engineers (2013). For a given Maximum Considered Tsunami, the runup elevation and inundation depth at the site are increased by at least the future projected relative sea level rise during the design life of the structure. The minimum design lifecycle of 50 years can be exceeded by the physical life span of many structures. Relative sea level change for a particular site may result from ground subsidence and long-term erosion, and change in local temperature of the ocean causing thermal expansion, as well as possible global effects by melting of ice sheets. Note that as the ASCE provisions for tsunami may later incorporate other areas with more significant relative sea level rise because of greater geologic (nonseismic) subsidence, this section on sea level rise would be more important. An example of this would be American Samoa and the other Pacific insular states and territories in the western Pacific Ocean associated with the United States, particularly where the habitable terrain elevation is low.

Historical sea level change trends over the past 150 years can be obtained from the NOAA Center for Operational Oceanographic Products and Services, which has been measuring sea level at tide stations along the U.S. coasts. Changes in local relative mean sea level (MSL) have been computed at 128 long-term water level stations using a minimum span of 30 years of observations at each location (http://tidesandcurrents.noaa.gov/sltrends/: "Tide gauge measurements are made with respect to a local fixed reference level on land; therefore, if there is some long-term vertical land motion occurring at that location, the relative MSL trend measured there is a combination of the global sea level rate and the local vertical land motion.")

C6.6 INUNDATION DEPTHS AND FLOW VELOCITIES BASED ON RUNUP

C6.6.1 Maximum Inundation Depth and Flow Velocities Based on Runup. Calculated flow velocities are subject to minimum and maximum limits based on observational data from past tsunamis of significant inundation depth. The upper limit of 50 ft/s (15.2 m/s) in Section 6.6 already includes a 1.5 factor on the velocity that is greater than any on-land flow velocity that the committee has examined from field or video evidence. The flow velocity upper limit is therefore applicable subsequent to any flow amplification factors caused by obstructions in the built environment, per Sections 6.6.5, 6.7.6.6, or 6.8.5.

In a Monte Carlo simulation of hundreds of variations of terrain profiles where the Energy Grade Line Analysis was compared to numerical models, it was found that cases of negative terrain slope gradient would cause the Energy Grade Line Analysis to fail to converge on a solution. To deal with the possibility of such cases in which local downslope acceleration of flow would occur, but to retain the simplicity of the analysis, a conservative adjustment of the runup elevation is prescribed in method 2 of Section 6.6.1, which essentially increases the energy grade line. Alternatively, a site-specific inundation analysis may be performed per method 1 of Section 6.6.1.

C6.6.2 Energy Grade Line Analysis of Maximum Inundation Depths and Flow Velocities. The coastline can be approximated in behavior by the use of one-dimensional linear transects of a composite bathymetric and/or topographic profile. The tsunami inundation design parameters of inundation depth and flow velocities are determined by an energy analysis approach. In the Energy Grade Line Analysis, hydraulic analysis using Manning's coefficient for equivalent terrain macroroughness is used to account for friction along with the profile comprised of a series of one-dimensional slopes, to determine the variation of depth and velocity-associated inundation depth across the inland profile. Velocity is assumed to be a function of inundation depth, calibrated to the prescribed Froude number $u/\sqrt{(gh)}$ at the shoreline and decreasing inland. The specified incremental horizontal distance maximum spacing of 100 ft (30.5 m) is necessary for accuracy of the hydraulic analysis.

The Energy Grade Line Analysis is based on rational hydraulic principles for a one-dimensional transect flow assumption, so it is not the same analysis technique as performed by a two-dimensional (2D) flow numerical simulation. The Energy Grade Line Analysis is simple and it is inherently less precise, but is statistically conservative for use as a practical design tool that can be performed by the engineer. One additional conservatism is to use the actual roughness of the terrain in the energy method rather than a "bare-earth" assumption. Also the parameters of "maximum" depth and flow speed are associated with the prescriptive Load Cases that define the combinations derived from these "maximum" flow parameters. When the actual roughness of the terrain is used in the energy method, it takes a greater initial energy at the shoreline in order to reach the runup. Therefore, conceptually, the energy at the shoreline in the prescriptive method is implicitly greater than the energy value used in the 2D inundation model run on "bare earth." The benefit of this approach is that it produces conservative design values and it implicitly accounts for flow amplification that occurs through the built environment (which the standard 2D inundation model technique does not consider). The 328-ft (100-m) depth is where the waveform based on Probabilistic Tsunami Hazard Analysis can be regionally characterized along the coastline. The Tsunami Design Zone Maps of this standard have been developed from a Probabilistic Tsunami Hazard Analysis (PTHA) using the NOAA Pacific Marine Environmental Laboratory (PMEL) short-term inundation forecasting for tsunamis (SIFT), method of splitting tsunamis (MOST), and community model interface for tsunamis (ComMIT) model for the 2D inundation analysis. The Tsunami Design Zone Maps have been produced to ensure that the inundation limit and runup elevation and the extent of the hazard zone are generally not underestimated. The Energy Grade Line Analysis uses this input so that the flow characteristics at the site are not severely underestimated for design purposes. The statistical level of conservatism has been verified by doing many thousands of simulations to compare the energy method with time history outputs from the primary types of numerical inundation models.

The Energy Grade Line Analysis stepwise procedure consists of the following steps:

1. Obtain the runup and inundation limit values from the Tsunami Design Zone Map generated by the ASCE Tsunami Design Geodatabase.
2. Approximate the principal topographic transect by a series of x–z grid coordinates defining a series of segmented slopes, in which x is the distance inland from the shoreline to the point and z is the ground elevation of the point. The horizontal spacing of transect points should be less than 100 ft (30.5 m), and the transect elevations should be obtained from a topographic Digital Elevation Model (DEM) of at least 33-ft (10-m) resolution.
3. Compute the topographic slope, ϕ_i, of each segment as the ratio of the increments of elevation and distance from point to point in the direction of the incoming flow.

4. Obtain the Manning's coefficient, n, from Table 6.6-1 for each segment based on terrain analysis.
5. Compute the Froude number at each point on the transect using Eq. (6.6-3).
6. Start at the point of runup with a boundary condition of $E_R = 0$ at the point of runup.
7. Select a nominally small value of inundation depth [~0.1 ft (0.03 m)] h_r at the point of runup.
8. Calculate the hydraulic friction slope, s_i, using Eq. (6.6-2).
9. Compute the hydraulic head, E_i, from Eq. (6.6-1) at successive points toward the shoreline.
10. Calculate the inundation depth, h_i, from the hydraulic head, E_i.
11. Using the definition of Froude number, determine the velocity u. Check against the minimum flow velocity required by Section 6.6.1.
12. Repeat through the transect until the h and u are calculated at the site. These are used as the maximum inundation depth, h_{max}, and the maximum velocity, u_{max}, at the site.

Load Cases based on these parameters are defined in Section 6.8.3.1. The inundation elevation is also determined from the inundation depth plus the ground elevation.

The Energy Grade Line Analysis algorithm and an example calculation for a portion of a transect follows in Fig. C6.6-1 and C6.6-2.

The effects of any sea level rise are to be added to the inundation depth results of the Energy Grade Line Analysis, per Section 6.5.3.

C6.6.3 Terrain Roughness. Terrain macroroughness is represented by Manning's coefficient (Kotani et al. 1998 and Shimada et al. 2003). Note that for the Energy Grade Line Analysis, a higher value of roughness is conservative because it requires a greater incident energy budget (of $h_0 + U_0^2/2g$) to overcome the frictional losses in reaching the given runup elevation. For site-specific inundation analysis to calculate the runup elevation and inundation limit, the opposite is true, in that lower friction allows the inundation to reach a higher point.

Table 6.6-1 provides values of Manning's coefficient, n, based on land use to represent equivalent roughness. Where an area is predominantly light-frame residential houses, which will not resist tsunami flow, it is advisable to use the coefficient of roughness for "all other cases." Where building obstructions are enclosed structures of concrete, masonry, or structural steel construction, Aburaya and Imamura (2002) and Imamura (2009) indicate that n varies with the coverage density of such buildings expressed as a coverage percentage, θ, their width, w, and the inundation depth, D, in accordance with the following equation, in which n_0 is 0.025 and C_D is 1.5:

$$n = \sqrt{n_0^2 + \frac{C_D}{2gw} \times \frac{\theta}{100 - \theta} \times D^{4/3}}$$

Generally, values of n greater than 0.040 result where the urban density θ is greater than 20%. For densities of 80% or more, values of n can approach 0.10. Studies by the Tsunami Loads and Effects Subcommittee indicate that it is not necessary to use values of Manning's roughness greater than 0.050 in the Energy Grade Line Analysis. Values of n greater than 0.050 result in much higher statistical conservatism in calculated momentum flux by the Energy Grade Line Analysis, compared to numerical solutions.

C6.6.4 Tsunami Bores. The criteria for the occurrence of bores consist of a complex interaction of the tsunami waveform and the coastal bathymetry, which can be determined through site-specific inundation numerical modeling. One such condition described in the literature arises where soliton fission waves are generated by shoaling over a long bathymetric slope of approximately 1/100 or milder (Murata et al. 2010; Madsen et al. 2008). These packets of shorter period waves on the front face of the long-period tsunami then break into a series of bores where the soliton amplitude over depth ratio reaches 0.78 to 0.83. Another case occurs where a shoaling tsunami encounters an abruptly rising seabed floor, such as a fringing shallow reef.

C6.7 INUNDATION DEPTHS AND FLOW VELOCITIES BASED ON SITE-SPECIFIC PROBABILISTIC TSUNAMI HAZARD ANALYSIS

A method of Probabilistic Tsunami Hazard Analysis has been established in the recognized literature that is generally consistent with Probabilistic Seismic Hazard Analysis in the treatment of uncertainty.

The basics of Probabilistic Tsunami Hazard Analysis for a region are as follows:

1. Tsunamigenic subduction zones and nonsubduction seismic thrust faults are discretized into a compiled system of rectangular subfaults, each with corresponding tectonic parameters.
2. Tsunami waveform generation is modeled as a linear combination of individual tsunami waveforms, each generated from a particular subfault, with a set sequence of subfaults used to describe the earthquake rupture in location, orientation, and rupture direction and sequence.
3. A statistically weighted logic tree approach is used to account for epistemic uncertainties in the tectonic parameters. The logic tree model parameters are developed from available tectonic, geodetic, historical, and paleotsunami data and estimated plate convergence rates. A probability distribution is used to account for aleatory uncertainties (i.e., intrinsic variability in each actual earthquake event).
4. Each individual tsunami waveform is propagated in deep water to 328-ft (100-m) depth using linear long-wave equations to take into account spatial variations in seafloor depth. This set of precomputed individual tsunami waveforms is combined in a linear fashion (i.e., Green's function approach) for subsequent analysis.
5. At the 328-ft (100-m) depth, determine the highest offshore wave heights and develop wave height, period, and waveform parameters for the design level exceedance of the 2,475-year tsunami event. The still-water level may be considered to be at the Reference Sea Level for this purpose.
6. Disaggregate (i.e., separate out for use in subsequent analysis) the seismic sources and associated moment magnitudes that together contribute at least 90% to the net offshore tsunami hazard at the site under consideration for the design level mean recurrence interval.
7. Use nonlinear shallow-water wave equations to transform each disaggregated tsunami event, or the Hazard-Consistent Tsunami event(s) (see below) from 328-ft (100-m) depth toward the shore to determine the associated maximum inundation.
8. Analyze each tsunami event from the disaggregated sample to determine flow parameters for the site of interest. Manning's coefficient for equivalent terrain

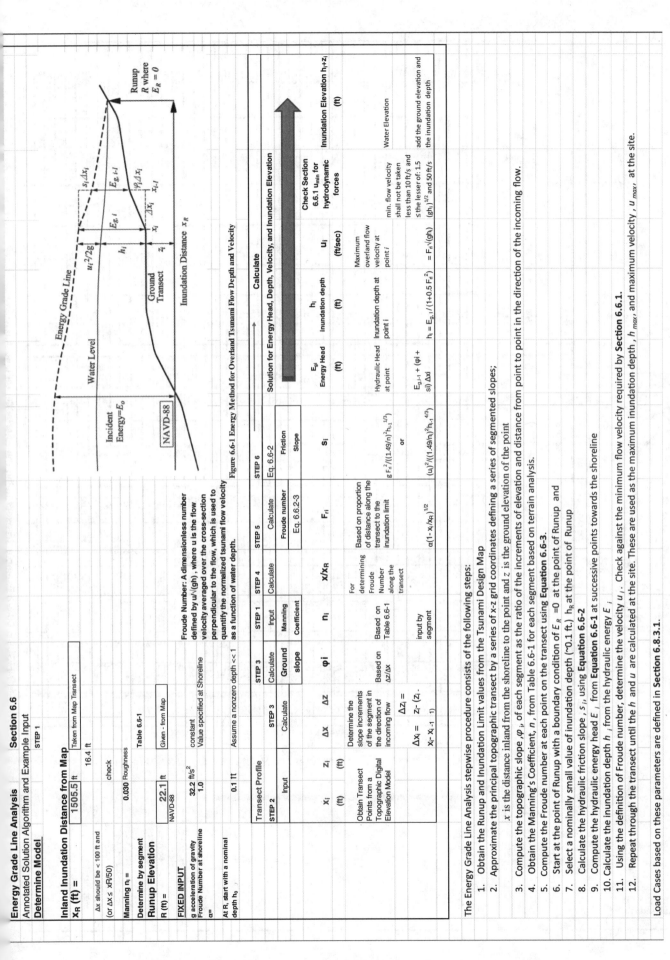

FIGURE C6.6-1 Energy Grade Line Analysis—Spreadsheet Algorithm

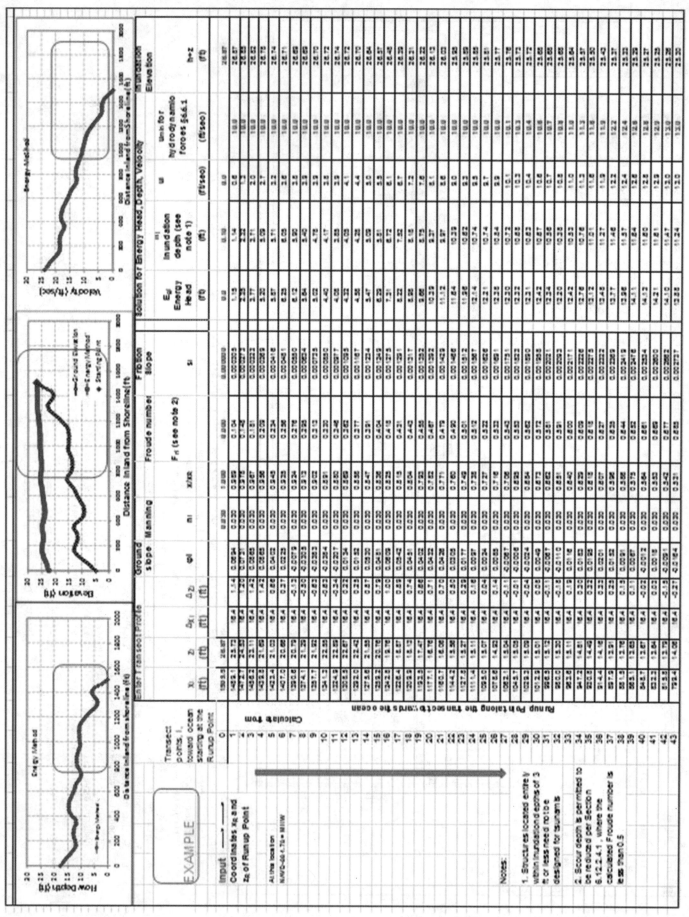

FIGURE C6.6-2 Example Transect using the Energy Grade Line Analysis

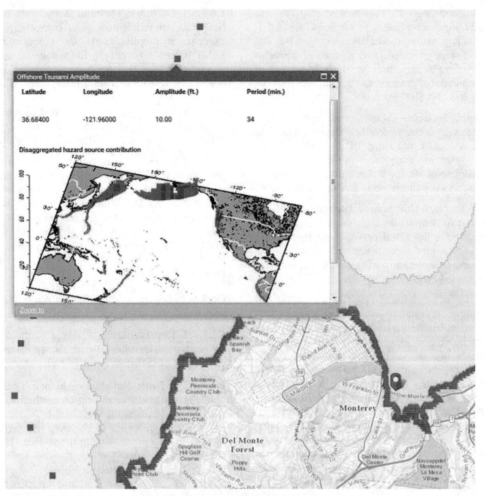

FIGURE C6.7-1 Example of the Offshore Tsunami Amplitude from the ASCE Tsunami Design Geodatabase (amplitude in feet, period in minutes)

macroroughness is used to account for friction. Maximum runup, inundation depth, flow velocity, and/or specific momentum flux at the site of interest is permitted to be evaluated by either of the following techniques:

a. Determine the Hazard-Consistent Tsunami as one or more surrogate events that replicate the weighted average waveform corresponding to the offshore wave height for the return period. The inundation limit for the Hazard-Consistent Tsunami shall match the area that is inundated by tsunami waves from all the disaggregated major source zones for that particular return period. Flow parameters at the site of interest shall be determined from the Hazard-Consistent Tsunami waveform.
b. Develop the probabilistic distributions of flow parameters from the disaggregated sample of computed tsunamis and construct the statistical distributions of flow parameters for at least three critical Load Cases.
9. From the probabilistic events, capture the design flow parameters of inundation depth, flow velocity, and/or specific momentum flux at the site of interest.

Data referenced in Section 6.7 are provided by ASCE. The ASCE Tsunami Design Geodatabase is version-specific for the application of each edition of ASCE 7, and the user of the standard should take care to select the proper data set as appropriate for the building code adopted by the jurisdiction.

The ASCE Tsunami Design Geodatabase of geocoded reference points of offshore 328-ft (100-m) depth; Tsunami Amplitude, H_T; and Predominant Period, T_{TSU}, of the Maximum Considered Tsunami are available at the ASCE website at http://asce7tsunami.online, including the following data sets:

- Offshore tsunami amplitude with predominant period and
- Disaggregated hazard source contribution figures.

The GIS digital map layers of probabilistic subsidence are available in the ASCE Tsunami Design Geodatabase at http://asce7tsunami.online.

An example of the ASCE Tsunami Design Geodatabase of Offshore Tsunami Amplitude, Predominant Wave Period, and geocoded location of a database point is shown in Fig. C6.7-1.

C6.7.1 Tsunami Waveform. The waveform prescribed here is a sum of two solitons, one with a positive and one with a negative amplitude. This pair reflects (1) the case where the leading wave has a positive amplitude, and (2) the case where the largest (positive) amplitude wave follows a large trough wave. Because tsunamis consist of multiple waves interacting with the coastal bathymetry, both cases of leading or trailing depression waveforms are to be considered; the range of parameter a_2 shall also be considered in developing the suite of waveforms to be analyzed by inundation modeling. The parameters a_1, a_2, and T_{TSU} can be obtained from the map (Fig. 6.7-1) or from Table 6.7-1, which provides a more regionalized model based on Fig. 6.7-1. The values of Table 6.7-1 were compiled from a series of scenarios modeled for each of the five western states, in which the maxima of the offshore wave amplitudes of the crest and trough were sampled at offshore locations at 328-ft (100-m) depth.

C6.7.2 Tsunamigenic Sources. Tsunamigenic sources that affect the Pacific states have been identified through historical records as well as systematic modeling of all circum-Pacific subduction zones. Maximum magnitudes were derived from perceived maximum dimensions for future earthquake ruptures and scaling relations between dimensions and magnitude.

C6.7.3 Earthquake Rupture Unit Source Tsunami Functions for Offshore Tsunami Amplitude. For the computation of offshore amplitudes (e.g., at the 328-ft or 100-m depth contour), the principle of linearity is used to compute tsunami waves as a summation of precomputed waves for smaller subsources, which in total make up the intended earthquake rupture.

C6.7.4 Treatment of Modeling and Natural Uncertainties. *Aleatory* uncertainty is the uncertainty that arises because of unpredictable variability in the performance of the natural system being modeled. *Epistemic* uncertainty is the uncertainty that arises because of lack of knowledge of the physical mechanism of the natural system being modeled. Typically, aleatory uncertainties are represented in probabilistic analysis in the form of probability distribution functions, whereas epistemic uncertainties are included in the form of (weighted) alternative logic tree branches. The technique for a one-sigma allowance is specified in order that numerical modeling provide a Hazard-Consistent Tsunami waveform that accounts for uncertainty in tsunami amplitude. It is a key component of PTHA to include a reliability factor.

C6.7.5 Offshore Tsunami Amplitude. The 328-ft (100-m) depth contour was chosen because tsunami waves behave linearly in deeper water, which allows for rapid calculation of tsunami waveforms, which enables a comprehensive probabilistic approach. The 328-ft (100-m) depth was selected to minimize the amount of nonlinear shoaling caused by such things as the Nearshore Profile, embayment amplification, and resonance effects. The Offshore Tsunami Amplitude at this reference depth represents with more clarity the probabilistic tsunami amplitude values for an area that do not vary dramatically parallel to the shoreline. To use a fully shoaled tsunami wave amplitude at the shoreline would result in more difficulty in verifying or calibrating whether other site-specific models have actually used a Hazard-Consistent Tsunami because the waveform has become more complex at the shoreline, and in certain cases, it may have broken into a series of bores; for these reasons, the tsunami amplitudes have large spatial variation in the nearshore regime. It would also not be possible at the shoreline to specify regional waveform parameters, such as in Table 6.7-1 at the 328-ft (100-m) depth that would be valid in all nearshore regimes. The Offshore Tsunami Amplitude and associated parameters of Section 6.7 are calibrated to the 2,475-year hazard level, and the primary purpose of this information is to support site-specific inundation analysis of Tsunami Risk Category IV structures, where more detailed spatial bathymetry and topography are used with a 2D inundation model software code to develop flow parameters at the site. It is called a 2D model because the third vertical dimension of flow speed is depth-averaged, but it uses a 3D spatial digital raster representation of the terrain.

C6.7.5.1 Offshore Tsunami Amplitude for Distant Seismic Sources. An efficient way to compute probabilistic tsunami exceedance amplitudes is by precomputing tsunami time series (Green's functions) for a suite of subsources with unit slip (e.g., 3.28 ft or 1 m) that together can be used to represent actual slip distribution on a larger fault. The resultant tsunami time series is then a summation of the individual Green's functions, which are each multiplied by the actual slip relative to the unit slip for each subsource. Examples of this process are given in Thio et al. (2010) and Gica et al. (2008).

C6.7.5.2 Direct Computation of Probabilistic Inundation and Runup. For local sources, it is necessary to compute a suite of scenarios, which represent a probabilistic distribution, taking into account aleatory uncertainties in such things as slip distribution, magnitudes, and tide levels. The ground subsidence should be taken into account in the actual tsunami calculations. The surface deformation can be computed using a variety of algorithms that compute the elastic response of a medium caused by slip on a fault.

C6.7.6 Procedures for Determining Tsunami Inundation and Runup

C6.7.6.1 Representative Design Inundation Parameters. These parameters have been shown to be relevant for the estimation of different types of tsunami impact.

C6.7.6.2 Seismic Subsidence before Tsunami Arrival. The uplift and subsidence during an earthquake that generates the tsunami also causes vertical changes to the coastal topography, and this change needs to be taken into account for modeling local tsunamis directly from the source. The effect of seismic subsidence follows the recognized deformation model of Okada (1985) illustrated in Fig. C6.7-2. The subsidence can be considered instantaneous, and it occurs before the arrival of the tsunami waves. Fig. 6.7-3 is the 2,475-year subsidence map, which also identifies which areas are governed by the local coseismic tsunami source. Southern California and Hawaii are not adjacent to interplate subduction zones, so they are not subject to regional seismic ground subsidence before a tsunami.

C6.7.6.3 Model Macroroughness Parameter. The effects of ground friction on tsunami inundation are significant and need to be considered. There are different techniques in which ground friction is incorporated in numerical schemes; the most common is the Manning's coefficient. If other approaches are used, care should be taken so that they closely follow the effect related to the appropriate Manning's coefficient.

C6.7.6.4 Nonlinear Modeling of Inundation. Typically in numerical tsunami modeling, the equations solved correspond to the shallow water approximation, with nonlinear effects such as inundation (moving boundary), bottom friction, and advection. Usually, these algorithms solve the equations of motion in two (horizontal) dimensions, with the vertical properties averaged over depth. Most of the effects mentioned here are taken into account, provided that an acceptable tsunami inundation model is used. Dispersion is taken into account in some approaches (Boussinesq-type); dispersion is required for the modeling of small dimensional sources, such as submarine landslides. Bore formation is more accurately modeled with algorithms that preserve momentum. In the case of local tsunamis, where the bore formation is particularly pronounced, these algorithms are preferred over approaches that do not explicitly conserve mass.

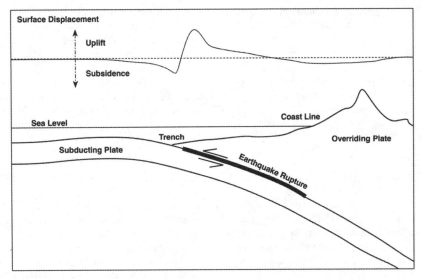

FIGURE C6.7-2 Subduction Zone Earthquake Mechanism for Tsunami Generation

C6.7.6.5 Model Spatial Resolution. Tsunami Digital Elevation Models for the United States are available from NOAA through the following website: http://www.ngdc.noaa.gov/mgg/inundation/. For select areas, models with resolutions of up to 32.8 ft (10 m), or 1/3 arcsec, are available as of 2015.

C6.7.6.6 Built Environment. The standard practice of all tsunami inundation models is to use a low Manning's roughness parameter for the topography, so that the lines for the inundation limit and runup elevation are not underestimated. The models have also been numerically benchmarked with that assumption against analytical and laboratory cases with smooth profiles. Digital Elevation Models (DEMs) for very high resolution modeling may be available in the form of lidar (light detection and ranging) models from federal, state, county, or city agencies, or from local site surveys. In some cases, the built environment has been removed from the lidar images, which negates their usefulness for detailed inundation modeling. When the built environment is modeled as fixed bluff bodies on the terrain, it is appropriate to include only those with sufficient structural resilience to resist tsunami forces.

C6.7.6.7 Inundation Model Validation. The National Tsunami Hazard Mitigation Program (NTHMP) (2012) has developed a number of benchmarking exercises that include analytical, laboratory, and real-world simulations, available at the Forecast Propagation Database: http://nctr.pmel.noaa.gov/propagation-database.html. Aida (1978) provides further information relating to the degree of fit criteria of K between 0.8 and 1.2 and κ less than 1.4.

C6.7.6.8 Determining Site-Specific Inundation Flow Parameters. The Hazard-Consistent Tsunami is a term of craft used to refer to a means to replicate the effects of the Maximum Considered Tsunami (MCT) hazard level by one or more surrogate scenarios. Hazard-Consistent Tsunamis are devised to incorporate the net effect of an explicit analysis of uncertainty in the parameters originally used in determining the MCT through conducting the Probabilistic Tsunami Hazard Analysis. To incorporate this aleatory uncertainty in a few surrogate Hazard-Consistent Tsunamis, the slip parameters of the generating rupture of the Hazard-Consistent Tsunami are increased. This is a procedural device to produce a convenient representation of input for a more limited number of inundation analyses. As a result, the Hazard-Consistent Tsunami appears greater than a single deterministic event generated by an earthquake of that mean recurrence interval at the same source.

In urban environments, the flow velocities determined by a site-specific inundation analysis for a given structure location may not be taken as less than 90% of those determined in accordance with the Energy Grade Line Analysis method. For other terrain roughness conditions, the site-specific inundation analysis flow velocities at a given structure location shall not be taken as less than 75% of those determined in accordance with the Energy Grade Line Analysis method. These restrictions apply before any velocity adjustments caused by flow amplification. The reason for these restrictions is that the tsunami model validation standard, Synolakis et al. (2007), presently does not validate flow velocities, and the committee evaluation indicated that in certain cases of field observations, some inundation models appear to underestimate flow velocities. The limiting values are based on a reliability analysis accounting for the reduced uncertainty of momentum flux associated with site-specific probabilistic tsunami hazard inundation analysis, compared to the prescriptive Energy Grade Line Analysis method based on the ASCE 7 Tsunami Design Zone Maps. (Chock et al. 2016)

C6.7.6.9 Tsunami Design Parameters for Flow over Land. The inundation numerical model should capture the time-correlated series of inundation depth and flow velocity (and therefore the representation of momentum flux) at the site for the three key Load Cases defined in Section 6.8.3.1. However, if the maximum momentum flux is found to occur at an inundation depth different than Load Case 2, that point in the time series should also be considered.

C6.8 STRUCTURAL DESIGN PROCEDURES FOR TSUNAMI EFFECTS

It is important to understand that building failure modes differ fundamentally between seismic loads (high-frequency dynamic effects generated on the inertial masses of a structure) and tsunami loads (externally and internally applied sustained fluid forces varying with inundation depth over long-period cycles of

Table C6.8-1 Tsunami Performance Levels per Tsunami Risk Category

Hazard Level and Tsunami Frequency	Tsunami Performance Level Objective		
	Immediate Occupancy	Damage Control	Collapse Prevention
Maximum Considered Tsunami (2,475-year mean recurrence interval)	Tsunami Vertical Evacuation Refuge Structures	Tsunami Risk Category IV and Tsunami Risk Category III Critical Facilities	Tsunami Risk Category III and Risk Category II

load reversal). Tsunami-induced failure modes of buildings have been examined in several detailed analyses of case studies taken from the Tohoku tsunami of March 11, 2011 (Chock et al. 2013a). Building components are subject simultaneously to internal forces generated by the external loading on the lateral-force-resisting system, together with high-intensity momentum pressure forces exerted on individual members. Performance objectives for structures subject to tsunami design requirements are given in Table C6.8-1.

C6.8.1 Performance of Tsunami Risk Category II and III Buildings and Other Structures. Based on analysis of prototypical buildings, a threshold of 65 ft (19.8 m) above grade plane is deemed the appropriate height for both reliable Life Safety and reasonable economic cost for Tsunami Risk Category II buildings to resist tsunami loads. Importantly, when evacuation from the inundation zone is not possible before tsunami arrival because of long evacuation distances, road congestion, or damaged infrastructure, the public will attempt to use taller buildings to escape the tsunami inundation and will inherently expect that such taller structures will not collapse during the tsunami. As just one example from the city of Ishinomaki, per a detailed investigation of the lessons for international preparedness after the 2011 Great East Japan Tsunami (Fraser et al. 2012), about 500 people sought refuge at three designated buildings in Ishinomaki. In addition to the few Tsunami Vertical Evacuation Refuge Structures available, "There was widespread use of buildings for informal (unplanned) vertical evacuation in Ishinomaki on March 11th, 2011. In addition to these three designated buildings, almost any building that is higher than a 2-storey residential structure was used for vertical evacuation in this event. About 260 official and unofficial evacuation places were used in total, providing refuge to around 50,000 people. These included schools, temples, shopping centres and housing" (Fraser et al. 2012).

Tsunami Risk Category III buildings and other structures are to be designed against collapse, since these structures include school buildings with mass public assembly occupancies, health-care facilities with 50 or more resident patients, critical infrastructure such as power and water treatment, and facilities that may store hazardous materials (when the quantities of hazardous material do not place them in Risk Category IV). Where the structure does not have an occupiable floor with an elevation exceeding 1.3 times the Maximum Considered Tsunami inundation elevation by more than 10 ft (3.0 m), the facility should implement a plan and procedure for evacuation to a location above and outside of the Tsunami Evacuation Zone or to a designated Tsunami Vertical Evacuation Refuge Structure.

C6.8.2 Performance of Tsunami Risk Category III Critical Facilities and Tsunami Risk Category IV Buildings and Other Structures. The state, local, or tribal government may specifically designate certain traditionally Risk Category III structures as more important Critical Facilities that have post-tsunami mission critical functions or have more critical roles in economic recovery. Critical Facilities and Risk Category IV Essential Facilities should be located outside of the Tsunami Design Zone whenever possible. For those structures that necessarily exist to serve critical and essential services to a community within a coastal zone subject to tsunami hazard, the design provisions target better than a Life Safety performance level (that is, what is called the Damage Control performance level) for the floor levels that are not inundated. Vertical Evacuation Refuge Structures have the highest performance level objective. The Damage Control performance level is an intermediate structural performance level between Life Safety and Immediate Occupancy.

C6.8.3 Structural Performance Evaluation

C6.8.3.1 Load Cases. The normalized inundation depth and depth-averaged velocity time history curves of Fig. 6.8-1 (and corresponding Table C6.8-2) are based on tsunami video analysis, and they are generally consistent with numerical modeling with respect to the Load Cases defining critical stages of structural loading for design purposes (Ngo and Robertson 2012). Load Case 1 is for calculating the maximum buoyant force on the structure with the associated hydrodynamic lateral force, the primary purpose of which is to check the overall stability of the structure and its foundation anchorage against net uplift. The uplift force, calculated in accordance with Section 6.9.1, depends on the differential inundation depth exterior to the structure versus the flooded depth within it. Load Case 2 is for calculating the maximum

Table C6.8-2 Values of Normalized Inundation and Flow Speed of Fig. 6.8-1

t/T_{tsu}	h/h_{max}	u/u_{max}
0.000	0.000	0.000
0.033	0.125	0.517
0.067	0.250	0.726
0.111	0.417	0.881
0.133	0.500	0.943
0.178	0.670	1.000
0.267	0.833	0.764
0.356	0.933	0.550
0.444	0.983	0.333
0.500	1.000	0.000
0.556	0.983	−0.333
0.644	0.933	−0.550
0.733	0.833	−0.764
0.822	0.670	−1.000
0.867	0.500	−0.943
0.889	0.417	−0.881
0.933	0.250	−0.726
0.967	0.125	−0.517
1.000	0.000	0.000

hydrodynamic forces on the structure. Load Case 3 is for calculating the hydrodynamic forces associated with the maximum inundation depth. The time history curves can be used to determine the inundation depth and velocities at other stages of inundation as a function of the maximum values determined by the Energy Grade Line Analysis. When a site-specific inundation analysis procedure is used, the local Authority Having Jurisdiction may approve a site-specific inundation and velocity time history curve, subject to the minimum values of 80% or 100% of the Energy Grade Line Analysis, as indicated.

C6.8.3.2 Tsunami Importance Factors. The values of Tsunami Importance Factors are derived from the target structural reliabilities, which were calculated using Monte Carlo simulation involving a million trial combinations of random variables independently occurring in proportion to their statistical distributions for the demand parameters of fluid density, closure ratio, Energy Grade Line Analysis momentum flux, inundation depth hazard, and the aleatory uncertainty of inundation depth. For capacity, the structural component analyzed is a beam-column member carrying gravity loads. The Importance Factors for each Risk Category, analyzed in combination with the other parameters discussed above, result in structural component reliabilities, given that the MCT has occurred, that are similar to seismic systemic performance given the MCE has occurred:

- Tsunami Risk Category II: Tsunami Importance Factor of 1.0 results in structural component limit state exceedance probability of 7.5%$_{(MCT)}$ vs. 10%$_{(MCE)}$ for collapse of the lateral-force-resisting system during an earthquake.
- Tsunami Risk Category III: Tsunami Importance Factor of 1.25 results in structural component limit state exceedance probability of 4.9%$_{(MCT)}$ vs. 5%$_{(MCE)}$ for collapse of the lateral-force-resisting system during an earthquake.
- Tsunami Risk Category IV: Tsunami Importance Factor of 1.25 results in structural component limit state exceedance probability of 2.7%$_{(MCT)}$ vs. 2.5%$_{(MCE)}$ for collapse of the lateral-force-resisting system during an earthquake.
- The exceedance probability for the limit state of structural components of a Tsunami Vertical Evacuation Refuge Structure given the occurrence of the MCT would be approximately 0.8%.

C6.8.3.3 Load Combinations. The structural load combinations given are consistent with the Extraordinary Load Combinations of Section 2.5 of this standard by applying F_{TSU} for incoming and receding directions for A_k in the load combinations given in Eq. (2.5-1), as modified to include the effect of tsunami-induced lateral earth pressure caused by water seepage through the soil, H_{TSU}, resulting in Eq. (6.8-1). The inundation depths and velocities for determining F_{TSU} and H_{TSU} must be consistent at the hazard level of the Maximum Considered Tsunami. The load factor for H_{TSU} is given as 1.0 because these load combinations occur during the defined submerged conditions of the Maximum Considered Tsunami (MCT). The Probabilistic Tsunami Hazard Analysis criteria for determining runup already include explicit mathematical consideration of uncertainty; therefore, no additional factor needs to be applied to the 2,475-year MCT flow characteristics for Risk Category II buildings and other structures.

No allowable stress design load combinations were deemed necessary. When foundation stability analysis is performed for tsunami-induced soil seepage pressures, the results are equivalent to the existing recognized U.S. Army Corps of Engineers (USACE 2005, 2011) specifications for geotechnical limit equilibrium analysis for both overturning and uplift. For foundation design, resistance factors are given in Section 6.12.1.

C6.8.3.4 Lateral-Force-Resisting System Acceptance Criteria. The use of the overstrength factor to evaluate the overall lateral-force-resisting system is permitted to be used where the Life Safety performance level is to be verified. If Life Safety criteria have been met, then the Collapse Prevention criteria have also been met as a consequence, which is the minimum requirement for Tsunami Risk Category II and Tsunami Risk Category III buildings and other structures. For Immediate Occupancy, the structural system should be explicitly analyzed.

The capacity of minimum-code seismically designed Risk Category II structures using this technique was demonstrated in the EERI Special Issue of March 2013 in the paper providing analysis of prototypical buildings for tsunami and seismic requirements (Chock et al. 2013a). In this paper, several prototypical buildings were investigated, with the following assumptions:

> Parameters for a comparison of overall systemic resistance to tsunami loading are as follows: (1) the prototypical Risk Category II buildings of increasing height selected for illustrative purposes were 120 ft (36.5 m) long and 90 ft (27.5 m) wide and are 25% open; (2) They were located in high seismic zones (in the United States, $S_s = 1.5$ and $S_1 = 0.6$); (3) Connections developed the inelastic capacities of the members; (4) Tsunami flow velocity was 26.2 ft/s (8 m/s); (5) Each tsunami inundation load curve represented the continuum of hydrodynamic loading cases inundation increases to the maximum depth; and (6) Seismic capacity is used based on the overstrength capacity. It was found by analysis that the structural systems of larger scaled and taller buildings are inherently less susceptible to tsunamis than this prototype, provided that there is adequate foundation anchorage for resistance to scour and uplift.

C6.8.3.5 Structural Component Acceptance Criteria. The design of structural components shall comply with Section 6.8.3.5.1 or in accordance with alternative performance-based criteria of Section 6.8.3.5.2 or 6.8.3.5.3, as applicable. The alternative Sections 6.8.3.5.2 and 6.8.3.5.3 are not prescriptive methods of analysis.

C6.8.3.5.1 Acceptability Criteria by Component Design Strength. The primary means of determining structural component acceptability is based on a linear elastic analysis and the evaluation of resultant actions of the load combinations compared with the design strength of the structural components and connections.

C6.8.3.5.2 Alternative Performance-Based Criteria. The alternative structural component acceptability criteria use acceptance criteria of ASCE 41-13 (ASCE 2014), *Seismic Evaluation and Retrofit of Existing Buildings*. With an adaptation of this method, strength and stability can be checked to determine that the design of the structural components is capable of withstanding the tsunami to achieve the Structural Performance level required. The tsunami adaption allows the techniques of the linear static and nonlinear static analysis. For the purposes of this chapter, it may be appropriate to generically adapt modeling parameters nonspecific to earthquake ground motion found in the Nonlinear Static Procedure or Simplified Nonlinear Static Procedure of ASCE 41-13 (ASCE 2014). For example, it should be acceptable to use the effective stiffness values of ASCE 41-13 (ASCE 2014) for this purpose. Structural components can then be checked using the ASCE 41-13 acceptability criteria for actual tsunami loads and depths that are correct from the standpoint of fluid mechanics without load factors. Results of the analysis procedure shall not exceed the numerical acceptance criteria for linear procedures of ASCE 41-13 (Chapters 9 through 11) for the Structural Performance criteria required for the building or structure Tsunami Risk Category. For nonlinear static analysis procedures, expected deformation capacities shall be greater than or equal to the maximum deformation demands calculated at the required tsunami forces and applied actions.

C6.8.3.5.3 Alternative Acceptability by Progressive Collapse Avoidance. Where tsunami loads or effects exceed acceptability criteria for a structural element, a recognized method of checking the residual structural system gravity-load-carrying capacity is the alternate load path procedure given by *Design of Structures to Resist Progressive Collapse* (DOD 2013). An ASCE standard for progressive collapse avoidance is not yet available.

C6.8.4 Minimum Fluid Density for Tsunami Loads. The tsunami inundation conditions relevant to structural design are by no means represented by pristine seawater. Overland flow is more similar to a debris flood with suspended soil and debris objects of various sizes and materials. The fluid density factor, k_s, is used to account for such soil and debris within equivalent fluid forces that are not accounted for in Section 6.11, which deals with debris impacts by larger objects. The value of $k_s = 1.1$ was selected to represent an equivalent sediment concentration of 7%, assuming a specific gravity of 2.5 for the suspended soil particles.

C6.8.5 Flow Velocity Amplification. Tsunami Design Zone Maps are based on so-called "bare-earth" conditions (i.e., without discrete objects) with an equivalent Manning's roughness to represent what exists above ground. When the Energy Grade Line Analysis is used with the Manning's roughness based on built environments, greater energy is required to reach the specified mapped runup. Urban environments are expected to have a higher probability of generating flow amplification effects through numerous diffraction scenarios that are not possible to specifically enumerate. Nevertheless, the Energy Grade Line Analysis implemented with an urban area's roughness produces sufficiently conservative depth and momentum flux, and additional flow amplification need not be considered. Therefore, to account for these possible effects, site flow velocities shall not be taken as less than 100% of those determined in accordance with the Energy Grade Line Analysis method of Section 6.6.

This section of the Commentary provides some examples of flow diffraction effects taken from available North American research. Experimental studies discussed in Thomas et al. (2015) and Nouri et al. (2010) relate to flow amplification. The example of channelized flow conditions is inferred from the University of Ottawa experiments, and the results indicated for symmetrically arranged objects is based on Oregon State University experiments conducted in the Network for Earthquake Engineering Simulation (NEES) tsunami wave basin. Note that these tests were for higher Froude numbers than are used in the provisions. The explicit flow amplification cases described here are for places where the geometry of nearby structures leads to higher than normal flows.

Building types deemed to be unable to resist the tsunami hydrodynamic forces are to be excluded because of the fact that they will not be able to redirect sustained flow to other buildings. Also, the provisions do not permit the shielding effect of neighboring structures to reduce flow velocity below the basic value derived from the Energy Grade Line Analysis method. The engineer should exercise judgment and conduct additional studies for other conditions where flow amplification would be expected based on the principles of fluid mechanics.

Flow Velocity Amplification for Channel Effect—University of Ottawa Research Discussed in Nouri et al. (2010). Where obstructing structures exist along each side of a street or open swath oriented parallel to the flow such that the ratio L/W is between 1 and 3 and W_c/W is between 0.6 and 0.8, as shown in Fig. C6.8-1, the incident flow velocity amplification on a structure located at the end of such a channelized flow was

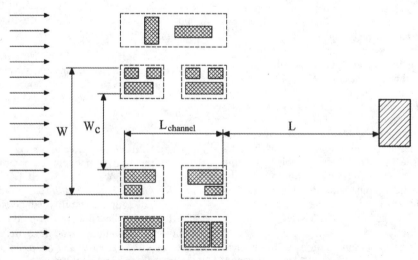

FIGURE C6.8-1 Flow Amplification for Channelized Conditions

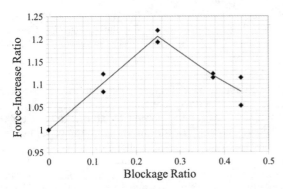

FIGURE C6.8-2 Flow Amplification Force-Increase Ratio for Channelized Conditions (Where the Blockage Ratio Is the Net Width of Constricting Obstructions Divided by the Total Flume Width)

found to be as shown in Fig. C6.8-2, in which the blockage ratio is $1 - W_c/W$.

Flow Amplification for Two Isolated Upstream Structures— Oregon State University Research Discussed in Thomas et al. (2015). Where tsunami bores flow through a layout of two upstream isolated buildings in a symmetric arrangement, the net flow amplification factor downstream was found to be related to the effective wake clearance angle, β. The effective wake angle is shown in Fig. C6.8-3, and a summary of results is shown in Fig. C6.8-4 and Table C6.8-3. This figure also illustrates what is the center third of the width of the downstream structure.

The approximate envelope of the flow amplification factors is given in Table C6.8-3. Linear interpolation is reasonable between the listed values.

C6.8.5.2 Flow Velocity Amplification by Physical or Numerical Modeling. The implementation of an alternative flow amplification analysis is expected to consist of a numerical model or an experimental model that is sufficiently documented to comply with Section 6.8.10. Park et al. (2013) provide a discussion of this approach.

C6.8.6 Directionality of Flow

C6.8.6.1 Flow Direction. Rather than assume that the flow is uniformly perpendicular to the shoreline, variation by ±22.5 degrees from the perpendicular transect is considered, in which the center of rotation of the possible transects is located at the site. The first objective is to find the highest runup elevation associated with the possible design transects. The second objective is to account for variation in the load application to the structure. The governing transect that produces the maximum runup elevation may be used to perform an Energy Grade Line Analysis, and the prescribed directional variation for the resulting flow parameters may be used to compute directional loads. Alternatively, an Energy Grade Line Analysis may be computed for each possible direction.

The determination of outflow direction may also require consideration that the outflow current may be affected by existing stream- and riverbeds and drainage canals, as well as additional scour and sediment transport, which may cause significant morphological changes in topography during a vigorous draw-down cycle after a large watershed is inundated.

C6.8.6.2 Site-Specific Directionality. Although a site-specific inundation analysis is permitted to establish flow vectoring at the site, some uncertainty in the accuracy of this estimation is reflected in the ±10 degree variability.

C6.8.7 Minimum-Closure-Ratio-for-Load-Determination. Loads on buildings shall be calculated assuming a minimum closure ratio of 70% of the pressure exposed surface area of the exterior enclosure. This assumption accounts for accumulated waterborne debris trapped against the side of the structure, as well as any internal blockage caused by building contents that cannot easily flow out of the structure. As a practical matter based on observations of buildings subjected to destructive tsunami, "breakaway" walls cannot be totally relied upon to relieve structural loading, primarily because of the copious amount of external debris. Also, studs and girts may be capable of entrapping contents within a building, thus generating hydrodynamic drag forces on the internal debris, which in turn transfer those loads to the structure, per Chock et al. (2013a, b).

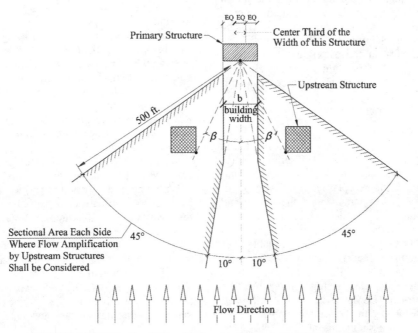

FIGURE C6.8-3 Effective Wake Clearance Angle, β, for Flow Amplification

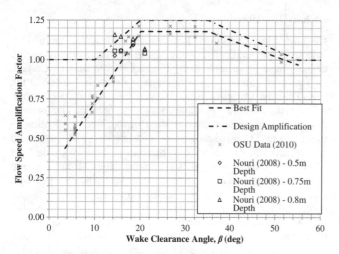

FIGURE C6.8-4 Amplification of Flow Speed vs. Wake Clearance Angle

Table C6.8-3 Flow Speed Amplification Factors

Effective Wake Clearance Angle (β) in degrees	0	10	20	35	≥55
Flow speed amplification factor from two symmetrically arranged buildings located upstream	1.0	1.0	1.25	1.25	1.0

C6.8.8 Minimum-Number-of-Tsunami-Flow-Cycles. Designers working on loading shall consider a minimum of two tsunami inflow and outflow cycles, the second of which shall be at the Maximum Considered Tsunami design level. This consideration is required because the condition of the building and its foundation may change in each tsunami inflow and outflow cycle. Therefore, building foundation designs shall consider changes in the site surface and the in situ soil properties during the multiwave tsunami event. Local scour effects caused by the first cycle shall be calculated as described in Section 6.12.2.5 but based on an inundation depth at 80% of the Maximum Considered Tsunami design level. The second tsunami cycle shall be considered to be at the Maximum Considered Tsunami design level, in which the scour of the first cycle is combined with the loads generated by the inflow of the second cycle.

C6.8.9 Seismic Effects on the Foundations Preceding Local Subduction Zone Maximum Considered Tsunami. Since the assumption in this standard for a local Maximum Considered Earthquake includes plastic deformation in structural performance, it is important that the postelastic strength of the structure not be degraded by excessive ductility demand. Tsunami loading (except for debris impacts) is of a sustained nature, so the strength capacity available after the earthquake should be maintained at a predictable level by limited ductility demand.

For Tsunami Vertical Evacuation Refuge Structures, the Commentary of FEMA P-646 (FEMA 2012) states, "Utilizing the approach in ASCE/SEI 41-13, the performance objective for code-defined essential facilities should be at least Immediate Occupancy performance for the Design Earthquake (DE) and Life Safety performance for the Maximum Considered Earthquake (MCE)." The ASCE design provisions only consider the MCE.

C6.8.10 Physical Modeling of Tsunami Flow, Loads, and Effects. The capacity to generate appropriately scaled flows means that a test facility should be able to model both the specific site geometry of the structure or structures, as well as the form and duration of the incident tsunami or multiple tsunamis with an accuracy and duration that are appropriate for the process being investigated. Testing of a building or multiple buildings, for example, may need to account for the relevant topography and bathymetry, as well as the scaled roughness of terrain. Testing for site-specific conditions may require separate simulations of inflow and drawdown outflow of a tsunami waveform or use of a bore generator or recirculating flume to ensure adequate duration of loading.

It is appropriate to use physical modeling together with numerical modeling to evaluate site-specific conditions. This combination includes tasks such as the calibration of numerical models with data obtained from physical models and the testing of loads on individual components in a site-specific numerical model. The spatial and temporal limitations of physical model testing may necessitate combination with numerical modeling for a site-specific analysis. For example, the required variation in wave directionality (Section 6.8.6) may not be achievable for the structure or structures under consideration, so results of physical flow tests at a single incident angle can be used to calibrate a numerical model that addresses the full variability in flow direction.

Physical models can be used to simulate the inundation depths and velocities in tsunami flows for use with the design procedures and equations in Sections 6.9, 6.10, 6.11, and 6.12. Physical modeling may also be indicated for cases where flows do not follow the assumptions used to develop the equations in Sections 6.8, 6.9, 6.10, 6.11, and 6.12 or cannot economically be addressed using a numerical model alone. This situation can be the case where the flow varies rapidly temporally or spatially and where there are dynamic pressure variations, rotational flow, or multiphase flows.

Dynamic and kinematic similarity of model and prototype should match as closely as possible. For tsunamis, the most important scaling parameter is usually taken to be the Froude number, but many other dimensionless parameters (such as Reynolds, Euler, and Cauchy numbers) should also be constrained in order to obtain useful results. For example, the Reynolds number should be maintained sufficiently high to ensure fully developed turbulent flow so that the model is insensitive to Reynolds number variation. For wave generation, the effects of turbulence and air entrainment generated in breaking waves and at the leading edge of turbulent bores are difficult to scale and may affect test results. Scaling concerns are also particularly difficult for problems involving water–soil–structure interaction, structural response, or debris impacts. For example, the limiting 1:10 scale provided for structural components is appropriate for modeling loads on a rigid element, but it may need to be increased in order to model the response of the structural element. In the case of debris impacts, experiments are often carried out at full scale because the nonlinear structural behavior and material properties are not adequately represented at smaller scales.

There is extensive guidance on scaling of physical models starting with Buckingham (1914) up to recent commentary on coastal engineering applications by Hughes (1993). Briggs et al. (2010) and Goseberg et al. (2013) address considerations on scaling and on tsunami wave and bore generation in the laboratory.

C6.9 HYDROSTATIC LOADS

C6.9.1 Buoyancy. Uplift caused by buoyancy has resulted in numerous structural failures during past tsunamis, including to concrete and steel framed buildings not designed for tsunami conditions (Chock et al. 2013a, b). High water table levels and rapid saturation of the ground surface during tsunami inundation may enable pressure to develop below the grade level of the building. The resulting uplift force is proportional to the volume of water displaced by the structural components and any enclosed spaces below the inundation level at the time buoyancy is being considered. The displaced volume should include, as a minimum, any structural components, enclosed spaces, floor soffits, and integrated structural slabs where air may be entrapped by beams. Nonstructural walls designed to break away under tsunami loads and standard windows can be assumed to fail and allow water into the interior of the building, thus relieving the buoyancy effect. However, windows designed for large missile wind-borne debris impact, such as those in hurricane zones, and those designed for blast loading, should be assumed to remain intact throughout the tsunami (unless analyzed to determine their breaking strength threshold). Enclosed spaces with openings or breakaway wall elements equal to or greater than 25% of the enclosure envelope below the inundation level can be assumed to fill with water, thus relieving the buoyancy effect.

That full hydrostatic pressure develops below the grade level of the building is a conservative assumption based on permeable foundation soils, such as silts, sands, and gravels, but may be too conservative for cohesive soils such as clay and clayey silt. Soil permeability should be evaluated in the context of the duration and pressure head of the tsunami inundation depth at the site. However, if the first-floor slab on grade has the typical isolation joints around the columns, uplift on the slab will lift the slab and yield it but will not lift the superstructure. Cohesive soils adjacent to foundation components and basement walls may provide resistance to the resulting buoyancy.

Load Case 1, defined in Section 6.8.3.1, requires that a minimum uplift condition be evaluated at an inundation depth of one story or the height of the top of the first-story windows. The exceptions to Load Case 1 of Section 6.8.3.1 also apply. In summary, net buoyancy may be avoided by preventing the buildup of hydrostatic pressure beneath structural slabs, allowing the interior space to become flooded or designing for pressure relief or structural yielding relief of the hydrostatic head pressure, sufficient deadweight, anchorage, or a combination of the above design considerations.

C6.9.2 Unbalanced Lateral Hydrostatic Force. Hydrostatic unbalanced lateral force develops on a wall element because of differences in water level on either side of the wall, irrespective of the wall orientation to the tsunami flow direction. Narrow walls or those with openings equal to or exceeding 10% of the wall area are assumed to allow water levels to equalize on opposite sides of the wall. However, for wide walls, or when perpendicular walls on either the front or back of the wall under consideration prevent water from getting to the other side of the wall, unbalanced hydrostatic loads should be considered. This condition need only be considered during inflow Load Cases 1 and 2.

C6.9.3 Residual Water Surcharge Load on Floors and Walls. During tsunami drawdown, water may not drain from elevated floor slabs that have perimeter structural components such as an upturn beam, perimeter masonry, or concrete wall or parapet. This lack of drainage results in surcharge loads on the floor slab that may exceed the slab capacity. The potential depth of water retained on the slab depends on the maximum inundation depth during the tsunami but would be limited to the height of any continuous perimeter structural components that have the capacity to survive the tsunami loads and that would retain water on an inundated floor. Nonstructural elements above this perimeter structural element are assumed to have failed during tsunami inflow so that they will not contribute to the retention of water on the slab during drawdown.

C6.9.4 Hydrostatic-Surcharge-Pressure-on-Foundation. During tsunami inundation and drawdown, it is possible for different water levels to exist on opposite sides of a wall, building, or other structure under consideration. The resulting differential in hydrostatic surcharge pressure on the foundations should therefore be considered in the foundation design.

C6.10 HYDRODYNAMIC LOADS

Hydrodynamic loads develop when fluid flows around objects in the flow path. Tsunami inundation may take the form of a rapidly rising tide or surge, or a broken bore. Both of these conditions are considered here. Because tsunami waves typically break offshore, no consideration is given to the wave breaking loads typically associated with wind-driven storm waves (FEMA 2011).

C6.10.1 Simplified Equivalent Uniform Lateral Static Pressure. It is anticipated that most buildings and other structures subject to the provisions of this chapter will be designed for other lateral load conditions, such as wind and seismic loads. For large or tall buildings, these other load conditions can result in greater loads on the lateral-force-resisting system than the tsunami loads, particularly in high seismic hazard regions. In such cases, it is therefore desirable to have a simplified but conservative approach to check whether or not tsunami load conditions will affect the structural system.

Eq. (6.10-1) is provided as a conservative alternative to more detailed tsunami loading analysis. This equation is based on the assumption that all of the most conservative provisions presented elsewhere in this section occur simultaneously on a rectangular building with no openings. The maximum hydrodynamic loads are assumed to occur during Load Case 2 (Section 6.8.3.1), when $h = 2/3 h_{max}$, assuming a conservative Froude number of $\sqrt{2}$ and a drag coefficient of $C_d = 2.0$. Based on the more severe condition of bore impact per Section 6.10.2.3, the resulting lateral load per unit width of the building is given by the following:

$$f_w = 1.5 * \frac{1}{2} k_s \rho_{sw} I_{tsu} C_d C_{cx} (hu^2)_{max}$$

$$= \frac{3}{4} * k_s \rho_{sw} I_{tsu} (2)(0.7) \left(\frac{2}{3} h_{max}\right) \left(\sqrt{2} \sqrt{g * \frac{2}{3} h_{max}}\right)^2$$

$$= 0.933 k_s \rho_{sw} g I_{tsu} h_{max}^2$$

Including the worst effects of flow focusing suggested in Commentary Section C6.8.6 as a 1.25 amplification on the flow velocity (Table C6.8-3) and a 1.1 factor to allow for additional uncertainty, results in Eq. (6.10-1):

$$f_w = 1.1(1.25)^2(0.933 k_s \rho_{sw} g I_{tsu} h_{max}^2) = 1.6 k_s \rho_{sw} g I_{tsu} h_{max}^2$$

To account for additional buildup of water level at the forward edge of the building, this load is distributed as a rectangular pressure distribution over a height of $1.3h_{max}$. The resulting pressure is therefore

$$P_{us} = 1.6 k_s \rho_{sw} g I_{tsu} h_{max}^2 / 1.3 h_{max} \approx 1.25 I_{tsu} \gamma_s h_{max}$$

The lateral-force-resisting system should be evaluated for this pressure distribution acting over the entire width of the building perpendicular to the flow direction for both incoming and outgoing flows. All structural members below $1.3h_{max}$ should be evaluated for the effects of this pressure on their tributary width of projected area. Although it is not possible for a bore to occur on drawdown outflow, topography and erosion may result in additional flow acceleration that is not accounted for by the 1.25 amplification factor of Commentary Section C6.8.5. Therefore, it was determined that the intent of this simplified equation was better met by using the same loading for both inflow and drawdown outflow cases.

C6.10.2.1 Overall Drag Force on Buildings and Other Structures. Once flow develops around the entire building or structure, the unbalanced lateral load caused by hydrodynamic effects can be estimated using Eq. (6.10-2), which is based on fluid mechanics (FEMA 2011). The values of drag coefficient, C_d, provided in Table 6.10-1 depend on the ratio between the building width perpendicular to the flow direction and the inundation depth (FEMA 2011). A wider structure results in a greater buildup of water level at the leading edge of the structure. The closure coefficient, C_{cx}, represents the vertical projected area of structural components, relative to the vertical projected area of the submerged section of the building. This ratio may not be taken as less than the value given in Section 6.8.7 so as to account for debris accumulation. Eq. (6.10-2) is evaluated for all three Load Cases defined in Section 6.8.3.1. Because the drag coefficient, C_d, depends on the inundation depth, and the inundation depth changes for each Load Case, the appropriate drag coefficient is determined from Table 6.10-1 for each of the specified Load Cases.

C6.10.2.2 Drag Force on Components. All structural components and exterior wall assemblies below the inundation depth are subjected to the hydrodynamic drag forces given by Eq. (6.10-3). This classical hydrodynamic drag expression depends on an empirically determined drag coefficient, C_d, based on the shape of the individual element. Typical values for C_d for common member cross-sectional shapes are given in Table 6.10-2 (OCADI 2009; Blevins 1984; Sarpkaya 2010; Newman 1977). Post-tsunami observations show that exterior elements are subject to debris accumulation that makes for an irregular shape, so for exterior elements, a C_d of 2.0 is used. The net force determined from Eq. (6.10-3) is to be applied as a distributed pressure load on the submerged portion of the component being designed. All three Load Cases defined in Section 6.8.3.1 should be considered. As the inundation depth increases for different Load Cases, the components that are inundated increase. Structural components that are a part of the lateral-force-resisting system may be subject to the net resultant of their participation in resisting the overall drag force on the structure of Section 6.10.2.1 and the pressure drag caused by local flow around the component defined in this section.

C6.10.2.3 Tsunami-Loads-on-Vertical-Structural-Components, F_w. Laboratory experiments have shown that when the leading edge of a tsunami surge, often taking the form of a broken bore, impinges on a wide wall element, a short-duration impulsive load develops that exceeds the hydrodynamic drag force obtained from Eq. (6.10-5) (Ramsden 1993; Arnason et al. 2009; Paczkowski 2011; Robertson et al. 2013). This increase is approximately 50%, resulting in the amplification factor of 1.5 applied to the steady-state drag expression in Eq. (6.10-5). Where the flow initiated as a bore but the Froude number has been reduced by topography to be less than 1.0 at a site, then the bore is considered dissipated and Eq. (6.10-5b) is not invoked.

This loading condition need only be checked when the inflow velocity is at a maximum, i.e., Load Case 2 in Section 6.8.3.1. It is applied to all wall elements that are wider than three times the inundation depth corresponding to Load Case 2. This aspect ratio is based on the bore height to specimen width ratio for which Arnason's experiments resulted in an impulsive force exceeding the subsequent steady-state drag force (Arnason et al. 2009). Because of the short duration of this impulsive load, windows and doors are assumed to be intact until the peak load is reached.

An alternative, more detailed approach to predicting the maximum impulsive force developed when a bore strikes a wall was developed by Robertson et al. (2013). Based on large-scale experiments at Oregon State University, it was determined that the lateral load per unit width of wall can be estimated using the following:

$$F_w = k_s \rho_{sw} \left(\frac{1}{2} g h_b^2 + h_j v_j^2 + g^{\frac{1}{3}}(h_j v_j)^{\frac{4}{3}} \right)$$

where

h_b is the bore height, equal to the sum of still water depth, d_s, and jump height, h_j; and

v_j is the bore velocity, which can be estimated using hydraulic jump theory as

$$v_j = \sqrt{\frac{1}{2} g h_b \left[\frac{h_b}{d_s} + 1 \right]}$$

This force acts as a triangular pressure distribution over a height of h_p with a base pressure of $2F_w/h_p$, where h_p is the instantaneous ponding height at maximum load given by

$$h_p = \left(0.25 \frac{h_j}{d_s} + 1 \right)(h_b + h_r) \leq 1.75(h_b + h_r)$$

where $h_r = \left(\dfrac{v_j h_j}{\sqrt{g}} \right)^{2/3}$

Application of this expression to a large structural wall damaged during the Tohoku tsunami showed excellent agreement with the observed damage (Chock et al. 2013a). The above equations can be appropriate when detailed information on a bore strike scenario is determined from field data or estimated from a site-specific inundation model analysis.

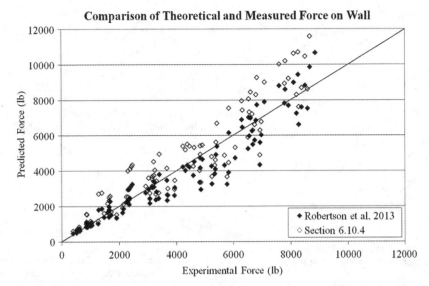

FIGURE C6.10-1 Predicted vs. Experimental Bore Loading on Wall (1:5 Model Scale)

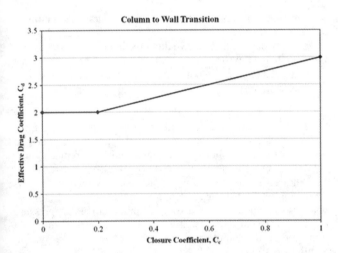

FIGURE C6.10-2 Hydrodynamic Drag Transition from Individual Column to Solid Wall

For design purposes, comparison between the more detailed approach and that used in Section 6.10.2.3 for all of the bore-on-wall experimental cases performed at Oregon State University shows good agreement between predicted and measured results (Fig. C6.10-1).

C6.10.2.4 Hydrodynamic Load on Perforated Walls, F_{pw}. The impulsive force on a solid wall obtained from Section 6.10.2.3 can be reduced if there are openings in the wall through which the flow can pass. When applied to Eq. (6.10-3), the 1.5 factor applied in Section 6.10.2.4 produces an effective drag coefficient, $C_d = 3.0$. Experiments performed at Oregon State University on perforated walls indicate that closure coefficients, C_{cx}, less than 20% have no effect on the force on individual wall piers (Santo and Robertson 2010). However, higher closure ratios resulted in increased loading on all wall piers. A linear transition is therefore assumed between the fully closed wall and a wall with 20% closure coefficient, as shown in Fig. C6.10-2.

C6.10.2.5 Walls Angled to the Flow. Eq. (6.10-7) provides for a reduction in hydrodynamic loads on a wall positioned oblique to the flow. This reduction is also the same that was provided for breaking waves in Chapter 5 of the 2010 edition of this standard.

C6.10.3.1 Flow Stagnation Pressure. Observations from the Tohoku tsunami indicate that structurally enclosed spaces created by structural walls on three sides and structural slabs can become pressurized by flow entering a wall-enclosed space without openings in the side or leeward walls (Chock et al. 2013b). Analysis of three of these conditions formed by reinforced concrete walls confirmed that the internal pressure reached the theoretical flow stagnation pressure given by Eq. (6.10-8) (Bernoulli 1738).

C6.10.3.2 Hydrodynamic Surge Uplift at Horizontal Slabs

C6.10.3.2.1 Slabs Submerged during Tsunami Inflow. Experiments performed at Oregon State University on horizontal slabs with no flow obstructions above or below the slab indicate that uplift pressures of 20 psf (0.958 kPa) can develop on the slab soffit (Ge and Robertson 2010). Blockage in the form of structural columns or walls below the slab results in larger uplift pressures given in Section 6.10.3.3.

C6.10.3.2.2 Slabs over Sloping Grade. A horizontal elevated slab located over sloping ground is subjected to upward hydrodynamic pressure if the flow reaches the slab soffit elevation. Because of the grade slope, the flow has a vertical velocity component of $u_v = u_{max} \tan \phi$, where ϕ is the average slope of the grade below the slab being considered. Eq. (6.10-9) estimates that this vertical velocity induces an upward pressure equivalent to the transient surge load on walls given by Section 6.10.2.3, assuming a C_d value of 2.0.

C6.10.3.3 Tsunami Bore Flow Entrapped in Structural Wall-Slab Recesses

C6.10.3.3.1 Pressure Load in Structural Wall-Slab Recesses. Evidence from past tsunamis shows that significant uplift pressures can develop below horizontal structural slabs, such as floors and piers, if the flow below the slab is blocked by a wall or other obstruction (Saatcioglu et al. 2005; Chock et al. 2013b). Incoming flow striking the wall is diverted upward but is blocked by the

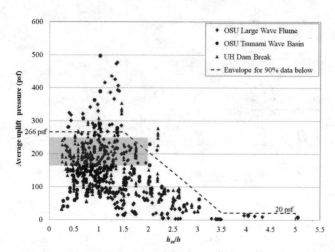

FIGURE C6.10-3 Average Uplift Pressure on Slab Soffit When Flow Is Blocked by a Solid Wall (1 psf = 0.048 kPa)

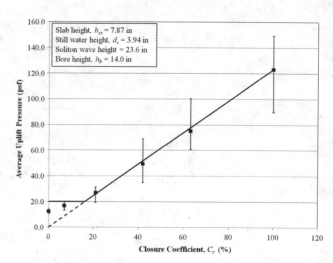

FIGURE C6.10-4 Mean Slab Uplift Pressure Reduction Caused by Perforated Wall (1 psf = 0.048 kPa; 1 in. = 25.4 mm)

Note: Vertical Bars Show Boxplot Range of Test Results at Each Closure Ratio.

slab, resulting in large pressures on both the wall and slab soffit close to the face of the wall. A series of experiments performed at small and large scales at Oregon State University and the University of Hawaii demonstrated that this effect is most severe when the inundation depth exceeds 2/3 of the slab elevation, or $h_s/h \leq 1.5$ (Ge and Robertson 2010). Fig. C6.10-3 shows the results for tests at small scale (OSU tsunami wave basin and UH dam break using identical test setup at approximately 1:10 model scale) and large scale (OSU large wave flume at approximately 1:5 model scale) with bore heights twice those in the small-scale experiments. An envelope was developed that enclosed 90% of the test data, as shown in Fig. C6.10-3.

Although there is considerable variability in the uplift pressures even for identical repeat wave conditions, there appeared to be consistency between the results for both small- and large-scale tests, indicating that the laboratory results also apply at full scale. Estimation of the uplift pressures required to cause uplift failure of pier access slabs during the Tohoku tsunami indicated that the pressures ranged from 180 to 250 psf (8.62 to 12.0 kPa), indicated by the shaded area in Fig. C6.10-3 (Chock et al. 2013b). Although the inundation depth at the time of slab failure is unknown, these uplift pressures observed at full scale confirm the appropriateness of the envelope in Fig. C6.10-3.

Pressure distributions measured at peak uplift indicated that the outward pressure on the wall and the uplift pressure under the slab adjacent to the wall were approximately a third greater than the average pressure, whereas the outward pressure away from the wall dropped to about half of this value. If the wall has finite width, l_w, and water is able to flow around the ends of the wall, then the uplift pressure on the slab beyond a distance $h_s + l_w$ is assumed to drop to the nominal uplift of 30 psf (1.436 kPa). The entire wall and the slab within h_s of the wall are therefore to be designed for an uplift pressure of 350 psf (16.76 kPa). Between a distance h_s and $h_s + l_w$, the slab is to be designed for an uplift pressure of 175 psf (8.38 kPa). Beyond $h_s + l_w$, the slab is to be designed for an uplift pressure of 30 psf (1.436 kPa).

C6.10.3.3.2 Reduction of Load with Inundation Depth. As shown in Fig. C6.10-4, when the inundation depth is less than 2/3 of the slab elevation, or $h_s/h > 1.5$, the uplift pressure envelope drops linearly. Eq. (6.10-10) provides the equivalent linear decrease for the 350-psf (16.76-kPa) pressure prescribed in Section 6.10.3.3.1. The slab uplift pressure outside a distance of h_s from the wall also decreases proportionately. A minimum average uplift pressure of 20 psf (0.958 kPa) is indicated by the test data for slabs as high as five times the inundation depth. Once the slab height exceeds five times the inundation depth, the upward directed flow did not reach the slab, so no uplift pressure need be considered.

C6.10.3.3.3 Reduction of Load for Wall Openings. Experiments were performed in the tsunami wave basin at Oregon State University using the same experimental setup as described in Section C6.10.3.3.1 but with a perforated wall replacing the solid wall behind the slab (Ge and Robertson 2010). These tests indicated that the slab uplift pressures decrease linearly as a function of the percentage closure provided by the perforated wall, as shown in Fig. C6.10-4. A minimum uplift pressure of 20 psf (0.958 kPa) was observed even when the wall was removed completely, leading to the provision in Section 6.10.3.2.1.

C6.10.3.3.4 Reduction in Load for Slab Openings. Experiments performed at the University of Hawaii demonstrated that the presence of an opening gap between the slab and the solid wall significantly relieves the uplift pressure on the slab Fig. C6.10-5 shows the test setup used for these experiments, performed at approximately 1:12 scale. Fig. C6.10-6 (Takakura and Robertson 2010) shows the reduction in uplift pressure, P_u, for a slab with an opening gap, compared with one that has no gap, as the gap width, w_g, changes relative to the slab soffit height, h_{ss}. Eqs. (6.10-12), (6.10-13), and (6.10-14) are based on these data and are represented by the dashed line in Fig. C6.10-6. The same effect is assumed to occur when the gap is created by means of a panel designed to break away at an uplift pressure less than 175 psf (8.38 kPa). Evidence from numerous piers inundated during the Tohoku tsunami indicate that pressure relief gratings and loose access panels were effective as breakaway panels for the purpose of relieving uplift pressures on the remaining slab (Chock et al. 2013b).

C6.10.3.3.5 Reduction in Load for Tsunami Breakaway Wall. If the wall restricting flow below the slab is designed as a tsunami breakaway wall, then it can be assumed to fail when the pressure on the wall exceeds that required to fail the connection between

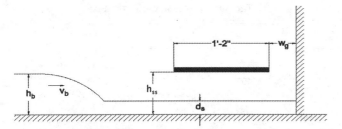

FIGURE C6.10-5 Test Setup to Study Effect of Opening Gap or Breakaway Slab on Uplift Pressure

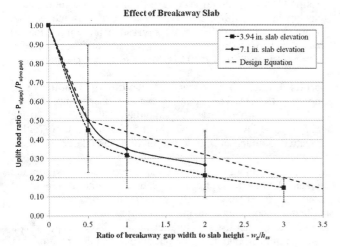

FIGURE C6.10-6 Reduction in Slab Uplift Pressure Caused by Presence of Opening Gap or Breakaway Slab (1 in. = 25.4 mm)

the wall and the slab. This pressure will be the highest that can be experienced by the slab before failure of the wall.

C6.11 DEBRIS IMPACT LOADS

Tsunamis can transport a large volume of debris. Virtually anything in the flow path that can float given the inundation depth and that cannot withstand the water flow becomes debris. Common examples are trees, wooden utility poles, cars, and wood-frame houses and portions thereof. Some nonfloating debris, such as boulders and pieces of concrete, can also be transported if the flow is strong enough. This section covers the specification of forces and duration of the impact on structures by such debris. Debris impact forces shall be determined for the location of the structure based on the potential debris in the surrounding area that would be expected to reach the site during the tsunami. Of particular concern are the perimeter structural components oriented perpendicular to the flow direction because they are at the greatest risk of impact and their loss may compromise the ability of the structure to support gravity loads.

The impact forces depend on the impact velocity, which is assumed to be equal to the flow velocity for floating debris. The points of application of the impact force, which is assumed to be a concentrated force, shall be chosen to give the worst case for shear and moment for each structural member that should be considered within the inundation depth and the corresponding flow velocity. Exceptions to this are specified in subsequent sections based on specific debris characteristics.

The ubiquity of (1) logs and/or poles; (2) passenger vehicles; and (3) boulders and concrete debris requires the assumption that these things will impact the structure if the inundation depth and velocity make it feasible.

Closed shipping containers float very easily, even if loaded. Therefore, for structures near a container yard, impact from floating containers should be considered. The site hazard assessment procedure in Section 6.11.5 is used to assess if impact from shipping containers should be considered at a particular location.

Ships (including ferries) and barges are also potential debris that may impact structures. The likelihood of such "extraordinary" debris impact is most significant for structures near ports and harbors that contain these vessels. Because impact from these objects is likely to place a demand that cannot be resisted economically for many structures, only Tsunami Risk Category III Critical Facilities and Tsunami Risk Category IV buildings and structures are required to consider such impact. The site hazard assessment procedure in Section 6.11.5 is used to assess if impact from marine vessels should be considered.

Table C6.11-1 summarizes the requirements for design, especially the threshold inundation depths at which level (or greater) it is required to consider each debris impact type.

C6.11.1 Alternative Simplified Debris Impact Static Load. Designing for a conservative, prescriptive load is allowed to replace specific consideration of impact by logs, poles, vehicles, boulders, concrete debris, and shipping containers. The value of 330 kips (1,470 kN) is based on the cap of 220 kips (980 kN) in Section 6.11.6 for shipping containers, multiplied by a dynamic amplification of 1.5. Note that the maximum dynamic amplification factor from Table 6.11-1 was not used in order to account for the reduction of peak forces caused by inelastic response of the impacted component. If it is shown that the site is not in the container or ship impact zone per Section 6.11.5, then the impact force is assumed to occur from a direct strike by a wood log. Lehigh University (Piran Aghl et al. 2014) tested a nominal wood log of 450-lb (204-kg) weight (see Section C6.11.2). A basic direct strike force of 165 kips (734 kN) can be used instead of the 330 kips (1,470 kN) based on the shipping container. The nominal design impact force for logs or poles is limited to the material crushing strength. This prescriptive 165-kip (734-kN) load includes a structural dynamic response factor of 1.5 and is associated with poles and logs with parallel to grain crushing strength of 5,000 psi (34.5 MPa) (approximately mean plus one standard deviation) for Southern Pine or Douglas Fir per ASTM D2555 (2011). The effective contact area is assumed to be 22 sq in. (142 cm^2), which represents about 20% of the

Table C6.11-1 Conditions for Which Design for Debris Impact Is Evaluated

Debris	Buildings and Other Structures	Threshold Inundation Depth
Poles, logs, and passenger vehicles	All[a]	3 ft (0.914 m)
Boulders and concrete debris	All[a]	6 ft (1.8 m)
Shipping containers	All[a]	3 ft (0.914 m)
Ships and/or barges	Tsunami Risk Category III Critical Facilities and Category IV[b]	12 ft (3.6 m)

[a]All buildings and other structures as specified in Section 6.1.1.
[b]Tsunami Risk Category III Critical Facilities and Category IV buildings and other structures in the debris impact hazard region as determined in Section 6.11.5.

end-on area of a 1-ft (30.5-cm) diameter pole, consistent with Piran Aghl et al. (2014). In all cases, the Orientation Factor, C_o, is applied. The net prescriptive simplified force is conservative compared with the laboratory test results, which are also considered conservative compared to likely field conditions.

C6.11.2 Wood Logs and Poles. Previous provisions for debris impact forces, such as in ASCE 7-10 (2013a), Section C5.4.5, have been based principally on an impulse-momentum formulation for rigid-body impact, which requires an assumption of the duration of impact. Eq. (6.11-2) is based on stress wave propagation in the debris and hence considers the flexibility of the debris and structural member. The assumptions are elastic impact and a longitudinal strike. That is, in the case of a pole or log, it hits the structure at its butt end, rather than transversely. Likewise, for a shipping container, per Section 6.11.6, the assumption is that the end of one bottom longitudinal rail of the container strikes the structural member. Full-scale testing at Lehigh University has validated the equation for a utility pole and a 20-ft (6.1-m) shipping container under these conditions (Piran Aghl et al. 2014; Riggs et al. 2014). Eq. (6.11-4) for the duration is also based on elastic impact. It assumes that the impact force is constant, resulting in a rectangular force–time history. Although the duration may be underestimated somewhat, the total impulse is conservative.

The nominal 1,000-lb (454-kg) log or pole is adopted from ASCE 7-10 (2013a) Section C5.4.5 for flooding. Wood properties vary widely, but this corresponds approximately to a 30-ft (9.15-m) log with a 1-ft (30.5-cm) diameter. However, much larger trees are possible in certain geographical areas, and design professionals should consider regional and local conditions. The minimum stiffness is based on these dimensions and a modulus of elasticity of approximately 1,100 ksi (7,580 MPa). It is obtained from the well-known EA/L relation for axial stiffness.

Debris impact is clearly a dynamic event, and the structural element responds dynamically. However, an equivalent static analysis is allowed, where the static displacement is multiplied by a dynamic factor. The scaling factors as a function of the ratio of impact duration to the natural period of the structural element represent the shock spectrum. Shock spectra also depend on the shape of the force–time curve. The shock spectrum specified in Table 6.11-1 is adopted from ASCE 7-10 (2013a), Section C5.4.5 and is a modified version of the shock spectrum for a half-sine wave. The difference is that the factor in Table 6.11-1 remains constant for $t_d/T > 1.4$, whereas the half-sine wave shock spectrum decreases. However, the Lehigh experimental results show better agreement with the values in Table 6.11-1 because the force–time history is not truly a half-sine wave (Piran Aghl et al. 2014).

The value of Orientation Factor, C_o, was derived from the data of Haehnel and Daly (2004), jointly sponsored by ASCE and FEMA. It is the mean plus one standard deviation value of the log debris impact force for trials that included glancing and direct impacts of freely floating logs.

C6.11.3 Impact by Vehicles. Passenger vehicles are ubiquitous, float, and are easily transported. This standard requires the assumption that impact occurs as long as the inundation depth is sufficient to float the vehicle, which is deemed to be 3 ft (0.914 m). NCAC (2011, 2012) describe an experimental and numerical analysis of the frontal crash impact against a wall of a 2,400-lb (1,090-kg) subcompact passenger vehicle traveling at 35 mph (15.6 m/s). Based on the results therein, the initial stiffness of the vehicle was estimated to be 5,700 lb/in. (1 kN/mm). With an assumed velocity of 9 mph (4 m/s), Eq. (6.11-2) results in an impact force of approximately 30 kips (133 kN) (Naito et al. 2014). Based on a more likely glancing impact with a smaller contact area, it is judged that 30 kips (133 kN) is a sufficiently conservative, prescriptive load to cover a range of possible vehicle impact scenarios. The impact can occur anywhere from 3 ft (0.914 m) up to the inundation depth.

C6.11.4 Impact by Submerged Tumbling Boulder and Concrete Debris. The tumbling boulder debris impact force has been established based on a simplified static approach (Chau and Bao 2010). A "boulder" weight of 5,000 lb (2,270 kg) (either an actual boulder at the lower end of the "large boulder" classification shown in Table C6.11-2 or similar-sized debris from failed structural components) is considered within the inundation zone tumbling at a relative maximum velocity of about 13.1 ft/s (4 m/s). A dynamic amplification factor of 2 is implicitly incorporated in the force. The tumbling boulder is assumed to impact the structural element at 2 ft (0.61 m) above grade to reflect that the motion of the boulder is tumbling (rolling) along the ground surface.

C6.11.5 Site Hazard Assessment for Shipping Containers, Ships, and Barges. A procedure is specified to determine if impact from these debris objects should be considered. The procedure is based on the assumption that the debris is disbursed from a point source and then the hazard region associated with the point source is identified. This assumption may require large container yards or ports to be split into several point sources, each with its own hazard region. For example, a port that is spread out longitudinally along the coast may need to be represented by several point sources.

The basic idea of the procedure is to find a 45° circular sector with an area equal to 50 times the combined plan area of the debris, such that the debris, once disbursed, would have an average concentration (i.e., "area density") of 2%. The 45° range about the perpendicular transect is adopted from Section 6.8.6.1. Because the debris can be transported toward the shoreline on drawdown, the circular sector is "flipped," as shown in Fig. 6.11-1 to account for risk of impact during drawdown. The combined areas of both circular segments define the hazard region for which impact from containers, barges, or ships needs to be considered. This standard allows for the directionality of flow, and hence impact directionality, to be considered. This basic procedure was applied to several regions after the 2011 Tohoku tsunami with reasonable success (Naito et al. 2014). In the cases considered therein, not all debris was within the hazard region, but most was, and it is assumed that debris that are distributed more laterally have a significantly smaller velocity than the flow velocity and hence represent a smaller threat.

Transport of the debris can be limited by tsunami-resilient structures, geography, and insufficient inundation depth. For example, containers in a yard that is ringed on the leeward side by structural steel and concrete structures will not disburse

Table C6.11-2 Boulder Class Size

Boulder Size		
in.	mm	Boulder Classification
160–80	4,000–2,000	Very large boulders
80–40	2,000–1,000	Large boulders
40–20	1,000–500	Medium boulders
20–10	500–250	Small boulders

Source: Data from Lane (1947).

beyond the structures, as long as they cannot float over the structures. Similarly, ships with, for example, a 4-ft (1.2-m) draft will not be transported far or at significant speed in a 4-ft (1.2-m) inundation depth.

C6.11.6 Shipping Containers. The impact force and duration equations in Section 6.11.2 are also valid for head-on (longitudinal) impact by the corner of a shipping container. See the discussion in Section C6.11.2.

The Lehigh test results (Piran Aghl et al. 2014) have shown that the mass of container contents does not significantly affect the impact force as long as the contents are not rigidly attached to the structural frame. Therefore, for shipping containers, the empty mass of the container is used in Eqs. (6.11-2) and (6.11-4). However, the contents may increase the duration somewhat, so Eq. (6.11-5) is used to obtain an alternative duration that should also be considered. Note that the force in that equation is the same as for an empty container. Eq. (6.11-5) is based on essentially plastic impact caused by plastic deformation of the container, i.e., the container is assumed to "stick" to the structural element and not rebound.

Shipping containers are standardized in terms of length, height, and width, but weight and structural details can vary somewhat by manufacturer. The values for weight and stiffness provided in Table 6.11-2 are considered to be reasonable approximations for most standard ISO shipping containers. The loaded weights assume that the containers are loaded to about 50%. Hence, these numbers converted to mass can be used directly in Eq. (6.11-5) for $m_d + m_{contents}$. The stiffness values are based on EA/L, where E is the modulus of elasticity of steel, A is the cross-sectional area of one bottom rail of the container, and L is the length of the rail, not including any cast end blocks.

Eq. (6.11-2) does not contain any factor to account for an increase in force caused by the fluid flow being affected by the sudden stoppage of the debris object, which some other formulations include. For longitudinal impact of a log, such an increase in force is not expected to be significant. Testing at the NEES tsunami wave flume at Oregon State University on scale-model shipping containers also showed that for longitudinal impacts, the impact force was not significantly affected by the fluid (Riggs et al. 2014). The force coming from Eq. (6.11-2) is considered to be sufficiently conservative to allow the transient fluid "added mass" effect to be ignored.

It should be noted that the maximum required impact force of 220 kip (980 kN) for a shipping container is not the maximum force that a container could apply. The value is based on the Lehigh tests at 8.5 mph (3.8 m/s). Recently published results for simulations have indicated that the maximum force may be higher (Madurapperuma and Wijeyewickrema 2013), depending on the impact scenario. The 220 kip (980 kN) force has been chosen as a reasonable value for design.

An Orientation Factor, C_o, value of 0.65 is used, assuming that there is similar randomness in the alignment of the lower corner steel chord of the shipping containers with the target structural component.

C6.11.7 Extraordinary Debris Impacts. Extraordinary debris impacts, defined as impact by 88,000-lb (39,916-kg) or larger marine vessels, should be considered for Tsunami Risk Category III Critical Facilities and Tsunami Risk Category IV buildings and structures that are in the debris hazard impact region of a port or harbor, as defined in Section 6.11.5 and for which the inundation depth is 12 ft (3.66 m) or larger. The size vessel to be used depends on the most probable size vessel typically present at the port or harbor. The harbormaster or port authority can be consulted to determine typical vessel sizes, ballasted drafts, and weight displacement under ballasted draft. Typical vessel sizes are also provided in PIANC (2014).

The nominal impact force is calculated with Eq. (6.11-3), with the assumption that the vessel stiffness is larger than the structural member stiffness. Hence, the transverse structural member stiffness is to be used. The calculated force may be larger than any economical capacity of the member. Hence, the member may be assumed to have failed, in which case progressive collapse should be prevented for these important structures.

C6.11.8 Alternative Methods of Response Analysis. It is also permitted to carry out a dynamic analysis. This standard specifies that a rectangular pulse be applied. For a linear elastic analysis of a single degree of freedom system, the peak response is higher than that obtained from an equivalent elastic analysis because the rectangular shock spectrum is larger than the half-sine wave spectrum; see, for example, Chopra (2012).

If inelastic response is considered, the ductility of the structure can reduce the force demands. Both a nonlinear time history analysis as well as a work-energy approach are allowed.

A work-energy method for impact by large, essentially rigid debris that considers the ductility of the impacted structure leads to Eq. (C6.11-1):

$$F_{cap} = u_{max} \left[\frac{(1+e)}{(1+M_m/m_d)\sqrt{2\mu - 1}} \right] \sqrt{k_e M_e} \quad (C6.11-1)$$

where

F_{cap} is the capacity of the structure;
  e is the coefficient of restitution between debris and structure (e shall be assumed to be 1 unless otherwise substantiated);
M_e is effective mass of structure at impact point;
k_e is the initial elastic stiffness of structure at impact point; and
μ is permissible ductility ratio (e.g., per ACI 349-13 (2013), Appendix C, Section C.3.7).

This equation is based on an elastic–perfectly plastic force–displacement relation in which k_e is the elastic stiffness, F_{cap} is the "yield" force, and μ is the ratio of maximum displacement to F_{cap}/k_e. The mass terms are calculated as $M_m = \sum m_i \Delta_i$ and $M_e = \sum m_i \Delta_i^2$, where Δ_i represents the static displacements of the structure, as a result of a force at the impact point, scaled to have a value of 1 at the impact point. This approach is a slightly modified version of the approach in Kuilanoff and Drake (1991).

Care should be taken in choosing what part of the structure is used to obtain k_e, M_m, and M_e. The duration of impact should be considered. For example, impact by a log or shipping container occurs most likely over a very short time frame, and it is unlikely for all of a large structure to respond during such a short duration. Therefore, a section of the structure, possibly down to the structural member, should be considered. However, impact by a ship may occur over long enough time as to allow the entire structure to respond, and it may be appropriate to consider the entire structure.

For other work-energy approaches, where forces are accommodated by inelastic behavior up to some permissible ductility, the structure's initial stiffness is modified to an effective stiffness that reflects that deformation. In lieu of a nonlinear time history analysis, the secant or effective stiffness is a recognized technique of linearizing the modeled response analysis based on a performance point in the inelastic range, as shown in Fig. C6.11-1. Sample analysis indicates that in order to reach consistent results over a broad range of inelastic behavior, the

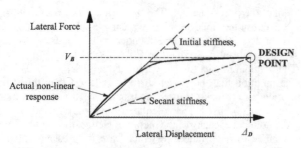

FIGURE C6.11-1 Stiffness Definitions for Initial Elastic and Effective Secant Stiffness

Source: Data from Sullivan et al. (2004).

velocity applied in the work-energy method of analysis shall be u_{max} multiplied by the product of importance factor, I_{tsu} and the Orientation Factor, C_o.

C6.12 FOUNDATION DESIGN

Design of structure foundations and tsunami barriers should consider changes in the site surface and in situ soil properties during the design tsunami. In addition to the site response and geologic site hazard considerations, similar to the seismic hazard provisions of this standard, the designer should consider both topographic changes caused by scour and erosion, and the effects of surrounding natural or design elements such as shielding and flow concentrations caused by other structures. Natural or designed countermeasures, such as barriers, berms, geotextile reinforcements, or ground improvements designed to protect the foundations and relieve them of direct loads, may be applied in the proximity of the foundation as well as in the foundation itself.

For deep foundations, the pile design procedure for tsunami loads relies upon determining the maximum anticipated loss of strength and scour depth from pore pressure softening while being subjected to the assigned inundation loading. The scour depths are based upon best available tsunami observation and analytical research. For comparison, the effects of general site erosion and local scour on deep foundations are also discussed for hurricane events producing similar magnitudes of general erosion and scour in FEMA 55 (2011), Section 10.5. Failures of deep foundations can result from either overloading the pile itself or from overloading at the pile–soil interface. Increasing a pile's embedment depth does not offset a pile with a cross section that is too small or pile material that is too weak; similarly, increasing a pile's cross section (or its material strength) does not compensate for inadequate pile embedment. The proposed approach provides a check on both cases for simple calculations. Advanced numerical modeling can assist in determining if load combinations can be reasonably isolated and may justify more efficient designs, as in the design of piles, to resist seismic liquefaction and shaking combinations.

The basic principle for foundation design is to apply the loads under pseudostatic conditions. As with seismic loads, for Critical Facilities subject to significant tsunami loads, it may be appropriate to apply time history loading, applying various combinations of loads and effects (Section 6.12.3).

This section is organized to provide a logical progression of analytical steps to define siting effects and to apply suitable countermeasures consistent with the direct loads. For in-water or over-water structures or barriers, which are beyond the intent of this standard, it is suggested that foundation design may be approached by applying the specified tsunami loads and using appropriate offshore design methods such as USACE CEM (2011), California MOTEMS (California State Lands Commission 2005), PIANC (2010), and API (2004).

In the tsunami load combination (as an extraordinary load in this standard Section 2.5.2), a 1.0 factor on H_{tsu} is used rather than the 1.6 factor in Section 2.3.2 since loading and unloading of bulk material storage structures would not be done during a tsunami.

A quotation from USACE (2005), *Stability Analysis of Concrete Structures*, is relevant to the necessary coordination between geotechnical and structural engineers:

> Even though stability analysis of concrete structures is a structural engineering responsibility, the analysis must be performed with input from other disciplines. It is necessary to determine hydrostatic loads consistent with water levels determined by hydraulic and hydrological engineers. Geotechnical engineers and geologists must provide information on properties of foundation materials, and must use experience and judgment to predict behavior of complex foundation conditions. *To ensure that the proper information is supplied, it is important that those supplying the information understand how it will be used by the structural engineer. To ensure that the information is applied appropriately, it is important that the structural engineer understand methods and assumptions used to develop this interdisciplinary data.*

Therefore, it is recommended that the report of the geotechnical investigation for a project in the Tsunami Design Zone include explanation of the derivation of design values for explicit use in strength design load combinations in accordance with Section 6.12.1.

C6.12.1 Resistance Factors for Foundation Stability Analyses. Typical failure mechanisms evaluated in foundation stability analyses are the following:

- Lateral sliding (due to tsunami forces with the added effects of any unbalanced lateral soil pressures caused by local scour on one side);
- Uplift or flotation;
- Piping conditions caused by excess seepage stresses reducing the strength and integrity of the soil fabric;
- Slope stability analysis (caused by saturation and pore pressure softening effects of inundation); and
- Bearing capacity (where soil strength properties may be affected by sustained pore pressures. However, lateral sliding failures typically occur before actual bearing stress failures, so this analysis is not expected to be governing, except where specifically cited in this chapter to be checked, e.g., hydrostatic fluid forces caused by differential water depth).

These effects are dominated by tsunami effects on the foundation and soil properties. As is typical with other geotechnical problems, soil loading analysis incorporates geotechnical judgment in selecting a reduced nominal strength that is valid across the variability of the soil deposit or load bearing strata and in recognition of the inherently nonlinear behavior of soil materials. For nonlinear materials, such as soils, a limit state is assumed to exist along some failure surface, and the resultant actions from an equilibrium analysis are compared to the reduced nominal strength for that material. Hence, this approach is commonly called limit equilibrium analysis. To ensure that the assumed failure does not occur, a resistance factor is applied to the material nominal strength. This factor does not imply an allowable stress design methodology of elastic analysis. However, the inverse of the resistance factor is often called a "factor of safety" in the recognized literature. So, equivalently,

Applied Load or Stress Resultant $\leq \phi$ Resistance

The "factor of safety" of 1.33 is a common minimum applicable to the analytical methods and practices for typical foundations, berms, geotextiles, and slope applications under a variety of design standards (USACE 2005, 2011). To recognize tsunami resistance as having progressive consequences to an initial failure, a ϕ of 0.67 is used as the corresponding resistance factor (i.e., a value lower than 1/1.33). The factor of safety for uplift of 1.5 has been adapted from USACE (2005) to allow a credit for the uplift resistance of piles and anchors since traditional water retention structures often rely solely on gravity for stability. Accordingly, the corresponding uplift resistance factor is 0.67. With a uniform resistance factor, combined effects of the tsunami on the foundation can be consistently evaluated.

C6.12.2 Load and Effect Characterization. Because of the successive cycles of tsunami inundation, foundation loads should be considered concurrent and cumulative and need to take into account landward flow of the tsunami waves, outflow of drawdown water, and the possibility that areas remain flooded between and after waves. Fig. C6.12-1 presents a schematic representation of the applicable loading on a foundation element for the design condition after local scour and general erosion have occurred and pore pressure and seepage effects are present.

In addition to the shear, axial, and bending forces that are transmitted to the foundation from a structure during a tsunami, the direct hydrostatic and hydrodynamic loading of the foundation or barrier because of exposure to flow from general erosion or local scour should be considered. Loading is affected by flow blockages and flow focusing as well as by dissipation of energy at the structure or foundation. Lateral soil pressure and pressure gradients are also affected by the tsunami inundation and may result in unbalanced loading because of flow blockage, excessive seepage, and unsteady flow effects.

C6.12.2.1 Uplift and Underseepage Forces. Hydrostatic uplift and seepage forces are traditionally considered for structures such as dams and levees that are designed to control or retain water. In the case of tsunamis, however, where there is sustained overland flow as well as trapping of water in low-lying areas and behind berms, roads, and foundations, the uplift force on the base of foundation elements should be evaluated. Guidance for design in the presence of uplift and seepage is available in

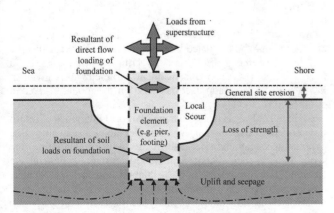

FIGURE C6.12-1 Schematic of Tsunami Loading Condition for a Foundation Element

USACE (1989, 2005). To evaluate the potential for soil saturation, seepage, and uplift, careful site investigations should be conducted to determine soil characteristics. Soil permeability and the potential for erosion, soil blowout, and piping during tsunami inundation should be assessed. Guidance for evaluating critical gradients with potential for internal erosion and piping caused by high-velocity inundation seepage stresses is available in Zhang et al. (2010) and Jantzer and Knutsson (2013).

To ensure that failure does not occur, a resistance factor is applied to the resistance of anchoring elements in recognition of the uncertainty and inherent variability in the soil resisting properties in uplift and/or underseepage conditions. Current load and resistant factors are provided in accordance with the unique dynamic loading magnitude and velocity conditions associated with tsunami inundation conditions. Future research is needed to determine if material-specific resistance factors are needed for conditions of excessive underseepage and resulting loss of strength. A methodology developing specific resistance factors for earthquake loading (Akbas and Tekin 2013) may provide an example to consider for adaptation to tsunami seepage conditions.

For foundations subject to tsunami flows and inundation, the weight of the structure and the soils that overlie the foundation act together with the foundation elements to resist uplift, as described in Eq. (C6.12-1):

$$0.9D + F_{tsu} \leq \phi R \qquad (C6.12\text{-}1)$$

where

- D = Counteracting downward weight of the structure, including deadweight and soil, above the bearing surface of the foundation exposed to uplift. The moist or saturated unit weight shall be used for soil above the saturation level, and the submerged unit weight shall be used for soil below the groundwater table.
- F_{tsu} = Net maximum uplift caused by the distribution of hydrostatic pressure around the building as determined from the analysis of tsunami inundation. The Load Cases during inflow and outflow of the tsunami are defined in Section 6.8.3.1.
- R = Upward design load resisting capacity for foundation structural elements, such as piles and anchors.
- ϕ = Resistance factor, which is 0.67 for these uplift resisting elements.

C6.12.2.2 Loss of Strength. Loss of strength is a critical design consideration that may require extensive mitigative countermeasures to be incorporated in the design. Loss of shear strength under tsunami loading can result from tsunami-induced pore pressure softening, piping, or seismic shaking. Pore pressure softening is a mechanism whereby increased pore-water pressure is generated during rapid tsunami loading and is released during drawdown. This increased pore-water pressure can soften the ground and decrease its effective shear strength. This decline decreases the shear stress required to initiate sediment transport and increases the scour depth. The primary differences between seismic liquefaction and tsunami-induced pore pressure softening are illustrated in Fig. C6.12-2.

The methods used to evaluate loss of strength caused by pore pressure softening should take into account the fundamentals of soil mechanics, including flow through porous media. The primary interest is estimated elevated pore pressures, uplift forces on soil grains, loss of confinement (decrease in effective stress), and associated loss of shear strength. The loss of shear strength is presumed to follow directly the percent loss of confinement.

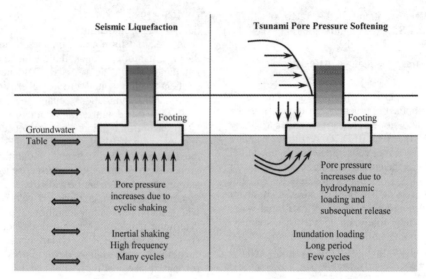

FIGURE C6.12-2 Schematic Diagram Showing Differences between Seismic Liquefaction and Tsunami-Induced Pore Pressure Softening

Scale modeling or numerical modeling of soil–structure–fluid interactions may be used. Alternatively, loss of strength caused by pore pressure softening may be evaluated by multiplying the shear strength by a factor $1 - \Lambda$, where Λ is a scour enhancement parameter (Tonkin et al. 2003). The scour enhancement parameter Λ approximates the fraction of the weight of soil grains that is supported by the excess pore water pressure. Equivalently, it is a measure of the loss of confinement. The loss of shear strength as it relates to both scour and structural design is presumed to follow directly this fractional loss of confinement. This strength reduction may be applied uniformly throughout the evaluation depth. The corresponding increased active and decreased passive earth pressures in this zone should be evaluated.

$$\Lambda = \min\left[0.5, \frac{2}{\sqrt{\pi}} \cdot \frac{h_{\max}\gamma_s}{\gamma_b\sqrt{c_V(T_{\text{draw}})}}\right] \quad \text{(C6.12-2)}$$

where

$h_{\max}$ is the maximum inundation depth;
γ_b is the buoyant weight density of the soils (see below);
γ_s is the fluid weight density for tsunami loads (from Section 6.8.5);
c_V is the consolidation coefficient of the soil (see below); and
T_{draw} is the drawdown timescale of the tsunami (see below).

The buoyant weight density of sediment, γ_b, is the difference between the bulk weight density of the saturated soil skeleton and the weight density of the pore water. When calculating γ_b, the pore water shall be treated as clean seawater with specific weight γ_{sw}.

Typical values of the consolidation coefficient c_V for sand and gravel are as follows (Tonkin et al. 2003; Hicher 1996; Francis 2008):

- Gravel: 10 ft²/s to 1,000 ft²/s (approximately 1 to 100 m²/s);
- Sand: 0.1 ft²/s to 1 ft²/s (approximately 0.01 to 0.1 m²/s).

When calculating Λ for finer materials, c_V may be taken as 0.1 ft²/s (approximately 0.01 m²/s). Although much lower values are commonly used in standard geotechnical practice, these very low values would lead to excessive loss of strength for fine materials. Field observations provide no evidence that loss of strength or scour is substantially greater in finer soils.

In the absence of a more detailed time series analysis, the drawdown timescale T_{draw} may be taken to be one-quarter of the tsunami wave period T_{tsu}. In areas where the maximum Froude number is less than 0.5, the value of Λ may be multiplied by an adjustment factor that linearly varies from 0 at the horizontal inundation limit to 1.0 at the point where the Froude number is 0.5.

C6.12.2.3 General Erosion. Both general erosion and local scour can contribute to the lowering of ground around a structure foundation. In the absence of countermeasures such as protective slabs on grade, the sum of general erosion and local scour is used in foundation design. Scour effects are generally discussed (nonspecific to tsunami) in USACE (1984, 1993), Simons and Senturk (1977), and FHWA (2012).

Evaluation of general site erosion may be based upon the standard literature and models that describe flood-induced general erosion, e.g., USACE 2010. However, these approaches do not include the effects of pore pressure softening. Pore pressure softening can increase the depth of general site erosion, as described in Yeh and Li (2008) and in Xiao et al. (2010). The effect of pore pressure softening on erosion and during drawdown may be evaluated using physical scale modeling or numerical modeling of soil–structure–fluid interactions similar to those described in these references.

Alternatively, the increase in general site erosion during drawdown may be evaluated by multiplying the buoyant specific weight γ_b of the sediment or the critical shear stress by a factor $1 - \Lambda$. The scour enhancement parameter Λ is given in Eq. (C6.12-2). This approach is based on the model of Tonkin et al. (2003).

Channelized scour occurs when significant quantities of return flow collect into a channel, for example, along a seawall or in a preexisting streambed. Because of this flow concentration, the scour depth in such channels can be greater than the general site erosion. If the geometry of the situation and lack of countermeasures suggest that channelized scour may be a factor, then this type of erosion should be analyzed. Pore pressure softening need not be factored into the depth of channelized scour because pore pressure softening is associated with rapid changes in hydrodynamic loading, whereas channelized scour is associated with a longer timescale of drawdown flow collected from a wider area.

C6.12.2.4 Scour. The geometry of the structure should be considered in the evaluation of local scour. Specifically, it should be determined whether tsunami flow is expected to be around the structure, causing flow acceleration around the obstruction, or whether the flow overtops the structure. Because of the very high stress gradients caused in tsunami flow, all levels of soil cohesion short of weathered rock or hard saprolite have been observed to rupture.

C6.12.2.4.1 Sustained Flow Scour. Sustained flow scour is caused by tsunami flow around a structure. Numerical, physical modeling or empirical methods may be used for sustained flow scour; the analysis method should consider the effects of pore pressure softening. The methodology provided here is described in Tonkin et al. (2013) and is based on a comparison of post-tsunami field observations of scour around structures with the model of Tonkin et al. (2003). The area extent is based solely on field observations in Francis (2008) and Chock et al. (2013b).

C6.12.2.4.2 Plunging Scour. Plunging scour is caused by tsunami flow over an overtopped structure. Numerical, physical modeling or empirical methods may be used for sustained flow scour; however, the analysis method is not required to include the effects of pore pressure softening. The methodology provided here is described in Tonkin et al. (2013) and is based on a comparison of post-tsunami field observations of scour around structures with the physical model results of Fahlbusch (1994) as described in Hoffmans and Verheij (1997).

C6.12.2.6 Displacements. Calculation of displacements is performed with the same procedures as other geotechnical displacement calculations recognized in literature for the identified cases of footing, slopes, walls, and piles. Each uses a different procedure. The calculations for tsunami cases are possible with no procedural modifications because the tsunami loads given in Section 6.12.2 are in a form consistent with other geotechnical loads.

C6.12.3 Alternative Foundation Performance-Based Design Criteria. In ASCE 41-13 (2014), foundation strength and stiffness characterization parameters are generally suitable for tsunami loading, but its procedures are not suitable for direct reference in the provisions. It may be also desirable for the foundation performance objectives for the facility to consider the role of the facility within community resilience and sustainability objectives. Tradeoffs that go beyond structural hardening include siting countermeasures, evacuation planning, and other emergency response planning provisions. In addition to Life Safety, loss of physical infrastructure and consequential damage and economic impacts to occupants and community services and commerce may be considered. These concepts are discussed in Presidential Policy Directive (2013) and TISP (2012). For Tsunami Risk Category IV buildings and structures, particularly in Site Classes D, E, and F, a tsunami–soil–structure interaction modeling analysis is recommended. Such modeling analysis is typically performed by geotechnical engineers.

C6.12.4 Foundation Countermeasures

C6.12.4.1 Fill. The uses of structural fill are discussed in FEMA 55 (2011), Section 10.3. Structural fill can be eroded during tsunamis, and it may not be feasible to provide adequate

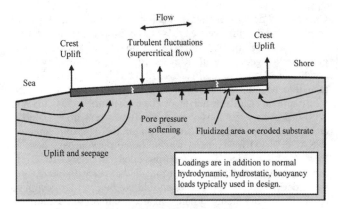

FIGURE C6.12-3 Schematic of Tsunami-Induced Loading on Exterior Slabs On Grade

countermeasures in some areas without additional ground improvement or reinforcement, such as geotextiles.

C6.12.4.2 Protective Slab on Grade. Exterior slab-on-grade uplift shall be assumed to occur as a preexisting condition to computation of local, sustained flow and plunging scour unless determined otherwise by a site-specific design analysis based upon recognized literature. The design of stable slabs on grade under tsunami loading relies on recognizing the potential for scour at slab edges and ensuring the stability of slab sections and substrate. At slab edges, grade changes often result in rapid changes in flow speed and depth, which can carry away material and substrate, while large-scale pressure fluctuations in high-speed flows over pavers or concrete slab sections can pry sections loose and cause further damage. This type of damage, as well as failure of substrate and loss of soil strength, has been observed both during overwash by tsunamis (Yeh and Li 2008; Yeh et al. 2012) and during coastal storms (Seed et al. 2008). A schematic showing the different loading conditions and consideration is shown in Fig. C6.12-3.

Guidance for protective slab-on-grade design is drawn from roadway design in the coastal environment and is discussed in Douglass and Krolak (2008), with specific recommendations for best practices at slab-on-grade transitions in Clopper and Chen (1988).

C6.12.4.3 Geotextiles and Reinforced Earth Systems. Use of geotextiles to provide foundation stability and erosion resistance under tsunami loading provides internal reinforcement to the soil mass through both high- and low-strength geotextiles. They are applied in various configurations, relying upon composite material behavior to a predetermined geometry of improved ground bearing on strata that remain stable through the event loading. Broad use in coastal environments has proven their effectiveness with varying levels of reinforcement used to address varying severity of water and wave loading. They can be effective for creating protective reinforcement to traditional shallow footings, slabs on grade, small retaining walls, berms, and larger structures up to tall mechanically stabilized earth walls as used in the transportation industry. Additional guidance for geotextile placement and design is available from the following:

1. FHWA (2010). Geotechnical Engineering Circular No. 11, "Design and construction of mechanically stabilized earth walls and reinforced soil slopes," FHWA-NHI-10-024.
2. AASHTO (2006). "Standard Specification for Geotextile Specification for Highway Applications," M288-06.

C6.12.4.4 Facing Systems. Facing materials in coastal structures and reinforced earth systems are critical to prevent raveling and erosion. AASHTO M288-06 (2006) provides design guidance for geotextile filter layers assuming high-energy wave conditions. Armor sizing in areas of high Froude number should take into account the high-velocity turbulent flows associated with tsunamis and the height of the incoming waves. FHWA (2009) provides methods appropriate for current flow. Esteban et al. (2014) provides an adaptation of the Hudson equation (USACE 2011) for tsunami waves. Some approaches, such as the Van der Meer equation provided in USACE 2011, recommend armor stone sizing that decreases with increasing wave periods; these approaches should not be used for design of tsunami-resistant facing systems. In areas of low Froude number, the tsunami acts more as current flow, and stone sizing may be treated accordingly using standard methods.

C6.12.4.5 Ground Improvement. Soil–cement ground improvement for foundations is effective under high-velocity turbulent flows such as tsunamis because it provides both strength and erosion resistance to the improved mass. The widely used methods of deep soil mixing and jet grouting can be applied in a variety of geometries and design strengths for particular tsunami loading conditions. These methods, when incorporated in the modeling and analysis methods in this section, can be used to determine the optimal limits of treatment for desired performance levels. Similar applications are used for bridge scour and foundations for levees, dikes, and coastal structures. Additional guidance for soil–cement ground improvement is available in the following:

1. FHWA (2000). "An Introduction to the Deep Soil Mixing Methods as Used in Geotechnical Applications," FHWA-RD-99-138.
2. USACE (2000). "Design and Construction of Levees," Appendix G: Soil–Cement for Protection, EM 1110-2-1913.
3. ASTM International (2007). "Standard Test Methods for Compressive Strength of Molded Soil–Cement Cylinders," D1633-00.

C6.13 STRUCTURAL COUNTERMEASURES FOR TSUNAMI LOADING

The potential extreme magnitude or severity of tsunami loading warrants use of robust or redundant structural countermeasures, including Open Structures, retrofitting and/or alterations, and use of tsunami mitigation barriers located exterior to buildings. The type of selected countermeasures and their strength and extent of protection are dependent upon both performance objectives of the structure to be protected and the extent of protection achievable by countermeasures applied to the structure itself. For most sites, an alternatives evaluation of these three types of countermeasures is needed to identify the optimal protection method or blend of methods. The application of structural countermeasures should be integrated with foundation countermeasures described in Section 6.12.4.

C6.13.2 Tsunami Barriers. Tsunami mitigation barriers consist of widely varying materials and designs, ranging from simple berms and engineered levees to advanced performance-based, instrumented, heavy coastal infrastructure systems of reinforced concrete barrier walls with active or passive floodgates. The design of large-scale coastal infrastructure barriers for extreme tsunami loading may involve other considerations beyond the scope of this section. Areas with existing tsunami mitigation barriers require coordination of performance objectives and interactions with new barrier designs, and design should include combined system scenario modeling.

Tsunami barriers are required in some cases to achieve reliable Life Safety performance under extreme tsunami loads, and this requirement highlights the need to consider siting design as well as traditional structure-level design. Tsunami barriers may also provide an opportunity for more cost-effective mitigation under modest to medium inundation loads.

Performance objectives of barriers may include the possibility of some overtopping. Overtopping may result in residual inundation of the protected structure. Design to account for the inundation caused by barrier overtopping is critical for those cases of intentional overtopping, and it is still prudent as a precaution for possible extreme events in excess of design levels.

This section focuses on adaptation of existing standards for modest-size levee design with geotextile reinforced earth and facing systems, using current best practices developed after Hurricane Katrina with consideration of the Tohoku tsunami (Kuwano et al. 2014). The standards are being evaluated and applied on a massive scale in the Gulf Coast and California (USACE 2000; California DWR 2012; CIRIA 2013). These methods incorporate specific failure mode analysis of fundamental performance requirements for stability, seepage control, and erosion and scour control, including overtopping conditions. The guidance also calls for checks of conventional levee design criteria.

C6.13.2.2 Site Layout. The simplified guidance for determining the layout (location and linear extent) of tsunami mitigation barriers uses simplified shoreline setback and tsunami incident angle shielding criteria, based upon general coastal engineering principles of wave inundation and barrier interaction (USACE 2011). This layout can be optimized with site-specific inundation scenario modeling. In this case, the designer should consider an analysis of various alternative barrier configurations to best address complex wave interactions during runup, drawdown, and channeling.

The requirement on radius of curvature for alignment changes is based on the avoidance of sharp corners, which may be vulnerable to scouring.

C6.14 TSUNAMI VERTICAL EVACUATION REFUGE STRUCTURES

Tsunami Vertical Evacuation Refuge Structures are a special classification of buildings and structures within the Tsunami Evacuation Zone designated as a means of alternative evacuation in communities where sufficiently high ground does not exist or where the time available after the tsunami warning is not deemed to be adequate for full evacuation before tsunami arrival. Such a building or structure should have the strength and resiliency needed to resist all effects of the Maximum Considered Tsunami. Despite the devastation of the March 11, 2011, Tohoku tsunami along the northeast coastline of Honshu Island of Japan, there were many tsunami evacuation buildings that provided safe refuge for thousands of survivors (Fraser et al. 2012; Chock et al. 2013b). In the United States, the Federal Emergency Management Agency published the P-646 *Guidelines for Design of Structures for Vertical Evacuation from Tsunamis* (FEMA 2012) as a set of guidelines for the siting, design, construction, and operation of vertical evacuation refuges. However, as a guideline, it is not written in mandatory language necessary for building code and design standards. Therefore, this standard includes the technical requirements for such structures, using FEMA (2012) as a pre-standard reference. The unreduced live load of 100 psf (4.8 kPa) for public assembly is deemed adequate because occupancy of the designated refuge areas should not be more densely packed than exits; this unreduced live load is consistent with the basic intent of FEMA (2012).

Particularly important considerations are the elevation and height of the refuge since the refuge should provide structural Life Safety for the occupants within a portion of the refuge that is not inundated. Therefore, additional conservatism is necessary in the estimation of inundation elevation. The minimum elevation for a tsunami refuge area is, therefore, the Maximum Considered Tsunami inundation elevation at the site, multiplied by 1.3, plus 10 ft (3.05 m). Section 6.14.1 states, "This same Maximum Considered Tsunami site-specific inundation elevation, factored by 1.3, shall also be used for design of the Tsunami Vertical Evacuation Refuge Structure in accordance with Sections 6.8 to 6.12." There have been extensive comparisons of predicted versus observed heights for historic tsunamis. A plus or minus 30% deviation is generally described as reasonable agreement between field-observed data and model-predicted values. For this reason, the additional 30% factor is consistent with the skill level of present-day tsunami inundation models, for example, as discussed in Tang et al. (2009, 2012).

In the event that it is discovered that the Tsunami Vertical Evacuation Refuge Structure is altered, damaged, or significantly deteriorated, the structure may need to be evaluated by a registered design professional to confirm that it still satisfies the requirements of this chapter.

C6.15 DESIGNATED NONSTRUCTURAL COMPONENTS AND SYSTEMS

"Designated nonstructural components and systems" is an explicitly defined term in Section 6.2; they are within certain buildings and structures of higher importance. Designated nonstructural components and systems are those that are assigned a component importance factor, I_p, equal to 1.5, per Section 13.1.3 of this standard. Designated nonstructural components and systems require special attention since they are needed to continue to perform their functions after both earthquake and tsunami events. For this reason, the same definition of what is considered a designated nonstructural component is used for tsunami effects as is used for earthquake effects. Nonstructural systems that are required for the continued operation of an essential building or structure in a Tsunami Design Zone need to be protected from tsunami inundation effects.

From a tsunami perspective, there are three approaches that can be used to better ensure that designated nonstructural components will perform as needed. One approach is to locate the components in the structure of concern above the Maximum Considered Tsunami inundation elevation. The second approach is to protect the components from inundation effects. Providing a tsunami barrier that surrounds the facility being protected and prevents water from reaching the component during tsunami inundation is one acceptable way of protecting the component(s) and the facility. The barrier height is set as 1.3 times the maximum inundation elevation, which is the same level of conservatism used for Tsunami Vertical Evacuation Refuge Structure design in Section 6.14. For large facilities where the tsunami barrier surrounds many structures, the height of the barrier may vary because the maximum inundation level may vary along the perimeter of the protective berm. The third approach that can be used is to allow the components to be designed directly for tsunami effects. The third approach may be suitable for pipes and vessels, which are inherently leak tight. However, it would not be suitable for mechanical or electrical equipment, where submersion in water (likely saltwater) would probably render the equipment inoperable. For the third approach, the designated nonstructural components and systems would need to be designed to resist flotation, collapse, and permanent lateral displacement caused by action of tsunami and debris loads in accordance with the earlier part of Chapter 6.

C6.16 NONBUILDING TSUNAMI RISK CATEGORY III AND IV STRUCTURES

The requirements of this section apply to nonbuilding structures that are required to be designed for tsunami effects. Risk Category II nonbuilding structures do not generally need to be designed for tsunami effects, and therefore requirements for Risk Category II nonbuilding structures are not provided. It should be noted, however, that some nonbuilding structures, such as tanks and vessels, could float if inundated, and it may be wise to tether or restrain them so that they would not cause damage to other nonbuilding critical structures in the vicinity (Naito et al. 2013). Requirements are provided for both Tsunami Risk Category III and IV nonbuilding structures.

From a tsunami perspective, there are four or more approaches that can be used to design nonbuilding structures to resist tsunami effects. The first is to design the structure and its foundation to resist the effects of tsunami forces directly per the requirements of Section 6.8. The second is to locate the nonbuilding structure safely above the Maximum Considered Tsunami inundation elevation. A safe height is deemed to be 1.3 times the maximum inundation elevation. The third approach is to protect the components from inundation effects. Providing a tsunami barrier that surrounds the facility being protected and prevents water from reaching the nonbuilding structures during tsunami inundation is one acceptable way of protecting the components and the facility. The barrier height is set as 1.3 times the maximum inundation elevation; this is the same level of conservatism used for Tsunami Vertical Evacuation Refuge Structure design in Section 6.14. For large facilities where a tsunami barrier surrounds many structures, the height of the barrier may vary because the maximum inundation level may vary along the perimeter of the protective berm. A fourth approach is to design a protective barrier to mitigate the inundation depth to a level sustainable by the structure (rather than keeping the structure entirely dry).

REFERENCES

Aburaya, T., and Imamura, F. (2002). "The proposal of a tsunami run-up simulation using combined equivalent roughness." *Proc., Coastal Eng. Conf. (JSCE)*, 49, 276–280 (in Japanese).

Aida, I. (1978). "Reliability of a tsunami source model derived from fault parameters." *J. Phys. Earth* 26, 57–73.

Akbas, S. O., and Tekin, E. (2013). "Estimation of resistance factors for reliability-based design of shallow foundations in cohesionless soils under earthquake loading." *Foundation engineering in the face of uncertainty: Honoring Fred H. Kulhawy*, Geotechnical Special Publication 229, J. L. Withiam, K.-K. Phoon, and M. Hussein, eds., ASCE, Reston, VA, 555–569.

American Association of State Highway and Transportation Officials (AASHTO). (2006). "Standard specification for geotextile specification for highway applications." *M288-06*. Washington, DC.

American Concrete Institute (ACI). (2013). "Code requirements for nuclear safety-related concrete structures and commentary." *ACI 349-13*. Farmington Hills, MI.

American Petroleum Institute (API). (2004). "Recommended practice 2A-WSD." Section 6, Foundation Design, 21st Ed., Washington, DC.

ASCE. (2013a). "Minimum design loads for buildings and other structures." *ASCE 7-10*, Reston, VA.

ASCE. (2013b). "Unified definitions for critical infrastructure resilience." *ASCE Policy Statement 518*, Reston, VA.

ASCE. (2014). "Seismic evaluation and retrofit of existing buildings." *ASCE 41-13*, Reston, VA.

Arnason, H., Petroff, C., and Yeh, H. (2009). "Tsunami bore impingement onto a vertical column." *J. Disaster Res.*, 4(6), 391–403.

ASTM International. (2007). "Standard test methods for compressive strength of molded soil–cement cylinders." D1633-00 (2007), West Conshohocken, PA.

ASTM International. (2011). "Standard practice for establishing clear wood strength values." *D2555-06*. West Conshohocken, PA.

Bernoulli, D. (1738). *Hydrodynamica, sive de viribus et motibus fluidorum commentarii*. English version, *Hydrodynamics, or commentaries on the forces and motions of fluids*, Trans. T. Carmody and H. Kobus (1968), Dover, New York, 342.

Blevins, R. D. (1984, republished 2003). *Applied fluid dynamics handbook*, Krieger Publishing, Malabar, FL.

Briggs, M. J., Yeh, H., and Cox, D. (2010). "Physical modeling of tsunami waves." *Handbook of coastal and ocean engineering*, Y. C. Kim, ed., World Scientific Publishing, Singapore, 1073–1106.

Buckingham, E. (1914). "On physically similar systems: Illustrations of the use of dimensional analysis." *Physics Review*, 4(4), 345.

California Department of Water Resources (California DWR). (2012). *Urban levee design criteria*, Sacramento, CA.

California Geological Survey. (2015). "Evaluation and application of probabilistic tsunami hazard analysis in California." *Special Report 237*, Sacramento, CA.

California State Lands Commission. (2005). "The marine oil terminal engineering and maintenance standard (MOTEMS)." Sacramento, CA.

California State Lands Commission (2016). 2016 CCR, Title 24, Part 2, California Building Code, Chapter 31F. "Marine oil terminals (effective January 1, 2017)", Sacramento, CA

Carden, L., Chock, G., Yu, G., and Robertson, I. (2015). "The New ASCE tsunami design standard applied to mitigate Tohoku tsunami building structural failure mechanisms." *Handbook of coastal disaster mitigation for engineers and planners*. Elsevier Science and Technology Books, Amsterdam, Netherlands.

Chau, K. T., and Bao, J. Q. (2010). "Hydrodynamic analysis of boulder transportation on Phi-Phi Island during the 2004 Indian Ocean tsunami." *7th International Conference on Urban Earthquake Engineering (7CUEE) and 5th International Conference on Earthquake Engineering (5ICEE), March 3–5, 2010*, Tokyo Institute of Technology, Tokyo.

Chock, G., Carden, L., Robertson, I., Olsen, M. J., and Yu, G. (2013a). "Tohoku tsunami-induced building failure analysis with implications for USA tsunami and seismic design codes." *Earthq. Spectra*, 29(S1), S99–S126.

Chock, G., Robertson, I., Kriebel, D., Francis, M., and Nistor, I. (2013b). *Tohoku, Japan, earthquake and tsunami of 2011: Performance of structures under tsunami loads*, ASCE, Reston, VA.

Chock, G. (2015). "The ASCE 7 tsunami loads and effects design standard for the United States," *Handbook of coastal disaster mitigation for engineers and planners*. Elsevier Science and Technology Books, Amsterdam, Netherlands.

Chock, G. (2016). "Design for tsunami loads and effects in the ASCE 7-16 standard." *J. Struct. Eng.* doi: 10.1061/(ASCE)ST.1943-541X.0001565, 04016093.

Chock, G., Yu, G., Thio, H. K., and Lynett, P. (2016). "Target structural reliability analysis for tsunami hydrodynamic loads of the ASCE 7 standard." *J. Struct. Eng.* doi: 10.1061/(ASCE)ST.1943-541X.0001499, 04016092.

Chopra, K. A. (2012). *Dynamics of structures, theory and applications to earthquake engineering*, 4th Ed., Prentice Hall, Upper Saddle River, NJ.

CIRIA, French Ministry of Ecology, and U.S. Army Corps of Engineers. (2013). *The international levee handbook, C731*. CIRIA, London.

Clopper, P., and Chen, Y. H. (1988). "Minimizing embankment damage during overtopping flow." *FHWA-RD-88-181*. Federal Highway Administration, Sterling, VA.

Department of Defense (DOD). (2013). "Design of structures to resist progressive collapse." *UFC 4-023-03*. July 2009, including Change 2–June 2013.

Didenkulova, I., and Pelinovsky, E. (2011). "Runup of tsunami waves in U-shaped bays." *Pure Applied Geophys.*, 168, 1239–1249.

Douglass, S., and Krolak, J. (2008). "Highways in the coastal environment." *Hydraulic Engineering Circular 25, FHWA NHI-07-096*, 2nd Ed., Federal Highway Administration, Sterling, VA.

Esteban, M., et al. (2014). "Stability of breakwater armor units against tsunami attacks." *J. Waterway, Port, Coastal, Ocean Eng.*, doi: 10.1061/(ASCE)WW.1943-5460.0000227, 188–198.

Fahlbusch, F. E. (1994). "Scour in rock riverbeds downstream of large dams." *Intl. J. Hydropower Dams*, 1, 30–32.

Federal Emergency Management Agency (FEMA). (2007). "Design guide for improving critical facility safety from flooding and high winds." *FEMA P-543*, Washington, DC.

FEMA. (2011). "Coastal construction manual," 4th Ed., *FEMA P-55*, Vol. II, Washington, DC.

FEMA. (2012). "Guidelines for design of structures for vertical evacuation from tsunamis." *P-646*. 2nd Ed., Washington, DC.

Federal Highway Administration (FHWA) (2000). "An introduction to the deep soil mixing methods as used in geotechnical applications." *FHWA-RD-99-138*, Maclean, VA.

FHWA. (2009). "Bridge scour and stream instability countermeasures, 3rd Ed." *Hydraul. Eng. Circ. 23, FHWA-NHI-09-111*, Washington, DC.

FHWA. (2010). "Design and construction of mechanically stabilized earth walls and reinforced soil slopes." *Geotech. Eng. Circ. No. 11, FHWA-NHI-10-024*, Washington, DC.

FHWA. (2012). "Evaluating scour at bridges, 5th Ed." *Hydraul. Eng. Circ. 18, FHWA-HIF-12-003*, Washington, DC.

Francis, M. (2008). "Tsunami inundation scour of roadways, bridges and foundations: Observations and technical guidance from the Great Sumatra Andaman tsunami." *EERI/FEMA NEHRP 2006 Prof. Fellowship Rep*.

Fraser, S., Leonard, G. S., Matsuo, I., and Murakami, H. (2012). "Tsunami evacuation: Lessons from the Great East Japan earthquake and tsunami of March 11th, 2011." *GNS Sci. Rep. 2012/17, Inst. Geol. Nucl. Sci.* Lower Hutt, Wellington, New Zealand.

Ge, M., and Robertson, I. N. (2010). "Uplift loading on elevated floor slab due to a tsunami bore." *Res. Rep. UHM/CEE/10-03*, University of Hawaii, Manoa.

Gica, E., Spillane, M. C., Titov, V. V., Chamberlin, C. D., and Newman, J. C. (2008). "Development of the forecast propagation database for NOAA's short-term inundation forecast for tsunamis (SIFT)." *NOAA Tech. Memo. OAR PMEL-139*.

Goseberg, N., Wurpts, A., and Schlurmann, T. (2013). "Laboratory-scale generation of tsunami and long waves." *Coastal Eng.*, 79, 57–74.

Haehnel, R. B., and Daly, S. F. (2004). "Maximum impact force of woody debris on floodplain structures." *J. Hydraul. Eng.*, 10.1061/(ASCE)0733-9429(2004)130:2(112), 112–120.

Hicher, P.-Y. (1996). "Elastic properties of soils." *J. Geo. Eng.*, 122, 641–648.

Hoffmans, G. J. C. M., and Verheij, H. J. (1997). *Scour manual*, Taylor & Francis, UK, CRC Press, Boca Raton, FL.

Hughes, S. A. (1993). *Physical models and laboratory techniques in coastal engineering*, Advanced Series on Ocean Engineering, 7, World Scientific, Singapore.

Hughes, S. A. (2004). "Estimation of wave run-up on smooth, impermeable slopes using the wave momentum flux parameter." *Coastal Eng.*, 51, 1085–1104.

Imamura, F. (2009). "Tsunami modeling: Calculating inundation and hazard maps." *The Sea*, 15, 321–332.

Intergovernmental Oceanographic Commission. (2013). *Tsunami glossary, 2013*, IOC Technical Series 85, United Nations Educational, Scientific, and Cultural Organization (UNESCO), Paris.

International Code Council, Inc. (2015). *International Building Code*, Country Club Hills, IL.

Jantzer, I., and Knutsson, S. (2013). *Critical gradients for tailings dam design*, Luleå University of Technology, Sweden.

Kotani, M., Imamura, F., and Shuto, N. (1998). "Tsunami run-up calculations and damage estimation method using GIS." *Coastal Eng. J.*, 45.

Kuilanoff, G., and Drake, R. M. (1991). "Design of DOE facilities for wind-generated missiles." *3rd DOE Natural Phenomena Hazards Mitigation Conference*, Washington, DC.

Kuwano, J., Koseki, J., and Miyata, Y. (2014). "Performance of reinforced soil walls during the 2011 Tohoku earthquake." *Geosynthet. Intl.* 21(3), 179–196.

Lane, E. W. (1947). "Report of the Subcommittee on Sediment Terminology." *Transactions of the American Geophysical Union*, 28(6), 936–938.

Li, Y. (2000). "Tsunamis: Non-breaking solitary wave run-up." *Report KH-R-60*, California Institute of Technology, Pasadena.

Li, Y., and Raichlen, F. (2001). "Solitary wave runup on plane slopes." *J. Waterway, Port, Coastal, Ocean Eng.*, 10.1061/(ASCE)0733-950X(2001) 127:1(33), 33–44.

Li, Y., and Raichlen, F. (2003). "Energy balance model for breaking solitary wave runup." *J. Waterway, Port, Coastal, Ocean Eng.*, 10.1061/(ASCE) 0733-950X(2003)129:2(47), 47–59.

Madsen, P. A., Fuhrman, D. R., and Schaffer, H. A. (2008). "On the solitary wave paradigm for tsunamis." *J. Geo. Res.* 113, C12012, doi: 10.1029/2008JC004932.

Madsen, P. A., and Schäffer, H. A. (2010). "Analytical solutions for tsunami runup on a plane beach: Single waves, N-waves and transient waves." *J. Fluid Mech.*, 645, 27–57.

Madurapperuma, M. A. K. M., and Wijeyewickrema, A. C. (2013). "Response of reinforced concrete columns impacted by tsunami dispersed 20′ and 40′ shipping containers." *Eng. Struct.* 56, doi: 10.1016/j.engstruct.2013.07.034, 1631–1644.

Murata, S., Imamura, F., Katoh, K., Kawata, Y., Takahashi, S., and Takayama, T. (2010). *Tsunami: To survive from tsunami*, Advanced Series on Ocean Engineering, 32, 116–120, 260–265.

Naito, C., Cercone, C., Riggs, H. R., and Cox, D. (2014). "Procedure for site assessment of the potential for tsunami debris impact." *J. Waterway, Port, Coastal, Ocean Eng.*, 10.1061/(ASCE)WW.1943-5460.0000222, 223–232.

Naito, C., Cox, D., Yu, Q. S., Brooker, H. (2013). "Fuel storage container performance during the 2011 Tohoku, Japan, tsunami." *J. Perform. Constr. Facil.*, 10.1061/(ASCE)CF.1943-5509.0000339, 373–380.

National Research Council (NRC). (2011). Tsunami warning and preparedness, An assessment of the U.S. tsunami program and the nation's preparedness effort. National Academies Press, Washington, DC.

National Tsunami Hazard Mitigation Program. (2012). "Proceedings and results of the 2011 NTHMP Model benchmarking workshop." *NOAA Special Report*, Boulder, CO, U.S. Department of Commerce/NOAA/NTHMP.

National Crash Analysis Center (NCAC). (2012). "Extended validation of the finite element model for the 2010 Toyota Yaris passenger sedan." *NCAC Working Paper NCAC 2012-W-005*, George Washington University, Washington, DC.

NCAC (2011). "Development and validation of a finite element model for the 2010 Toyota Yaris passenger sedan." George Washington University, *NCAC 2011-T-001* prepared for the Federal Highway Administration.

Newman, J. H. (1977). *Marine hydrodynamics*, MIT Press, Boston.

Ngo, N., and Robertson, I. N. (2012). "Video analysis of the March 2011 tsunami in Japan's coastal cities." *Univ. of Hawaii Research Report UHM/CEE/12-11*.

Nouri, Y., Nistor, I., Palermo, D., and Cornett, A. (2010). "Experimental investigation of the tsunami impact on free standing structures." *Coastal Eng. J.* 52(1), 43–70.

Okada, Y. (1985). "Surface deformation due to shear and tensile faults in a half-space." *Bull. Seis. Soc. of Am.*, 75, 1135–1154.

Overseas Coastal Area Development Institute of Japan (OCADI). (2009). "Technical standards and commentaries for port and harbour facilities in Japan." Ports and Harbours Bureau, Tokyo.

Paczkowski, K. (2011). "Bore impact upon vertical wall and water-driven, high-mass, low-velocity debris impact." Ph.D. Dissertation, Dept. of Civil and Environmental Engineering, Univ. of Hawaii, Honolulu.

Park, H., Cox, D. T., Lynett, P. J., Wiebe, D. M., and Shin, S. (2013). "Tsunami inundation modeling in constructed environments: A physical and numerical comparison of free-surface elevation, velocity, and momentum flux." *Coastal Eng.*, 79, 9–21.

Permanent International Association of Navigation Congresses (PIANC). (2010). "Mitigation of tsunami disasters in ports." PIANC Working Group 53, Brussels, Belgium.

PIANC. (2014). "Harbour approach channels: Design guidelines." Appendix C, Typical Ship Dimensions, *Report No. 121-2014*. Brussels, Belgium.

Piran Aghl, P., Naito, C. J., and Riggs, H. R. (2014). "Full-scale experimental study of impact demands resulting from high mass, low velocity debris." *J. Struct. Eng.*, doi: 10.1061/(ASCE)ST.1943-541X.0000948, 04014006.

Presidential Policy Directive. (2013). "Critical infrastructure security and resilience." *PPD-21*, U.S. Executive Office of the President, Washington, DC.

Ramsden, J. D. (1993). "Tsunamis: Forces on a vertical wall caused by long waves, bores, and surges on a dry bed." *Report KH-R-54*, W. M. Keck Laboratory, California Institute of Technology, Pasadena.

Riggs, H. R., Cox, D. T., Naito, C. J., Kobayashi, M. H., Piran Aghl, P., Ko, H. T.-S., et al. (2014). "Experimental and analytical study of water-driven debris impact forces on structures." *J. Offshore Mech. Arct. Eng.*, doi: 10.1115/1.4028338 OMAE-13-1042.

Robertson, I. N., Pacskowski, K., Riggs, H. R., and Mohamed, A. (2013). "Experimental investigation of tsunami bore forces on vertical walls." *J. Offshore Mech. Arct. Eng.*, 135(2), 021601-1-021601-8.

Saatcioglu, M., Ghobarah, A., Nistor, I. (2005). *Reconnaissance Report on the 26 Dec 2004 Sumatra Earthquake and Tsunami*, Canadian Association for Earthquake Engineering, Ottawa.

Santo, J., and Robertson, I. N. (2010). "Lateral loading on vertical structural elements due to a tsunami bore." *Research Report UHM/CEE/10-02*, University of Hawaii, Manoa.

Sarpkaya, T. (2010). *Wave forces on offshore structures*, Cambridge University Press, Cambridge, MA.

Seed, R., et al. (2008). "New Orleans and Hurricane Katrina. II: The central region and the lower Ninth Ward." *J. Geotech. Geoenviron. Eng.*, doi: 10.1061/(ASCE)1090-0241(2008)134:5(718), 718–738.

Shimada, Y., Kurachi, Y., Toyota, M., and Tomidokoro, G. (2003). "Flood inundation simulation considering fine land categories by means of unstructured grids." *Proc., Intl. Symp. Disaster Mitigation and Basin-Wide Water Management*, ISDB 2003, JSCE, Niigata, Japan, 70–77.

Simons, D. B., and Senturk, F. (1977). *Sediment transport technology*, Water Resources Publications, Fort Collins, CO.

Sullivan, T. J., Calvi, G. M., and Priestley, M. J. N. (2004). "Initial stiffness versus secant stiffness in displacement based design." *13th World Conference on Earthquake Engineering*, Vancouver, BC, Canada.

Synolakis, C. E. (1986). "The runup of long waves." *Report KH-R-61*, California Institute of Technology, Pasadena.

Synolakis, C. E., Bernard, E. N., Titov, V. V., Kanoglu, U., Gonzalez, F. I. (2007). "Standards, criteria, and procedures for NOAA evaluation of tsunami numerical models." *NOAA Technical Memorandum OAR PMEL-135*, as modified by the National Tsunami Hazard Mitigation Program.

Takakura, R., and Robertson, I. N. (2010). "Reducing tsunami bore uplift forces by providing a breakaway panel." *Research Report UHM/CEE/10-07*, University of Hawaii, Manoa.

Tang, L., Titov, V. V., Bernard, E., Wei, Y., Chamberlin, C., Newman, J. C., et al. (2012). "Direct energy estimation of the 2011 Japan tsunami using deep-ocean pressure measurements." *J. Geophys. Res.*, 117, C08008, doi: 10.1029/2011JC007635.

Tang, L., Titov, V. V., and Chamberlin, C. D. (2009). "Development, testing, and applications of site-specific tsunami inundation models for real-time forecasting," *J. Geophys. Res.*, 114, C12025, doi: 10.1029/2009JC005476.

The Infrastructure Security Partnership (TISP). (2012). *A guide to regional resilience planning*, 2nd Ed., Society of Military Engineers Press, Arlington, VA.

Thio, H. K., Somerville, P. G., and Polet, J. (2010). "Probabilistic tsunami hazard in California." *Pacific Earthquake Engineering Research Center Report.* 108, 331, Berkeley, CA.

Thomas, S., Killian, J., and Bridges, K. (2014). "Influence of macroroughness on tsunami loading of coastal structures." *J. Waterway, Port, Coastal, Ocean Eng.* 10.1061/(ASCE)WW.1943-5460.0000268, 04014028.

Tonkin, S. P., Francis, M., and Bricker, J. D. (2013). "Limits on coastal scour depths due to tsunami." *Intl. Efforts in Lifeline Earthq. Eng.*, C. Davis, X. Du, M. Miyajima, and L. Yan, eds. *TCLEE Monograph 38*, ASCE, Reston, VA.

Tonkin, S. P., Yeh, H., Kato, F., and Sato, S. (2003). "Tsunami scour around a cylinder." *J. Fluid Meh.*, 496, 165–192.

U.S. Army Corps of Engineers (USACE). (1984). *Shore protection manual*, Washington, DC.

USACE. (1989). "Retaining and flood walls." *EM 1110-2-2502*.

USACE. (1993). "Coastal scour problems and prediction of maximum scour." *Technical Report CERC-93-8*, Washington, DC.

USACE. (2000). "Design and construction of levees." *EM 1110-2-1913*.

USACE. (2005). "Stability analysis of concrete structures." *EM 1110-2-2100*.

USACE. (2010). "Hydrologic engineering centers, computer programs." *HEC-RAS*, River Analysis System, v. 4.1.

USACE. (2011). *Coastal engineering manual (CEM)*, Chapter VI-6, *EM 1110-2-1100*.

USACE. (2013). "Incorporating sea-level change considerations for civil works programs." *Eng. Reg. 1100-2-8162*.

Wood, N. (2007). "Variations in city exposure and sensitivity to tsunami hazards in Oregon." *Geological Survey (U.S.) Scientific Investigations Report 2007-5283*, iv.

Wood, N., Church, A., Frazier, T., and Yarnal, B. (2007). "Variations in community exposure and sensitivity to tsunami hazards in the state of Hawai'i." *Geological Survey (U.S.) Scientific Investigations Report 2007-5208*, iv.

Wood, N., and Peters, J. (2015). "Variations in population vulnerability to tectonic and landslide-related tsunami hazards in Alaska." *Nat. Haz.*, 75(2), 1811–1831.

Wood, N., and Soulard, C. (2008). "Variations in community exposure and sensitivity to tsunami hazards on the open-ocean and Strait of Juan de Fuca Coasts of Washington." *Geological Survey (U.S.) Scientific Investigations Report 2008-5004*, vi.

Wood, N., Ratliff, J., and Peters, J. (2013). "Community exposure to tsunami hazards in California." *Geological Survey (U.S.) Scientific Investigations Report 2012-5222*, iv.

Xiao, H., Young, Y. L., and Prévost, J. H. (2010). "Parametric study of breaking solitary wave induced liquefaction of coastal sandy slopes." *Ocean Eng.*, 37, 1546–1553.

Yeh, H., and Li, W. (2008). "Tsunami scour and sedimentation." *Proc., 4th Intl. Conf. on Scour and Erosion*, Japanese Geotechnical Society, Tokyo, 95–106.

Yeh, H., Sato, S., and Tajima, Y. (2012). "The 11 March 2011 East Japan earthquake and tsunami: Tsunami effects on coastal infrastructure and buildings." *Pure Applied Geophys.* 170(6), 1019–1031.

Zhang, J., Jiang, S., Wang, Q., Hou, Y., and Chen, Z. (2010). "Critical hydraulic gradient of piping in sand." *Proc., 20th Intl. Soc. of Offshore and Polar Engineers (ISOPE) Conference*, Beijing.

Zilkoski, D. B., Richards, J. H., and Young, G. M. (1992). "Results of the general adjustment of the North American vertical datum of 1988." American Congress on Surveying and Mapping, *Survey. Land Inform. Sys.* 52(3), 133–149.

OTHER REFERENCES (NOT CITED)

U.S. Army Corps of Engineers (USACE). (2014). "Procedures to evaluate sea level change: Impacts, responses, and adaptation." *Eng. Tech. Lett. 1100-2-1*.

CHAPTER C7
SNOW LOADS

Methodology. The procedure established for determining design snow loads is as follows:

1. Determine the ground snow load for the geographic location (Sections 7.2 and C7.2).
2. Generate a flat roof snow load from the ground load with consideration given to (1) roof exposure (Sections 7.3.1, C7.3, and C7.3.1); (2) roof thermal - condition (Sections 7.3.2, C7.3, and C7.3.2); and (3) occupancy and function of structure (Sections 7.3.3 and C7.3.3).
3. Consider roof slope (Sections 7.4 through 7.4.5 and C7.4).
4. Consider partial loading (Sections 7.5 and C7.5).
5. Consider unbalanced loads (Sections 7.6 through 7.6.4 and C7.6).
6. Consider snow drifts: (1) on lower roofs (Sections 7.7 through 7.7.2 and C7.7) and (2) from projections (Sections 7.8 and C7.8).
7. Consider sliding snow (Sections 7.9 and C7.9).
8. Consider extra loads from rain on snow (Sections 7.10 and C7.10).
9. Consider ponding loads (Section 7.11 and C7.11).
10. Consider existing roofs (Sections 7.12 and C7.12).
11. Consider other roofs and sites (Section C7.13).
12. Consider the consequences of loads in excess of the design value (see the following text).

Loads in Excess of the Design Value. The philosophy of the probabilistic approach used in this standard is to establish a design value that reduces the risk of a snow load–induced failure to an acceptably low level. Because snow loads in excess of the design value may occur, the implications of such "excess" loads should be considered. For example, if a roof is deflected at the design snow load so that slope to drain is eliminated, "excess" snow load might cause ponding (Section C7.11) and perhaps progressive failure.

The snow load/dead load ratio of a roof structure is an important consideration when assessing the implications of "excess" loads. If the design snow load is exceeded, the percentage increase in total load would be greater for a lightweight structure (i.e., one with a high snow load/dead load ratio) than for a heavy structure (i.e., one with a low snow load/dead load ratio). For example, if a 40 lb/ft^2 (1.92 kN/m^2) roof snow load is exceeded by 20 lb/ft^2 (0.96 kN/m^2) for a roof that has a 25 lb/ft^2 (1.19 kN/m^2) dead load, the total load increases by 31% from 65 to 85 lb/ft^2 (3.11 to 4.07 kN/m^2). If the roof had a 60 lb/ft^2 (2.87 kN/m^2) dead load, the total load would increase only by 20% from 100 to 120 lb/ft^2 (4.79 to 5.75 kN/m^2).

C7.2 GROUND SNOW LOADS, p_g

The mapped snow load provisions in Fig. 7.2-1 were developed from an extreme-value statistical analysis of weather records of snow on the ground (Ellingwood and Redfield 1983). The map was produced by the Corps of Engineers, Cold Regions Research and Engineering Laboratory (CRREL).The lognormal distribution was selected to estimate ground snow loads, which have a 2% annual probability of being exceeded (50-yr mean recurrence interval).

Maximum measured ground snow loads and ground snow loads with a 2% annual probability of being exceeded are presented in Table C7.2-1 for National Weather Service (NWS) "first-order" stations at which ground snow loads have been measured for at least 11 years during the period 1952–1992.

Concurrent records of the depth and load of snow on the ground at 204 NWS first-order stations were used to estimate the ground snow load and the ground snow depth that has a 2% annual probability of being exceeded for each of these locations. The period of record for these 204 locations, where both snow depth and snow load have been measured, averages 33 years up through the winter of 1991–1992. A mathematical relationship was developed between the 2% depths and the 2% loads. The nonlinear best-fit relationship between these extreme values was used to estimate 2% (50-yr mean recurrence interval) ground snow loads at about 9,200 other locations at which only snow depths were measured. These loads, as well as the extreme-value loads developed directly from snow load measurements at 204 first-order locations, were used to construct the maps.

In general, loads from these two sources were in agreement. In areas where there were differences, loads from the 204 first-order locations were considered to be more valuable when the map was constructed. This procedure ensures that the map is referenced to the NWS observed loads and contains spatial detail provided by snow depth measurements at about 9,200 other locations.

The maps were generated from data current through the 1991–1992 winter. Where statistical studies using more recent information are available, they may be used to produce improved design guidance.

However, adding a big snow year to data developed from periods of record exceeding 20 years will usually not change 50-yr values much. As examples, the databases for Boston and Chattanooga, Tennessee, were updated to include the winters of 1992–1993 and 1993–1994 because record snows occurred there during that period. In Boston, 50-yr loads based on water equivalent measurements only increased from 34 to 35 lb/ft^2 (1.63 to 1.68 kN/m^2), and loads generated from snow depth measurements remained at 25 lb/ft^2 (1.20 kN/m^2). In Chattanooga, loads generated from water equivalent measurements increased from 6 to 7 lb/ft^2 (0.29 to 0.34 kN/m^2), and loads

Table C7.2-1 Ground Snow Loads at National Weather Service Locations Where Load Measurements Are Made

Location	Ground Snow Load (lb/ft²)		
	Years of Record	Maximum Observed	2% Annual Probability[a]
ALABAMA			
Birmingham	40	4	3
Huntsville	33	7	5
Mobile	40	1	1
ARIZONA			
Flagstaff	38	88	48
Tucson	40	3	3
Winslow	39	12	7
ARKANSAS			
Fort Smith	37	6	5
Little Rock	24	6	6
CALIFORNIA			
Bishop	31	6	8
Blue Canyon	26	213	242
Mt. Shasta	32	62	62
Red Bluff	34	3	3
COLORADO			
All cities	NA	NA	See Table 7.2-2
CONNECTICUT			
Bridgeport	39	21	24
Hartford	40	23	33
New Haven	17	11	15
DELAWARE			
Wilmington	39	12	16
GEORGIA			
Athens	40	6	5
Atlanta	39	4	3
Augusta	40	8	7
Columbus	39	1	1
Macon	40	8	7
Rome	28	3	3
IDAHO			
All cities	NA	NA	See Table 7.2-3
ILLINOIS			
Chicago	26	37	22
Chicago-O'Hare	32	25	17
Moline	39	21	19
Peoria	39	27	15
Rockford	26	31	19
Springfield	40	20	21
INDIANA			
Evansville	40	12	17
Fort Wayne	40	23	20
Indianapolis	40	19	22
South Bend	39	58	41
IOWA			
Burlington	11	15	17
Des Moines	40	22	22
Dubuque	39	34	32
Sioux City	38	28	28
Waterloo	33	25	32
KANSAS			
Concordia	30	12	17
Dodge City	40	10	14
Goodland	39	12	15
Topeka	40	18	17
Wichita	40	10	14
KENTUCKY			
Covington	40	22	13
Jackson	11	12	18
Lexington	40	15	13
Louisville	39	11	12
LOUISIANA			
Alexandria	17	2	2
Shreveport	40	4	3
MAINE			
Caribou	34	68	95
Portland	39	51	60
MARYLAND			
Baltimore	40	20	22
MASSACHUSETTS			
Boston	39	25	34
Nantucket	16	14	24
Worcester	33	29	44
MICHIGAN			
Alpena	31	34	48
Detroit Airport	34	27	18
Detroit City	14	6	10
Detroit—Willow	12	11	22
Flint	37	20	24
Grand Rapids	40	32	36
Houghton Lake	28	33	48
Lansing	35	34	36
Marquette	16	44	53
Muskegon	40	40	51
Sault Ste. Marie	40	68	77
MINNESOTA			
Duluth	40	55	63
International Falls	40	43	44
Minneapolis-St. Paul	40	34	51
Rochester	40	30	47
St. Cloud	40	40	53
MISSISSIPPI			
Jackson	40	3	3
Meridian	39	2	2
MISSOURI			
Columbia	39	19	20
Kansas City	40	18	18
St. Louis	37	28	21
Springfield	39	14	14
MONTANA			
All cities	NA	NA	See Table 7.2-4
NEBRASKA			
Grand Island	40	24	23
Lincoln	20	15	22
Norfolk	40	28	25
North Platte	39	16	13
Omaha	25	23	20
Scottsbluff	40	10	12
Valentine	26	26	22
NEVADA			
Elko	12	12	20
Ely	40	10	9
Las Vegas	39	3	3
Reno	39	12	11
Winnemucca	39	7	7
NEW HAMPSHIRE			
ll cities	NA	NA	See Table 7.2-8
NEW JERSEY			
Atlantic City	35	12	15
Newark	39	18	15

continues

Table C7.2-1 (Continued)

Location	Years of Record	Maximum Observed	2% Annual Probability[a]
		Ground Snow Load (lb/ft²)	
NEW MEXICO			
All cities	NA	NA	See Table 7.2-6
NEW YORK			
Albany	40	26	27
Binghamton	40	30	35
Buffalo	40	41	39
NYC–Kennedy	18	8	15
NYC–LaGuardia	40	23	16
Rochester	40	33	38
Syracuse	40	32	32
NORTH CAROLINA			
Asheville	28	7	14
Cape Hatteras	34	5	5
Charlotte	40	8	11
Greensboro	40	14	11
Raleigh-Durham	36	13	14
Wilmington	39	14	7
Winston-Salem	12	14	20
NORTH DAKOTA			
Bismarck	40	27	27
Fargo	39	27	41
Williston	40	28	27
OHIO			
Akron-Canton	40	16	14
Cleveland	40	27	19
Columbus	40	11	11
Dayton	40	18	11
Mansfield	30	31	17
Toledo Express	36	10	10
Youngstown	40	14	10
OKLAHOMA			
Oklahoma City	40	10	8
Tulsa	40	5	8
OREGON			
All cities	NA	NA	See Table 7.2-7
PENNSYLVANIA			
Allentown	40	16	23
Erie	32	20	18
Harrisburg	19	21	23
Philadelphia	39	13	14
Pittsburgh	40	27	20
Scranton	37	13	18
Williamsport	40	18	21
RHODE ISLAND			
Providence	39	22	23
SOUTH CAROLINA			
Charleston	39	2	2
Columbia	38	9	8
Florence	23	3	3
Greenville-Spartanburg	24	6	7
SOUTH DAKOTA			
Aberdeen	27	23	43
Huron	40	41	46
Rapid City	40	14	15
Sioux Falls	39	40	40
TENNESSEE			
Bristol	40	7	9
Chattanooga	40	6	6
Knoxville	40	10	9

Table C7.2-1 (Continued)

Location	Years of Record	Maximum Observed	2% Annual Probability[a]
		Ground Snow Load (lb/ft²)	
Memphis	40	7	6
Nashville	40	6	9
TEXAS			
Abilene	40	6	6
Amarillo	39	15	10
Austin	39	2	2
Dallas	23	3	3
El Paso	38	8	8
Fort Worth	39	5	4
Lubbock	40	9	11
Midland	38	4	4
San Angelo	40	3	3
San Antonio	40	9	4
Waco	40	3	2
Wichita Falls	40	4	5
UTAH			
Milford	23	23	14
Salt Lake City	40	11	11
Wendover	13	2	3
VERMONT			
Burlington	40	43	36
VIRGINIA			
Dulles Airport	29	15	23
Lynchburg	40	13	18
National Airport	40	16	22
Norfolk	38	9	10
Richmond	40	11	16
Roanoke	40	14	20
WASHINGTON			
All cities	NA	NA	Table 7.2-5
WEST VIRGINIA			
Beckley	20	20	30
Charleston	38	21	18
Elkins	32	22	18
Huntington	30	15	19
WISCONSIN			
Green Bay	40	37	36
La Crosse	16	23	32
Madison	40	32	35
Milwaukee	40	34	29
WYOMING			
Casper	40	9	10
Cheyenne	40	18	18
Lander	39	26	24
Sheridan	40	20	23

[a]It is not appropriate to use only the site-specific information in this table for design purposes. Reasons are given in Section C7.2.
Note: To convert lb/ft² to kN/m², multiply by 0.0479.

generated from snow depth measurements remained at 6 lb/ft² (0.29 kN/m²).

The following additional information was also considered when establishing the snow load zones on the map of the United States (Fig. 7.2-1).

1. The number of years of record available at each location;
2. Additional meteorological information available from NWS, National Resources Conservation Service (NRCS) snow surveys (SNOTEL) and other sources. NRCS was formerly known as Soil Conservation Service;

3. Maximum snow loads observed;
4. Regional topography; and
5. The elevation of each location.

The map was updated in the 1995 edition of this standard and was changed in the 2016 edition.

In much of the south, infrequent but severe snowstorms disrupted life in the area to the point that meteorological observations were missed. In these and similar circumstances, more value was given to the statistical values for stations with complete records. Year-by-year checks were made to verify the significance of data gaps.

The mapped snow loads cannot be expected to represent all the local differences that may occur within each zone. Because local differences exist, each zone has been positioned so as to encompass essentially all the statistical values associated with normal sites in that zone. Although the zones represent statistical values, not maximum observed values, the maximum observed values were helpful in establishing the position of each zone.

For sites not covered in Fig. 7.2-1, design values should be established from meteorological information, with consideration given to the orientation, elevation, and records available at each location. The same method can also be used to improve upon the values presented in Fig. 7.2-1. Detailed study of a specific site may generate a design value lower than that indicated by the generalized national map. It is appropriate in such a situation to use the lower value established by the detailed study. Occasionally a detailed study may indicate that a higher design value should be used than the national map indicates. Again, results of the detailed study should be followed.

Using the database used to establish the ground snow loads in Fig. 7.2-1, additional meteorological data, and a methodology that meets the requirements of Section 7.2 (Tobiasson and Greatorex 1996), ground snow loads have been determined for every town in New Hampshire (Tobiasson et al. 2000, 2002).

The area covered by a site-specific case study varies depending on local climate and topography. In some places, a single case study suffices for an entire community, but in others, varying local conditions limit a "site" to a much smaller area. The area of applicability usually becomes clear as information in the vicinity is examined for the case study.

As suggested by the footnote, it is not appropriate to use only the site-specific information in Table C7.2-1 for design purposes. It lacks an appreciation for surrounding station information and, in a few cases, is based on rather short periods of record. The map or a site-specific case study provides more valuable information.

The importance of conducting detailed studies for locations not covered in Fig. 7.2-1 is shown in Table C7.2-2.

For some locations within the case study (CS) areas of the northeast (Fig. 7.2-1), ground snow loads exceed 100 lb/ft^2 (4.79 kN/m^2). Even in the southern portion of the Appalachian Mountains, not far from sites where a 15-lb/ft^2 (0.72-kN/m^2) ground snow load is appropriate, ground loads exceeding 50 lb/ft^2 (2.39 kN/m^2) may be required. Lake-effect storms create requirements for ground loads in excess of 75 lb/ft^2 (3.59 kN/m^2) along portions of the Great Lakes. In some areas of the Rocky Mountains, ground snow loads exceed 200 lb/ft^2 (9.58 kN/m^2).

Local records and experience should also be considered when establishing design values.

The values in Table 7.2-1 are for specific Alaskan locations only and generally do not represent appropriate design values for other nearby locations. They are presented to illustrate the extreme variability of snow loads within Alaska. This variability precludes statewide mapping of ground snow loads there. Site-specific case studies were conducted to provide the Alaskan values in Table 7.2-1.

Valuable information on snow loads for the western states is contained in Structural Engineers Association of Northern California (1981), MacKinlay and Willis (1965), Brown (1970), U.S. Department of Agriculture, Soil Conservation Service (1970), Structural Engineers Association of Colorado (2016), Structural Engineers Association of Oregon (2013), Structural Engineers Association of Arizona (1981), Theisen et al. (2004), Structural Engineers Association of Washington (1995), Structural Engineers Association of Utah (1992), Placer County Building Division (1985), Al Hatailah et al. (2015), Curtis and Grimes (2004), and Maji (1999).

Most of these references for the western United States use annual probabilities of being exceeded of 2% (50-year mean recurrence interval). Reasonable, but not exact, factors for converting from other annual probabilities of being exceeded to the value herein are presented in Table C7.2-3. For example, a ground snow load based on a 3.3% annual probability of being exceeded (30-yr mean recurrence interval) should be multiplied by 1.15 to generate a value of p_g for use in Eq. (7.3-1).

The design loads in Table 7.2-2 are based upon a reliability analysis, targeted at a reliability index ß = 3.0, which is consistent with Table 1.3-1. Therefore, unlike the ground snow loads for other states, the mean recurrence interval for the tabulated Colorado snow loads is not necessarily 50 years. However, like all snow loads in the standard, the Colorado snow loads are designed to be used with the load factor of 1.6 for strength-based design per Chapter 2 of this standard. Refer to the cited report from the Structural Engineers Association of Colorado (2016).

Regardless of the methodology used to obtain ground snow loads (e.g., Fig. 7.2-1, a case study or a state study), the ASCE 7 snow load provisions should be used to obtain the ground-to-roof conversion, unbalanced loads, drift loads, and related items.

Table C7.2-2 Comparison of Some Site-Specific Values and Zoned Values in Fig. 7.2-1

State	Location	Elevation ft (m)	Zoned Value lb/ft^2 (kN/m^2)	Case Study Value[a] lb/ft^2 (kN/m^2)
California	Mount Hamilton	4,210 (1,283)	0 to 2,400 ft (732 m)	30 (1.44)
Arizona	Palisade Ranger Station	7,950 (2,423)	0 to 3,500 ft (1,067 m) 5 to 4,600 ft (0.24 to 1,402 m) 10 to 5,000 ft (0.48 to 1,524 m)	120 (5.75)
Tennessee	Monteagle	1,940 (591)	10 to 1,800 ft (0.48 to 549 m)	15 (0.72)
Maine	Sunday River Ski Area	900 (274)	90 to 700 ft (4.31 to 213 m)	100 (4.79)

[a]Based on a detailed study of information in the vicinity of each location.

Table C7.2-3 Factors for Converting from Other Annual Probabilities of Being Exceeded and Other Mean Recurrence Intervals to That Used in This Standard

Annual Probability of Being Exceeded (%)	Mean Recurrence Interval (yr)	Multiplication Factor
10	10	1.82
4	25	1.20
3.3	30	1.15
1	100	0.82

The snow load provisions of several editions of the National Building Code of Canada served as a guide in preparing the snow load provisions in this standard. However, there are some important differences between the Canadian and the United States databases. They include the following:

1. The Canadian ground snow loads are based on a 3.3% annual probability of being exceeded (30-yr mean recurrence interval) generated by using the extreme-value, Type-I (Gumbel) distribution, while the normal-risk values in this standard are based on a 2% annual probability of being exceeded (50-yr mean recurrence interval) generated by a lognormal distribution.
2. The Canadian loads are based on measured depths and regionalized densities based on four or fewer measurements per month. Because of the infrequency of density measurements, an additional weight of rain is added (Newark 1984). In this standard, the weight of the snow is based on many years of frequently measured weights obtained at 204 locations across the United States. Those measurements contain many rain-on-snow events, and thus a separate rain-on-snow surcharge load is not needed except for some roofs with a slope less than $W/50$ as per Section 7.10.

The Importance Factor times the ground snow load is the required balanced snow load for snow accumulation surfaces such as decks, balconies, and subterranean spaces located below the depth of the ground snow. Such snow accumulation surfaces are not subject to the same level of wind erosion or building thermal effect as typical "aboveground" roofs. As such, the flat roof snow load in Eq. (7.3-1) would underestimate the snow load. Note that sliding snow loads and snow drift loads on such surfaces are also possible.

In the 2016 version of the standard, ground snow load contours for selected western states were eliminated from the ground load map in Fig. 7.2-1. They were replaced with tables listing the ground snow load and elevation for select locations in each state. These new western state tables (Tables 7.2-2 through 7.2-8) were developed from detailed state ground snow load studies that satisfy the requirements of this section. Finally, one must also consider any statutory requirements of the Authority Having Jurisdiction. Any such Authority Having Jurisdiction statutory requirements are not included in the state ground snow load tables.

C7.3 FLAT ROOF SNOW LOADS, p_f

The live load reductions in Section 4.8 should not be applied to snow loads. The minimum allowable values of p_f presented in Section 7.3 acknowledge that in some areas a single major storm can generate loads that exceed those developed from an analysis of weather records and snow load case studies.

The factors in this standard that account for the thermal, aerodynamic, and geometric characteristics of the structure in its particular setting were developed using the National Building Code of Canada (National Research Council of Canada 1990) as a point of reference. The case study reports in Peter et al. (1963), Schriever et al. (1967), Lorenzen (1970), Lutes and Schriever (1971), Elliott (1975), Mitchell (1978), Meehan (1979), and Taylor (1979, 1980) were examined in detail.

In addition to these published references, an extensive program of snow load case studies was conducted by eight universities in the United States, the U.S. Army Corps of Engineers' Alaska District, and the U.S. Army Cold Regions Research and Engineering Laboratory (CRREL) for the Corps of Engineers. The results of this program were used to modify the Canadian methodology to better fit U.S. conditions. Measurements obtained during the severe winters of 1976–1977 and 1977–1978 are included. A statistical analysis of some of that information is presented in O'Rourke et al. (1983). The experience and perspective of many design professionals, including several with expertise in building failure analysis, have also been incorporated.

C7.3.1 Exposure Factor, C_e. Except in areas of "aerodynamic shade," where loads are often increased by snow drifting, less snow is present on most roofs than on the ground. Loads in unobstructed areas of conventional flat roofs average less than 50% of ground loads in some parts of the country. The values in this standard are above-average values, chosen to reduce the risk of snow load–induced failures to an acceptably low level. Because of the variability of wind action, a conservative approach has been taken when considering load reductions by wind.

The effects of exposure are handled on two scales. First, Eq. (7.3-1) contains a basic exposure factor of 0.7. Second, the type of surface roughness and the exposure of the roof are handled by exposure factor C_e. This two-step procedure generates ground-to-roof load reductions as a function of exposure that range from 0.49 to 0.84.

Table 7.3-1 has been changed from what appeared in a prior version of this standard to separate regional wind issues associated with surface roughness from local wind issues associated with roof exposure. This change was made to better define categories without significantly changing the values of C_e.

Although there is a single "regional" surface roughness category for a specific site, different roofs of a structure may have different exposure factors caused by obstruction provided by higher portions of the structure or by objects on the roof. For example, in surface roughness category C, an upper level roof could be fully exposed ($C_e = 0.9$) while a lower level roof would be partially exposed ($C_e = 1.0$) because of the presence of the upper level roof, as shown in Example 3, this chapter.

The adjective "windswept" is used in the "mountainous areas" surface roughness category to preclude use of this category in those high mountain valleys that receive little wind.

The normal, combined exposure reduction in this standard is 0.70 as compared with a normal value of 0.80 for the ground-to-roof conversion factor in the 1990 National Building Code of Canada. The decrease from 0.80 to 0.70 does not represent decreased safety but arises because of increased choices of exposure and thermal classification of roofs (i.e., five surface roughness categories, three roof exposure categories, and four thermal categories in this standard vs. three exposure categories and no thermal distinctions in the Canadian code).

It is virtually impossible to establish exposure definitions that clearly encompass all possible exposures that exist across the country. Because individuals may interpret exposure categories somewhat differently, the range in exposure has been divided into several categories rather than just two or three. A difference of opinion of one category results in about a 10% "error" using these several categories and an "error" of 25% or more if only three categories are used.

C7.3.2 Thermal Factor, C_t. Usually, more snow will be present on cold roofs than on warm roofs. An exception to this is discussed in the following text. The thermal condition selected from Table 7.3-2 should represent that which is likely to exist during the life of the structure. Although it is possible that a brief power interruption will cause temporary cooling of a heated structure, the joint probability of this event and a simultaneous peak snow load event is very small. Brief power interruptions and loss of heat are acknowledged in the $C_t = 1.0$ category. Although it is possible that a heated structure will subsequently be used as an unheated structure, the probability of this is rather low. Consequently, heated structures need not be designed for this unlikely event.

Some dwellings are not used during the winter. Although their thermal factor may increase to 1.2 at that time, they are unoccupied, so their Importance Factor reduces to 0.8. The net effect is to require the same design load as for a heated, occupied dwelling.

Discontinuous heating of structures may cause thawing of snow on the roof and subsequent refreezing in lower areas. Drainage systems of such roofs have become clogged with ice, and extra loads associated with layers of ice several inches thick have built up in these undrained lower areas. The possibility of similar occurrences should be investigated for any intermittently heated structure.

Similar icings may build up on cold roofs subjected to meltwater from warmer roofs above. Exhaust fans and other mechanical equipment on roofs may also generate meltwater and icings.

Icicles and ice dams are a common occurrence on cold eaves of sloped roofs. They introduce problems related to leakage and to loads. Large ice dams that can prevent snow from sliding off roofs are generally produced by heat losses from within buildings. Icings associated with solar melting of snow during the day and refreezing along eaves at night are often small and transient. Although icings can occur on cold or warm roofs, roofs that are well insulated and ventilated are not commonly subjected to serious icings at their eaves. Methods of minimizing eave icings are discussed in Grange and Hendricks (1976), Klinge (1978), de Marne (1988), Mackinlay (1988), Tobiasson (1988), and Tobiasson and Buska (1993). Ventilation guidelines to prevent problematic icings at eaves have been developed for attics (Tobiasson et al. 1998) and for cathedral ceilings (Tobiasson et al. 1999).

Because ice dams can prevent load reductions by sliding on some warm ($C_t \leq 1.0$) roofs, the "unobstructed slippery surface" curve in Fig. 7.4-1a now only applies to unventilated roofs with a thermal resistance equal to or greater than 30 ft² h °F/Btu (5.3°C m²/W) and to ventilated roofs with a thermal resistance equal to or greater than 20 ft² h °F/Btu (3.5°C m²/W). For roofs that are well insulated and ventilated, see $C_t = 1.1$ in Table 7.3-2.

Glass, plastic, and fabric roofs of continuously heated structures are seldom subjected to much snow load because their high heat losses cause snowmelt and sliding. For such specialty roofs, knowledgeable manufacturers and designers should be consulted. The National Greenhouse Manufacturers Association (1988) recommends use of $C_t = 0.83$ for continuously heated greenhouses and $C_t = 1.00$ for unheated or intermittently heated greenhouses. They suggest a value of $I_s = 1.0$ for retail greenhouses and $I_s = 0.8$ for all other greenhouses. To qualify as a continuously heated greenhouse, a production or retail greenhouse must have a constantly maintained temperature of 50°F (10°C) or higher during winter months. In addition, it must also have a maintenance attendant on duty at all times or an adequate temperature alarm system to provide warning in the event of a heating system failure. Finally, the greenhouse roof material must have a thermal resistance, R-value, less than 2 ft² × h × °F/Btu (0.4°C m²/W). In this standard, the C_t factor for such continuously heated greenhouses is set at 0.85. An unheated or intermittently heated greenhouse is any greenhouse that does not meet the requirements of a continuously heated single- or double-glazed greenhouse. Greenhouses should be designed so that the structural supporting members are stronger than the glazing. If this approach is used, any failure caused by heavy snow loads will be localized and in the glazing. This should avert progressive collapse of the structural frame. Higher design values should be used where drifting or sliding snow is expected.

Little snow accumulates on warm air-supported fabric roofs because of their geometry and slippery surface. However, the snow that does accumulate is a significant load for such structures and should be considered.

The combined consideration of exposure and thermal conditions generates ground-to-roof factors that range from a low of 0.49 to a high of 1.01. The equivalent ground-to-roof factors in the 1990 National Building Code of Canada (National Research of Canada) are 0.8 for sheltered roofs, 0.6 for exposed roofs, and 0.4 for exposed roofs in exposed areas north of the tree line, all regardless of their thermal condition.

Sack (1988) and case history experience indicate that the roof snow load on open-air structures (e.g., parking structures and roofs over loading docks) and on buildings intentionally kept below freezing (e.g., freezer buildings) can be larger than the nearby ground snow load. It is thought that this effect is caused by the lack of heat flow up from the "warm" earth for these select groups of structures. Open-air structures are explicitly included with unheated structures. For freezer buildings, the thermal factor is specified to be 1.3 to account for this effect. This value is intended specifically for structures constructed to act as freezer buildings and not those that contain freezer enclosures inside.

C7.3.3 Importance Factor, I_s. The Importance Factor, I_s, has been included to account for the need to relate design loads to the consequences of failure. Roofs of most structures that have normal occupancies and functions are designed with an Importance Factor of 1.0, which corresponds to unmodified use of the statistically determined ground snow load for a 2% annual probability of being exceeded (50-yr mean recurrence interval).

A study of the locations in Table C7.2-1 showed that the ratio of the values for 4% and 2% annual probabilities of being exceeded (the ratio of the 25-yr to 50-yr mean recurrence interval values) averaged 0.80 and had a standard deviation of 0.06. The ratio of the values for 1% and 2% annual probabilities of being exceeded (the ratio of the 100-yr to 50-yr mean recurrence interval values) averaged 1.22 and had a standard deviation of 0.08. On the basis of the nationwide consistency of these values, it was decided that only one snow load map need be prepared for design purposes and that values for lower and higher risk situations could be generated using that map and constant factors.

Lower and higher risk situations are established using the Importance Factors for snow loads in Table 1.5-2. These factors range from 0.8 to 1.2. The factor 0.8 bases the average design value for that situation on an annual probability of being exceeded of about 4% (about a 25-year mean recurrence interval). The factor 1.2 is nearly that for a 1% annual probability of being exceeded (about a 100-year mean recurrence interval). In the 2016 version of the standard, the Importance Factor is now appropriately applied to the drift height in Fig. 7.6-1. As a result of the functional form, the drift surcharge in pounds per unit cross-wind width ($h_d(4h_d)\gamma/2$) is proportional to the Importance Factor.

C7.3.4 Minimum Snow Load for Low-Slope Roofs, p_m.
These minimums account for a number of situations that develop on low-slope roofs. They are particularly important considerations for regions where p_g is 20 lb/ft² (0.96 kN/m²) or less. In such areas, single storm events can result in loading for which the basic ground-to-roof conversion factor of 0.7, as well as the C_e and C_t factors, are not applicable.

The unbalanced load for hip and gable roofs, with an eave to ridge distance W of 20 ft (6.1 m) or less that have simply supported prismatic members spanning from ridge to eave, is greater than or equal to the minimum roof snow load, p_m. Hence, if such a hip and gable roof has a slope that requires unbalanced loading, the minimum snow load would not control and need not be checked for the roof.

C7.4 SLOPED ROOF SNOW LOADS, p_s

Snow loads decrease as the slopes of roofs increase. Generally, less snow accumulates on a sloped roof because of wind action. Also, such roofs may shed some of the snow that accumulates on them by sliding and improved drainage of meltwater. The ability of a sloped roof to shed snow load by sliding is related to the absence of obstructions not only on the roof but also below it, the temperature of the roof, and the slipperiness of its surface. It is difficult to define "slippery" in quantitative terms. For that reason, a list of roof surfaces that qualify as slippery and others that do not is presented in the standard. Most common roof surfaces are on that list. The slipperiness of other surfaces is best determined by comparisons with those surfaces. Some tile roofs contain built-in protrusions or have a rough surface that prevents snow from sliding. However, snow does slide off other smooth-surfaced tile roofs. When a surface may or may not be slippery, the implications of treating it either as a slippery or nonslippery surface should be determined. Because valleys obstruct sliding on slippery surfaced roofs, the dashed lines in Figs. 7.4-1a, b, and c should not be used in such roof areas.

Discontinuous heating of a building may reduce the ability of a sloped roof to shed snow by sliding because meltwater created during heated periods may refreeze on the roof's surface during periods when the building is not heated, thereby "locking" the snow to the roof.

All these factors are considered in the slope reduction factors presented in Fig. 7.4-1 and are supported by Taylor (1983, 1985), Sack et al. (1987), and Sack (1988). The thermal resistance requirements have been added to the "unobstructed slippery surfaces" curve in Fig. 7.4-1a to prevent its use for roofs on which ice dams often form because ice dams prevent snow from sliding. Mathematically the information in Fig. 7.4-1 can be represented as follows:

1. Warm roofs ($C_t = 1.0$ or less):
   (a) Unobstructed slippery surfaces with $R \geq$ 30 ft²h °F/Btu (5.3°C m²/W) if unventilated and $R \geq$ 20 ft² h °F/Btu (3.5°C m²/W) if ventilated:
       0°–5° slope $C_s = 1.0$
       5°–70° slope $C_s = 1.0 - (\text{slope} - 5°)/65$
       >70° slope $C_s = 0$
   (b) All other surfaces:
       0°–30° slope $C_s = 1.0$
       30°–70° slope $C_s = 1.0 - (\text{slope} - 30°)/40°$
       >70° slope $C_s = 0$
2. Cold roofs with $C_t = 1.1$
   (a) Unobstructed slippery surfaces:
       0°–10° slope $C_s = 1.0$
       10°–70° slope $C_s = 1.0 - (\text{slope} - 10°)/60$
       >70° slope $C_s = 0$
   (b) All other surfaces:
       0°–37.5° slope $C_s = 1.0$
       37.5°–70° slope $C_s = 1.0 - (\text{slope} - 37.5°)/32.5°$
       >70° slope $C_s = 0$
3. Cold roofs ($C_t = 1.2$):
   (a) Unobstructed slippery surfaces:
       0°–15° slope $C_s = 1.0$
       15°–70° slope $C_s = 1.0 - (\text{slope} - 15°)/55$
       >70° slope $C_s = 0$
   (b) All other surfaces:
       0°–45° slope $C_s = 1.0$
       45°–70° slope $C_s = 1.0 - (\text{slope} - 45°)/25$
       >70° slope $C_s = 0$

If the ground (or another roof of less slope) exists near the eave of a sloped roof, snow may not be able to slide completely off the sloped roof. This may result in the elimination of snow loads on upper portions of the roof and their concentration on lower portions. Steep A-frame roofs that nearly reach the ground are subject to such conditions. Lateral and vertical loads induced by such snow should be considered for such roofs.

If the roof has snow retention devices (installed to prevent snow and ice from sliding off the roof), it should be considered an obstructed roof and the slope factor C_s should be based on the "All Other Surfaces" curves in Fig. 7.4-1.

C7.4.3 Roof Slope Factor for Curved Roofs.
These provisions were changed from those in the 1993 edition of this standard to cause the load to diminish along the roof as the slope increases.

C7.4.4 Roof Slope Factor for Multiple Folded Plate, Sawtooth, and Barrel Vault Roofs.
Because these types of roofs collect extra snow in their valleys by wind drifting and snow creep and sliding, no reduction in snow load should be applied because of slope.

C7.4.5 Ice Dams and Icicles along Eaves.
The intent is to consider heavy loads from ice that forms along eaves only for structures where such loads are likely to form. It is also not considered necessary to analyze the entire structure for such loads, just the eaves themselves. Eave ice dam loads with various return periods on roofs with overhangs of 4 ft (1.2 m) or less are presented in O'Rourke et al. (2007).

This provision is intended for short roof overhangs and projections, with a horizontal extent less than 5 ft (1.5 m). In instances where the horizontal extent is greater than 5 ft (1.5 m), the surcharge that accounts for eave ice damming need only extend for a maximum of 5 ft (1.5 m) from the eave of the heated structure (Fig. C7.4-1).

C7.5 PARTIAL LOADING

In many situations, a reduction in snow load on a portion of a roof by wind scour, melting, or snow-removal operations

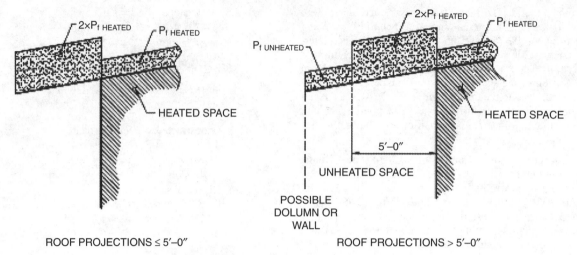

Figure C7.4-1 Eave Ice Dam Loading

simply reduces the stresses in the supporting members. However, in some cases a reduction in snow load from an area induces higher stresses in the roof structure than occur when the entire roof is loaded. Cantilevered roof joists are a good example; removing half the snow load from the cantilevered portion increases the bending stress and deflection of the adjacent continuous span. In continuous beam roof systems, problems can occur during snow-removal operations. The nonuniform loading imposed by the removal of snow load in an indiscriminate manner can significantly alter the load distribution throughout the roof system and can result in marked increases in stresses and deflections over those experienced during uniform loading. In other situations, adverse stress reversals may result.

The simplified provisions offered for continuous beam roof systems have been adopted to mimic real loadings experienced by this common structural system. The Case 1 load scenario simulates a critical condition encountered when, for example, the process of removing a portion of the snow in each span starts at one building end and proceeds toward the other end. The Case 2 load scenario is intended to model loading nonuniformity caused by wind scour or local heat loss and snowmelt at the building edges. The Case 3 load group is intended to encompass conditions that might be encountered, for example, when removal operations must be discontinued and are restarted in a different location, or when multiple snow removal crews start at different roof locations.

The intent is not to require consideration of multiple "checkerboard" loadings.

Members that span perpendicular to the ridge in gable roofs with slopes between 2.38° (½ on 12) and 30.3° (7 on 12) are exempt from partial load provisions because the unbalanced load provisions of Section 7.6.1 provide for this situation.

C7.6 UNBALANCED ROOF SNOW LOADS

Unbalanced snow loads may develop on sloped roofs because of sunlight and wind. Winds tend to reduce snow loads on windward portions and increase snow loads on leeward portions. Because it is not possible to define wind direction with assurance, winds from all directions should generally be considered when establishing unbalanced roof loads.

C7.6.1 Unbalanced Snow Loads for Hip and Gable Roofs. The expected shape of a gable roof drift is nominally a triangle located close to the ridgeline. Recent research suggests that the size of this nominally triangular gable roof drift is comparable to a leeward roof step drift with the same fetch. For certain simple structural systems, for example, wood or light-gauge roof rafter systems with either a ridge board or a supporting ridge beam, with small eave to ridge distances, the drift is represented by a uniform load of $I_s \times p_g$ from eave to ridge. For all other gable roofs, the drift is represented by a rectangular distribution located adjacent to the ridge. The location of the centroid for the rectangular distribution is identical to that for the expected triangular distribution. The intensity is the average of that for the expected triangular distribution.

The design snow load on the windward side for the unbalanced case, $0.3p_s$, is based upon case histories presented in Taylor (1979) and O'Rourke and Auren (1997) and discussed in Tobiasson (1999). The lower limit of $\theta = 2.38°$ is intended to exclude low-slope roofs, such as membrane roofs, on which significant unbalanced loads have not been observed. The upper bound of $\theta > 7$ on 12 (30.2°) is intended to exclude high-slope roofs on which significant unbalanced loads have not been observed. That is, although an upper bound for the angle of repose for fresh-fallen snow is about 70° as given in Fig. 7.4-1, the upper bound for the angle of repose of drifted snow is about 30°.

As noted, observed gable roof drifts are nominally triangular in shape. The surcharge is essentially zero at the ridge, and the top surface of the surcharge is nominally horizontal. As such, an upper bound for an actual surcharge atop the sloped roof snow load, p_s, would be a triangular distribution: zero at the ridge and a height at the eave equal to the elevation difference between the eave and the ridge.

For intersecting gable roofs and similar roof geometries, some codes and standards have required a valley drift load. Such valley drift loads are not required in ASCE 7. However, valley locations are subject to unbalanced or gable roof drifts, as described in Section 7.6.1. An example of unbalanced loading on an L-shaped gable roof is presented in O'Rourke (2007).

For intersecting monoslope roofs and intersecting gable roofs with slopes greater than 7 on 12, unbalanced loads are not required in ASCE 7. However, at such valleys, snow on each side of the valley is prevented from sliding by the presence of roof snow on the other side of the valley. As such, the valley portion of the roof (drainage area upslope of the reentrant corner) is obstructed and the slope factor C_s should be based on the "All Other Surfaces" lines in Fig. 7.4-1.

C7.6.2 Unbalanced Snow Loads for Curved Roofs.
The method of determining roof slope is the same as in the 1995 edition of this standard. C_s is based on the actual slope, not an equivalent slope. These provisions do not apply to roofs that are concave upward. For such roofs, see Section C7.13.

C7.6.3 Unbalanced Snow Loads for Multiple Folded Plate, Sawtooth, and Barrel Vault Roofs.
A minimum slope of 3/8 in./ft (1.79°) has been established to preclude the need to determine unbalanced loads for most internally drained membrane roofs that slope to internal drains. Case studies indicate that significant unbalanced loads can occur when the slope of multiple gable roofs is as low as 1/2 in./ft (2.38°).

The unbalanced snow load in the valley is $2p_f/C_e$ to create a total unbalanced load that does not exceed a uniformly distributed ground snow load in most situations.

Sawtooth roofs and other "up-and-down" roofs with significant slopes tend to be vulnerable in areas of heavy snowfall for the following reasons:

1. They accumulate heavy snow loads and are therefore expensive to build.
2. Windows and ventilation features on the steeply sloped faces of such roofs may become blocked with drifting snow and be rendered useless.
3. Meltwater infiltration is likely through gaps in the steeply sloped faces if they are built as walls because slush may accumulate in the valley during warm weather. This accumulation can promote progressive deterioration of the structure.
4. Lateral pressure from snow drifted against clerestory windows may break the glass.
5. The requirement that snow above the valley not be at an elevation higher than the snow above the ridge may limit the unbalanced load to less than $2p_f/C_e$.

C7.6.4 Unbalanced Snow Loads for Dome Roofs.
This provision is based on a similar provision in the 1990 National Building Code of Canada.

C7.7 DRIFTS ON LOWER ROOFS (AERODYNAMIC SHADE)

When a rash of snow load failures occurs during a particularly severe winter, there is a natural tendency for concerned parties to initiate across-the-board increases in design snow loads. This is generally a technically ineffective and expensive way of attempting to solve such problems because most failures associated with snow loads on roofs are caused not by moderate overloads on every square foot (square meter) of the roof, but rather by localized significant overloads caused by drifted snow.

Drifts accumulate on roofs (even on sloped roofs) in the wind shadow of higher roofs or terrain features. Parapets have the same effect. The affected roof may be influenced by a higher portion of the same structure or by another structure or terrain feature nearby if the separation is 20 ft (6.1 m) or less. When a new structure is built within 20 ft (6.1 m) of an existing structure, drifting possibilities should also be investigated for the existing structure (see Sections C7.7.2 and C7.12). The snow that forms drifts may come from the roof on which the drift forms, from higher or lower roofs, or, on occasion, from the ground.

The leeward drift load provisions are based on studies of snow drifts on roofs (Speck 1984; Taylor 1984; and O'Rourke et al. 1985, 1986). Drift size is related to the amount of driftable snow as quantified by the upwind roof length and the ground snow load. Drift loads are considered for ground snow loads as low as 5 lb/ft² (0.24 kN/m²). Case studies show that, in regions with low ground snow loads, drifts 3 to 4 ft (0.9 to 1.2 m) high can be caused by a single storm accompanied by high winds. The leeward drift height limit of 60% of the lower roof length applies to canopies and other lower level roofs with small horizontal projection from the building wall. It is based upon a 30° angle of repose for drifted snow (tan (30°) = drift height/drift length = 0.577 ~ 0.6) and is consistent with provisions in Section 7.6.1, which exclude roofs steeper than 7 on 12 from the gable roof drift (unbalanced load) provisions.

A change from a prior (1988) edition of this standard involves the width w when the drift height h_d from Fig. 7.6-1 exceeds the clear height h_c. In this situation, the width of the drift is taken as $4h_d^2/h_c$ with a maximum value of $8h_c$. This drift width relation is based upon equating the cross-sectional area of this drift (i.e., $1/2h_c \times w$) with the cross-sectional area of a triangular drift where the drift height is not limited by h_c (i.e., $1/2h_d \times 4h_d$) as suggested by Zallen (1988). The upper limit of drift width is based on studies by Finney (1939) and Tabler (1975) that suggest that a "full" drift has a rise-to-run of about 1:6.5, and case studies (Zallen 1988) that show observed drifts with a rise-to-run greater than 1:10.

The drift height relationship in Fig. 7.6-1 is based on snow blowing off a high roof upwind of a lower roof. The change in elevation where the drift forms is called a "leeward step." Drifts can also form at "windward steps." An example is the drift that forms at the downwind end of a roof that abuts a higher structure there. Fig. 7.7-1 shows "windward step" and "leeward step" drifts.

For situations that have the same amount of available snow (i.e., upper and lower roofs of the same length), the drifts that form in leeward steps are larger than those that form in windward steps. In previous versions of the standard, the windward drift height was given as $1/2h_d$ from Fig. 7.6-1 using the length of the lower roof for l_u. Based upon an analysis of case histories in O'Rourke and De Angelis (2002), a value of 3/4 is now prescribed.

Depending on wind direction, any change in elevation between roofs can be either a windward or leeward step. Thus the height of a drift is determined for each wind direction as shown in Example 3, this chapter, and the larger of the two heights is used as the design drift. The drift height relation in Fig. 7.6-1 was based upon a data set of leeward roof step drifts for which the average upwind fetch distance was about 170 ft (52 m). Unfortunately, the empirical relation provides unrealistic results for very short upwind fetch distances and very low ground snow loads. For example, one would calculate a negative drift height for $l_u = 5$ ft (1.5 m) and $p_g = 5$ psf (0.96 kPa). In prior versions of this load standard, this shortcoming was handled by specifying a minimum l_u of 20 ft (6.1 m). In the 2016 edition, this requirement was modified by noting that h_d need not be taken as greater than $\sqrt{(I_s p_g l_n/4\gamma)}$. This new limiting relation was determined by assuming that all the upwind snow is transported by wind and the trapping efficiency (percentage of transported snow that remains in the drift) is 50%.

The drift load provisions cover most, but not all, situations. Finney (1939) and O'Rourke (1989) document a larger drift than would have been expected based on the length of the upper roof. The larger drift was caused when snow on a somewhat lower roof, upwind of the upper roof, formed a drift between those two roofs, allowing snow from the upwind lower roof to be carried up onto the upper roof then into the drift on its downwind side. It was suggested that the sum of the lengths of both roofs could be used to calculate the size of the leeward drift. The situation of two

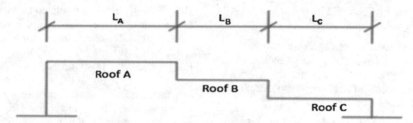

Figure C7.7-1 Roof Steps in Series

roof steps in series was studied by O'Rourke and Kuskowski (2005). For the roof geometry sketched in Fig. C7.7-1, the effective upwind fetch for the leeward drift atop roof C caused by wind from left to right was shown to be $L_B + 0.75L_A$. That is, some of the snow originally on roof A ends up in the leeward drift atop roof B, thereby reducing the amount of snow available for drift formation atop roof C. For the windward snowdrift atop roof B caused by wind from right to left, the effective upwind fetch was shown to be $L_B + 0.85L_C$. The analysis for windward drifts assumed that the elevation difference between roofs B and C was small enough to allow snow initially atop roof C to be blown up onto roof B.

Generally, the addition of a parapet wall on a high roof cannot be relied upon to substantially reduce the leeward snowdrift loading on an adjacent or adjoining lower roof. This is particularly true for the case of a single parapet wall of typical height located at the roof step. Also, the addition of a parapet wall at a roof step would increase the space available for windward drift formation on the lower roof. The issue of potential reduction in leeward drift size at a roof step caused by a parapet wall is discussed in more detail in O'Rourke (2007).

In another situation (Kennedy et al. 1992), a long "spike" drift was created at the end of a long skylight with the wind about 30° off the long axis of the skylight. The skylight acted as a guide or deflector that concentrated drifting snow. This action caused a large drift to accumulate in the lee of the skylight. This drift was replicated in a wind tunnel.

As shown in Fig. 7.7-2, the clear height, h_c, is determined based on the assumption that the upper roof is blown clear of snow in the vicinity of the drift. This assumption is reasonable for windward drifting but does not necessarily hold for leeward drifting. For leeward drifting, the last portion of the upper level roof that would become blown clear of snow is the portion adjacent to the roof step. That is, there may still be snow on the upper level roof when the roof step drift has stopped growing. Nevertheless, for simplicity, the same assumption regarding clear height is used for both leeward and windward drifts.

Tests in wind tunnels (Irwin et al. 1992; Isyumov and Mikitiuk 1992) and flumes (O'Rourke and Weitman 1992) have proven quite valuable in determining patterns of snow drifting and drift loads. For roofs of unusual shape or configuration, wind tunnel or water-flume tests may be needed to help define drift loads. An ASCE standard for wind tunnel testing including procedures to assist in the determination of snow loads on roofs is currently under development.

C7.7.2 Adjacent Structures. One expects a leeward drift to form on an adjacent lower roof only if the lower roof is low enough and close enough to be in the wind shadow (aerodynamic shade) region of the upper roof as sketched in Fig. C7.7-2. The provisions in Section 7.7.2 are based upon a wind shadow region that trails from the upper roof at a 1 downward to 6 horizontal slope.

For windward drifts, the requirements of Section 7.7.1 are to be used. However, the resulting drift may be truncated by eliminating the drift in the horizontal separation region as sketched in Fig. C7.7-3.

C7.7.3 Intersecting Drifts at Low Roofs. The accumulation of drifting snow from perpendicular wind directions can occur concurrently from single or multiple wind and snow events to create an intersecting snowdrift load at reentrant corners. The two resulting drifts are combined as shown in Fig. 7.7-3, where the snowdrift load at the drift intersection is based on the larger drift, not the additive effect of the two drifts (i.e., design h_d at the intersection is h_{d1} or h_{d2}, but not h_{d1} added to h_{d2}). Wind direction at a specific site can change during a single snow event.

C7.8 ROOF PROJECTIONS AND PARAPETS

Drifts around penthouses, roof obstructions, and parapet walls are also of the "windward step" type because the length of the upper roof is small or no upper roof exists. Solar panels, mechanical equipment, parapet walls, and penthouses are examples of roof projections that may cause "windward" drifts on the roof around them. The drift load provisions in Sections 7.7 and 7.8 cover most of these situations adequately, but flat-plate solar collectors may warrant some additional attention. Roofs equipped with several rows of them are subjected to additional snow loads. Before the collectors were installed, these roofs may have sustained minimal snow loads, especially if they were windswept. First, because a roof with collectors is apt to be somewhat "sheltered" by the collectors, it seems appropriate to assume that the roof is partially exposed and calculate a uniform snow load for the entire area as though the collectors did not exist. Second, the extra snow that might fall on the collectors and then slide onto the roof should be computed using the "All Other Surfaces" curve in Fig. 7.4-1b. This value should be applied as a uniform load on the roof at the base of each collector over an area about 2 ft (0.6 m) wide along the length of the collector. The uniform load combined with the load at the base of each collector probably represents a reasonable design load for such situations, except in very windy areas, where extensive snow drifting is to be expected among the collectors. By elevating collectors at least 2 ft (0.61 m) above the height of the balanced snow on an open system of structural supports, the potential for drifting will be diminished significantly. Finally, the collectors should be designed to sustain a load calculated by using the "Unobstructed Slippery Surfaces" curve in Fig. 7.4-1a. This last load should not be used in the design of the roof because the heavier load of sliding snow from the collectors has already been considered. The influence of solar collectors on snow accumulation is discussed in Corotis et al. (1979) and O'Rourke (1979).

Refer to Section C7.7 for more description of the effects that a parapet wall at a high roof can have on the snowdrift loading at an adjacent or adjoining lower roof.

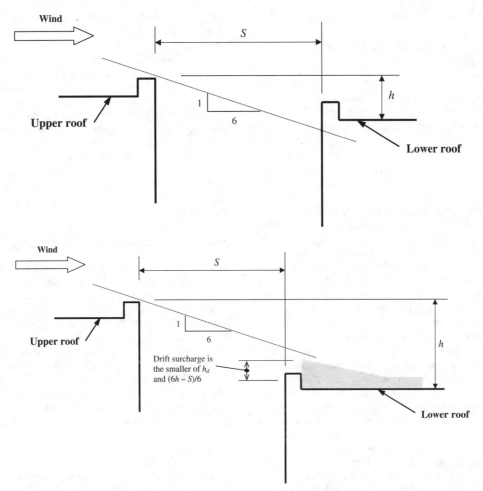

FIGURE C7.7-2 Leeward Snow Drift on Adjacent Roof, Separation $S < 20$ ft (top) Elevation View, $S \geq 6H$; Lower Roof above Wind Shadow (Aerodynamic Shade) Region, No Leeward Drift on Lower Roof. (bottom) Elevation View, $S < 6H$; Lower Roof within Wind Shadow (Aerodynamic Shade) Region, Leeward Drift on Lower Roof; Drift Length Is the Smaller of $(6H - S)$ and $6H_D$

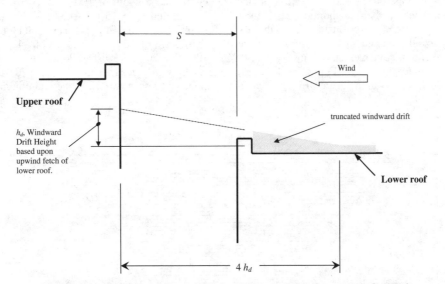

FIGURE C7.7-3 Windward Snow Drift on Adjacent Roof, Separation $S < 20$ ft

The accumulation of drifting snow from perpendicular wind directions can occur concurrently from single or multiple wind and snow events to create an intersecting snowdrift load at parapet wall corners and intersecting roof projections. Where the two drifts intersect, the drift loads need not be superimposed to create a combined (additive) drift height at the drift intersection. The individual drift loads are based on windward or leeward drifting, as indicated in Section 7.8, and the only difference with intersecting drifts is the determination of the governing drift load where the two drifts come together.

C7.9 SLIDING SNOW

Situations that permit snow to slide onto lower roofs should be avoided (Paine 1988). Where this is not possible, the extra load of the sliding snow should be considered. Roofs with little slope have been observed to shed snow loads by sliding.

The final resting place of any snow that slides off a higher roof onto a lower roof depends on the size, position, and orientation of each roof (Taylor 1983). Distribution of sliding loads might vary from a uniform 5-ft (1.5-m) wide load, if a significant vertical offset exists between the two roofs, to a 20-ft (6.1-m) wide uniform load, where a low-slope upper roof slides its load onto a second roof that is only a few feet (about a meter) lower or where snowdrifts on the lower roof create a sloped surface that promotes lateral movement of the sliding snow.

In some instances, a portion of the sliding snow may be expected to slide clear of the lower roof. Nevertheless, it is prudent to design the lower roof for a substantial portion of the sliding load to account for any dynamic effects that might be associated with sliding snow.

Snow retention devices are needed on some roofs to prevent roof damage and eliminate hazards associated with sliding snow (Tobiasson et al. 1996). When snow retention devices are added to a sloping roof, snow loads on the roof can be expected to increase. Thus, it may be necessary to strengthen a roof before adding snow retention devices. When designing a roof that will likely need snow retention devices in the future, it may be appropriate to use the "All Other Surfaces" curves in Fig. 7.4-1, not the "Unobstructed Slippery Surfaces" curves. The design of snow retention devices, their anchorages, and their supporting elements should consider the total maximum design roof snow load upslope of the snow retention device (including sloped roof snow load, drift surcharge, unbalanced loads, and sliding snow loads) and the resultant force caused by the roof slope. See Tobiasson et al. (1996) for further information on the effects of roof slope on the snow retention device force. Snow retention devices, their anchorages, and their supporting structural elements should consider the tributary snow loads from a trapezoidal-shaped area with a boundary at the edge of the snow retention device extending upslope at 45 degrees from the vertical to the ridge (Fig. C7.9-1) or to where the tributary area meets an adjacent tributary area.

Friction between the snow and the roof surface is typically taken as zero when calculating the resultant load on the snow retention device.

Sliding snow should also be considered at changes in roof slope where a higher and steeper roof meets a lower, less steep roof (where the difference in slope is greater than 2 on 12). Where this condition occurs, the low roof prevents the high roof from completely unloading because the low roof slides more gradually or not at all. In this case, the accumulated snow at the change in roof slope exceeds p_f of either roof.

When designing the structural elements of the building at the change in roof slope, a rational approach would be to take the sliding snow $(0.4 p_f W)$ and distribute it a distance of $W/4$ each side of the change in roof slope, accumulating it with the distributed design snow p_f.

C7.10 RAIN-ON-SNOW SURCHARGE LOAD

The ground snow load measurements on which this standard is based contain the load effects of light rain on snow. However, because heavy rains percolate down through snow packs and may drain away, they might not be included in measured values. Where p_g is greater than 20 lb/ft² (0.96 kN/m²), it is assumed that the full rain-on-snow effect has been measured and a separate rain-on-snow surcharge is not needed. The temporary roof load contributed by a heavy rain may be significant. Its magnitude depends on the duration and intensity of the design rainstorm, the drainage characteristics of the snow on the roof, the geometry of the roof, and the type of drainage provided. Loads associated with rain on snow are discussed in Colbeck (1977a, b) and O'Rourke and Downey (2001).

O'Rourke and Downey (2001) show that the surcharge from rain-on-snow loading is an increasing function of eave to ridge distance and a decreasing function of roof slope. That is, rain-on-snow surcharges are largest for wide, low-sloped roofs. The minimum slope reflects that functional relationship.

The following example illustrates the evaluation of the rain-on-snow surcharge. Consider a monoslope roof with slope of 1/4 on 12 and a width of 100 ft (30.5 m) with $C_e = 1.0$, $C_t = 1.1$, $I = 1.2$, and $p_g = 15$ psf (0.72 kN/m²). Because $C_s = 1.0$ for a slope of 1/4 on 12, $p_s = 0.7(1.0)(1.1)(1.0)(1.2)(15) = 14$ psf (0.67 kN/m²). Because the roof slope 1.19° is less than $100/50 = 2.0$, the 5 psf (0.24 kN/m²) surcharge is added to p_s, resulting in a design load of 19 psf (0.91 kN/m²). Because the slope is less than 15°, the minimum load from 7.34 is

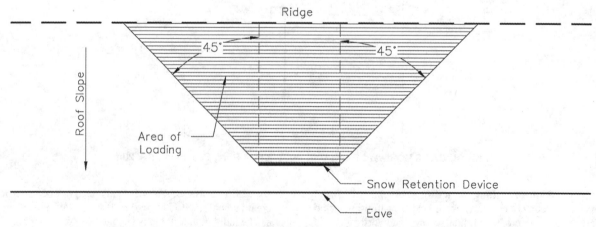

FIGURE C7.9-1 Plan View of Tributary Snow Loads for Snow Retention Device

$I \times p_g = 1.2(15) = 18$ psf (0.86 kN/m^2). Hence, the rain on snow modified load controls.

C7.11 PONDING INSTABILITY

Where adequate slope to drain does not exist, or where drains are blocked by ice, snow meltwater and rain may pond in low areas. Intermittently heated structures in very cold regions are particularly susceptible to blockages of drains by ice. A roof designed without slope or one sloped with only 1/8 in./ft (0.6°) to internal drains probably contains low spots away from drains by the time it is constructed. When a heavy snow load is added to such a roof, it is even more likely that undrained low spots exist. As rainwater or snow meltwater flows to such low areas, these areas tend to deflect increasingly, allowing a deeper pond to form. If the structure does not possess enough stiffness to resist this progression, failure by localized overloading can result. This mechanism has been responsible for several roof failures under combined rain and snow loads.

It is very important to consider roof deflections caused by snow loads when determining the likelihood of ponding instability from rain-on-snow or snow meltwater.

Internally drained roofs should have a slope of at least 1/4 in./ft (1.19°) to provide positive drainage and to minimize the chance of ponding. Slopes of 1/4 in./ft (1.19°) or more are also effective in reducing peak loads generated by heavy spring rain on snow. Further incentive to build positive drainage into roofs is provided by significant improvements in the performance of waterproofing membranes when they are sloped to drain.

Rain loads and ponding instability are discussed in detail in Chapter 8.

C7.12 EXISTING ROOFS

Numerous existing roofs have failed when additions or new buildings nearby caused snow loads to increase on the existing roof. A prior (1988) edition of this standard mentioned this issue only in its commentary where it was not a mandatory provision. The 1995 edition moved this issue to the standard.

The addition of a gable roof alongside an existing gable roof as shown in Fig. C7.12-1 most likely explains why some such metal buildings failed in the south during the winter of 1992–1993. The change from a simple gable roof to a multiple folded plate roof increased loads on the original roof, as would be expected from Section 7.6.3. Unfortunately, the original roofs were not strengthened to account for these extra loads, and they collapsed.

If the eaves of the new roof in Fig. C7.12-1 had been somewhat higher than the eaves of the existing roof, the exposure factor C_e for the original roof may have increased, thereby increasing snow loads on it. In addition, drift loads and loads from sliding snow would also have to be considered.

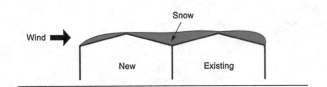

FIGURE C7.12-1 Valley in Which Snow Will Drift Is Created When New Gable Roof Is Added alongside Existing Gable Roof

C7.13 SNOW ON OPEN-FRAME EQUIPMENT STRUCTURES

Snow loads should be considered on all levels of the open-frame structures that can retain snow. The snow accumulations on the flooring, equipment, cable trays, and pipes should be considered. Snow cornicing effects impacting the accumulation of snow on surfaces such as grating, pipes, cable trays, and equipment should also be considered. Grating is considered to retain snow because of the cornicing effect of the snow between the grates.

C7.13.2 Snow at Levels below the Top Level. In the absence of site-specific information, the length of the loaded zone is approximated for convenience as the vertical distance to the covering level above.

The drift snow at levels below the top levels caused by wind walls may be ignored because of the limited snow accumulation on such levels.

C.7.13.3 Snow Loads on Pipes and Cable Trays. The snow loading on any pipe rack or pipe bridge occurs because of the snow loading on the pipes and cable trays at each level. For pipe racks and pipe bridges wider than 12 ft (4 m) where the spaces between individual pipes are less than the pipe diameter (including insulation), Section 7.13.1 and 7.13.2 should be applied.

C7.13.4 Snow Loads on Equipment and Equipment Platforms. Extended out of service conditions for equipment shall be considered for the loading criteria, particularly equipment that is often out of service for the whole winter.

C7.14 OTHER ROOFS AND SITES

Wind tunnel model studies, similar tests using fluids other than air, for example, water flumes, and other special experimental and computational methods have been used with success to establish design snow loads for other roof geometries and complicated sites (Irwin et al. 1992; Isyumov and Mikitiuk 1992; O'Rourke and Weitman 1992). To be reliable, such methods must reproduce the mean and turbulent characteristics of the wind and the manner in which snow particles are deposited on roofs and then redistributed by wind action. Reliability should be demonstrated through comparisons with situations for which full-scale experience is available.

Examples. The following three examples illustrate the method used to establish design snow loads for some of the situations discussed in this standard. Additional examples are found in O'Rourke and Wrenn (2004).

Example 1: Determine balanced and unbalanced design snow loads for an apartment complex in a suburb of Hartford, Connecticut. Each unit has an 8-on-12 slope unventilated gable roof. The building length is 100 ft (30.5 m), and the eave to ridge distance, W, is 30 ft (9.1 m). Composition shingles clad the roofs. Trees will be planted among the buildings.

Flat Roof Snow Load:

$$p_f = 0.7 C_e C_t I_s p_g$$

where

$p_g = 30$ lb/ft^2 (1.44 kN/m^2) (from Fig. 7.2-1),
$C_e = 1.0$ (from Table 7.3-1 for Surface Roughness Category B and a partially exposed roof),
$C_t = 1.0$ (from Table 7.3-2), and
$I_s = 1.0$ (from Table 1.5-2).

Thus, $p_f = (0.7)(1.0)(1.0)(1.0)(30) = 21$ lb/ft^2 (balanced load), (in SI, $p_f = (0.7)(1.0)(1.0)(1.0)(1.44) = 1.01$ kN/m^2).

Because the roof slope is greater than 15°, the minimum roof snow load, p_m, does not apply (see Section 7.3.4).

Sloped-Roof Snow Load:

$$p_s = C_s p_f \quad \text{where } C_s = 1.0$$

(using the "All Other Surfaces" [or solid] line, Fig. 7.4-1a). Thus, $p_s = 1.00(21) = 19$ lb/ft² (in SI, $p_s = 1.00(1.01) = 0.92$ kN/m²).

Unbalanced Snow Load: Because the roof slope is greater than 1/2 on 12 (2.38°), unbalanced loads must be considered. For $p_g = 30$ psf (1.44 kN/m²) and $W = l_u = 30$ ft (9.14 m), $h_d = 1.86$ ft (0.56 m) from Fig. 7.6-1 and $\gamma = 17.9$ pcf (2.80 kN/m³) from Eq. (7.7-1). For a 6 on 12 roof, $S = 2.0$ and hence the intensity of the drift surcharge, $h_d\gamma/\sqrt{S}$, is 23.5 psf (1.31 kN/m²), and its horizontal extent $8\sqrt{S}h_d/3$ is 7.0 ft (2.14 m).

Rain-on-Snow Surcharge: A rain-on-snow surcharge load need not be considered because $p_g > 20$ psf (0.96 kN/m²) (see Section 7.10). See Fig. C7.14-1 for both loading conditions.

Example 2: Determine the roof snow load for a vaulted theater that can seat 450 people, planned for a suburb of Chicago, Illinois. The building is the tallest structure in a recreation-shopping complex surrounded by a parking lot. Two large deciduous trees are located in an area near the entrance. The building has an 80-ft (24.4-m) span and 15-ft (4.6-m) rise circular arc structural concrete roof covered with insulation and aggregate surfaced built-up roofing. The unventilated roofing system has a thermal resistance of 20 ft² hr °F/Btu (3.5 Km²/W). It is expected that the structure will be exposed to winds during its useful life.

Flat Roof Snow Load:

$$p_f = 0.7 C_e C_t I_s p_g$$

where

$p_g = 25$ lb/ft² (1.20 kN/m²) (from Fig. 7.2-1),
$C_e = 0.9$ (from Table 7.3-1 for Surface Roughness Category B and a fully exposed roof),
$C_t = 1.0$ (from Table 7.3-2), and
$I_s = 1.1$ (from Table 1.5-2).

Thus, $p_f = (0.7)(0.9)(1.0)(1.1)(25) = 17$ lb/ft². In SI, $p_f = (0.7)(0.9)(1.0)(1.1)(1.19) = 0.83$ kN/m².

Tangent of vertical angle from eaves to crown = 15/40 = 0.375.
Angle = 21°.

Because the vertical angle exceeds 10°, the minimum roof snow load, p_m, does not apply. See Section 7.3.4.

Sloped-Roof Snow Load:

$$p_s = C_s p_f$$

From Fig. 7.4-1a, $C_s = 1.0$ until slope exceeds 30°, which (by geometry) is 30 ft (9.1 m) from the centerline. In this area, $p_s = 17(1) = 17$ lb/ft² (in SI $p_s = 0.83(1) = 0.83$ kN/m²). At the eaves, where the slope is (by geometry) 41°, $C_s = 0.72$ and $p_s = 17(0.72) = 12$ lb/ft² (in SI $p_s = 0.83(0.72) = 0.60$ kN/m²). Because slope at the eaves is 41°, Case II loading applies.

Unbalanced Snow Load: Because the vertical angle from the eaves to the crown is greater than 10° and less than 60°, unbalanced snow loads must be considered.

Unbalanced load at crown = $0.5 p_f = 0.5(17) = 9$ lb/ft² (in SI, = $0.5(0.83) = 0.41$ kN/m²).
Unbalanced load at 30° point = $2 p_f C_s / C_e = 2(17)(1.0)/0.9 = 38$ lb/ft² (in SI, = $2(0.83)(1.0)/0.9 = 1.84$ kN/m²).
Unbalanced load at eaves = $2(17)(0.72)/0.9 = 27$ lb/ft² (in SI, = $2(0.83)(0.72)/0.9 = 1.33$ kN/m²).

Rain-on-Snow Surcharge: A rain-on-snow surcharge load need not be considered because $p_g > 20$ psf (0.96 kN/m²) (see Section 7.10). See Fig. C7.14-2 for both loading conditions.

Example 3: Determine design snow loads for the upper and lower flat roofs of a building located where $p_g = 40$ lb/ft² (1.92 kN/m²). The elevation difference between the roofs is 10 ft (3 m). The 100 ft × 100 ft (30.5 m × 30.5 m) unventilated high portion is heated, and the 170-ft (51.8-m) wide, 100-ft (30.5-m) long low portion is an unheated storage area. The building is in an industrial park in flat open country with no trees or other structures offering shelter.

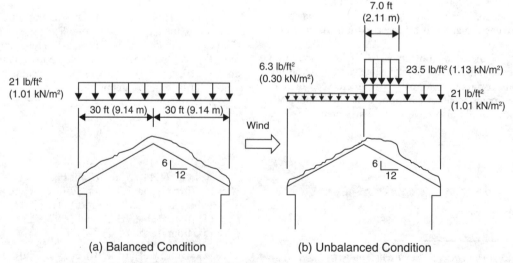

FIGURE C7.14-1 Design Snow Loads for Example 1

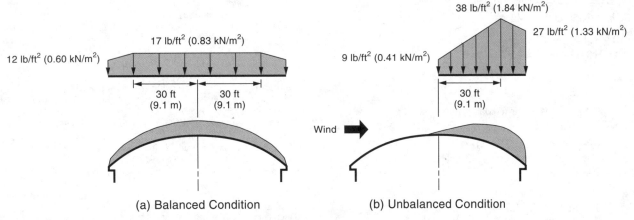

FIGURE C7.14-2 Design Snow Loads for Example 2

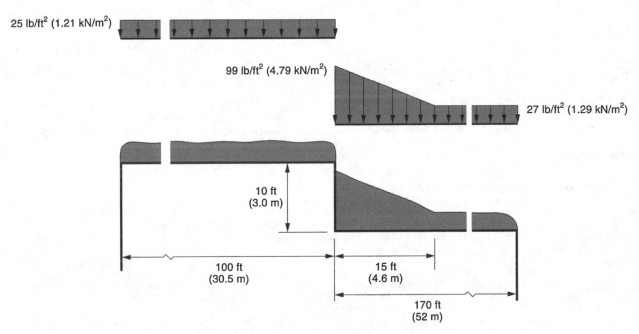

FIGURE C7.14-3 Design Snow Loads for Example 3

High Roof:

$$p_f = 0.7 C_e C_t I_s p_g$$

where

$p_g = 40$ lb/ft^2 (1.92 kN/m^2) (given),
$C_e = 0.9$ (from Table 7.3-1),
$C_t = 1.0$ (from Table 7.3-2), and
$I_s = 1.0$ (from Table 1.5-2).

Thus, $p_f = 0.7(0.9)(1.0)(1.0)(40) = 25$ lb/ft^2
In SI, $p_f = 0.7(0.9)(1.0)(1.0)(1.92) = 1.21$ kN/m^2.

Because $p_g = 40$ lb/ft^2 (1.92 kN/m^2) and $I_s = 1.0$, the minimum roof snow load value of $p_m = 20(1.0) = 20$ lb/ft^2 (0.96 kN/m^2) and hence does not control (see Section 7.3.4).

Low Roof:

$$p_f = 0.7 C_e C_t I_s p_g$$

where

$p_g = 40$ lb/ft^2 (1.92 kN/m^2) (given),
$C_e = 1.0$ (from Table 7.3-1) partially exposed because of the presence of a high roof,
$C_t = 1.2$ (from Table 7.3-2), and
$I_s = 0.8$ (from Table 1.5-2).

Thus, $p_f = 0.7(1.0)(1.2)(0.8)(40) = 27$ lb/ft^2
In SI, $p_f = 0.7(1.0)(1.2)(0.8)(1.92) = 1.29$ kN/m^2.

Because $p = 40$ lb/ft^2 (1.92 kN/m^2) and $I_s = 0.8$, the minimum roof snow load value of $p_m = 20(0.8) = 16$ lb/ft^2 (0.77 kN/m^2) and hence does not control (see Section 7.3.4).

Drift Load Calculation:

$\gamma = 0.13(40) + 14 = 19$ lb/ft^3
In SI, $\gamma = 0.426(1.92) + 2.2 = 3.02$ kN/m^3.

$h_b = p_f/19 = 27/19 = 1.4$ ft
In SI, $h_b = 1.29/3.02 = 0.43$ m.

$h_c = 10 - 1.4 = 8.6$ ft
In SI, $h_c = 3.05 - 0.43 = 2.62$ m.

$h_c/h_b = 8.6/1.4 = 6.1$
In SI, $h_c/h_b = 2.62/0.43 = 6.1$).

Because $h_c/h_b \geq 0.2$, drift loads must be considered (see Section 7.7.1).

h_d (leeward step) = 3.8 ft (1.16 m)
(Fig. 7.6-1 with $p_g = 40$ lb/ft^2 (1.92 kN/m^2)
and $l_u = 100$ ft [30.5 m])

h_d (windward step) = $3/4 \times 4.8$ ft (1.5 m)
= 3.6 ft (1.1 m) (4.8 ft [1.5 m])
from Fig. 7.6-1 with $p = 40$ lb/ft^2 [1.92 kN/m^2]
and l_u = length of lower roof = 170 ft [52 m])

Leeward drift governs, use $h_d = 3.8$ ft (1.16 m)
Because $h_d < h_c$,

$h_d = 3.8$ ft (1.16 m)
$w = 4h_d = 15.2$ ft (4.64 m), say, 15 ft (4.6 m)
$p_d = h_d \gamma = 3.8(19) = 72$ lb/ft^2
In SI, $p_d = 1.16(3.02) = 3.50$ kN/m^2)

Rain-on-Snow Surcharge: A rain-on-snow surcharge load need not be considered because p_g is greater than 20 lb/ft^2 (0.96 kN/m^2). See Fig. C7.14-3 for snow loads on both roofs.

REFERENCES

Al Hatailah, H., Godfrey, B. L., Nielsen, R. J., and Sack, R. L. (2015). "Ground snow loads for Idaho." *Univ. Idaho Civil Engineering Department Report*, Moscow, ID. http://www.lib.uidaho.edu/digital/idahosnow.

Brown, J. (1970). "An approach to snow load evaluation." *Proc., 38th Western Snow Conf.*, Western Snow Conference, Brush Prairie, WA. http://www.westernsnowconference.org/node/984.

Colbeck, S. C. (1977a). "Roof loads resulting from rain-on-snow: Results of a physical model." *Can. J. Civil Eng.*, 4, 482–490.

Colbeck, S. C. (1977b). "Snow loads resulting from rain-on-snow." U.S. Army, *CRREL Report 77-12*, Cold Regions Research and Engineering Laboratory, Hanover, NH.

Corotis, R. B., Dowding, C. H., and Rossow, E. C. (1979). "Snow and ice accumulation at solar collector installations in the Chicago metropolitan area." *NBS-GCR-79 181*, U.S. Dept. of Commerce, National Bureau of Standards, Washington, DC.

Curtis, J., and Grimes, G. (2004). *Wyoming climate atlas*, Dept. of Civil and Arch. Eng., Univ. of Wyoming, Chapter 5, 81–92. http:///www.wrds.uwyo.edu/wrds/wsc/climateatlas/snow.html; and http://www.wy.nrcs.usda.gov/snotel/Wyoming/wyoming.html.

de Marne, H. (1988). "Field experience in control and prevention of leaking from ice dams in New England." *Proc. 1st Intl. Conf. on Snow Eng., CRREL Special Report 89-67*, U.S. Army Cold Regions Research and Engineering Laboratory, Hanover, NH.

Ellingwood, B., and Redfield, R. (1983). "Ground snow loads for structural design." *J. Struct. Eng.*, 109(4), 950–964.

Elliott, M. (1975). "Snow load criteria for western United States, case histories and state-of-the-art." *Proc. 1st Western States Conf. of Structural Engineer Associations*. Sun River, OR.

Finney, E. (1939). "Snow drift control by highway design." *Bull. 86*, Michigan State College Engineering Station, Lansing, MI.

Grange, H. L., and Hendricks, L. T. (1976). "Roof-snow behavior and ice-dam prevention in residential housing," *Ext. Bull. 399*, University of Minnesota, Agricultural Extension Service, St. Paul, MN.

Irwin, P., William, C., Gamle, S., and Retziaff, R. (1992). "Snow prediction in Toronto and the Andes Mountains; FAE simulation capabilities." *Proc., 2nd Intl. Conf. on Snow Engineering*, CRREL Special Report 92-27, U. S. Army Cold Regions Research and Engineering Laboratory, Hanover, NH.

Isyumov, N., and Mikitiuk, M. (1992). "Wind tunnel modeling of snow accumulation on large roofs." *Proc., 2nd Intl. Conf. on Snow Engineering, CRREL Special Report 92-27*, U. S. Army Cold Regions Research and Engineering Laboratory, Hanover, NH.

Kennedy, D., Isyumov, M., and Mikitiuk, M. (1992). "The effectiveness of code provisions for snow accumulations on stepped roofs." *Proc., 2nd Intl. Conf. on Snow Engineering, CRREL Special Report 92-27*, U. S. Army Cold Regions Research and Engineering Laboratory, Hanover, NH.

Klinge, A. F. (1978). "Ice dams." *Pop. Sci.*, 119–120.

Lorenzen, R. T. (1970). "Observations of snow and wind loads precipitant to building failures in New York State, 1969–1970." *American Society of Agricultural Engineers North Atlantic Region meeting*, Paper NA 70-305, American Society of Agricultural Engineers, St. Joseph, MO.

Lutes, D. A., and Schriever, W. R. (1971). "Snow accumulation in Canada: Case histories: II. Ottawa." National Research Council of Canada, *DBR Technical Paper 339*, NRCC 11915.

Mackinlay, I. (1988). "Architectural design in regions of snow and ice." *Proc. 1st Intl. Conf. on Snow Engineering*, CRREL Special Report 89-67, U.S. Army Cold Regions Research and Engineering Laboratory, Hanover, NH.

Mackinlay, I., and Willis, W. E. (1965). *Snow country design*, National Endowment for the Arts, Washington, DC.

Maji, A. K. (1999). *Ground snow load database for New Mexico*, Univ. of New Mexico, Albuquerque.

Meehan, J. F. (1979). "Snow loads and roof failures." *Proc., Structural Engineers Association of California*. Structural Engineers Association of California, San Francisco.

Mitchell, G. R. (1978). "Snow loads on roofs—An interim report on a survey." *Wind and snow loading*, Construction Press, Lancaster, UK, 177–190.

National Greenhouse Manufacturers Association. (1988). "Design loads in greenhouse structures." Taylors, SC.

National Research Council of Canada. (1990). *National building code of Canada 1990*, Ottawa.

Newark, M. (1984). "A new look at ground snow loads in Canada." *Proc., 41st Eastern Snow Conf.* June 7–8, 1984, New Carrollton, MD, 37–48.

O'Rourke, M. (1989). "Discussion of 'Roof collapse under snowdrift loading and snow drift design criteria.'" *J. Perform. Constr. Facil.*, 3(4), 266–268.

O'Rourke, M. (2007). "Snow loads: A guide to the snow load provisions of ASCE 7-05." ASCE, Reston, VA.

O'Rourke, M. J. (1979). "Snow and ice accumulation around solar collector installations." *U.S. Department of Commerce*, NBS-GCR-79 180, National Bureau of Standards, Washington, DC.

O'Rourke, M., and Auren, M. (1997). "Snow loads on gable roofs." *J. Struct. Eng.*, 123(12), 1645–1651.

O'Rourke, M., and De Angelis, C. (2002). "Snow drifts at windward roof steps." *J. Struct. Eng.*, 128(10), 1330–1336.

O'Rourke, M., and Downey, C. (2001). "Rain-on-snow surcharge for roof design." *J. Struct. Eng.*, 127(1), 74–79.

O'Rourke, M., and Kuskowski, N. (2005). "Snow drifts at roof steps in series." *J. Struct. Eng.*, 131(10), 1637–1640.

O'Rourke, M., and Weitman, N. (1992). "Laboratory studies of snow drifts on multilevel roofs." *Proc., 2nd Intl. Conf. on Snow Engineering, CRREL Special Report 92-27*, U. S. Army Cold Regions Research and Engineering Laboratory, Hanover, NH.

O'Rourke, M., and Wrenn, P. D. (2004). *Snow loads: A guide to the use and understanding of the snow load provisions of ASCE 7-02*, ASCE, Reston, VA.

O'Rourke, M., Ganguly, M., and Thompson, L. (2007). "Eave ice dams." *Civil and Environmental Engineering Department Report*, Rensselaer Polytechnic Institute, Troy, NY.

O'Rourke, M., Koch, P., and Redfield, R. (1983). "Analysis of roof snow load case studies: Uniform loads." U.S. Army, *CRREL Report No. 83-1*, Cold Regions Research and Engineering Laboratory, Hanover, NH.

O'Rourke, M. J., Speck, R. S., Jr., and Stiefel, U. (1985). "Drift snow loads on multilevel roofs." *J. Struct. Eng.*, 111(2), 290–306.

O'Rourke, M., Tobiasson, W., and Wood, E. (1986). "Proposed code provisions for drifted snow loads." *J. Struct. Eng.*, 112(9), 2080–2092.

Paine, J. C. (1988). "Building design for heavy snow areas." *Proc., 1st Intl. Conf. on Snow Engineering, CRREL Special Report 89-67*, U.S. Army Cold Regions Research and Engineering Laboratory, Hanover, NH.

Peter, B. G. W., Dalgliesh, W. A., and Schriever, W. R. (1963). "Variations of snow loads on roofs." *Trans. Engrg. Inst. Can.*, 6(A-1), 8.

Placer County Building Division. (1985). "Snow load design." *Placer County Code*, Chapter 4, Sec. 4.20(V), Auburn, CA.

Sack, R. L. (1988). "Snow loads on sloped roofs." *J. Struct. Eng.*, 114(3), 501–517.

Sack, R. L., Arnholtz, D. A., and Haldeman, J. S. (1987). "Sloped roof snow loads using simulation." *J. Struct. Eng.*, 113(8), 1820–1833.

Schriever, W. R., Faucher, Y., and Lutes, D. A. (1967). "Snow accumulation in Canada: Case histories: I. Ottawa, Ontario, Canada." *Technical Paper NRCC 9287*, Issue 237, National Research Council of Canada, Division of Building Research, Ottawa.

Speck, R., Jr. (1984). "Analysis of snow loads due to drifting on multilevel roofs." M.S. Thesis, Dept. of Civil Eng., Rensselaer Polytechnic Institute, Troy, NY.

Structural Engineers Association of Arizona. (1981). *Snow load data for Arizona*, Tempe.

Structural Engineers Association of Colorado. (2016). *Colorado design snow loads*, Structural Engineers Association of Colorado, Denver.

Structural Engineers Association of Northern California. (1981). *Snow load design data for the Lake Tahoe area*, Structural Engineers Association of Northern California, San Francisco.

Structural Engineers Association of Oregon. (2013). *Snow load analysis for Oregon*, Oregon Department of Commerce, Building Codes Division, Salem.

Structural Engineers Association of Utah. (1992). *Utah snow load study*, Structural Engineers Association of Utah, Salt Lake City.

Structural Engineers Association of Washington. (SEAW). (1995). *Snow loads analysis for Washington*, Structural Engineers Association of Washington, Seattle.

Tabler, R. (1975). "Predicting profiles of snowdrifts in topographic catchments." *Proc., 43rd Annual Western Snow Conf.*, Western Snow Conference, Brush Prairie, WA. http://www.westernsnowconference.org/node/1073.

Taylor, D. (1983). "Sliding snow on sloping roofs." *Can. Bldg. Digest 228*. National Research Council of Canada, Ottawa.

Taylor, D. (1985). "Snow loads on sloping roofs: Two pilot studies in the Ottawa area." *Can. J. Civ. Eng.*, Division of Building Research Paper 1282, 12(2), 334–343.

Taylor, D. A. (1979). "A survey of snow loads on roofs of arena-type buildings in Canada." *Can. J. Civ. Eng.* 6(1), 85–96.

Taylor, D. A. (1980). "Roof snow loads in Canada." *Can. J. Civ. Eng.* 7(1), 1–18.

Taylor, D. A. (1984). "Snow loads on two-level flat roofs." *Proc. 41st Eastern Snow Conf.*, June 7–8, 1984, New Carrollton, MD.

Theisen, G. P., Keller, M. J., Stephens, J. E., Videon, F. F., and Schilke, J. P. (2004). "Snow loads for structural design in Montana." *Dept. of Civil Engineering*, Montana State Univ., Bozeman. http://www.coe.montana.edu/matlabwebserver/snowloadinput.html and http://www.ce.montana.edu/ce/snowloads/home.html.

Tobiasson, W. (1988). "Roof design in cold regions." *Proc.., 1st Intl. Conf. on Snow Eng., CRREL Special Report 89-67*, U.S. Army Cold Regions Research and Engineering Laboratory, Hanover, NH.

Tobiasson, W. (1999). "Discussion of 'Snow loads on gable roofs.'" *J. Struct. Eng.*, 125(4), 470–471.

Tobiasson, W., and Buska, J. (1993). "Standing seam metal roofs in cold regions." *Proc., 10th Conf. on Roofing Technology*, National Roofing Contractors Association, Rosemont, IL, 34–44. http://docserver.nrca.net/pdfs/technical/1818.pdf.

Tobiasson, W., and Greatorex, A. (1996). "Database and methodology for conducting site specific snow load case studies for the United States." *Proc., 3rd Intl. Conf. on Snow Eng.*, Sendai, Japan, 249–256.

Tobiasson, W., Buska, J., and Greatorex, A. (1996). "Snow guards for metal roofs." In *Cold regions engineering: The cold regions infrastructure—An international imperative for the 21st century*, R. F. Carlson, ed., ASCE, New York.

Tobiasson, W., Buska, J., and Greatorex, A. (1998). "Attic ventilation guidelines to minimize icings at eaves." *Interface* 16(1), Roof Consultants Institute, Raleigh, NC.

Tobiasson, W., Buska, J., Greatorex, A., Tirey, J., Fisher, J., and Johnson, S. (2000). "Developing ground snow loads for New Hampshire." *Snow engineering: Recent advances and developments*, A. A. Balkema, Brookfield, VT.

Tobiasson, W., Buska, J., Greatorex, A., Tirey, J., Fisher, J., and Johnson, S. (2002). "Ground snow loads for New Hampshire." U.S. Army Corps of Engineers, *Technical Report ERDC/CRREL TR-02-6*, Engineer Research and Development Center (ERDC), Cold Regions Research and Engineering Laboratory (CRREL), Hanover, NH.

Tobiasson, W., Tantillo, T., and Buska, J. (1999). "Ventilating cathedral ceilings to prevent problematic icings at their eaves." *Proc., North American Conf. on Roofing Technology*, National Roofing Contractors Association, Rosemont, IL, 34–44.

U.S. Department of Agriculture, Soil Conservation Service. (1970). *Lake Tahoe basin snow load zones*, U.S. Dept. of Agriculture, Soil Conservation Service, Reno, NV.

Zallen, R. M. (1988). "Roof collapse under snowdrift loading and snow drift design criteria." *J. Perform. Constr. Facil.* 2(2), 80–98.

OTHER REFERENCES (NOT CITED)

Sack, R. L., and Sheikh-Taheri, A. (1986). *Ground and roof snow loads for Idaho*, Department of Civil Engineering, University of Idaho, Moscow. http://www.lib.uidaho.edu/digital/idahosnow/index.html.

CHAPTER C8
RAIN LOADS

C8.1 DEFINITIONS AND SYMBOLS

C8.1.2 Symbols

A = Tributary roof area, plus one-half the wall area that diverts rainwater onto the roof, serviced by a single drain outlet in the secondary drainage system, in ft^2 (m^2).

D = Drain bowl diameter for a primary roof drain, or overflow dam or standpipe diameter for a secondary roof drain, in in. (mm).

i = Design rainfall intensity, in./h (mm/h).

Q = Flow rate out of a single drainage system, in gal./min (m^3/s).

L_r = Length of level roof edge that allows for free overflow drainage of rainwater when the roof edge is acting as the secondary drainage system, in ft (m).

C8.2 ROOF DRAINAGE

Roof drainage systems are not always designed to handle all the flow associated with intense, short-duration rainfall events. For example, the International Plumbing Code (ICC 2012) uses a 1-h duration event with a 100-yr return period for the design of both the primary and secondary drainage systems. An adequate secondary (overflow) drainage system, which is used to limit the depth of water on the roof in the event of clogging of the primary drains, must be designed for an adequately short-duration rainfall event. Some plumbing codes use an arbitrary 1-hour duration storm event for the design of roof drain systems; however, the critical duration for a roof is generally closer to 15 min (the critical duration depends on the roof geometry and drain sizes), and therefore the plumbing codes do not appropriately account for the coincidence of both blocked primary drains and short-duration rainfall events at the design mean recurrence interval (return period or frequency). Graber (2009) provides guidance for determining the critical durations for different types of roof configurations. A very severe local storm or thunderstorm in excess of the 100-year return period storm may produce a deluge of such intensity and duration that properly designed primary drainage systems are temporarily overloaded. Such temporary loads are typically covered in design when blocked drains (see Section 8.3) and a rainfall duration of 15 min are considered. The use of a 60-min duration/100-year return period rainfall event for the design of the primary drainage system and the 15-min duration/100-year return period rainfall event for the secondary drainage system (assuming the primary drainage system is completely blocked) is consistent with the NFPA 5000 Building Construction and Safety Code (2012). Internal gutters are typically designed for 2-min to 5-min duration rainfall events since their critical duration is very short due to their limited storage volume and inability to attenuate a rainfall event.

The National Oceanic and Atmospheric Administration (NOAA's) National Weather Service Precipitation Frequency Data Server, Hydrometeorological Design Studies Center provides rainfall intensity data in inches per hour for the 15-min duration/100-year mean recurrence interval (http://hdsc.nws.noaa.gov/hdsc/pfds/index.html). Precipitation intensity (i in Eq. [C8.3-1]) is in the units of inches per hour; if precipitation depth is provided, a conversion to intensity is required.

The following roof conditions adversely affect the critical duration, or increase the peak flow rate, and should be avoided or appropriately considered by the designer when determining the design rain load:

1. Roofs with internal gutters that have limited storage capacity and quickly fill with rainwater. Gutters are typically designed for 2- to 5-min duration storms since their critical duration is much shorter than the critical duration for typical roofs with scuppers or internal drains.

2. Architecturally complex roofs with internal gutters with significant gutter slopes. Significant gutter slopes allow water to flow at high velocities, which need to be considered when designing the gutter outlets and determining rain loads.

3. Areas susceptible to a concentration of flow, for example, when an addition is added to a low-sloped gable roof (Fig. C8.2-1). In this case, rainwater at the edge of the main roof cannot build over the primary or secondary drains to attain the design flow rate, and the water flows into and inundates the small roof extension.

4. Small roofs adjacent to large walls, where the wall is capable of contributing substantial wind-driven rain flow (sheet flow down the wall) to the roof.

5. High roof areas that drain onto a low roof, increasing the tributary roof area and decreasing the critical rainfall duration.

Item Nos. 4 and 5 can occur when building additions occur, i.e., the new construction imposes an unfavorable condition on the existing construction conditions.

Roof drainage is a structural, architectural, and mechanical (plumbing) issue. The type and location of secondary drains and the hydraulic head above their inlets at the design flow must be known in order to determine rain loads. Design team coordination is particularly important when establishing rain loads.

C8.3 DESIGN RAIN LOADS

The amount of water that could accumulate on a roof from blockage of the primary drainage system is determined, and the roof is designed to withstand the load created by that water plus the uniform load caused by water that rises above the inlet of the

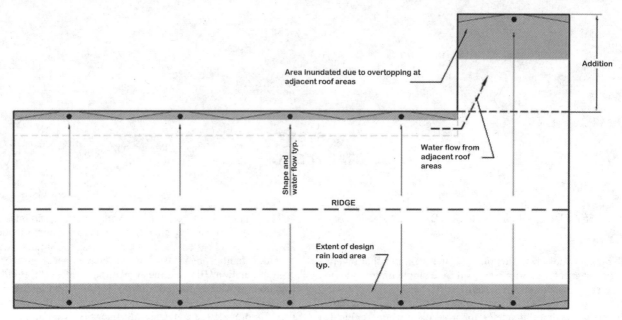

FIGURE C8.2-1 Low-Slope Gable Roof with Drainage Condition at Building Extension

secondary drainage systems at its design flow. If parapet walls, cant strips, expansion joints, and other features create the potential for deep water in an area, it may be advisable to install in that area secondary (overflow) drains with separate drain lines rather than overflow scuppers to reduce the magnitude of the design rain load. Where geometry permits, free discharge is the preferred form of emergency drainage.

When determining these water loads, it is assumed that the roof does not deflect. This assumption eliminates complexities associated with determining the distribution of water loads within deflection depressions. However, it is quite important to consider this water when assessing ponding instability in Section 8.4.

The depth of water, d_h, above the inlet of the secondary drainage system (i.e., the hydraulic head) is a function of the rainfall intensity, i, at the site, the area of roof serviced by that drainage system, and the size of the drainage system.

The flow rate through a single drainage system is as follows:

$$Q = 0.0104 Ai \qquad (C8.3\text{-}1)$$

$$Q = 0.278 \times 10^{-6} Ai \qquad (C8.3\text{-}1.\text{si})$$

The hydraulic head, d_h, is related to flow rate, Q, for various drainage systems in Table C8.3-1. This table indicates that d_h can vary considerably depending on the type and size of each drainage system and the flow rate it must handle. For this reason, the single value of 1 in. (25 mm) (i.e., 5 lb/ft² [0.24 kN/m²]) used in ASCE 7-93 was eliminated.

The hydraulic head, d_h, can generally be assumed to be negligible for design purposes when the secondary drainage system is free to overflow along a roof edge where the length of the level roof edge (L_r) providing free drainage is

$$L_r \geq Ai/400 \qquad (C8.3\text{-}2)$$

$$L_r \geq Ai/3{,}100 \qquad (C8.3\text{-}2.\text{si})$$

Eq. (C8.3-2) is based on the assumption that hydraulic head (d_h) of approximately 0.25 in. (6 mm) above the level roof edge, which represents a rain load of 1.3 lb/ft² (6.3 kg/m²), is negligible in most circumstances.

Flow rates and corresponding hydraulic heads for roof drains are often not available in industry codes, standards, or drain manufacturers' literature for many commonly specified drain types and sizes. Since the hydraulic characteristics and performance of roof drains can depend not only on the size of the drain outlet but also on the geometry of the drain body (e.g., the diameter of the drain dam and depth of the drain bowl), determining the flow rate and corresponding hydraulic head for a drain can be difficult based only on hydraulic calculations. This is particularly true when considering the difficulty in predicting the flow regime (i.e., weir flow, orifice flow, or transition between the two) and the significant effect that flow regime has on the relationship between flow rate and corresponding hydraulic head for a drain.

Based on a drain flow testing program completed by FM Global (2012), the hydraulic heads corresponding to a given range of drain flow rates are provided in Tables C8.3-1 and C8.3-2. This drain testing program included six sizes of primary roof drains and seven sizes and types of secondary (overflow) roof drains. The drains were tested with debris guards (strainers) in place and in a test basin with a relatively smooth bottom surface (waterproofing membrane) to simulate typical smooth-surface roofing material. Measurements of water depth in the test basin were made at a distance of 2 ft (0.6 m) or more from the drain, which ensured that the head measurements were not significantly affected by surface water velocity, and therefore were made where the velocity head was negligible, which was confirmed when comparing water depth based on direct depth measurements to hydraulic head based on pressure taps embedded in the bottom surface of the test basin.

Refer to Fig. C8.3-1 for a schematic view of a secondary drain and the relationship between the drain, the roof surface, and the head.

The following method can be used to approximate hydraulic head for differing drain body dimensions:

(a) For weir flow and transition flow regime designations (cells that are not shaded) in Tables C8.3-1 and C8.3-2:

Table C8.3-1 Flow Rate (Q) in Gallons per Minute for Secondary (Overflow) Roof Drains at Various Hydraulic Heads (d_h) above the Dam or Standpipe, in Inches

	Hydraulic Head (in.) above Dam or Standpipe						
	Overflow Dam 8 in. Diameter			Overflow Dam 12.75 in. Diameter		Overflow Dam 17 in. Diameter	Overflow Standpipe 6 in. Diameter
	Drain Outlet Size (in.)						
	3	4	6	6	8	10	4
	Drain Bowl Depth (in.)						
Flow rate (gal./min)	2	2	2	2	3.25	4.25	2
50	0.5	0.5	0.5	0.5	0.5	—	1.0
75	1.0	—	—	—	—	—	—
100	1.5	1.0	1.0	1.0	0.5	1.0	1.5
125	2.0	—	—	—	—	—	—
150	2.0	1.5	1.5	1.0	—	—	2.5
175	3.0	—	—	—	—	—	—
200	—	2.0	2.0	1.5	1.5	1.5	2.5
225	—	—	—	—	—	—	—
250	—	2.5	2.5	1.5	—	—	2.5
300	—	3.0	3.0	2.0	2.0	1.5	3.0
350	—	3.5	3.5	2.5	—	—	3.5
400	—	5.5	3.5	3.0	2.5	2.0	—
450	—	—	4.0	3.0	—	—	—
500	—	—	5.0	3.5	3.0	2.5	—
550	—	—	5.5	4.0	—	—	—
600	—	—	6.0	5.5	3.5	2.5	—
650	—	—	—	—	—	—	—
700	—	—	—	—	3.5	3.0	—
800	—	—	—	—	4.5	3.0	—
900	—	—	—	—	5.0	3.5	—
1,000	—	—	—	—	5.5	3.5	—
1,100	—	—	—	—	—	4.0	—
1,200	—	—	—	—	—	4.5	—

Notes:
1. Assume that the flow regime is either weir flow or transition flow, except where the hydraulic head values are in shaded cells below the heavy line that designates orifice flow.
2. To determine total head, add the depth of water (static head, d_s) above the roof surface to the secondary drain inlet (which is the height of the dam or standpipe above the roof surface) to the hydraulic head listed in this table.
3. Linear interpolation for flow rate and hydraulic head is appropriate for approximations.
4. Extrapolation is not appropriate.

Source: Adapted from FM Global (2012).

Where the specified secondary (overflow) drain dam or standpipe diameter differs from what is provided in Tables C8.3-1 and C8.3-2, the hydraulic head can be adjusted based on Eq. (C8.3-3) while holding flow rate constant; however, it is advisable not to use an adjusted design hydraulic head less than 80% of the hydraulic head indicated in the tables (for a given flow rate) unless flow test results are provided to justify the hydraulic head values.

$$d_{h2} = [(D_1/D_2)^{0.67}](d_{h1}) \quad (C8.3\text{-}3)$$

where

d_{h1} = known hydraulic head from Tables C8.3-1 and C8.3-2.
D_1 = overflow dam or standpipe diameter for secondary (overflow) drain, corresponding to d_{h1} for a given flow rate, as shown in Tables C8.3-1 and C8.3-2.
d_{h2} = hydraulic head to be determined for the specified secondary drain.
D_2 = specified overflow dam or standpipe diameter for secondary (overflow) drain corresponding to d_{h2} for a given flow rate.

Example 1: Determine the total head for an 8-in. secondary drain (8-in. outlet diameter) with a 10-in. diameter ×2 in. high overflow dam (d_s) at a flow rate (Q) of 300 gal./min.
From Table C8.3-1:
$D_1 = 12.75$ in. (dam diameter).
$d_{h1} = 2.0$ in. for 300 gal./min, 8-in. outlet.
For the specified 10-in. diameter overflow dam on an 8-in. drain outlet:
$D_2 = 10$ in. (dam diameter).
Therefore,
$d_{h2} = [(D_1/D_2)^{0.67}](d_{h1})$
$d_{h2} = [(12.75 \text{ in.}/10 \text{ in.})^{0.67}](2.0 \text{ in.}) = 2.4$ in. at $Q = 300$ gal./min
Total head $= d_{h2} + d_s = 2.4$ in. $+ 2$ in. $= 4.4$ in.
(b) For orifice flow regime designations for roof drains, as shown in the shaded cells in Tables C8.3-1 and C8.3-2:

Table C8.3-2 Flow Rate (Q) in Cubic Meters per Second for Secondary (Overflow) Roof Drains at Various Hydraulic Heads (d_h) above the Dam or Standpipe, in Millimeters

Flow Rate (m^3/s)	Hydraulic Head (mm) above Dam or Standpipe						
	Overflow Dam 203 mm Diameter			Overflow Dam 329 mm Diameter		Overflow Dam 432 mm Diameter	Overflow Standpipe 152 mm Diameter
	Drain Outlet Size (mm)						
	76	102	152	152	203	254	102
	Drain Bowl Depth (mm)						
	51	51	51	51	83	108	51
0.0032	13	13	13	13	13	—	25
0.0047	25	—	—	—	—	—	—
0.0063	38	25	25	25	13	25	38
0.0079	51	—	—	—	—	—	—
0.0095	51	38	38	25	—	—	64
0.0110	76	—	—	—	—	—	—
0.0126	—	51	51	38	38	38	64
0.0142	—	—	—	—	—	—	—
0.0158	—	64	64	38	—	—	64
0.0189	—	76	76	51	51	38	76
0.0221	—	89	89	64	—	—	89
0.0252	—	140	89	76	64	51	—
0.0284	—	—	102	76	—	—	—
0.0315	—	—	127	89	76	64	—
0.0347	—	—	140	102	—	—	—
0.0379	—	—	152	140	89	64	—
0.0410	—	—	—	—	—	—	—
0.0442	—	—	—	—	89	76	—
0.0505	—	—	—	—	114	76	—
0.0568	—	—	—	—	127	89	—
0.0631	—	—	—	—	140	89	—
0.0694	—	—	—	—	—	102	—
0.0757	—	—	—	—	—	114	—

Notes:
1. Assume that the flow regime is either weir flow or transition flow, except where the hydraulic head values are in shaded cells below the heavy line that designates orifice flow.
2. To determine total head, add the depth of water (static head, d_s) above the roof surface to the secondary drain inlet (which is the height of the dam or standpipe above the roof surface) to the hydraulic head listed in this table.
3. Linear interpolation for flow rate and hydraulic head is appropriate for approximations.
4. Extrapolation is not appropriate.

Source: Adapted from FM Global (2012).

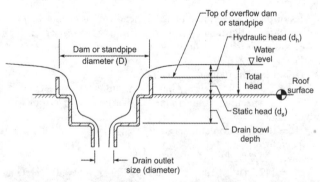

FIGURE C8.3-1 Schematic Cross Section of Secondary (Overflow) Roof Drain and Total Head (d_s+d_h). Drain Debris Guard (Strainer) and Ring Clamp (Gravel Stop) Not Shown for Clarity

The depth of the drain bowl can affect the hydraulic head acting on the drain outlet for a given flow rate; therefore, where the depth of the specified drain bowl is less than the depth of the tested drain bowl (indicated in the tables), the difference in drain bowl depth should be added to the hydraulic head from the tables to determine the design hydraulic head and total head. Where the depth of the specified drain bowl is greater than that indicated in the tables, the difference in drain bowl depth can be subtracted from hydraulic head in the tables to determine the design hydraulic head and total head; however, it is advisable not to use an adjusted design hydraulic head less than 80% of the hydraulic head provided in the tables (for a given flow rate) unless flow test results are provided to justify the hydraulic head values.

Example 2: Determine the total head for a 4 in. secondary drain (4-in. outlet diameter) with a drain bowl depth of 1.5 in. and a 2.5-in.-high overflow dam (d_s), at a flow rate (Q) of 350 gal./min.

Table C8.3-3 Flow Rate, Q, in Gallons Per Minute for Scuppers at Various Hydraulic Heads (d_h) in Inches

Drainage System	Hydraulic Head, d_h, in.									
	1	2	2.5	3	3.5	4	4.5	5	7	8
6-in. wide channel scupper[a]	18	50	b	90	b	140	b	194	321	393
24-in. wide channel scupper	72	200	b	360	b	560	b	776	1,284	1,572
6-in. wide, 4-in. high, closed scupper[a]	18	50	b	90	b	140	b	177	231	253
24-in. wide, 4-in. high, closed scupper	72	200	b	360	b	560	b	708	924	1,012
6-in. wide, 6-in. high, closed scupper	18	50	b	90	b	140	b	194	303	343
24-in. wide, 6-in. high, closed scupper	72	200	b	360	b	560	b	776	1,212	1,372

[a]Channel scuppers are open-topped (i.e., three-sided). Closed scuppers are four-sided.
[b]Interpolation is appropriate, including between widths of each scupper.
Source: Adapted from FM Global (2012).

Table C8.3-4 In SI, Flow Rate, Q, in Cubic Meters Per Second for Scuppers at Various Hydraulic Heads (d_h) in Millimeters

Drainage System	Hydraulic Head d_h, mm									
	25	51	64	76	89	102	114	127	178	203
152-mm wide channel scupper[a]	0.0011	0.0032	b	0.0057	b	0.0088	b	0.0122	0.0202	0.0248
610-mm wide channel scupper	0.0045	0.0126	b	0.0227	b	0.0353	b	0.0490	0.0810	0.0992
152-mm wide, 102-mm high, closed scupper[a]	0.0011	0.0032	b	0.0057	b	0.0088	b	0.0112	0.0146	0.0160
610-mm wide, 102-mm high, closed scupper	0.0045	0.0126	b	0.0227	b	0.0353	b	0.0447	0.0583	0.0638
152-mm wide, 152-mm high, closed scupper	0.0011	0.0032	b	0.0057	b	0.0088	b	0.0122	0.0191	0.0216
610-mm wide, 152-mm high, closed scupper	0.0045	0.0126	b	0.0227	b	0.0353	b	0.0490	0.0765	0.0866

[a]Channel scuppers are open-topped (i.e., three-sided). Closed scuppers are four-sided.
[b]Interpolation is appropriate, including between widths of each scupper.
Source: Adapted from FM Global (2012).

Table C8.3-5 Flow Rate (Q) in Gallons per Minute, for Circular Scuppers at Various Hydraulic Heads (d_h) in Inches

	Scupper Flow (gal./min)						
	Scupper Diameter (in.)						
d_h (in.)	5	6	8	10	12	14	16
1	6	7	8	8	10	10	10
2	25	25	30	35	40	40	45
3	50	55	65	75	75	90	95
4	80	90	110	130	140	155	160
5	115	135	165	190	220	240	260
6	155	185	230	270	300	325	360
7	190	230	300	350	410	440	480
8	220	280	375	445	510	570	610

Notes:
1. Hydraulic head (d_h) is taken above the scupper invert (design water level above base of scupper opening).
2. Linear interpolation is appropriate for approximations.
3. Extrapolation is not appropriate.
Source: Data from Carter (1957) and Bodhaine (1968).

Table C8.3-6 Flow Rate (Q) in Cubic Meters per Second for Circular Scuppers at Various Hydraulic Heads (d_h) in Millimeters

	Scupper Flow Rate (m³/s)						
	Scupper Diameter (mm)						
d_h (mm)	127	152	203	254	305	356	406
25	0.0004	0.0004	0.0005	0.0005	0.0006	0.0006	0.0006
51	0.0016	0.0016	0.0019	0.0022	0.0025	0.0025	0.0028
76	0.0032	0.0035	0.0041	0.0047	0.0047	0.0057	0.0060
100	0.0050	0.0057	0.0069	0.0082	0.0088	0.0098	0.0101
127	0.0073	0.0085	0.0104	0.0120	0.0139	0.0151	0.0164
152	0.0098	0.0117	0.0145	0.0170	0.0189	0.0205	0.0227
178	0.0120	0.0145	0.0189	0.0221	0.0259	0.0278	0.0303
203	0.0139	0.0177	0.0237	0.0281	0.0322	0.0360	0.0385

Notes:
1. Hydraulic head (d_h) is taken above the scupper invert (design water level above base of scupper opening).
2. Linear interpolation is appropriate for approximations.
3. Extrapolation is not appropriate.
Source: Data from Carter (1957) and Bodhaine (1968).

From Table C8.3-1: When $Q = 350$ gal./min, for a 4-in. drain with an 8-in. dam, $d_h = 3.5$ in., the flow regime is orifice flow (shaded portion of the table), and the drain bowl depth is 2 in.

The specified drain bowl depth is 1.5 in., and since this is 0.5 in. less than the drain bowl depth referenced in the table, the hydraulic head from the table is increased by 0.5 in.

Therefore, for the proposed drain: $d_h = 3.5$ in. $+ 0.5$ in. $= 4.0$ in.
Total head $= d_h + d_s = 4.0$ in. $+ 2.5$ in. $= 6.5$ in.

Drain outlet sizes are generally standard in the industry, so it is unlikely that adjustments to hydraulic head values in Tables C8.3-1 and C8.3-2 based on differing drain outlet sizes will be needed.

Where a roof drain is installed in a sump pan located below the adjoining roof surface, reductions in hydraulic head and rain load on the adjoining roof surface should only be credited when based on hydraulic analysis from a qualified plumbing engineer.

Refer to Tables C8.3-3, C8.3-4, C8.3-5, and C8.3-6 for flow rates of rectangular and circular (pipe) roof scuppers at various hydraulic heads. Note that these tables are based on the assumption that no backwater is present (i.e., free outfall) at the discharge end of the scupper. If backwater is present, then the hydraulic head can be expected to increase for the same flow rate.

C8.4 PONDING INSTABILITY AND PONDING LOAD

As water accumulates on roofs, roof deflection allows additional water flows to such areas, and the roof tends to deflect more, allowing a deeper pond to form there. If the structure does not possess enough stiffness to resist this progression, failure by localized overloading may result. Haussler (1962), Chinn (1965), Marino (1966), Salama and Moody (1967), Sawyer (1967, 1968), Chinn et al. (1969), Heinzerling (1971), Burgett (1973), AITC (1978), Associate Committee on the National Building Code (1990), FM Global (2012), SBCCI (1991), BOCA (1993), AISC 360 (2016), and SJI (2007) contain information on ponding and its importance in the design of flexible roofs. Rational design methods to preclude instability from ponding are presented in AISC 360 (2016) and SJI (2007). In determining ponding loads, the primary and secondary members in two-way roof framing systems are to be analyzed concurrently and iteratively since the deflection of primary members (girders) will cause deflection and increase the ponding loads on secondary members (e.g., joists, purlins, or rafters) and the deflection of secondary members will increase the deflection and ponding loads on girders. The deflection of decking can also add additional ponding loads. AISC 360 (2016), Appendix 2, provides methods to evaluate ponding stability for structural steel roof framing. Note, however, that all framing systems can be impacted by ponding.

Regardless of roof slope, if water is impounded on the roof to reach a secondary drainage system, ponding instability can occur. Where such impounded water situations exist, the bay is considered a susceptible bay. Shown in Fig. C8.4-1 are typical susceptible bays for a roof with any slope. For the same structure with secondary members perpendicular to the free draining edge and a roof slope less than 1/4 in./ft, all bays are susceptible. For the same structure with secondary members parallel to the free draining edge and a roof slope less than 1 in./ft, all bays are susceptible. Fig. C8.4-2 shows a roof with perimeter overflow (secondary) drains and interior primary drains. Irrespective of the roof slope, all bays are susceptible. Susceptible bays must be checked to preclude ponding instability and confirmed to have adequate strength with the ponding load.

The limits of 1/4 in. per ft and 1 in. per ft are based upon a maximum deflection to span ration of 1/240. It can be shown that for secondary members parallel to the free draining edge, the minimum rise (β in inches) for a run of 1 ft for no impounded water/free drainage is

$$\beta = (L_s/S + \pi)/20$$

where L_s is the span of the secondary members and S is the spacing of the secondary members ($\beta = 0.76$ for $L_s = 60$ ft and $S = 5$ ft). For secondary members perpendicular to the free

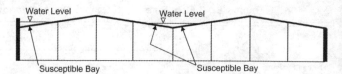

FIGURE C8.4-1 Susceptible Bays for Ponding Evaluation

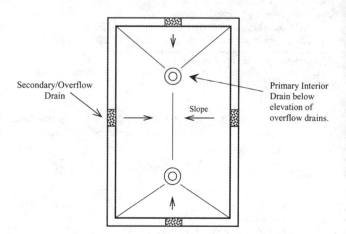

FIGURE C8.4-2 All Bays Susceptible to Ponding

draining edge, the minimum rise (β in inches) for a run of 1 ft for no impounded water/free drainage is

$$\beta = (1 + (L_p/2L_s))/10$$

where L_p is the span of the primary members ($\beta = 0.175$ for $L_s = 40$ ft and $L_p = 60$ ft).

C8.5 CONTROLLED DRAINAGE

In some areas of the country, ordinances are in effect that either limit the rate or delay the release of rainwater flow from roofs into storm drains. Controlled-flow drains are often used on such roofs. Those roofs must be capable of supporting the stormwater temporarily stored on them, similar to traditional roof drainage systems.

Examples

The following two examples illustrate the method used to establish design rain loads based on Chapter 8 of this standard.

Example 3: Determine the design rain load, R, at the secondary drainage for the roof plan shown in Fig. C8.5-1, located at a site in Birmingham, Alabama. Assume that the primary drains are blocked. The design rainfall intensity, i, based on the NOAA Precipitation Frequency Data Server (http://hdsc.nws.noaa.gov/hdsc/pfds/index.html) for a 100-yr, 15-min duration rainfall is 7.23 in./h (184 mm/h). The inlet of the 4-in.(102-mm) diameter secondary roof drain with an 8-in. diameter overflow dam is set 2 in. (51 mm) above the roof surface.

Flow rate, Q, for the secondary drainage 4-in. (102-mm) diameter roof drain:

$$Q = 0.0104 Ai$$

$Q = 0.0104(2,500)(7.23 \text{ in./h}) = 188 \text{ gal./min } (0.0119 \text{ m}^3/\text{s})$

Hydraulic head, d_h:

Using Table C8.3-1 and C8.3-2 for a 4-in. (102-mm) diameter secondary roof drain with a flow rate of 188 gal./min

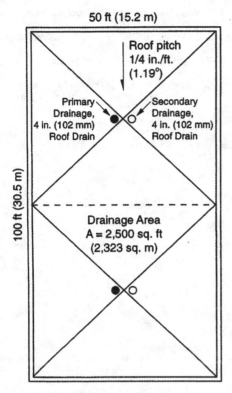

FIGURE C8.5-1 Example 3 Roof Plan
Note: Dashed line indicates the boundary between separate drainage areas.

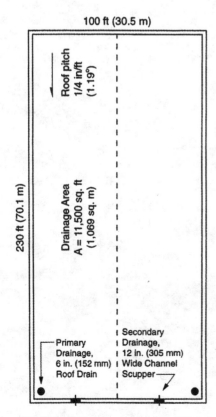

FIGURE C8.5-2 Example 4 Roof Plan
Note: Dashed line indicates the boundary between separate drainage areas.

(0.0119 m^3/s), interpolate between a hydraulic head of 1.5 in. and 2 in. (38 mm and 51 mm) as follows:

$$d_h = 1.5 \text{ in.} + [188 \text{ gal./min} - 150 \text{ gal./min}]$$
$$\div (200 \text{ gal./min} - 150 \text{ gal./min})] \times [2.0 \text{ in.}$$
$$-1.5 \text{ in.}] = 1.9 \text{ in.}$$

In S.I.

$$d_h = 38 \text{ mm} + [0.0119 \text{ m}^3/\text{s} - 0.0095 \text{ m}^3/\text{s}]$$
$$\div (0.0126 \text{ m}^3/\text{s} - 0.0095 \text{ m}^3/\text{s})] \times [51 \text{ mm}$$
$$- 38 \text{ mm}] = 48.2 \text{ mm}$$

Static head $d_s = 2$ in. (51 mm), the water depth from drain inlet to the roof surface.

Design rain load, R, adjacent to the drains:

$$R = 5.2(d_s + d_h) \qquad (8.3\text{-}1)$$

$$R = 5.2(2 + 1.9) = 20.3 \text{ psf}$$

$$R = 0.0098(d_s + d_h) \qquad (8.3\text{-}1.\text{si})$$

$$R = 0.0098(51 + 48.2) = 0.97 \text{ kN/m}^2$$

Example 4: Determine the design rain load, R, at the secondary drainage for the roof plan shown in Fig. C8.5-2, located at a site in Los Angeles, California. Assume that the primary drains are blocked. The design rainfall intensity, i, based on the NOAA Precipitation Frequency Data Server (http://hdsc.nws.noaa.gov/hdsc/pfds/index.html) for a 100-yr, 15-min duration rainfall is 3.28 in./h (83 mm/h). The inlet of the 12-in. (305-mm) secondary roof scuppers is set 2 in. (51 mm) above the roof surface.

Flow rate, Q, for the secondary drainage, 12-in. (305-mm) wide channel scupper:

$$Q = 0.0104 A i$$

$Q = 0.0104(11,500)(3.28) = 392$ gal./min (0.0247 m^3/s).

Hydraulic head, d_h:

Using Tables C8.3-3 and C8.3-4, by interpolation, the flow rate for a 12 in. (305 mm) wide channel scupper is twice that of a 6 in. (152 mm) wide channel scupper. Using Tables C8.3-3 and C8.3-4, the hydraulic head, d_h, for one-half the flow rate, Q, or 196 gal./min (0.0124 m^3/s), through a 6 in. (152 mm) wide channel scupper is 3 in. (127 mm).

$d_h = 5$ in. (127 mm) for a 12-in. (305-mm) wide channel scupper with a flow rate, Q, of 392 gal./min (0.0247 m^3/s).

Static head, $d_s = 2$ in. (51 mm), depth of water from the scupper inlet to the roof surface.

Design rain load, R, adjacent to the scuppers:

$$R = 5.2(2 + 5) = 36.4 \text{ psf}$$

In S.I.

$$R = 0.0098(51 + 127) = 1.7 \text{ kN/m}^2$$

REFERENCES

American Institute of Steel Construction (AISC) (2016). "Specifications for structural steel buildings," (*AISC 360*) American Institute of Steel Construction, Chicago.

American Institute of Timber Construction (AITC). (1978). "Roof slope and drainage for flat or nearly flat roofs." *AITC Technical Note No. 5*, Englewood, CO.

Associate Committee on the National Building Code. (1990). "National building code of Canada 1990," National Research Council of Canada, Ottawa.

Building Officials and Code Administrators International (BOCA). (1993). "The BOCA national plumbing code/1993," BOCA Inc., Country Club Hills, IL.

Bodhaine, G. L. (1968). "Measurement of peak discharge at culverts by indirect methods." *Techniques of water-resources investigations of the United States Geological Survey: Book 3 Application of hydraulics*, U.S. Geological Survey, Reston, VA.

Burgett, L. B. (1973). "Fast check for ponding." *Eng. J.* 10(1), 26–28.

Carter, R. W. (1957). "Computation of peak discharge at culverts." *Geological Survey Circular* 376, U.S. Geological Survey, Washington, DC.

Chinn, J. (1965). "Failure of simply supported flat roofs by ponding of rain." *Eng. J.* 3(2), 38–41.

Chinn, J., Mansouri, A. H., and Adams, S. F. (1969). "Ponding of liquids on flat roofs." *J. Struct. Div.*, 95(5), 797–807.

Factory Mutual Global (FM Global). (2012). *Loss prevention data 1–54, roof loads for new construction*, Factory Mutual Engineering Corp., Norwood, MA.

Graber, S. D. (2009). "Rain loads and flow attenuation on roofs." *J. Arch. Eng.* 15(3), 91–101.

Haussler, R. W. (1962). "Roof deflection caused by rainwater pools." *Civil Eng.* 32, 58–59.

Heinzerling, J. E. (1971). "Structural design of steel joist roofs to resist ponding loads." *Technical Digest No. 3*. Steel Joist Institute, Arlington, VA.

International Code Council (ICC) (2012) "International plumbing code." Washington, DC.

Marino, F. J. (1966). "Ponding of two-way roof systems." *Eng. J.* 3(3), 93–100.

National Fire Protection Association (NFPA). (2012). *NFPA 5000 building construction and safety code*. NFPA, Quincy, MA.

Salama, A. E., and Moody, M. L. (1967). "Analysis of beams and plates for ponding loads." *J. Struct. Div.*, 93(1), 109–126.

Sawyer, D. A. (1967). "Ponding of rainwater on flexible roof systems." *J. Struct. Div.*, 93(1), 127–148.

Sawyer, D. A. (1968). "Roof-structure roof-drainage interaction." *J. Struct. Div.*, 94(1), 175–198.

Southern Building Code Congress International (SBCCI). (1991). "Standard plumbing code," SBCCI Inc., Birmingham, AL.

Steel Joist Institute (SJI). (2007). "Structural design of steel roofs to resist ponding loads." *Technical Digest No. 3*. Myrtle Beach, SC.

CHAPTER C9
RESERVED FOR FUTURE COMMENTARY

CHAPTER C10
ICE LOADS—ATMOSPHERIC ICING

C10.1 GENERAL

In most of the contiguous United States, freezing rain is considered the cause of the most severe ice loads. Values for ice thicknesses caused by in-cloud icing and snow suitable for inclusion in this standard are not currently available.

Very few sources of direct information or observations of naturally occurring ice accretions (of any type) are available. Bennett (1959) presents the geographical distribution of the occurrence of ice on utility wires from data compiled by various railroad, electric power, and telephone associations in the nine-year period from the winter of 1928–1929 to the winter of 1936–1937. The data include measurements of all forms of ice accretion on wires, including glaze ice, rime ice, and accreted snow, but do not differentiate among them. Ice thicknesses were measured on wires of various diameters, heights above ground, and exposures. No standardized technique was used in measuring the thickness. The maximum ice thickness observed during the nine-year period in each of 975 squares, 60 mi (97 km) on a side, in a grid covering the contiguous United States is reported. In every state except Florida, thickness measurements of accretions with unknown densities of approximately one radial inch were reported. Information on the geographical distribution of the number of storms in this nine-year period with ice accretions greater than specified thicknesses is also included.

Tattelman and Gringorten (1973) reviewed ice load data, storm descriptions, and damage estimates in several meteorological publications to estimate maximum ice thicknesses with a 50-year mean recurrence interval in each of seven regions in the United States. *Storm Data* (NOAA 1959–Present) is a monthly publication that describes damage from storms of all sorts throughout the United States. The compilation of this qualitative information on storms causing damaging ice accretions in a particular region can be used to estimate the severity of ice and wind-on-ice loads. The Electric Power Research Institute has compiled a database of icing events from the reports in *Storm Data* (Shan and Marr 1996). Damage severity maps were also prepared.

Bernstein and Brown (1997) and Robbins and Cortinas (1996) provide information on freezing rain climatology for the 48 contiguous states based on recent meteorological data.

C10.1.1 Site-Specific Studies. In-cloud icing may cause significant loadings on ice-sensitive structures in mountainous regions and for very tall structures in other areas. Mulherin (1996) reports that of 120 communications tower failures in the United States caused by atmospheric icing, 38 were caused by in-cloud icing, and in-cloud icing combined with freezing rain caused an additional 26 failures. In-cloud ice accretion is very sensitive to the degree of exposure to moisture-laden clouds, which is related to terrain, elevation, and wind direction and velocity. Large differences in accretion size can occur over a few hundred feet and can cause severe load unbalances in overhead wire systems. Advice from a meteorologist familiar with the area is particularly valuable in these circumstances. In Arizona, New Mexico, and the panhandles of Texas and Oklahoma, the United States Forest Service specifies ice loads caused by in-cloud icing for towers constructed at specific mountaintop sites (U.S. Forest Service 1994). Severe in-cloud icing has been observed in southern California (Mallory and Leavengood 1983a, 1983b), eastern Colorado (NOAA Feb. 1978), the Pacific Northwest (Winkleman 1974; Richmond et al. 1977; Sinclair and Thorkildson 1980), Alaska (Ryerson and Claffey 1991), and the Appalachians (Ryerson 1987, 1988a, 1988b, 1990; Govoni 1990).

Snow accretions also can result in severe structural loads and may occur anywhere snow falls, even in localities that may experience only one or two snow events per year. Some examples of locations where snow accretion events resulted in significant damage to structures are Nebraska (NPPD 1976), Maryland (Mozer and West 1983), Pennsylvania (Goodwin et al. 1983), Georgia and North Carolina (Lott 1993), Colorado (McCormick and Pohlman 1993), Alaska (Peabody and Wyman 2005), and the Pacific Northwest (Hall 1977; Richmond et al. 1977).

For Alaska, available information indicates that moderate to severe snow and in-cloud icing can be expected. The measurements made by Golden Valley Electric Association (Jones et al. 2002) are consistent in magnitude with visual observations across a broad area of central Alaska (Peabody 1993). Several meteorological studies using an ice accretion model to estimate ice loads have been performed for high-voltage transmission lines in Alaska (Gouze and Richmond 1982a, 1982b; Richmond 1985, 1991, 1992; Peterka et al. 1996). Estimated 50-year mean recurrence interval accretion thicknesses from snow range from 1.0 to 5.5 in. (25 to 140 mm), and in-cloud ice accretions range from 0.5 to 6.0 in. (12 to 150 mm). The assumed accretion densities for snow and in-cloud ice accretions, respectively, were 5 to 31 lb/ft^3 (80 to 500 kg/m^3) and 25 lb/ft^3 (400 kg/m^3). These loads are valid only for the particular regions studied and are highly dependent on the elevation and local terrain features.

In Hawaii, for areas where freezing rain (Wylie 1958), snow, and in-cloud icing are known to occur at higher elevations, site-specific meteorological investigations are needed.

Local records and experience should be considered when establishing the design ice thickness, concurrent wind speed, and concurrent temperature. In determining equivalent radial ice thicknesses from historical weather data, the quality, completeness, and accuracy of the data should be considered along with the robustness of the ice accretion algorithm. Meteorological stations may be closed by ice storms because of power outages, anemometers may be iced over, and hourly precipitation data recorded only after the storm when the ice in the rain gauge melts. These problems are likely to be more severe at automatic weather stations where observers are not available to estimate the weather parameters or correct erroneous readings. Note also that (1) air temperatures are recorded only to the nearest 1°F, at best, and may vary significantly from the recorded value in the region around the weather station; (2) the wind speed during freezing rain has a significant effect on the accreted ice load on objects oriented perpendicular to the wind direction; (3) wind speed and direction vary with terrain and exposure; (4) enhanced precipitation may occur on the windward side of mountainous terrain; and (5) ice may remain on the structure for days or weeks after freezing rain ends, subjecting the iced structure to wind speeds that may be significantly higher than those that accompanied the freezing rain. These factors should be considered both in estimating the accreted ice thickness at a weather station in past storms and in extrapolating those thicknesses to a specific site.

In using local data, it must also be emphasized that sampling errors can lead to large uncertainties in the specification of the 500-year ice thickness. Sampling errors are the errors associated with the limited size of the climatological data samples (years of record). When local records of limited extent are used to determine extreme ice thicknesses, care should be exercised in their use.

A robust ice accretion algorithm is not sensitive to small changes in input variables. For example, because temperatures are normally recorded in whole degrees, the calculated amount of ice accreted should not be sensitive to temperature changes of fractions of a degree.

C10.1.2 Dynamic Loads. While design for dynamic loads is not specifically addressed in this edition of the standard, the effects of dynamic loads are an important consideration for some ice-sensitive structures and should be considered in the design when they are anticipated to be significant. For example, large-amplitude galloping (Rawlins 1979; Section 6.2 of Simiu and Scanlan 1996) of guys and overhead cable systems occurs in many areas. The motion of the cables can cause damage because of direct impact of the cables on other cables or structures and can also cause damage because of wear and fatigue of the cables and other components of the structure (White 1999). Ice shedding from the guys on guyed masts can cause substantial dynamic loads in the mast.

C10.1.3 Exclusions. Additional guidance is available in Committee on Electrical Transmission Structures (1982) and CSA (1987, 1994).

C10.2 DEFINITIONS

FREEZING RAIN: Freezing rain occurs when warm, moist air is forced over a layer of subfreezing air at the Earth's surface. The precipitation usually begins as snow that melts as it falls through the layer of warm air aloft. The drops then cool as they fall through the cold surface air layer and freeze on contact with structures or the ground. Upper air data indicate that the cold surface air layer is typically between 1,000 and 3,900 ft (300 and 1,200 m) thick (Young 1978), averaging 1,600 ft (500 m) (Bocchieri 1980). The warm air layer aloft averages 5,000 ft (1,500 m) thick in freezing rain, but in freezing drizzle the entire temperature profile may be below 32°F (0°C) (Bocchieri 1980).

Precipitation rates and wind speeds are typically low to moderate in freezing rainstorms. In freezing rain, the water impingement rate is often greater than the freezing rate. The excess water drips off and may freeze as icicles, resulting in a variety of accretion shapes that range from a smooth cylindrical

FIGURE C10.2-1 Glaze Ice Accretion Caused by Freezing Rain

sheath, through a crescent on the windward side with icicles hanging on the bottom, to large irregular protuberances, see Fig. C10.2-1. The shape of an accretion depends on a combination of varying meteorological factors and the cross-sectional shape of the structural member, its spatial orientation, and flexibility.

Note that the theoretical maximum density of ice (917 kg/m^3 or 57 lb/ft^3) is never reached in naturally formed accretions because of the presence of air bubbles.

HOARFROST: Hoarfrost, which is often confused with rime, forms by a completely different process. Hoarfrost is an accumulation of ice crystals formed by direct deposition of water vapor from the air on an exposed object. Because it forms on objects with surface temperatures that have fallen below the frost point (a dew point temperature below freezing) of the surrounding air because of strong radiational cooling, hoarfrost is often found early in the morning after a clear, cold night. It is feathery in appearance and typically accretes up to about 1 in. (25 mm) in thickness with very little weight. Hoarfrost does not constitute a significant loading problem; however, it is a very good collector of supercooled fog droplets. In light winds, a hoarfrost-coated wire may accrete rime faster than a bare wire (Power 1983).

ICE-SENSITIVE STRUCTURES: Ice-sensitive structures are structures for which the load effects from atmospheric icing control the design of part or all of the structural system. Many open structures are efficient ice collectors, so ice accretions can have a significant load effect. The sensitivity of an open structure to ice loads depends on the size and number of structural members, components, and appurtenances and also on the other loads for which the structure is designed. For example, the additional weight of ice that may accrete on a heavy wide-flange member is smaller in proportion to the dead load than the same ice thickness on a light angle member. Also, the percentage increase in projected area for wind loads is smaller for the wide-flange member than for the angle member. For some open structures, other design loads, for example, snow loads and live loads on a catwalk floor, may be larger than the design ice load.

IN-CLOUD ICING: This icing condition occurs when a cloud or fog (consisting of supercooled water droplets 100 microns (100 mm) or less in diameter) encounters a surface that is at or below freezing temperature. It occurs in mountainous areas where adiabatic cooling causes saturation of the atmosphere to occur at temperatures below freezing, in free air in supercooled clouds, and in supercooled fogs produced by a stable air mass with a strong temperature inversion. In-cloud ice accretions can reach thicknesses of 1 ft (0.30 m) or more because the icing conditions can include high winds and typically persist or recur episodically during long periods of subfreezing temperatures. Large concentrations of supercooled droplets are not common at air temperatures below about 0°F (−18°C).

In-cloud ice accretions have densities ranging from that of low-density rime to glaze. When convective and evaporative cooling removes the heat of fusion as fast as it is released by the freezing droplets, the drops freeze on impact. When the cooling rate is lower, the droplets do not completely freeze on impact. The unfrozen water then spreads out on the object and may flow completely around it and even drip off to form icicles. The degree to which the droplets spread as they collide with the structure and freeze governs how much air is incorporated in the accretion, and thus its density. The density of ice accretions caused by in-cloud icing varies over a wide range from 5 to 56 pcf (80 to 900 kg/m^3) (Macklin 1962; Jones 1990). The resulting accretion can be either white or clear, possibly with attached icicles; see Fig. C10.2-2.

The amount of ice accreted during in-cloud icing depends on the size of the accreting object, the duration of the icing condition, and the wind speed. If, as often occurs, wind speed increases and air temperature decreases with height above ground, larger amounts of ice accrete on taller structures. The accretion shape depends on the flexibility of the structural member, component, or appurtenance. If it is free to rotate, such as a long guy or a long span of a single conductor or wire, the ice accretes with a roughly circular cross section. On more rigid structural members, components, and appurtenances, the ice forms in irregular pennant shapes extending into the wind.

SNOW: Under certain conditions, snow falling on objects may adhere because of capillary forces, interparticle freezing (Colbeck and Ackley 1982), and/or sintering (Kuroiwa 1962). On objects with circular cross section, such as a wire, cable,

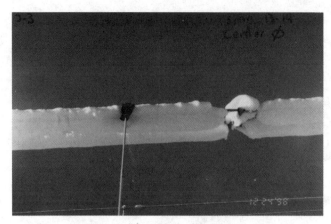

FIGURE C10.2-3 Snow Accretion on Wires

conductor, or guy, sliding, deformation, and/or torsional rotation of the underlying cable may occur, resulting in the formation of a cylindrical sleeve, even around bundled conductors and wires; see Fig. C10.2-3. Because accreting snow is often accompanied by high winds, the density of accretions may be much higher than the density of the same snowfall on the ground.

Damaging snow accretions have been observed at surface air temperatures ranging from about 23 to 36°F (−5 to 2°C). Snow with a high moisture content appears to stick more readily than drier snow. Snow falling at a surface air temperature above freezing may accrete even at wind speeds above 25 mi/h (10 m/s), producing dense 37 to 50 pcf (600 to 800 kg/m^3) accretions. Snow with a lower moisture content is not as sticky, blowing off the structure in high winds. These accreted snow densities are typically between 2.5 and 16 pcf (40 and 250 kg/m^3) (Kuroiwa 1965).

Even apparently dry snow can accrete on structures (Gland and Admirat 1986). The cohesive strength of the dry snow is initially supplied by the interlocking of the flakes and ultimately by sintering, as molecular diffusion increases the bond area between adjacent snowflakes. These dry snow accretions appear to form only in very low winds and have densities estimated at between 5 and 10 pcf (80 and 150 kg/m^3) (Sakamoto et al. 1990; Peabody 1993).

C10.4 ICE LOADS CAUSED BY FREEZING RAIN

C10.4.1 Ice Weight. The ice thicknesses shown in Figs. 10.4-2 through 10.4-6 were determined for a horizontal cylinder oriented perpendicular to the wind. These ice thicknesses cannot be applied directly to cross sections that are not round, such as channels and angles. However, the ice area from Eq. (10.4-1) is the same for all shapes for which the circumscribed circles have equal diameters (Peabody and Jones 2002; Jones and Peabody 2006). It is assumed that the maximum dimension of the cross section is perpendicular to the trajectory of the raindrops. Similarly, the ice volume in Eq. (10.4-2) is for a flat plate perpendicular to the trajectory of the raindrops. The constant π in Eq. (10.4-2) corrects the thickness from that on a cylinder to the thickness on a flat plate. For vertical cylinders and horizontal cylinders parallel to the wind direction, the ice area given by Eq. (10.4-1) is conservative.

C10.4.2 Nominal Ice Thickness. The 500-year mean recurrence interval ice thicknesses shown in Figs. 10.4-2 to 10.4-6 are based on studies using an ice accretion model and local data.

FIGURE C10.2-2 Rime Ice Accretion Caused by In-Cloud Icing

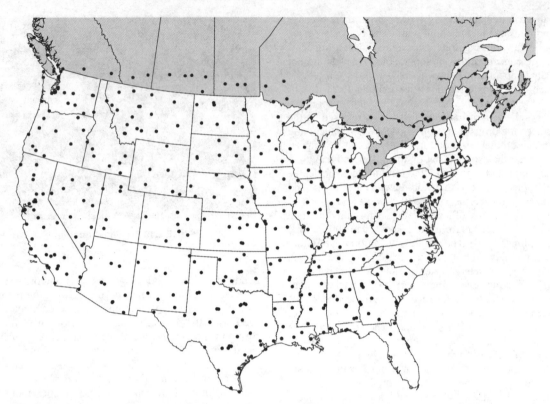

FIGURE C10.4-1 Locations of Weather Stations Used in Preparation of Figures 10.4-2 through 10.4-6

Historical weather data from 540 National Weather Service (NWS), military, Federal Aviation Administration (FAA), and Environment Canada weather stations were used with the U.S. Army's Cold Regions Research and Engineering Laboratory (CRREL) and simple ice accretion models (Jones 1996, 1998) to estimate uniform radial glaze ice thicknesses in past freezing rainstorms. The models and algorithms have been applied to additional stations in Canada along the border of the lower 48 states. The station locations are shown in Fig. C10.4-1 for the 48 contiguous states and in Fig. 10.4-6 for Alaska. The period of record of the meteorological data at any station is typically 20 to 50 years. The ice accretion models use weather and precipitation data to simulate the accretion of ice on cylinders 33 ft (10 m) above the ground, oriented perpendicular to the wind direction in freezing rainstorms. Accreted ice is assumed to remain on the cylinder until after freezing rain ceases and the air temperature increases to at least 33°F (0.6°C). At each station, the maximum ice thickness and the maximum wind-on-ice load were determined for each storm. Severe storms, those with significant ice or wind-on-ice loads at one or more weather stations, were researched in *Storm Data* (NOAA 1959–Present), newspapers, and utility reports to obtain corroborating qualitative information on the extent of and damage from the storm. Yet very little corroborating information was obtained about damaging freezing rainstorms in Alaska, perhaps because of the low population density and relatively sparse newspaper coverage in the state.

Extreme ice thicknesses were determined from an extreme value analysis using the peaks-over-threshold method and the generalized Pareto distribution (Hoskings and Wallis 1987; Wang 1991; Abild et al. 1992). To reduce sampling error, weather stations were grouped into superstations (Peterka 1992) based on the incidence of severe storms, the frequency of freezing rainstorms, latitude, proximity to large bodies of water, elevation, and terrain. Concurrent wind-on-ice speeds were back-calculated from the extreme wind-on-ice load and the extreme ice thickness. The analysis of the weather data and the calculation of extreme ice thicknesses are described in more detail in Jones et al. (2002).

The maps in Figs. 10.4-2 to 10.4-6 represents the most consistent and best available nationwide map for nominal design ice thicknesses and wind-on-ice speeds. The icing model used to produce the map has not, however, been verified with a large set of collocated measurements of meteorological data and uniform radial ice thicknesses. Furthermore, the weather stations used to develop this map are almost all at airports. Structures in more exposed locations at higher elevations or in valleys or gorges, for example, Signal and Lookout Mountains in Tennessee, the Pontotoc Ridge and the edge of the Yazoo Basin in Mississippi, the Shenandoah Valley and Poor Mountain in Virginia, Mount Washington in New Hampshire, and Buffalo Ridge in Minnesota and South Dakota, may be subject to larger ice thicknesses and higher concurrent wind speeds. However, structures in more sheltered locations, for example, along the north shore of Lake Superior within 300 vertical ft (90 m) of the lake, may be subject to smaller ice thicknesses and lower concurrent wind speeds. Loads from snow or in-cloud icing may be more severe than those from freezing rain (see Section C10.1.1).

Special Icing Regions. Special icing regions are identified on the map. As described previously, freezing rain occurs only under special conditions when a cold, relatively shallow layer of air at the surface is overrun by warm, moist air aloft. For this reason, severe freezing rainstorms at high elevations in mountainous terrain typically do not occur in the same weather systems that cause severe freezing rainstorms at the nearest airport with a weather station. Furthermore, in these regions, ice thicknesses and wind-on-ice loads may vary significantly over short distances because of local variations in elevation, topography, and exposure. In these mountainous regions, the values given in

Fig. 10.4-1 should be adjusted, based on local historical records and experience, to account for possibly higher ice loads from both freezing rain and in-cloud icing (see Section C10.1.1).

C10.4.4 Importance Factors. The importance factors for ice and concurrent wind adjust the nominal ice thickness and concurrent wind pressure for Risk Category I structures from a 500-year mean recurrence interval to a 250-year mean recurrence interval. For Risk Category III and IV structures, they are adjusted to 1,000-year and 1,400-year mean recurrence intervals, respectively. The importance factor is multiplied times the ice thickness rather than the ice load because the ice load from Eq. (10.4-1) depends on the diameter of the circumscribing cylinder as well as the design ice thickness. The concurrent wind speed used with the nominal ice thickness is based on both the winds that occur during the freezing rainstorm and those that occur between the time the freezing rain stops and the time the temperature rises to above freezing. When the temperature rises above freezing, it is assumed that the ice melts enough to fall from the structure. In the colder northern regions, the ice generally stays on structures for a longer period of time after the end of a storm resulting in higher concurrent wind speeds. The results of the extreme value analysis show that the concurrent wind speed does not change significantly with mean recurrence interval. The lateral wind-on-ice load does, however, increase with mean recurrence interval because the ice thickness increases. The importance factors differ from those used for both the wind loads in Chapter 6 and the snow loads in Chapter 7 because the extreme value distribution used for the ice thickness is different from the distributions used to determine the extreme wind speeds in Chapter 6 and snow loads in Chapter 7. See also Table C10.4-1 and the discussion under Section C10.4.6.

C10.4.6 Design Ice Thickness for Freezing Rain. The design load on the structure is a product of the nominal design load and the load factors specified in Chapter 2. The load factors for load and resistance factor design (LRFD) for atmospheric icing are 1.0. Table C10.4-1 shows the multipliers on the 500-year mean recurrence interval ice thickness and concurrent wind speed used to adjust to other mean recurrence intervals.

The studies of ice accretion on which the maps are based indicate that the concurrent wind speed on ice does not increase with mean recurrence interval (see Section C10.4.4).

The 2002 Edition of ASCE 7 was the first edition to include atmospheric icing maps. Fifty-year mean recurrence interval (MRI) maps were provided at that time in order to match the approach used for the wind and snow load maps. An ice-thickness multiplier equal to 2 was included (see Eq. (10.4-5) in ASCE 7-02 to ASCE 7-10) to adjust the mapped values to a 500-year MRI for design. Thus, the adjusted 500-year MRI value would be appropriate for use with the LRFD load factor of 1.0 shown for ice loads in Chapter 2. A 500-year MRI load was consistent at that time with those historically used for seismic and wind loads. The 2016 edition of ASCE 7 includes atmospheric icing maps based on a 500-year MRI load with no ice-thickness multiplier in Eq. (10.4-5). The 2016 maps have been redrawn directly from the original extreme value analysis. Design load changes from the 2010 edition to the 2016 edition are caused by the maps being redrawn, not caused by changing the map MRI.

When the reliability of a system of structures or one interconnected structure of large extent is important, spatial effects should also be considered. All of the cellular telephone antenna structures that serve a state or a metropolitan area could be considered to be a system of structures. Long overhead electric transmission lines and communications lines are examples of large, interconnected structures. Figs. 10.4-2 through 10.4-6 are for ice loads appropriate for a single structure of small areal extent. Large, interconnected structures and systems of structures are hit by icing storms more frequently than a single structure. The frequency of occurrence increases with the area encompassed or the linear extent. To obtain equal risks of exceeding the design load in the same icing climate, the individual structures forming the system or the large, interconnected structure should be designed for a larger ice load than a single structure (CEATI 2003, 2005; Chouinard and Erfani 2006; Golikova 1982; Golikova et al. 1982; Jones 2010).

C10.5 WIND ON ICE-COVERED STRUCTURES

Ice accretions on structures change the structure's wind drag coefficients. The ice accretions tend to round sharp edges, reducing the drag coefficient for such members as angles and bars. Natural ice accretions can be irregular in shape with an uneven distribution of ice around the object on which the ice has accreted. The shape varies from storm to storm and from place to place within a storm. The actual projected area of a glaze ice accretion may be larger than that obtained by assuming a uniform ice thickness.

C10.5.5 Wind on Ice-Covered Guys and Cables. There are practically no published experimental data giving the force coefficients for ice-covered guys and cables. There have been many studies of the force coefficient for cylinders without ice. The force coefficient varies with the surface roughness and the Reynolds number. At subcritical Reynolds numbers, both smooth and rough cylinders have force coefficients of approximately 1.2, as do square sections with rounded edges (Fig. 4.5.5 in Simiu and Scanlan 1996). For a wide variety of stranded electrical transmission cables, the supercritical force coefficients are approximately 1.0 with subcritical values as high as 1.3 (Fig. 5-2 in Shan 1997). The transition from subcritical to supercritical depends on the surface characteristics and takes place over a wide range of Reynolds numbers. For the stranded cables described in Shan (1997), the range is from approximately 25,000 to 150,000. For a square section with rounded edges, the transition takes place at a Reynolds number of approximately 800,000 (White 1999). The concurrent 3-s gust wind speed in Figs. 10.4-2 through 10.4-5 for the contiguous 48 states varies from 30 to 60 mi/h (13.4 to 26.8 m/s), with speeds in Fig. 10.4-6 for Alaska up to 80 mi/h (35.8 m/s). Table C10.5-1 shows the Reynolds numbers (using U.S. standard atmosphere) for a range of iced guys and cables. In practice, the Reynolds numbers range from subcritical through critical to supercritical depending on the roughness of the ice accretion. Considering that

Table C10.4-1 Mean Recurrence Interval Factors

Mean Recurrence Interval	Multiplier on Ice Thickness	Multiplier on Wind Pressure
25	0.40	1.0
50	0.50	1.0
100	0.625	1.0
200	0.75	1.0
250	0.80	1.0
300	0.85	1.0
400	0.90	1.0
500	1.00	1.0
1,000	1.15	1.0
1,400	1.25	1.0

Table C10.5-1 Typical Reynolds Numbers for Iced Guys and Cables

Guy or Cable Diameter (in.)	Ice Thickness t (in.)	Importance Factor I_w	Design Ice Thickness t_d (in.)	Iced Diameter (in.)	Concurrent 3-s Gust Wind Speed (mi/h)	Reynolds Number
Contiguous 48 States						
0.250	0.25	0.80	0.20	0.650	30	15,200
0.375	0.25	0.80	0.20	0.775	30	18,100
0.375	1.25	1.25	1.563	3.500	60	163,000
1.000	0.25	0.80	0.20	1.400	30	32,700
1.000	1.25	1.25	1.563	4.125	60	192,000
2.000	1.25	1.25	1.563	5.125	60	239,000
Alaska						
0.250	0.25	0.80	0.20	0.650	50	27,000
2.000	0.50	1.25	0.625	3.250	80	202,000

Note: To convert in. to mm, multiply by 25.4. To convert mi/h to km/h, multiply by 1.6.

the shape of ice accretions is highly variable from relatively smooth cylindrical shapes to accretions with long icicles with projected areas greater than the equivalent radial thickness used in the maps, a single force coefficient of 1.2 has been chosen.

C10.6 DESIGN TEMPERATURES FOR FREEZING RAIN

Some ice-sensitive structures, particularly those using overhead cable systems, are also sensitive to changes in temperature. In some cases, the maximum load effect occurs around the melting point of ice (32°F or 0°C) and in others at the lowest temperature that occurs while the structure is loaded with ice. Figs. 10.6-1 and 10.6-2 show the low temperatures to be used for design in addition to the melting temperature of ice.

The freezing rain model described in Section C10.4.2 tracked the temperature during each modeled icing event. For each event, the minimum temperature that occurred with the maximum ice thickness was recorded. The minimum temperatures for all the freezing rain events used in the extreme value analysis of ice thickness were analyzed to determine the 10th percentile temperature at each superstation (i.e., the temperature that was exceeded during 90% of the extreme icing events). These temperatures were used to make the maps shown in Figs. 10.6-1 and 10.6-2. In areas where the temperature contours were close to the wind or ice thickness contours, they were moved to coincide with, first, the concurrent wind boundaries, and, second, the ice zone boundaries.

C10.7 PARTIAL LOADING

Variations in ice thickness caused by freezing rain on objects at a given elevation are small over distances of about 1,000 ft (300 m). Therefore, partial loading of a structure from freezing rain is usually not significant (Cluts and Angelos 1977).

In-cloud icing is more strongly affected by wind speed, thus partial loading caused by differences in exposure to in-cloud icing may be significant. Differences in ice thickness over several structures or components of a single structure are associated with differences in the exposure. The exposure is a function of shielding by other parts of the structure and by the upwind terrain.

Partial loading associated with ice shedding may be significant for snow or in-cloud ice accretions and for guyed structures when ice is shed from some guys before others.

REFERENCES

Abild, J., Andersen, E. Y., and Rosbjerg, L. (1992). "The climate of extreme winds at the Great Belt, Denmark." *J. Wind Engrg. Indust. Aerodyn.* 41–44, 521–532.

Bennett, I. (1959). "Glaze: Its meteorology and climatology, geographical distribution and economic effects." *Environmental Protection Research Division Technical Report EP-105*, U.S. Army Quartermaster, Research and Engineering Center, Natick, MA.

Bernstein, B. C., and Brown, B. G. (1997). "A climatology of supercooled large drop conditions based upon surface observations and pilot reports of icing." *Proc., 7th Conf. on Aviation, Range and Aerospace Meteorology*, Feb. 2–7, Long Beach, CA.

Bocchieri, J. R. (1980). "The objective use of upper air soundings to specify precipitation type." *Mon. Weather Rev.* 108, 596–603.

Canadian Standards Association (CSA). (1987). "Overhead lines." *CAN/CSA-C22.3 No. 1-M87*. Rexdale, ON.

CSA. (1994). "Antennas, towers and antenna-supporting structures." *CSA-S37-94*. CSA, Rexdale, ON.

Centre for Energy Advancement through Technological Innovation. (CEATI). (2003). "Spatial factors for extreme ice and extreme wind: Task 1 literature review on the determination of spatial factors." *T033700-3316B*. Montreal.

CEATI. (2005). "Spatial factors for extreme ice and extreme wind: Task 2 calculation of spatial factors from ice and wind data." *T033700-3316B-2*. Montreal.

Chouinard, L., and Erfani, R. (2006). "Spatial modeling of atmospheric icing hazards." *Intl. Forum on Engineering Decision Making*, April 26–29, Lake Louise, Canada.

Cluts, S., and Angelos, A. (1977). "Unbalanced forces on tangent transmission structures." *IEEE Winter Power Meeting, Paper No.A77-220-7*. IEEE, Los Alamitos, CA.

Colbeck, S. C., and Ackley, S. F. (1982). "Mechanisms for ice bonding in wet snow accretions on power lines." *Proc., 1st Intl. Workshop on Atmospheric Icing of Structures*, L. D. Minsk, ed. *U.S. Army CRREL Special Report*, 25–30, 83–17. Hanover, NH.

Committee on Electrical Transmission Structures. (1982). "Loadings for electrical transmission structures by the committee on electrical transmission structures." *J. Struct. Div.*, 108(5), 1088–1105.

Gland, H., and Admirat, P. (1986). "Meteorological conditions for wet snow occurrence in France—Calculated and measured results in a recent case study on 5 March 1985." *Proc., 3rd Intl. Workshop on Atmospheric Icing of Structures*, L. E. Welsh and D. J. Armstrong, eds., Canadian Climate Program, Vancouver, 91–96.

Golikova, T. N. (1982). "Probability of increased ice loads on overhead lines depending on their length." *Soviet Power No. 10*, Ralph McElroy Co., 888–894.

Golikova, T. N., Golikov, B. F., and Savvaitov, D. S. (1982). "Methods of calculating icing loads on overhead lines as spatial constructions." *Proc., 1st Intl. Workshop on Atmospheric Icing of Structures*, L. D. Minsk, ed. *U.S. Army CRREL Special Report No. 83-17*, Cold Regions Research and Engineering Laboratory, Hanover, NH, 341–345.

Goodwin, E. J., Mozer, J. D., DiGioia, A. M., Jr., and Power, B. A. (1983). "Predicting ice and snow loads for transmission line design." *Proc., 3rd*

Intl. Workshop on Atmospheric Icing of Structures, L. E. Welsh, and D. J. Armstrong, eds., Canadian Climate Program, 1991, Vancouver, 267–275.

Gouze, S. C., and Richmond, M. C. (1982a). *Meteorological evaluation of the proposed Alaska transmission line routes*, Meteorology Research, Altadena, CA.

Gouze, S. C., and Richmond, M. C. (1982b). *Meteorological evaluation of the proposed Palmer to Glennallen transmission line route*, Meteorology Research, Inc., Altadena, CA.

Govoni, J. W. (1990). "A comparative study of icing rates in the White Mountains of New Hampshire, Paper A1-9." *Proc., 5th Intl. Workshop on Atmospheric Icing of Structures*, Tokyo.

Hall, E. K. (1977). "Ice and wind loading analysis of Bonneville Power Administration's transmission lines and test spans." *IEEE Power Engineering Society Summer Meeting*, July 20, Mexico City.

Hoskings, J. R. M., and Wallis, J. R. (1987). "Parameter and quantile estimation for the generalized Pareto distribution." *Technometrics*, 29(3), 339–349.

Jones, K. (1998). "A simple model for freezing rain loads." *Atmos. Res.*, 46, 87–97.

Jones, K. F. (1990). "The density of natural ice accretions related to nondimensional icing parameters." *Q. J. Royal Meteorolog. Soc.*, 116, 477–496.

Jones, K. F. (1996). "Ice accretion in freezing rain" *U.S. Army CRREL Report No. 96-2*. Cold Regions Research and Engineering Laboratory, Hanover, NH.

Jones, K. F. (2010). *Evaluation of extreme ice loads from freezing rain for Newfoundland and Labrador Hydro, Final Report, Muskrat Falls Project Exhibit 96*. Terrestrial and Cryospheric Sciences Branch, Cold Regions Research and Engineering Laboratory, Hanover, NH.

Jones, K. F., and Peabody, A. B. (2006). "The application of a uniform radial ice thickness to structural sections." *Cold Reg. Sci. Tech.*, 44(2), 145–148.

Jones, K. F., Thorkildson, R., and Lott, J. N. (2002). "The development of the map of extreme ice loads for ASCE Manual 74." *Electrical Transmission in a New Age*. D. E. Jackman, ed., ASCE, Reston, VA, 9–31.

Kuroiwa, D. (1962). "A study of ice sintering." *U.S. Army CRREL, Research Report No. 86*, Hanover, NH.

Kuroiwa, D. (1965). "Icing and snow accretion on electric wires." *U.S. Army CRREL Research Paper 123*, Hanover, NH.

Lott, N. (1993). *NCDC Technical Report Nos. 93-01 and 93-03*, National Climatic Data Center, Asheville, NC.

Macklin, W. C. (1962). "The density and structure of ice formed by accretion." *Q. J. Royal Meteorolog. Soc.*, 88, 30–50.

Mallory, J. H., and Leavengood, D. C. (1983a). "Extreme glaze and rime ice loads in Southern California: Part I—Rime." *Proc., 1st Intl. Workshop on Atmospheric Icing of Structures*, L. D. Minsk, ed., *U.S. Army CRREL Special Report No. 83-17*. Hanover, NH, 299–308.

Mallory, J. H., and Leavengood, D. C. (1983b). "Extreme glaze and rime ice loads in Southern California: Part II—Glaze." *Proc., 1st Intl. Workshop on Atmospheric Icing of Structures*, L. D. Minsk, ed., *U.S. Army CRREL Special Report No. 83-17*. Hanover, NH, 309–318.

McCormick, T., and Pohlman, J. C. (1993). "Study of compact 220 kV line system indicates need for micro-scale meteorological information." *Proc., 6th Intl. Workshop on Atmospheric Icing of Structures*. Budapest, Hungary. 155–159.

Mozer, J. D., and West, R. J. (1983). "Analysis of 500 kV tower failures." *Meeting of the Pennsylvania Electric Association*, Harrisburg, PA.

Mulherin, N. D. (1996). "Atmospheric icing and tower collapse in the United States." *7th Intl. Workshop on Atmospheric Icing of Structures*. M. Farzaneh and J. Laflamme, eds. Chicoutimi, Quebec, June 3–6.

National Oceanic and Atmospheric Administration (NOAA). (1959–Present). *Storm data*, National Oceanic and Atmospheric Administration, Washington, DC.

Nebraska Public Power District (NPPD). (1976). *The storm of March 29, 1976*, Public Relations Department, Columbus, NE.

Peabody, A. B. (1993). "Snow loads on transmission and distribution lines in Alaska." *Proc., 6th Intl. Workshop on Atmospheric Icing of Structures*. Budapest, Hungary. 201–205.

Peabody, A. B., and Jones, K. F. (2002). "Effect of wind on the variation of ice thickness from freezing rain." *Proc., 10th Intl. Workshop on Atmospheric Icing of Structures*. June 17–20, Brno, Czech Republic.

Peabody, A. B., and Wyman, G. (2005). "Atmospheric icing measurements in Fairbanks, Alaska." *Proc., 11th Intl. Workshop on Atmospheric Icing of Structures*, Masoud Farzaneh and Anand P. Goel, eds., Oct. 9–13, Yokohama, Japan.

Peterka, J. A. (1992). "Improved extreme wind prediction for the United States." *J. Wind Engrg. Indust. Aerodyn.*, 41–44, 533–541.

Peterka, J. A., Finstad, K., and Pandy, A. K. (1996). *Snow and wind loads for Tyee transmission line*. Cermak Peterka Petersen, Fort Collins, CO.

Power, B. A. (1983). "Estimation of climatic loads for transmission line design." *CEA No. ST 198*. Canadian Electric Association, Montreal.

Rawlins, C. B. (1979). "Galloping conductors." *Transmission line reference book, wind-induced conductor motion*, prepared by Gilbert/Commonwealth. Electric Power Research Institute, Palo Alto, CA, 113–168.

Richmond, M. C. (1985). *Meteorological evaluation of Bradley Lake hydroelectric project 115kV transmission line route*, Richmond Meteorological Consultant, Torrance, CA.

Richmond, M. C. (1991). *Meteorological evaluation of Tyee Lake hydroelectric project transmission line route, Wrangell to Petersburg*, Richmond Meteorological Consulting, Torrance, CA.

Richmond, M. C. (1992). *Meteorological evaluation of Tyee Lake hydroelectric project transmission line route*, Tyee power plant to Wrangell, Richmond Meteorological Consulting, Torrance, CA.

Richmond, M. C., Gouze, S. C., and Anderson, R. S. (1977). *Pacific Northwest icing study*, Meteorology Research, Altadena, CA.

Robbins, C. C., and Cortinas, J. V., Jr. (1996). "A climatology of freezing rain in the contiguous United States: Preliminary results." Preprints, 15th AMS Conference on Weather Analysis and Forecasting, Norfolk, VA, Aug. 19–23.

Ryerson, C. (1987). "Rime meteorology in the Green Mountains." *U.S. Army CRREL Report No. 87-1*. Cold Regions Research and Engineering Laboratory, Hanover, NH.

Ryerson, C. (1988a). "Atmospheric icing climatologies of two New England mountains." *J. Appl. Meteorol.*, 27(11), 1261–1281.

Ryerson, C. (1988b). "New England mountain icing climatology." *CRREL Report No. 88-12*. Cold Regions Research and Engineering Laboratory, Hanover, NH.

Ryerson, C. (1990). "Atmospheric icing rates with elevation on northern New England mountains, U.S.A." *Arctic Alpine Res.*, 22(1), 90–97.

Ryerson, C., and Claffey, K. (1991). "High latitude, West Coast mountaintop icing." *Proc., Eastern Snow Conference*, Guelph, ON, 221–232.

Sakamoto, Y., Mizushima, K., and Kawanishi, S. (1990). "Dry snow type accretion on overhead wires: Growing mechanism, meteorological conditions under which it occurs and effect on power lines." *Proc. 5th International Workshop on Atmospheric Icing of Structures*, Tokyo, Paper 5–9.

Shan, L. (1997). "Wind tunnel study of drag coefficients of single and bundled conductors." *EPRI TR-108969*. Electric Power Research Institute, Palo Alto, CA.

Shan, L., and Marr, L. (1996). *Ice storm data base and ice severity maps*, Electric Power Research Institute, Palo Alto, CA.

Simiu, E., and Scanlan, R. H. (1996). *Wind effects on structures: Fundamentals and applications to design*. John Wiley & Sons, New York.

Sinclair, R. E., and Thorkildson, R. M. (1980). "In-cloud moisture droplet impingement angles and track clearances at the Moro UHV test site." *BPA 1200 kV Project Report No. ME-80-7*. Bonneville Power Administration, Portland, OR.

Tattelman, P., and Gringorten, I. (1973). "Estimated glaze ice and wind loads at the earth's surface for the contiguous United States." Report AFCRL-TR-73-0646. U.S. Air Force Cambridge Research Laboratories, Bedford, MA.

U.S. Forest Service (USFS). (1994). *Forest Service handbook FSH6609.14, Telecommunications handbook*, R3 Supplement 6609.14-94-2. USFS, Washington, DC.

Wang, Q. J. (1991). "The POT model described by the generalized Pareto distribution with Poisson arrival rate." *J. Hydrol.*, 129, 263–280.

White, H. B. (1999). "Galloping of ice covered wires." *Proc., 10th International Conf. on Cold Regions Engineering*, Hanover, NH, 799–804.

Winkleman, P. F. (1974). *Investigation of ice and wind loads: Galloping, vibrations and subconductor oscillations*, Bonneville Power Administration, Portland, OR.

Wylie, W. G. (1958). "Tropical ice storms—Winter invades Hawaii." *Weatherwise* (June), 84–90.

Young, W. R. (1978). "Freezing precipitation in the Southeastern United States." M.S. Thesis, Texas A&M University, College Station, TX.

CHAPTER C11
SEISMIC DESIGN CRITERIA

C11.1 GENERAL

Many of the technical changes made to the seismic provisions of the 2010 edition of this standard are primarily based on Part 1 of the 2009 edition of the *NEHRP Recommended Seismic Provisions for New Buildings and Other Structures* (FEMA 2009), which was prepared by the Building Seismic Safety Council (BSSC) under sponsorship of the Federal Emergency Management Agency (FEMA) as part of its contribution to the National Earthquake Hazards Reduction Program (NEHRP). The National Institute of Standards and Technology (NIST) is the lead agency for NEHRP, the federal government's long-term program to reduce the risks to life and property posed by earthquakes in the United States. Since 1985, the NEHRP provisions have been updated every three to five years. The efforts by BSSC to produce the NEHRP provisions were preceded by work performed by the Applied Technology Council (ATC) under sponsorship of the National Bureau of Standards (NBS)—now NIST—which originated after the 1971 San Fernando Valley earthquake. These early efforts demonstrated the design rules of that time for seismic resistance but had some serious shortcomings. Each subsequent major earthquake has taught new lessons. The NEHRP agencies (FEMA, NIST, the National Science Foundation [NSF], and the U.S. Geological Survey [USGS]), ATC, BSSC, ASCE, and others have endeavored to work individually and collectively to improve each succeeding document to provide state-of-the-art earthquake engineering design and construction provisions and to ensure that the provisions have nationwide applicability.

Content of Commentary. This commentary is updated from the enhanced commentary to ASCE/SEI 7-10 that was based substantially on Part 2, Commentary, of the 2009 *NEHRP Recommended Seismic Provisions for New Buildings and Other Structures* (FEMA 2009). For additional background on the earthquake provisions contained in Chapters 11 through 23 of ASCE/SEI 7-10, the reader is referred to *Recommended Lateral Force Requirements and Commentary* (SEAOC 1999).

Nature of Earthquake "Loads." Earthquakes load structures indirectly through ground motion. As the ground shakes, a structure responds. The response vibration produces structural deformations with associated strains and stresses. The computation of dynamic response to earthquake ground shaking is complex. The design forces prescribed in this standard are intended only as approximations to generate internal forces suitable for proportioning the strength and stiffness of structural elements and for estimating the deformations (when multiplied by the deflection amplification factor, C_d) that would occur in the same structure in the event of the design-level earthquake ground motion (not MCE_R).

The basic methods of analysis in the standard use the common simplification of a response spectrum. A response spectrum for a specific earthquake ground motion provides the maximum value of response for elastic single-degree-of-freedom oscillators as a function of period without the need to reflect the total response history for every period of interest. The design response spectrum specified in Section 11.4 and used in the basic methods of analysis in Chapter 12 is a smoothed and normalized approximation for many different recorded ground motions.

The design limit state for resistance to an earthquake is unlike that for any other load within the scope of ASCE 7. The earthquake limit state is based upon system performance, not member performance, and considerable energy dissipation through repeated cycles of inelastic straining is assumed. The reason is the large demand exerted by the earthquake and the associated high cost of providing enough strength to maintain linear elastic response in ordinary buildings. This unusual limit state means that several conveniences of elastic behavior, such as the principle of superposition, are not applicable and makes it difficult to separate design provisions for loads from those for resistance. This difficulty is the reason Chapter 14 of the standard contains so many provisions that modify customary requirements for proportioning and detailing structural members and systems. It is also the reason for the construction quality assurance requirements.

Use of Allowable Stress Design Standards. The conventional design of almost all masonry structures and many wood and steel structures has been accomplished using allowable stress design (ASD). Although the fundamental basis for the earthquake loads in Chapters 11 through 23 is a strength limit state beyond the first yield of the structure, the provisions are written such that conventional ASD methods can be used by the design engineer. Conventional ASD methods may be used in one of two ways:

1. The earthquake load as defined in Chapters 11 through 23 may be used directly in allowable stress load combinations of Section 2.4, and the resulting stresses may be compared directly with conventional allowable stresses.
2. The earthquake load may be used in strength design load combinations, and resulting stresses may be compared with amplified allowable stresses (for those materials for which the design standard gives the amplified allowable stresses, e.g., masonry).

Federal Government Construction. The Interagency Committee on Seismic Safety in Construction has prepared an order executed by the president (Executive Order 12699 2016) that all federally owned or leased building construction, as well as federally regulated and assisted construction, should be constructed to mitigate seismic hazards and that the NEHRP provisions are deemed to be the suitable standard. It is expected that this standard would be deemed equivalent, but the reader should bear in mind that there are certain differences.

C11.1.1 Purpose. The purpose of Section 11.1.1 is to clarify that the detailing requirements and limitations prescribed in this section and referenced standards are still required even when the design load combinations involving the wind forces of Chapters 26 through 29 produce greater effects than the design load combinations involving the earthquake forces of Chapters 11 through 23. This detailing is required so that the structure resists, in a ductile manner, potential seismic loads in excess of the prescribed wind loads. A proper, continuous load path is an obvious design requirement, but experience has shown that it is often overlooked and that significant damage and collapse can result. The basis for this design requirement is twofold:

1. To ensure that the design has fully identified the seismic force-resisting system and its appropriate design level and
2. To ensure that the design basis is fully identified for the purpose of future modifications or changes in the structure.

Detailed requirements for analyzing and designing this load path are given in the appropriate design and materials chapters.

C11.1.2 Scope. Certain structures are exempt for the following reasons:

Exemption 1—Detached wood-frame dwellings not exceeding two stories above grade plane constructed in accordance with the prescriptive provisions of the International Residential Code (IRC) for light-frame wood construction, including all applicable IRC seismic provisions and limitations, are deemed capable of resisting the anticipated seismic forces. Detached one- and two-story wood-frame dwellings generally have performed well even in regions of higher seismicity. Therefore, within its scope, the IRC adequately provides the level of safety required for buildings. The structures that do not meet the prescriptive limitations of the IRC are required to be designed and constructed in accordance with the International Building Code (IBC) and the ASCE 7 provisions adopted therein.

Exemption 2—Agricultural storage structures generally are exempt from most code requirements because such structures are intended only for incidental human occupancy and represent an exceptionally low risk to human life.

Exemption 3—Bridges, transmission towers, nuclear reactors, and other structures with special configurations and uses are not covered. The regulations for buildings and buildinglike structures presented in this document do not adequately address the design and performance of such special structures.

ASCE 7 is not retroactive and usually applies to existing structures only when there is an addition, change of use, or alteration. Minimum acceptable seismic resistance of existing buildings is a policy issue normally set by the authority having jurisdiction. Appendix 11B of the standard contains rules of application for basic conditions. ASCE 41 (2014) provides technical guidance but does not contain policy recommendations. A chapter in the *International Building Code* (IBC) applies to alteration, repair, addition, and change of occupancy of existing buildings, and the International Code Council maintains the *International Existing Building Code* (IEBC) and associated commentary.

C11.1.3 Applicability. Industrial buildings may be classified as nonbuilding structures in certain situations for the purposes of determining seismic design coefficients and factors, system limitations, height limits, and associated detailing requirements. Many industrial building structures have geometries and/or framing systems that are different from the broader class of occupied structures addressed by Chapter 12, and the limited nature of the occupancy associated with these buildings reduces the hazard associated with their performance in earthquakes. Therefore, when the occupancy is limited primarily to maintenance and monitoring operations, these structures may be designed in accordance with the provisions of Section 15.5 for nonbuilding structures similar to buildings. Examples of such structures include, but are not limited to, boiler buildings, aircraft hangars, steel mills, aluminum smelting facilities, and other automated manufacturing facilities, whereby the occupancy restrictions for such facilities should be uniquely reviewed in each case. These structures may be clad or open structures.

C11.1.4 Alternate Materials and Methods of Construction. It is not possible for a design standard to provide criteria for the use of all possible materials and their combinations and methods of construction, either existing or anticipated. This section serves to emphasize that the evaluation and approval of alternate materials and methods require a recognized and accepted approval system. The requirements for materials and methods of construction contained within the document represent the judgment of the best use of the materials and methods based on well-established expertise and historical seismic performance. It is important that any replacement or substitute be evaluated with an understanding of all the ramifications of performance, strength, and durability implied by the standard.

Until needed standards and agencies are created, authorities that have jurisdiction need to operate on the basis of the best evidence available to substantiate any application for alternates. If accepted standards are lacking, applications for alternative materials or methods should be supported by test data obtained from test data requirements in Section 1.3.1. The tests should simulate expected load and deformation conditions to which the system, component, or assembly may be subjected during the service life of the structure. These conditions, when applicable, should include several cycles of full reversals of loads and deformations in the inelastic range.

C11.1.5 Quality Assurance. Quality assurance (QA) requirements are essential for satisfactory performance of structures in earthquakes. QA requirements are usually incorporated in building codes as special inspections and tests or as structural observation, and they are enforced through the authorities having jurisdiction. Many building code requirements parallel or reference the requirements found in standards adopted by ASCE 7. Where special inspections and testing, or structural observations are not specifically required by the building code, a level of quality assurance is usually provided by inspectors employed by the authority having jurisdiction.

Where building codes are not in force or where code requirements do not apply to or are inadequate for a unique structure or system, the registered design professional for the structure or system should develop a QA program to verify that the structure or system is constructed as designed. A QA program could be modeled on similar provisions in the building code or applicable standards.

The quality assurance plan is used to describe the QA program to the owner, the authority having jurisdiction, and to all other participants in the QA program. As such, the QA plan should include definitions of roles and responsibilities of the participants. It is anticipated that in most cases the owner of the project would be responsible for implementing the QA plan.

C11.2 DEFINITIONS

ATTACHMENTS, COMPONENTS, AND SUPPORTS: The distinction among attachments, components, and supports is necessary to the understanding of the requirements for nonstructural components and nonbuilding structures. Common cases associated with nonstructural elements are illustrated in Fig. C11.2-1. The

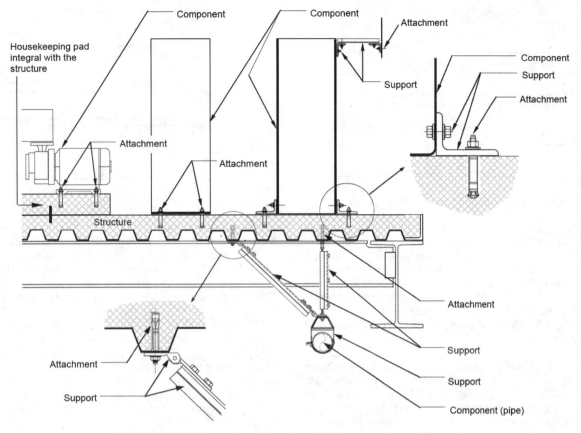

FIGURE C11.2-1 Examples of Attachments, Components, and Supports

definitions of attachments, components, and supports are generally applicable to components with a defined envelope in the as-manufactured condition and for which additional supports and attachments are required to provide support in the as-built condition. This distinction may not always be clear, particularly when the component is equipped with prefabricated supports; therefore, judgment must be used in the assignment of forces to specific elements in accordance with the provisions of Chapter 13.

BASE: The following factors affect the location of the seismic base:

- location of the grade relative to floor levels,
- soil conditions adjacent to the building,
- openings in the basement walls,
- location and stiffness of vertical elements of the seismic force-resisting system,
- location and extent of seismic separations,
- depth of basement,
- manner in which basement walls are supported,
- proximity to adjacent buildings, and
- slope of grade.

For typical buildings on level sites with competent soils, the base is generally close to the grade plane. For a building without a basement, the base is generally established near the ground-level slab elevation, as shown in Fig. C11.2-2. Where the vertical elements of the seismic force-resisting system are supported on interior footings or pile caps, the base is the top of these elements. Where the vertical elements of the seismic force-resisting system are supported on top of perimeter foundation walls, the base is typically established at the top of the foundation walls. Often vertical elements are supported at various elevations on the top of

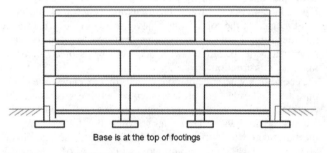

FIGURE C11.2-2 Base for a Level Site

footings, pile caps, and perimeter foundation walls. Where this occurs, the base is generally established as the lowest elevation of the tops of elements supporting the vertical elements of the seismic force-resisting system.

For a building with a basement located on a level site, it is often appropriate to locate the base at the floor closest to grade, as shown in Fig. C11.2-3. If the base is to be established at the level located closest to grade, the soil profile over the depth of the basement should not be liquefiable in the MCE_G ground motion. The soil profile over the depth of the basement also should not include quick and highly sensitive clays or weakly cemented soils prone to collapse in the MCE_G ground motion. Where liquefiable soils or soils susceptible to failure or collapse in an MCE_G ground motion are located within the depth of the basement, the base may need to be located below these soils rather than close to grade. Stiff soils are required over the depth of the basement because seismic forces are transmitted to and

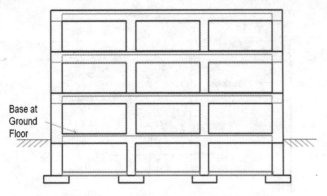

FIGURE C11.2-3 Base at Ground Floor Level

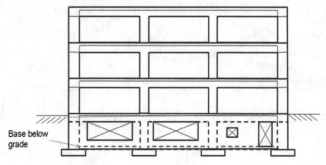

FIGURE C11.2-5 Base Below Substantial Openings in Basement Wall

from the building at this level and over the height of the basement walls. The engineer of record is responsible for establishing whether the soils are stiff enough to transmit seismic forces near grade. For tall or heavy buildings or where soft soils are present within the depth of the basement, the soils may compress laterally too much during an earthquake to transmit seismic forces near grade. For these cases, the base should be located at a level below grade.

In some cases, the base may be at a floor level above grade. For the base to be located at a floor level above grade, stiff foundation walls on all sides of the building should extend to the underside of the elevated level considered the base. Locating the base above grade is based on the principles for the two-stage equivalent lateral force procedure for a flexible upper portion of a building with one-tenth the stiffness of the lower portion of the building, as permitted in Section 12.2.3.2. For a floor level above grade to be considered the base, it generally should not be above grade more than one-half the height of the basement story, as shown in Fig. C11.2-4. Fig. C11.2-4 illustrates the concept of the base level located at the top of a floor level above grade, which also includes light-frame floor systems that rest on top of stiff basement walls or stiff crawl space stem walls of concrete or masonry construction.

A condition where the basement walls that extend above grade on a level site may not provide adequate stiffness is where the basement walls have many openings for items such as light wells, areaways, windows, and doors, as shown in Fig. C11.2-5. Where the basement wall stiffness is inadequate, the base should be taken as the level close to but below grade. If all of the vertical elements of the seismic force-resisting system are located on top of basement walls and there are many openings in the basement walls, it may be appropriate to establish the base at the bottom of the openings. Another condition where the basement walls may not be stiff enough is where the vertical elements of the seismic force-resisting system are long concrete shear walls extending over the full height and length of the building, as shown in Fig. C11.2-6. For this case, the appropriate location for the base is the foundation level of the basement walls.

Where the base is established below grade, the weight of the portion of the story above the base that is partially above and below grade must be included as part of the effective seismic weight. If the equivalent lateral force procedure is used, this procedure can result in disproportionately high forces in upper levels because of a large mass at this lowest level above the base. The magnitude of these forces can often be mitigated by using the two-stage equivalent lateral force procedure where it is allowed or by using dynamic analysis to determine force distribution over the height of the building. If dynamic analysis is used, it may be necessary to include multiple modes to capture the required mass participation, unless soil springs are incorporated into the model. Incorporation of soil springs into the model generally reduces seismic forces in the upper levels. With one or more stiff stories below more flexible stories, the dynamic behavior of the structure may result in the portion of the base shear from the first mode being less than the portion of base shear from higher modes.

Other conditions may also necessitate establishing the base below grade for a building with a basement that is located on a level site. Such conditions include those where seismic separations extend through all floors, including those located close to and below grade; those where the floor diaphragms close to and below grade are not tied to the foundation wall; those where the floor diaphragms, including the diaphragm for the floor close to grade, are flexible; and those where other buildings are located nearby.

For a building with seismic separations extending through the height of the building including levels close to and below grade, the separate structures are not supported by the soil against a

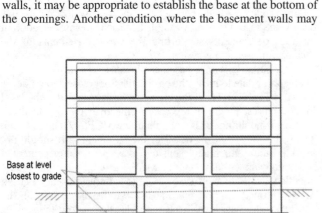

FIGURE C11.2-4 Base at Level Closest to Grade Elevation

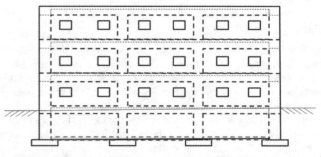

FIGURE C11.2-6 Base at Foundation Level Where There Are Full-Length Exterior Shear Walls

basement wall on all sides in all directions. If there is only one joint through the building, assigning the base to the level close to grade may still be appropriate if the soils over the depth of the basement walls are stiff and the diaphragm is rigid. Stiff soils are required so that the seismic forces can be transferred between the soils and basement walls in both bearing and side friction. If the soils are not stiff, adequate side friction may not develop for movement in the direction perpendicular to the joint.

For large footprint buildings, seismic separation joints may extend through the building in two directions and there may be multiple parallel joints in a given direction. For individual structures within these buildings, substantial differences in the location of the center of rigidity for the levels below grade relative to levels above grade can lead to torsional response. For such buildings, the base should usually be at the foundation elements below the basement or the highest basement slab level where the separations are no longer provided.

Where floor levels are not tied to foundation walls, the base may need to be located well below grade at the foundation level. An example is a building with tieback walls and posttensioned floor slabs. For such a structure, the slabs may not be tied to the wall to allow relative movement between them. In other cases, a soft joint may be provided. If shear forces cannot be transferred between the wall and a ground level or basement floor, the location of the base depends on whether forces can be transferred through bearings between the floor diaphragm and basement wall and between the basement wall and the surrounding soils. Floor diaphragms bearing against the basement walls must resist the compressive stress from earthquake forces without buckling. If a seismic or expansion joint is provided in one of these buildings, the base almost certainly needs to be located at the foundation level or a level below grade where the joint no longer exists.

If the diaphragm at grade is flexible and does not have substantial compressive strength, the base of the building may need to be located below grade. This condition is more common with existing buildings. Newer buildings with flexible diaphragms should be designed for compression to avoid the damage that can otherwise occur.

Proximity to other structures can also affect where the base should be located. If other buildings with basements are located adjacent to one or more sides of a building, it may be appropriate to locate the base at the bottom of the basement. The closer the adjacent building is to the building, the more likely it is that the base should be below grade.

For sites with sloping grade, many of the same considerations for a level site are applicable. For example, on steeply sloped sites, the earth may be retained by a tieback wall so that the building does not have to resist the lateral soil pressures. For such a case, the building is independent of the wall, so the base should be located at a level close to the elevation of grade on the side of the building where it is lowest, as shown in Fig. C11.2-7. Where

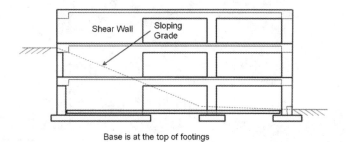

FIGURE C11.2-8 Building with Vertical Elements of the Seismic Force-Resisting System Supporting Lateral Earth Pressures

the building's vertical elements of the seismic force-resisting system also resist lateral soil pressures, as shown in Fig. C11.2-8, the base should also be located at a level close to the elevation of grade on the side of the building where grade is low. For these buildings, the seismic force-resisting system below highest grade is often much stiffer than the system used above it, as shown in Fig. C11.2-9, and the seismic weights for levels close to and below highest grade are greater than for levels above highest grade. Use of a two-stage equivalent lateral force procedure can be useful for these buildings.

Where the site is moderately sloped such that it does not vary in height by more than a story, stiff walls often extend to the underside of the level close to the elevation of high grade, and the seismic force-resisting system above grade is much more flexible above grade than it is below grade. If the stiff walls extend to the underside of the level close to high grade on all sides of the building, locating the base at the level closest to high grade may be appropriate. If the stiff lower walls do not extend to the underside of the level located closest to high grade on all sides of the building, the base should be assigned to the level closest to low grade. If there is doubt as to where to locate the base, it should conservatively be taken at the lower elevation.

DISTRIBUTION SYSTEM: For the purposes of determining the anchorage of components in Chapter 13, a distribution system is characterized as a series of individual in-line mechanical or electrical components that have been physically attached together to function as an interconnected system. In general, the individual in-line components of a distribution system are comparable to those of the pipe, duct, or electrical raceway so that the overall seismic behavior of the system is relatively uniform along its length. For example, a damper in a duct or a valve in a pipe is sufficiently similar to the weight of the duct or pipe itself, as opposed to a large fan or large heat exchanger. If a component is large enough to require support that is independent of the piping,

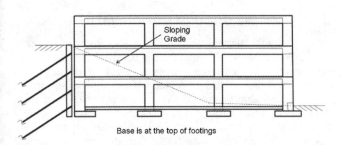

FIGURE C11.2-7 Building with Tie-Back or Cantilevered Retaining Wall That Is Separate from the Building

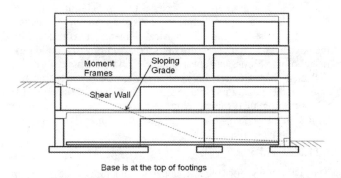

FIGURE C11.2-9 Building with Vertical Elements of the Seismic Force-Resisting System Supporting Lateral Earth Pressures

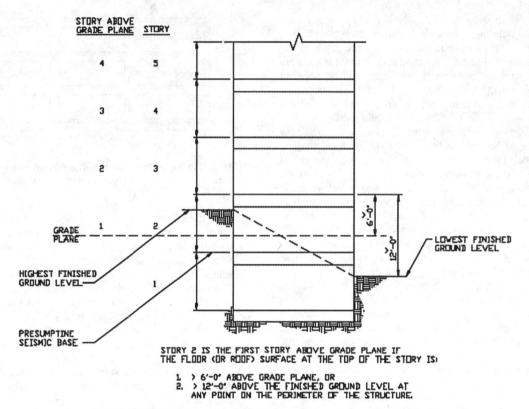

FIGURE C11.2-10 Illustration of Definition of Story above Grade Plane

duct, or conduit to which it is attached, it should likely be treated as a discrete component with regard to both exemptions and general design requirements. Representative distribution systems are listed in Table 13.6-1.

FLEXURE-CONTROLLED DIAPHRAGM: An example of a flexure-controlled diaphragm is a cast-in-place concrete diaphragm, where the flexural yielding mechanism would typically be yielding of the chord tension reinforcement.

SHEAR-CONTROLLED DIAPHRAGM: Shear-controlled diaphragms fall into two main categories. The first category is diaphragms that cannot develop a flexural mechanism because of aspect ratio, chord member strength, or other constraints. The second category is diaphragms that are intended to yield in shear rather than in flexure. Wood-sheathed diaphragms, for example, typically fall in the second category.

STORY ABOVE GRADE PLANE: Fig. C11.2-10 illustrates this definition.

TRANSFER FORCES: Transfer forces are diaphragm forces that are not caused by the acceleration of the diaphragm inertial mass. Transfer forces occur because of discontinuities in the vertical elements of the seismic force-resisting system or because of changes in stiffness in these vertical elements from one story to the next, even if there is no discontinuity. Additionally, buildings that combine frames and shear walls, which would have different deflected shapes under the same loading, also develop transfer forces in the diaphragms that constrain the frames and shear walls to deform together; this development is especially significant in dual systems.

C11.3 SYMBOLS

The provisions for precast concrete diaphragm design are intended to ensure that yielding, when it occurs, is ductile. Since yielding in shear is generally brittle at precast concrete connections, an additional overstrength factor, Ω_v, has been introduced; the required shear strength for a precast diaphragm is required to be amplified by this factor. This term is added to the symbols.

δ_{MDD} = This symbol refers to in-plane diaphragm deflection and is therefore designated with a lower-case delta. Note that the definition for δ_{MDD} refers to "lateral load" without any qualification, and the definition for Δ_{ADVE} refers to "tributary lateral load equivalent to that used in the computation of δ_{MDD}." This equivalency is an important concept that was part of the 1997 Uniform Building Code (UBC) (ICBO 1997) definition for a flexible diaphragm.

Ω_v = The provisions for precast concrete diaphragm design are intended to ensure that yielding, when it occurs, is ductile. Since yielding in shear is generally brittle at precast concrete connections, an additional overstrength factor, Ω_v, has been introduced; the required shear strength for a precast diaphragm is required to be amplified by this factor. This term is added to the symbols.

C11.4 SEISMIC GROUND MOTION VALUES

The basis for the mapped values of the MCE_R ground motions in ASCE 7-16 is identical to that in ASCE 7-10. Both of these are significantly different from mapped values of MCE ground motions in earlier editions of ASCE 7. These differences include use of (1) probabilistic ground motions that are based on uniform risk, rather than uniform hazard, (2) deterministic ground motions that are based on the 84th percentile (approximately 1.8 times median), rather than 1.5 times median response spectral acceleration for sites near active faults, and (3) ground motion intensity that is based on maximum rather than average (geometric mean) response spectral acceleration in the horizontal plane. These differences are explained in detail in the Commentary of the 2009 *NEHRP Recommended Provisions*. Except for determining the MCE_G PGA values in Chapters 11 and 21, the mapped values are given as MCE_R spectral values.

C11.4.1 Near-Fault Sites. In addition to very large accelerations, ground motions on sites located close to the zone of fault rupture of large-magnitude earthquakes can exhibit impulsive characteristics as well as unique directionality not typically recorded at sites located more distant from the zone of rupture. In past earthquakes, these characteristics have been observed to be particularly destructive. Accordingly, this standard establishes more restrictive design criteria for structures located on sites where such ground motions may occur. The standard also requires direct consideration of these unique characteristics in selection and scaling of ground motions used in nonlinear response history analysis and for the design of structures using seismic isolation or energy-dissipation devices when located on such sites.

The distance from the zone of fault rupture at which these effects can be experienced is dependent on a number of factors, including the rupture type, depth of fault, magnitude, and direction of fault rupture. Therefore, a precise definition of what constitutes a near-fault site is difficult to establish on a general basis. This standard uses two categorizations of near-fault conditions, both based on the distance of a site from a known active fault, capable of producing earthquakes of a defined magnitude or greater, and having average annual slip rates of nonnegligible amounts. These definitions were first introduced in the 1997 UBC (ICBO 1997). Fig. C11.4-1 illustrates the means of determining the distance of a site from a fault, where the fault plane dips at an angle relative to the ground surface.

C11.4.2 Mapped Acceleration Parameters. Mapped response spectral accelerations (5% damping) are provided on U.S. Geological Survey (USGS) maps for short periods, S_S, and at 1 s, S_1, for sites at the boundary of Site Classes B and C, which is $\bar{v}_s = 760$ m/s (2,500 ft/s). The USGS maps have been applicable to this site condition since 1996 (Frankel et al. 1996), but now are more clearly marked as being applicable to the reference value of $\bar{v}_s$. USGS ground motion maps are available at https://doi.org/10.5066/F7NK3C76.

C11.4.3 Site Class. The new site coefficients, F_a and F_v, necessitated a revision to the default site class when the site is known to be soil not classified as Site Class E or F. The F_a and F_v values for Site Class D in ASCE 7-10 were always equal to or greater than F_a and F_v values for Site Class C. Thus, specifying Site Class D as the default site class ensured that the response spectral accelerations would not be underestimated. However, the F_a values for Site Class C are greater than those for Site Class D for $S_S \geq 1.0$. Thus, a minor modification was required, consisting of adding the sentence "Where Site Class D is selected as the default site class per Section 11.4.3, the value of Fa shall not be less than1.2." in the last paragraph of Section 11.4.4 to ensure that the larger of the site coefficients for Site Classes C and D would be selected when the soil properties are not sufficiently known to determine the site class, and the authority having jurisdiction or geotechnical data have determined that Site Class E or F soils are not present.

Because the site coefficients, F_a and F_v, are less than unity (1.0) for Site Class B, a new paragraph was added to this section that requires the measurement of shear-wave velocity to demonstrate that the site is Site Class B according to the definition in Section 20.3. Furthermore, when $S_1 \geq 1.0$, the values of F_a for Site Class C are now greater than those for Site Class D.

C11.4.4 Site Coefficients and Risk-Targeted Maximum Considered Earthquake (MCE_R) Spectral Response Acceleration Parameters. Acceleration response parameters obtained from the maps (figures) cited in Section 11.4.2 are applicable for sites that have $\bar{v}_s = 760$ m/s ($\bar{v}_s = 2{,}500$ ft/s). For other site conditions, the S_S and S_1 values are computed as indicated in Section 11.4.4. This section has been revised from ASCE 7-10 to adjust the site factors to a reference site condition of $\bar{v}_s = 760$ m/s ($\bar{v}_s = 2{,}500$ ft/s) (instead of Site Class B) and to reflect more recent knowledge and data pertaining to site response.

The site coefficients, F_a and F_v, presented, respectively, in Tables 11.4-1 and 11.4-2 for the various site classes are based on analysis of strong-motion data and on numerical simulations of nonlinear site response. The development of the factors that were in place from the 1994 *NEHRP Provisions* through ASCE 7-10 is described in Dobry et al. (2000) and the references therein. Motivation for the revisions to these site factors includes the following (Seyhan and Stewart 2012): (1) updating the reference site condition used for the factors to match the condition on the national maps, which is $\bar{v}_s = 760$ m/s ($\bar{v}_s = 2{,}500$ ft/s); and (2) incorporating into the factors the substantial knowledge gains (stemming in large part from an enormous increase in available data) on site response over the past two decades.

The work undertaken to develop the revised factors is described in two PEER reports (Boore et al. 2013, Stewart and Seyhan 2013). They develop a semiempirical site amplification model for shallow crustal regions with two components: (1) a component to account for the change in ground motion with $\bar{v}_s$ for weak shaking conditions (referred to as the $\bar{v}_s$-scaling component); and (2) a component accounting for the effect of nonlinearity. The $\bar{v}_s$-scaling component was derived from strong ground motion data compiled and analyzed as part of the NGA-West 2 project (http://peer.berkeley.edu/ngawest2/). Whereas Stewart and Seyhan (2013) describe some regional variations in $\bar{v}_s$-scaling, such variations were not judged to be sufficiently robust for inclusion in the model, and therefore the model's $\bar{v}_s$-scaling reflects the average of the full international data set. The nonlinear component of the model is designed to jointly capture nonlinear effects revealed by strong-motion data analysis and the results of numerical simulations by Kamai et al. (2013).

The complete model (incorporating both $\bar{v}_s$-scaling and nonlinearity) is used to derive the recommended values of F_a and F_v for Site Classes B–D. The reference velocity used in the

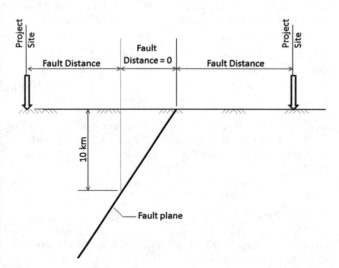

FIGURE C11.4-1 Fault Distance for Various Project Site Locations

computations was 760 m/s (2,500 ft/s). The values of $\bar{v}_s$ used to compute the tabulated factors for Classes B, C, and D were 913 m/s (2,995 ft/s), 489 m/s (1,604 ft/s), and 266 m/s (873 ft/s), respectively. These are average values of $\bar{v}_s$ for sites in the respective classes based on the NGA-West 2 data set. For Site Class E, median estimates of site amplification were computed using the complete model (applied at 155 m/s (509 ft/s)) as with the other classes. However, the recommended factors for Site Class E are increased above the median by half of the within-event standard deviation derived from the data, which increases site factors by approximately a factor of 1.3 to 1.4. This increase introduces a conservative bias to the Class E factors that is considered desirable because of the relatively modest amount of data for this site condition. A conservative bias was applied in the original site factors for Class E as well.

Fig. C11.4-2 shows the recommended site factors as a function of $\bar{v}_s$ for the levels of excitation (specified as values of S_S and S_1) given in Tables 11.4-1 and 11.4-2. The revised site factors for Site Class B (rock) are smaller than earlier values because of the change in reference velocity from 1,050 to 760 m/s (3,445 to 2,500 ft/s). The revised values for soil sites are generally similar to the prior values. However, for stronger shaking levels and Site Class C and D soils, the revised site factors are the same as or greater than the earlier values because of reduced levels of nonlinearity, especially at long periods (i.e., in the F_v parameter). For Class E, the relative levels of revised and prior site factors are strongly influenced by the amount of conservative bias adopted in their selection.

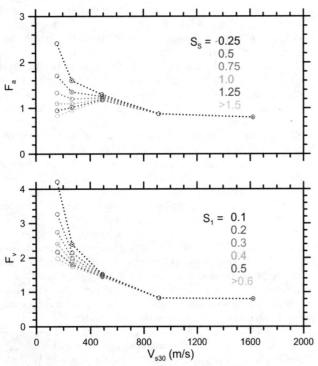

Note: In the top frame of this figure, the values of F_a for V_{S30} = 155 m/s (Site Class E) are 1.1, 1.0, and 0.8 for S_S = 1.0, 1.25, and >1.5, respectively. These values were originally approved by the Provisions Update Committee but were later replaced by the note, "See Sec. 11.4.8," which was necessitated by the results of more recent research. See Section C11.4.8 for details.

FIGURE C11.4-2 Site Factors F_a and F_v as Function of $\bar{v}_s$ (Shown as V_{s30} in Figure) for Various Amplitudes of Reference Rock Shaking.

Source: Stewart and Seyhan 2013.

Whereas the overall levels of Class E site amplification remain about the same, the degree of nonlinearity has been reduced somewhat for F_a and increased somewhat for F_v.

The revised factors are applicable for average site conditions in tectonically active regions (e.g., the West Coast of the United States). Because of different average site conditions in stable continental regions (such as the central and eastern United States), differences in average site response relative to the factors in Tables 11.4-1 and 11.4-2 should be anticipated for such regions. This difference can be addressed through site-specific analysis performed in accordance with Section 11.4.8.

For Site Classes B–D, site coefficients F_a and F_v may be computed from the following equations in lieu of using the site factors in Tables 11.4-1 and 11.4-2:

$$F_a = \exp\left[-0.727 \ln\left(\frac{V_{s30}}{760}\right)\right.$$
$$-0.2298 \left[\begin{array}{l} \exp\{-0.00638(\min(V_{s30}, 760) - 360)\} - \\ \exp\{-0.00638 \times 400\} \end{array}\right]$$
$$\left. \times \ln\left(\frac{(S_s/2.3) + 0.1}{0.1}\right)\right] \quad \text{(C11.4-1)}$$

$$F_v = \exp\left[-1.03 \ln\left(\frac{V_{s30}}{760}\right)\right.$$
$$-0.118 \left[\begin{array}{l} \exp\{-0.00756(\min(V_{s30}, 760) - 360)\} - \\ \exp\{-0.00756 \times 400\} \end{array}\right]$$
$$\left. \times \ln\left(\frac{(S_1/0.7) + 0.1}{0.1}\right)\right] \quad \text{(C11.4-2)}$$

In Eqs. (C11.4-1) and (C11.4-2), $\bar{v}_s$ (shown as V_{s30} in the equations) is in units of m/s and S_s and S_1 are in units of g. The equations are considered useful for $\bar{v}_s$ = 150 to 1,000 m/s ($\bar{v}_s$ = 492 to 3,281 ft/s), S_s = 0 – 1.8 g, and S_1 = 0 – 0.6 g (gravity). To obtain the F_a and F_v for $\bar{v}_s$ < 180 m/s ($\bar{v}_s$ < 590 ft/s), the +1/2 standard-deviation correction for Site Class E described earlier in this section would need to be applied to the natural logarithm of F_a and F_v resulting from both equations. The standard deviations are 0.67 for F_a and 0.58 for F_v. One half of these standard deviations are to be added to the natural logarithms of F_a and F_v; the antilogs of the resulting values yield F_a and F_v for Site Class E. Equations applicable for specific periods, and that allow use of $\bar{v}_s$ > 1,000 m/s ($\bar{v}_s$ > 3,281 ft/s), are given in Boore et al. (2013).

C11.4.5 Design Spectral Acceleration Parameters. As described in Section C11.4, structural design in ASCE 7 is performed for earthquake demands that are 2/3 of the MCE_R response spectra. As set forth in Section 11.4.5, two additional parameters, S_{DS} and S_{D1}, are used to define the acceleration response spectrum for this design level event. These parameters are 2/3 of the respective S_{MS} and S_{M1} values and define a design response spectrum for sites of any characteristics and for natural periods of vibration less than the transition period, T_L. Values of S_{MS}, S_{M1}, S_{DS}, and S_{D1} can also be obtained from the USGS website cited previously.

C11.4.6 Design Response Spectrum. The design response spectrum (Fig. 11.4-1) consists of several segments. The constant-acceleration segment covers the period band from T_0 to T_s; response accelerations in this band are constant and equal to S_{DS}. The constant-velocity segment covers the period band from T_s to T_L, and the response accelerations in this band are proportional to $1/T$ with the response acceleration at a 1-s period equal to S_{D1}. The long-period portion of the design response spectrum is defined on the basis of the parameter, T_L, the period that marks the transition from the constant-velocity segment to the constant-displacement segment of the design response spectrum. Response accelerations in the constant-displacement segment, where $T \geq T_L$, are proportional to $1/T^2$. Values of T_L are provided on maps in Figs. 22-14 through 22-17.

The T_L maps were prepared following a two-step procedure. First, a correlation between earthquake magnitude and T_L was established. Then, the modal magnitude from deaggregation of the ground-motion seismic hazard at a 2-s period (a 1-s period for Hawaii) was mapped. Details of the procedure and the rationale for it are found in Crouse et al. (2006).

C11.4.8 Site-Specific Ground Motion Procedures. Site-specific ground motions are permitted for design of any structure and are required for design of certain structures and certain site soil conditions. The objective of a site-specific ground motion analysis is to determine ground motions for local seismic and site conditions with higher confidence than is possible using the general procedure of Section 11.4.

As noted earlier, the site-specific procedures of Chapter 21 are the same as those used by the U.S. Geological Survey to develop the mapped values of MCE_R ground motions shown in Figs. 22-1 through 22-8 of Chapter 22. Unless significant differences in local seismic and site conditions are determined by a site-specific analysis of earthquake hazard, site-specific ground motions would not be expected to differ significantly from those of the mapped values of MCE_R ground motions prepared by the USGS.

Site-specific ground motions are required for design of structures at softer soil sites and stronger ground motion intensities for which the two domains of constant acceleration and constant velocity (e.g., of the design response spectrum) do not adequately characterize site response and MCE_R response spectral acceleration cannot be reliably calculated using procedures and formulas of Section 11.4. Softer soil sites requiring site-specific ground motions were identified by a study that investigated and developed solutions to potential shortcomings in equivalent lateral force (ELF) (and modal response spectrum analysis, MRSA) design procedures (Kircher & Associates 2015). The impetus for the ELF study came from a BSSC Provisions Update Committee effort (late in the 2015 cycle) to define seismic design forces at additional response periods beyond 1.0 s; a first step toward ultimately basing seismic design forces on multiperiod MCE_R response spectra.

Multiperiod MCE_R response spectra would eliminate potential shortcomings associated with the use of seismic forces based on only two response periods by directly providing reliable values of seismic demand at all design periods of interest. Unfortunately, multiperiod hazard and associated design methods are not yet mature enough for incorporation in seismic codes, and the site-specific requirements of Section 11.4.8 for softer sites and stronger ground motions provide a short-term solution to a problem that will ultimately be resolved by adoption of design methods based on multiperiod response spectra.

The value of parameter S_{MS} is based on response at a period of 0.2 s, and the value of the parameter S_{M1} is based on response at a period of 1 s. The domain of constant acceleration defined by the parameter (S_{MS}) and the domain of constant velocity (S_{M1}/T) are crude approximations to the actual shape of response spectral accelerations of MCE_R ground motions, such as those calculated using the site-specific procedures of Chapter 21 for a number of different periods of response (so-called multiperiod MCE_R response spectra).

Although approximate, the two domains of constant acceleration and velocity provide reasonably accurate and conservative representation of the frequency content of design ground motions when peak response spectral acceleration occurs at or near $T = 0.2$ s, the period used to define S_{MS}, and peak response spectral velocity (i.e., peak response spectral acceleration divided by response period) occurs at or near $T = 1.0$ s, the period used to define S_{M1}. Such is the case for response at stiffer sites governed by smaller magnitude earthquakes, but generally it is inaccurate and potentially unconservative at softer sites (e.g., Site Classes D and E), in particular sites for which seismic hazard is dominated by large-magnitude earthquakes. In the latter case, values of S_{MS} and S_{M1} would be more accurately calculated if based on response at periods that better represent peak response spectral acceleration and peak response spectral velocity, and hence the frequency content, of MCE_R ground motions of the site of interest.

The site-specific requirements of Section 21.4 of ASCE 7-10 recognized that periods of peak response are not always at 0.2 s and 1.0 s and require that S_{D1} be taken as equal to 2 times the response spectral acceleration at 2 s with greater than 1-s response spectral acceleration, and that S_{DS} be taken as equal to 0.2-s response spectral acceleration, but not less than 90% of response at any period to better represent the frequency content of site-specific ground motions. For softer sites governed by large-magnitude events, the peak value of response spectral velocity can occur at a period beyond 2 s, and ASCE 7-16 extends the S_{D1} criteria of Section 21.4 to a period range of 1 s to 5 s for Site Class D and E sites (i.e., $V_{s30} \leq 1,200$ ft/s [$V_{s30} \leq 365.76$ m/s]).

Potential shortcomings in ELF seismic design forces are illustrated in Figs. C11.4-3, C11.4-4, and C11.4-5, each of which shows plots of a multiperiod MCE_R response spectrum for Site Class BC, multiperiod MCE_R, design response spectra for the site class of interest (Site Class C, D, or E), and the two-domain ELF design spectrum, defined by the product $C_s \times (R/I_e)$. In these figures, the MCE_R ground motions represent a magnitude 7.0 earthquake at $R_x = 6.75$ km ($R_x = 4.19$ mi), which has values of the parameters $S_S = 1.5$ g and $S_1 = 0.6$ g for Site Class BC conditions [$V_s v_{s30} = 2,500$ ft/s ($V_s v_{s30} = 762$ m/s)]. The ELF design spectrum is based on these values of S_S and S_1 and values of the site coefficients F_a and F_v for the site class of

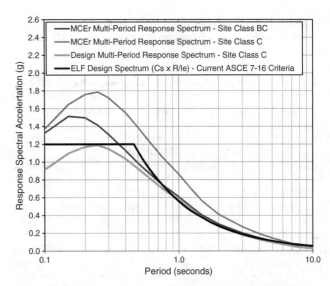

FIGURE C11.4-3 Comparison of ELF and Multiperiod Design Spectra—Site Class C Ground Motions ($V_s v_{s30} = 1,600$ ft/s)

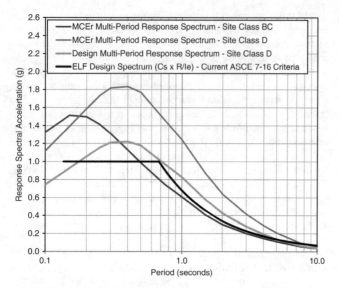

FIGURE C11.4-4 Comparison of ELF and Multiperiod Design Spectra—Site Class D Ground Motions (V_{s30} = 870 ft/s)

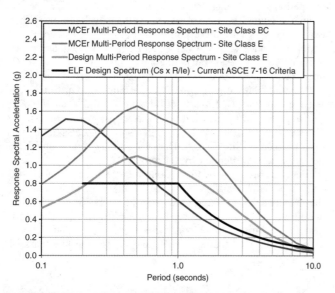

FIGURE C11.4-5 Comparison of ELF and Multiperiod Design Spectra—Site Class E Ground Motions ($V_s v_{s30}$ = 510 ft/s)

interest. For example, the domain of constant acceleration is defined by the value of the parameter $S_{DS} = 2/3 \times 0.8 \times 1.5$ g = 0.8 g, and the domain of constant velocity is defined by the value of the parameter $S_{D1} = 2/3 \times 2.0 \times 0.6$ g = 0.8 g for the ELF design spectrum shown in Fig. C11.4-5 for Site Class E conditions.

Comparisons of multiperiod and ELF design spectra in Figs. C11.4-3, C11.4-4, and C11.4-5 show varying degrees of similarity. For Site Class C (Fig. C11.4-3), the ELF design spectrum is similar to the multiperiod design spectrum. The ELF domain of constant acceleration matches the peak of the multiperiod design spectrum at period of 0.25 s, and the ELF domain of constant velocity ($1/T$) matches the multiperiod design spectrum at periods of 1 s and greater. For Site Class D (Fig. C11.4-4), the ELF design spectrum is moderately unconservative at most periods (e.g., by about 20% at a period of 1 s to 2 s), and for Site Class E (Fig. C11.4-5), the design spectrum is significantly unconservative at all periods (e.g., by about a factor of 1.65 at periods of 2 s to 3 s). These figures are based on multiperiod response spectra whose shape corresponds to a magnitude 7.0 earthquake. Spectral shape is a function of magnitude, and smaller magnitude events would show greater similarity between the multiperiod design spectrum and ELF spectrum of site class of interest, whereas larger magnitude events would show more significant differences.

In general, Section 11.4.8 requires site-specific hazard analysis for structures on Site Class E with values of S_S greater than or equal to 1.0 g, and for structures on Site Class D or Site Class E for values of S_1 greater than or equal to 0.2 g. These requirements significantly limit the use of practical ELF and MRSA design methods, which is of particular significance for Site Class D sites. To lessen the effect of these requirements on design practice, three exceptions permit the use of conservative values of design parameters for certain conditions for which conservative values of design were identified by the ELF study. These exceptions do not apply to seismically isolated structures and structures with damping systems for which site-specific analysis is required in all cases at sites with S_1 greater than or equal to 0.6.

The first exception permits use of the value of the site coefficient F_a of Site Class C ($F_a = 1.2$) for Site Class E sites (for values of S_S greater than or equal to 1.0 g) in lieu of site-specific hazard analysis. The ELF study found that while values of the site coefficient F_a tend to decrease with intensity for softer sites, values of spectrum shape adjustment factor C_a tend to increase such that the net effect is approximately the same intensity of MCE_R ground motions for Site Classes C, D, and E when MCE_R ground motion intensity is strong (i.e., $S_{MS} \geq 1.0$). Site Class C was found to not require spectrum shape adjustment, and the value of site coefficient F_a for Site Class C ($F_a = 1.2$) is large enough to represent both site class and spectrum shape effects for Site Class E (and Site Class D).

The second exception permits both ELF and MRSA design of structures at Site Class D sites for values of S_1 greater than or equal to 0.2 g, provided that the value of the seismic response coefficient C_s is conservatively calculated using Eq. (12.8-2) for $T \leq 1.5T_s$ and using 1.5 times the value computed in accordance with either Eq. (12.8-3) for $T_L \geq T > 1.5T_s$ or Eq. (12.8-4) for $T > T_L$. This exception recognizes that structures are conservatively designed for the response spectral acceleration defined by the domain of constant acceleration (S_{DS}) or by a 50% increase in the value of seismic response coefficient C_s for structures with longer periods ($T \geq 1.5T_s$). The underlying presumption of this exception for MRSA design of structures is that the shape of the design response spectrum (Fig. 11.4-1) is sufficiently representative of the frequency content of Site Class D ground motions to permit use of MRSA and that the potential underestimation of fundamental-mode response using the design response spectrum shape of Fig. 11.4-1 is accounted for by scaling MRSA design values (Section 12.9.1.4) with a conservative value of the seismic response coefficient C_s. In general, this exception effectively limits the requirements for site-specific hazard analysis to very tall and or flexible structures at Site Class D sites ($S_1 \geq 0.2$ g).

The third exception permits ELF design of short-period structures ($T \leq T_s$) at Site Class E sites for values of S_S greater than or equal to 0.2 g. This exception recognizes that short-period structures are conservatively designed using the ELF procedure for values of seismic response coefficient C_s based on the domain of constant acceleration (S_{DS}), which is, in all cases, greater than or equal to response spectral accelerations of the domain of constant velocity and therefore need not consider the effects of spectrum shape at periods $T > T_s$. In general, the shape of the design response spectrum (Fig. 11.4-1) is not representative of the frequency content of Site Class E ground motions, and

MRSA is not permitted for design unless the design spectrum is calculated using the site-specific procedures of Section 21.2.

C11.5 IMPORTANCE FACTOR AND RISK CATEGORY

Large earthquakes are rare events that include severe ground motions. Such events are expected to result in damage to structures even if they were designed and built in accordance with the minimum requirements of the standard. The consequence of structural damage or failure is not the same for the various types of structures located within a given community. Serious damage to certain classes of structures, such as critical facilities (e.g., hospitals), disproportionately affects a community. The fundamental purpose of this section and of subsequent requirements that depend on this section is to improve the ability of a community to recover from a damaging earthquake by tailoring the seismic protection requirements to the relative importance of a structure. That purpose is achieved by requiring improved performance for structures that

1. Are necessary to response and recovery efforts immediately after an earthquake,
2. Present the potential for catastrophic loss in the event of an earthquake, or
3. House a large number of occupants or occupants less able to care for themselves than the average.

The first basis for seismic design in the standard is that structures should have a suitably low likelihood of collapse in the rare events defined as the maximum considered earthquake (MCE) ground motion. A second basis is that life-threatening damage, primarily from failure of nonstructural components in and on structures, is unlikely in a design earthquake ground motion (defined as two-thirds of the MCE). Given the occurrence of ground motion equivalent to the MCE, a population of structures built to meet these design objectives probably still experiences substantial damage in many structures, rendering these structures unfit for occupancy or use. Experience in past earthquakes around the world has demonstrated that there is an immediate need to treat injured people, to extinguish fires and prevent conflagration, to rescue people from severely damaged or collapsed structures, and to provide shelter and sustenance to a population deprived of its normal means. These needs are best met when structures essential to response and recovery activities remain functional.

This standard addresses these objectives by requiring that each structure be assigned to one of the four Risk Categories presented in Chapter 1 and by assigning an Importance Factor, I_e, to the structure based on that Risk Category. (The two lowest categories, I and II, are combined for all purposes within the seismic provisions.) The Risk Category is then used as one of two components in determining the Seismic Design Category (see Section C11.6) and is a primary factor in setting drift limits for building structures under the design earthquake ground motion (see Section C12.12).

Fig. C11.5-1 shows the combined intent of these requirements for design. The vertical scale is the likelihood of the ground motion; the MCE is the rarest considered. The horizontal scale is the level of performance intended for the structure and attached nonstructural components, which range from collapse to operational. The basic objective of collapse prevention at the MCE for ordinary structures (Risk Category II) is shown at the lower right by the solid triangle; protection from life-threatening damage at the design earthquake ground motion (defined by the standard as two-thirds of the MCE) is shown by the hatched triangle. The performance implied for higher Risk Categories III and IV is shown by squares and circles, respectively. The performance anticipated for less severe ground motion is shown by open symbols.

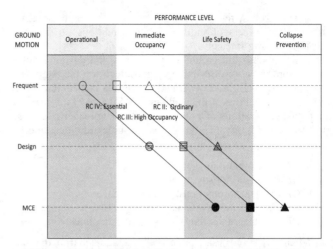

FIGURE C11.5-1 Expected Performance as Related to Risk Category and Level of Ground Motion

C11.5.1 Importance Factor. The Importance Factor, I_e, is used throughout the standard in quantitative criteria for strength. In most of those quantitative criteria, the Importance Factor is shown as a divisor on the factor R or R_p to reduce damage for important structures in addition to preventing collapse in larger ground motions. The R and R_p factors adjust the computed linear elastic response to a value appropriate for design; in many structures, the largest component of that adjustment is ductility (the ability of the structure to undergo repeated cycles of inelastic strain in opposing directions). For a given strength demand, reducing the effective R factor (by means of the Importance Factor) increases the required yield strength, thus reducing ductility demand and related damage.

C11.5.2 Protected Access for Risk Category IV. Those structures considered essential facilities for response and recovery efforts must be accessible to carry out their purpose. For example, if the collapse of a simple canopy at a hospital could block ambulances from the emergency room admittance area, then the canopy must meet the same structural standard as the hospital. The protected access requirement must be considered in the siting of essential facilities in densely built urban areas.

C11.6 SEISMIC DESIGN CATEGORY

Seismic Design Categories (SDCs) provide a means to step progressively from simple, easily performed design and construction procedures and minimums to more sophisticated, detailed, and costly requirements as both the level of seismic hazard and the consequence of failure escalate. The SDCs are used to trigger requirements that are not scalable; such requirements are either on or off. For example, the basic amplitude of ground motion for design is scalable—the quantity simply increases in a continuous fashion as one moves from a low hazard area to a high hazard area. However, a requirement to avoid weak stories is not particularly scalable. Requirements such as this create step functions. There are many such requirements in the standard, and the SDCs are used systematically to group these step functions. (Further examples include whether seismic anchorage of nonstructural components is required or not, whether particular inspections will be required or not, and structural height limits applied to various seismic force-resisting systems.)

In this regard, SDCs perform one of the functions of the seismic zones used in earlier U.S. building. However, SDCs also depend on a building's occupancy and, therefore, its desired performance. Furthermore, unlike the traditional implementation of seismic zones, the ground motions used to define the SDCs include the effects of individual site conditions on probable ground-shaking intensity.

In developing the ground-motion limits and design requirements for the various Seismic Design Categories, the equivalent modified Mercalli intensity (MMI) scale was considered. There are now correlations of the qualitative MMI scale with quantitative characterizations of ground motions. The reader is encouraged to consult any of a great many sources that describe the MMIs. The following list is a coarse generalization:

MMI V   No real damage
MMI VI Light nonstructural damage
MMI VII Hazardous nonstructural damage
MMI VIII Hazardous damage to susceptible structures
MMI IX Hazardous damage to robust structures

When the current design philosophy was adopted from the 1997 NEHRP provisions and Commentary (FEMA 1997a and FEMA 1997b), the upper limit for SDC A was set at roughly one-half of the lower threshold for MMI VII, and the lower limit for SDC D was set at roughly the lower threshold for MMI VIII. However, the lower limit for SDC D was more consciously established by equating that design value (two-thirds of the MCE) to one-half of what had been the maximum design value in building codes over the period of 1975 to 1995. As more correlations between MMI and numerical representations of ground motion have been created, it is reasonable to make the following correlation between the MMI at MCE ground motion and the Seismic Design Category (all this discussion is for ordinary occupancies):

MMI V   SDC A
MMI VI SDC B
MMI VII SDC C
MMI VIII SDC D
MMI IX SDC E

An important change was made to the determination of SDC when the current design philosophy was adopted. Earlier editions of the *NEHRP Provisions* used the peak velocity-related acceleration, A_v, to determine a building's seismic performance category. However, this coefficient does not adequately represent the damage potential of earthquakes on sites with soil conditions other than rock. Consequently, the 1997 NEHRP provisions (FEMA 1997a) adopted the use of response spectral acceleration parameters S_{DS} and S_{D1}, which include site soil effects for this purpose.

Except for the lowest level of hazard (SDC A), the SDC also depends on the Risk Categories. For a given level of ground motion, the SDC is one category higher for Risk Category IV structures than for lower risk structures. This rating has the effect of increasing the confidence that the design and construction requirements can deliver the intended performance in the extreme event.

Note that the tables in the standard are at the design level, defined as two-thirds of the MCE level. Also recall that the MMIs are qualitative by their nature and that the above correlation will be more or less valid, depending on which numerical correlation for MMI is used. The numerical correlations for MMI roughly double with each step, so correlation between design earthquake ground motion and MMI is not as simple or convenient.

In sum, at the MCE level, SDC A structures should not see motions that are normally destructive to structural systems, whereas the MCE level motions for SDC D structures can destroy vulnerable structures. The grouping of step function requirements by SDC is such that there are a few basic structural integrity requirements imposed at SDC A, graduating to a suite of requirements at SDC D based on observed performance in past earthquakes, analysis, and laboratory research.

The nature of ground motions within a few kilometers of a fault can be different from more distant motions. For example, some near-fault motions have strong velocity pulses, associated with forward rupture directivity, that tend to be highly destructive to irregular structures, even if they are well detailed. For ordinary occupancies, the boundary between SDCs D and E is set to define sites likely to be close enough to a fault that these unusual ground motions may be present. Note that this boundary is defined in terms of mapped bedrock outcrop motions affecting response at 1 s, not site-adjusted values, to better discriminate between sites near and far from faults. Short-period response is not normally as affected as the longer period response. The additional design criteria imposed on structures in SDCs E and F specifically are intended to provide acceptable performance under these very intense near-fault ground motions.

For most buildings, the SDC is determined without consideration of the building's period. Structures are assigned to an SDC based on the more severe condition determined from 1-s acceleration and short-period acceleration. This assigning is done for several reasons. Perhaps the most important of these is that it is often difficult to estimate precisely the period of a structure using default procedures contained in the standard. Consider, for example, the case of rigid wall–flexible diaphragm buildings, including low-rise reinforced masonry and concrete tilt-up buildings with either untopped metal deck or wood diaphragms. The formula in the standard for determining the period of vibration of such buildings is based solely on the structural height, h_n, and the length of wall present. These formulas typically indicate very short periods for such structures, often on the order of 0.2 s or less. However, the actual dynamic behavior of these buildings often is dominated by the flexibility of the diaphragm—a factor neglected by the formula for approximate fundamental period. Large buildings of this type can have actual periods on the order of 1 s or more. To avoid misclassifying a building's SDC by inaccurately estimating the fundamental period, the standard generally requires that the more severe SDC determined on the basis of short- and long-period shaking be used.

Another reason for this requirement is a desire to simplify building regulation by requiring all buildings on a given soil profile in a particular region to be assigned to the same SDC, regardless of the structural type. This assignment has the advantage of permitting uniform regulation in the selection of seismic force-resisting systems, inspection and testing requirements, seismic design requirements for nonstructural components, and similar aspects of the design process regulated on the basis of SDC, within a community.

Notwithstanding the above, it is recognized that classification of a building as SDC C instead of B or D can have a significant impact on the cost of construction. Therefore, the standard includes an exception permitting the classification of buildings that can reliably be classified as having short structural periods on the basis of short-period shaking alone.

Local or regional jurisdictions enforcing building regulations may desire to consider the effect of the maps, typical soil conditions, and Seismic Design Categories on the practices in their jurisdictional areas. For reasons of uniformity of practice or reduction of potential errors, adopting ordinances could stipulate particular values of ground motion, particular site classes, or particular Seismic Design Categories for all or part of the area of their jurisdiction. For example,

1. An area with a historical practice of high seismic zone detailing might mandate a minimum SDC of D regardless of ground motion or site class.

2. A jurisdiction with low variation in ground motion across the area might stipulate particular values of ground motion rather than requiring the use of maps.
3. An area with unusual soils might require use of a particular site class unless a geotechnical investigation proves a better site class.

C11.7 DESIGN REQUIREMENTS FOR SEISMIC DESIGN CATEGORY A

The 2002 edition of the standard included a new provision of minimum lateral force for Seismic Design Category A structures. The minimum load is a structural integrity issue related to the load path. It is intended to specify design forces in excess of wind loads in heavy low-rise construction. The design calculation in Section 1.4.2 of the standard is simple and easily done to ascertain if the seismic load or the wind load governs. This provision requires a minimum lateral force of 1% of the total gravity load assigned to a story to ensure general structural integrity.

Seismic Design Category A is assigned when the MCE ground motions are below those normally associated with hazardous damage. Damaging earthquakes are not unknown or impossible in such regions, however, and ground motions close to such events may be large enough to produce serious damage. Providing a minimum level of resistance reduces both the radius over which the ground motion exceeds structural capacities and resulting damage in such rare events. There are reasons beyond seismic risk for minimum levels of structural integrity.

The requirements for SDC A in Section 1.4 are all minimum strengths for structural elements stated as forces at the level appropriate for direct use in the strength design load combinations of Section. 2.3. The two fundamental requirements are a minimum strength for a structural system to resist lateral forces (Section 1.4.2) and a minimum strength for connections of structural members (Section 1.4.3).

For many buildings, the wind force controls the strength of the lateral-force-resisting system, but for low-rise buildings of heavy construction with large plan aspect ratios, the minimum lateral force specified in Section 1.4.2 may control. Note that the requirement is for strength and not for toughness, energy-dissipation capacity, or some measure of ductility. The force level is not tied to any postulated seismic ground motion. The boundary between SDCs A and B is based on a spectral response acceleration of 25% of gravity (MCE level) for short-period structures; clearly the 1% acceleration level (from Eq. (1.4-1)) is far smaller. For ground motions below the A/B boundary, the spectral displacements generally are on the order of a few inches or less depending on period. Experience has shown that even a minimal strength is beneficial in providing resistance to small ground motions, and it is an easy provision to implement in design. The low probability of motions greater than the MCE is a factor in taking the simple approach without requiring details that would produce a ductile response. Another factor is that larger design forces are specified in Section 1.4.3 for connections between main elements of the lateral force load path.

The minimum connection force is specified in three ways: a general minimum horizontal capacity for all connections; a special minimum for horizontal restraint of in-line beams and trusses, which also includes the live load on the member; and a special minimum for horizontal restraint of concrete and masonry walls perpendicular to their plane (Section 1.4.4). The 5% coefficient used for the first two is a simple and convenient value that provides some margin over the minimum strength of the system as a whole.

C11.8 GEOLOGIC HAZARDS AND GEOTECHNICAL INVESTIGATION

In addition to this commentary, Part 3 of the 2009 NEHRP recommended provisions (FEMA 2009) includes additional and more detailed discussion and guidance on evaluation of geologic hazards and determination of seismic lateral pressures.

C11.8.1 Site Limitation for Seismic Design Categories E and F. Because of the difficulty of designing a structure for the direct shearing displacement of fault rupture and the relatively high seismic activity of SDCs E and F, locating a structure on an active fault that has the potential to cause rupture of the ground surface at the structure is prohibited.

C11.8.2 Geotechnical Investigation Report Requirements for Seismic Design Categories C through F. Earthquake motion is only one factor in assessing potential for geologic and seismic hazards. All of the listed hazards can lead to surface ground displacements with potential adverse consequences to structures. Finally, hazard identification alone has little value unless mitigation options are also identified.

C11.8.3 Additional Geotechnical Investigation Report Requirements for Seismic Design Categories D through F. New provisions for computing peak ground acceleration for soil liquefaction and stability evaluations have been introduced in this section. Of particular note in this section is the explicitly stated requirement that liquefaction must now be evaluated for the MCE_G ground motion. These provisions include maps of the maximum considered earthquake geometric mean (MCE_G) peak ground acceleration (PGA) for Site Class B bedrock plus a site-coefficient (F_{PGA}) table to convert the PGA value to one adjusted for site class effects (PGA_M).

A requirement, similar to the one in Section 11.4.4, has been added to the provisions to take the larger of the F_{PGA} for Site Classes C and D to conservatively account for the amplification when the site is known to consist of soil that is not in Site Class E or F.

The equation used to derive the F_{PGA} values is similar to Eqs. (C11.4-1) and (C11.4-2) for F_a and F_v; it is as follows:

$$F_{PGA} = \exp\left[-0.604 \ln\left(\frac{\bar{v}_s}{760}\right)\right.$$
$$-0.150 \left[\begin{array}{l}\exp\{-0.00701(\min(\bar{v}_s, 760) - 360)\}-\\ \exp\{-0.00701 \times 400\}\end{array}\right]$$
$$\left.\times \ln\left(\frac{PGA + 0.1}{0.1}\right)\right] \quad (C11.8-1)$$

In Eq. (C11.8-1), $\bar{v}_s$ is in units of m/s and PGA is in units of g. Velocities measured in ft/s can be converted to m/s by multiplying by 0.3048. To obtain the F_{PGA} for $\bar{v}_s < 180$ m/s ($\bar{v}_s < 590$ ft/s), the $+1/2$ standard-deviation correction described for Site Class E in Section C11.4.4 would need to be applied to the natural logarithm of F_{PGA}. The standard deviation is 0.70.

PGA Provisions. Item 2 of Section 11.8.3 states that peak ground acceleration shall be determined based on either a site-specific study, taking into account soil amplification effects, or using Eq. (11.8-1), for which MCE_G peak ground acceleration is obtained from national maps of PGA for bedrock Site Class B multiplied by a site coefficient (F_{PGA}) to obtain peak ground acceleration for other site classes (PGA_M). This methodology for determining peak ground acceleration for liquefaction evaluations improved the methodology in ASCE 7-05 by using mapped

PGA rather than the approximation for PGA by the ratio $S_s/2.5$. Furthermore, in the central and eastern United States, the ratio $S_s/2.5$ tends to underestimate PGA. $S_s/2.5$ is applicable for bedrock Site Class B and thus could be used as input at depth to a site response analysis under the provisions of ASCE 7-05. The use of Eq. (11.8-1) provides an alternative to conducting site response analysis using rock PGA by providing a site-adjusted ground surface acceleration (PGA_M) that can directly be applied in the widely used empirical correlations for assessing liquefaction potential. Correlations for evaluating liquefaction potential are elaborated on in Resource Paper RP 12, "Evaluation of Geologic Hazards and Determination of Seismic Lateral Earth Pressures," published in the 2009 NEHRP provisions (FEMA 2009).

Maps of MCE_G PGA for Site Class B bedrock, similar to maps of S_s and S_1, are shown in Figs. 22-9 to 22-13 in Chapter 22. Similar to adjustments for the bedrock spectral response accelerations for site response through the F_a and F_v coefficients, bedrock motions for PGA are adjusted for these same site effects using a site coefficient, F_{PGA}, that depends on the level of ground shaking in terms of PGA and the stiffness of the soil, typically defined in terms of the shear-wave velocity in the upper 30 m (98.4 ft) of geologic profile, V_{s30}. Values of F_{PGA} are presented in Table 11.8-1, and the adjustment is made through Eq. (11.8-1), i.e., $PGA_M = F_{PGA}$ PGA, where PGA_M is peak ground acceleration adjusted for site class. The method of determining site class, used in the determination of F_a and F_v, is also identical to that in the present and previous ASCE 7 documents.

There is an important difference in the derivation of the PGA maps and the maps of S_s and S_1 in ASCE 7-10. Unlike previous editions of ASCE 7, the S_s and S_1 maps in ASCE 7-10 were derived for the "maximum direction shaking" and are risk based rather than hazard based. However, the PGA maps have been derived based on the geometric mean of the two horizontal components of motion. The geometric mean was used in the PGA maps rather than the PGA for the maximum direction shaking to ensure that there is consistency between the determination of PGA and the basis of the simplified empirical field procedure for estimating liquefaction potential based on results of standard penetration tests (SPTs), cone penetrometer tests (CPTs), and other similar field investigative methods. When these correlations were originally derived, the geomean (or a similar metric) of peak ground acceleration at the ground surface was used to identify the cyclic stress ratio for sites with or without liquefaction. The resulting envelopes of data define the liquefaction cyclic resistance ratio (CRR). Rather than reevaluating these case histories for the "maximum direction shaking," it was decided to develop maps of the geomean PGA and to continue using the existing empirical methods.

Liquefaction Evaluation Requirements. Beginning with ASCE 7-02, it has been the intent that liquefaction potential be evaluated at MCE ground motion levels. There was ambiguity in the previous requirement in ASCE 7-05 as to whether liquefaction potential should be evaluated for the MCE or for the design earthquake. Paragraph 2 of Section 11.8.3 of ASCE 7-05 stated that liquefaction potential would be evaluated for the design earthquake; it also stated that in the absence of a site-specific study, peak ground acceleration shall be assumed to be equal to $S_s/2.5$ (S_s is the MCE short-period response spectral acceleration on Site Class B rock). There has also been a difference in provisions between ASCE 7-05 and the 2006 edition of the IBC, in which Section 1802.2.7 stated that liquefaction shall be evaluated for the design earthquake ground motions and the default value of peak ground acceleration in the absence of a site-specific study was given as $S_{DS}/2.5$ (S_{DS} is the short-period site-adjusted design response spectral acceleration). ASCE 7-10, in item 2 of Section 11.8.3 and Eq. (11.8-1), requires explicitly that liquefaction potential be evaluated based on the MCE_G peak ground acceleration.

The explicit requirement in ASCE 7-10 to evaluate liquefaction for MCE ground motion rather than to design earthquake ground motion ensures that the full potential for liquefaction is addressed during the evaluation of structure stability, rather than a lesser level when the design earthquake is used. This change also ensures that, for the MCE ground motion, the performance of the structure is considered under a consistent hazard level for the effects of liquefaction, such as collapse prevention or life safety, depending on the risk category for the structure (Fig. C11.5-1). By evaluating liquefaction for the MCE rather than the design earthquake peak ground acceleration, the ground motion for the liquefaction assessment increases by a factor of 1.5. This increase in peak ground acceleration to the MCE level means that sites that previously were nonliquefiable could now be liquefiable, and sites where liquefaction occurred to a limited extent under the design earthquake could undergo more liquefaction, in terms of depth and lateral extent. Some mechanisms that are directly related to the development of liquefaction, such as lateral spreading and flow or ground settlement, could also increase in severity.

This change in peak ground acceleration level for the liquefaction evaluation addressed an issue that has existed and has periodically been discussed since the design earthquake concept was first suggested in the 1990s. The design earthquake ground motion was obtained by multiplying the MCE ground motion by a factor of 2/3 to account for a margin in capacity in most buildings. Various calibration studies at the time of code development concluded that for the design earthquake, most buildings had a reserve capacity of more than 1.5 relative to collapse. This reserve capacity allowed the spectral accelerations for the MCE to be reduced using a factor of 2/3, while still achieving safety from collapse. However, liquefaction potential is evaluated at the selected MCE_G peak ground acceleration and is typically determined to be acceptable if the factor of safety is greater than 1.0, meaning that there is no implicit safety margin on liquefaction potential. By multiplying peak ground acceleration by a factor of 2/3, liquefaction would be assessed at an effective return period or probability of exceedance different than that for the MCE. However, ASCE 7-10 requires that liquefaction be evaluated for the MCE.

Item 3 of Section 11.8.3 of the ASCE 7-10 standard lists the various potential consequences of liquefaction that must be assessed; soil downdrag and loss in lateral soil reaction for pile foundations are additional consequences that have been included in this paragraph. This section of the new provisions, as in previous editions, does not present specific seismic criteria for the design of the foundation or substructure, but item 4 does state that the geotechnical report must include discussion of possible measures to mitigate these consequences.

A liquefaction resource document has been prepared in support of these revisions to Section 11.8.3. The resource document "Evaluation of Geologic Hazards and Determination of Seismic Lateral Earth Pressures," includes a summary of methods that are currently being used to evaluate liquefaction potential and the limitations of these methods. This summary appears as Resource Paper RP 12 in the 2009 NEHRP provisions (FEMA 2009). The resource document summarizes alternatives for evaluating liquefaction potential, methods for evaluating the possible consequences of liquefaction (e.g., loss of ground support and increased lateral earth pressures) and methods of mitigating the liquefaction hazard. The resource document also identifies alternate methods of evaluating liquefaction hazards, such as analytical and physical modeling. Reference is made to the use of nonlinear effective stress methods for modeling the buildup in pore water pressure during seismic events at liquefiable sites.

Evaluation of Dynamic Seismic Lateral Earth Pressures. The dynamic lateral earth pressure on basement and retaining walls during earthquake ground shaking is considered to be an earthquake load, E, for use in design load combinations. This dynamic earth pressure is superimposed on the preexisting static lateral earth pressure during ground shaking. The preexisting static lateral earth pressure is considered to be an H load.

C11.9 VERTICAL GROUND MOTIONS FOR SEISMIC DESIGN

C11.9.2 MCE$_R$ Vertical Response Spectrum. Previous editions of ASCE 7 do not provide adequate guidance regarding procedures for estimating vertical ground motion levels for use in earthquake-resistant design. Historically, the amplitude of vertical ground motion has been inferred to be two-thirds (2/3) the amplitude of the horizontal ground motion. However, studies of horizontal and vertical ground motions over the past 25 years have shown that such a simple approach is not valid in many situations (e.g., Bozorgnia and Campbell 2004, and references therein) for the following main reasons: (1) vertical ground motion has a larger proportion of short-period (high-frequency) spectral content than horizontal ground motion, and this difference increases with decreasing soil stiffness, and (2) vertical ground motion attenuates at a higher rate than horizontal ground motion, and this difference increases with decreasing distance from the earthquake. The observed differences in the spectral content and attenuation rate of vertical and horizontal ground motion lead to the following observations regarding the vertical/horizontal (V/H) spectral ratio (Bozorgnia and Campbell 2004):

1. The V/H spectral ratio is sensitive to spectral period, distance from the earthquake, local site conditions, and earthquake magnitude and is insensitive to earthquake mechanism and sediment depth;
2. The V/H spectral ratio has a distinct peak at short periods that generally exceeds 2/3 in the near-source region of an earthquake; and
3. The V/H spectral ratio is generally less than 2/3 at mid-to-long periods.

Therefore, depending on the period, the distance to the fault, and the local site conditions of interest, use of the traditional 2/3 V/H spectral ratio can result in either an under- or overestimation of the expected vertical ground motions.

The procedure for defining the MCE$_R$ vertical response spectrum in ASCE 7 is a modified version of the procedure taken from the 2009 *NEHRP Provisions*. Unlike the procedure contained in the 2009 *NEHRP Provisions*, the procedure provided in Section 11.9 is keyed to the MCE$_R$ spectral response acceleration parameter at short periods, S_{MS}. The procedure is based on the studies of horizontal and vertical ground motions conducted by Campbell and Bozorgnia (2003) and Bozorgnia and Campbell (2004). These procedures are also generally compatible with the general observations of Abrahamson and Silva (1997) and Silva (1997) and the proposed design procedures of Elnashai (1997). The procedure has been modified to express the vertical ground motions in terms of MCE$_R$ ground motions instead of design ground motions.

To be consistent with the shape of the horizontal design response spectrum, the vertical design response spectrum has four regions defined by the vertical period of vibration (T_v).

Based on the study of Bozorgnia and Campbell (2004), the periods that define these regions are approximately constant with respect to the magnitude of the earthquake, the distance from the earthquake, and the local site conditions. In this respect, the shape of the vertical response spectrum is simpler than that of the horizontal response spectrum.

The equations that are used to define the design vertical response spectrum are based on three observations made by Bozorgnia and Campbell (2004):

1. The short-period part of the 5% damped vertical response spectrum is controlled by the spectral acceleration at $T_v = 0.1$ s;
2. The mid-period part of the vertical response spectrum is controlled by a spectral acceleration that decays as the inverse of the 0.75 power of the vertical period of vibration ($T_v^{-0.75}$); and
3. The short-period part of the V/H spectral ratio is a function of the local site conditions, the distance from the earthquake (for sites located within about 30 mi (60 km) of the fault), and the earthquake magnitude (for soft sites).

ASCE 7 does not include seismic design maps for the vertical spectral acceleration at $T_v = 0.1$ s and does not preserve any information on the earthquake magnitudes or the source-to-site distances that contribute to the horizontal spectral accelerations that are mapped. Therefore, the general procedure recommended by Bozorgnia and Campbell (2004) was modified to use only those horizontal spectral accelerations that are available from the seismic design maps, as follows:

1. Estimate the vertical spectral acceleration at $T_v = 0.1$ s from the ratio of this spectral acceleration to the horizontal spectral acceleration at $T = 0.2$ s for the Site Class B/C boundary (i.e., the boundary between Site Classes B and C $\bar{v}_s = 2,500$ ft/s ($\bar{v}_s = 760$ m/s), the reference site condition for the 2008 U.S. Geological Survey National Seismic Hazard Maps). For earthquakes and distances for which the vertical spectrum might be of engineering interest (magnitudes greater than 6.5 and distances less than 30 mi (60 km), this ratio is approximately 0.8 for all site conditions (Campbell and Bozorgnia 2003).
2. Estimate the horizontal spectral acceleration at $T = 0.2$ s from the Next Generation Attenuation (NGA) relationship of Campbell and Bozorgnia (2008) for magnitudes greater than 6.5 and distances ranging between 1 and 30 mi (1 and 60 km) for the Site Class B/C boundary $\bar{v}_s = 2,500$ ft/s ($\bar{v}_s = 760$ m/s). The relationship of Campbell and Bozorgnia (2008), rather than that of Campbell and Bozorgnia (2003), was used for this purpose to be consistent with the development of the 2008 U.S. Geological Survey National Seismic Hazard Maps, which use the NGA attenuation relationships to estimate horizontal ground motions in the western United States. Similar results were found for the other two NGA relationships that were used to develop the seismic hazard and design maps (Boore and Atkinson 2008; Chiou and Youngs 2008).
3. Use the dependence between the horizontal spectral acceleration at $T = 0.2$ s and source-site distance estimated in Item 2 and the relationship between the V/H spectral ratio, source-site distance, and local site conditions in Bozorgnia and Campbell (2004) to derive a relationship between the vertical spectral acceleration and the mapped MCE$_R$ spectral response acceleration parameter at short periods, S_S.
4. Use the dependence between the vertical spectral acceleration and the mapped MCE$_R$ spectral response acceleration

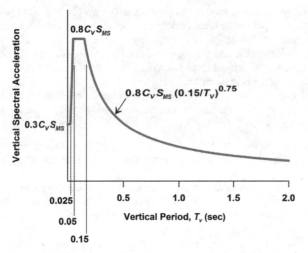

FIGURE C11.9-1 Illustrative Example of the Vertical Response Spectrum

parameter at short periods, S_S, in Item 3 to derive a vertical coefficient, C_v, that when multiplied by 0.8 and the MCE_R horizontal response acceleration at short periods, S_{MS}, results in an estimate of the design vertical spectral acceleration at $T_v = 0.1$ s.

The following description of the detailed procedure listed in Section 11.9.2 refers to the illustrated MCE_R vertical response spectrum in Fig. C11.9-1.

Vertical Periods Less Than or Equal to 0.025 s. Eq. (11.9-1) defines that part of the MCE_R vertical response spectrum that is controlled by the vertical peak ground acceleration. The 0.3 factor was approximated by dividing the 0.8 factor that represents the ratio between the vertical spectral acceleration at $T_v = 0.1$ s and the horizontal spectral acceleration at $T = 0.2$ s by 2.5, the factor that represents the ratio between the MCE_R horizontal spectral acceleration at $T = 0.2$ s, S_{MS}, and the zero-period acceleration used in the development of the MCE_R horizontal response spectrum. The vertical coefficient, C_v, in Table 11.9-1 accounts for the dependence of the vertical spectral acceleration on the amplitude of the horizontal spectral acceleration and the site dependence of the V/H spectral ratio as determined in Items 3 and 4 above. The factors are applied to S_{MS} rather than to S_S because S_{MS} already includes the effects of local site conditions.

Vertical Periods Greater Than 0.025 s and Less Than or Equal to 0.05 s. Eq. (11.9-2) defines that part of the MCE_R vertical response spectrum that represents the linear transition from the part of the spectrum that is controlled by the vertical peak ground acceleration and the part of the spectrum that is controlled by the dynamically amplified short-period spectral plateau. The factor of 20 is the factor that is required to make this transition continuous and piecewise linear between these two adjacent parts of the spectrum.

Vertical Periods Greater Than 0.05 s and Less Than or Equal to 0.15 s. Eq. (11.9-3) defines that part of the MCE_R vertical response spectrum that represents the dynamically amplified short-period spectral plateau.

Vertical Periods Greater Than 0.15 s and Less Than or Equal to 2.0 s. Eq. (11.9-4) defines that part of the MCE_R vertical response spectrum that decays with the inverse of the vertical period of vibration raised to the 0.75 power.

Two limits are imposed on the MCE_R vertical response spectrum defined by Eqs. (11.9-1) through (11.9-4) and illustrated in Fig. C11.9-1. The first limit restricts the applicability of the vertical response spectrum to a maximum vertical period of vibration of 2 s. This limit accounts for the fact that such large vertical periods are rare (structures are inherently stiff in the vertical direction) and that the vertical spectrum might decay differently with period at longer periods. There is an allowance for developing a site-specific MCE_R vertical response spectrum if this limit is exceeded (see Section 11.4 or Chapter 21 for guidance on applying site-specific methods). The second limit restricts the MCE_R vertical response spectrum to be no less than 50% of the MCE_R horizontal response spectrum. This limit accounts for the fact that a V/H spectral ratio of one-half ($1/2$) is a reasonable, but somewhat conservative, lower bound over the period range of interest, based on the results of Campbell and Bozorgnia (2003) and Bozorgnia and Campbell (2004).

REFERENCES

Abrahamson, N. A., and Silva, W. J. (1997). "Empirical response spectral attenuation relations for shallow crustal earthquakes." *Seismol. Res. Lett.*, 68, 94–127.

ASCE. (2014). "Seismic evaluation and retrofit of existing buildings." *ASCE/SEI 41-13*. Reston, VA.

Boore, D. M., and Atkinson, G. M. (2008). "Ground-motion prediction equations for the average horizontal component of PGA, PGV, and 5%-damped PSA at spectral periods between 0.01 s and 10.0 s." *Earthq. Spectra*, 24, 99–138.

Boore, D. M., Stewart, J. P., Seyhan, E., and Atkinson, G. M. (2013). "NGA-West 2 equations for predicting response spectral accelerations for shallow crustal earthquakes." *PEER Report 2013/05*. Pacific Earthquake Engineering Research Center, University of California, Berkeley, CA.

Bozorgnia, Y., and Campbell, K. W. (2004). "The vertical-to-horizontal response spectral ratio and tentative procedures for developing simplified V/H and vertical design spectra." *J. Earthq. Eng.*, 8, 175–207.

Campbell, K. W., and Bozorgnia, Y. (2003). "Updated near-source ground motion (attenuation) relations for the horizontal and vertical components of peak ground acceleration and acceleration response spectra." *Bull. Seismol. Society of Am.*, 93, 314–331.

Campbell, K. W., and Bozorgnia, Y. (2008). "NGA ground motion model for the geometric mean horizontal component of PGA, PGV, PGD, and 5% damped linear elastic response spectra for periods ranging from 0.01 to 10 s." *Earthq. Spectra*, 24, 139–171.

Chiou, B. S.-J., and Youngs, R. R. (2008). "An NGA model for the average horizontal component of peak ground motion and response spectra." *Earthq. Spectra*, 24, 173–215.

Crouse, C. B., Leyendecker, E. V., Somerville, P. G., Power, M., and Silva, W. J. (2006). "Development of seismic ground-motion criteria for the ASCE/SEI 7 standard," Paper 533. *Proc. 8th U.S. National Conference on Earthquake Engineering*.

Dobry, R., Borcherdt, R., Crouse, C. B., Idriss, I. M., Joyner, W. B., Martin, G. R., et al. (2000). "New site coefficients and site classification system used in recent building seismic code provisions." *Earthq. Spectra*, 16(1), 41–67.

Elnashai, A. S. (1997). "Seismic design with vertical earthquake motion." *Seismic design for the next generation of codes*, P. Fajfar, and H. Krawinkler, eds., Balkema, Rotterdam, Netherlands, 91–100.

Executive Order 12699. (2016). *Establishing a Federal Earthquake Risk Management Standard*. http://www.whitehouse.gov/the-press-office/2016/02/02/executive-order-establishing-federal-earthquake-risk-management-standard.

Federal Emergency Management Agency (FEMA). (1997a). "NEHRP recommended provisions for seismic regulations in new buildings and other structures." *FEMA 302*. Washington, DC.

FEMA. (1997b). "NEHRP recommended provisions for seismic regulations in new buildings and other structures: Part 2, commentary." *FEMA 303*. Washington, DC.

FEMA. (2009). "NEHRP recommended seismic provisions for new buildings and other structures." *FEMA P-750*. Washington, DC.

Frankel, A., Mueller, C., Barnhard, T., Perkins, D., Leyendecker, E. V., Dickman, N., et al. (1996). "National seismic hazard maps." *Open File Report 96-532*, U.S. Geological Survey, Denver, CO.

International Conference of Building Officials (ICBO). (1997). *Uniform building code*. Whittier, CA.

Kamai, R., Abrahamson, N. A., and Silva, W. J. (2013). "Nonlinear horizontal site response for the NGA-West 2 project." *PEER Report 2013/12*, Pacific Earthquake Engineering Research Center, Berkeley, CA.

Kircher & Associates. (2015). "Investigation of an identified shortcoming in the seismic design procedures of ASCE 7-10 and development of recommended improvements for ASCE 7-16." Prepared for Building Seismic Safety Council, National Institute of Building Sciences, Washington, DC, Palo Alto, CA ⟨https://c.ymcdn.com/sites/www.nibs.org/resource/resmgr/BSSC2/Seismic_Factor_Study.pdf⟩.

Seyhan, E., and Stewart, J. P. (2012). "Site response in NEHRP provisions and NGA models." *Geotechnical engineering state of the art and practice: Volume of keynote lectures from GeoCongress 2012*," ASCE Geotechnical Special Publication No. 226, K. Rollins and D. Zekkos, eds., ASCE, Reston, VA, 359–379.

Silva, W. (1997). "Characteristics of vertical strong ground motions for applications to engineering design. FHWA/NCEER workshop on the national representation of seismic ground motion for new and existing highway facilities." *Technical Report NCEER-97-0010*, National Center for Earthquake Engineering Research, Buffalo, NY.

Stewart, J. P., and Seyhan, E. (2013). "Semi-empirical nonlinear site amplification and its application in NEHRP site factors." *PEER Report 2013*. Pacific Earthquake Engineering Research Center, University of California, Berkeley, CA.

Structural Engineers Association of California (SEAOC). (1999). *Recommended lateral force requirements and commentary*, Seismology Committee, Sacramento, CA.

OTHER REFERENCES (NOT CITED)

Abrahamson, N. A. (2000). "Effects of rupture directivity on probabilistic seismic hazard analysis." *Proc. 6th Intl. Conference on Seismic Zonation*, Palm Springs, CA.

ASCE. (2003). "Seismic evaluation of existing buildings." *ASCE/SEI 31-03*. Reston, VA.

ASCE. (2007). "Seismic rehabilitation of existing buildings." *ASCE/SEI 41-06*. Reston, VA.

Borcherdt, R. D. (2002). "Empirical evidence for site coefficients in building-code provisions." *Earthq. Spectra*, 18(2), 189–217.

Crouse, C. B., and McGuire, J. W. (1996). "Site response studies for purposes of revising NEHRP seismic provisions." *Earthq. Spectra*, 12(3), 129–143.

Dobry, R., Ramos, R., and Power, M. S. (1999). "Site factors and site categories in seismic codes." *Technical Report MCEER-99-0010*. Multidisciplinary Center for Earthquake Engineering Research, University of Buffalo, NY.

Field, E. H. (2000). "A modified ground motion attenuation relationship for Southern California that accounts for detailed site classification and a basin depth effect." *Bull. Seismol. Soc. of Am.*, 90, S209–S221.

Harmsen, S. C. (1997). "Determination of site amplification in the Los Angeles urban area from inversion of strong motion records." *Bull. Seismol. Soc. of Am.*, 87, 866–887.

Huang, Y.-N., Whittaker, A. S., and Luco, N. (2008). "Orientation of maximum spectral demand in the near-fault region." *Earthq. Spectra*, 24, 319–341.

Joyner, W. B., and Boore, D. M. (2000). "Recent developments in earthquake ground motion estimation." *Proc. 6th Intl. Conference on Seismic Zonation*, Palm Springs, CA.

Luco, N., Ellingwood, B. R., Hamburger, R. O., Hooper, J. D., Kimball, J. K., and Kircher, C. A. (2007). "Risk-targeted vs. current seismic design maps for the conterminous United States." *Proc. SEAOC 76th Annual Convention*, Sacramento, CA.

Petersen, M. D., Frankel, A. D., Harmsen, S. C., Mueller, C. S., Haller, K. M., Wheeler, R. L., et al. (2008). "Documentation for the 2008 update of the United States national seismic hazard maps." *USGS Open File Report 2008-1128*, U.S. Geological Survey, Reston, VA.

Rodriguez-Marek, A., Bray, J. D., and Abrahamson, N. (2001). "An empirical geotechnical site response procedure." *Earthq. Spectra*, 17(1), 65–87.

Seyhan, E. (2014). *Weighted average of 2014 NGA West-2 GMPEs*, Excel file: NGAW2_GMPE_Spreadsheets_v5.6_070514, Pacific Earthquake Engineering Center, Berkeley, CA ⟨http://peer.berkeley.edu/ngawest2/databases/⟩.

Silva, W., Darragh, R., Gregor, N., Martin, G., Abrahamson, N., and Kircher, C. (2000). "Reassessment of site coefficients and near-fault factors for building code provisions." Building Code Provisions, Program Element II, *Report 98-HQ-GR-1010*, U.S. Geological Survey, Reston, VA.

Somerville, P. G., Smith, N. F., Graves, R. W., and Abrahamson, N. A. (1997). "Modification of empirical strong ground motion attenuation relations to include the amplitude and duration effects of rupture directivity." *Seismol. Res. Lett.*, 68, 199–222.

Steidl, J. H. (2000). "Site response in Southern California for probabilistic seismic hazard analysis." *Bull. Seismol. Soc. of Am.*, 90, S149–S169.

Stewart, J. P., Liu, A. H., and Choi, Y. (2003). "Amplification factors for spectral acceleration in tectonically active regions." *Bull. Seismol. Soc. of Am.*, 93(1), 332–352.

CHAPTER C12
SEISMIC DESIGN REQUIREMENTS FOR BUILDING STRUCTURES

C12.1 STRUCTURAL DESIGN BASIS

The performance expectations for structures designed in accordance with this standard are described in Sections C11.1 and C11.5. Structures designed in accordance with the standard are likely to have a low probability of collapse but may suffer serious structural damage if subjected to the risk-targeted maximum considered earthquake (MCE_R) or stronger ground motion.

Although the seismic requirements of the standard are stated in terms of forces and loads, there are no external forces applied to the structure during an earthquake as, for example, is the case during a windstorm. The design forces are intended only as approximations to generate internal forces suitable for proportioning the strength and stiffness of structural elements and for estimating the deformations (when multiplied by the deflection amplification factor, C_d) that would occur in the same structure in the event of design earthquake (not MCE_R) ground motion.

C12.1.1 Basic Requirements. Chapter 12 of the standard sets forth a set of coordinated requirements that must be used together. The basic steps in structural design of a building structure for acceptable seismic performance are as follows:

1. Select gravity- and seismic force-resisting systems appropriate to the anticipated intensity of ground shaking. Section 12.2 sets forth limitations depending on the Seismic Design Category.
2. Configure these systems to produce a continuous, regular, and redundant load path so that the structure acts as an integral unit in responding to ground shaking. Section 12.3 addresses configuration and redundancy issues.
3. Analyze a mathematical model of the structure subjected to lateral seismic motions and gravity forces. Sections 12.6 and 12.7 set forth requirements for the method of analysis and for construction of the mathematical model. Sections 12.5, 12.8, and 12.9 set forth requirements for conducting a structural analysis to obtain internal forces and displacements.
4. Proportion members and connections to have adequate lateral and vertical strength and stiffness. Section 12.4 specifies how the effects of gravity and seismic loads are to be combined to establish required strengths, and Section 12.12 specifies deformation limits for the structure.

One- to three-story structures with shear wall or braced frame systems of simple configuration may be eligible for design under the simplified alternative procedure contained in Section 12.14. Any other deviations from the requirements of Chapter 12 are subject to approval by the authority having jurisdiction (AHJ) and must be rigorously justified, as specified in Section 11.1.4.

The baseline seismic forces used for proportioning structural elements (individual members, connections, and supports) are static horizontal forces derived from an elastic response spectrum procedure. A basic requirement is that horizontal motion can come from any direction relative to the structure, with detailed requirements for evaluating the response of the structure provided in Section 12.5. For most structures, the effect of vertical ground motions is not analyzed explicitly; it is implicitly included by adjusting the load factors (up and down) for permanent dead loads, as specified in Section 12.4. Certain conditions requiring more detailed analysis of vertical response are defined in Chapters 13 and 15 for nonstructural components and nonbuilding structures, respectively.

The basic seismic analysis procedure uses response spectra that are representative of, but substantially reduced from, the anticipated ground motions. As a result, at the MCE_R level of ground shaking, structural elements are expected to yield, buckle, or otherwise behave inelastically. This approach has substantial historical precedent. In past earthquakes, structures with appropriately ductile, regular, and continuous systems that were designed using *reduced* design forces have performed acceptably. In the standard, such design forces are computed by dividing the forces that would be generated in a structure behaving elastically when subjected to the design earthquake ground motion by the response modification coefficient, R, and this design ground motion is taken as two-thirds of the MCE_R ground motion.

The intent of R is to reduce the demand determined, assuming that the structure remains elastic at the design earthquake, to target the development of the first significant yield. This reduction accounts for the displacement ductility demand, R_d, required by the system and the inherent overstrength, Ω, of the seismic force-resisting system (SFRS) (Fig. C12.1-1). Significant yield is the point where complete plastification of a critical region of the SFRS first occurs (e.g., formation of the first plastic hinge in a moment frame), and the stiffness of the SFRS to further increases in lateral forces decreases as continued inelastic behavior spreads within the SFRS. This approach is consistent with member-level ultimate strength design practices. As such, first significant yield should not be misinterpreted as the point where first yield occurs in any member (e.g., 0.7 times the yield moment of a steel beam or either initial cracking or initiation of yielding in a reinforcing bar in a reinforced concrete beam or wall).

Fig. C12.1-1 shows the lateral force versus deformation relation for an archetypal moment frame used as an SFRS. First significant yield is shown as the lowest plastic hinge on the force–deformation diagram. Because of particular design rules and limits, including material strengths in excess of nominal or project-specific design requirements, structural elements are stronger by some degree than the strength required by analysis. The SFRS is therefore expected to reach first significant yield for forces in excess of design forces. With increased lateral loading,

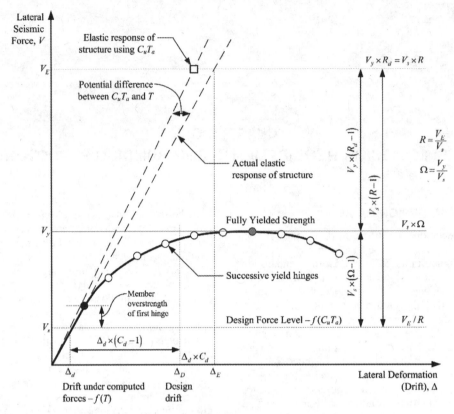

FIGURE C12.1-1 Inelastic Force–Deformation Curve

additional plastic hinges form and the resistance increases at a reduced rate (following the solid curve) until the maximum strength is reached, representing a fully yielded system. The maximum strength developed along the curve is substantially higher than that at first significant yield, and this margin is referred to as the system overstrength capacity. The ratio of these strengths is denoted as Ω. Furthermore, the figure illustrates the potential variation that can exist between the actual elastic response of a system and that considered using the limits on the fundamental period (assuming 100% mass participation in the fundamental mode—see Section C12.8.6). Although not a concern for strength design, this variation can have an effect on the expected drifts.

The system overstrength described above is the direct result of overstrength of the elements that form the SFRS and, to a lesser extent, the lateral force distribution used to evaluate the inelastic force–deformation curve. These two effects interact with applied gravity loads to produce sequential plastic hinges, as illustrated in the figure. This member overstrength is the consequence of several sources. First, material overstrength (i.e., actual material strengths higher than the nominal material strengths specified in the design) may increase the member overstrength significantly. For example, a recent survey shows that the mean yield strength of ASTM A36 steel is about 30% to 40% higher than the specified yield strength used in design calculations. Second, member design strengths usually incorporate a strength reduction or resistance factor, ϕ, to produce a low probability of failure under design loading. It is common to not include this factor in the member load-deformation relation when evaluating the seismic response of a structure in a nonlinear structural analysis. Third, designers can introduce additional strength by selecting sections or specifying reinforcing patterns that exceed those required by the computations. Similar situations occur where prescriptive minimums of the standard, or of the referenced design standards, control the design. Finally, the design of many flexible structural systems (e.g., moment-resisting frames) can be controlled by the drift rather than strength, with sections selected to control lateral deformations rather than to provide the specified strength.

The result is that structures typically have a much higher lateral strength than that specified as the minimum by the standard, and the first significant yielding of structures may occur at lateral load levels that are 30% to 100% higher than the prescribed design seismic forces. If provided with adequate ductile detailing, redundancy, and regularity, full yielding of structures may occur at load levels that are two to four times the prescribed design force levels.

Most structural systems have some elements whose action cannot provide reliable inelastic response or energy dissipation. Similarly, some elements are required to remain essentially elastic to maintain the structural integrity of the structure (e.g., columns supporting a discontinuous SFRS). Such elements and actions must be protected from undesirable behavior by considering that the actual forces within the structure can be significantly larger than those at first significant yield. The standard specifies an overstrength factor, Ω_0, to amplify the prescribed seismic forces for use in design of such elements and for such actions. This approach is a simplification to determining the maximum forces that could be developed in a system and the distribution of these forces within the structure. Thus, this specified overstrength factor is neither an upper nor a lower bound; it is simply an approximation specified to provide a nominal degree of protection against undesirable behavior.

The elastic deformations calculated under these reduced forces (see Section C12.8.6) are multiplied by the deflection amplification factor, C_d, to estimate the deformations likely to result from the design earthquake ground motion. This factor was first introduced in ATC 3-06 (ATC 1978). For a vast majority of systems, C_d is less than R, with a few notable exceptions, where inelastic drift is strongly coupled with an increased risk of collapse (e.g., reinforced concrete bearing walls). Research over the past 30 years has illustrated that inelastic displacements may be significantly greater than Δ_E for many structures and less than Δ_E for others. Where C_d is substantially less than R, the system is considered to have damping greater than the nominal 5% of critical damping. As set forth in Section 12.12 and Chapter 13, the amplified deformations are used to assess story drifts and to determine seismic demands on elements of the structure that are not part of the seismic force-resisting system and on nonstructural components within structures.

Fig. C12.1-1 illustrates the significance of seismic design parameters contained in the standard, including the response modification coefficient, R; the deflection amplification factor, C_d; and the overstrength factor, Ω_0. The values of these parameters, provided in Table 12.2-1, as well as the criteria for story drift and P-delta effects, have been established considering the characteristics of typical properly designed structures. The provisions of the standard anticipate an SFRS with redundant characteristics wherein significant system strength above the level of first significant yield can be obtained by plastification at other critical locations in the structure before the formation of a collapse mechanism. If excessive "optimization" of a structural design is performed with lateral resistance provided by only a few elements, the successive yield hinge behavior depicted in Fig. C12.1-1 is not able to form, the actual overstrength (Ω) is small, and use of the seismic design parameters in the standard may not provide the intended seismic performance.

The response modification coefficient, R, represents the ratio of the forces that would develop under the specified ground motion if the structure had an entirely linear-elastic response to the prescribed design forces (Fig. C12.1-1). The structure must be designed so that the level of significant yield exceeds the prescribed design force. The ratio R_d, expressed as $R_d = V_E/V_S$, where V_E is the elastic seismic force demand and V_S is the prescribed seismic force demand, is always larger than 1.0; thus, all structures are designed for forces smaller than those the design ground motion would produce in a structure with a completely linear-elastic response. This reduction is possible for a number of reasons. As the structure begins to yield and deform inelastically, the effective period of response of the structure lengthens, which results in a reduction in strength demand for most structures. Furthermore, the inelastic action results in a significant amount of energy dissipation (hysteretic damping) in addition to other sources of damping present below significant yield. The combined effect, which is known as the ductility reduction, explains why a properly designed structure with a fully yielded strength (V_y in Fig. C12.1-1) that is significantly lower than V_E can be capable of providing satisfactory performance under the design ground motion excitations.

The energy dissipation resulting from hysteretic behavior can be measured as the area enclosed by the force–deformation curve of the structure as it experiences several cycles of excitation. Some structures have far more energy dissipation capacity than others. The extent of energy dissipation capacity available depends largely on the amount of stiffness and strength degradation the structure undergoes as it experiences repeated cycles of inelastic deformation. Fig. C12.1-2 shows representative load deformation curves for two simple substructures, such as a beam–column assembly in a frame. Hysteretic curve (a) in the figure represents the behavior of substructures that have been detailed for ductile behavior. The substructure can maintain almost all of its strength and stiffness over several large cycles of inelastic deformation. The resulting force–deformation "loops" are quite wide and open, resulting in a large amount of energy dissipation. Hysteretic curve (b) represents the behavior of a substructure that has much less energy dissipation than that for the substructure (a) but has a greater change in response period. The structural response is determined by a combination of energy dissipation and period modification.

The principles of this section outline the conceptual intent behind the seismic design parameters used by the standard. However, these parameters are based largely on engineering judgment of the various materials and performance of structural systems in past earthquakes and cannot be directly computed using the relationships presented in Fig. C12.1-1. The seismic design parameters chosen for a specific project or system should be chosen with care. For example, lower values should be used for structures possessing a low degree of redundancy wherein all the plastic hinges required for the formation of a mechanism may be formed essentially simultaneously and at a force level close to the specified design strength. This situation can result in considerably more detrimental P-delta effects. Because it is difficult for individual designers to judge the extent to which the value of R should be adjusted based on the inherent redundancy of their designs, Section 12.3.4 provides the redundancy factor, ρ, that is typically determined by being based on the removal of individual seismic force-resisting elements.

Higher order seismic analyses are permitted for any structure and are required for some structures (*see* Section 12.6); lower limits based on the equivalent lateral force procedure may, however, still apply.

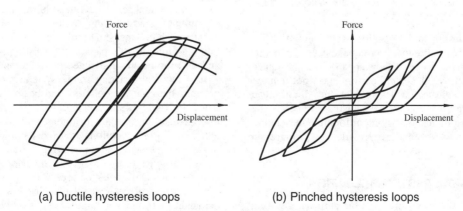

FIGURE C12.1-2 Typical Hysteretic Curves

C12.1.2 Member Design, Connection Design, and Deformation Limit. Given that key elements of the seismic force-resisting system are likely to yield in response to ground motions, as discussed in Section C12.1.1, it might be expected that structural connections would be required to develop the strength of connected members. Although that is a logical procedure, it is not a general requirement. The actual requirement varies by system and generally is specified in the standards for design of the various structural materials cited by reference in Chapter 14. Good seismic design requires careful consideration of this issue.

C12.1.3 Continuous Load Path and Interconnection. In effect, Section 12.1.3 calls for the seismic design to be complete and in accordance with the principles of structural mechanics. The loads must be transferred rationally from their point of origin to the final point of resistance. This requirement should be obvious, but it often is overlooked by those inexperienced in earthquake engineering. Design consideration should be given to potentially adverse effects where there is a lack of redundancy. Given the many unknowns and uncertainties in the magnitude and characteristics of earthquake loading, in the materials and systems of construction for resisting earthquake loadings, and in the methods of analysis, good earthquake engineering practice has been to provide as much redundancy as possible in the seismic force-resisting system of buildings. Redundancy plays an important role in determining the ability of the building to resist earthquake forces. In a structural system without redundant elements, every element must remain operative to preserve the integrity of the building structure. However, in a highly redundant system, one or more redundant elements may fail and still leave a structural system that retains its integrity and can continue to resist lateral forces, albeit with diminished effectiveness.

Although a redundancy requirement is included in Section 12.3.4, overall system redundancy can be improved by making all joints of the vertical load-carrying frame moment resisting and incorporating them into the seismic force-resisting system. These multiple points of resistance can prevent a catastrophic collapse caused by distress or failure of a member or joint. (The overstrength characteristics of this type of frame are discussed in Section C12.1.1.)

The minimum connection forces are not intended to be applied simultaneously to the entire seismic force-resisting system.

C12.1.4 Connection to Supports. The requirement is similar to that given in Section 1.4 on connections to supports for general structural integrity. See Section C1.4.

C12.1.5 Foundation Design. Most foundation design criteria are still stated in terms of allowable stresses, and the forces computed in the standard are all based on the strength level of response. When developing strength-based criteria for foundations, all the factors cited in Section 12.1.5 require careful consideration. Section C12.13 provides specific guidance.

C12.1.6 Material Design and Detailing Requirements. The design limit state for resistance to an earthquake is unlike that for any other load within the scope of the standard. The earthquake limit state is based on overall system performance, not member performance, where repeated cycles of inelastic straining are accepted as an energy-dissipating mechanism. Provisions that modify customary requirements for proportioning and detailing structural members and systems are provided to produce the desired performance.

C12.2 STRUCTURAL SYSTEM SELECTION

C12.2.1 Selection and Limitations. For the purpose of seismic analysis and design requirements, seismic force-resisting systems are grouped into categories as shown in Table 12.2-1. These categories are subdivided further for various types of vertical elements used to resist seismic forces. In addition, the sections for detailing requirements are specified.

Specification of response modification coefficients, R, requires considerable judgment based on knowledge of actual earthquake performance and research studies. The coefficients and factors in Table 12.2-1 continue to be reviewed in light of recent research results. The values of R for the various systems were selected considering observed performance during past earthquakes, the toughness (ability to dissipate energy without serious degradation) of the system, and the amount of damping typically present in the system when it undergoes inelastic response. FEMA P-695 (2009b) has been developed with the purpose of establishing and documenting a methodology for quantifying seismic force-resisting system performance and response parameters for use in seismic design. Whereas R is a key parameter being addressed, related design parameters such as the overstrength factor, Ω_0, and the deflection amplification factor, C_d, also are addressed. Collectively, these terms are referred to as "seismic design coefficients (or factors)." Future systems are likely to derive their seismic design coefficients (or factors) using this methodology, and existing system coefficients (or factors) also may be reviewed in light of this new procedure.

Height limits have been specified in codes and standards for more than 50 years. The structural system limitations and limits on structural height, h_n, specified in Table 12.2-1, evolved from these initial limitations and were further modified by the collective expert judgment of the NEHRP Provisions Update Committee (PUC) and the ATC-3 project team (the forerunners of the PUC). They have continued to evolve over the past 30 years based on observations and testing, but the specific values are based on subjective judgment.

In a bearing wall system, major load-carrying columns are omitted and the walls carry a major portion of the gravity (dead and live) loads. The walls supply in-plane lateral stiffness and strength to resist wind and earthquake loads and other lateral loads. In some cases, vertical trusses are used to augment lateral stiffness. In general, lack of redundancy for support of vertical and horizontal loads causes values of R to be lower for this system compared with R values of other systems.

In a building frame system, gravity loads are carried primarily by a frame supported on columns rather than by bearing walls. Some portions of the gravity load may be carried on bearing walls, but the amount carried should represent a relatively small percentage of the floor or roof area. Lateral resistance is provided by shear walls or braced frames. Light-framed walls with shear panels are intended for use only with wood and steel building frames. Although gravity load-resisting systems are not required to provide lateral resistance, most of them do. To the extent that the gravity load-resisting system provides additional lateral resistance, it enhances the building's seismic performance capability, so long as it is capable of resisting the resulting stresses and undergoing the associated deformations.

In a moment-resisting frame system, moment-resisting connections between the columns and beams provide lateral resistance. In Table 12.2-1, such frames are classified as ordinary, intermediate, or special. In high seismic design categories, the anticipated ground motions are expected to produce large inelastic demands, so special moment frames designed and detailed for ductile response in accordance with Chapter 14 are required. In low Seismic Design Categories, the inherent overstrength in typical structural designs is such that the anticipated inelastic demands are somewhat reduced, and less ductile systems may be used safely. Because these less ductile ordinary framing systems do not possess as much toughness, lower values of R are specified.

The values for R, Ω_0, and C_d at the composite systems in Table 12.2-1 are similar to those for comparable systems of structural steel and reinforced concrete. Use of the tabulated values is allowed only when the design and detailing requirements in Section 14.3 are followed.

In a dual system, a three-dimensional space frame made up of columns and beams provides primary support for gravity loads. Primary lateral resistance is supplied by shear walls or braced frames, and secondary lateral resistance is provided by a moment frame complying with the requirements of Chapter 14.

Where a beam–column frame or slab–column frame lacks special detailing, it cannot act as an effective backup to a shear wall subsystem, so there are no dual systems with ordinary moment frames. Instead, Table 12.2-1 permits the use of a shear wall–frame interactive system with ordinary reinforced concrete moment frames and ordinary reinforced concrete shear walls. Use of this defined system, which requires compliance with Section 12.2.5.8, offers a significant advantage over a simple combination of the two constituent ordinary reinforced concrete systems. Where those systems are simply combined, Section 12.2.3.3 would require use of seismic design parameters for an ordinary reinforced concrete moment frame.

In a cantilevered column system, stability of mass at the top is provided by one or more columns with base fixity acting as a single-degree-of-freedom system.

Cantilever column systems are essentially a special class of moment-resisting frame, except that they do not possess the redundancy and overstrength that most moment-resisting frames derive from sequential formation of yield or plastic hinges. Where a typical moment-resisting frame must form multiple plastic hinges in members to develop a yield mechanism, a cantilever column system develops hinges only at the base of the columns to form a mechanism. As a result, their overstrength is limited to that provided by material overstrength and by design conservatism.

It is permitted to construct cantilever column structures using any of the systems that can be used to develop moment frames, including ordinary and special steel; ordinary, intermediate, and special concrete; and timber frames. The system limitations for cantilever column systems reflect the type of moment frame detailing provided but with a limit on structural height, h_n, of 35 ft (10.7 m).

The value of R for cantilever column systems is derived from moment-resisting frame values where R is divided by Ω_0 but is not taken as less than 1 or greater than 2 1/2. This range accounts for the lack of sequential yielding in such systems. C_d is taken as equal to R, recognizing that damping is quite low in these systems and inelastic displacement of these systems is not less than the elastic displacement.

C12.2.1.1 Alternative Structural Systems. Historically, this standard has permitted the use of alternative seismic force-resisting systems subject to satisfactory demonstration that the proposed systems' lateral force resistance and energy dissipation capacity is equivalent to structural systems listed in Table 12.2-1, for equivalent values of the response modification coefficient, R, overstrength factor, Ω_0, and deflection amplification coefficient, C_d. These design factors were established based on limited analytical and laboratory data and the engineering judgment of the developers of the standard.

Under funding from the Federal Emergency Management Agency, the Applied Technology Council developed a rational methodology for validation of design criteria for seismic force-resisting systems under its ATC-63 project. Published as *FEMA P-695* (2009b), this methodology recognizes that the fundamental goal of seismic design rules contained in the standard is to limit collapse probability to acceptable levels. The *FEMA P-695* methodology uses nonlinear response history analysis to predict an adjusted collapse margin ratio (ACMR) for a suite of archetypical structures designed in accordance with a proposed set of system-specific design criteria and subjected to a standard series of ground motion accelerograms. The suite of archetypical structures is intended to represent the typical types and sizes of structures that are likely to incorporate the system. The ACMR relates to the conditional probability of collapse given MCE_R shaking and considers uncertainties associated with the record-to-record variability of ground motions, the quality of the design procedure, the comprehensiveness and quality of the laboratory data upon which the analytical modeling is based; and uncertainties associated with the analytical modeling. Subsequent studies have been used to benchmark this methodology against selected systems contained in Table 12.2-1 and have demonstrated that the methodology provides rational results consistent with past engineering judgment for many systems. The *FEMA P-695* methodology is therefore deemed to constitute the preferred procedure for demonstrating adequate collapse resistance for new structural systems not currently contained in Table 12.2-1.

Under the *FEMA P-695* methodology, the archetypes used to evaluate seismic force-resisting systems are designed using the criteria for Risk Category II structures and are evaluated to demonstrate that the conditional probability of collapse of such structures conforms to the 10% probability of collapse goal stated in this section and also described in Section C1.3.1 of the commentary to this standard. It is assumed that application of the seismic importance factors and more restrictive drift limits associated with the design requirements for structures assigned to Risk Categories III and IV will provide such structures with the improved resistance to collapse described in Section C1.3.1 for those Risk Categories.

In addition to providing a basis for establishment of design criteria for structural systems that can be used for design of a wide range of structures, the *FEMA P-695* methodology also contains a building-specific methodology intended for application to individual structures. The rigor associated with application of the *FEMA P-695* methodology may not be appropriate to the design of individual structures that conform with limited and clearly defined exceptions to the criteria contained in the standard for a defined structural system. Nothing contained in this section is intended to require the use of *FEMA P-695* or similar methodologies for such cases.

C12.2.1.2 Elements of Seismic Force-Resisting Systems. This standard and its referenced standards specify design and detailing criteria for members and their connections (elements) of seismic force-resisting systems defined in Table 12.2-1. Substitute elements replace portions of the defined seismic force-resisting systems. Examples include proprietary products made up of special steel moment-resisting connections or proprietary shear walls for use in light-frame construction. Requirements for qualification of substitute elements of seismic force-resisting systems are intended to ensure equivalent seismic performance of the element and the system as a whole. The evaluation of suitability for substitution is based on comparison of key performance parameters of the code-defined (conforming) element and the substitute element.

FEMA P-795, *Quantification of Building Seismic Performance Factors: Component Equivalency Methodology* (2011) is an acceptable methodology to demonstrate equivalence of substitute elements and their connections and provides methods for component testing, calculation of parameter statistics from test data, and acceptance criteria for evaluating equivalency. Key performance parameters include strength ratio, stiffness ratio, deformation capacity, and cyclic strength and stiffness characteristics.

Section 12.2.1.2, item f, requires independent design review as a condition of approval of the use of substitute elements. It is not the intent that design review be provided for every project that incorporates a substitute component, but rather that such review would be performed one time, as part of the general qualification of such substitute components. When used on individual projects, evidence of such review could include an evaluation service report or review letter indicating the conditions under which use of the substitute component is acceptable.

C12.2.2 Combinations of Framing Systems in Different Directions. Different seismic force-resisting systems can be used along each of the two orthogonal axes of the structure, as long as the respective values of R, Ω_0, and C_d are used. Depending on the combination selected, it is possible that one of the two systems may limit the extent of the overall system with regard to structural system limitations or structural height, h_n; the more restrictive of these would govern.

C12.2.3 Combinations of Framing Systems in the Same Direction. The intent of the provision requiring use of the most stringent seismic design parameters (R, Ω_0, and C_d) is to prevent mixed seismic force-resisting systems that could concentrate inelastic behavior in the lower stories.

C12.2.3.1 R, C_d, and Ω_0 Values for Vertical Combinations. This section expands upon Section 12.2.3 by specifying the requirements specific to the cases where (a) the value of R for the lower seismic force-resisting system is lower than that for the upper system, and (b) the value of R for the upper seismic force-resisting system is lower than that for the lower system.

The two cases are intended to account for all possibilities of vertical combinations of seismic force-resisting systems in the same direction. For a structure with a vertical combination of three or more seismic force-resisting systems in the same direction, Section 12.2.3.1 must be applied to the adjoining pairs of systems until the vertical combinations meet the requirements therein.

There are also exceptions to these requirements for conditions that do not affect the dynamic characteristics of the structure or that do not result in concentration of inelastic demand in critical areas.

C12.2.3.2 Two-Stage Analysis Procedure. A two-stage equivalent lateral force procedure is permitted where the lower portion of the structure has a minimum of 10 times the stiffness of the upper portion of the structure. In addition, the period of the entire structure is not permitted to be greater than 1.1 times the period of the upper portion considered as a separate structure supported at the transition from the upper to the lower portion. An example would be a concrete podium under a wood- or steel-framed upper portion of a structure. The upper portion may be analyzed for seismic forces and drifts using the values of R, Ω_0, and C_d for the upper portion as a separate structure. The seismic forces (e.g., shear and overturning) at the base of the upper portion are applied to the top of the lower portion and scaled up by the ratio of $(R/\rho)_{upper}$ to $(R/\rho)_{lower}$. The lower portion, which now includes the seismic forces from the upper portion, may then be analyzed using the values of R, Ω_0, and C_d for the lower portion of the structure.

C12.2.3.3 R, C_d, and Ω_0 Values for Horizontal Combinations. For almost all conditions, the least value of R of different seismic force-resisting systems in the same direction must be used in design. This requirement reflects the expectation that the entire system will undergo the same deformation with its behavior controlled by the least ductile system. However, for light-frame construction or flexible diaphragms meeting the listed conditions, the value of R for each independent line of resistance can be used. This exceptional condition is consistent with light-frame construction that uses the ground for parking with residential use above.

C12.2.4 Combination Framing Detailing Requirements. This requirement is provided so that the seismic force-resisting system with the highest value of R has the necessary ductile detailing throughout. The intent is that details common to both systems be designed to remain functional throughout the response to earthquake load effects to preserve the integrity of the seismic force-resisting system.

C12.2.5 System-Specific Requirements

C12.2.5.1 Dual System. The moment frame of a dual system must be capable of resisting at least 25% of the design seismic forces; this percentage is based on judgment. The purpose of the 25% frame is to provide a secondary seismic force-resisting system with higher degrees of redundancy and ductility to improve the ability of the building to support the service loads (or at least the effect of gravity loads) after strong earthquake shaking. The primary system (walls or bracing) acting together with the moment frame must be capable of resisting all of the design seismic forces. The following analyses are required for dual systems:

1. The moment frame and shear walls or braced frames must resist the design seismic forces, considering fully the force and deformation interaction of the walls or braced frames and the moment frames as a single system. This analysis must be made in accordance with the principles of structural mechanics that consider the relative rigidities of the elements and torsion in the system. Deformations imposed upon members of the moment frame by their interaction with the shear walls or braced frames must be considered in this analysis.
2. The moment frame must be designed with sufficient strength to resist at least 25% of the design seismic forces.

C12.2.5.2 Cantilever Column Systems. Cantilever column systems are singled out for special consideration because of their unique characteristics. These structures often have limited redundancy and overstrength and concentrate inelastic behavior at their bases. As a result, they have substantially less energy dissipation capacity than other systems. A number of apartment buildings incorporating this system experienced severe damage and, in some cases, collapsed in the 1994 Northridge (California) earthquake. Because the ductility of columns that have large axial stress is limited, cantilever column systems may not be used where individual column axial demands from seismic load effects exceed 15% of their available axial strength, including slenderness effects.

Elements providing restraint at the base of cantilever columns must be designed for seismic load effects, including overstrength, so that the strength of the cantilever columns is developed.

C12.2.5.3 Inverted Pendulum-Type Structures. Inverted pendulum-type structures do not have a unique entry in Table 12.2-1 because they can be formed from many structural systems. Inverted pendulum-type structures have more than half of their mass concentrated near the top (producing one degree of freedom in horizontal translation) and rotational compatibility of the mass with the column (producing vertical accelerations acting in opposite directions). Dynamic response amplifies this rotation; hence, the bending moment induced at the top of the column can exceed that computed using the procedures of Section 12.8. The requirement to design for a top moment that is one-half of the base moment calculated in accordance with Section 12.8 is based on analyses of inverted pendulums covering a wide range of practical conditions.

C12.2.5.4 Increased Structural Height Limit for Steel Eccentrically Braced Frames, Steel Special Concentrically Braced Frames, Steel Buckling-Restrained Braced Frames, Steel Special Plate Shear Walls, and Special Reinforced Concrete Shear Walls. The first criterion for an increased limit on structural height, h_n, precludes extreme torsional irregularity because premature failure of one of the shear walls or braced frames could lead to excessive inelastic torsional response. The second criterion, which is similar to the redundancy requirements, is to limit the structural height of systems that are too strongly dependent on any single line of shear walls or braced frames. The inherent torsion resulting from the distance between the center of mass and the center of rigidity must be included, but accidental torsional effects are neglected for ease of implementation.

C12.2.5.5 Special Moment Frames in Structures Assigned to Seismic Design Categories D through F. Special moment frames, either alone or as part of a dual system, are required to be used in Seismic Design Categories D through F where the structural height, h_n, exceeds 160 ft (48.8 m) (or 240 ft [73.2 m]) for buildings that meet the provisions of Section 12.2.5.4) as indicated in Table 12.2-1. In shorter buildings where special moment frames are not required to be used, the special moment frames may be discontinued and supported on less ductile systems as long as the requirements of Section 12.2.3 for framing system combinations are followed.

For the situation where special moment frames are required, they should be continuous to the foundation. In cases where the foundation is located below the building's base, provisions for discontinuing the moment frames can be made as long as the seismic forces are properly accounted for and transferred to the supporting structure.

C12.2.5.6 Steel Ordinary Moment Frames. Steel ordinary moment frames (OMFs) are less ductile than steel special moment frames; consequently, their use is prohibited in structures assigned to Seismic Design Categories D, E, and F (Table 12.2-1). Structures with steel OMFs, however, have exhibited acceptable behavior in past earthquakes where the structures were sufficiently limited in their structural height, number of stories, and seismic mass. The provisions in the standard reflect these observations. The exception is discussed separately below. Table C12.2-1 summarizes the provisions.

C12.2.5.6.1 Seismic Design Category D or E. Single-story steel OMFs are permitted, provided that (a) the structural height, h_n, is a maximum of 65 ft (20 m), (b) the dead load supported by and tributary to the roof is a maximum of 20 lb/ft² (0.96 kN/m²), and (c) the dead load of the exterior walls more than 35 ft (10.6 m) above the seismic base tributary to the moment frames is a maximum of 20 lb/ft² (0.96 kN/m²).

In structures of light-frame construction, multistory steel OMFs are permitted, provided that (a) the structural height, h_n, is a maximum of 35 ft (10.6 m), (b) the dead load of the roof and each floor above the seismic base supported by and tributary to the moment frames are each a maximum of 35 lb/ft² (1.68 kN/m²), and (c) the dead load of the exterior walls tributary to the moment frames is a maximum of 20 lb/ft² (0.96 kN/m²).

EXCEPTION: Industrial structures, such as aircraft maintenance hangars and assembly buildings, with steel OMFs have performed well in past earthquakes with strong ground motions (EQE Inc. 1983, 1985, 1986a, 1986b, 1986c, and 1987); the exception permits single-story steel OMFs to be unlimited in height provided that (a) the structure is limited to the enclosure of equipment or machinery; (b) its occupants are limited to maintaining and monitoring the equipment, machinery, and their associated processes; (c) the sum of the dead load and equipment loads supported by and tributary to the roof is a maximum of 20 lb/ft² (0.96 kN/m²); and (d) the dead load of the exterior wall system, including exterior columns more than 35 ft (10.6 m) above the seismic base is a maximum of 20 lb/ft² (0.96 kN/m²). Though the latter two load limits (Items C and D) are similar to those described in this section, there are meaningful differences.

The exception further recognizes that these facilities often require large equipment or machinery, and associated systems, not supported by or considered tributary to the roof, that support the intended operational functions of the structure, such as top running bridge cranes, jib cranes, and liquid storage containment and distribution systems. To limit the seismic interaction between the seismic force-resisting systems and these components, the exception requires the weight of equipment or machinery that is not self-supporting (i.e., not freestanding) for all loads (e.g., dead, live, or seismic) to be included when determining compliance with the roof or exterior wall load limits. This *equivalent* equipment load shall be in addition to the loads listed above.

To determine the equivalent equipment load, the exception requires the weight to be considered fully (100%) tributary to an area not exceeding 600 ft² (55.8 m²). This limiting area can be taken either to an adjacent exterior wall for cases where the weight is supported by an exterior column (which may also span to the first interior column) or to the adjacent roof for cases where the weight is supported entirely by an interior column or columns, but not both; nor can a fraction of the weight be allocated to each zone. Equipment loads within overlapping tributary areas should be combined in the same limiting area. Other provisions in the standard, as well as in past editions, require satisfying wall load limits tributary to the moment frame, but this requirement is not included in the exception in that it is based on a component-level approach that does not consider the interaction between systems in the structure. As such, the limiting area is considered to be a reasonable approximation of the tributary area of a moment frame segment for the purpose of this conversion. Although this weight allocation procedure may not represent an accurate physical distribution, it is considered to be a reasonable method for verifying compliance with the specified load limits to limit seismic interactions. The engineer must still be attentive to actual mass distributions when computing seismic loads. Further information is discussed in Section C11.1.3.

C12.2.5.6.2 Seismic Design Category F. Single-story steel OMFs are permitted, provided that they meet conditions (a) and (b) described in Section C12.2.5.6.1 for single-story frames and (c) the dead load of the exterior walls tributary to the moment frames is a maximum of 20 lb/ft² (0.96 kN/m²).

C12.2.5.7 Steel Intermediate Moment Frames. Steel intermediate moment frames (IMFs) are more ductile than steel ordinary moment frames (OMFs) but less ductile than steel special moment frames; consequently, restrictions are placed

Table C12.2-1 Summary of Conditions for OMFs and IMFs in Structures Assigned to Seismic Design Category D, E, or F
(Refer to the Standard for Additional Requirements)

Section	Frame	SDC	Max. Number Stories	Light-Frame Construction	Max. h_n, ft	Max. roof/floor D_L (lb/ft^2)	Exterior Wall DL Max. (lb/ft^2)	Wall[a] Height (ft)
12.2.5.6.1(a)	OMF	D, E	1	NA	65	20	20	35
12.2.5.6.1(a)-Exc	OMF	D, E	1	NA	NL	20	20	35
12.2.5.6.1(b)	OMF	D, E	NL	Required	35	35	20	0
12.2.5.6.2	OMF	F	1	NA	65	20	20	0
12.2.5.7.1(a)	IMF	D	1	NA	65	20	20	35
12.2.5.7.1(a)-Exc	IMF	D	1	NA	NL	20	20	35
12.2.5.7.1(b)	IMF	D	NL	NA	35	NL	NL	NA
12.2.5.7.2(a)	IMF	E	1	NA	65	20	20	35
12.2.5.7.2(a)-Exc	IMF	E	1	NA	NL	20	20	35
12.2.5.7.2(b)	IMF	E	NL	NA	35	35	20	0
12.2.5.7.3(a)	IMF	F	1	NA	65	20	20	0
12.2.5.7.3(b)	IMF	F	NL	Required	35	35	20	0

Note: NL means No Limit; NA means Not Applicable. For metric units, use 20 m for 65 ft and use 10.6 m for 35 ft. For 20 lb/ft^2, use 0.96 kN/m^2 and for 30 lb/ft^2, use 1.68 kN/m^2.
[a]Applies to portion of wall above listed wall height.

on their use in structures assigned to Seismic Design Category D and their use is prohibited in structures assigned to Seismic Design Categories E and F (Table 12.2-1). As with steel OMFs, steel IMFs have also exhibited acceptable behavior in past earthquakes where the structures were sufficiently limited in their structural height, number of stories, and seismic mass. The provisions in the standard reflect these observations. The exceptions are discussed separately (following). Table C12.2-1 summarizes the provisions.

C12.2.5.7.1 Seismic Design Category D. Single-story steel IMFs are permitted without limitations on dead load of the roof and exterior walls, provided that the structural height, h_n, is a maximum of 35 ft (10.6 m). An increase to 65 ft (20 m) is permitted for h_n, provided that (a) the dead load supported by and tributary to the roof is a maximum of 20 lb/ft^2 (0.96 kN/m^2), and (b) the dead load of the exterior walls more than 35 ft (10.6 m) above the seismic base tributary to the moment frames is a maximum of 20 lb/ft^2 (0.96 kN/m^2).

The exception permits single-story steel IMFs to be unlimited in height, provided that they meet all of the conditions described in the exception to Section C12.2.5.6.1 for the same structures.

C12.2.5.7.2 Seismic Design Category E. Single-story steel IMFs are permitted, provided that they meet all of the conditions described in Section C12.2.5.6.1 for single-story OMFs.

The exception permits single-story steel IMFs to be unlimited in height, provided that they meet all of the conditions described in Section C12.2.5.6.1 for the same structures.

Multistory steel IMFs are permitted, provided that they meet all of the conditions described in Section C12.2.5.6.1 for multistory OMFs, except that the structure is not required to be of light-frame construction.

C12.2.5.7.3 Seismic Design Category F. Single-story steel IMFs are permitted, provided that (a) the structural height, h_n, is a maximum of 65 ft (20 m), (b) the dead load supported by and tributary to the roof is a maximum of 20 lb/ft^2 (0.96 kN/m^2), and (c) the dead load of the exterior walls tributary to the moment frames is a maximum of 20 lb/ft^2 (0.96 kN/m^2).

Multistory steel IMFs are permitted, provided that they meet all of the conditions described in the exception to Section C12.2.5.6.1 for multistory OMFs in structures of light-frame construction.

C12.2.5.8 Shear Wall–Frame Interactive Systems. For structures assigned to Seismic Design Category A or B (where seismic hazard is low), it is usual practice to design shear walls and frames of a shear wall–frame structure to resist lateral forces in proportion to their relative rigidities, considering interaction between the two subsystems at all levels. As discussed in Section C12.2.1, this typical approach would require use of a lower response modification coefficient, R, than that defined for shear wall–frame interactive systems. Where the special requirements of this section are satisfied, more reliable performance is expected, justifying a higher value of R.

C12.3 DIAPHRAGM FLEXIBILITY, CONFIGURATION IRREGULARITIES, AND REDUNDANCY

C12.3.1 Diaphragm Flexibility. Most seismic force-resisting systems have two distinct parts: the horizontal system, which distributes lateral forces to the vertical elements and the vertical system, which transmits lateral forces between the floor levels and the base of the structure.

The horizontal system may consist of diaphragms or a horizontal bracing system. For the majority of buildings, diaphragms offer the most economical and positive method of resisting and distributing seismic forces in the horizontal plane. Typically, diaphragms consist of a metal deck (with or without concrete), concrete slabs, and wood sheathing and/or decking. Although most diaphragms are flat, consisting of the floors of buildings, they also may be inclined, curved, warped, or folded configurations, and most diaphragms have openings.

The diaphragm stiffness relative to the stiffness of the supporting vertical seismic force-resisting system ranges from flexible to rigid and is important to define. Provisions defining diaphragm flexibility are given in Sections 12.3.1.1 through 12.3.1.3. If a diaphragm cannot be idealized as either flexible or rigid, explicit consideration of its stiffness must be included in the analysis.

The diaphragms in most buildings braced by wood light-frame shear walls are semirigid. Because semirigid diaphragm modeling is beyond the capability of available software for wood light-frame buildings, it is anticipated that this requirement will be met by evaluating force distribution using both rigid and flexible diaphragm models and taking the worse case of the two. Although this procedure is in conflict with common design practice, which typically includes only flexible diaphragm force distribution for wood light-frame buildings, it is one method of capturing the effect of the diaphragm stiffness.

C12.3.1.1 Flexible Diaphragm Condition. Section 12.3.1.1 defines broad categories of diaphragms that may be idealized as flexible, regardless of whether the diaphragm meets the calculated conditions of Section 12.3.1.3. These categories include the following:

a. Construction with relatively stiff vertical framing elements, such as steel-braced frames and concrete or masonry shear walls;
b. One- and two-family dwellings; and
c. Light-frame construction (e.g., construction consisting of light-frame walls and diaphragms) with or without nonstructural toppings of limited stiffness.

For item c above, compliance with story drift limits along each line of shear walls is intended as an indicator that the shear walls are substantial enough to share load on a tributary area basis and not require torsional force redistribution.

C12.3.1.2 Rigid Diaphragm Condition. Span-to-depth ratio limits are included in the deemed-to-comply condition as an indirect measure of the flexural contribution to diaphragm stiffness.

C12.3.1.3 Calculated Flexible Diaphragm Condition. A diaphragm is permitted to be idealized as flexible if the calculated diaphragm deflection (typically at midspan) between supports (lines of vertical elements) is greater than two times the average story drift of the vertical lateral force-resisting elements located at the supports of the diaphragm span.

Fig. 12.3-1 depicts a distributed load, conveying the intent that the tributary lateral load be used to compute δ_{MDD}, consistent with the Section 11.3 symbols. A diaphragm opening is illustrated, and the shorter arrows in the portion of the diaphragm with the opening indicate lower load intensity because of lower tributary seismic mass.

C12.3.2 Irregular and Regular Classification. The configuration of a structure can significantly affect its performance during a strong earthquake that produces the ground motion contemplated in the standard. Structural configuration can be divided into two aspects: horizontal and vertical. Most seismic design provisions were derived for buildings that have regular configurations, but earthquakes have shown repeatedly that buildings that have irregular configurations suffer greater damage. This situation prevails even with good design and construction.

There are several reasons for the poor behavior of irregular structures. In a regular structure, the inelastic response, including energy dissipation and damage, produced by strong ground shaking tends to be well distributed throughout the structure. However, in irregular structures, inelastic behavior can be concentrated by irregularities and can result in rapid failure of structural elements in these areas. In addition, some irregularities introduce unanticipated demands into the structure, which designers frequently overlook when detailing the structural system. Finally, the elastic analysis methods typically used in the design of structures often cannot predict the distribution of earthquake demands in an irregular structure very well, leading to inadequate design in the areas associated with the irregularity. For these reasons, the standard encourages regular structural configurations and prohibits gross irregularity in buildings located on sites close to major active faults, where very strong ground motion and extreme inelastic demands are anticipated. The termination of seismic force-resisting elements at the foundation, however, is not considered to be a discontinuity.

C12.3.2.1 Horizontal Irregularity. A building may have a symmetric geometric shape without reentrant corners or wings but still be classified as irregular in plan because of its distribution of mass or vertical seismic force-resisting elements. Torsional effects in earthquakes can occur even where the centers of mass and rigidity coincide. For example, ground motion waves acting on a skew with respect to the building axis can cause torsion. Cracking or yielding in an asymmetric fashion also can cause torsion. These effects also can magnify the torsion caused by eccentricity between the centers of mass and rigidity. Torsional structural irregularities (Types 1a and 1b) are defined to address this concern.

A square or rectangular building with minor reentrant corners would still be considered regular, but large reentrant corners creating a crucifix form would produce an irregular structural configuration (Type 2). The response of the wings of this type of building generally differs from the response of the building as a whole, and this difference produces higher local forces than would be determined by application of the standard without modification. Other winged plan configurations (e.g., H-shapes) are classified as irregular even if they are symmetric because of the response of the wings.

Significant differences in stiffness between portions of a diaphragm at a level are classified as Type 3 structural irregularities because they may cause a change in the distribution of seismic forces to the vertical components and may create torsional forces not accounted for in the distribution normally considered for a regular building.

Where there are discontinuities in the path of lateral force resistance, the structure cannot be considered regular. The most critical discontinuity defined is the out-of-plane offset of vertical elements of the seismic force-resisting system (Type 4). Such offsets impose vertical and lateral load effects on horizontal elements that are difficult to provide for adequately.

Where vertical lateral force-resisting elements are not parallel to the major orthogonal axes of the seismic force-resisting system, the equivalent lateral force procedure of the standard cannot be applied appropriately, so the structure is considered to have an irregular structural configuration (Type 5).

Fig. C12.3-1 illustrates horizontal structural irregularities.

C12.3.2.2 Vertical Irregularity. Vertical irregularities in structural configuration affect the responses at various levels and induce loads at these levels that differ significantly from the distribution assumed in the equivalent lateral force procedure given in Section 12.8.

A moment-resisting frame building might be classified as having a soft story irregularity (Type 1) if one story is much taller than the adjoining stories and the design did not compensate for the resulting decrease in stiffness that normally would occur.

A building is classified as having a weight (mass) irregularity (Type 2) where the ratio of mass to stiffness in adjacent stories differs significantly. This difference typically occurs where a heavy mass (e.g., an interstitial mechanical floor) is placed at one level.

A vertical geometric irregularity (Type 3) applies regardless of whether the larger dimension is above or below the smaller one.

Vertical lateral force-resisting elements at adjoining stories that are offset from each other in the vertical plane of the elements and impose overturning demands on supporting structural elements, such as beams, columns, trusses, walls, or slabs, are classified as in-plane discontinuity irregularities (Type 4).

Buildings with a weak-story irregularity (Type 5) tend to develop all of their inelastic behavior and consequent damage at the weak story, possibly leading to collapse.

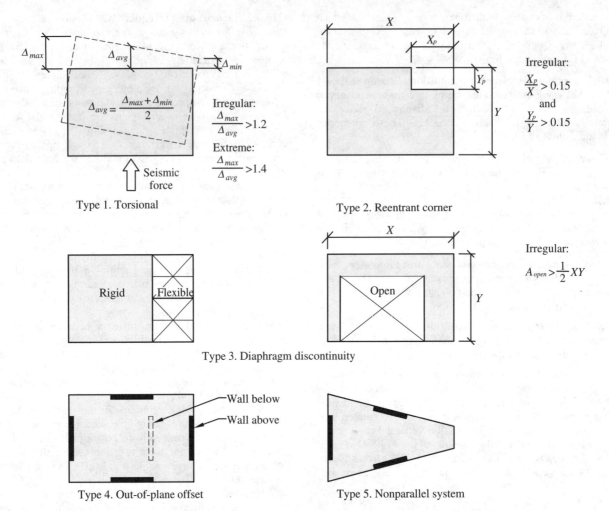

FIGURE C12.3-1 Horizontal Structural Irregularity Examples

Fig. C12.3-2 illustrates examples of vertical structural irregularities.

C12.3.3 Limitations and Additional Requirements for Systems with Structural Irregularities

C12.3.3.1 Prohibited Horizontal and Vertical Irregularities for Seismic Design Categories D through F. The prohibitions and limits caused by structural irregularities in this section stem from poor performance in past earthquakes and the potential to concentrate large inelastic demands in certain portions of the structure. Even where such irregularities are permitted, they should be avoided whenever possible in all structures.

C12.3.3.2 Extreme Weak Stories. Because extreme weak story irregularities are prohibited in Section 12.3.3.1 for buildings located in Seismic Design Categories D, E, and F, the limitations and exceptions in this section apply only to buildings assigned to Seismic Design Category B or C. Weak stories of structures assigned to Seismic Design Category B or C that are designed for seismic load effects including overstrength are exempted because reliable inelastic response is expected.

C12.3.3.3 Elements Supporting Discontinuous Walls or Frames. The purpose of requiring elements (e.g., beams, columns, trusses, slabs, and walls) that support discontinuous walls or frames to be designed to resist seismic load effects including overstrength is to protect the gravity load-carrying system against possible overloads caused by overstrength of the seismic force-resisting system. Either columns or beams may be subject to such failure; therefore, both should include this design requirement. Beams may be subject to failure caused by overloads in either the downward or upward directions of force. Examples include reinforced concrete beams, the weaker top laminations of glued laminated beams, or unbraced flanges of steel beams or trusses. Hence, the provision has not been limited simply to downward force, but instead to the larger context of "vertical load." Additionally, walls that support isolated point loads from frame columns or discontinuous perpendicular walls or walls with significant vertical offsets, as shown in Figs. C12.3-3 and C12.3-4, can be subject to the same type of failure caused by overload.

The connection between the discontinuous element and the supporting member must be adequate to transmit the forces required for the design of the discontinuous element. For example, where the discontinuous element is required to comply with the seismic load effects, including overstrength in Section 12.4.3, as is the case for a steel column in a braced frame or a moment frame, its connection to the supporting member is required to be designed to transmit the same forces. These same seismic load effects are not required for shear walls, and thus, the connection between the shear wall and the supporting member would only need to be designed to transmit the loads associated with the shear wall.

For wood light-frame shear wall construction, the final sentence of Section 12.3.3.3 results in the shear and overturning connections at the base of a discontinued shear wall (i.e., shear

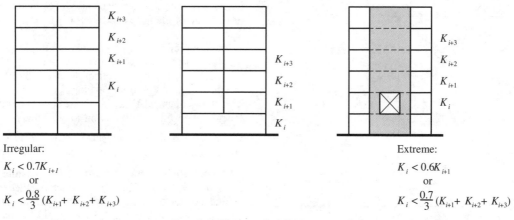

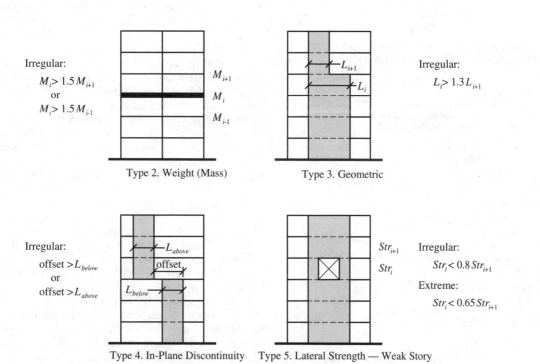

FIGURE C12.3-2 Vertical Structural Irregularities

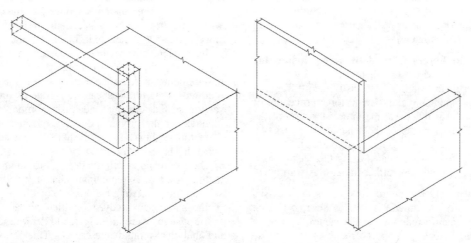

FIGURE C12.3-3 Vertical In-Plane-Discontinuity Irregularity from Columns or Perpendicular Walls (Type 4)

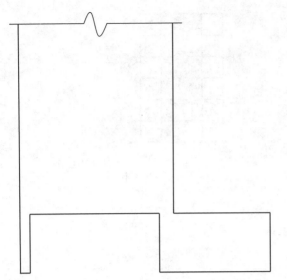

FIGURE C12.3-4 Vertical In-Plane-Discontinuity Irregularity from Walls with Significant Offsets (Type 4)

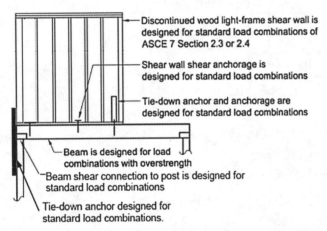

FIGURE C12.3-5 Discontinued Wood Light-Frame Shear Wall

fasteners and tie-downs) being designed using the load combinations of Section 2.3 or 2.4 rather than the load combinations with overstrength of Section 12.4.3 (Fig. C12.3-5). The intent of the first sentence of Section 12.3.3.3 is to protect the system providing resistance to forces transferred from the shear wall by designing the system for seismic load effects including overstrength; strengthening of the shear wall anchorage to this system is not required to meet this intent.

C12.3.3.4 Increase in Forces Caused by Irregularities for Seismic Design Categories D through F. The listed irregularities may result in loads that are distributed differently than those assumed in the equivalent lateral force procedure of Section 12.8, especially as related to the interconnection of the diaphragm with vertical elements of the seismic force-resisting system. The 25% increase in force is intended to account for this difference. Where the force is calculated using the seismic load effects including overstrength, no further increase is warranted.

C12.3.4 Redundancy. The standard introduces a revised redundancy factor, ρ, for structures assigned to Seismic Design Category D, E, or F to quantify redundancy. The value of this factor is either 1.0 or 1.3. This factor has the effect of reducing the response modification coefficient, R, for less redundant structures, thereby increasing the seismic demand. The factor is specified in recognition of the need to address the issue of redundancy in the design.

The desirability of redundancy, or multiple lateral force-resisting load paths, has long been recognized. The redundancy provisions of this section reflect the belief that an excessive loss of story shear strength or development of an extreme torsional irregularity (Type 1b) may lead to structural failure. The value of ρ determined for each direction may differ.

C12.3.4.1 Conditions Where Value of ρ is 1.0. This section provides a convenient list of conditions where ρ is 1.0.

C12.3.4.2 Redundancy Factor, ρ, for Seismic Design Categories D through F. There are two approaches to establishing a redundancy factor, ρ, of 1.0. Where neither condition is satisfied, ρ is taken as equal to 1.3. It is permitted to take ρ equal to 1.3 without checking either condition. A reduction in the value of ρ from 1.3 is not permitted for structures assigned to Seismic Design Category D that have an extreme torsional irregularity (Type 1b). Seismic Design Categories E and F are not also specified because extreme torsional irregularities are prohibited (see Section 12.3.3.1).

The first approach is a check of the elements outlined in Table 12.3-3 for cases where the seismic design story shear exceeds 35% of the base shear. Parametric studies (conducted by Building Seismic Safety Council Technical Subcommittee 2 but unpublished) were used to select the 35% value. Those studies indicated that stories with story shears of at least 35% of the base shear include all stories of low-rise buildings (buildings up to five to six stories) and about 87% of the stories of tall buildings. The intent of this limit is to exclude penthouses of most buildings and the uppermost stories of tall buildings from the redundancy requirements.

This approach requires the removal (or loss of moment resistance) of an individual lateral force-resisting element to determine its effect on the remaining structure. If the removal of elements, one by one, does not result in more than a 33% reduction in story strength or an extreme torsional irregularity, ρ may be taken as 1.0. For this evaluation, the determination of story strength requires an in-depth calculation. The intent of the check is to use a simple measure (elastic or plastic) to determine whether an individual member has a significant effect on the overall system. If the original structure has an extreme torsional irregularity to begin with, the resulting ρ is 1.3. Fig. C12.3-6 presents a flowchart for implementing the redundancy requirements.

As indicated in Table 12.3-3, braced frame, moment frame, shear wall, and cantilever column systems must conform to redundancy requirements. Dual systems also are included but, in most cases, are inherently redundant. Shear walls or wall piers with a height-to-length aspect ratio greater than 1.0 within any story have been included; however, the required design of collector elements and their connections for Ω_0 times the design force may address the key issues. To satisfy the collector force requirements, a reasonable number of shear walls usually is required. Regardless, shear wall systems are addressed in this section so that either an adequate number of wall elements is included or the proper redundancy factor is applied. For wall piers, the height is taken as the height of the adjacent opening and generally is less than the story height.

The second approach is a deemed-to-comply condition wherein the structure is regular and has a specified arrangement of seismic force-resisting elements to qualify for a ρ of 1.0. As part of the parametric study, simplified braced frame and moment frame systems were investigated to determine their sensitivity to the analytical redundancy criteria. This simple deemed-to-comply condition is consistent with the results of the study.

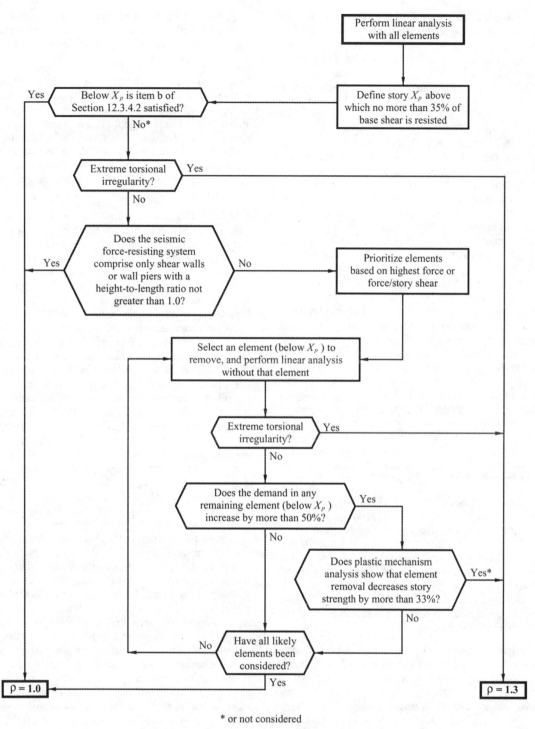

FIGURE C12.3-6 Calculation of the Redundancy Factor, ρ

C12.4 SEISMIC LOAD EFFECTS AND COMBINATIONS

C12.4.1 Applicability. Structural elements designated by the engineer as part of the seismic force-resisting system typically are designed directly for seismic load effects. None of the seismic forces associated with the design base shear are formally assigned to structural elements that are not designated as part of the seismic force-resisting system, but such elements must be designed using the load conditions of Section 12.4 and must accommodate the deformations resulting from application of seismic loads.

C12.4.2 Seismic Load Effect. The seismic load effect includes horizontal and vertical components. The horizontal seismic load effects, E_h, are caused by the response of the structure to horizontal seismic ground motions, whereas the vertical seismic load effects are caused by the response of the structure to vertical seismic ground motions. The basic load combinations in Chapter 2 were duplicated and reformulated in Section 12.4 to clarify the intent of the provisions for the vertical seismic load effect term, E_v.

The concept of using an equivalent static load coefficient applied to the dead load to represent vertical seismic load effects was first introduced in ATC 3-06 (1978), where it was defined as simply $\pm 0.2D$. The load combinations where the vertical seismic load coefficient was to be applied assumed strength design load combinations. Neither ATC 3-06 (1978) nor the early versions of the NEHRP provisions (FEMA 2009a) clearly explained how the values of 0.2 were determined, but it is reasonable to assume that it was based on the judgment of the writers of those documents. It is accepted by the writers of this standard that vertical ground motions do occur and that the value of $\pm 0.2S_{DS}$ was determined based on consensus judgment. Many issues enter into the development of the vertical coefficient, including phasing of vertical ground motion and appropriate R factors, which make determination of a more precise value difficult. Although no specific rationale or logic is provided in editions of the NEHRP provisions (FEMA 2009a) on how the value of $0.2S_{DS}$ was determined, one possible way to rationalize the selection of the $0.2S_{DS}$ value is to recognize that it is equivalent to $(2/3)(0.3)S_{DS}$, where the 2/3 factor represents the often-assumed ratio between the vertical and horizontal components of motion, and the 0.3 factor represents the 30% in the 100% to 30% orthogonal load combination rule used for horizontal motions.

For situations where the vertical component of ground motion is explicitly included in design analysis, the vertical ground motion spectra definition that is provided in Section 11.9 should be used. Following the rationale described above, the alternate vertical ground motion component determined in Section 11.9, S_{av}, is combined with the horizontal component of ground motion by using the 100%–30% orthogonal load combination rule used for horizontal motions resulting in the vertical seismic load effect determined with Eq. (12.4-4b), $E_v = 0.3S_{av}D$.

C12.4.2.1 Horizontal Seismic Load Effect. Horizontal seismic load effects, E_h, are determined in accordance with Eq. (12.4-3) as $E_h = \rho Q_E$. Q_E is the seismic load effect of horizontal seismic forces from V or F_p. The purpose of E_h is to approximate the horizontal seismic load effect from the design basis earthquake to be used in load combinations including E for the design of lateral force-resisting elements including diaphragms, vertical elements of seismic force-resisting systems as defined in Table 12.2-1, the design and anchorage of elements such as structural walls, and the design of nonstructural components.

C12.4.2.2 Vertical Seismic Load Effect. The vertical seismic load effect, E_v, is determined with Eq. (12.4-4a) as $E_v = 0.2S_{DS}D$ or with Eq. (12.4-4b) as $E_v = 0.3S_{av}D$. E_v is permitted to be taken as zero in Eqs. (12.4-1), (12.4-2), (12.4-5), and (12.4-6) for structures assigned to Seismic Design Category B and in Eq. (12.4-2) for determining demands on the soil–structure interface of foundations. E_v increases the load on beams and columns supporting seismic elements and increases the axial load in the P–M interaction of walls resisting seismic load effects.

C12.4.3 Seismic Load Effects Including Overstrength. Some elements of properly detailed structures are not capable of safely resisting ground-shaking demands through inelastic behavior. To ensure safety, these elements must be designed with sufficient strength to remain elastic.

The horizontal load effect including overstrength may be calculated in either of two ways. The load effect may be approximated by use of an overstrength factor, Ω_0, which approximates the inherent overstrength in typical structures based on the structure's seismic force-resisting systems. This approach is addressed in Section 12.4.3.1. Alternatively, the expected system strength may be directly calculated based on actual member sizes and expected material properties, as addressed in Section 12.4.3.2.

C12.4.3.1 Horizontal Seismic Load Effect Including Overstrength. Horizontal seismic load effects including overstrength, E_{mh}, are determined in accordance with Eq. (12.4-7) as $E_{mh} = \Omega_0 Q_E$. Q_E is the effect of horizontal seismic forces from V, F_{px}, or F_p. The purpose for E_{mh} is to approximate the maximum seismic load for the design of critical elements, including discontinuous systems, transfer beams and columns supporting discontinuous systems, and collectors. Forces calculated using this approximate method need not be used if a more rigorous evaluation as permitted in Section 12.4.3.2 is used.

C12.4.3.2 Capacity-Limited Horizontal Seismic Load Effect. The standard permits the horizontal seismic load effect including overstrength to be calculated directly using actual member sizes and expected material properties where it can be determined that yielding of other elements in the structure limits the force that can be delivered to the element in question. When calculated this way, the horizontal seismic load effect including overstrength is termed the capacity-limited seismic load effect, E_{cl}.

As an example, the axial force in a column of a moment-resisting frame results from the shear forces in the beams that connect to this column. The axial forces caused by seismic loads need never be taken as greater than the sum of the shear forces in these beams at the development of a full structural mechanism, considering the probable strength of the materials and strain-hardening effects. For frames controlled by beam hinge-type mechanisms, these shear forces would typically be calculated as $2M_{pr}/L_h$, where M_{pr} is the probable flexural strength of the beam considering expected material properties and strain hardening, and L_h is the distance between plastic hinge locations. Both ACI 318 and AISC 341 require that beams in special moment frames be designed for shear calculated in this manner, and both standards include many other requirements that represent the capacity-limited seismic load effect instead of the use of a factor approximating overstrength. This design approach is sometimes termed "capacity design." In this design method, the capacity (expected strength) of one or more elements is used to generate the demand (required strength) for other elements, because the yielding of the former limits the forces delivered to the latter. In this context, the capacity of the yielding element is its expected or mean anticipated strength, considering potential variation in material yield strength and strain-hardening effects. When calculating the capacity of elements for this purpose, expected member strengths should not be reduced by strength reduction or resistance factors, ϕ.

The capacity-limited design is not restricted to yielding limit states (axial, flexural, or shear); other examples include flexural buckling (axial compression) used in steel special concentrically braced frames, or lateral-torsional buckling in steel ordinary moment frame beams, as confirmed by testing.

C12.4.4 Minimum Upward Force for Horizontal Cantilevers for Seismic Design Categories D through F. In Seismic Design Categories D, E, and F, horizontal cantilevers are designed for an upward force that results from an effective vertical acceleration of 1.2 times gravity. This design requirement is meant to provide some minimum strength in the upward direction and to account for possible dynamic amplification of vertical ground motions resulting from the vertical flexibility of the cantilever. The requirement is not applied to downward forces on cantilevers, for which the typical load combinations are used.

C12.5 DIRECTION OF LOADING

Seismic forces are delivered to a building through ground accelerations that may approach from any direction relative to the orthogonal directions of the building; therefore, seismic effects are expected to develop in both directions simultaneously. The standard requires structures to be designed for the most critical loading effects from seismic forces applied in any direction. The procedures outlined in this section are deemed to satisfy this requirement.

For horizontal structural elements such as beams and slabs, orthogonal effects may be minimal; however, design of vertical elements of the seismic force-resisting system that participate in both orthogonal directions is likely to be governed by these effects.

C12.5.1 Direction of Loading Criteria. For structures with orthogonal seismic force-resisting systems, the most critical load effects can typically be computed using a pair of orthogonal directions that coincide with the principal axes of the structure. Structures with nonparallel or nonorthogonal systems may require a set of orthogonal direction pairs to determine the most critical load effects. If a three-dimensional mathematical model is used, the analyst must be attentive to the orientation of the global axes in the model in relation to the principal axes of the structure.

C12.5.2 Seismic Design Category B. Recognizing that design of structures assigned to Seismic Design Category (SDC) B is often controlled by nonseismic load effects and, therefore, is not sensitive to orthogonal loadings regardless of any horizontal structural irregularities, it is permitted to determine the most critical load effects by considering that the maximum response can occur in any single direction; simultaneous application of response in the orthogonal direction is not required. Typically, the two directions used for analysis coincide with the principal axes of the structure.

C12.5.3 Seismic Design Category C. Design of structures assigned to SDC C often parallels the design of structures assigned to SDC B and, therefore, as a minimum conforms to Section 12.5.2. Although it is not likely that design of the seismic force-resisting systems in regular structures assigned to SDC C would be sensitive to orthogonal loadings, special consideration must be given to structures with nonparallel or nonorthogonal systems (Type 5 horizontal structural irregularity) to avoid overstressing by different directional loadings. In this case, the standard provides two methods to approximate simultaneous orthogonal loadings and requires a three-dimensional mathematical model of the structure for analysis in accordance with Section 12.7.3.

The orthogonal combination procedure in item (a) of Section 12.5.3.1 combines the effects from 100% of the seismic load applied in one direction with 30% of the seismic load applied in the perpendicular direction. This general approximation—the "30% rule"—was introduced by Rosenblueth and Contreras (1977) based on earlier work by A. S. Veletsos and also N. M. Newmark (cited in Rosenblueth and Contreras 1977) as an alternative to performing the more rational, yet computationally demanding, response history analysis, and is applicable to any elastic structure. Combining effects for seismic loads in each direction, and accidental torsion in accordance with Sections 12.8.4.2 and 12.8.4.3, results in the following 16 load combinations:

- $Q_E = \pm Q_{E.X+AT} \pm 0.3 Q_{E.Y}$ where $Q_{E.Y}$ = effect of Y-direction load at the center of mass (Section 12.8.4.2);
- $Q_E = \pm Q_{E.X-AT} \pm 0.3 Q_{E.Y}$ where $Q_{E.X}$ = effect of X-direction load at the center of mass (Section 12.8.4.2);
- $Q_E = \pm Q_{E.Y+AT} \pm 0.3 Q_{E.X}$ where AT = accidental torsion computed in accordance with Sections 12.8.4.2 and 12.8.4.3; and
- $Q_E = \pm Q_{E.Y-AT} \pm 0.3 Q_{E.X}$.

Though the standard permits combining effects from forces applied independently in any pair of orthogonal directions (to approximate the effects of concurrent loading), accidental torsion need not be considered in the direction that produces the lesser effect, per Section 12.8.4.2. This provision is sometimes disregarded when using a mathematical model for three-dimensional analysis that can automatically include accidental torsion, which then results in 32 load combinations.

The maximum effect of seismic forces, Q_E, from orthogonal load combinations is modified by the redundancy factor, ρ, or the overstrength factor, Ω_0, where required, and the effects of vertical seismic forces, E_v, are considered in accordance with Section 12.4 to obtain the seismic load effect, E.

These orthogonal combinations should not be confused with uniaxial modal combination rules, such as the square root of the sum of the squares (SRSS) or the complete quadratic combination (CQC) method. In past standards, an acceptable alternative to the above was to use the SRSS method to combine effects of the two orthogonal directions, where each term computed is assigned the sign that resulted in the most conservative result. This method is no longer in common use. Although both approaches described for considering orthogonal effects are approximations, it is important to note that they were developed with consideration of results for a square building.

Orthogonal effects can alternatively be considered by performing three-dimensional response history analyses (see Chapter 16) with application of orthogonal ground motion pairs applied simultaneously in any two orthogonal directions. If the structure is located within 3 mi (5 km) of an active fault, the ground motion pair should be rotated to the fault-normal and fault-parallel directions of the causative fault.

C12.5.4 Seismic Design Categories D through F. The direction of loading for structures assigned to SDCs D, E, or F conforms to Section 12.5.3 for structures assigned to SDC C. If a Type 5 horizontal structural irregularity exists, then orthogonal effects are similarly included in design. Recognizing the higher seismic risk associated with structures assigned to SDCs D, E, or F, the standard provides additional requirements for vertical members coupled between intersecting seismic force-resisting systems.

C12.6 ANALYSIS PROCEDURE SELECTION

Table 12.6-1 provides the permitted analysis procedures for all Seismic Design Categories. The table is applicable only to buildings without seismic isolation (Chapter 17) or passive energy devices (Chapter 18) for which there are additional requirements in Sections 17.4 and 18.2.4, respectively.

The four basic procedures provided in Table 12.6-1 are the equivalent lateral force (ELF, Section 12.8), the modal response spectrum (MRS analysis, Section 12.9.1), the linear response history (LRH, Section 12.9.2), and the nonlinear response history (NRH, Section 16.1) analysis procedures. Nonlinear static pushover analysis is not provided as an "approved" analysis procedure in the standard.

The ELF procedure is allowed for all buildings assigned to Seismic Design Category B or C and for all buildings assigned

Table C12.6-1 Values of $3.5T_s$ for Various Cities and Site Classes

Location	S_s (g)	S_1 (g)	$3.5T_s$ (s) for Site Class			
			A & B	C	D	E
Denver	0.219	0.057	0.91	1.29	1.37	1.07
Boston	0.275	0.067	0.85	1.21	1.30	1.03
New York City	0.359	0.070	0.68	0.97	1.08	0.93
Las Vegas	0.582	0.179	1.08	1.50	1.68	1.89
St. Louis	0.590	0.169	1.00	1.40	1.60	1.81
San Diego	1.128	0.479	1.31	1.73	1.99	2.91
Memphis	1.341	0.368	0.96	1.38	1.59	2.25
Charleston	1.414	0.348	0.86	1.25	1.47	2.08
Seattle	1.448	0.489	1.18	1.55	1.78	2.63
San Jose	1.500	0.600	1.40	1.82	2.10	2.12
Salt Lake City	1.672	0.665	1.39	1.81	2.09	3.10

to Seismic Design Category D, E, or F, except for the following:

- Structures with structural height, $h_n > 160$ ft ($h_n > 48.8$ m) and $T > 3.5T_s$;
- Structures with structural height, $h_n > 160$ ft ($h_n > 48.8$ m) and $T \leq 3.5T_s$ but with one or more of the structural irregularities in Table 12.3-1 or 12.3-2; and
- Structures with structural height, $h_n < 160$ ft ($h_n < 48.8$ m) and with one or more of the following structural irregularities: torsion or extreme torsion (Table 12.3-1); or soft story, extreme soft story, weight (mass), or vertical geometric (Table 12.3-2).

$T_s = S_{D1}/S_{DS}$ is the period at which the horizontal and descending parts of the design response spectrum intersect (Fig. 11.4-1). The value of T_s depends on the site class because S_{DS} and S_{D1} include such effects. Where the ELF procedure is not allowed, the analysis must be performed using modal response spectrum or response history analysis.

The use of the ELF procedure is limited to buildings with the listed structural irregularities because the procedure is based on an assumption of a gradually varying distribution of mass and stiffness along the height and negligible torsional response. The basis for the $3.5T_s$ limitation is that the higher modes become more dominant in taller buildings (Lopez and Cruz 1996, Chopra 2007a,b), and as a result, the ELF procedure may underestimate the seismic base shear and may not correctly predict the vertical distribution of seismic forces in taller buildings.

As Table C12.6-1 demonstrates, the value of $3.5T_s$ generally increases as ground motion intensity increases and as soils become softer. Assuming a fundamental period of approximately 0.1 times the number of stories, the maximum structural height, h_n, for which the ELF procedure applies ranges from about 10 stories for low seismic hazard sites with firm soil to 30 stories for high seismic hazard sites with soft soil. Because this trend was not intended, the 160-ft (48.8-m) height limit is introduced.

C12.7 MODELING CRITERIA

C12.7.1 Foundation Modeling. Structural systems consist of three interacting subsystems: the structural framing (girders, columns, walls, and diaphragms), the foundation (footings, piles, and caissons), and the supporting soil. The ground motion that a structure experiences, as well as the response to that ground motion, depends on the complex interaction among these subsystems.

Those aspects of ground motion that are affected by site characteristics are assumed to be independent of the structure-foundation system because these effects would occur in the free field in the absence of the structure. Hence, site effects are considered separately (Sections 11.4.3 through 11.4.5 and Chapters 20 and 21).

Given a site-specific ground motion or response spectrum, the dynamic response of the structure depends on the foundation system and on the characteristics of the soil that support the system. The dependence of the response on the structure–foundation–soil system is referred to as soil–structure interaction (SSI). Such interactions usually, but not always, result in a reduction of seismic base shear. This reduction is caused by the flexibility of the foundation–soil system and an associated lengthening of the fundamental period of vibration of the structure. In addition, the soil system may provide an additional source of damping. However, that total displacement typically increases with soil–structure interaction.

If the foundation is considered to be rigid, the computed base shears are usually conservative, and it is for this reason that rigid foundation analysis is permitted. The designer may neglect soil-structure interaction or may consider it explicitly in accordance with Section 12.13.3 or implicitly in accordance with Chapter 19.

As an example, consider a moment-frame building without a basement and with moment-frame columns supported on footings designed to support shear and axial loads (i.e., pinned column bases). If foundation flexibility is not considered, the columns should be restrained horizontally and vertically, but not rotationally. Consider a moment-frame building with a basement. For this building, horizontal restraint may be provided at the level closest to grade, as long as the diaphragm is designed to transfer the shear out of the moment frame. Because the columns extend through the basement, they may also be restrained rotationally and vertically at this level. However, it is often preferable to extend the model through the basement and provide the vertical and rotational restraints at the foundation elements, which is more consistent with the actual building geometry.

C12.7.2 Effective Seismic Weight. During an earthquake, the structure accelerates laterally, and these accelerations of the structural mass produce inertial forces. These inertial forces, accumulated over the height of the structure, produce the seismic base shear.

When a building vibrates during an earthquake, only that portion of the mass or weight that is physically tied to the structure needs to be considered as effective. Hence, live loads (e.g., loose furniture, loose equipment, and human occupants) need not be included. However, certain types of live loads, such as storage loads, may develop inertial forces, particularly where they are densely packed.

Also considered as contributing to effective seismic weight are the following:

1. All permanent equipment (e.g., air conditioners, elevator equipment, and mechanical systems);
2. Partitions to be erected or rearranged as specified in Section 4.3.2 (greater of actual partition weight and 10 lb/ft² (0.5 kN/m²) of floor area);
3. 20% of significant snow load, $p_f > 30$ lb/ft² ($p_f > 1.4$ kN/m²) and
4. The weight of landscaping and similar materials.

The full snow load need not be considered because maximum snow load and maximum earthquake load are unlikely to occur simultaneously and loose snow does not move with the roof.

C12.7.3 Structural Modeling. The development of a mathematical model of a structure is always required because the story drifts and the design forces in the structural members cannot be determined without such a model. In some cases, the mathematical model can be as simple as a free-body diagram as long as the model can appropriately capture the strength and stiffness of the structure.

The most realistic analytical model is three-dimensional, includes all sources of stiffness in the structure and the soil–foundation system as well as P-delta effects, and allows for nonlinear inelastic behavior in all parts of the structure–foundation–soil system. Development of such an analytical model is time-consuming, and such analysis is rarely warranted for typical building designs performed in accordance with the standard. Instead of performing a nonlinear analysis, inelastic effects are accounted for indirectly in the linear analysis methods by means of the response modification coefficient, R, and the deflection amplification factor, C_d.

Using modern software, it often is more difficult to decompose a structure into planar models than it is to develop a full three-dimensional model, so three-dimensional models are now commonplace. Increased computational efficiency also allows efficient modeling of diaphragm flexibility. Three-dimensional models are required where the structure has horizontal torsional (Type 1), out-of-plane offset (Type 4), or nonparallel system (Type 5) irregularities.

Analysis using a three-dimensional model is not required for structures with flexible diaphragms that have horizontal out-of-plane offset irregularities. It is not required because the irregularity imposes seismic load effects in a direction other than the direction under consideration (orthogonal effects) because of eccentricity in the vertical load path caused by horizontal offsets of the vertical lateral force-resisting elements from story to story. This situation is not likely to occur, however, with flexible diaphragms to an extent that warrants such modeling. The eccentricity in the vertical load path causes a redistribution of seismic design forces from the vertical elements in the story above to the vertical elements in the story below in essentially the same direction. The effect on the vertical elements in the orthogonal direction in the story below is minimal. Three-dimensional modeling may still be required for structures with flexible diaphragms caused by other types of horizontal irregularities (e.g., nonparallel system).

In general, the same three-dimensional model may be used for the equivalent lateral force, the modal response spectrum, and the linear response history analysis procedures. Modal response spectrum and linear response history analyses require a realistic modeling of structural mass; the response history method also requires an explicit representation of inherent damping. Five percent of critical damping is automatically included in the modal response spectrum approach. Chapter 16 and the related commentary have additional information on linear and nonlinear response history analysis procedures.

It is well known that deformations in the panel zones of the beam–column joints of steel moment frames are a significant source of flexibility. Two different mechanical models for including such deformations are summarized in Charney and Marshall (2006). These methods apply to both elastic and inelastic systems. For elastic structures, centerline analysis provides reasonable, but not always conservative, estimates of frame flexibility. Fully rigid end zones should not be used because this method always results in an overestimation of lateral stiffness in steel moment-resisting frames. Partially rigid end zones may be justified in certain cases, such as where doubler plates are used to reinforce the panel zone.

Including the effect of composite slabs in the stiffness of beams and girders may be warranted in some circumstances. Where composite behavior is included, due consideration should be paid to the reduction in effective composite stiffness for portions of the slab in tension (Schaffhausen and Wegmuller 1977, Liew et al. 2001).

For reinforced concrete buildings, it is important to address the effects of axial, flexural, and shear cracking in modeling the effective stiffness of the structural elements. Determining appropriate effective stiffness of the structural elements should take into consideration the anticipated demands on the elements, their geometry, and the complexity of the model. Recommendations for computing cracked section properties may be found in Paulay and Priestley (1992) and similar texts.

When dynamic analysis is performed, at least three dynamic degrees of freedom must be present at each level consistent with language in Section 16.2.2. Depending on the analysis software and modal extraction technique used, dynamic degrees of freedom and static degrees of freedom are not identical. It is possible to develop an analytical model that has many static degrees of freedom but only one or two dynamic degrees of freedom. Such a model does not capture response properly.

C12.7.4 Interaction Effects. The interaction requirements are intended to prevent unexpected failures in members of moment-resisting frames. Fig. C12.7-1 illustrates a typical situation where masonry infill is used and this masonry is fitted tightly against reinforced concrete columns. Because the masonry is much stiffer than the columns, hinges in a column form at the top of the column and at the top of the masonry rather than at the top

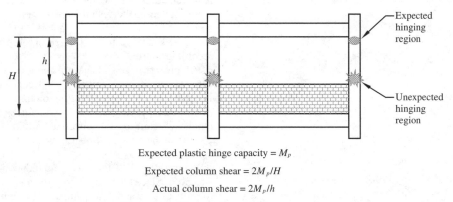

Expected plastic hinge capacity = M_p

Expected column shear = $2M_p/H$

Actual column shear = $2M_p/h$

FIGURE C12.7-1 Undesired Interaction Effects

and bottom of the column. If the column flexural capacity is M_p, the shear in the columns increases by the factor H/h, and this increase may cause an unexpected nonductile shear failure in the columns. Many building collapses have been attributed to this effect.

C12.8 EQUIVALENT LATERAL FORCE PROCEDURE

The equivalent lateral force (ELF) procedure provides a simple way to incorporate the effects of inelastic dynamic response into a linear static analysis. This procedure is useful in preliminary design of all structures and is allowed for final design of the vast majority of structures. The procedure is valid only for structures without significant discontinuities in mass and stiffness along the height, where the dominant response to ground motions is in the horizontal direction without significant torsion.

The ELF procedure has three basic steps:

1. Determine the seismic base shear, V;
2. Distribute V vertically along the height of the structure; and
3. Distribute V horizontally across the width and breadth of the structure.

Each of these steps is based on a number of simplifying assumptions. A broader understanding of these assumptions may be obtained from any structural dynamics textbook that emphasizes seismic applications.

C12.8.1 Seismic Base Shear. Treating the structure as a single-degree-of-freedom system with 100% mass participation in the fundamental mode, Eq. (12.8-1) simply expresses V as the product of the effective seismic weight, W, and the seismic response coefficient, C_s, which is a period-dependent, spectral pseudoacceleration, in g units. C_s is modified by the response modification coefficient, R, and the Importance Factor, I_e, as appropriate, to account for inelastic behavior and to provide for improved performance for high-occupancy or essential structures.

C12.8.1.1 Calculation of Seismic Response Coefficient. The standard prescribes five equations for determining C_s. Eqs. (12.8-2), (12.8-3), and (12.8-4) are illustrated in Fig. C12.8-1.

Eq. (12.8-2) controls where $0.0 < T < T_s$ and represents the constant acceleration part of the design response spectrum (Section 11.4.5). In this region, C_s is independent of period. Although the theoretical design response spectrum shown in Fig. 11.4-1 illustrates a transition in pseudoacceleration to the peak ground acceleration as the fundamental period, T, approaches zero from T_0, this transition is not used in the ELF procedure. One reason is that simple reduction of the response spectrum by $(1/R)$ in the short-period region would exaggerate inelastic effects.

Eq. (12.8-3), representing the constant velocity part of the spectrum, controls where $T_s < T < T_L$. In this region, the seismic response coefficient is inversely proportional to period, and the pseudovelocity (pseudoacceleration divided by circular frequency, ω, assuming steady-state response) is constant. T_L, the long-period transition period, represents the transition to constant displacement and is provided in Figs. 22-12 through 22-16. T_L ranges from 4 s in the north-central conterminous states and western Hawaii to 16 s in the Pacific Northwest and in western Alaska.

Eq. (12.8-4), representing the constant displacement part of the spectrum, controls where $T > T_L$. Given the current mapped values of T_L, this equation only affects long-period structures. The transition period has recently received increased attention because displacement response spectra from the 2010 magnitude 8.8 Chilean earthquake indicate that a considerably lower transition period is possible in locations controlled by subduction zone earthquakes.

The final two equations represent minimum base shear levels for design. Eq. (12.8-5) is the minimum base shear and primarily affects sites in the far field. This equation provides an allowable strength of approximately 3% of the weight of the structure. This minimum base shear was originally enacted in 1933 by the state of California (Riley Act). Based on research conducted in the ATC-63 project (FEMA 2009b), it was determined that this equation provides an adequate level of collapse resistance for long-period structures when used in conjunction with other provisions of the standard.

Eq. (12.8-6) applies to sites near major active faults (as reflected by values of S_1) where pulse-type effects can increase long-period demands.

C12.8.1.2 Soil–Structure Interaction Reduction. Soil–structure interaction, which can significantly influence the dynamic response of a structure during an earthquake, is addressed in Chapter 19.

C12.8.1.3 Maximum S_{DS} Value in Determination of C_s and E_v. This cap on the maximum value of S_{DS} reflects engineering judgment about performance of code-complying, regular, low-rise buildings in past earthquakes. It was created during the update from the 1994 UBC to the 1997 UBC and has been carried through to this standard. At that time, near-source factors were introduced, which increased the design force for buildings in Zone 4, which is similar to Seismic Design Categories D through F in this standard. The near-source factor was based on observations of instrument recording during the 1994 Northridge earthquake and new developments in seismic hazard and ground motion science. The cap placed on S_{DS} for design reflected engineering judgment by the SEAOC Seismology Committee about performance of code-complying low-rise structures based on anecdotal evidence from past California earthquakes, specifically the 1971 San Fernando, 1989 Loma Prieta, and 1994 Northridge earthquakes.

In the 1997 UBC, the maximum reduction of the cap provided was 30%. Since the change from seismic zones in the 1997 UBC to probabilistic and deterministic seismic hazard in ASCE 7-02 (2003) and subsequent editions, S_{DS} values in some parts of the

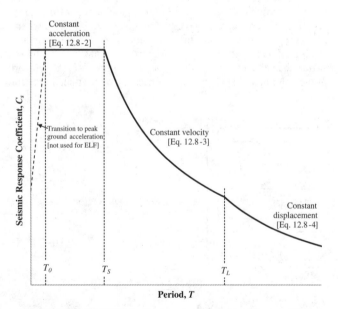

FIGURE C12.8-1 Seismic Response Coefficient Versus Period

country can exceed $S_{DS} = 2.0$, creating reductions well beyond the original permitted reduction. That is the rationale for this provision providing a maximum reduction in design force of 30%.

The structural height, period, redundancy, and regularity conditions required for use of the limit are important qualifiers. Additionally, the observations of acceptable performance have been with respect to collapse and life safety, not damage control or preservation of function, so this cap on the design force is limited to Risk Category I and II structures, not Risk Category III and IV structures, where higher performance is expected. Also, because past earthquake experience has indicated that buildings on very soft soils, Site Classes E and F, have performed noticeably more poorly than buildings on more competent ground, this cap cannot be used on those sites.

C12.8.2 Period Determination. The fundamental period, T, for an elastic structure is used to determine the design base shear, V, as well as the exponent, k, that establishes the distribution of V along the height of the structure (see Section 12.8.3). T may be computed using a mathematical model of the structure that incorporates the requirements of Section 12.7 in a properly substantiated analysis. Generally, this type of analysis is performed using a computer program that incorporates all deformational effects (e.g., flexural, shear, and axial) and accounts for the effect of gravity load on the stiffness of the structure. For many structures, however, the sizes of the primary structural members are not known at the outset of design. For preliminary design, as well as instances where a substantiated analysis is not used, the standard provides formulas to compute an approximate fundamental period, T_a (see Section 12.8.2.1). These periods represent lower-bound estimates of T for different structure types. Period determination is typically computed for a mathematical model that is fixed at the base. That is, the base where seismic effects are imparted into the structure is globally restrained (e.g., horizontally, vertically, and rotationally). Column base modeling (i.e., pinned or fixed) for frame-type seismic force-resisting systems is a function of frame mechanics, detailing, and foundation (soil) rigidity; attention should be given to the adopted assumption. However, this conceptual restraint is not the same for the structure as is stated above. Soil flexibility may be considered for computing T (typically assuming a rigid foundation element). The engineer should be attentive to the equivalent linear soil-spring stiffness used to represent the deformational characteristics of the soil at the base (see Section 12.13.3). Similarly, pinned column bases in frame-type structures are sometimes used to conservatively account for soil flexibility under an assumed rigid foundation element. Period shifting of a fixed-base model of a structure caused by soil–structure interaction is permitted in accordance with Chapter 19.

The fundamental mode of a structure with a geometrically complex arrangement of seismic force-resisting systems determined with a three-dimensional model may be associated with the torsional mode of response of the system, with mass participating in both horizontal directions (orthogonal) concurrently. The analyst must be attentive to this mass participation and recognize that the period used to compute the design base shear should be associated to the mode with the largest mass participation in the direction being considered. Often in this situation, these periods are close to each other. Significant separation between the torsional mode period (when fundamental) and the shortest translational mode period may be an indicator of an ill-conceived structural system or potential modeling error. The standard requires that the fundamental period, T, used to determine the design base shear, V, does not exceed the approximate fundamental period, T_a, times the upper limit coefficient, C_u, provided in Table 12.8-1. This period limit prevents the use of an unusually low base shear for design of a structure that is, analytically, overly flexible because of mass and stiffness inaccuracies in the analytical model. C_u has two effects on T_a. First, recognizing that project-specific design requirements and design assumptions can influence T, C_u lessens the conservatism inherent in the empirical formulas for T_a to more closely follow the mean curve (Fig. C12.8-2). Second, the values for C_u recognize that the formulas for T_a are targeted to structures in high seismic hazard locations. The stiffness of a structure is most likely to decrease in areas of lower seismicity, and this decrease is accounted for in the values of C_u. The response modification coefficient, R, typically decreases to account for reduced ductility demands, and the relative wind effects increase in lower seismic hazard locations. The design engineer must therefore be attentive to the value used for design of seismic force-resisting systems in structures that are controlled by wind effects. Although the value for C_u is most likely to be independent of the governing design forces in high wind areas, project-specific serviceability requirements may add considerable stiffness to a structure and decrease the value of C_u from considering seismic effects alone. This effect should be assessed where design forces for seismic and wind effects are almost equal. Lastly, if T from a properly substantiated analysis (Section 12.8.2) is less than $C_u T_a$, then the lower value of T and $C_u T_a$ should be used for the design of the structure.

C12.8.2.1 Approximate Fundamental Period. Eq. (12.8-7) is an empirical relationship determined through statistical analysis of the measured response of building structures in small- to moderate-sized earthquakes, including response to wind effects (Goel and Chopra 1997, 1998). Fig. C12.8-2 illustrates such data

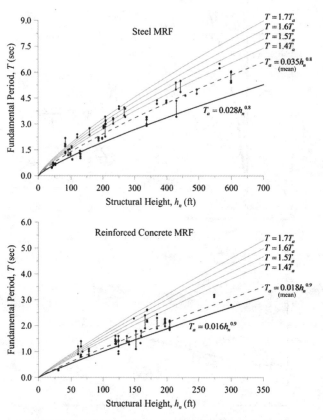

FIGURE C12.8-2 Variation of Fundamental Period with Structural Height

for various building structures with steel and reinforced concrete moment-resisting frames. Historically, the exponent, x, in Eq. (12.8-7) has been taken as 0.75 and was based on the assumption of a linearly varying mode shape while using Rayleigh's method. The exponents provided in the standard, however, are based on actual response data from building structures, thus more accurately reflecting the influence of mode shape on the exponent. Because the empirical expression is based on the lower bound of the data, it produces a lower bound estimate of the period for a building structure of a given height. This lower bound period, when used in Eqs. (12.8-3) and (12.8-4) to compute the seismic response coefficient, C_s, provides a conservative estimate of the seismic base shear, V.

C12.8.3 Vertical Distribution of Seismic Forces. Eq. (12.8-12) is based on the simplified first mode shape shown in Fig. C12.8-3. In the figure, F_x is the inertial force at level x, which is simply the absolute acceleration at level x times the mass at level x. The base shear is the sum of these inertial forces, and Eq. (12.8-11) simply gives the ratio of the lateral seismic force at level x, F_x, to the total design lateral force or shear at the base, V.

The deformed shape of the structure in Fig. C12.8-3 is a function of the exponent k, which is related to the fundamental period of the structure, T. The variation of k with T is illustrated in Fig. C12.8-4. The exponent k is intended to approximate the effect of higher modes, which are generally more dominant in structures with a longer fundamental period of vibration. Lopez and Cruz (1996) discuss the factors that influence higher modes of response. Although the actual first mode shape for a structure is also a function of the type of seismic force-resisting system, that effect is not reflected in these equations. Also, because T is limited to $C_u T_a$ for design, this mode shape may differ from that corresponding to the statistically based empirical formula for the approximate fundamental period, T_a. A drift analysis in accordance with Section 12.8.6 can be conducted using the actual period (see Section C12.8.6). As such, k changes to account for the variation between T and the actual period.

The horizontal forces computed using Eq. (12.8-11) do not reflect the actual inertial forces imparted on a structure at any particular point in time. Instead, they are intended to provide lateral seismic forces at individual levels that are consistent with enveloped results from more accurate analyses (Chopra and Newmark 1980).

C12.8.4 Horizontal Distribution of Forces. Within the context of an ELF analysis, the horizontal distribution of lateral forces in a given story to various seismic force-resisting elements in that story depends on the type, geometric arrangement, and vertical extents of the structural elements and on the shape and flexibility of the floor or roof diaphragm. Because some elements of the seismic force-resisting system are expected to respond inelastically to the design ground motion, the distribution of forces to the various structural elements and other systems also depends on the strength of the yielding elements and their sequence of yielding (see Section C12.1.1). Such effects cannot be captured accurately by a linear elastic static analysis (Paulay 1997), and a nonlinear dynamic analysis is too computationally cumbersome to be applied to the design of most buildings. As such, approximate methods are used to account for uncertainties in horizontal distribution in an elastic static analysis, and to a lesser extent in elastic dynamic analysis.

Of particular concern in regard to the horizontal distribution of lateral forces is the torsional response of the structure during the earthquake. The standard requires that the inherent torsional moment be evaluated for every structure with diaphragms that are not flexible (see Section C12.8.4.1). Although primarily a factor for torsionally irregular structures, this mode of response has also been observed in structures that are designed to be symmetric in plan and layout of seismic force-resisting systems (De La Llera and Chopra 1994). This torsional response in the case of a torsionally regular structure is caused by a variety of "accidental" torsional moments caused by increased eccentricities between the centers of rigidity and mass that exist because of uncertainties in quantifying the mass and stiffness distribution of the structure, as well as torsional components of earthquake ground motion that are not included explicitly in code-based designs (Newmark and Rosenblueth 1971). Consequently, the accidental torsional moment can affect any structure, and potentially more so for a torsionally irregular structure. The standard requires that the accidental torsional moment be considered for every structure (see Section C12.8.4.2) as well as the amplification of this torsion for structures with torsional irregularity (see Section C12.8.4.3).

C12.8.4.1 Inherent Torsion. Where a rigid diaphragm is in the analytical model, the mass tributary to that floor or roof can be idealized as a lumped mass located at the resultant location on the floor or roof—termed the center of mass (CoM). This point represents the resultant of the inertial forces on the floor or roof. This diaphragm model simplifies structural analysis by reducing what would be many degrees of freedom in the two principal directions of a structure to three degrees of freedom (two horizontal and one rotational about the vertical axis). Similarly, the resultant stiffness of the structural members

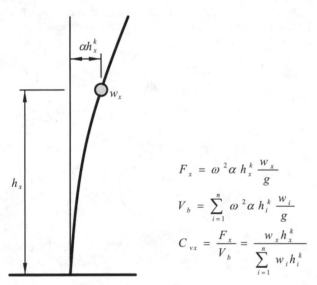

FIGURE C12.8-3 Basis of Eq. (12.8-12)

$$F_x = \omega^2 \alpha h_x^k \frac{w_x}{g}$$

$$V_b = \sum_{i=1}^{n} \omega^2 \alpha h_i^k \frac{w_i}{g}$$

$$C_{vx} = \frac{F_x}{V_b} = \frac{w_x h_x^k}{\sum_{i=1}^{n} w_i h_i^k}$$

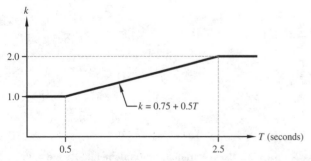

FIGURE C12.8-4 Variation of Exponent k with Period T

providing lateral stiffness to the structure tributary to a given floor or roof can be idealized as the center of rigidity (CoR).

It is difficult to accurately determine the center of rigidity for a multistory building because the center of rigidity for a particular story depends on the configuration of the seismic force-resisting elements above and below that story and may be load dependent (Chopra and Goel 1991). Furthermore, the location of the CoR is more sensitive to inelastic behavior than the CoM. If the CoM of a given floor or roof does not coincide with the CoR of that floor or roof, an inherent torsional moment, M_t, is created by the eccentricity between the resultant seismic force and the CoR. In addition to this *idealized* inherent torsional moment, the standard requires that an accidental torsional moment, M_{ta}, be considered (see Section C12.8.4.2).

Similar principles can be applied to models of semirigid diaphragms that explicitly model the in-plane stiffness of the diaphragm, except that the deformation of the diaphragm needs to be included in computing the distribution of the resultant seismic force and inherent torsional moment to the seismic force-resisting system.

This inherent torsion is included automatically when performing a three-dimensional analysis using either a rigid or semirigid diaphragm. If a two-dimensional planar analysis is used, where permitted, the CoR and CoM for each story must be determined explicitly and the applied seismic forces must be adjusted accordingly.

For structures with flexible diaphragms (as defined in Section 12.3), vertical elements of the seismic force-resisting system are assumed to resist inertial forces from the mass that is tributary to the elements with no explicitly computed torsion. No diaphragm is perfectly flexible; therefore some torsional forces develop even when they are neglected.

C12.8.4.2 Accidental Torsion. The locations of the centers of mass and rigidity for a given floor or roof typically cannot be established with a high degree of accuracy because of mass and stiffness uncertainty and deviations in design, construction, and loading from the idealized case. To account for this inaccuracy, the standard requires the consideration of a minimum eccentricity of 5% of the width of a structure perpendicular to the direction being considered to any static eccentricity computed using idealized locations of the centers of mass and rigidity. Where a structure has a geometrically complex or nonrectangular floor plan, the eccentricity is computed using the diaphragm extents perpendicular to the direction of loading (see Section C12.5).

One approach to account for this variation in eccentricity is to shift the CoM each way from its calculated location and apply the seismic lateral force at each shifted location as separate seismic load cases. It is typically conservative to assume that the CoM offsets at all floors and roof occur simultaneously and in the same direction. This offset produces an "accidental" static torsional moment, M_{ta}, at each story. Most computer programs can automate this offset for three-dimensional analysis by automatically applying these static moments in the autogenerated seismic load case (along the global coordinate axes used in the computer model—see Section C12.5). Alternatively, user-defined torsional moments can be applied as separate load cases and then added to the seismic lateral force load case. For two-dimensional analysis, the accidental torsional moment is distributed to each seismic force-resisting system as an applied static lateral force in proportion to its relative elastic lateral stiffness and distance from the CoR.

Shifting the CoM is a static approximation and thus does not affect the dynamic characteristics of the structure, as would be the case were the CoM to be physically moved by, for example, altering the horizontal mass distribution and mass moment of inertia. Although this "dynamic" approach can be used to adjust the eccentricity, it can be too computationally cumbersome for static analysis and therefore is reserved for dynamic analysis (see Section C12.9.1.5).

The previous discussion is applicable only to a rigid diaphragm model. A similar approach can be used for a semirigid diaphragm model except that the accidental torsional moment is decoupled into nodal moments or forces that are placed throughout the diaphragm. The amount of nodal action depends on how sensitive the diaphragm is to in-plane deformation. As the in-plane stiffness of the diaphragm decreases, tending toward a flexible diaphragm, the nodal inputs decrease proportionally.

The physical significance of this mass eccentricity should not be confused with the physical meaning of the eccentricity required for representing nonuniform wind pressures acting on a structure. However, this accidental torsion also incorporates to a lesser extent the potential torsional motion input into structures with large footprints from differences in ground motion within the footprint of the structure.

Torsionally irregular structures whose fundamental mode is potentially dominated by the torsional mode of response can be more sensitive to dynamic amplification of this accidental torsional moment. Consequently, the 5% minimum can underestimate the accidental torsional moment. In these cases, the standard requires the amplification of this moment for design when using an elastic static analysis procedure, including satisfying the drift limitations (see Section C12.8.4.3).

Accidental torsion results in forces that are combined with those obtained from the application of the seismic design story shears, V_x, including inherent torsional moments. All elements are designed for the maximum effects determined, considering positive accidental torsion, negative accidental torsion, and no accidental torsion (see Section C12.5). Where consideration of earthquake forces applied concurrently in any two orthogonal directions is required by the standard, it is permitted to apply the 5% eccentricity of the center of mass along the single orthogonal direction that produces the greater effect, but it need not be applied simultaneously in the orthogonal direction.

The exception in this section provides relief from accidental torsion requirements for buildings that are deemed to be relatively insensitive to torsion. It is supported by research (Debock et al. 2014) that compared the collapse probability (using nonlinear dynamic response history analysis) of buildings designed with and without accidental torsion requirements. The research indicated that, while accidental torsion requirements are important for most torsionally sensitive buildings (i.e., those with plan torsional irregularities arising from torsional flexibility or irregular plan layout), and especially for buildings in Seismic Design Category D, E or F, the implementation of accidental torsion provisions has little effect on collapse probability for Seismic Design Category B buildings without Type 1b horizontal structural irregularity and for Seismic Design Category D buildings without Type1a or 1b irregularity.

C12.8.4.3 Amplification of Accidental Torsional Moment. For structures with torsional or extreme torsional irregularity (Type 1a or 1b horizontal structural irregularity) analyzed using the equivalent lateral force procedure, the standard requires amplification of the accidental torsional moment to account for increases in the torsional moment caused by potential yielding of the perimeter seismic force-resisting systems (i.e., shifting of the center of rigidity), as well as other factors potentially leading to dynamic torsional instability. For verifying torsional irregularity requirements in Table 12.3-1, story drifts

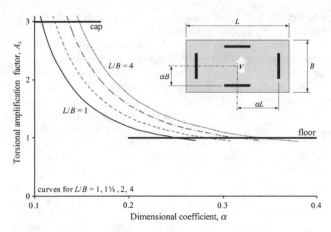

FIGURE C12.8-5 Torsional Amplification Factor for Symmetric Rectangular Buildings

resulting from the applied loads, which include both the inherent and accidental torsional moments, are used with no amplification of the accidental torsional moment ($A_x = 1$). The same process is used when computing the amplification factor, A_x, except that displacements (relative to the base) at the level being evaluated are used in lieu of story drifts. Displacements are used here to indicate that amplification of the accidental torsional moment is primarily a system-level phenomenon, proportional to the increase in acceleration at the extreme edge of the structure, and not explicitly related to an individual story and the components of the seismic force-resisting system contained therein.

Eq. (12.8-14) was developed by the SEAOC Seismology Committee to encourage engineers to design buildings with good torsional stiffness; it was first introduced in the UBC (1988). Fig. C12.8-5 illustrates the effect of Eq. (12.8-14) for a symmetric rectangular building with various aspect ratios (L/B) where the seismic force-resisting elements are positioned at a variable distance (defined by α) from the center of mass in each direction. Each element is assumed to have the same stiffness. The structure is loaded parallel to the short direction with an eccentricity of $0.05L$.

For α equal to 0.5, these elements are at the perimeter of the building, and for α equal to 0.0, they are at the center (providing no torsional resistance). For a square building ($L/B = 1.00$), A_x is greater than 1.0 where α is less than 0.25 and increases to its maximum value of 3.0 where α is equal to 0.11. For a rectangular building with L/B equal to 4.00, A_x is greater than 1.0 where α is less than 0.34 and increases to its maximum value of 3.0 where α is equal to 0.15.

C12.8.5 Overturning. The overturning effect on a vertical lateral force-resisting element is computed based on the calculation of lateral seismic force, F_x, times the height from the base to the level of the horizontal lateral force-resisting element that transfers F_x to the vertical element, summed over each story. Each vertical lateral force-resisting element resists its portion of overturning based on its relative stiffness with respect to all vertical lateral force-resisting elements in a building or structure. The seismic forces used are those from the equivalent lateral force procedure determined in Section 12.8.3 or based on a dynamic analysis of the building or structure. The overturning forces may be resisted by dead loads and can be combined with dead and live loads or other loads, in accordance with the load combinations of Section 2.3.7.

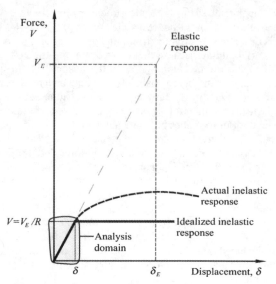

FIGURE C12.8-6 Displacements Used to Compute Drift

C12.8.6 Story Drift Determination. Eq. (12.8-15) is used to estimate inelastic deflections (δ_x), which are then used to calculate design story drifts, Δ. These story drifts must be less than the allowable story drifts, Δ_a, of Table 12.12-1. For structures without torsional irregularity, computations are performed using deflections of the centers of mass of the floors bounding the story. If the eccentricity between the centers of mass of two adjacent floors, or a floor and a roof, is more than 5% of the width of the diaphragm extents, it is permitted to compute the deflection for the bottom of the story at the point on the floor that is vertically aligned with the location of the center of mass of the top floor or roof. This situation can arise where a building has story offsets and the diaphragm extents of the top of the story are smaller than the extents of the bottom of the story. For structures assigned to Seismic Design Category C, D, E, or F that are torsionally irregular, the standard requires that deflections be computed along the edges of the diaphragm extents using two vertically aligned points.

Fig. C12.8-6 illustrates the force-displacement relationships between elastic response, response to reduced design-level forces, and the expected inelastic response. If the structure remained elastic during an earthquake, the force developed would be V_E, and the corresponding displacement would be δ_E. V_E does not include R, which accounts primarily for ductility and system overstrength. According to the equal displacement approximation rule of seismic response, the maximum displacement of an inelastic system is approximately equal to that of an elastic system with the same initial stiffness. This condition has been observed for structures idealized with bilinear inelastic response and a fundamental period, T, greater than T_s (see Section 11.4.6). For shorter period structures, peak displacement of an inelastic system tends to exceed that of the corresponding elastic system. Because the forces are reduced by R, the resulting displacements are representative of an elastic system and need to be amplified to account for inelastic response.

The deflection amplification factor, C_d, in Eq. (12.8-15) amplifies the displacements computed from an elastic analysis using prescribed forces to represent the expected inelastic displacement for the design-level earthquake and is typically less than R (Section C12.1.1). It is important to note that C_d is a story-level amplification factor and does not represent displacement amplification of the elastic response of a structure, either modeled as an *effective* single-degree-of-freedom structure (fundamental mode) or a constant amplification to represent the deflected shape of a multiple-degree-of-freedom structure, in effect, implying that the mode shapes do not change during inelastic response. Furthermore, drift-level forces are different than design-level forces used for strength compliance of the structural elements. Drift forces are typically lower because the computed fundamental period can be used to compute the base shear (see Section C12.8.6.2).

When conducting a drift analysis, the analyst should be attentive to the applied gravity loads used in combination with the strength-level earthquake forces so that consistency between the forces used in the drift analysis and those used for stability verification (P-Δ) in Section 12.8.7 is maintained, including consistency in computing the fundamental period if a second-order analysis is used. Further discussion is provided in Section C12.8.7.

The design forces used to compute the elastic deflection (δ_{xe}) include the Importance Factor, I_e, so Eq. (12.8-15) includes I_e in the denominator. This inclusion is appropriate because the allowable story drifts (except for masonry shear wall structures) in Table 12.12-1 are more stringent for higher Risk Categories.

C12.8.6.1 Minimum Base Shear for Computing Drift.
Except for period limits (as described in Section C12.8.6.2), all of the requirements of Section 12.8 must be satisfied when computing drift for an ELF analysis, except that the minimum base shear determined from applying Eq. (12.8-5) does not need to be considered. This equation represents a minimum strength that needs to be provided to a system (see Section C12.8.1.1). Eq. (12.8-6) needs to be considered, when triggered, because it represents the increase in the response spectrum in the long-period range from near-fault effects.

C12.8.6.2 Period for Computing Drift.
Where the design response spectrum of Section 11.4.6 or the corresponding equations of Section 12.8.1 are used and the fundamental period of the structure, T, is less than the long-period transition period, T_L, displacements increase with increasing period (even though forces may decrease). Section 12.8.2 applies an upper limit on T so that design forces are not underestimated, but if the lateral forces used to compute drifts are inconsistent with the forces corresponding to T, then displacements can be overestimated. To account for this variation in dynamic response, the standard allows the determination of displacements using forces that are consistent with the computed fundamental period of the structure without the upper limit of Section 12.8.2.

The analyst must still be attentive to the period used to compute drift forces. The same analytical representation (see Section C12.7.3) of the structure used for strength design must also be used for computing displacements. Similarly, the same analysis method (Table 12.6-1) used to compute design forces must also be used to compute drift forces. It is generally appropriate to use 85% of the computed fundamental period to account for mass and stiffness inaccuracies as a precaution against overly flexible structures, but it need not be taken as less than that used for strength design. The more flexible the structure, the more likely it is that P-delta effects ultimately control the design (see Section C12.8.7). Computed values of T that are significantly greater than (perhaps more than 1.5 times in high seismic areas) $C_u T_a$ may indicate a modeling error. Similar to the discussion in Section C12.8.2, the analyst should assess the value of C_u used where serviceability constraints from wind effects add significant stiffness to the structure.

C12.8.7 P-Delta Effects.
Fig. C12.8-7 shows an idealized static force-displacement response for a simple one-story structure

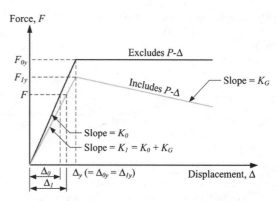

FIGURE C12.8-7 Idealized Response of a One-Story Structure with and without P-Δ

(e.g., idealized as an inverted pendulum-type structure). As the top of the structure displaces laterally, the gravity load, P, supported by the structure acts through that displacement and produces an increase in overturning moment by P times the story drift, Δ, that must be resisted by the structure—the so-called "P-delta (P-Δ) effect." This effect also influences the lateral displacement response of the structure from an applied lateral force, F.

The response of the structure not considering the P-Δ effect is depicted by Condition 0 in the figure with a slope of K_0 and lateral first-order yield force F_{0y}. This condition characterizes the first-order response of the structure (the response of the structure from an analysis not including P-delta effects). Where the P-Δ effect is included (depicted by Condition 1 in the figure), the related quantities are K_1 and F_{1y}. This condition characterizes the second-order response of the structure (the response of the structure from an analysis including P-delta effects).

The geometric stiffness of the structure, K_G, in this example is equal to the gravity load, P, divided by the story height, h_{sx}. K_G is used to represent the change in lateral response by analytically reducing the elastic stiffness, K_0. K_G is negative where gravity loads cause compression in the structure. Because the two response conditions in the figure are for the same structure, the inherent yield displacement of the structure is the same ($\Delta_{0y} = \Delta_{1y} = \Delta_y$).

Two consequential points taken from the figure are (1) the increase in required strength and stiffness of the seismic force-resisting system where the P-Δ effect influences the lateral response of the structure must be accounted for in design, and (2) the P-Δ effect can create a negative stiffness condition during postyield response, which could initiate instability of the structure. Where the postyield stiffness of the structure may become negative, dynamic displacement demands can increase significantly (Gupta and Krawinkler 2000).

One approach that can be used to assess the influence of the P-Δ effect on the lateral response of a structure is to compare the first-order response to the second-order response, which can be done using an elastic stability coefficient, θ, defined as the absolute value of K_G divided by K_0.

$$\theta = \frac{|K_G|}{K_0} = \left| \frac{P\Delta_{0y}}{F_{0y}h_{sx}} \right| \qquad (C12.8\text{-}1)$$

Given the above, and the geometric relationships shown in Fig. C12.8-7, it can be shown that the force producing yield in condition 1 (with P-Δ effects) is

$$F_{1y} = F_{0y}(1 - \theta) \qquad (C12.8\text{-}2)$$

and that for a force, F, less than or equal to F_{1y}

$$\Delta_1 = \frac{\Delta_0}{1-\theta} \qquad \text{(C12.8-3)}$$

Therefore, the stiffness ratio, K_0/K_1, is

$$\frac{K_0}{K_1} = \frac{1}{1-\theta} \qquad \text{(C12.8-4)}$$

In the previous equations,

F_{0y} = the lateral first-order yield force;
F_{1y} = the lateral second-order yield force;
h_{sx} = the story height (or structure height in this example);
K_G = the geometric stiffness;
K_0 = the elastic first-order stiffness;
K_1 = the elastic second-order stiffness;
P = the total gravity load supported by the structure;
Δ_0 = the lateral first-order drift;
Δ_{0y} = the lateral first-order yield drift;
Δ_1 = the lateral second-order drift;
Δ_{1y} = the lateral second-order yield drift; and
θ = the elastic stability coefficient.

A physical interpretation of this effect is that to achieve the second-order response depicted in the figure, the seismic force-resisting system must be designed to have the increased stiffness and strength depicted by the first-order response. As θ approaches unity, Δ_1 approaches infinity and F_1 approaches zero, defining a state of static instability.

The intent of Section 12.8.7 is to determine whether P-Δ effects are significant when considering the first-order response of a structure and, if so, to increase the strength and stiffness of the structure to account for P-Δ effects. Some material-specific design standards require P-Δ effects to always be included in the elastic analysis of a structure and strength design of its members. The amplification of first-order member forces in accordance with Section 12.8.7 should not be misinterpreted to mean that these other requirements can be disregarded; nor should they be applied concurrently. Therefore, Section 12.8.7 is primarily used to verify compliance with the allowable drifts and check against potential postearthquake instability of the structure, while provisions in material-specific design standards are used to increase member forces for design, if provided. In so doing, the analyst should be attentive to the stiffness of each member used in the mathematical model so that synergy between standards is maintained.

Eq. (12.8-16) is used to determine the elastic stability coefficient, θ, of each story of a structure.

$$\theta = \left| \frac{P\Delta_0}{F_0 h_{sx}} \right| = \frac{P\Delta I_e}{V_x h_{sx} C_d} \qquad \text{(C12.8-5)}$$

where

h_{sx}, I_e, and V_x are the same as defined in the standard and
F_0 = the force in a story causing $\Delta_0 = \sum F_x = V_x$;
Δ_0 = the elastic lateral story drift = $\Delta I_e/C_d$;
Δ = the inelastic story drift determined in accordance with Section 12.8.6; and
P = the total *point-in-time* gravity load supported by the structure.

Structures with θ less than 0.10 generally are expected to have a positive monotonic postyield stiffness. Where θ for any story exceeds 0.10, P-Δ effects must be considered for the entire structure using one of the two approaches in the standard. Either first-order displacements and member forces are multiplied by $1/(1-\theta)$ or the P-Δ effect is explicitly included in the structural analysis and the resulting θ is multiplied by $1/(1+\theta)$ to verify compliance with the first-order stability limit. Most commercial computer programs can perform second-order analysis. The analyst must therefore be attentive to the algorithm incorporated in the software and cognizant of any limitations, including suitability of iterative and noniterative methods, inclusion of second-order effects (P-Δ and P-δ) in automated modal analyses, and appropriateness of superposition of design forces.

Gravity load drives the increase in lateral displacements from the equivalent lateral forces. The standard requires the total vertical design load, and the largest vertical design load for combination with earthquake loads is given by combination 6 from Section 2.3.6, which is transformed to

$$(1.2 + 0.2S_{DS})D + 1.0L + 0.2S + 1.0E$$

where the 1.0 factor on L is actually 0.5 for many common occupancies. The provision of Section 12.8.7 allows the factor on dead load D to be reduced to 1.0 for the purpose of P-delta analysis under seismic loads. The vertical seismic component need not be considered for checking $\theta_{\max}$.

As explained in the commentary for Chapter 2, the 0.5 and 0.2 factors on L and S, respectively, are intended to capture the arbitrary point-in-time values of those loads. The factor 1.0 results in the dead load effect being fairly close to best estimates of the arbitrary point-in-time value for dead load. L is defined in Chapter 4 of the standard to include the reduction in live load based on floor area. Many commercially available computer programs do not include live load reduction in the basic structural analysis. In such programs, live reduction is applied only in the checking of design criteria; this difference results in a conservative calculation with regard to the requirement of the standard.

The seismic story shear, V_x (in accordance with Section 12.8.4), used to compute θ includes the Importance Factor, I_e. Furthermore, the design story drift, Δ (in accordance with Section 12.8.6), does not include this factor. Therefore, I_e has been added to Eq. (12.8-16) to correct an apparent omission in previous editions of the standard. Nevertheless, the standard has always required V_x and Δ used in this equation to be those occurring simultaneously.

Eq. (12.8-17) establishes the maximum stability coefficient, $\theta_{\max}$, permitted. The intent of this requirement is to protect structures from the possibility of instability triggered by postearthquake residual deformation. The danger of such failures is real and may not be eliminated by apparently available overstrength. This problem is particularly true of structures designed in regions of lower seismicity.

For the idealized system shown in Fig. C12.8-7, assume that the maximum displacement is $C_d\Delta_0$. Assuming that the unloading stiffness, K_u, is equal to the elastic stiffness, K_0, the residual displacement is

$$\left(C_d - \frac{1}{\beta}\right)\Delta_0 \qquad \text{(C12.8-6)}$$

Additionally, assume that there is a factor of safety, FS, of 2 against instability at the maximum residual drift, $\Delta_{r,\max}$. Evaluating the overturning and resisting moments ($F_0 = V_0$ in this example),

$$P\Delta_{r,\max} \leq \frac{V_0}{\beta FS}h \quad \text{where} \quad \beta = \frac{V_0}{V_{0y}} \leq 1.0 \qquad \text{(C12.8-7)}$$

Therefore,

$$\frac{P[\Delta_0(\beta C_d - 1)]}{V_0 h} \leq 0.5 \to \theta_{max}(\beta C_d - 1)$$
$$= 0.5 \to \theta_{max} = \frac{0.5}{\beta C_d - 1} \quad \text{(C12.8-8)}$$

Conservatively assume that $\beta C_d - 1 \approx \beta C_d$

$$\theta_{max} = \frac{0.5}{\beta C_d} \leq 0.25 \quad \text{(C12.8-9)}$$

In the previous equations,

C_d = the displacement amplification factor;
FS = the factor of safety;
h_{sx} = the story height (or height of the structure in this example);
P = the total point-in-time gravity load supported by the structure;
V_0 = the first-order story shear demand;
V_{0y} = the first-order yield strength of the story;
β = the ratio of shear demand to shear capacity;
Δ_0 = the elastic lateral story drift;
$\Delta_{r,max}$ = the maximum residual drift at $V_0 = 0$; and
θ_{max} = the maximum elastic stability coefficient.

The standard requires that the computed stability coefficient, θ, not exceed 0.25 or $0.5/\beta C_d$, where βC_d is an adjusted ductility demand that takes into account the variation between the story strength demand and the story strength supplied. The story strength demand is simply V_x. The story strength supplied may be computed as the shear in the story that occurs simultaneously with the attainment of the development of first significant yield of the overall structure. To compute first significant yield, the structure should be loaded with a seismic force pattern similar to that used to compute story strength demand and iteratively increased until first yield. Alternatively, a simple and conservative procedure is to compute the ratio of demand to strength for each member of the seismic force-resisting system in a particular story and then use the largest such ratio as β.

The principal reason for inclusion of β is to allow for a more equitable analysis of those structures in which substantial extra strength is provided, whether as a result of added stiffness for drift control, code-required wind resistance, or simply a feature of other aspects of the design. Some structures inherently possess more strength than required, but instability is not typically a concern. For many flexible structures, the proportions of the structural members are controlled by drift requirements rather than strength requirements; consequently, β is less than 1.0 because the members provided are larger and stronger than required. This method has the effect of reducing the inelastic component of total seismic drift, and thus, β is placed as a factor on C_d.

Accurate evaluation of β would require consideration of all pertinent load combinations to find the maximum ratio of demand to capacity caused by seismic load effects in each member. A conservative simplification is to divide the total demand with seismic load effects included by the total capacity; this simplification covers all load combinations in which dead and live load effects add to seismic load effects. If a member is controlled by a load combination where dead load counteracts seismic load effects, to be correctly computed, β must be based only on the seismic component, not the total. The gravity load, P, in the P-Δ computation would be less in such a circumstance and, therefore, θ would be less. The importance of the counteracting load combination does have to be considered, but it rarely controls instability.

Although the P-Δ procedure in the standard reflects a simple static idealization as shown in Fig. C12.8-7, the real issue is one of dynamic stability. To adequately evaluate second-order effects during an earthquake, a nonlinear response history analysis should be performed that reflects variability of ground motions and system properties, including initial stiffness, strain hardening stiffness, initial strength, hysteretic behavior, and magnitude of point-in-time gravity load, P. Unfortunately, the dynamic response of structures is highly sensitive to such parameters, causing considerable dispersion to appear in the results (Vamvatsikos 2002). This dispersion, which increases dramatically with stability coefficient θ, is caused primarily by the incrementally increasing residual deformations (ratcheting) that occur during the response. Residual deformations may be controlled by increasing either the initial strength or the secondary stiffness. Gupta and Krawinkler (2000) give additional information.

C12.9 LINEAR DYNAMIC ANALYSIS

C12.9.1 Modal Response Spectrum Analysis. In the modal response spectrum analysis method, the structure is decomposed into a number of single-degree-of-freedom systems, each having its own mode shape and natural period of vibration. The number of modes available is equal to the number of mass degrees of freedom of the structure, so the number of modes can be reduced by eliminating mass degrees of freedom. For example, rigid diaphragm constraints may be used to reduce the number of mass degrees of freedom to one per story for planar models and to three per story (two translations and rotation about the vertical axis) for three-dimensional structures. However, where the vertical elements of the seismic force-resisting system have significant differences in lateral stiffness, rigid diaphragm models should be used with caution because relatively small in-plane diaphragm deformations can have a significant effect on the distribution of forces.

For a given direction of loading, the displacement in each mode is determined from the corresponding spectral acceleration, modal participation, and mode shape. Because the sign (positive or negative) and the time of occurrence of the maximum acceleration are lost in creating a response spectrum, there is no way to recombine modal responses exactly. However, statistical combination of modal responses produces reasonably accurate estimates of displacements and component forces. The loss of signs for computed quantities leads to problems in interpreting force results where seismic effects are combined with gravity effects, produce forces that are not in equilibrium, and make it impossible to plot deflected shapes of the structure.

C12.9.1.1 Number of Modes. The key motivation to perform modal response spectrum analysis is to determine how the actual distribution of mass and stiffness of a structure affects the elastic displacements and member forces. Where at least 90% of the modal mass participates in the response, the distribution of forces and displacements is sufficient for design. The scaling required by Section 12.9.1.4 controls the overall magnitude of design values so that incomplete mass participation does not produce nonconservative results.

The number of modes required to achieve 90% modal mass participation is usually a small fraction of the total number of modes. Lopez and Cruz (1996) contribute further discussion of the number of modes to use for modal response spectrum analysis.

In general, the provisions require modal analysis to determine all individual modes of vibration, but permit modes with periods less than or equal to 0.05 s to be collectively treated as a single, rigid mode of response with an assumed period of 0.05 s. In general, structural modes of interest to building design have periods greater than 0.05 s (frequencies greater than 20 Hz), and earthquake records tend to have little, if any, energy, at frequencies greater than 20 Hz. Thus, only "rigid" response is expected for modes with frequencies above 20 Hz. Although not responding dynamically, the "residual mass" of modes with frequencies greater than 20 Hz should be included in the analysis to avoid underestimation of earthquake design forces.

Section 4.3 of ASCE 4 (ASCE 2000) provides formulas that may be used to calculate the modal properties of the residual-mass mode. When using the formulas of ASCE 4 to calculate residual-mass mode properties, the "cut-off" frequency should be taken as 20 Hz and the response spectral acceleration at 20 Hz (0.05 s) should be assumed to govern response of the residual-mass mode. It may be noted that the properties of residual-mass mode are derived from the properties of modes with frequencies less than or equal to 20 Hz, such that modal analysis need only determine properties of modes of vibration with periods greater than 0.05 s (when the residual-mass mode is included in the modal analysis). The design response spectral acceleration at 0.05 s (20 Hz) should be determined using Eq. (11.4-5) of this standard where the design response spectrum shown in Fig. 11.4-1 is being used for the design analysis. Substituting 0.05 s for T and $0.2 T_s$ for T_0 in Eq. (11.4-5), one obtains the residual-mode response spectral acceleration as $S_a = S_{DS}[0.4 + 0.15/T_s]$. Most general-purpose linear structural analysis software has the capacity to consider residual mass modes in order to meet the existing requirements ASCE 4 (ASCE 2000).

The exception permits excluding modes of vibration when such would result in a modal mass in each orthogonal direction of at least 90% of the actual mass. This approach has been included in ASCE 7 (2003, 2010) for many years and is still considered adequate for most building structures that typically do not have significant modal mass in the very short period range.

C12.9.1.2 Modal Response Parameters. The design response spectrum (whether the general spectrum from Section 11.4.6 or a site-specific spectrum determined in accordance with Section 21.2) is representative of linear elastic structures. Division of the spectral ordinates by the response modification coefficient, R, accounts for inelastic behavior, and multiplication of spectral ordinates by the Importance Factor, I_e, provides the additional strength needed to improve the performance of important structures. The displacements that are computed using the response spectrum that has been modified by R and I_e (for strength) must be amplified by C_d and reduced by I_e to produce the expected inelastic displacements (see Section C12.8.6.)

C12.9.1.3 Combined Response Parameters. Most computer programs provide for either the SRSS or the CQC method (Wilson et al. 1981) of modal combination. The two methods are identical where applied to planar structures, or where zero damping is specified for the computation of the cross-modal coefficients in the CQC method. The modal damping specified in each mode for the CQC method should be equal to the damping level that was used in the development of the design response spectrum. For the spectrum in Section 11.4.6, the damping ratio is 0.05.

The SRSS or CQC method is applied to loading in one direction at a time. Where Section 12.5 requires explicit consideration of orthogonal loading effects, the results from one direction of loading may be added to 30% of the results from loading in an orthogonal direction. Wilson (2000) suggests that a more accurate approach is to use the SRSS method to combine 100% of the results from each of two orthogonal directions where the individual directional results have been combined by SRSS or CQC, as appropriate.

The CQC4 method, as modified by ASCE 4 (1998), is specified and is an alternative to the required use of the CQC method where there are closely spaced modes with significant cross-correlation of translational and torsional response. The CQC4 method varies slightly from the CQC method through the use of a parameter that forces a correlation in modal responses where they are partially or completely in phase with the input motion. This difference primarily affects structures with short fundamental periods, T, that have significant components of response that are in phase with the ground motion. In these cases, using the CQC method can be nonconservative. A general overview of the various modal response combination methods can be found in U.S Nuclear Regulatory Commission (2012).

The SRSS or CQC method is applied to loading in one direction at a time. Where Section 12.5 requires explicit consideration of orthogonal loading effects, the results from one direction of loading may be added to 30% of the results from loading in an orthogonal direction. Wilson (2000) suggests that a more accurate approach is to use the SRSS method to combine 100% of the results from each of two orthogonal directions where the individual directional results have been combined by SRSS or CQC, as appropriate. Menun and Der Kiureghian (1998) propose an alternate method, referred to as CQC3, which provides the critical orientation of the earthquake relative to the structure. Wilson (2000) now endorses the CQC3 method for combining the results from multiple component analyses.

C12.9.1.4 Scaling Design Values of Combined Response. The modal base shear, V_t, may be less than the ELF base shear, V, because: (a) the calculated fundamental period, T, may be longer than that used in computing V, (b) the response is not characterized by a single mode, or (c) the ELF base shear assumes 100% mass participation in the first mode, which is always an overestimate.

C12.9.1.4.1 Scaling of Forces. The scaling required by Section 12.9.1.4.1 provides, in effect, a minimum base shear for design. This minimum base shear is provided because the computed fundamental period may be the result of an overly flexible (incorrect) analytical model. Recent studies of building collapse performance, such as those of FEMA P-695 (the ATC-63 project, 2009b), NIST GCR 10-917-8 (the ATC-76 project) and NIST GCR 12-917-20 (the ATC-84 project) show that designs based on the ELF procedure generally result in better collapse performance than those based on modal response spectrum analysis (MRSA) with the 15% reduction in base shear included. In addition, many of the designs using scaled MRSA did not achieve the targeted 10% probability of collapse given MCE ground shaking. Whereas scaling to 100% of the ELF base shear and to 100% of the drifts associated with Eq. (12.8-6) does not necessarily achieve the intended collapse performance, it does result in performance that is closer to the stated goals of this standard.

C12.9.1.4.2 Scaling of Drifts. Displacements from the modal response spectrum are only scaled to the ELF base shear where V_t is less than $C_s W$ and C_s is determined based on Eq. (12.8-6). For all other situations, the displacements need not be scaled because the use of an overly flexible model will result in conservative estimates of displacement that need not be further scaled. The reason for requiring scaling when Eq. (12.8-6) controls the minimum base shear is to be consistent with the requirements for designs based on the ELF procedure.

C12.9.1.5 Horizontal Shear Distribution. Torsion effects in accordance with Section 12.8.4 must be included in the modal response spectrum analysis (MRSA) as specified in Section 12.9 by requiring use of the procedures in Section 12.8 for the determination of the seismic base shear, V. There are two basic approaches for consideration of accidental torsion.

The first approach follows the static procedure discussed in Section C12.8.4.2, where the total seismic lateral forces obtained from MRSA—using the computed locations of the centers of mass and rigidity—are statically applied at an artificial point offset from the center of mass to compute the accidental torsional moments. Most computer programs can automate this procedure for three-dimensional analysis. Alternatively, the torsional moments can be statically applied as separate load cases and added to the results obtained from MRSA.

Because this approach is a static approximation, amplification of the accidental torsion in accordance with Section 12.8.4.3 is required. MRSA results in a single, positive response, thus inhibiting direct assessment of torsional response. One method to circumvent this problem is to determine the maximum and average displacements for each mode participating in the direction being considered and then apply modal combination rules (primarily the CQC method) to obtain the total displacements used to check torsional irregularity and compute the amplification factor, A_x. The analyst should be attentive about how accidental torsion is included for individual modal responses.

The second approach, which applies primarily to three-dimensional analysis, is to modify the dynamic characteristics of the structure so that dynamic amplification of the accidental torsion is directly considered. This modification can be done, for example, by either reassigning the lumped mass for each floor and roof (rigid diaphragm) to alternate points offset from the initially calculated center of mass and modifying the mass moment of inertia, or physically relocating the initially calculated center of mass on each floor and roof by modifying the horizontal mass distribution (typically presumed to be uniformly distributed). This approach increases the computational demand significantly because all possible configurations would have to be analyzed, primarily two additional analyses for each principal axis of the structure. The advantage of this approach is that the dynamic effects of direct loading and accidental torsion are assessed automatically. Practical disadvantages are the increased bookkeeping required to track multiple analyses and the cumbersome calculations of the mass properties.

Where this "dynamic" approach is used, amplification of the accidental torsion in accordance with Section 12.8.4.3 is not required because repositioning the center of mass increases the coupling between the torsional and lateral modal responses, directly capturing the amplification of the accidental torsion.

Most computer programs that include accidental torsion in a MRSA do so statically (first approach discussed above) and do not physically shift the center of mass. The designer should be aware of the methodology used for consideration of accidental torsion in the selected computer program.

C12.9.1.6 P-Delta Effects. The requirements of Section 12.8.7, including the stability coefficient limit, θ_{max}, apply to modal response spectrum analysis.

C12.9.1.7 Soil–Structure Interaction Reduction. The standard permits including soil–structure interaction (SSI) effects in a modal response spectrum analysis in accordance with Chapter 19. The increased use of modal analysis for design stems from computer analysis programs automatically performing such an analysis. However, common commercial programs do not give analysts the ability to customize modal response parameters. This problem hinders the ability to include SSI effects in an automated modal analysis.

C12.9.1.8 Structural Modeling. Using modern software, it often is more difficult to decompose a structure into planar models than it is to develop a full three-dimensional model. As a result, three-dimensional models are now commonplace. Increased computational efficiency also allows efficient modeling of diaphragm flexibility. As a result, when modal response spectrum analysis is used, a three-dimensional model is required for all structures, including those with diaphragms that can be designated as flexible.

C12.9.2 Linear Response History Analysis

C12.9.2.1 General Requirements. The linear response history (LRH) analysis method provided in this section is intended as an alternate to the modal response spectrum (MRS) analysis method. The principal motivation for providing the LRH analysis method is that signs (positive–negative bending moments, tension–compression brace forces) are preserved, whereas they are lost in forming the SRSS and CQC combination in MRS analysis.

It is important to note that, like the ELF procedure and the MRS analysis method, the LRH analysis method is used as a basis for structural design, and not to predict how the structure will respond to a given ground motion. Thus, in the method provided in this section, spectrum-matched ground motions are used in lieu of amplitude-scaled motions. The analysis may be performed using modal superposition, or by analysis of the fully coupled equations of motion (often referred to as direct integration response history analysis).

As discussed in Section 12.9.2.3, the LRH analysis method requires the use of three sets of ground motions, with two orthogonal components in each set. These motions are then modified such that the response spectra of the modified motions closely match the shape of the target response spectrum. Thus, the maximum computed response in each mode is virtually identical to the value obtained from the target response spectrum. The only difference between the MRS analysis method and the LRH analysis method (as developed in this section using the spectrum-matched ground motions) is that in the MRS analysis method the system response is computed by statistical combination (SRSS or CQC) of the modal responses and in the LRH analysis method, the system response is obtained by direct addition of modal responses or by simultaneous solution of the full set of equations of motion.

C12.9.2.2 General Modeling Requirements. Three-dimensional (3D) modeling is required for conformance with the inherent and accidental torsion requirements of Section 12.9.2.2.2.

C12.9.2.2.1 P-Delta Effects. A static analysis is required to determine the stability coefficients using Eq. (12.8-17). Typically, the mathematical model used to compute the quantity Δ in Eq. (12.8-16) does not directly include P-delta effects. However, Section 12.8.7 provides a methodology for checking compliance with the θ_{max} limit where P-delta effects are directly included in the model. For dynamic analysis, an ex post facto modification of results from an analysis that does not include P-delta effects to one that does (approximately) include such effects is not rational.

Given that virtually all software that performs linear response history analysis has the capability to directly include P-delta effects, it is required that P-delta effects be included in all analyses, even when the maximum stability ratio at any level is less than 0.1. The inclusion of such effects causes a lengthening of the period of vibration of the structure, and this period should be used for establishing the range of periods for spectrum matching (Section 12.9.2.3.1) and for selecting the number of modes to include in the response (Section 12.9.2.2.4).

While the P-delta effect is essentially a nonlinear phenomenon (stiffness depends on displacements and displacements depend on stiffness), such effects are often "linearized" by forming a constant geometric stiffness matrix that is created from member forces generated from an initial gravity load analysis (Wilson and Habibullah 1987; Wilson 2004). This approach works for both the modal superposition method and the direct analysis method. It is noted, however, that there are some approximations in this method, principally the way the global torsional component of P-delta effects is handled. The method is of sufficient accuracy in analysis for which materials remain elastic. Where direct integration is used, a more accurate response can be computed by iteratively updating the geometric stiffness at each time step or by iteratively satisfying equilibrium about the deformed configuration. In either case, the analysis is in fact "nonlinear," but it is considered as a linear analysis in Section 12.9.2 because material properties remain linear.

For 3D models, it is important to use a realistic spatial distribution of gravity loads because such a distribution is necessary to capture torsional P-delta effects.

C12.9.2.2.2 Accidental Torsion. The required 5% offset of the center of mass need not be applied in both orthogonal directions at the same time. Direct modeling of accidental torsion by offsetting the center of mass is required to retain the signs (positive–negative bending moments, tension–compression forces in braces). In addition to the four mathematical models with mass offsets, a fifth model without accidental torsion (including only inherent torsion) must also be prepared. The model without accidental torsion is needed as the basis for scaling results as required in Section 12.9.2.5. Though not a requirement of the LRH analysis method, the analyst may also compare the modal characteristics (periods, mode shapes) to the systems with and without accidental mass eccentricity to gauge the sensitivity of the structure to accidental torsional response.

C12.9.2.2.3 Foundation Modeling. Foundation flexibility may be included in the analysis. Where such modeling is used, the requirements of Section 12.13.3 should be satisfied. Additional guidance on modeling foundation effects may be found in *Nonlinear Structural Analysis for Seismic Design: A Guide for Practicing Engineers* (NIST 2010).

C12.9.2.2.4 Number of Modes to Include in Modal Response History Analysis. Where modal response history analysis is used, it is common to analyze only a subset of the modes. In the past, the number of modes to analyze has been determined such that a minimum of 90% of the effective mass in each direction is captured in the response. An alternate procedure that produces participation of 100% of the effective mass is to represent all modes with periods less than 0.05 s in a single rigid body mode having a period of 0.05 s. In direct analysis, the question of the number of modes to include does not arise because the system response is computed without modal decomposition.

An example of a situation where it would be difficult to obtain 90% of the mass in a reasonable number of modes is reported in Chapter 4 of FEMA P-751 (2013), which presents the dynamic analysis of a 12-story building over a 1-story basement. When the basement walls and grade-level diaphragm were excluded from the model, 12 modes were sufficient to capture 90% of the effective mass. When the basement was modeled as a stiff first story, it took more than 120 modes to capture 90% of the total mass (including the basement and the ground-level diaphragm). It is noted in the Chapter 4 discussion that when the full structure was modeled and only 12 modes were used, the member forces and system deformations obtained were virtually identical to those obtained when 12 modes were used for the fixed-base system (modeled without the podium).

If modal response history analysis is used and it is desired to use a mathematical model that includes a stiff podium, it might be beneficial to use Ritz vectors in lieu of eigenvectors (Wilson 2004). Another approach is the use of the "static correction method," in which the responses of the higher modes are determined by a static analysis instead of a dynamic analysis (Chopra 2007). The requirement in Section 12.9.2.2.4 of including all modes with periods of less than 0.05 s as a rigid body mode is in fact an implementation of the static correction method.

C12.9.2.2.5 Damping. Where modal superposition analysis is used, 5% damping should be specified for each mode because it is equal to the damping used in the development of the response spectrum specified in Section 11.4.6 and in Section 21.1.3. Where direct analysis is used, it is possible but not common to form a damping matrix that provides uniform damping across all modes (Wilson and Penzien 1972). It is more common to use a mass and stiffness proportional damping matrix (i.e., Rayleigh damping), but when this is done, the damping ratio may be specified at only two periods. Damping ratios at other periods depend on the mass and stiffness proportionality constants. At periods associated with higher modes, the damping ratios may become excessive, effectively damping out important modes of response. To control this effect, Section 12.9.2.2.5 requires the damping in all included modes (with periods as low as T_{lower}) be less than or equal to 5% critical.

C12.9.2.3 Ground Motion Selection and Modification. Response spectrum matching (also called spectral matching) is the nonuniform scaling of an actual or artificial ground motion such that its pseudoacceleration response spectrum closely matches a target spectrum. In most cases, the target spectrum is the same spectrum used for scaling actual recorded ground motions (i.e., the ASCE 7 design spectrum). Spectral matching can be contrasted with amplitude scaling, in which a uniform scale factor is applied to the ground motion. The principal advantage of spectral matching is that fewer ground motions, compared to amplitude scaling, can be used to arrive at an acceptable estimate of the mean response as recommended in NIST GCR 11-918-15 (NIST 2011). Fig. C12.9-1(a) shows the response spectra of two ground motions that have been spectral matched, and Fig. C12.9-1(b) shows the response spectra of the original ground motions. In both cases, the ground motions are normalized to match the target response spectrum at a period of 1.10 s. Clearly, the two amplitude-scaled records will result in significantly different responses, whereas analysis using the spectrum-matched records will be similar. As described later, however, there is enough variation in the response using spectrum-matched records to require the use of more than one record in the response history analysis.

A variety of methods is available for spectrum matching, and the reader is referred to Hancock et al. (2006) for details. Additional information on use of spectrum-matched ground motions in response history analysis is provided by Grant and Diaferia (2012).

C12.9.2.3.1 Procedure for Spectrum Matching. Experience with spectrum matching has indicated that it is easier to get a good match when the matching period extends beyond the period range of interest. It is for this reason that spectrum matching is required over the range $0.8T_{\text{lower}}$ to $1.2T_{\text{upper}}$. For the purposes of this section, a good match is defined when the ordinates of the average (arithmetic mean) of the computed acceleration spectrum from the matched records in each direction does not fall above or below the target spectrum by more than 10% over the period range of interest.

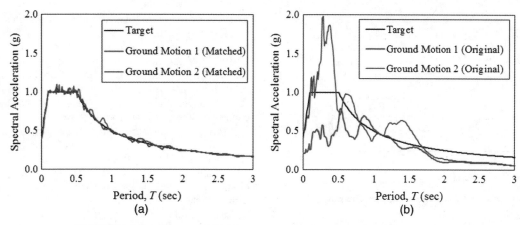

FIGURE C12.9-1. Spectral Matching vs. Amplitude-Scaled Response Spectra

C12.9.2.4 Application of Ground Acceleration Histories. One of the advantages of linear response history analysis is that analyses for gravity loads and for ground shaking may be computed separately and then combined in accordance with Section 12.4.2. Where linear response history analysis is performed in accordance with Section 12.9.2, it is required that each direction of response for each ground motion be computed independently. This requirement is based on the need to apply different scaling factors in the two orthogonal directions. Analyses with and without accidental torsion are required to be run for each ground motion. Thus, the total number of response histories that need to be computed is 18. (For each ground motion, one analysis is needed in each direction without mass eccentricity, and two analyses are needed in each direction to account for accidental torsion. These six cases times three ground motions give 18 required analyses.)

C12.9.2.5 Modification of Response for Design. The dynamic responses computed using spectrum-matched motions are elastic responses and must be modified for inelastic behavior.

For force-based quantities, the design base shear computed from the dynamic analysis must not be less than the base shear computed using the equivalent lateral force procedure. The factors η_X and η_Y, computed in Section 12.9.2.5.2, serve that purpose. Next, the force responses must be multiplied by I_e and divided by R. This modification, together with the application of the ELF scale factors, is accomplished in Section 12.9.2.5.3.

For displacement base quantities, it is not required to normalize to ELF, and computed response history quantities need be multiplied only by the appropriate C_d/R in the direction of interest. This step is accomplished in Section 12.9.2.5.4.

Whereas accidental torsion is not required for determining the maximum elastic base shear, which is used only for determining the required base shear scaling, it is required for all analyses that are used to determine design displacements and member forces.

C12.9.2.6 Enveloping of Force Response Quantities. Forces used in design are the envelope of forces computed from all analyses. Thus, for a brace, the maximum tension and the maximum compression forces are obtained. For a beam-column, envelope values of axial force and envelope values of bending moment are obtained, but these actions do not likely occur at the same time, and using these values in checking member capacity is not rational. The preferred approach is to record the histories of axial forces and bending moments, and to plot their traces together with the interaction diagram of the member. If all points of the force trace fall inside the interaction diagram, for all ground motions analyzed, the design is sufficient.

An alternate is to record member demand to capacity ratio histories (also called usage ratio histories), and to base the design check on the envelope of these values.

C12.10 DIAPHRAGMS, CHORDS, AND COLLECTORS

This section permits choice of diaphragm design in accordance with either Sections 12.10.1 and 12.10.2 provisions or the new provisions of Section 12.10.3. Section 12.10.3 is mandatory for precast concrete diaphragms in buildings assigned to SDC C, D, E, or F and is optional for precast concrete diaphragms in SDC B buildings, cast-in-place concrete diaphragms, and wood diaphragms. The required mandatory use of Section 12.10.3 for precast diaphragm systems in SDC C through F buildings is based on recent research that indicates that improved earthquake performance can thus be attained. Many conventional diaphragm systems designed in accordance with Sections 12.10.1 and 12.10.2 have performed adequately. Continued use of Sections 12.10.1 and 12.10.2 is considered reasonable for diaphragm systems other than those for which Section 12.10.3 is mandated.

C12.10.1 Diaphragm Design. Diaphragms are generally treated as horizontal deep beams or trusses that distribute lateral forces to the vertical elements of the seismic force-resisting system. As deep beams, diaphragms must be designed to resist the resultant shear and bending stresses. Diaphragms are commonly compared to girders, with the roof or floor deck analogous to the girder web in resisting shear, and the boundary elements (chords) analogous to the flanges of the girder in resisting flexural tension and compression. As in girder design, the chord members (flanges) must be sufficiently connected to the body of the diaphragm (web) to prevent separation and to force the diaphragm to work as a single unit.

Diaphragms may be considered flexible, semirigid, or rigid. The flexibility or rigidity of the diaphragm determines how lateral forces are distributed to the vertical elements of the seismic force-resisting system (see Section C12.3.1). Once the distribution of lateral forces is determined, shear and moment diagrams are used to compute the diaphragm shear and chord forces. Where diaphragms are not flexible, inherent and accidental torsion must be considered in accordance with Section 12.8.4.

Diaphragm openings may require additional localized reinforcement (subchords and collectors) to resist the subdiaphragm chord forces above and below the opening and to collect shear forces where the diaphragm depth is reduced (Fig. C12.10-1). Collectors on each side of the opening drag shear into the subdiaphragms above and below the opening. The subchord and collector reinforcement must extend far enough into the adjacent diaphragm to develop the axial force through shear transfer. The

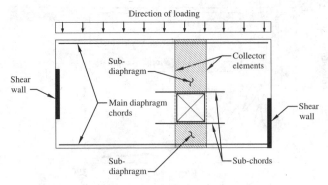

FIGURE C12.10-1 Diaphragm with an Opening

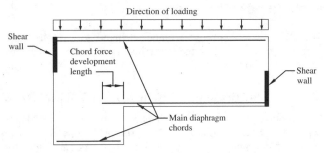

FIGURE C12.10-2 Diaphragm with a Reentrant Corner

required development length is determined by dividing the axial force in the subchord by the shear capacity (in force/unit length) of the main diaphragm.

Chord reinforcement at reentrant corners must extend far enough into the main diaphragm to develop the chord force through shear transfer (Fig. C12.10-2). Continuity of the chord members also must be considered where the depth of the diaphragm is not constant.

In wood and metal deck diaphragm design, framing members are often used as continuity elements, serving as subchords and collector elements at discontinuities. These continuity members also are often used to transfer wall out-of-plane forces to the main diaphragm, where the diaphragm itself does not have the capacity to resist the anchorage force directly. For additional discussion, see Sections C12.11.2.2.3 and C12.11.2.2.4.

C12.10.1.1 Diaphragm Design Forces. Diaphragms must be designed to resist inertial forces, as specified in Eq. (12.10-1), and to transfer design seismic forces caused by horizontal offsets or changes in stiffness of the vertical resisting elements. Inertial forces are those seismic forces that originate at the specified diaphragm level, whereas the transfer forces originate above the specified diaphragm level. The redundancy factor, ρ, used for design of the seismic force-resisting elements also applies to diaphragm transfer forces, thus completing the load path.

C12.10.2.1 Collector Elements Requiring Load Combinations Including Overstrength for Seismic Design Categories C through F. The overstrength requirement of this section is intended to keep inelastic behavior in the ductile elements of the seismic force-resisting system (consistent with the response modification coefficient, R) rather than in collector elements.

C12.10.3 Alternative Design Provisions for Diaphragms, Including Chords and Collectors. The provisions of Section 12.10.3 are being mandated for precast concrete diaphragms in buildings assigned to SDC C, D, E, or F and are being offered as an alternative to those of Sections 12.10.1 and 12.10.2 for other precast concrete diaphragms, cast-in-place concrete diaphragms, and wood-sheathed diaphragms supported by wood framing. Diaphragms designed by Sections 12.10.1 and 12.10.2 have generally performed adequately in past earthquakes. The level of diaphragm design force from Sections 12.10.1 and 12.10.2 may not ensure, however, that diaphragms have sufficient strength and ductility to mobilize the inelastic behavior of vertical elements of the seismic force-resisting system. Analytical and experimental results show that actual diaphragm forces over much of the height of a structure during the design-level earthquake may be significantly greater than those from Sections 12.10.1 and 12.10.2, particularly when diaphragm response is near-elastic. There are material-specific factors that are related to overstrength and deformation capacity that may account for the adequate diaphragm performance in past earthquakes. The provisions of Section 12.10.3 consider both the significantly greater forces observed in near-elastic diaphragms and the anticipated overstrength and deformation capacity of diaphragms, resulting in an improved distribution of diaphragm strength over the height of buildings and among buildings with different types of seismic force-resisting systems.

Based on experimental and analytical data and observations of building performance in past earthquakes, changes are warranted to the procedures of Sections 12.10.1 and 12.10.2 for some types of diaphragms and for some locations within structures. Examples include the large diaphragms in some parking garages.

Section 12.10.3, Item 1, footnote b to Table 12.2-1 permits reduction in the value of Ω_0 for structures with flexible diaphragms. The lowered Ω_0 results in lower diaphragm forces, which is not consistent with experimental and analytical observations. Justification for footnote b is not apparent; therefore, to avoid the inconsistency, the reduction is eliminated when using the Section 12.10.3 design provisions.

Section 12.10.3, item 2: The ASCE 7-10 Section 12.3.3.4 provision requiring a 25% increase in design forces for certain diaphragm elements in buildings with several listed irregularities is eliminated when using the Section 12.10.3 design provisions because the diaphragm design force level in this section is based on realistic assessment of anticipated diaphragm behavior. Under the Sections 12.10.1 and 12.10.2 design provisions, the 25% increase is invariably superseded by the requirement to amplify seismic design forces for certain diaphragm elements by Ω_0; the only exception is wood diaphragms, which are exempt from the Ω_0 multiplier.

Section 12.10.3, items 3 and 4: Section 12.10.3.2 provides realistic seismic design forces for diaphragms. Section 12.10.3.4 requires that diaphragm collectors be designed for 1.5 times the force level used for diaphragm in-plane shear and flexure. Based on these forces, the use of a ρ factor greater than one for collector design is not necessary and would overly penalize designs. The unit value of the redundancy factor is retained for diaphragms designed by the force level given in Sections 12.10.1 and 12.10.2. This value is reflected in the deletion of item 7 and the addition of diaphragms to item 5. For transfer diaphragms, see Section 12.10.3.3.

C12.10.3.1 Design. This provision is a rewrite of ASCE 7-10, Sections 12.10.1 and 12.10.2. The phrase "diaphragms including chords, collectors, and their connections to the vertical elements" is used consistently throughout the added or modified provisions, to emphasize that its provisions apply to all portions of a diaphragm. It is also emphasized that the diaphragm is to be designed for motions in two orthogonal directions.

C12.10.3.2 Seismic Design Forces for Diaphragms, Including Chords and Collectors. Eq. (12.10-4) makes the diaphragm seismic design force equal to the weight tributary to the diaphragm, w_{px}, times a diaphragm design acceleration coefficient, C_{px}, divided by a diaphragm design force reduction factor, R_s, which is material-dependent and whose background is given in Section C12.10.3.5. The background to the diaphragm design acceleration coefficient, C_{px}, is given below.

The diaphragm design acceleration coefficient at any height of the building can be determined from linear interpolation, as indicated in Fig. 12.10-2.

The diaphragm design acceleration coefficient at the building base, C_{p0}, equals the peak ground acceleration consistent with the design response spectrum in ASCE 7-10, Section 11.4.5, times the Importance Factor I_e. Note that the term $0.4S_{DS}$ can be calculated from Eq. (11.4-5) by making $T = 0$.

The diaphragm design acceleration coefficient at 80% of the structural height, C_{pi}, given by Eqs. (12.10-8) and (12.10-9), reflects the observation that at about this height, floor accelerations are largely, but not solely, contributed by the first mode of response. In an attempt to provide a simple design equation, coefficient C_{pi} was formulated as a function of the design base shear coefficient, C_s, of ASCE 7-10, which may be determined from equivalent static analysis or modal response spectrum analysis of the structure. Note that C_s includes a reduction by the response modification factor, R, of the seismic force-resisting system. It is magnified back up by the overstrength factor, Ω_0, of the seismic force-resisting system because overstrength will generate higher first-mode forces in the diaphragm. In many lateral systems, at 80% of the building height, the contribution of the second mode is negligible during linear response, and during nonlinear response it is typically small, though nonnegligible. In recognition of this observation, the diaphragm seismic design coefficient at this height has been made a function of the first mode of response only, and the contribution of this mode has been factored by $0.9\Gamma_{m1}$ as a weighed value between contributions at the first-mode effective height (approximately 2/3 of the building height) and the building height.

Systems that make use of high R-factors, such as buckling-restrained braced frames (BRBFs) and moment-resisting frames (MRFs), show that in the lower floors the higher modes add to the accelerations, whereas the contribution of the first mode is minimal. For this reason, the coefficient C_{pi} needs to have a lower bound. A limit of C_{p0} has been chosen; it makes the lower floor acceleration coefficients independent of R. Wall systems are unlikely to be affected by this lower limit on C_{pi}.

At the structural height, h_n, the diaphragm design acceleration coefficient, C_{pn}, given by Eq. (12.10-7), reflects the influence of the first mode, amplified by system overstrength, and of the higher modes without amplification on the floor acceleration at this height. The individual terms are combined using the square root of the sum of the squares. The overstrength amplification of the first mode recognizes that the occurrence of an inelastic mechanism in the first mode is an anticipated event under the design earthquake, whereas inelastic mechanisms caused by higher mode behavior are not anticipated. The higher mode seismic response coefficient, C_{s2}, is computed as the smallest of the values given by Eqs. (12.10-10), (12.10-11), and (12.10-12a) or (12.10-12b). These four equations consider that the periods of the higher modes contributing to the floor acceleration can lie on the ascending, constant, or first descending branch of the design response spectrum shown in ASCE 7-10, Fig. 11.4-1. Users are warned against extracting higher modes from their modal analysis of buildings and using them in lieu of the procedure presented in Section 12.10.3.2.1 because the higher mode contribution to floor accelerations can come from a number of modes, particularly when there is lateral-torsional coupling of the modes.

Note that Eq. (12.10-7) makes use of the modal contribution factor defined here as the mode shape ordinate at the building

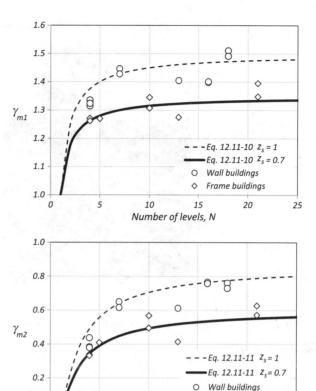

FIGURE C12.10-3 Comparison of Factors Γ_{m1} and Γ_{m2} Obtained from Analytical Models and Actual Structures with Those Predicted by Eqs. (12.10-13) and (12.10-14)

height times the modal participation factor and is uniquely defined for each mode of response (Chopra 1995). A building database was compiled to obtain approximate equations for the first mode and higher mode contribution factors. The first and second translational modes, as understood in the context of two-dimensional modal analysis, were extracted from the mode shapes obtained from three-dimensional modal analysis by considering modal ordinates at the center of mass. These buildings had diverse lateral systems, and the number of stories ranged from 3 to 23. Eqs. (12.10-13) and (12.10-14) were empirically calibrated from simple two-dimensional models of realistic frame-type and wall-type buildings and then compared with data extracted from the database (Fig. C12.10-3). In Eq. (12.10-7), C_{pn} is required to be no less than C_{pi}, based on judgment, in order to eliminate instances where the design acceleration at roof level might be lower than that at $0.8h_n$. This cap will particularly affect low-z_s systems such as BRBFs.

To validate Eq. (12.10-4), coefficients C_{px} were calculated for various buildings tested on a shake table. Figs. C12.10-4 and C12.10-5 plot the floor acceleration envelopes and the floor accelerations predicted from Eq. (12.10-4) with $R_s = 1$ for two buildings built at full scale and tested on a shake table (Panagiotou et al. 2011, Chen et al. 2015), with C_{p0} defined as the diaphragm design acceleration coefficient at the structure base and C_{px} defined as the diaphragm design acceleration coefficient at level x. Measured floor accelerations are reasonably predicted by Eq. (12.10-4). Research work by Choi et al. (2008) concluded that buckling-restrained braced frames are very effective in limiting floor accelerations in buildings arising from higher mode effects.

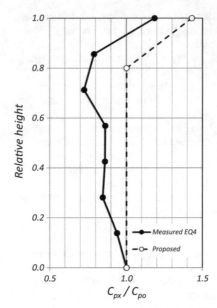

FIGURE C12.10-4 Comparison of Measured Floor Accelerations and Accelerations Predicted by Eq. (12.10-4) for a Seven-Story Bearing Wall Building
Source: Panagiotou et al. 2011.

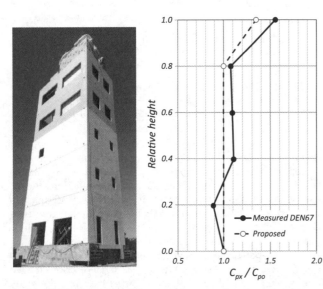

FIGURE C12.10-5 Comparison of Measured Floor Accelerations and Accelerations Predicted by Eq. (12.10-4) for a Five-Story Special Moment-Resisting Frame Building
Source: (left) Courtesy of Michelle Chen; (right) Adapted from Chen et al. (2015).

This finding is reflected in this proposal, where the mode shape factor z_s has been made the smallest for buckling-restrained braced frame systems. Fig. C12.10-6 compares average floor accelerations obtained from the nonlinear time history analyses of four buildings (two steel buckling-restrained braced frame systems and two steel special moment frame systems) when subjected to an ensemble of spectrum-compatible earthquakes with floor accelerations computed from Eqs. (12.10-4) and (12.10-5). The proposed design equations predict the accelerations in the uppermost part of the building and in the lowest levels reasonably well.

The significant difference between a low-z_s system such as the BRBF and a high-z_s system such as a bearing wall system is that inelastic deformations are distributed throughout the height of the structure in a low-z_s system, whereas they are concentrated at the base of the structure in a high-z_s system. If rational analysis can be performed to demonstrate that inelastic deformations are in fact distributed along the height of the structure, as is often the case with eccentrically braced frame or coupled shear wall systems, then the use of a low z_s value, as has been assigned to the BRBF for such a system, would be justified.

During the calibration of the design procedure leading to Eq. (12.10-4), it was found that at intermediate levels in lateral systems designed using large response modification coefficients, diaphragm design forces given by this equation could be rather low. There was consensus within the BSSC PUC Issue Team that developed Section 12.10.3 that diaphragm design forces should not be taken as less than the minimum force currently prescribed by ASCE 7-10, and hence they developed Eq. (12.10-5).

The procedure presented in Section 12.10.3 is based on consideration of buildings and structures whose mass distribution is reasonably uniform along the building height. Buildings or structures with tapered mass distribution along their height or with setbacks in their upper levels may experience diaphragm forces in the upper levels that are greater than those derived from Eq. (12.10-4). In such buildings and structures, it is preferable to define an effective building height, h_{ne}, and a corresponding level, n_e, the level to which the structural effective height is measured. The effective number of levels in a building, n_e, is defined as level x where the ratio $\sum_{i=1}^{x} w_i / \sum_{i=1}^{n} w_i$ first exceeds 0.95. Level 1 is defined as the first level above the base. The effective structural height, h_{ne}, is the height of the building measured from the base to level n_e. In buildings with tapered mass distribution or setbacks, the diaphragm design acceleration coefficient, C_{pn}, is calculated by interpolation and extrapolation, as shown in Fig. C12.10-7, with n replaced by n_e in Eqs. (12.10-10) through (12.10-14).

C12.10.3.3 Transfer Forces in Diaphragms. All diaphragms are subject to inertial forces caused by the weight tributary to the diaphragm. Where the relative lateral stiffnesses of vertical seismic force-resisting elements vary from story to story, or the vertical seismic force-resisting elements have out-of-plane offsets, lateral forces in the vertical elements need to be transferred through the diaphragms as part of the load path between vertical elements

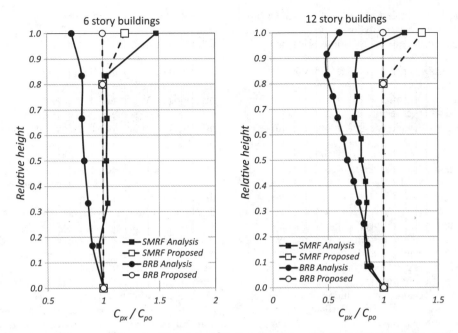

FIGURE C12.10-6 Comparison of Measured Floor Accelerations with Proposed Eqs. (12.10-4) and (12.10-5) for Steel Buckling-Restrained Braced Frame and Special Moment-Resisting Frame Buildings
Source: Adapted from Choi et al. 2008

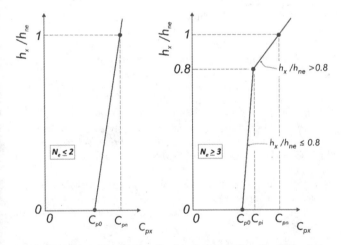

FIGURE C12.10-7 Diaphragm Design Acceleration Coefficient C_{px} for Buildings with Nonuniform Mass Distribution

above and below the diaphragm. These transfer forces are in addition to the inertial forces and can at times be quite large.

For structures that have a horizontal structural irregularity of Type 4 in Table 12.3-1, the magnitude of the transfer forces is largely dependent upon the overstrength in the offset vertical elements of the seismic force-resisting system. Therefore, the transfer force caused by the offset is required to be amplified by the overstrength factor, Ω_0, of the seismic force-resisting system. The amplified transfer force is to be added to the inertial force for the design of this portion of the diaphragm.

Transfer forces can develop in many other diaphragms, even within regular buildings; the design of diaphragms with such transfer forces can be for the sum of the transfer forces, unamplified, and the inertial forces.

C12.10.3.4 Collectors—Seismic Design Categories C through F. For structures in Seismic Design Categories C through F, ASCE 7-10, Section 12.10.2.1 specifies the use of forces including the overstrength factor, Ω_0, for design of diaphragm collectors and their connections to vertical elements of the seismic force-resisting system. The intent of this requirement is to increase collector forces in order to help ensure that collectors will not be the weak links in the seismic force-resisting system.

In this section the collector force is instead differentiated by using a multiplier of 1.5. This is a smaller multiplier than has been used in the past, but it is justified because the diaphragm forces are more accurately determined by Eq. (12.10-4). For collector elements of diaphragms that carry transfer forces caused by out-of-plane offsets of the vertical elements of the seismic force-resisting system, only the inertia force is amplified by 1.5; the transfer forces, already amplified by Ω_0, are not further amplified by 1.5.

Some seismic force-resisting systems, such as braced frames and moment frames, have a fairly well defined lateral strength corresponding to a well-defined yield mechanism. When collectors deliver seismic forces to such systems, it is not sensible to have to design the collectors for forces higher than those corresponding to the lateral strength of the supporting elements in the story below. This is why the cap on collector design forces is included. The lateral strength of a braced frame or moment frame may be calculated using the same methods as are used for determining whether a weak-story irregularity is present (Table 12.3-2). It should be noted that only the moment frames or braced frames below the collector are to be considered in calculating the upper-bound collector design force. The shear strength of the gravity columns and the lateral strength of the frames above are not included.

The design forces in diaphragms that deliver forces to collectors can also be limited by the maximum forces that can be generated in those collectors by the moment frames or braced frames below.

C12.10.3.5 Diaphragm Design Force Reduction Factor. Despite the fact that analytical and shake table studies indicate higher diaphragm accelerations than currently used in diaphragm design, many commonly used diaphragm systems, including diaphragms designed under a number of U.S. building codes and editions, have a history of excellent earthquake performance. With limited exceptions, diaphragms have not been reported to have performed below the life-safety intent of building code seismic design provisions in past earthquakes. Based on this history, it is felt that, for many diaphragm systems, no broad revision is required to the balance between demand and capacity used for design of diaphragms under current ASCE 7 provisions. In view of this observation, it was recognized that the analytical studies and diaphragm testing from which the higher accelerations and design forces were being estimated used diaphragms that were elastic or near-elastic in their response. Commonly used diaphragm systems are recognized to have a wide range of overstrength and inelastic displacement capacity (ductility). It was recognized that the effect of the varying diaphragm systems on seismic demand required evaluation and incorporation into the proposed diaphragm design forces. Eq. (12.10-4) incorporates the diaphragm overstrength and inelastic displacement capacity through the use of the diaphragm force reduction factor, R_s. This factor is most directly based on the global ductility capacity of the diaphragm system; however, the derivation of the global ductility capacity inherently also captures the effect of diaphragm overstrength.

For diaphragm systems with inelastic deformation capacity sufficient to permit inelastic response under the design earthquake, the diaphragm design force reduction factor, R_s, is typically greater than 1.0, so that the design force demand, F_{px}, is reduced relative to the force demand for a diaphragm that remains linear elastic under the design earthquake. For diaphragm systems that do not have sufficient inelastic deformation capacity, R_s should be less than 1.0, or even 0.7, so that linear-elastic force-deformation response can be expected under the risk-targeted maximum considered earthquake (MCE_R).

Diaphragms with R_s values greater than 1.0 shall have the following characteristics: (1) a well-defined, specified yield mechanism, (2) global ductility capacity for the specified yield mechanism, which exceeds anticipated ductility demand for the risk-targeted maximum considered earthquake, and (3) sufficient local ductility capacity to provide for the intended global ductility capacity, considering that the specified yield mechanism may require concentrated local inelastic deformations to occur. The following discussion addresses these characteristics and the development of R_s-factors in detail.

A diaphragm system with an R_s value greater than 1.0 should have a specified, well-defined yield mechanism, for which both the global strength and the global deformation capacity can be estimated. For some diaphragm systems, a shear-yield mechanism may be appropriate, whereas for other diaphragm systems, a flexural-yield mechanism may be appropriate.

Fig. C12.10-8(a) shows schematically the force-deformation ($F_{dia} - \Delta_{dia}$) response of a diaphragm with significant inelastic deformation capacity. The figure illustrates the response of a diaphragm system, such as a wood diaphragm or a steel deck diaphragm, which is not expected to exhibit a distinct yield point, so that an effective yield point ($F_{Y\text{-eff}}$ and $\Delta_{Y\text{-eff}}$) needs to be defined. For wood diaphragms and steel deck diaphragms, the figure illustrates one way to define the effective yield point. The stiffness of a test specimen is defined by the secant stiffness through a point corresponding to 40% of the peak strength (F_{peak}). The effective yield point ($F_{Y\text{-eff}}$ and $\Delta_{Y\text{-eff}}$) for a diaphragm is defined by the secant stiffness through $0.4F_{peak}$ and the nominal diaphragm strength reduced by a strength reduction factor (ϕF_n), as shown in the figure. The $F_{dia} - \Delta_{dia}$ response is then idealized with a bilinear model, using the effective yield point ($F_{Y\text{-eff}}$ and $\Delta_{Y\text{-eff}}$) and F_{peak} and the corresponding deformation Δ_{peak} as shown in the figure.

Fig. C12.10-8(b) shows schematically the force-deformation ($F_{dia} - \Delta_{dia}$) response of a diaphragm with significant inelastic deformation capacity, which is expected to have nearly linear $F_{dia} - \Delta_{dia}$ response up to a distinct yield point, such as a cast-in-place reinforced concrete diaphragm. For this type of diaphragm system, the effective yield point can be taken as the actual yield point ($F_{Y\text{-actual}}$ and $\Delta_{Y\text{-actual}}$) of the diaphragm (accounting for diaphragm material overstrength and not including a strength reduction factor (ϕ)).

The global (or system) deformation capacity of a diaphragm system (Δ_{cap}) should be estimated from analyses of test data. The force-deformation ($F_{dia} - \Delta_{dia}$) response shown schematically in Figs. C12.10-8(a) and C12.10-8(b) is the global force-deformation behavior.

In some cases, tests provide directly the global deformation capacity, but more often, tests provide only the local response, including the strength and deformation capacity, of diaphragm

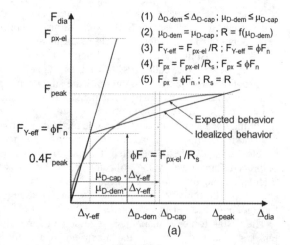
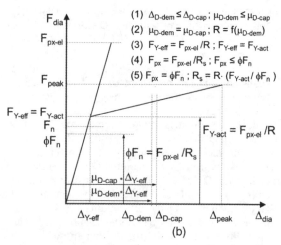

FIGURE C12.10-8. Diaphragm Inelastic Response Models for (a) a Diaphragm System That Is Not Expected to Exhibit a Distinct Yield Point and (b) a Diaphragm System That Does Exhibit a Distinct Yield Point

components and connections. When tests provide only the local deformation capacity, analyses of typical diaphragms should be made to estimate the global deformation capacity of these diaphragms. These analyses should consider: (1) the specified yield mechanism, (2) the local force-deformation response data from tests, (3) the typical distributions of design strength and internal force demands across the diaphragm, and (4) any other factors that may require concentrated local inelastic deformation to occur when the intended yield mechanism forms.

After the global force-deformation ($F_{dia} - \Delta_{dia}$) response of a diaphragm has been estimated, the global deformation capacity (Δ_{cap}) can be determined. In Fig. C12.10-8(a), for example, Δ_{cap} can be taken as Δ_{peak}, which is the deformation corresponding to the strength (F_{peak}). For some diaphragm systems, it may be acceptable to take the deformation corresponding to 80% of F_{peak} (i.e., postpeak) as Δ_{cap}.

Only a selected portion of the deformation capacity of a diaphragm (Δ_{cap}) should be used under the design earthquake in recognition of two major concerns: (1) the diaphragm must perform adequately under the MCE_R, which has a design spectrum 50% more intense than the design earthquake design spectrum, and (2) significant inelastic deformation under the design earthquake may result in undesirable damage to the diaphragm. As a rough estimate, the diaphragm deformation capacity under the design earthquake ($\Delta_{D\text{-cap}}$) should be limited to approximately one-half to two-thirds of the deformation capacity Δ_{cap}.

To develop the diaphragm force reduction factor, R_s, the diaphragm global deformation capacity should be expressed as a global ductility capacity (μ_{cap}), which equals the deformation capacity (Δ_{cap}) divided by the effective yield deformation ($\Delta_{Y\text{-eff}}$). The corresponding diaphragm design ductility capacity ($\mu_{D\text{-cap}}$) equals $\Delta_{D\text{-cap}}/\Delta_{Y\text{-eff}}$.

From the diaphragm global deformation capacity and corresponding ductility capacity, an appropriate R_s factor can be estimated. Use of the estimated R_s factor in design should result in diaphragm ductility demands that do not exceed the ductility capacity that was used to estimate R_s. The force reduction factor is ideally derived from system-specific studies. Where such studies are unavailable, however, some guidance on the conversion from global ductility to force reduction is available from past studies.

Expressions that provide the force reduction factor, R, for the seismic force-resisting system of a building corresponding to an expected ductility demand (μ_{dem}) have been proposed by numerous research teams. Numerous factors, including vibration period, inherent damping, deformation hardening (stiffness after the effective yield point), and hysteretic energy dissipation under cyclic loading, have been considered in developing these expressions. Two such expressions, which are based on elastoplastic force-deformation response under cyclic loading (Newmark and Hall 1982), are as follows: (1) $R = (2\mu_{dem} - 1)^{0.5}$, applicable to short-period systems, and (2) $R = \mu_{dem}$, applicable to systems with longer periods. The first function, known as the equal energy rule, gives a smaller value of R for a given value of μ_{dem}; the second function, known as the equal displacement rule, is also widely used.

Figs. C12.10-8(a) and C12.10-8(b) summarize an approach to estimating R_s as follows:

1. For the selected value of R_s, the diaphragm deformation demand under the design earthquake ($\Delta_{D\text{-dem}}$) should not exceed the diaphragm design deformation capacity ($\Delta_{D\text{-cap}}$). This design constraint, expressed in terms of diaphragm ductility, requires that the diaphragm ductility demand under the design earthquake ($\mu_{D\text{-dem}}$) should not exceed the diaphragm design ductility capacity ($\mu_{D\text{-cap}}$).

2. The largest value of R that can be justified for a given diaphragm design deformation capacity is obtained by setting the ductility demand ($\mu_{D\text{-dem}}$) equal to the design ductility capacity ($\mu_{D\text{-cap}}$) and determining R from a function that provides R for a given μ_{dem}. For example, if $\mu_{D\text{-cap}} = 2.5$, then $\mu_{D\text{-dem}}$ is set equal to 2.5 and the corresponding $R = 2$ from the equal energy rule or $R = 2.5$ from the equal displacement rule.

3. R from step (2) is the ratio of the force demand for a linear elastic diaphragm ($F_{px\text{-el}}$) to the effective yield strength of the diaphragm ($F_{Y\text{-eff}}$). For a diaphragm system that is not expected to exhibit a distinct yield point (Fig. C12.10-8a), $F_{Y\text{-eff}}$ equals the factored nominal diaphragm strength (ϕF_n). For a diaphragm system that is expected to exhibit a distinct yield point (Fig. C12.10-8b), $F_{Y\text{-eff}}$ equals the actual yield strength ($F_{Y\text{-actual}}$), accounting for diaphragm material overstrength and not including the strength reduction factor (ϕ).

4. R_s is, however, defined as the ratio of the force demand for a linear elastic diaphragm ($F_{px\text{-el}}$) to the design force demand (F_{px}). The diaphragm must be designed such that the design force demand (F_{px}) is less than or equal to the factored nominal diaphragm strength (ϕF_n).

5. For a diaphragm system without a distinct yield point (Fig. C12.10-8(a)) that has the minimum strength ($F_{px} = \phi F_n$), R_s equals R from step (2). For a diaphragm system with a distinct yield point (Fig. C12.10-8(b)), which has the minimum strength ($F_{px} = \phi F_n$), R_s equals R from step (2) multiplied by the ratio $F_{Y\text{-eff}}/\phi F_n$.

Diaphragm force reduction factors, R_s, have been developed for some commonly used diaphragm systems. The derivation of factors for each of these systems is explained in detail in the following commentary sections. For each, the specific design standard considered in the development of the R_s factor is specified. The resulting R_s factors are specifically tied to the design and detailing requirements of the noted standard because these play a significant role in setting the ductility and overstrength of the diaphragm system. For this reason, the applicability of the R_s factor to diaphragms designed using other standards must be specifically considered and justified.

Cast-in-Place Concrete Diaphragms. The R_s values in Table 12.10-1 address cast-in-place concrete diaphragms designed in accordance with ACI 318.

Intended Mechanism. Flexural yielding is the intended yield mechanism for a reinforced concrete diaphragm. Where this can be achieved, designation as a flexure-controlled diaphragm and use of the corresponding R_s factor in Table 12.10-1 is appropriate. There are many circumstances, however, where the development of a well-defined yielding mechanism is not possible because of diaphragm geometry (aspect ratio or complex diaphragm configuration), in which case, designation as a shear-controlled diaphragm and use of the lower R_s factor is required.

Derivation of Diaphragm Force Reduction Factor. Test results for reinforced concrete diaphragms are not available in the literature. Test results for shear walls under cyclic lateral loading were considered. The critical regions of shear wall test specimens usually have high levels of shear force, moment, and flexural deformation demands; high levels of shear force are known to degrade the flexural ductility capacity. The flexural ductility capacity of shear wall test specimens under cyclic lateral loading was used to estimate the flexural ductility capacity of reinforced concrete diaphragms, using the previously described method based on Newmark and Hall (1982).

Based on shear wall test results, the estimated global flexural ductility capacity of a reinforced concrete diaphragm is 3, based on the actual yield displacement ($\Delta_{Y\text{-actual}}$) of the test specimens. The design ductility capacity is taken as 2/3 of the ductility capacity; the design ductility capacity ($\mu_{D\text{-cap}}$) is 2.

Setting the ductility demand (μ_{dem}) equal to the design ductility capacity ($\mu_{D\text{-cap}}$) and using the equal energy rule, the force reduction factor R is $R = (2\mu_{dem} - 1)^{0.5} = 1.73$.

R_s equals R multiplied by the ratio $F_{Y\text{-eff}}/\phi F_n$. $F_{Y\text{-eff}}$ is taken equal to $F_{Y\text{-actual}}$, which is assumed to be $1.1 F_n$ and ϕ equals 0.9. Therefore, $R_s = 2.11$, which is rounded to 2.

Because of the geometric characteristics of a building or other factors, such as minimum reinforcement requirements, it is not possible to design some reinforced concrete diaphragms to yield in flexure. Such diaphragms are termed "shear controlled" to indicate that they are expected to yield in shear. Shear-controlled reinforced concrete diaphragms should be designed to remain essentially elastic under the design earthquake, with their available global ductility held in reserve for safety under the MCE_R.

Based on the following considerations, R_s is specified as 1.5 for shear-controlled reinforced concrete diaphragms: Reinforced-concrete diaphragms have performed well in past earthquakes. ACI-318 specifies ϕ of 0.75 or 0.6 for diaphragm shear strength and limits the concrete contribution to the shear strength to only $2(f'_c)^{0.5}$. In addition, reinforced concrete floor slabs often have gravity load reinforcement that is not considered in determining the diaphragm shear strength. Therefore, shear-controlled reinforced concrete diaphragms are expected to have significant overstrength. The ratio $F_{Y\text{-eff}}/\phi F_n$ for a reinforced concrete diaphragm, where $F_{Y\text{-eff}}$ is taken equal to $F_{Y\text{-actual}}$, is expected to exceed 1.5, which is the rationale for $R_s = 1.5$, even though μ_{dem} is assumed to be 1 for the design earthquake.

Precast Concrete Diaphragms. The R_s values in Table 12.10-1 address precast concrete diaphragms designed in accordance with ACI 318.

Derivation of Diaphragm Force Reduction Factors. The diaphragm force reduction factors, R_s, in Table 12.10-1 for precast concrete diaphragms were established based on the results of analytical earthquake simulation studies conducted within a multiple-university project: Diaphragm Seismic Design Methodology (DSDM) for Precast Concrete Diaphragms (Fleischman et al. 2013). In this research effort, diaphragm design force levels have been aligned with the diaphragm deformation capacities specifically for precast concrete diaphragms. Three different design options were proposed according to different design performance targets, as indicated in Table C12.10-1. The relationships between diaphragm design force levels and diaphragm local/global ductility demands have been established in the DSDM research project. These relationships have been used to derive the R_s for precast concrete diaphragms in Table 12.10-1.

Diaphragm R_{dia}-μ_{global}-μ_{local} Relationships. Extensive analytical studies have been performed (Fleischman et al. 2013) to develop the relationship of R_{dia}-μ_{global}-μ_{local}. R_{dia} is the diaphragm force reduction factor (similar to the R_s in Table 12.10-1) measured from the required elastic diaphragm design force at MCE_R level. μ_{global} is the diaphragm global ductility demand, and μ_{local} is the diaphragm local connector ductility demand measured at MCE level. Fig. C12.10-9 shows the μ_{global}-μ_{local} and R_{dia}-μ_{global} analytical results for different diaphragm aspect ratios (AR) and proposed linear equations derived from the data.

Diaphragm Force Reduction Factor (R_s). Using the equations in Fig. C12.10-9, the R_s can be calculated for different diaphragm design options provided that the diaphragm local reinforcement ductility capacity is known. In the DSDM research, precast diaphragm connectors have been extensively tested (Fleischman et al. 2013) and have been qualified into three categories: high deformability elements (HDEs), moderate deformability elements (MDEs), and low deformability elements (LDEs), which are required as a minimum for designs using the reduced design objective (RDO), the basic design objective (BDO), and the elastic design objective (EDO), respectively. The local deformation and ductility capacities for diaphragm connector categories are shown in Table C12.10-2. Considering that the proposed diaphragm design force level [Eq. (12.10-1)] targets elastic diaphragm response at the design earthquake, which is equivalent to design using BDO where $\mu_{local} = 3.5$ at MCE_R (see Table C12.10-2), the available diaphragm global

Table C12.10-1. Diaphragm Design Performance Targets

Options	Flexure		Shear
	DE	MCE_R	DE and MCE_R
EDO	Elastic	Elastic	Elastic
BDO	Elastic	Inelastic	Elastic
RDO	Inelastic	Inelastic	Elastic

Note: DE, design earthquake, MCE_R, risk-targeted maximum considered earthquake, EDO, elastic design objective; BDO, basic design objective; and RDO, reduced design objective.

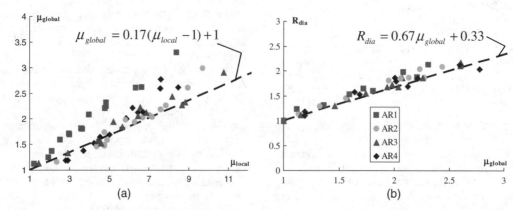

FIGURE C12.10-9. Relationships: (a) μ_{global}-μ_{local} and (b) R_{dia}-μ_{global}

Table C12.10-2 Diaphragm Force Reduction Factors, R_s

Options	Diaphragm Connector Category	δ_{local} (in.)	μ_{local}	μ_{global}	$\mu_{global,red}$	R_s
EDO	LDE	0.06	1.0	1.0	0.58	0.7
BDO	MDE	0.2	3.5	1.4	1.0	1.0
RDO	HDE	0.4	7.0	2.0	1.6	1.4

Note: EDO, elastic design objective; LDE, low deformability elements; BDO, basic design objective; MDE, moderate deformability elements; RDO reduced design objective, and HDE, high deformability elements.

ductility capacity has to be reduced from Fig. C12.10-9(a), acknowledging more severe demands in the MCE_R,

$$\mu_{global,red} = 0.17(\mu_{local} - 3.5) + 1 \qquad (C12.10\text{-}1)$$

Accordingly, the R_s-factor can be modified from Fig. C12.10-9(b) (see Table C12.10-2):

$$R_s = 0.67\mu_{global,red} + 0.33 \qquad (C12.10\text{-}2)$$

Diaphragm Shear Overstrength Factor. Precast diaphragms typically exhibit ductile flexural response but brittle shear response. In order to avoid brittle shear failure, elastic shear response targets are required for both flexure-controlled and shear-controlled systems at design earthquake and MCE_R levels. Thus, a shear overstrength factor, Ω_v, is required for diaphragm shear design. For EDO design, since the diaphragm is expected to remain elastic under the MCE_R, no shear overstrength is needed. Fig. C12.10-10 shows the analytical results for required shear overstrength factors for BDO and RDO (shown as marks). A simplified conservative equation is proposed as (see black lines in Fig. C12.10-10):

$$\Omega_v = 1.4 R_s \qquad (C12.10\text{-}3)$$

Wood-Sheathed Diaphragms. The R_s values given in Table 12.10-1 are for wood-sheathed diaphragms designed in accordance with *Special Design Provisions for Wind and Seismic* (AWC 2008).

Intended Mechanism. Wood-sheathed diaphragms are shear-controlled, with design strength determined in accordance with AWC (2008) and the shear behavior based on the sheathing-to-framing connections. Wood diaphragm chord members are unlikely to form flexural mechanisms (ductile or otherwise) because of the overstrength inherent in axially loaded members designed in accordance with applicable standards.

Derivation of Diaphragm Design Force Reduction Factor. An R_s factor of 3 is assigned in Table 12.10-1, based on diaphragm test data (APA 1966, 2000, DFPA 1954, 1963) and analytical studies. The available testing includes diaphragm spans (loaded as simple-span beams) ranging from 24 to 48 ft (7.3 to 14.6 m), aspect ratios ranging between 1 and 3.3, and diaphragm construction covering a range of construction types including blocked and unblocked construction, and regular and high-load diaphragms. The loading was applied with a series of point loads at varying spacing; however, the loading was reasonably close to uniform. Whereas available diaphragm testing was monotonic, based on shear wall loading protocol studies (Gatto and Uang 2002), it is believed that the monotonic load-deflection behavior is reasonably representative of the cyclic load-deflection envelope, suggesting that it is appropriate to use monotonic load-deflection behavior in the estimation of overstrength, ductility, and displacement capacity.

Analytical studies using nonlinear response history analysis evaluated the relationship between global ductility and diaphragm force reduction factor for a model wood building. The analysis identified the resulting diaphragm force reduction factor as ranging from just below 3 to significantly in excess of 5. A force reduction factor of 3 was selected so that diaphragm design force levels would generally not be less than determined in accordance with provisions of Section 12.10.

The calibration approach for selection of R_s of 3 was considered appropriate to limit conditions where diaphragm force levels would drop below those determined in accordance with Section 12.10. This was due in part to historical experience of good diaphragm performance across a range of wood diaphragm types, even though test data showed varying levels of ductility and deformation capacity. Tests of nailed wood diaphragms showed significant but varying levels of overstrength. It is recognized that even further variation of overstrength will result from

- Presence of floor coverings or toppings and their attachment or bond to diaphragm sheathing,
- Presence of wall to floor framing nailing through diaphragm sheathing, and
- Presence of adhesives in combination with required sheathing nailing (commonly used for purposes of mitigating floor vibration, increasing floor stiffness for gravity loading, and reducing the potential for squeaking).

These sources of overstrength are not considered to be detrimental to overall diaphragm performance.

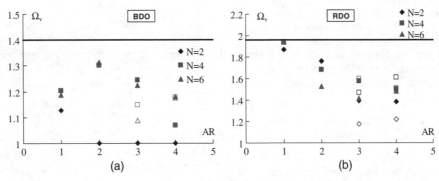

FIGURE C12.10-10 Diaphragm Shear Overstrength Factor, Ω_v vs. Aspect Ratio, AR, for Different Numbers of Stories, N: (a) BDO; (b) RDO
Source: Fleischman et al. 2013.

C12.11 STRUCTURAL WALLS AND THEIR ANCHORAGE

As discussed in Section C1.4, structural integrity is important not only in earthquake-resistant design but also in resisting high winds, floods, explosion, progressive failure, and even such ordinary hazards as foundation settlement. The detailed requirements of this section address wall-to-diaphragm integrity.

C12.11.1 Design for Out-of-Plane Forces. Because they are often subjected to local deformations caused by material shrinkage, temperature changes, and foundation movements, wall connections require some degree of ductility to accommodate slight movements while providing the required strength.

Although nonstructural walls are not subject to this requirement, they must be designed in accordance with Chapter 13.

C12.11.2 Anchorage of Structural Walls and Transfer of Design Forces into Diaphragms or Other Supporting Structural Elements. There are numerous instances in U.S. earthquakes of tall, single-story, and heavy walls becoming detached from supporting roofs, resulting in collapse of walls and supported bays of roof framing (Hamburger and McCormick 2004). The response involves dynamic amplification of ground motion by response of the vertical system and further dynamic amplification from flexible diaphragms. The design forces for Seismic Design Category D and higher have been developed over the years in response to studies of specific failures. It is generally accepted that the rigid diaphragm value is reasonable for structures subjected to high ground motions. For a simple idealization of the dynamic response, these values imply that the combined effects of inelastic action in the main framing system supporting the wall, the wall (acting out of plane), and the anchor itself correspond to a reduction factor of 4.5 from elastic response to an MCE_R motion, and therefore the value of the response modification coefficient, R, associated with nonlinear action in the wall or the anchor itself is 3.0. Such reduction is generally not achievable in the anchorage itself; thus, it must come from yielding elsewhere in the structure, for example, the vertical elements of the seismic force-resisting system, the diaphragm, or walls acting out of plane. The minimum forces are based on the concept that less yielding occurs with smaller ground motions and less yielding is achievable for systems with smaller values of R, which are permitted in structures assigned to Seismic Design Categories B and C. The minimum value of R in structures assigned to Seismic Design Category D, except cantilever column systems and light-frame walls sheathed with materials other than wood structural panels, is 3.25, whereas the minimum values of R for Categories B and C are 1.5 and 2.0, respectively.

Where the roof framing is not perpendicular to anchored walls, provision needs to be made to transfer both the tension and sliding components of the anchorage force into the roof diaphragm. Where a wall cantilevers above its highest attachment to, or near, a higher level of the structure, the reduction factor based on the height within the structure, $(1 + 2z/h)/3$, may result in a lower anchorage force than appropriate. In such an instance, using a value of 1.0 for the reduction factor may be more appropriate.

C12.11.2.1 Wall Anchorage Forces. Diaphragm flexibility can amplify out-of-plane accelerations so that the wall anchorage forces in this condition are twice those defined in Section 12.11.1.

C12.11.2.2 Additional Requirements for Anchorage of Concrete or Masonry Structural Walls to Diaphragms in Structures Assigned to Seismic Design Categories C through F

C12.11.2.2.1 Transfer of Anchorage Forces into Diaphragm. This requirement, which aims to prevent the diaphragm from tearing apart during strong shaking by requiring transfer of anchorage forces across the complete depth of the diaphragm, was prompted by failures of connections between tilt-up concrete walls and wood panelized roof systems in the 1971 San Fernando earthquake.

Depending on diaphragm shape and member spacing, numerous suitable combinations of subdiaphragms, continuous tie elements, and smaller sub-subdiaphragms connecting to larger subdiaphragms and continuous ties are possible. The configurations of each subdiaphragm (or sub-subdiaphragm) provided must comply with the simple 2.5-to-1 length-to-width ratio, and the continuous tie must have adequate member and connection strength to carry the accumulated wall anchorage forces. The 2.5-to-1 aspect ratio is applicable to subdiaphragms of all materials, but only when they serve as part of the continuous tie system.

C12.11.2.2.2 Steel Elements of Structural Wall Anchorage System. A multiplier of 1.4 has been specified for strength design of steel elements to obtain a fracture strength of almost 2 times the specified design force (where ϕ_t is 0.75 for tensile rupture).

C12.11.2.2.3 Wood Diaphragms. Material standards for wood structural panel diaphragms permit the sheathing to resist shear forces only; use of diaphragm sheathing to resist direct tension or compression forces is not permitted. Therefore, seismic out-of-plane anchorage forces from structural walls must be transferred into framing members (such as beams, purlins, or subpurlins) using suitable straps or anchors. For wood diaphragms, it is common to use local framing and sheathing elements as subdiaphragms to transfer the anchorage forces into more concentrated lines of drag or continuity framing that carry the forces across the diaphragm and hold the building together. Fig. C12.11-1 shows a schematic plan of typical roof framing using subdiaphragms.

Fasteners that attach wood ledgers to structural walls are intended to resist shear forces from diaphragm sheathing attached to the ledger that act longitudinally along the length of the ledger but not shear forces that act transversely to the ledger, which tend to induce splitting in the ledger caused by cross-grain bending. Separate straps or anchors generally are provided to transfer out-of-plane wall forces into perpendicular framing members.

Requirements of Section 12.11.2.2.3 are consistent with requirements of AWC SDPWS, SDPWS-15 (2014) Section 4.1.5.1 but also apply to wood use in diaphragms that may fall outside

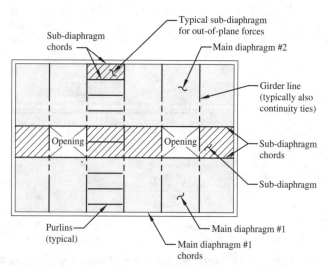

FIGURE C12.11-1 Typical Subdiaphragm Framing

the scope of AWC SDPWS. Examples include use of wood structural panels attached to steel bar joists or metal deck attached to wood nailers.

C12.11.2.2.4 Metal Deck Diaphragms. In addition to transferring shear forces, metal deck diaphragms often can resist direct axial forces in at least one direction. However, corrugated metal decks cannot transfer axial forces in the direction perpendicular to the corrugations and are prone to buckling if the unbraced length of the deck as a compression element is large. To manage diaphragm forces perpendicular to the deck corrugations, it is common for metal decks to be supported at 8- to 10-ft (2.4- to 3.0-m) intervals by joists that are connected to walls in a manner suitable to resist the full wall anchorage design force and to carry that force across the diaphragm. In the direction parallel to the deck corrugations, subdiaphragm systems are considered near the walls; if the compression forces in the deck become large relative to the joist spacing, small compression reinforcing elements are provided to transfer the forces into the subdiaphragms.

C12.11.2.2.5 Embedded Straps. Steel straps may be used in systems where heavy structural walls are connected to wood or steel diaphragms as the wall-to-diaphragm connection system. In systems where steel straps are embedded in concrete or masonry walls, the straps are required to be bent around reinforcing bars in the walls, which improve their ductile performance in resisting earthquake load effects (e.g., the straps pull the bars out of the wall before the straps fail by pulling out without pulling the reinforcing bars out). Consideration should be given to the probability that light steel straps have been used in past earthquakes and have developed cracks or fractures at the wall-to-diaphragm framing interface because of gaps in the framing adjacent to the walls.

C12.11.2.2.6 Eccentrically Loaded Anchorage System. Wall anchors often are loaded eccentrically, either because the anchorage mechanism allows eccentricity or because of anchor bolt or strap misalignment. This eccentricity reduces the anchorage connection capacity and hence must be considered explicitly in design of the anchorage. Fig. C12.11-2 shows a one-sided roof-to-wall anchor that is subjected to severe eccentricity because of a misplaced anchor rod. If the detail were designed as a concentric two-sided connection, this condition would be easier to correct.

C12.11.2.2.7 Walls with Pilasters. The anchorage force at pilasters must be calculated considering two-way bending in wall panels. It is customary to anchor the walls to the diaphragms

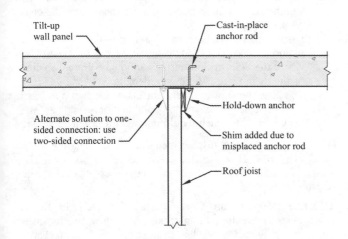

FIGURE C12.11-2 Plan View of Wall Anchor with Misplaced Anchor Rod

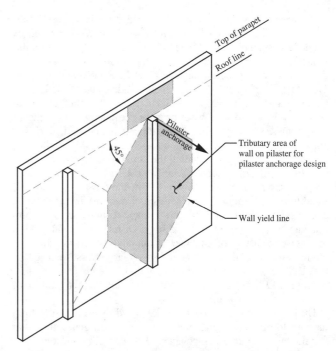

FIGURE C12.11-3 Tributary Area Used to Determine Anchorage Force at Pilaster

assuming one-way bending and simple supports at the top and bottom of the wall. However, where pilasters are present in the walls, their stiffening effect must be taken into account. The panels between pilasters are typically supported along all panel edges. Where this support occurs, the reaction at the top of the pilaster is the result of two-way action of the panel and is applied directly to the anchorage supporting the top of the pilaster. The anchor load at the pilaster generally is larger than the typical uniformly distributed anchor load between pilasters. Fig. C12.11-3 shows the tributary area typically used to determine the anchorage force for a pilaster.

Anchor points adjacent to the pilaster must be designed for the full tributary loading, conservatively ignoring the effect of the adjacent pilaster.

C12.12 DRIFT AND DEFORMATION

As used in the standard, deflection is the absolute lateral displacement of any point in a structure relative to its base, and design story drift, Δ, is the difference in deflection along the height of a story (i.e., the deflection of a floor relative to that of the floor below). The drift, Δ, is calculated according to the procedures of Section 12.8.6. (Sections 12.9.2 and 16.1 give procedures for calculating displacements for modal response spectrum and linear response history analysis procedures, respectively; the definition of Δ in Section 11.3 should be used).

Calculated story drifts generally include torsional contributions to deflection (i.e., additional deflection at locations of the center of rigidity at other than the center of mass caused by diaphragm rotation in the horizontal plane). The provisions allow these contributions to be neglected where they are not significant, such as in cases where the calculated drifts are much less than the allowable story drifts, Δ_a, no torsional irregularities exist, and more precise calculations are not required for structural separations (see Sections C12.12.3 and C12.12.4).

The deflections and design story drifts are calculated using the design earthquake ground motion, which is two-thirds of the risk-targeted maximum considered earthquake (MCE_R) ground motion. The resulting drifts are therefore likely to be underestimated.

The design base shear, V, used to calculate Δ is reduced by the response modification coefficient, R. Multiplying displacements by the deflection amplification factor, C_d, is intended to correct for this reduction and approximate inelastic drifts corresponding to the design response spectrum unreduced by R. However, it is recognized that use of values of C_d less than R underestimates deflections (Uang and Maarouf 1994). Also Sections C12.8.6.2 and C12.9.1.4 deal with the appropriate base shear for computing displacements.

For these reasons, the displacements calculated may not correspond well to MCE_R ground motions. However, they are appropriate for use in evaluating the structure's compliance with the story drift limits put forth in Table 12.12-1 of the standard.

There are many reasons to limit drift; the most significant are to address the structural performance of member inelastic strain and system stability and to limit damage to nonstructural components, which can be life-threatening. Drifts provide a direct but imprecise measure of member strain and structural stability. Under small lateral deformations, secondary stresses caused by the P-delta effect are normally within tolerable limits (see Section C12.8.7). The drift limits provide indirect control of structural performance.

Buildings subjected to earthquakes need drift control to limit damage to partitions, shaft and stair enclosures, glass, and other fragile nonstructural components. The drift limits have been established without regard to economic considerations such as a comparison of present worth of future repairs with additional structural costs to limit drift. These are matters for building owners and designers to address.

The allowable story drifts, Δ_a, of Table 12.12-1 reflect the consensus opinion of the ASCE 7 Committee taking into account the life-safety and damage control objectives described in the aforementioned commentary. Because the displacements induced in a structure include inelastic effects, structural damage as the result of a design-level earthquake is likely. This notion may be seen from the values of Δ_a stated in Table 12.12-1. For other structures assigned to Risk Category I or II, the value of Δ_a is $0.02h_{sx}$, which is about 10 times the drift ordinarily allowed under wind loads. If deformations well in excess of Δ_a were to occur repeatedly, structural elements of the seismic force-resisting system could lose so much stiffness or strength that they would compromise the safety and stability of the structure.

To provide better performance for structures assigned to Risk Category III or IV, their allowable story drifts, Δ_a, generally are more stringent than for those assigned to Risk Category I or II. However, those limits are still greater than the damage thresholds for most nonstructural components. Therefore, though the performance of structures assigned to Risk Category III or IV should be improved, there may be considerable damage from a design-level earthquake.

The allowable story drifts, Δ_a, for structures a maximum of four stories above the base are relaxed somewhat, provided that the interior walls, partitions, ceilings, and exterior wall systems have been designed to accommodate story drifts. The type of structure envisioned by footnote d in Table 12.12-1 would be similar to a prefabricated steel structure with metal skin.

The values of Δ_a set forth in Table 12.12-1 apply to each story. For some structures, satisfying strength requirements may produce a system with adequate drift control. However, the design of moment-resisting frames and of tall, narrow shear walls or braced frames often is governed by drift considerations. Where design spectral response accelerations are large, seismic drift considerations are expected to control the design of midrise buildings.

C12.12.3 Structural Separation. This section addresses the potential for impact from adjacent structures during an earthquake. Such conditions may arise because of construction on or near a property line or because of the introduction of separations within a structure (typically called "seismic joints") for the purpose of permitting their independent response to earthquake ground motion. Such joints may effectively eliminate irregularities and large force transfers between portions of the building with different dynamic properties.

The standard requires the distance to be "sufficient to avoid damaging contact under total deflection." It is recommended that the distance be computed using the square root of the sum of the squares of the lateral deflections. Such a combination method treats the deformations as linearly independent variables. The deflections used are the expected displacements (e.g., the anticipated maximum inelastic deflections including the effects of torsion and diaphragm deformation). Just as these displacements increase with height, so does the required separation. If the effects of impact can be shown not to be detrimental, the required separation distances can be reduced.

For rigid shear wall structures with rigid diaphragms whose lateral deflections cannot be reasonably estimated, the NEHRP provisions (FEMA 2009a) suggest that older code requirements for structural separations of at least 1 in. (25 mm) plus 1/2 in. (13 mm) for each 10 ft (3 m) of height above 20 ft (6 m) be followed.

C12.12.4 Members Spanning between Structures. Where a portion of the structure is seismically separated from its support, the design of the support requires attention to ensure that support is maintained as the two portions move independently during earthquake ground motions. To prevent loss of gravity support for members that bridge between the two portions, the relative displacement must not be underestimated. Displacements computed for verifying compliance with drift limits [Eq. (12.8-15)] and structural separations [Eq. (12.12-1)] may be insufficient for this purpose.

The provision gives four requirements to ensure that displacement is not underestimated:

1. The deflections calculated using Eq. (12.8-15) are multiplied by $1.5R/C_d$ to correct for likely underestimation of displacement by the equation. The factor of 1.5 corrects for the 2/3 factor that is used in the calculation of seismic base shear, V, by reducing the base shear from the value based on the MCE_R ground motion (Section 11.4.4). Multiplying by R/C_d corrects for the fact that values of C_d less than R underestimate deflections (Uang and Maarouf 1994).
2. The deflections are calculated for torsional effects, including amplification factors. Diaphragm rotation can add significantly to the center-of-mass displacements calculated using Eq. (12.8-15).
3. Displacements caused by diaphragm deformations are required to be calculated, as in some types of construction where the deformation during earthquake ground motions of the diaphragm can be considerable.
4. The absolute sum of displacements of the two portions is used instead of a modal combination, such as with Eq. (12.12-2), which would represent a probable value.

It is recognized that displacements so calculated are likely to be conservative. However, the consequences of loss of gravity support are likely to be severe, and some conservatism is deemed appropriate.

C12.12.5 Deformation Compatibility for Seismic Design Categories D through F. In regions of high seismicity, many designers apply ductile detailing requirements to elements that are intended to resist seismic forces but neglect such practices for nonstructural components, or for structural components that are designed to resist only gravity forces but must undergo the same lateral deformations as the designated seismic force-resisting system. Even where elements of the structure are not intended to resist seismic forces and are not detailed for such resistance, they can participate in the response and may suffer severe damage as a result. This provision requires the designer to provide a level of ductile detailing or proportioning to all elements of the structure appropriate to the calculated deformation demands at the design story drift, Δ. This provision may be accomplished by applying details in gravity members similar to those used in members of the seismic force-resisting system or by providing sufficient strength in those members, or by providing sufficient stiffness in the overall structure to preclude ductility demands in those members.

In the 1994 Northridge earthquake, such participation was a cause of several failures. A preliminary reconnaissance report of that earthquake (EERI 1994) states the following:

> Of much significance is the observation that six of the seven partial collapses (in modern precast concrete parking structures) seem to have been precipitated by damage to the gravity load system. Possibly, the combination of large lateral deformation and vertical load caused crushing in poorly confined columns that were not detailed to be part of the lateral load resisting system. Punching shear failures were observed in some structures at slab-to-column connections, such as at the Four Seasons building in Sherman Oaks. The primary lateral load resisting system was a perimeter ductile frame that performed quite well. However, the interior slab–column system was incapable of undergoing the same lateral deflections and experienced punching failures.

This section addresses such concerns. Rather than relying on designers to assume appropriate levels of stiffness, this section explicitly requires that the stiffening effects of adjoining rigid structural and nonstructural elements be considered and that a rational value of member and restraint stiffness be used for the design of structural components that are not part of the seismic force-resisting system.

This section also includes a requirement to address shears that can be induced in structural components that are not part of the seismic force-resisting system because sudden shear failures have been catastrophic in past earthquakes.

The exception is intended to encourage the use of intermediate or special detailing in beams and columns that are not part of the seismic force-resisting system. In return for better detailing, such beams and columns are permitted to be designed to resist moments and shears from unamplified deflections. This design approach reflects observations and experimental evidence that well-detailed structural components can accommodate large drifts by responding inelastically without losing significant vertical load-carrying capacity.

C12.13 FOUNDATION DESIGN

C12.13.1 Design Basis. In traditional geotechnical engineering practice, foundation design is based on allowable stresses, with allowable foundation load capacities, Q_{as}, for dead plus live loads based on limiting static settlements, which provides a large factor of safety against exceeding ultimate capacities. In this practice, allowable soil stresses for dead plus live loads often are increased arbitrarily by one-third for load combinations that include wind or seismic forces. That approach is overly conservative and not entirely consistent with the design basis prescribed in Section 12.1.5, since it is not based on explicit consideration of the expected strength and dynamic properties of the site soils. Strength design of foundations in accordance with Section 12.13.5 facilitates more direct satisfaction of the design basis.

Section 12.13.1.1 provides horizontal load effect, E_h, values that are used in Section 12.4.2 to determine foundation load combinations that include seismic effects. Vertical seismic load effects are still determined in accordance with Section 12.4.2.2.

Foundation horizontal seismic load effect values specified in Section 12.13.1.1 are intended to be used with horizontal seismic forces, Q_E, defined in Section 12.4.2.1.

C12.13.3 Foundation Load-Deformation Characteristics. For linear static and dynamic analysis methods, where foundation flexibility is included in the analysis, the load-deformation behavior of the supporting soil should be represented by an equivalent elastic stiffness using soil properties that are compatible with the soil strain levels associated with the design earthquake motion. The strain-compatible shear modulus, G, and the associated strain-compatible shear wave velocity, v_s, needed for the evaluation of equivalent elastic stiffness are specified in Chapter 19 of the standard or can be based on a site-specific study. Although inclusion of soil flexibility tends to lengthen the fundamental period of the structure, it should not change the maximum period limitations applied when calculating the required base shear of a structure.

A mathematical model incorporating a combined superstructure and foundation system is necessary to assess the effect of foundation and soil deformations on the superstructure elements. Typically, frequency-independent linear springs are included in the mathematical model to represent the load-deformation characteristics of the soil, and the foundation components are either explicitly modeled (e.g., mat foundation supporting a configuration of structural walls) or are assumed to be rigid (e.g., spread footing supporting a column). In specific cases, a spring may be used to model both the soil and the foundation component (e.g., grade beams or individual piles).

For dynamic analysis, the standard requires a parametric evaluation with upper and lower bound soil parameters to account for the uncertainty in as-modeled soil stiffness and in situ soil variability and to evaluate the sensitivity of these variations on the superstructure. Sources of uncertainty include variability in the rate of loading, including the cyclic nature of building response, level of strain associated with loading at the design earthquake (or stronger), idealization of potentially non-linear soil properties as elastic, and variability in the estimated soil properties. To a lesser extent, this variation accounts for variability in the performance of the foundation components, primarily when a rigid foundation is assumed or distribution of cracking of concrete elements is not explicitly modeled.

Commonly used analysis procedures tend to segregate the "structural" components of the foundation (e.g., footing, grade beam, pile, and pile cap) from the supporting (e.g., soil) components. The "structural" components are typically analyzed using standard strength design load combinations and methodologies, whereas the adjacent soil components are analyzed using allowable stress design (ASD) practices, in which earthquake forces (that have been reduced by R) are considered using ASD load combinations, to make comparisons of design forces versus allowable capacities. These "allowable" soil capacities are typically based on expected strength divided by a factor of safety, for a given level of potential deformations.

When design of the superstructure and foundation components is performed using strength-level load combinations, this traditional practice of using allowable stress design to verify soil compliance can become problematic for assessing the behavior of foundation components. The 2009 NEHRP provisions (FEMA 2009a) contain two resource papers (RP 4 and RP 8) that provide guidance on the application of ultimate strength design procedures in the geotechnical design of foundations and the development of foundation load-deformation characterizations for both linear and nonlinear analysis methods. Additional guidance on these topics is contained in ASCE 41 (2014b).

C12.13.4 Reduction of Foundation Overturning. Since the vertical distribution of horizontal seismic forces prescribed for use with the equivalent lateral force procedure is intended to envelope story shears, the resulting base overturning forces can be exaggerated in some cases. (See Section C12.13.3.) Such overturning will be over-estimated where multiple vibration modes are excited, so a 25 percent reduction in overturning effects is permitted for verification of soil stability. This reduction is not permitted for inverted pendulum or cantilevered column type structures, which typically have a single mode of response.

Since the modal response spectrum analysis procedure more accurately reflects the actual distribution of base shear and overturning moment, the permitted reduction is reduced to 10 percent.

C12.13.5 Strength Design for Foundation Geotechnical Capacity. This section provides guidance for determination of nominal strengths, resistance factors, and acceptance criteria when the strength design load combinations of Section 12.4.2 are used, instead of allowable stress load combinations, to check stresses at the soil–foundation interface.

C12.13.5.1.1 Soil Strength Parameters. If soils are saturated or anticipated to become so, undrained soil properties might be used for transient seismic loading, even though drained strengths may have been used for static or more sustained loading. For competent soils that are not expected to degrade in strength during seismic loading (e.g., due to partial or total liquefaction of cohesionless soils or strength reduction of sensitive clays), use of static soil strengths is recommended for determining the nominal foundation geotechnical capacity, Q_{ns}, of foundations. Use of static strengths is somewhat conservative for such soils because rate-of-loading effects tend to increase soil strengths for transient loading. Such rate effects are neglected because they may not result in significant strength increase for some soil types and are difficult to estimate confidently without special dynamic testing programs. The assessment of the potential for soil liquefaction or other mechanisms for reducing soil strengths is critical, because these effects may reduce soil strengths greatly below static strengths for susceptible soils.

The best estimated nominal strength of footings, Q_{ns}, should be determined using accepted foundation engineering practice. In the absence of moment loading, the ultimate vertical load capacity of a rectangular footing of width B and length L may be written as $Q_{ns} = q_c(BL)$, where q_c = ultimate soil bearing pressure.

For rigid footings subject to moment and vertical load, contact stresses become concentrated at footing edges, particularly as footing uplift occurs. Although the nonlinear behavior of soils causes the actual soil pressure beneath a footing to become nonlinear, resulting in an ultimate foundation strength that is slightly greater than the strength that is determined by assuming a simplified trapezoidal or triangular soil pressure distribution with a maximum soil pressure equal to the ultimate soil pressure, q_c, the difference between the nominal ultimate foundation strength and the effective ultimate strength calculated using these simplified assumptions is not significant.

Lateral resistance may be determined from test data, or by a combination of lateral bearing, lateral friction, and cohesion values. The lateral bearing values may represent values determined from the passive strength values of soil or rock, or they may represent a reduced "allowable" value determined to meet a defined deformation limit. Lateral friction values may represent side-friction values caused by uplift or movement of a foundation against soils, such as for pile uplift or a side friction caused by lateral foundation movement, or they may represent the lateral friction resistance that may be present beneath a foundation caused by the gravity weight of loads that is bearing upon the supporting material.

The lateral foundation geotechnical capacity of a footing may be assumed to be equal to the sum of the best estimated soil passive resistance against the vertical face of the footing plus the best estimated soil friction force on the footing base. The determination of passive resistance should consider the potential contribution of friction on the vertical face.

For piles, the best estimated vertical strength (for both axial compression and axial tensile loading) should be determined using accepted foundation engineering practice. The moment capacity of a pile group should be determined assuming a rigid pile cap, leading to an initial triangular distribution of axial pile loading from applied overturning moments. However, the full expected axial capacity of piles may be mobilized when computing moment capacity, in a manner analogous to that described for a footing. The strength provided in pile caps and intermediate connections should be capable of transmitting the best estimated pile forces to the supported structure. When evaluating axial tensile strength, consideration should be given to the capability of pile cap and splice connections to resist the factored tensile loads.

The lateral foundation geotechnical capacity of a pile group may be assumed to be equal to the best estimated passive resistance acting against the face of the pile cap plus the additional resistance provided by piles.

When the nominal foundation geotechnical capacity, Q_{ns}, is determined by in situ testing of prototype foundations, the test program, including the appropriate number and location of test specimens, should be provided to the authority having jurisdiction by a registered design professional, based on the scope and variability of geotechnical conditions present at the site.

C12.13.5.2 Resistance Factors. Resistance factors, ϕ, are provided to reduce nominal foundation geotechnical capacities, Q_{ns}, to design foundation geotechnical capacities, ϕQ_{ns}, to verify foundation acceptance criteria. The values of ϕ recommended here have been based on the values presented in the AASHTO *LRFD Bridge Design Specifications* (2010). The AASHTO values have been further simplified by using the lesser values when multiple values are presented. These resistance factors account not only for unavoidable variations in design, fabrication, and erection, but also for the variability that often is found in site conditions and test methods (AASHTO 2010).

C12.13.5.3 Acceptance Criteria. The design foundation geotechnical capacity, ϕQ_{ns}, is used to assess acceptability for the linear analysis procedures. The mobilization of ultimate capacity in nonlinear analysis procedures does not necessarily lead to unacceptable performance because structural deformations caused by foundation displacements may be tolerable. For the nonlinear analysis procedures, Section 12.13.3 also requires evaluation of structural behavior using parametric variation of foundation strength to identify potential changes in structural ductility demands.

C12.13.6 Allowable Stress Design for Foundation Geotechnical Capacity. In traditional geotechnical engineering practice, foundation design is based on allowable stresses, with allowable foundation load capacities, Q_{as}, for dead plus live loads based on limiting static settlements, which provides a large factor of safety against exceeding ultimate capacities. In this practice, allowable soil stresses for dead plus live loads often are increased arbitrarily by one-third for load combinations that include wind or seismic forces. That approach may be both more conservative and less consistent than the strength design basis prescribed in Section 12.1.5, since it is not based on explicit consideration of the expected strength and dynamic properties of the site soils.

C12.13.7 Requirements for Structures Assigned to Seismic Design Category C

C12.13.7.1 Pole-Type Structures. The high contact pressures that develop between an embedded pole and soil as a result of lateral loads make pole-type structures sensitive to earthquake motions. Pole-bending strength and stiffness, the soil lateral bearing capacity, and the permissible deformation at grade level are key considerations in the design. For further discussion of pole–soil interaction, see Section C12.13.8.7.

C12.13.7.2 Foundation Ties. One important aspect of adequate seismic performance is that the foundation system acts as an integral unit, not permitting one column or wall to move appreciably to another. To attain this performance, the standard requires that pile caps be tied together. This requirement is especially important where the use of deep foundations is driven by the existence of soft surface soils.

Multistory buildings often have major columns that run the full height of the building adjacent to smaller columns that support only one level; the calculated tie force should be based on the heavier column load.

The standard permits alternate methods of tying foundations together when appropriate. Relying on lateral soil pressure on pile caps to provide the required restraint is not a recommended method because ground motions are highly dynamic and may occasionally vary between structure support points during a design-level seismic event.

C12.13.7.3 Pile Anchorage Requirements. The pile anchorage requirements are intended to prevent brittle failures of the connection to the pile cap under moderate ground motions. Moderate ground motions can result in pile tension forces or bending moments that could compromise shallow anchorage embedment. Loss of pile anchorage could result in increased structural displacements from rocking, overturning instability, and loss of shearing resistance at the ground surface. A concrete bond to a bare steel pile section usually is unreliable, but connection by means of deformed bars properly developed from the pile cap into concrete confined by a circular pile section is permitted.

C12.13.8 Requirements for Structures Assigned to Seismic Design Categories D through F

C12.13.8.1 Pole-Type Structures. See Section C12.13.7.1.

C12.13.8.2 Foundation Ties. See Section C12.13.7.2. For Seismic Design Categories D through F, the requirement is extended to spread footings on soft soils (Site Class E or F).

C12.13.8.3 General Pile Design Requirement. Design of piles is based on the same response modification coefficient, R, used in design of the superstructure; because inelastic behavior results, piles should be designed with ductility similar to that of the superstructure. When strong ground motions occur, inertial pile–soil interaction may produce plastic hinging in piles near the bottom of the pile cap, and kinematic soil–pile interaction results in bending moments and shearing forces throughout the length of the pile, being higher at interfaces between stiff and soft soil strata. These effects are particularly severe in soft soils and liquefiable soils, so Section 14.2.3.2.1 requires special detailing in areas of concern.

The shears and curvatures in piles caused by inertial and kinematic interaction may exceed the bending capacity of conventionally designed piles, resulting in severe damage. Analysis techniques to evaluate pile bending are discussed by Margason and Holloway (1977) and Mylonakis (2001), and these effects on concrete piles are further discussed by Sheppard (1983). For homogeneous, elastic media and assuming that the pile follows the soil, the free-field curvature (soil strains without a pile present) can be estimated by dividing the peak ground acceleration by the square of the shear wave velocity of the soil. Considerable judgment is necessary in using this simple relationship for a layered, inelastic profile with pile–soil interaction effects. Norris (1994) discusses methods to assess pile–soil interaction.

Where determining the extent of special detailing, the designer must consider variation in soil conditions and driven pile lengths, so that adequate ductility is provided at potentially high curvature interfaces. Confinement of concrete piles to provide ductility and maintain functionality of the confined core pile during and after the earthquake may be obtained by use of heavy spiral reinforcement or exterior steel liners.

C12.13.8.4 Batter Piles. Partially embedded batter piles have a history of poor performance in strong ground shaking, as shown by Gerwick and Fotinos (1992). Failure of battered piles has been attributed to design that neglects loading on the piles from ground deformation or assumes that lateral loads are resisted by axial response of piles without regard to moments induced in the pile at the pile cap (Lam and Bertero 1990). Because batter piles are considered to have limited ductility, they must be designed using the load combinations including overstrength. Moment-resisting connections between pile and pile cap must resolve the eccentricities inherent in batter pile configurations. This concept is illustrated clearly by EQE Engineering (1991).

C12.13.8.5 Pile Anchorage Requirements. Piles should be anchored to the pile cap to permit energy-dissipating mechanisms, such as pile slip at the pile–soil interface, while maintaining a competent connection. This section of the standard sets forth a capacity design approach to achieve that objective. Anchorages occurring at pile cap corners and edges should be reinforced to preclude local failure of plain concrete sections caused by pile shears, axial loads, and moments.

C12.13.8.6 Splices of Pile Segments. A capacity design approach, similar to that for pile anchorage, is applied to pile splices.

C12.13.8.7 Pile–Soil Interaction. Short piles and long slender piles embedded in the earth behave differently when subjected to lateral forces and displacements. The response of a long slender pile depends on its interaction with the soil considering the nonlinear response of the soil. Numerous design aid curves and computer programs are available for this type of analysis, which is necessary to obtain realistic pile moments, forces, and deflections and is common in practice (Ensoft 2004b). More sophisticated models, which also consider inelastic behavior of the pile itself, can be analyzed using general-purpose nonlinear analysis computer programs or closely approximated using the pile–soil limit state methodology and procedure given by Song et al. (2005).

Each short pile (with length-to-diameter ratios no more than 6) can be treated as a rigid body, simplifying the analysis. A method assuming a rigid body and linear soil response for lateral bearing is given in the current building codes. A more accurate and comprehensive approach using this method is given in a study by Czerniak (1957).

C12.13.8.8 Pile Group Effects. The effects of groups of piles, where closely spaced, must be taken into account for vertical and horizontal response. As groups of closely spaced piles move laterally, failure zones for individual piles overlap and horizontal strength and stiffness response of the pile–soil system is reduced. Reduction factors or "p-multipliers" are used to account for these groups of closely spaced piles. For a pile center-to-center spacing of three pile diameters, reduction factors of 0.6 for the leading pile row and 0.4 for the trailing pile rows are recommended by Rollins et al. (1999). Computer programs are available to analyze group effects assuming nonlinear soil and elastic piles (Ensoft 2004a).

C12.13.9 Requirements for Foundations on Liquefiable Sites. This new section provides requirements for foundations of structures that are located on sites that have been determined to have the potential to liquefy when subjected to Geomean Maximum Considered Earthquake ground motions. This section complements the requirements of Section 11.8, which provides requirements for geotechnical investigations in areas with significant seismic ground motion hazard with specific requirements for additional geotechnical information and recommendations for sites that have the potential to liquefy when subjected to the Geomean Maximum Considered Earthquake ground motion.

Before the 2010 edition of ASCE 7 (which was based on the *2009 NEHRP Recommended Seismic Provisions for New Buildings and Other Structures*, FEMA 2009a), the governing building code requirements for foundations where potentially liquefiable soil conditions were present was Chapter 18 of the International Building Code (ICC 2009). Chapter 18 of the IBC (ICC 2009) specified the use of the design earthquake (DE) ground motions for all structural and geotechnical evaluations for buildings. Chapter 18 of IBC (ICC 2012) references ASCE 7-10 (2010) and deletes reference to the DE. Chapter 11 of ASCE 7-10 (2010) has new requirements that specify that Maximum Considered Earthquake (MCE) rather than the DE ground motions should be used for geotechnical (liquefaction-related) evaluations that are specified in IBC (ICC 2009).

The reason that the change to MCE ground motions for liquefaction evaluations was made in ASCE 7-10 (2010) was to make the ground motions used in the evaluations consistent with the ground motions used as the basis for the design of structures. Starting with the 2000 edition of the IBC (ICC 2000), the ground motion maps provided in the code for seismic design were MCE mapped values and not DE values. Although design values for structures in the IBC are based on DE ground motions, which are two-thirds of the MCE, studies (FEMA 2009b) have indicated that structures designed for DE motions had a low probability of collapse at MCE level motions. However, these studies presumed nonliquefiable soil conditions. It should also be noted that most essential structures, such as hospitals, are required to be explicitly designed for MCE motions. Whereas ASCE 7-10 has specific requirements for MCE-level liquefaction evaluations, it has no specific requirements for foundation design when these conditions exist. This lack of clear direction was the primary reason for the development of this new section.

The requirements of this section, along with the seismic requirements of this standard, are intended to result in structure foundation systems that satisfy the performance goals stated in Section 1.1 of the 2009 *NEHRP Recommended Seismic Provisions for New Buildings and Other Structures* for structure sites that have been determined to be liquefiable per Section 11.8. They require mitigation of significant liquefaction-induced risks, either through ground improvement or structural measures, aimed at preventing liquefaction-induced building collapse and permitting the structure and its nonstructural system to satisfy the Section 1.1 performance goals. With the exception of Risk Category IV Essential Facilities, the provisions do not seek to control non-life-threatening damage to buildings that may occur as a result of liquefaction-induced settlement. For Risk Category IV Essential Facilities, the provisions seek to limit damage attributable to liquefaction to levels that would permit post-earthquake use. For example, settlement is controlled to levels that would be expected to allow for continued operation of doors.

There is nothing in these provisions that is intended to preclude the Authority Having Jurisdiction from enacting more stringent planning regulations for building on sites susceptible to potential geologic hazards, in recognition of losses that may occur in the event of an earthquake that triggers liquefaction.

In the first paragraph of Section 12.13.9, it is stated that the foundation must also be designed to resist the effects of design earthquake seismic load effects assuming that liquefaction does not occur. This additional requirement is imposed since maximum seismic loads on a foundation during an earthquake can occur before liquefaction. This additional requirement provides assurance that the foundation will be adequate regardless of when liquefaction occurs during the seismic event.

Observed Liquefaction-Related Structural Damage in Past Earthquakes

Damage to structures from liquefaction-related settlement, punching failure of footings, and lateral spreading has been common in past earthquakes. Whereas total postliquefaction settlement values have varied from several inches to several feet (depending on the relative density and thickness of saturated sand deposits), differential settlements depend on the uniformity of site conditions and the depth of liquefied strata. For example, in the 1995 Kobe, Japan, earthquake, total settlements of 1.5 to 2.5 ft (0.46 to 0.76 m) were observed but with relatively small differential settlements.

In the 1989 Loma Prieta, California, earthquake, settlements of as much as 2 ft (0.61 m) and lateral spreading that ranged between 0.25 and 5 ft (0.08 and 1.5 m) were observed on the Moss Landing spit. The Monterey Bay Aquarium Research Institute's (MBARI's) technology building was supported on shallow foundation with ties and located some 30 ft (9.14 m) away from the edge of the Moss Landing South Harbor. Whereas 0.25 ft (0.76 m) of lateral spreading was measured at the MBARI building, it suffered only minor cracks. On the other hand, the Moss Landing Marine Lab (MLML) building was located on a different part of the spit where between 4 and 5 ft (1.22 and 1.52 m) of lateral spreading was measured. The MLML building, which was supported on shallow foundations without ties, collapsed as the building footings were pulled apart. The MBARI research pier, located at the harbor, across the street from the Technology Building, suffered no damage except for minor spalling at the underside of the concrete deck, where the 16-in. (406.4-mm) diameter cylindrical driven piles for the pier interfaced with the overlying concrete deck.

The 1999 Kocaeli, Turkey, earthquake provided numerous examples of the relationship between liquefaction-induced soil deformations and building and foundation damage in the city of Adapazari. Examples include a five-story reinforced concrete frame building on a mat foundation that settled about 0.5 ft (0.15 m) at one corner and 5 ft (1.5 m) at the opposite corner with related tilting associated with rigid body motion. Essentially no foundation or structural damage was observed. In contrast, several buildings on mat foundations underwent bearing capacity failures and overturned. The foundation soil strength loss, evidenced by bulging around the building perimeter, initiated the failures, as opposed to differential settlement caused by post-liquefaction volume change in the former case history. In addition, lateral movements of building foundations were also observed. Movements were essentially rigid body for buildings on stiff mat foundations, and they led to no significant building damage. For example, a five-story building experienced about 1.5 ft (0.46 m) of settlement and 3 ft (0.91 m) of lateral displacement.

In the 2011 and 2012 Christchurch, New Zealand, earthquakes, significant differential settlement occurred for several buildings on spread footings. Values of differential settlement of 1 to 1.5 ft (0.31 to 0.46 m) were measured for three- to five-story buildings, resulting in building tilt of 2 to 3 deg. Structural damage was less for cases where relatively strong reinforced concrete ties between footings were used to minimize differential settlement. Footing punching failures also occurred leading to significant damage. For taller buildings on relatively rigid raft foundations, differential ground settlement resulted in building tilt, but less structural damage. In contrast, structures on pile foundations performed relatively well.

C12.13.9.1 Foundation Design. Foundations are not allowed to lose the strength capacity to support vertical reactions after liquefaction. This requirement is intended to prevent bearing capacity failure of shallow foundations and axial load failure of deep foundations. Settlement in the event of such failures cannot be accurately estimated and has potentially catastrophic consequences. Such failures can be prevented by using ground improvement or adequately designed deep foundations.

Liquefaction-induced differential settlement can result from variations in the thickness, relative density, or fines content of potentially liquefiable layers that occur across the footprint of the structure. When planning a field exploration program for a potentially liquefiable site, where it is anticipated that shallow foundations may be used, the geotechnical engineer must have information on the proposed layout of the building(s) on the site. This information is essential to properly locating and spacing exploratory holes to obtain an appropriate estimate of anticipated differential settlement. One acceptable method for dealing with unacceptable liquefaction-induced settlements is by performing ground improvement. There are many acceptable methods for ground improvement.

C12.13.9.2 Shallow Foundations. Shallow foundations are permitted where individual footings are tied together so that they have the same horizontal displacements, and differential settlements are limited or where the expected differential settlements can be accommodated by the structure and the foundation system. The lateral spreading limits provided in Table 12.13-2 are based on engineering judgment and are the judged upper limits of lateral spreading displacements that can be tolerated while still achieving the desired performance for each Risk Category, presuming that the foundation is well tied together. Differential settlement is defined as δ_v/L, where δ_v and L are illustrated for an example structure in Fig. C12.13-1. The differential settlement limits specified in Table 12.13-3 are intended to provide collapse resistance for Risk Category II and III structures.

The limit for one-story Risk Category II structures with concrete or masonry structural walls is consistent with the drift limit in ASCE 41 (2014b) for concrete shear walls to maintain collapse prevention. The limit for taller structures is more restrictive because of the effects that the tilt would have on the floors of upper levels. This more restrictive limit is consistent with the "moderate to severe damage" for multistory masonry structures, as indicated in Boscardin and Cording (1989).

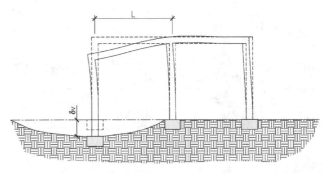

FIGURE C12.13-1 Example Showing Differential Settlement Terms δ_v and L

The limits for structures without concrete or masonry structural walls are less restrictive and are consistent with the drift limits in ASCE 41 (2014b) for high-ductility concrete frames to maintain collapse prevention. Frames of lower ductility are not permitted in Seismic Design Categories C and above, which are the only categories where liquefaction hazards need to be assessed.

The limits for Risk Category III structures are two-thirds of those specified for Risk Category II.

The limits for Risk Category IV are intended to maintain differential settlements less than the distortion that will cause doors to jam in the design earthquake. The numerical value is based on the median value of drift (0.0023) at the onset of the damage state for jammed doors developed for the ATC-58 project (ATC 2012), multiplied by 1.5 to account for the dispersion and scaled to account for the higher level of shaking in the MCE relative to the DE.

Shallow foundations are required to be interconnected by ties, regardless of the effects of liquefaction. The additional detailing requirements in this section are intended to provide moderate ductility in the behavior of the ties because the adjacent foundations may settle differentially. The tie force required to accommodate lateral ground displacement is intended to be a conservative assessment to overcome the maximum frictional resistance that could occur between footings along each column or wall line. The tie force assumes that the lateral spreading displacement occurs abruptly midway along the column or wall line. The coefficient of friction between the footings and underlying soils may be taken conservatively as 0.50. This requirement is intended to maintain continuity throughout the substructure in the event of lateral ground displacement affecting a portion of the structure. The required tie force should be added to the force determined from the lateral loads for the design earthquake in accordance with Sections 12.8, 12.9, 12.14, or Chapter 16.

C12.13.9.3 Deep Foundations. Pile foundations are intended to remain elastic under axial loadings, including those from gravity, seismic, and downdrag loads. Since geotechnical design is most frequently performed using allowable stress design (ASD) methods, and liquefaction-induced downdrag is assessed at an ultimate level, the requirements state that the downdrag is considered as a reduction in the ultimate capacity. Since structural design is most frequently performed using load and resistance factor design (LRFD) methods, and the downdrag is considered as a load for the pile structure to resist, the requirements clarify that the downdrag is considered as a seismic axial load, to which a factor of 1.0 would be applied for design.

The ultimate geotechnical capacity of the pile should be determined using only the contribution from the soil below the liquefiable layer. The net ultimate capacity is the ultimate capacity reduced by the downdrag load (Fig. C12.13-2).

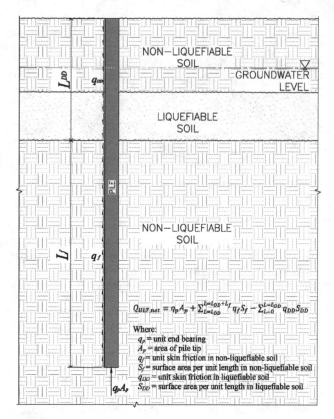

FIGURE C12.13-2 Determination of Ultimate Pile Capacity in Liquefiable Soils

Lateral resistance of the foundation system includes resistance of the piles as well as passive pressure acting on walls, pile caps, and grade beams. Analysis of the lateral resistance provided by these disparate elements is usually accomplished separately. In order for these analyses to be applicable, the displacements used must be compatible. Lateral pile analyses commonly use nonlinear soil properties. Geotechnical recommendations for passive pressure should include the displacement at which the pressure is applicable, or they should provide a nonlinear mobilization curve. Liquefaction occurring in near-surface layers may substantially reduce the ability to transfer lateral inertial forces from foundations to the subgrade, potentially resulting in damaging lateral deformations to piles. Ground improvement of surface soils may be considered for pile-supported structures to provide additional passive resistance to be mobilized on the sides of embedded pile caps and grade beams, as well as to increase the lateral resistance of piles. Otherwise, the check for transfer of lateral inertial forces is the same as for structures on nonliquefiable sites.

IBC (ICC 2012), Section 1810.2.1, requires that deep foundation elements in fluid (liquefied) soil be considered unsupported for lateral resistance until a point 5 ft (1.5 m) into stiff soil or 10 ft (3.1 m) into soft soil unless otherwise approved by the authority having jurisdiction on the basis of a geotechnical investigation by a registered design professional. Where liquefaction is predicted to occur, the geotechnical engineer should provide the dimensions (depth and length) of the unsupported length of the pile or should indicate if the liquefied soil will provide adequate resistance such that the length is considered laterally supported in this soil. The geotechnical engineer should develop these dimensions by performing an analysis of the nonlinear resistance of the soil to lateral displacement of the pile (i.e., p-y analysis).

Concrete pile detailing includes transverse reinforcing requirements for columns in ACI 318-14 (2014). This is intended to provide ductility within the pile similar to that required for columns.

Where permanent ground displacement is indicated, piles are not required to remain elastic when subjected to this displacement. The provisions are intended to provide ductility and maintain vertical capacity, including flexure-critical behavior of concrete piles.

The required tie force specified in Section 12.13.9.3.5 should be added to the force determined from the lateral loads for the design earthquake in accordance with Sections 12.8, 12.9, 12.14, or Chapter 16.

C12.14 SIMPLIFIED ALTERNATIVE STRUCTURAL DESIGN CRITERIA FOR SIMPLE BEARING WALL OR BUILDING FRAME SYSTEMS

C12.14.1 General. In recent years, engineers and building officials have become concerned that the seismic design requirements in codes and standards, though intended to make structures perform more reliably, have become so complex and difficult to understand and implement that they may be counterproductive. Because the response of buildings to earthquake ground shaking is complex (especially for irregular structural systems), realistically accounting for these effects can lead to complex requirements. There is a concern that the typical designers of small, simple buildings, which may represent more than 90% of construction in the United States, have difficulty understanding and applying the general seismic requirements of the standard.

The simplified procedure presented in this section of the standard applies to low-rise, stiff buildings. The procedure, which was refined and tested over a five-year period, was developed to be used for a defined set of buildings deemed to be sufficiently regular in structural configuration to allow a reduction of prescriptive requirements. For some design elements, such as foundations and anchorage of nonstructural components, other sections of the standard must be followed, as referenced within Section 12.14.

C12.14.1.1 Simplified Design Procedure. Reasons for the limitations of the simplified design procedure of Section 12.14 are as follows:

1. The procedure was developed to address adequate seismic performance for standard occupancies. Because it was not developed for higher levels of performance associated with structures assigned to Risk Categories III and IV, no Importance Factor (I_e) is used.
2. Site Class E and F soils require specialized procedures that are beyond the scope of the procedure.
3. The procedure was developed for stiff, low-rise buildings, where higher mode effects are negligible.
4. Only stiff systems where drift is not a controlling design criterion may use the procedure. Because of this limitation, drifts are not computed. The response modification coefficient, R, and the associated system limitations are consistent with those found in the general Chapter 12 requirements.
5. To achieve a balanced design and a reasonable level of redundancy, two lines of resistance are required in each of the two major axis directions. Because of this stipulation, no redundancy factor (ρ) is applied.
6. When combined with the requirements in items 7 and 8, this requirement reduces the potential for dominant torsional response.

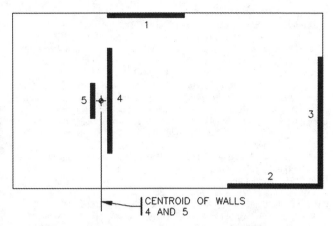

FIGURE C12.14-1 Treatment of Closely Spaced Walls

7. Although concrete diaphragms may be designed for even larger overhangs, the torsional response of the system would be inconsistent with the behavior assumed in development of Section 12.14. Large overhangs for flexible diaphragm buildings can also produce a response that is inconsistent with the assumptions associated with the procedure.
8. Linear analysis shows a significant difference in response between flexible and rigid diaphragm behavior. However, nonlinear response history analysis of systems with the level of ductility present in the systems permitted in Table 12.14-1 for the higher Seismic Design Categories has shown that a system that satisfies these layout and proportioning requirements provides essentially the same probability of collapse as a system with the same layout but proportioned based on rigid diaphragm behavior (BSSC 2015). This procedure avoids the need to check for torsional irregularity, and calculation of accidental torsional moments is not required. Fig. C12.14-1 shows a plan with closely spaced walls in which the method permitted in subparagraph (c) should be implemented. In that circumstance, the flexible diaphragm analysis would first be performed as if there were one wall at the location of the centroid of walls 4 and 5, then the force computed for that group would be distributed to walls 4 and 5 based on an assessment of their relative stiffnesses.
9. An essentially orthogonal orientation of lines of resistance effectively uncouples response along the two major axis directions, so orthogonal effects may be neglected.
10. Where the simplified design procedure is chosen, it must be used for the entire design in both major axis directions.
11. Because in-plane and out-of-plane offsets generally create large demands on diaphragms, collectors, and discontinuous elements, which are not addressed by the procedure, these irregularities are prohibited.
12. Buildings that exhibit weak-story behavior violate the assumptions used to develop the procedure.

C12.14.3 Seismic Load Effects and Combinations. The equations for seismic load effects in the simplified design procedure are consistent with those for the general procedure, with one notable exception: The overstrength factor (corresponding to Ω_0 in the general procedure) is set at 2.5 for all systems, as indicated in Section 12.14.3.2.1. Given the limited systems that can use the simplified design procedure, specifying unique overstrength factors was deemed unnecessary.

C12.14.7 Design and Detailing Requirements. The design and detailing requirements outlined in this section are similar to those for the general procedure. The few differences include the following:

1. Forces used to connect smaller portions of a structure to the remainder of the structures are taken as 0.20 times the short-period design spectral response acceleration, S_{DS}, rather than the general procedure value of 0.133 (Section 12.14.7.1).
2. Anchorage forces for concrete or masonry structural walls for structures with diaphragms that are not flexible are computed using the requirements for nonstructural walls (Section 12.14.7.5).

C12.14.8 Simplified Lateral Force Analysis Procedure

C12.14.8.1 Seismic Base Shear. The seismic base shear in the simplified design procedure, as given by Eq. (12.14-11), is a function of the short-period design spectral response acceleration, S_{DS}. The value for F in the base shear equation addresses changes in dynamic response for buildings that are two or three stories above grade plane (see Section 11.2 for definitions of "grade plane" and "story above grade plane"). As in the general procedure (Section 12.8.1.3), S_{DS} may be computed for short, regular structures with S_S taken as no greater than 1.5.

C12.14.8.2 Vertical Distribution. The seismic forces for multistory buildings are distributed vertically in proportion to the weight of the respective floor. Given the slightly amplified base shear for multistory buildings, this assumption, along with the limit of three stories above grade plane for use of the procedure, produces results consistent with the more traditional triangular distribution without introducing that more sophisticated approach.

C12.14.8.5 Drift Limits and Building Separation. For the simplified design procedure, which is restricted to stiff shear wall and braced frame buildings, drift need not be calculated. Where drifts are required (such as for structural separations and cladding design) a conservative drift value of 1% is specified.

REFERENCES

American Association of State Highway and Transportation Officials (AASHTO). (2010). *LRFD bridge design specifications*, Washington, DC.
ACI. (2014). "Building code requirements for structural concrete and commentary." *ACI 318-14*, Farmington Hills, MI.
American Institute of Steel Construction (AISC). (2016). "Seismic provisions for structural steel buildings." *AISC 341-10*, Chicago.
American Plywood Association (APA). (1966). "1966 horizontal plywood diaphragm tests." *Laboratory Report 106*, APA, Tacoma, WA.
APA. (2000). "Plywood diaphragms." *Research Report 138*, APA, Tacoma, WA.
Applied Technology Council (ATC). (1984), "Tentative provisions for the development of seismic regulations for buildings." *ATC 3-06*, Redwood City, CA.
Applied Technology Council (ATC). (2012), "Seismic performance assessment of buildings." *ATC 58* ATC Redwood City, CA.
ASCE. (2000). "Seismic analysis of safety-related nuclear structures and commentary." *ASCE 4-98*, ASCE Reston, VA.
ASCE. (2003). "Minimum design loads for buildings and other structures." *ASCE 7-02*, ASCE, Reston, VA.
ASCE. (2010). "Minimum design loads for buildings and other structures." *ASCE 7-10*, ASCE, Reston, VA.
ASCE. (2014b). "Seismic evaluation and retrofit of existing buildings." *ASCE 41-13*, ASCE, Reston, VA.
ASCE/SEI. (2007). "Seismic rehabilitation of existing buildings." *ASCE/SEI 41-06*, ASCE, Reston, VA.
American Wood Council (AWC). (2008). "Special design provisions for wind and seismic." *AWC SDPWS-2008*, AWC, Leesburg, VA.

AWC (2014). "Special design provisions for wind and seismic." *AWC SDPWS-15*, AW, Leesburg, VA.

Boscardin, M. D., and Cording, E. J. (1989). "Building response to excavation-induced settlement." *J. Geotech. Eng.*, ASCE, 115(1), 1–21.

Building Seismic Safety Council (BSSC). (2015). "Development of simplified seismic design provisions, Working Group 1 Report: Simplifying Section 12.14." *National Institute of Building Sciences*, Washington, DC.

Charney, F. A., and Marshall, J. (2006). "A comparison of the Krawinkler and scissors models for including beam-column joint deformations in the analysis of moment-resisting frames." *AISC Engrg. J.*, 43(1), 31–48.

Chen, M., Pantoli, E., Wang, X., Astroza, R., Ebrahimian, H., Hutchinson, T., et al. (2015). "Full-scale structural and nonstructural building system performance during earthquakes part I – specimen description, test protocol and structural response." *Earthquake Spectra*, 32(2), 737–770.

Choi, H., Erochko, J., Christopoulos, C., and Tremblay, R. (2008). "Comparison of the seismic response of steel buildings incorporating self-centering energy-dissipative braces, buckling restrained braces and moment-resisting frames." *Research Report 05-2008*, University of Toronto, Canada.

Chopra, A. K. (1995). *Structural dynamics*, Prentice Hall International Series in Civil Engineering, New York.

Chopra, A. K. (2007a). *Dynamics of structures*, 4th Ed. Prentice Hall International Series in Civil Engineering, Upper Saddle River, NJ.

Chopra, A. K. (2007b). *Structural dynamics*, Prentice Hall International Series in Civil Engineering, Upper Saddle River, N.J.

Chopra, A. K., and Goel, R. K. (1991). "Evaluation of torsional provisions in seismic codes." *J. Struct. Engrg.*, 117(12), 3762–3782.

Chopra, A. K., and Newmark, N. M. (1980). "Analysis." *Design of earthquake resistant structures*, E. Rosenblueth, ed., John Wiley & Sons, New York.

Czerniak, E. (1957). "Resistance to overturning of single, short piles." *J. Struct. Div.*, 83(2), 1–25.

Debock, D. J., Liel, A. B., Haselton, C. B., Hopper, J. D., and Henige, R. (2014). "Importance of Seismic Design Accidental Torsion Requirements for Building Collapse," *Earthq. Engrg. Struct. Dyn.*, 43(6).

De La Llera, J. C., and Chopra, A. K. (1994). "Evaluation of code accidental-torsion provisions from building records." *J. Struct. Engrg.*, 120(2), 597–616.

Douglas Fir Plywood Association (DFPA). (1954). "Horizontal plywood diaphragm tests." *Laboratory Report 63a*. DFPA, Tacoma, WA.

DFPA. (1963). "Lateral tests on plywood sheathed diaphragms." *Laboratory Report 55*. DFPA, Tacoma, WA.

Earthquake Engineering Research Institute (EERI). (1994). "Northridge Earthquake, January 17, 1994: Preliminary reconnaissance report," *John F. Hall*, ed., EERI, Oakland, CA, 56–57.

Ensoft, Inc. (2004a). Computer program GROUP, Version 6.0: A program for the analysis of a group of piles subjected to axial and lateral loading, user's manual and technical manual. Ensoft, Austin, TX.

Ensoft, Inc. (2004b). Computer program LPILE Plus, Version 5.0: A program for the analysis of piles and drilled shafts under lateral loads, user's manual and technical manual. Ensoft, Austin, TX.

EQE Engineering. (1991). "Structural concepts and details for seismic design." *UCRL-CR-106554*. Department of Energy, Washington, DC.

EQE Incorporated. (1983). "The effects of the May 2, 1983, Coalinga, California, earthquake on industrial facilities." EQE Inc., Newport Beach, CA.

EQE. (1985). "Summary of the September 19, 1985, Mexico earthquake." EQE Inc., Newport Beach, CA.

EQE. (1986a). "Summary of the March 3, 1985, Chile earthquake." EQE Inc., Newport Beach, CA.

EQE. (1986b). "The effects of the March 3, 1985, Chile earthquake on power and industrial facilities." EQE Inc., Newport Beach, CA.

EQE. (1986c). "Power and industrial facilities in the epicentral area of the 1985 Mexico earthquake." *EQE Inc.*, Newport Beach, CA.

EQE. (1987). "Summary of the 1987 Bay of Plenty, New Zealand, earthquake." *EQE Inc.*, Newport Beach, CA.

Federal Emergency Management Agency (FEMA). (2009a). "NEHRP recommended seismic provisions for new buildings and other structures." *FEMA P-750*, Building Seismic Safety Council, FEMA, Washington, DC.

FEMA. (2009b). "Quantification of building seismic performance factors." *FEMA P-695*, Applied Technology Council, FEMA, Washington, DC.

FEMA. (2011). "Quantification of building seismic performance factors: Component equivalency methodology." *FEMA P-795*, Washington, DC.

FEMA (2012) "Seismic Performance Assessment of Buildings." FEMA P-58, Applied Technology Council, FEMA, Washington, DC.

FEMA. (2013). "2009 NEHRP recommended seismic provisions: Design examples." *FEMA P-751CD*, Washington, DC

Fleischman, R. B., Restrepo, J. I., Naito, C. J., Sause, R., Zhang, D., and Schoettler, M. (2013). "Integrated analytical and experimental research to develop a new seismic design methodology for precast concrete diaphragms," *J. Struct. Eng.*, 139(7), 1192–1204.

Gatto, K., and Uang, C.-M. (2002). *Cyclic response of woodframe shearwalls: Loading protocol and rate of loading effects*. Consortium of Universities for Research in Earthquake Engineering, Richmond, CA.

Gerwick, B., Jr., and Fotinos, G. (1992). "Drilled piers and driven piles for foundations in areas of high seismicity." SEAONC Fall Seminar, October 29, San Francisco.

Goel, R. K., and Chopra, A. K. (1997). "Period formulas for moment-resisting frame buildings." *J. Struct. Engrg.*, 123(11), 1454–1461.

Goel, R. K., and Chopra, A. K. (1998). "Period formulas for concrete shear wall buildings." *J. Struct. Engrg.*, 124(4), 426–433.

Grant, D., and Diaferia, R. (2012). "Assessing adequacy of spectrum matched ground motions for response history analysis." *Earthq. Eng. Struct. Dyn.*, 42(9), 1265–1280.

Gupta, A., and Krawinkler, H. (2000). "Dynamic P-delta effects for flexible inelastic steel structures." *J. Struct. Engrg.*, 126(1), 145–154.

Hamburger, R. O., and McCormick, D. L. (2004). "Implications of the January 17, 1994, Northridge earthquake on tilt-up and masonry buildings with wood roofs." *Proc. 63rd Annual Convention*, Structural Engineers Association of California, Lake Tahoe, CA, 243–255.

Hancock, J., Watson-Lamprey, J., Abrahamson, N. A., Bommer, J. J., Markatis, A., McCoy, E., et al. (2006). "An improved method of matching response spectra of recorded earthquake ground motion using wavelets." *J. Earthq. Eng.*, 10(1), 67–89.

International Conference of Building Officials (ICBO). (1988). *Uniform Building Code*, Whittier, CA.

ICBO. (1994). *Uniform Building Code*, Whittier, CA.

ICBO. (1997). *Uniform Building Code*, Whittier, CA.

International Code Council (ICC). (2000). *International Building Code*, Country Club Hills, IL.

ICC. (2009). *International Building Code*, Country Club Hills, IL.

ICC. (2012). *International Building Code*, Country Club Hills, IL.

ICC. (2015). *International Building Code*, Country Club Hills, IL.

Lam, I., and Bertero, V. (1990). "A seismic design of pile foundations for port facilities." *Proc. POLA Seismic Workshop on Seismic Engineering*, March 21–23, San Pedro, CA, Port of Los Angeles.

Liew, J. Y. R., Chen, H., and Shanmugam, N. E. (2001). "Inelastic analysis of steel frames with composite beams." *J. Struct. Engrg.*, 127(2), 194–202.

Lopez, O. A., and Cruz, M. (1996). "Number of modes for the seismic design of buildings." *Earthq. Engrg. Struct. Dyn.*, 25(8), 837–856.

Margason, E., and Holloway, M. (1977). "Pile bending during earthquakes." *Proc. 6th World Conference on Earthquake Engineering*, New Delhi.

Menun, C., and Der Kiureghian, A. (1998). "A replacement for the 30%, 40%, and SRSS rules for multicomponent seismic analysis." *Earthq. Spectra*, 14(1).

Mylonakis, G. (2001). "Seismic pile bending at soil-layer interfaces." *Soils and Found.*, 41(4), 47–58.

National Institute of Standards and Technology (NIST). (2010). "Nonlinear structural analysis for seismic design: A guide for practicing engineers." *Report NIST GCR 10-917-5*, Gaithersburg, MD.

NIST. (2011). "Selecting and scaling earthquake ground motions for performing response-history analyses." *NIST GCR 11-918-15*. Prepared by the NEHRP Consultants, Joint Venture for the National Institute of Standards and Technology, Gaithersburg, MD.

Newmark, N. M., and Hall, W. J. (1982). *Earthquake spectra and design*, EERI Monograph Series, Earthquake Engineering Research Institute, Oakland, CA.

Newmark, N. M., and Rosenblueth, E. (1971). *Fundamentals of earthquake engineering*, Prentice Hall, Englewood Cliff, NJ.

Norris, G. M. (1994). "Seismic bridge pile foundation behavior." *Proc. Intl. Conf. on Design and Construction of Deep Foundations*, Federal Highway Administration, Vol. 1.

Paulay, T. (1997). "Are existing seismic torsion provisions achieving design aims?" *Earthq. Spectra*, 13(2), 259–280.

Paulay, T., and Priestley, M. J. N. (1992). *Seismic design of reinforced concrete and masonry structures*. John Wiley & Sons, New York.

Panagiotou, M., Restrepo, J. I., and Conte, J. P. (2011). "Shake-table test of a full-scale 7-story building slice. Phase I: Rectangular wall." *J. Struct. Eng.*, 137(6), 691–704.

Rollins, K. M., Peterson, K. T., Weaver, T. J., and Sparks, A. E. (1999). "Static and dynamic lateral load behavior on a full-scale pile group in clay." *Brigham Young University*, Provo, UT, and Utah Department of Transportation, Salt Lake City, June 23.

Rosenblueth, E., and Contreras, H. (1977). "Approximate design for multicomponent earthquakes." *J. Engrg. Mech. Div.*, 103(5), 881–893.

Schaffhausen, R., and Wegmuller, A. (1977). "Multistory rigid frames with composite girders under gravity and lateral forces." *AISC Engrg. J.*, 2nd Quarter.

Sheppard, D. A. (1983). "Seismic design of prestressed concrete piling." *PCI J.*, 28(2), 20–49.

Song, S. T., Chai, Y. H., and Hale, T. H. (2005). "Analytical model for ductility assessment of fixed-head concrete piles." *J. Struct. Engrg.*, 131(7), 1051–1059.

Uang, C.-M., and Maarouf, A. (1994). "Deflection amplification factor for seismic design provisions." *J. Struct. Engrg.*, 120(8), 2423–2436.

U.S. Nuclear Regulatory Commission. (2012). "Combining modal responses and spatial components in seismic response analysis," *Regulatory Guide 1.92*, Brookhaven National Laboratory, Upton, NY.

Vamvatsikos, D. (2002). "Seismic performance, capacity and reliability of structures as seen through incremental dynamic analysis." Ph.D. Dissertation, Stanford University, Palo Alto, CA.

Wilson, E. L. (2000). "Three-dimensional static and dynamic analysis of structures." *Computers and Structures*, Inc., Berkeley, CA.

Wilson, E. L. (2004). *Static and dynamic analysis of structures*, Computers and Structures, Berkeley, CA.

Wilson, E. L., Der Kiureghian, A., and Bayo, E. P. (1981). "A replacement for the SRSS method in seismic analysis." *Earthq. Engrg. Struct. Dyn.* 9(2), 187–194.

Wilson, E. L., and Habibullah, A. (1987). "Static and dynamic analysis of multistory buildings, including P-delta effects," *Earthq. Spectra* 3(2), 289–298.

Wilson, E. L., and Penzien, J. (1972). "Evaluation of orthogonal damping matrices." *Int. J. Numer. Meth.*, 4, 5–10.

OTHER REFERENCES (NOT CITED)

American Plywood Association (APA). (1966). "1966 horizontal plywood diaphragm tests." *Laboratory Report 106*. APA, Tacoma, WA.

APA. (2000). "Plywood diaphragms" *Research Report 138*. American Plywood Association, Tacoma, WA.

American Wood Council (AWC). (2008). "Special design provisions for wind and seismic." *AWC SDPWS-2008*. AWC, Leesburg, VA.

"Assessing adequacy of spectrum matched ground motions for response history analysis," *Earthq. Eng. Struct. Dyn.*, 42(9), 1265–1280.

Bernal, D. (1987). "Amplification factors for inelastic dynamic P-delta effects in earthquake analysis." *Earthq. Engrg. Struct. Dyn.*, 18, 635–681.

California Geological Survey. (2008). "Special publication SP 117A–Guidelines for evaluation and mitigation of seismic hazards in California."

Charney, F. A. (1990). "Wind drift serviceability limit state design of multistory buildings." *J. Wind Engrg. Indust. Aerodyn.*, 36, 203–212.

City of Newport Beach. (2012). "Minimum liquefaction mitigation measures," April 24.

Degenkolb, H. J. (1987). "A study of the P-delta effect." *Earthq. Spectra*, 3(1).

"Geotechnical guidelines for buildings on liquefiable sites in accordance with the NBC 2005 for Greater Vancouver." (2007). *Task Force Report*, May 8 ⟨https://static1.squarespace.com/static/523c951be4b0728273e73d94/t/53234518e4b0556272c33257/1394820376117/2007+Task+Force+Report.pdf⟩.

Griffis, L. (1993). "Serviceability limit states under wind load." *Engrg. J., Am. Inst. Steel Construction*, First Quarter.

International Conference of Building Officials (ICBO). (1988). *Uniform building code, UBC-88*.

Newmark, N. M., and Hall, W. J. (1978). "Development of criteria for seismic review of selected nuclear power plants," *NUREG/CR-0098*, U.S. Nuclear Regulatory Commission.

Newmark, N. M., and Hall, W. J. (1982). *Earthquake Spectra and Design*, EERI Monograph Series, Earthquake Engineering Research Institute, Oakland, CA.

NIST. (2010). Nonlinear Structural Analysis for Seismic Design, A Guide for Practicing Structural Engineers, Report Number *NIST GCR 10-917-5*, National Institute of Standards and Technology, Gaithersburg, MD.

NIST. (2011). Selecting and Scaling Earthquake Ground Motions for Performing Response-History Analyses, *NIST GCR 11-918-15*. Prepared by the NEHRP Consultants Joint Venture for the National Institute of Standards and Technology, Gaithersburg, MD.

Pantoli, E., Chen, M., Wang, X., Astroza, R., Mintz, S., Ebrahimian, H., et al. (2013). "BNCS Report #2: Full-scale structural and nonstructural building system performance during earthquakes and postearthquake fire–test results." *Structural Systems Research Project Report Series, SSRP 13/10*. University of California San Diego, La Jolla, CA.

Post Tensioning Institute, *PTI DC10.5-Standard Requirements of the Design and Analysis of Post-Tensioned Concrete Foundations on Expansive Soils, 2012*.

U.S. Nuclear Regulatory Commission. (1999). "Reevaluation of regulatory guidance on modal response combination methods for seismic response spectrum analysis." *NUREG/CR-6645*, Brookhaven National Laboratory, Upton, NY.

Wilson, E. L., Suhawardy, I., and Habibullah, A. (1995). "A clarification of the orthogonal effects in a three-dimensional seismic analysis." *Earthq. Spectra*, 11(4).

CHAPTER C13
SEISMIC DESIGN REQUIREMENTS FOR NONSTRUCTURAL COMPONENTS

C13.1 GENERAL

Chapter 13 defines minimum design criteria for architectural, mechanical, electrical, and other nonstructural systems and components, recognizing structure use, occupant load, the need for operational continuity, and the interrelation of structural, architectural, mechanical, electrical, and other nonstructural components. Nonstructural components are designed for design earthquake ground motions as defined in Section 11.2 and determined in Section 11.4.5 of the standard. In contrast to structures, which are implicitly designed for a low probability of collapse when subjected to the risk-targeted maximum considered earthquake (MCE_R) ground motions, there are no implicit performance goals associated with the MCE_R for nonstructural components. Performance goals associated with the design earthquake are discussed in Section C13.1.3.

Suspended or attached nonstructural components that could detach either in full or in part from the structure during an earthquake are referred to as falling hazards and may represent a serious threat to property and life safety. Critical attributes that influence the hazards posed by these components include their weight, their attachment to the structure, their failure or breakage characteristics (e.g., nonshatterproof glass), and their location relative to occupied areas (e.g., over an entry or exit, a public walkway, an atrium, or a lower adjacent structure). Architectural components that pose potential falling hazards include parapets, cornices, canopies, marquees, glass, large ornamental elements (e.g., chandeliers), and building cladding. In addition, suspended mechanical and electrical components (e.g., mixing boxes, piping, and ductwork) may represent serious falling hazards. Figs. C13.1-1 through C13.1-4 show damage to nonstructural components in past earthquakes.

Components whose collapse during an earthquake could result in blockage of the means of egress deserve special consideration. The term "means of egress" is used commonly in building codes with respect to fire hazard. Egress paths may include intervening aisles, doors, doorways, gates, corridors, exterior exit balconies, ramps, stairways, pressurized enclosures, horizontal exits, exit passageways, exit courts, and yards. Items whose failure could jeopardize the means of egress include walls around stairs and corridors, veneers, cornices, canopies, heavy partition systems, ceilings, architectural soffits, light fixtures, and other ornaments above building exits or near fire escapes. Examples of components that generally do not pose a significant falling hazard include fabric awnings and canopies. Architectural, mechanical, and electrical components that, if separated from the structure, fall in areas that are not accessible to the public (e.g., into a mechanical shaft or light well) also pose little risk to egress routes.

For some architectural components, such as exterior cladding elements, wind design forces may exceed the calculated seismic design forces. Nevertheless, seismic detailing requirements may still govern the overall structural design. Where this is a possibility, it must be investigated early in the structural design process.

The seismic design of nonstructural components may involve consideration of nonseismic requirements that are affected by seismic bracing. For example, accommodation of thermal expansion in pressure piping systems often is a critical design consideration, and seismic bracing for these systems must be arranged in a manner that accommodates thermal movements. Particularly in the case of mechanical and electrical systems, the design for seismic loads should not compromise the functionality, durability, or safety of the overall design; this method requires collaboration among the various disciplines of the design and construction team.

For various reasons (e.g., business continuity), it may be desirable to consider higher performance than that required by the building code. For example, to achieve continued operability of a piping system, it is necessary to prevent unintended operation of valves or other in-line components in addition to preventing collapse and providing leak tightness. Higher performance also is required for components containing substantial quantities of hazardous contents (as defined in Section 11.2). These components must be designed to prevent uncontrolled release of those materials.

The requirements of Chapter 13 are intended to apply to new construction and tenant improvements installed at any time during the life of the structure, provided that they are listed in Table 13.5-1 or 13.6-1. Furthermore, they are intended to reduce (not eliminate) the risk to occupants and to improve the likelihood that essential facilities remain functional. Although property protection (in the sense of investment preservation) is a possible consequence of implementation of the standard, it is not currently a stated or implied goal; a higher level of protection may be advisable if such protection is desired or required.

C13.1.1 Scope. The requirements for seismic design of nonstructural components apply to the nonstructural component and to its supports and attachments to the main structure. In some cases, as defined in Section 13.2, it is necessary to consider explicitly the performance characteristics of the component. The requirements are intended to apply only to permanently attached components, not to furniture, temporary items, or mobile units. Furniture, such as tables, chairs, and desks, may shift during strong ground shaking but generally poses minimal hazards provided that it does not obstruct emergency egress routes. Storage cabinets, tall bookshelves, and other items of significant mass do not fall into this category and should be anchored or braced in accordance with this chapter.

Temporary items are those that remain in place for short periods of time (months, not years). Components that are

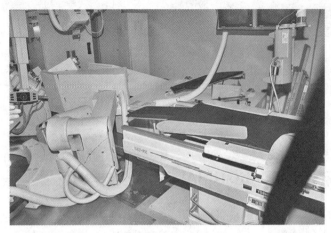

FIGURE C13.1-1. Hospital Imaging Equipment That Fell from Overhead Mounts

FIGURE C13.1-2 Damaged Ceiling System

FIGURE C13.1-3 Collapsed Light Fixtures

FIGURE C13.1-4 Toppled Storage Cabinets

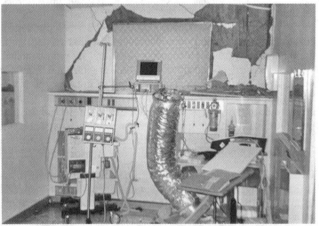

FIGURE C13.1-5 Collapsed Duct and HVAC Diffuser

FIGURE C13.1-6 Skid-Mounted Components

expected to remain in place for periods of a year or longer, even if they are designed to be movable, should be considered permanent for the purposes of this section. Modular office systems are considered permanent because they ordinarily remain in place for long periods. In addition, they often include storage units that have significant capacity and may topple in an earthquake. They are subject to the provisions of Section 13.5.8 for partitions if they exceed 6 ft (1.8 m) high. Mobile units include components that are moved from one point in the structure to another during ordinary use. Examples include desktop computers, office equipment, and other components that are not permanently attached to the building utility systems (Fig. C13.1-5). Components that are mounted on wheels to facilitate periodic maintenance or cleaning but that otherwise remain in the same location (e.g., server racks) are not considered movable for the purposes of anchorage and bracing. Likewise, skid-mounted components (as shown in Fig. C13.1-6), as well as the skids themselves, are considered permanent equipment.

With the exception of solar panels satisfying the provisions of Section 13.6.12, equipment must be anchored if it is permanently attached to utility services (electricity, gas, and water). For the purposes of this requirement, "permanently attached" should be understood to include all electrical connections except NEMA 5-15 and 5-20 straight-blade connectors (duplex receptacles).

C13.1.2 Seismic Design Category. The requirements for nonstructural components are based in part on the Seismic Design Category (SDC) to which they are assigned. As the SDC is established considering factors not unique to specific nonstructural components, all nonstructural components occupying or attached to a structure are assigned to the same SDC as the structure.

C13.1.3 Component Importance Factor. Performance expectations for nonstructural components often are defined in terms of the functional requirements of the structure to which the components are attached. Although specific performance goals for nonstructural components have yet to be defined in building codes, the component Importance Factor (I_p) implies performance levels for specific cases. For noncritical nonstructural components (those with a component Importance Factor, I_p, of 1.0), the following behaviors are anticipated for shaking of different levels of intensity:

1. Minor earthquake ground motions—minimal damage; not likely to affect functionality;
2. Moderate earthquake ground motions—some damage that may affect functionality; and
3. Design earthquake ground motions—major damage but significant falling hazards are avoided; likely loss of functionality.

Components with Importance Factors greater than 1.0 are expected to remain in place, sustain limited damage, and when necessary, function after an earthquake (see Section C13.2.2). These components can be located in structures that are not assigned to Risk Category IV. For example, fire sprinkler piping systems have an Importance Factor, I_p, of 1.5 in all structures because these essential systems should function after an earthquake. Egress stairways are assigned an I_p of 1.5 as well, although in many cases the design of these stairways is dictated by differential displacements, not inertial force demands.

The component Importance Factor is intended to represent the greater of the life-safety importance of the component and the hazard-exposure importance of the structure. It indirectly influences the survivability of the component via required design forces and displacement levels, as well as component attachments and detailing. Although this approach provides some degree of confidence in the seismic performance of a component, it may not be sufficient in all cases. For example, individual ceiling tiles may fall from a ceiling grid that has been designed for larger forces. This problem may not represent a serious falling hazard if the ceiling tiles are made of lightweight materials, but it may lead to blockage of critical egress paths or disruption of the facility function. When higher levels of confidence in performance are required, the component is classified as a designated seismic system (Section 11.2), and in certain cases, seismic qualification of the component or system is necessary. Seismic qualification approaches are provided in Sections 13.2.5 and 13.2.6. In addition, seismic qualification approaches presently in use by the Department of Energy (DOE) can be applied.

Risk Category IV structures are intended to be functional after a design earthquake; critical nonstructural components and equipment in such structures are designed with I_p equal to 1.5. This requirement applies to most components and equipment because damage to vulnerable unbraced systems or equipment may disrupt operations after an earthquake even if they are not directly classified as essential to life safety. The nonessential and nonhazardous components are themselves not affected by this requirement. Instead, requirements focus on the supports and attachments. UFC 3-310-04 (DOD 2007) has additional guidance for improved performance.

C13.1.4 Exemptions. Several classes of nonstructural components are exempted from the requirements of Chapter 13. The exemptions are made on the assumption that, either because of their inherent strength and stability or the lower level of earthquake demand (accelerations and relative displacements), or both, these nonstructural components and systems can achieve the performance goals described earlier in this commentary without explicitly satisfying the requirements of this chapter.

The requirements are intended to apply only to permanent components, not furniture and temporary or mobile equipment. Furniture (with the exception of more massive elements like storage cabinets) may shift during strong ground shaking but poses minimal hazards. With the exception of solar panels satisfying the provisions of Section 13.6.12, equipment must be anchored if it is permanently attached to the structure utility services, such as electricity, gas, or water. For the purposes of this requirement, "permanently attached" includes all electrical connections except plugs for duplex receptacles.

Temporary items are those that remain in place for six months or less. Modular office systems are considered permanent since they ordinarily remain in place for long periods. In addition, they often include storage units of significant capacity, which may topple in earthquakes. Mobile units include components that are moved from one point in the structure to another during ordinary use. Examples include desktop computers, office equipment, and other components that are not permanently attached to the building utility systems. Components mounted on wheels to facilitate periodic maintenance or cleaning but that otherwise remain in the same location are not considered movable for the purposes of anchorage and bracing.

Furniture resting on floors, such as tables, chairs, and desks, may shift during strong ground shaking, but they generally pose minimal hazards, provided that they do not obstruct emergency egress routes. Examples also include desktop computers, office equipment, and other components that are not permanently attached to the building utility systems.

With the exception of parapets supported by bearing walls or shear walls, all components in Seismic Design Categories A and B are exempt because of the low levels of ground shaking expected. Parapets are not exempt because experience has shown that these items can fail and pose a significant falling hazard, even at low-level shaking levels.

Discrete components are generally understood to be stand-alone items such as cabinets, pumps, electrical boxes, lighting, and signage. Discrete components, architectural or mechanical, weighing 20 lb (89 N) or less generally do not pose a risk and are exempted provided that they are positively attached to the structure, regardless of whether they carry an Importance Factor, I_p, of 1.5 or not. Larger items up to 400 lb (1,780 N) in weight with $I_p = 1.0$ have historically been exempted provided that they are positively attached and have flexible connections. The exemption for mechanical and electrical components in Seismic Design Categories D, E, or F based on weight and location of the center of mass is particularly applicable to vertical equipment racks and similar components. Where detailed information regarding the center of mass of the intended installation is unavailable, a conservative estimate based on potential equipment configurations

FIGURE C13.1-7 Premanufactured Modular Mechanical Systems
Source: Courtesy of Matthew Tobolski.

should be used. The exemption for components weighing 400 lb (1,780 N) or less has existed in provisions for nonstructural components for many years and corresponds roughly to the weight of a 40-gal. (150-L) hot water tank. Coupled with this and the other exemptions in SDC D, E, and F is a requirement that the component be positively attached to the structure. Positive attachment is provided when the attachment is carried out using appropriate structural-grade materials whereby explicit design calculations for the anchorage are not required.

Although the exemptions listed in Section 13.1.4 are intended to waive bracing requirements for nonstructural components that are judged to pose negligible life-safety hazard, in some cases it may nevertheless be advisable to consider bracing (in consultation with the owner) for exempted components to minimize repair costs and/or disproportionate loss (e.g., art works of high value).

The bracing exemptions for short hangers have been moved to the respective sections in which they apply. These exemptions are based on the assumption that the hangers have sufficient ductility to undergo plastic deformations without failure while at the same time providing sufficient stiffness to limit lateral displacement to a reasonable level. This assumption extends to the anchors, and as such the design and detailing of the connections to the structure should take this into account. Raceways, ducts, and piping systems must be able to accommodate the relative displacement demands calculated in Section 13.3.2, since these displacements can be substantially greater than those that occur at connections to equipment. At seismic separation joints between structures, large displacements may occur over a short distance.

Short hangers fabricated from threaded rods resist lateral force primarily through bending and are prone to failure through cyclic fatigue. Tests conducted by Soulages and Weir (2011) suggest that low cycle fatigue is not an issue when the ductility ratios for the rods are less than about 4. The testing also indicated that swivel connections are not required, provided that the load and rod length limitations are observed. The limits on unbraced trapezes and hangers are based on limiting the ductility ratios to reasonable levels, when subject to the maximum force demands in the highest seismic risk regions. It should be noted that in areas of lower seismic risk, less restrictive criteria could be used.

The exemption for short hangers is limited to the case where every hanger in the raceway run is less than 12 in. (305 mm) because of the need to carefully consider the seismic loads and compatible displacement limits for the portions of raceways with longer hanger supports.

The historical exemption for trapeze-supported conduit less than 2.5 in. (64 mm) trade size has been removed, since its application to specific cases, such as a trapeze supporting multiple conduit runs, was unclear.

The exemption for trapezes with short rod hangers applies only to trapezes configured with the rod hangers attached directly to the trapeze and the structural framing. Where one or more rod hangers for a trapeze are supported from another trapeze, the bracing exemption does not apply.

C13.1.5 Premanufactured Modular Mechanical and Electrical Systems. Large premanufactured modular mechanical and electrical systems (as shown in Fig. C13.1-7) should be considered nonbuilding structures for the purposes of the enveloping structural system design, unless the module has been prequalified in accordance with Section 13.2.2. However, where the premanufactured module has not been prequalified, the nonstructural components contained within the module should be addressed through the requirements of Chapter 13. Note that this provision is not intended to address skid-mounted equipment assemblies not equipped with an enclosure, nor does is address single large components, such as air handlers, cooling towers, chillers, and boilers.

C13.1.6 Application of Nonstructural Component Requirements to Nonbuilding Structures. At times, a nonstructural component should be treated as a nonbuilding structure. When the physical characteristics associated with a given class of nonstructural components vary widely, judgment is needed to select the appropriate design procedure and coefficients. For example, cooling towers vary from small packaged units with an operating weight of 2,000 lb (8.9 kN) or less to structures the size of buildings. Consequently, design coefficients for the design of "cooling towers" are found both in Tables 13.6-1 and 15.4-2. Small cooling towers are best designed as nonstructural components using the provisions of Chapter 13, whereas large ones are clearly nonbuilding structures that are more appropriately designed using the provisions of Chapter 15. Similar issues arise for other classes of nonstructural component (e.g., boilers and bins). Guidance on determining whether an item should be treated as a nonbuilding structure or nonstructural component for the purpose of seismic design is provided in Bachman and Dowty (2008).

The specified weight limit for nonstructural components (25% relative to the combined weight of the structure and component) relates to the condition at which dynamic interaction between the component and the supporting structural system is potentially significant. Section 15.3.2 contains requirements for addressing this interaction in design.

C13.1.7 Reference Documents. Professional and trade organizations have developed nationally recognized codes and standards for the design and construction of specific mechanical and electrical components. These documents provide design guidance for normal and upset (abnormal) operating conditions and for various environmental conditions. Some of these documents include earthquake design requirements in the context of the overall mechanical or electrical design. It is the intent of the standard that seismic requirements in referenced documents be used. The developers of these documents are familiar with the expected performance and failure modes of the components; however, the documents may be based on design considerations not immediately obvious to a structural design professional. For example, in the design of industrial piping, stresses caused by seismic inertia forces typically are not added to those caused by thermal expansion.

Where reference documents have been adopted specifically by this standard as meeting the force and displacement requirements of this chapter with or without modification, they are considered to be a part of the standard.

There is a potential for misunderstanding and misapplication of reference documents for the design of mechanical and electrical systems. A registered design professional familiar with both the standard and the reference documents used should be involved in the review and acceptance of the seismic design.

Even when reference documents for nonstructural components lack specific earthquake design requirements, mechanical and electrical equipment constructed in accordance with industry-standard reference documents have performed well historically when properly anchored. Nevertheless, manufacturers of mechanical and electrical equipment are expected to consider seismic loads in the design of the equipment itself, even when such consideration is not explicitly required by this chapter.

Although some reference documents provide requirements for seismic capacity appropriate to the component being designed, the seismic demands used in design may not be less than those specified in the standard.

Specific guidance for selected mechanical and electrical components and conditions is provided in Section 13.6.

Unless exempted in Section 13.1.4, components should be anchored to the structure and, to promote coordination, required supports and attachments should be detailed in the construction documents. Reference documents may contain explicit instruction for anchorage of nonstructural components. The anchorage requirements of Section 13.4 must be satisfied in all cases, however, to ensure a consistent level of robustness in the attachments to the structure.

C13.1.8 Reference Documents Using Allowable Stress Design. Many nonstructural components are designed using specifically developed reference documents that are based on allowable stress loads and load combinations and generally permit increases in allowable stresses for seismic loading. Although Section 2.4.1 of the standard does not permit increases in allowable stresses, Section 13.1.8 explicitly defines the conditions for stress increases in the design of nonstructural components where reference documents provide a basis for earthquake-resistant design.

C13.2 GENERAL DESIGN REQUIREMENTS

C13.2.1 Applicable Requirements for Architectural, Mechanical, and Electrical Components, Supports, and Attachments. Compliance with the requirements of Chapter 13 may be accomplished by project-specific design or by a manufacturer's certification of seismic qualification of a system or component. When compliance is by manufacturer's certification, the items must be installed in accordance with the manufacturer's requirements. Evidence of compliance may be provided in the form of a signed statement from a representative of the manufacturer or from the registered design professional indicating that the component or system is seismically qualified. One or more of the following options for evidence of compliance may be applicable:

1. An analysis (e.g., of a distributed system such as piping) that includes derivation of the forces used for the design of the system, the derivation of displacements and reactions, and the design of the supports and anchorages;
2. A test report, including the testing configuration and boundary conditions used (where testing is intended to address a class of components, the range of items covered by the testing performed should also include the justification of similarities of the items that make this certification valid); and/or
3. An experience data report.

Components addressed by the standard include individual simple units and assemblies of simple units for which reference documents establish seismic analysis or qualification requirements. Also addressed by the standard are complex architectural, mechanical, and electrical systems for which reference documents either do not exist or exist for only elements of the system. In the design and analysis of both simple components and complex systems, the concepts of flexibility and ruggedness often can assist the designer in determining the necessity for analysis and, when analysis is necessary, the extent and methods by which seismic adequacy may be determined. These concepts are discussed in Section C13.6.1.

C13.2.2 Special Certification Requirements for Designated Seismic Systems. This section addresses the qualification of active designated seismic equipment, its supports, and attachments with the goals of improving survivability and achieving a high level of confidence that a facility will be functional after a design earthquake. Where components are interconnected, the qualification should provide the permissible forces (e.g., nozzle loads) and, as applicable, anticipated displacements of the component at the connection points to facilitate assessment for consequential damage, in accordance with Section 13.2.3. Active equipment has parts that rotate, move mechanically, or are energized during operation. Active designated seismic equipment constitutes a limited subset of designated seismic systems. Failure of active designated seismic equipment itself may pose a significant hazard. For active designated seismic equipment, failure of structural integrity and loss of function are to be avoided.

Examples of active designated seismic equipment include mechanical (components of HVACR systems and piping systems) or electrical (power supply distribution) equipment, medical equipment, fire pump equipment, and uninterruptible power supplies for hospitals. It is generally understood that fire protection sprinkler piping systems designed and installed per NFPA 13 are deemed to comply with the special certification requirements of Section 13.2.2. See Section 13.6.7.2.

There are practical limits on the size of a component that can be qualified via shake table testing. Components too large to be qualified by shake table testing need to be qualified by a combination of structural analysis and qualification testing or empirical evaluation through a subsystem approach. Subsystems of large, complex components (e.g., large chillers, skid-mounted equipment assemblies, and boilers) can be qualified individually, and the overall structural frame of the component can be evaluated by structural analysis.

Evaluating postearthquake operational performance for active equipment by analysis generally involves sophisticated modeling with experimental validation and may not be reliable. Therefore, the use of analysis alone for active or energized components is not permitted unless a comparison can be made to components that have been otherwise deemed as rugged. As an example, a transformer is energized but contains components that can be shown to remain linearly elastic and are inherently rugged. However, switch equipment that contains fragile components is similarly energized but not inherently rugged, and it therefore cannot be certified solely by analysis. For complex components, testing or experience may therefore be the only practical way to ensure that the equipment will be operable after a design earthquake. Past earthquake experience has shown that much active equipment is inherently rugged. Therefore, evaluation of experience data, together with analysis of anchorage, is adequate to demonstrate compliance of active equipment such as pumps, compressors, and electric motors. In other cases, such as for motor control centers and switching equipment, shake table testing may be required.

With some exceptions (e.g., elevator motors), experience indicates that active mechanical and electrical components that contains electric motors of greater than 10 hp (7.4 kW) or that have a thermal exchange capacity greater than 200 MBH are unlikely to merit the exemption from shake table testing on the basis of inherent ruggedness. Components with lesser motor horsepower and thermal exchange capacity are generally considered to be small active components and are deemed rugged. Exceptions to this rule may be appropriate for specific cases, such as elevator motors that have higher horsepower but have been shown by experience to be rugged. Analysis is still required to ensure the structural integrity of the nonactive components. For example, a 15-ton condenser would require analysis of the load path between the condenser fan and the coil to the building structure attachment.

Where certification is accomplished by analysis, the type and sophistication of the required analysis varies by specific equipment type and construction. Static analysis using the total force specified in Section 13.3 considering applicable load combinations may be appropriate for single components where the structural frame is the only item to be certified and where internal dynamic effects are shown to be negligible. For single components where dynamic effects may be significant, or for assemblages of components, dynamic analysis is strongly suggested. Either modal analysis or response history procedures may be used, but care should be exercised when using modal analysis to ensure that the significant interactions between individual components are properly captured. In all analyses, it is essential that the stiffness, mass, and applied load be distributed in accordance with the component properties, and in sufficient detail (number of degrees of freedom) to allow for the desired forces, deformations, and accelerations to be accurately determined. Input motions for dynamic procedures should reflect the expected motion at the attachment points of the component. Nonlinear behavior of the component is typically not advisable in the certification analysis in the absence of well-documented test results for specific components. Generally, the input motion is (a) a generic floor response spectrum such as that provided in the ICC-ES AC 156, (b) location- and structure-specific floor spectra generated using the procedures of Section 13.3.1, or (c) acceleration time histories developed using dynamic analysis procedures similar to those specified in Chapter 16 or Section 12.9. Horizontal and vertical inputs are usually applied simultaneously when performing these types of dynamic analyses. As with all structural analysis, judgment is required to ensure that the results are applicable and representative of the behavior anticipated for the input motions.

C13.2.3 Consequential Damage. Although the components identified in Tables 13.5-1 and 13.6-1 are listed separately, significant interrelationships exist and must be considered. Consequential damage occurs because of interaction between components and systems. Even "braced" components displace, and the displacement between lateral supports can be significant in the case of distributed systems such as piping systems, cable and conduit systems, and other linear systems. It is the intent of the standard that the seismic displacements considered include both relative displacement between multiple points of support (addressed in Section 13.3.2) and, for mechanical and electrical components, displacement within the component assemblies. Impact of components must be avoided, unless the components are fabricated of ductile materials that have been shown to be capable of accommodating the expected impact loads. With protective coverings, ductile mechanical and electrical components and many more fragile components are expected to survive all but the most severe impact loads. Flexibility and ductility of the connections between distribution systems and the equipment to which they attach is essential to the seismic performance of the system.

The determination of the displacements that generate these interactions is not addressed explicitly in Section 13.3.2.1. That section concerns relative displacement of support points. Consequential damage may occur because of displacement of components and systems between support points. For example, in older suspended ceiling installations, excessive lateral displacement of a ceiling system may fracture sprinkler heads that project through the ceiling. A similar situation may arise if sprinkler heads projecting from a small-diameter branch line pass through a rigid ceiling system. Although the branch line may be properly restrained, it may still displace sufficiently between lateral support points to affect other components or systems. Similar interactions occur where a relatively flexible distributed system connects to a braced or rigid component.

The potential for impact between components that are in contact with or close to other structural or nonstructural components must be considered. However, where considering these potential interactions, the designer must determine if the potential interaction is both credible and significant. For example, the fall of a ceiling panel located above a motor control center is a credible interaction because the falling panel in older suspended ceiling installations can reach and impact the motor control center. An interaction is significant if it can result in damage to the target. Impact of a ceiling panel on a motor control center may not be significant because of the light weight of the ceiling panel. Special design consideration is appropriate where the failure of a nonstructural element could adversely influence the performance of an adjacent critical nonstructural component, such as an emergency generator.

C13.2.4 Flexibility. In many cases, flexibility is more important than strength in the performance of distributed systems, such as piping and ductwork. A good understanding of the displacement demand on the system, as well as its displacement capacity, is required. Components or their supports and attachments must be flexible enough to accommodate the full range of expected differential movements; some localized inelasticity is permitted in accommodating the movements. Relative movements in all directions must be considered. For example, even a braced branch line of a piping system may displace, so it needs to be connected to other braced or rigid components in a manner that accommodates the displacements without failure (Fig. C13.2-1). A further example is provided by cladding units (such as precast concrete

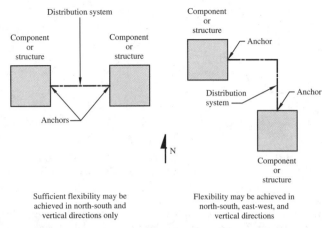

FIGURE C13.2-1 Schematic Plans Illustrating Branch Line Flexibility

wall units). Often very rigid in plane, cladding units require connections capable of accommodating story drift if attached at more than one level. (See Fig. C13.3-4 for an illustration.)

If component analysis assumes rigid anchors or supports, the predicted loads and local stresses can be unrealistically large, so it may be necessary to consider anchor and/or support stiffness.

C13.2.5 Testing Alternative for Seismic Capacity Determination. Testing is a well-established alternative method of seismic qualification for small- to medium-size equipment. Several national reference documents have testing requirements adaptable for seismic qualification. One such reference document (ICC-ES AC 156) is a shake table testing protocol that has been adopted by the International Code Council Evaluation Service. It was developed specifically to be consistent with acceleration demands (that is, force requirements) of the standard.

The development or selection of testing and qualification protocols should at a minimum include the following:

1. Description of how the protocol meets the intent for the project-specific requirements and relevant interpretations of the standard;
2. Definition of a test input motion with a response spectrum that meets or exceeds the design earthquake spectrum for the site;
3. Accounting for dynamic amplification caused by above-grade equipment installations (consideration of the actual dynamic characteristics of the primary support structure is permitted, but not required);
4. Definition of how shake table input demands were derived;
5. Definition and establishment of a verifiable pass/fail acceptance criterion for the seismic qualification based on the equipment Importance Factor and consistent with the building code and project-specific design intent; and
6. Development of criteria that can be used to rationalize test unit configuration requirements for highly variable equipment product lines.

To aid the design professional in assessing the adequacy of the manufacturer's certificate of compliance, it is recommended that certificates of compliance include the following:

1. Product family or group covered;
2. Building code(s) and standard(s) for which compliance was evaluated;
3. Testing standard used;
4. Performance objective and corresponding Importance Factor ($I_p = 1.0$ or $I_p = 1.5$);
5. Seismic demand for which the component is certified, including code and/or standard design parameters used to calculate seismic demand (such as values used for a_p, R_p, and site class); and
6. Installation restrictions, if any (grade, floor, or roof level).

Without a test protocol recognized by the building code, qualification testing is inconsistent and difficult to verify. The use of ICC-ES AC 156 simplifies the task of compliance verification because it was developed to address directly the testing alternative for nonstructural components, as specified in the standard. It also sets forth minimum test plan and report deliverables.

Use of other standards or ad hoc protocols to verify compliance of nonstructural components with the requirement of the standard should be considered carefully and used only where project-specific requirements cannot be met otherwise.

Where other qualification test standards are used, in whole or in part, it is necessary to verify compliance with this standard. For example, IEEE 693 (2005) indicates that it is to be used for the sole purpose of qualifying electrical equipment (specifically listed in the standard) for use in utility substations. Where equipment testing has been conducted to other standards (for instance, testing done in compliance with IEEE 693), a straightforward approach would be to permit evaluation, by the manufacturer, of the test plan and data to validate compliance with the requirements of ICC-ES AC 156 because it was developed specifically to comply with the seismic demands of this standard.

The qualification of mechanical and electrical components for seismic loads alone may not be sufficient to achieve high-performance objectives. Establishing a high confidence that performance goals will be met requires consideration of the performance of structures, systems (e.g., fluid, mechanical, electrical, and instrumentation), and their interactions (e.g., interaction of seismic and other loads), as well as compliance with installation requirements.

C13.2.6 Experience Data Alternative for Seismic Capacity Determination. An established method of seismic qualification for certain types of nonstructural components is the assessment of data for the performance of similar components in past earthquakes. The seismic capacity of the component in question is extrapolated based on estimates of the demands (e.g., force or displacement) to which the components in the database were subjected. Procedures for such qualification have been developed for use in nuclear facility applications by the Seismic Qualification Utility Group (SQUG) of the Electric Power Research Institute.

The SQUG rules for implementing the use of experience data are described in a proprietary Generic Implementation Procedure database. It is a collection of findings from detailed engineering studies by experts for equipment from a variety of utility and industrial facilities.

Valid use of experience data requires satisfaction of rules that address physical characteristics; manufacturer's classification and standards; and findings from testing, analysis, and expert consensus opinion.

Four criteria are used to establish seismic qualification by experience, as follows:

1. Seismic capacity versus demand (a comparison with a bounding spectrum);
2. Earthquake experience database cautions and inclusion rules;

3. Evaluation of anchorage; and
4. Evaluation of seismic interaction.

Experience data should be used with care because the design and manufacture of components may have changed considerably in the intervening years. The use of this procedure is also limited by the relative rarity of strong-motion instrument records associated with corresponding equipment experience data.

C13.2.7 Construction Documents. Where the standard requires seismic design of components or their supports and attachments, appropriate construction documents defining the required construction and installation must be prepared. These documents facilitate the special inspection and testing needed to provide a reasonable level of quality assurance. Of particular concern are large nonstructural components (such as rooftop chillers) whose manufacture and installation involve multiple trades and suppliers and which impose significant loads on the supporting structure. In these cases, it is important that the construction documents used by the various trades and suppliers be prepared by a registered design professional to satisfy the seismic design requirements.

The information required to prepare construction documents for component installation includes the dimensions of the component, the locations of attachment points, the operating weight, and the location of the center of mass. For instance, if an anchorage angle is attached to the side of a metal chassis, the gauge and material of the chassis must be known so that the number and size of required fasteners can be determined. Or when a piece of equipment has a base plate that is anchored to a concrete slab with expansion anchors, the drawings must show the base plate's material and thickness, the diameter of the bolt holes in the plate, and the size and depth of embedment of the anchor bolts. If the plate is elevated above the slab for leveling, the construction documents must also show the maximum gap permitted between the plate and the slab.

C13.3 SEISMIC DEMANDS ON NONSTRUCTURAL COMPONENTS

The seismic demands on nonstructural components, as defined in this section, are acceleration demands and relative displacement demands. Acceleration demands are represented by equivalent static forces. Relative displacement demands are provided directly and are based on either the actual displacements computed for the structure or the maximum allowable drifts that are permitted for the structure.

C13.3.1 Seismic Design Force. The seismic design force for a component depends on the weight of the component, the component Importance Factor, the component response modification factor, the component amplification factor, and the component acceleration at a point of attachment to the structure. The forces prescribed in this section of the standard reflect the dynamic and structural characteristics of nonstructural components. As a result of these characteristics, forces used for verification of component integrity and design of connections to the supporting structure typically are larger than those used for design of the overall seismic force-resisting system.

Certain nonstructural components lack the desirable attributes of structures (such as ductility, toughness, and redundancy) that permit the use of greatly reduced lateral design forces. Thus values for the response modification factor, R_p, in Tables 13.5-1 and 13.6-1 generally are smaller than R values for structures. These R_p values, used to represent the energy absorption capability of a component and its attachments, depend on both

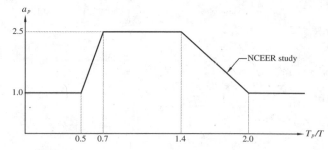

FIGURE C13.3-1 NCEER Formulation for a_p as Function of Structural and Component Periods

overstrength and deformability. At present, these potentially separate considerations are combined in a single factor. The tabulated values are based on the collective judgment of the responsible committee.

Beginning with the 2005 edition of ASCE 7, significant adjustments have been made to tabulated R_p values for certain mechanical and electrical systems. For example, the value of R_p for welded steel piping systems is increased from 3.5 to 9. The a_p value increased from 1.0 to 2.5, so although it might appear that forces on such piping systems have been reduced greatly, the net change is negligible because R_p/a_p changes from 3.5 to 3.6. The minimum seismic design force of Eq. (13.3-3), which governs in many cases, is unchanged.

The component amplification factor (a_p) represents the dynamic amplification of component responses as a function of the fundamental periods of the structure (T) and component (T_p). When components are designed or selected, the structural fundamental period is not always defined or readily available. The component fundamental period (T_p) is usually only accurately obtained by shake table or pull-back tests and is not available for the majority of components. Tabulated a_p values are based on component behavior that is assumed to be either rigid or flexible. Where the fundamental period of the component is less than 0.06 s, dynamic amplification is not expected and the component is considered rigid. The tabulation of assumed a_p values is not meant to preclude more precise determination of the component amplification factor where the fundamental periods of both structure and component are available. The National Center for Earthquake Engineering Research formulation shown in Fig. C13.3-1 may be used to compute a_p as a function of T_p/T.

Dynamic amplification occurs where the period of a nonstructural component closely matches that of any mode of the supporting structure, although this effect may not be significant depending on the ground motion. For most buildings, the primary mode of vibration in each direction has the most influence on the dynamic amplification for nonstructural components. For long-period structures (such as tall buildings), where the period of vibration of the fundamental mode is greater than 3.5 times T_s, higher modes of vibration may have periods that more closely match the period of nonstructural components. For this case, it is recommended that amplification be considered using such higher mode periods in lieu of the higher fundamental period. This approach may be generalized by computing floor response spectra for various levels that reflect the dynamic characteristics of the supporting structure to determine how amplification varies as a function of component period. Calculation of floor response spectra is described in Section 13.3.1.1. This procedure can be complex, but a simplified procedure is presented in Section 13.3.1.2. Consideration of nonlinear behavior of the structure greatly complicates the analysis.

In ASCE 7-10, values for Ω_0 were added to Tables 13.5-1 and 13.6-1. Ω_0 is included in the load combinations for the anchorage of nonstructural components to concrete or masonry in cases where the anchorage is deemed nonductile. ACI 318-14, Section 17.2.3.4.3 (d) requires the inclusion of Ω_0 for anchors that do not otherwise satisfy ductility requirements. This factor closely resembles the 1.3 factor and R_p penalty in ASCE 7-05 for nonductile anchorages. The upper value for Ω_0 was originally selected based on similar provisions in prior editions of the IBC and ACI codes. Research at San Diego State University indicates that this value need not exceed 2.

Eq. (13.3-1) represents a trapezoidal distribution of floor accelerations within a structure, varying linearly from the acceleration at the ground (taken as $0.4S_{DS}$) to the acceleration at the roof (taken as $1.2S_{DS}$). The ground acceleration ($0.4S_{DS}$) is intended to be the same acceleration used as design input for the structure itself, including site effects. The roof acceleration is established as three times the input ground acceleration based on examination of recorded in-structure acceleration data for short and moderate-height structures in response to large California earthquakes. Work by Dowell and Johnson (2013) suggests that, for taller structures, the amplification with height may vary significantly because of higher mode effects. Where more information is available, Eq. (13.3-4) permits an alternate determination of the component design forces based on the dynamic properties of the structure. When using a modal analysis procedure, as described in Section 12.9.1, the maximum floor accelerations are used.

Eq. (13.3-3) establishes a minimum seismic design force, F_p, which is consistent with current practice. Eq. (13.3-2) provides a simple maximum value of F_p that prevents multiplication of the individual factors from producing a design force that would be unreasonably high, considering the expected nonlinear response of support and component. Fig. C13.3-2 illustrates the distribution of the specified lateral design forces.

For elements with points of attachment at more than one height, it is recommended that design be based on the average of values of F_p determined individually at each point of attachment (but with the entire component weight, W_p) using Eqs. (13.3-1) through (13.3-3).

Alternatively, for each point of attachment, a force F_p may be determined using Eqs. (13.3-1) through (13.3-3), with the portion of the component weight, W_p, tributary to the point of attachment. For design of the component, the attachment force F_p must be distributed relative to the component's mass distribution over the area used to establish the tributary weight. To illustrate these options, consider a solid exterior nonstructural wall panel, supported top and bottom, for a one-story building with a rigid diaphragm. The values of F_p computed, respectively, for the top and bottom attachments using Eqs. (13.3-1) through (13.3-3) are $0.48S_{DS}I_pW_p$ and $0.30S_{DS}I_pW_p$. In the recommended method, a uniform load is applied to the entire panel based on $0.39S_{DS}I_pW_p$. In the alternative method, a trapezoidal load varying from $0.48S_{DS}I_pW_p$ at the top to $0.30S_{DS}I_pW_p$ at the bottom is applied. Each anchorage force is then determined considering static equilibrium of the complete component subject to all the distributed loads.

Cantilever parapets that are part of a continuous element should be checked separately for parapet forces. The seismic force on any component must be applied at the center of gravity of the component and must be assumed to act in any horizontal direction. Vertical forces on nonstructural components equal to $\pm 0.2S_{DS}W_p$ are specified in Section 13.3.1 and are intended to be applied to all nonstructural components and not just cantilevered elements. Nonstructural concrete or masonry walls laterally supported by flexible diaphragms must be anchored out of plane in accordance with Section 12.11.2.

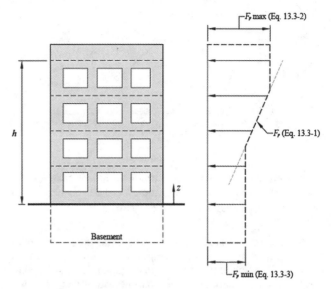

FIGURE C13.3-2 Lateral Force Magnitude over Height

C13.3.1.4 Dynamic Analysis

C13.3.1.4.1 Floor Response Spectra. The response of nonstructural components to earthquake shaking is influenced by the dynamic interaction of the nonstructural component with the response of the structure. Dynamic amplification occurs where the period of a nonstructural component closely matches that of any mode of the supporting structure, although this effect may not be significant depending on the ground motion. For most buildings, the primary mode of vibration in each direction will have the most influence on the dynamic amplification for nonstructural components. For long-period structures (such as tall buildings) higher modes of vibration may have periods that more closely match the period of nonstructural components. For this case, it is recommended that amplification be considered using such higher mode periods in lieu of the higher fundamental period. The approach may be generalized by computing floor response spectra for various levels that reflect the dynamic characteristics of the supporting structure to determine how amplification will vary as a function of component period. In order to properly consider this dynamic amplification, the vibrational characteristics of the building and the component must be known. The characteristics of the building are typically determined from a structural model of the building. The vibrational characteristics of the nonstructural components can be determined either by calculation or by testing. It should be noted that many types of nonstructural components, such as mechanical and electrical equipment, have multiple modes of vibration. The floor response spectra approach may not be applicable to nonstructural components that cannot be characterized as having a predominant mode of vibration.

C13.3.1.4.2 Alternate Floor Response Spectra. Calculation of floor response spectra can be complex since it requires a response history analysis. An alternative method of calculating a floor response history has been presented in Kehoe and Hachem (2003). The procedure described is based on a method of calculating floor response spectra initially developed for the U.S. Army Tri-Service manual (1986). The referenced paper compares the alternate floor response spectra to floor response spectra generated for three example buildings using ground motion records from several California earthquakes.

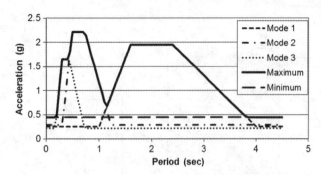

FIGURE C13.3-3 Approximate Floor Response Spectrum
Source: Kehoe and Hachem (2003); reproduced with permission from the Applied Technology Council.

This method considers the dynamic amplification contribution to the nonstructural component response based on the first three modes of vibration of the structure in calculating a floor response spectrum. It is important to note that the use of this procedure requires that the dynamic properties of both the building and the nonstructural components be known. A dynamic amplification factor is applied to each of the modal floor acceleration values for each of the first three modes of vibration, which is calculated based on linear dynamic response spectrum analysis procedures as described in Section 12.9. The dynamic amplification factor is approximated as a function of the ratio of the component period to the period of vibration of a mode of vibration of the building. The magnitude of amplification can vary depending on the ground motion and the building's dynamic characteristics. A peak amplification of 5 has been assumed for the cases where the ratio of component period to building period is between 0.8 and 1.2 using 5% damping for the nonstructural component based on review of results from previous studies (Kehoe and Hachem 2003). Where the ratio of component period to building period is less than 0.5 or more than 2.0, there is no dynamic amplification (Bachman et al. 1993).

The method allows calculation of floor response spectra in each direction at each floor level of the building using the mode shapes and the modal participation factors to amplify the ground motion to each floor level. The method considers linear building response. An example of a floor response spectrum is shown in Fig. C13.3-3. As shown in this figure, the floor response spectrum is taken as the maximum acceleration at each component period from each of the first three modes of the building in each orthogonal direction. The floor response spectrum can be used as a basis for determining the design horizontal force for the nonstructural component by multiplying the acceleration by the component weight and the Importance Factor and dividing by the response factor, R_p.

C13.3.2 Seismic Relative Displacements. The equations of this section are for use in design of cladding, stairways, windows, piping systems, sprinkler components, and other components connected to one structure at multiple levels or to multiple structures. Two equations are given for each situation. Eqs. (13.3-7) and (13.3-9) produce structural displacements as determined by elastic analysis, unreduced by the structural response modification factor (R). Because the actual displacements may not be known when a component is designed or procured, Eqs. (13.3-8) and (13.3-10) provide upper-bound displacements based on structural drift limits. Use of upper-bound equations may facilitate timely design and procurement of components, but may also result in costly added conservatism.

The value of seismic relative displacements is taken as the calculated displacement, D_p, times the Importance Factor, I_e, because the elastic displacement calculated in accordance with Eq. (12.8-15) to establish δ_x (and thus D_p) is adjusted for I_e in keeping with the philosophy of displacement demand for the structure. For component design, the unreduced elastic displacement is appropriate.

The standard does not provide explicit acceptance criteria for the effects of seismic relative displacements, except for glazing. Damage to nonstructural components caused by relative displacement is acceptable, provided that the performance goals defined elsewhere in the chapter are achieved.

The design of some nonstructural components that span vertically in the structure can be complicated when supports for the element do not occur at horizontal diaphragms. The language in Section 13.3.2 was previously amended to clarify that story drift must be accommodated in the elements that actually distort. For example, a glazing system supported by precast concrete spandrels must be designed to accommodate the full story drift, even though the height of the glazing system is only a fraction of the floor-to-floor height. This condition arises because the precast spandrels behave as rigid bodies relative to the glazing system and therefore all the drift must be accommodated by anchorage of the glazing unit, the joint between the precast spandrel and the glazing unit, or some combination of the two.

C13.3.2.1 Displacements within Structures. Seismic relative displacements can subject components or systems to unacceptable stresses. The potential for interaction resulting from component displacements (in particular for distributed systems) and the resulting impact effects should also be considered (see Section 13.2.3).

These interrelationships may govern the clearance requirements between components or between components and the surrounding structure. Where sufficient clearance cannot be provided, consideration should be given to the ductility and strength of the components and associated supports and attachments to accommodate the potential impact.

Where nonstructural components are supported between, rather than at, structural levels, as frequently occurs for glazing systems, partitions, stairs, veneers, and mechanical and electrical distributed systems, the height over which the displacement demand, D_p, must be accommodated may be less than the story height, h_{sx}, and should be considered carefully. For example, consider the glazing system supported by rigid precast concrete spandrels shown in Fig. C13.3-4. The glazing system may be subjected to full story drift, D_p, although its height ($h_x - h_y$) is only a fraction of the story height. The design drift must be accommodated by anchorage of the glazing unit, the joint between the precast spandrel and the glazing unit, or some combination of the two. Similar displacement demands arise where pipes, ducts, or conduits that are braced to the floor or roof above are connected to the top of a tall, rigid, floor-mounted component.

For ductile components, such as steel piping fabricated with welded connections, the relative seismic displacements between support points can be more significant than inertial forces. Ductile piping can accommodate relative displacements by local yielding with strain accumulations well below failure levels. However, for components fabricated using less ductile materials, where local yielding must be avoided to prevent unacceptable failure consequences, relative displacements must be accommodated by flexible connections.

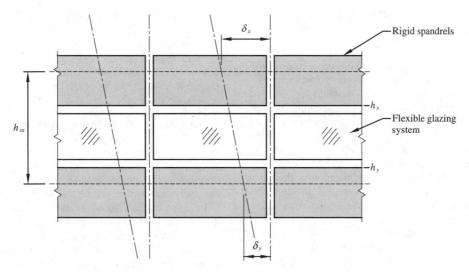

FIGURE C13.3-4 Displacements over Less than Story Height

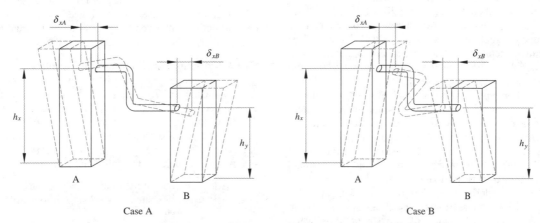

FIGURE C13.3-5 Displacements between Structures

C13.3.2.2 Displacements between Structures. A component or system connected to two structures must accommodate horizontal movements in any direction, as illustrated in Fig. C13.3-5.

C13.3.3 Component Period. Component period is used to classify components as rigid ($T \leq 0.06$ s) or flexible ($T > 0.06$ s). Determination of the fundamental period of an architectural, mechanical, or electrical component using analytical or test methods is often difficult and, if not properly performed, may yield incorrect results. In the case of mechanical and electrical equipment, the flexibility of component supports and attachments typically dominates component response and fundamental component period, and analytical determinations of component period should consider those sources of flexibility. Where testing is used, the dominant mode of vibration of concern for seismic evaluation must be excited and captured by the test setup. The dominant mode of vibration for these types of components cannot generally be acquired through in situ tests that measure only ambient vibrations. To excite the mode of vibration with the highest fundamental period by in situ testing, relatively significant input levels of motion may be required to activate the flexibility of the base and attachment. A resonant frequency search procedure, such as that given in ICC-ES acceptance criteria (AC156 2010), may be used to identify the dominant modes of vibration of a component.

Many mechanical components have fundamental periods below 0.06 s and may be considered rigid. Examples include horizontal pumps, engine generators, motor generators, air compressors, and motor-driven centrifugal blowers. Other types of mechanical equipment, while relatively stiff, have fundamental periods (up to about 0.13 s) that do not permit automatic classification as rigid. Examples include belt-driven and vane axial fans, heaters, air handlers, chillers, boilers, heat exchangers, filters, and evaporators. Where such equipment is mounted on vibration isolators, the fundamental period is substantially increased.

Electrical equipment cabinets can have fundamental periods ranging from 0.06 to 0.3 s, depending upon the supported weight and its distribution, the stiffness of the enclosure assembly, the flexibility of the enclosure base, and the load path through to the attachment points. Tall, narrow motor control centers and switchboards lie at the upper end of this period range. Low- and medium-voltage switchgear, transformers, battery chargers, inverters, instrumentation cabinets, and instrumentation racks usually have fundamental periods ranging from 0.1 to 0.2 s. Braced battery racks, stiffened vertical control panels, bench boards, electrical cabinets with top bracing, and wall-mounted panelboards generally have fundamental periods ranging from 0.06 to 0.1 s.

C13.4 NONSTRUCTURAL COMPONENT ANCHORAGE

Unless exempted in Section 13.1.4 or 13.6.9, components must be anchored to the structure, and all required supports and attachments must be detailed in the construction documents. To satisfy the load path requirement of this section, the detailed information described in Section C13.2.7 must be communicated during the design phase to the registered design professional responsible for the design of the supporting structure. The load path includes housekeeping slabs and curbs, which must be adequately reinforced and positively fastened to the supporting structure. Because the exact magnitude and location of the loads imposed on the structure may not be known until nonstructural components are ordered, the initial design of supporting structural elements should be based on conservative assumptions. The design of the supporting structural elements must be verified once the final magnitude and location of the design loads have been established. The limited exception for ballasted rooftop solar panels meeting the requirements of Section 13.6.12 is intended to accommodate the increasing use of such arrays on roof systems where positive attachment is difficult.

Design documents should provide details with sufficient information so that compliance with these provisions can be verified. Parameters such as a_p, R_p, I_p, S_{DS}, and W_p should be noted. Attachment details may include, as appropriate, dimensions and material properties of the connecting material, weld sizes, bolt sizes and material types for steel-to-steel connections, postinstalled anchor types, diameters, embedments, installation requirements, sheet metal screw diameters and material thicknesses of the connected parts, wood fastener types, and minimum requirements for specific gravity of the base materials.

Seismic design forces are determined using the provisions of Section 13.3.1. Specific reference standards should be consulted for additional adjustments to loads or strengths. Refer, for example, to the anchor design provisions of ACI 318, Chapter 17, for specific provisions related to seismic design of anchors in concrete. Unanchored components often rock or slide when subjected to earthquake motions. Because this behavior may have serious consequences, is difficult to predict, and is exacerbated by vertical ground motions, positive restraint must be provided for each component.

The effective seismic weight used in design of the seismic force-resisting system must include the weight of supported components. To satisfy the load path requirements of this section, localized component demand must also be considered. This satisfaction may be accomplished by checking the capacity of the first structural element in the load path (for example, a floor beam directly under a component) for combined dead, live, operating, and seismic loads, using the horizontal and vertical loads from Section 13.3.1 for the seismic demand, and repeating this procedure for each structural element or connection in the load path until the load case, including horizontal and vertical loads from Section 13.3.1, no longer governs design of the element. The load path includes housekeeping slabs and curbs, which must be adequately reinforced and positively fastened to the supporting structure.

Because the exact magnitude and location of loads imposed on the structure may not be known until nonstructural components are ordered, the initial design of supporting structural elements should be based on conservative assumptions. The design of the supporting structural elements may need to be verified once the final magnitude and location of the design loads have been established.

Tests have shown that there are consistent shear ductility variations between bolts installed in drilled or punched plates with nuts and connections using welded shear studs. The need for reductions in allowable loads for particular anchor types to account for loss of stiffness and strength may be determined through appropriate dynamic testing. Although comprehensive design recommendations are not available at present, this issue should be considered for critical connections subject to dynamic or seismic loading.

C13.4.1 Design Force in the Attachment. Previous editions of ASCE/SEI 7 included provisions for the amplification of forces to design the component anchorage. These provisions were intended to ensure that the anchorage either (a) would respond to overload in a ductile manner or (b) would be designed so that the anchorage would not be the weakest link in the load path.

Because of the difficulties associated with the application of the anchorage provisions in Section 13.4 in conjunction with anchorage provisions in other reference standards, the provisions for anchorage in ASCE/SEI 7-10 are substantially simplified. Adjustments on the R_p value used for the anchorage calculation have been eliminated, with the exception of the upper limit on R_p of 6, which is intended primarily to address the anchorage of ductile piping systems that are assigned higher R_p values. These higher component response modification factors reflect the inherent ductility and overstrength of ductile piping but may result in an underprediction of the forces on the anchorage.

C13.4.2 Anchors in Concrete or Masonry. Design capacity for anchors in concrete must be determined in accordance with ACI 318, Chapter 17. Design capacity for anchors in masonry is determined in accordance with TMS 402. Anchors must be designed to have ductile behavior or to provide a specified degree of excess strength. Depending on the specifics of the design condition, ductile design of anchors in concrete may satisfy one or more of the following objectives:

1. Adequate load redistribution between anchors in a group;
2. Allowance for anchor overload without brittle failure; or
3. Energy dissipation.

Achieving deformable, energy-absorbing behavior in the anchor itself is often difficult. Unless the design specifically addresses the conditions influencing desirable hysteretic response (e.g., adequate gauge length, anchor spacing, edge distance, and steel properties), anchors cannot be relied on for energy dissipation. Simple geometric rules, such as restrictions on the ratio of anchor embedment length to depth, may not be adequate to produce reliable ductile behavior. For example, a single anchor with sufficient embedment to force ductile tension failure in the steel body of the anchor bolt may still experience concrete fracture (a nonductile failure mode) if the edge distance is small, the anchor is placed in a group of tension-loaded anchors with reduced spacing, or the anchor is loaded in shear instead of tension. In the common case where anchors are subject primarily to shear, response governed by the steel element may be nonductile if the deformation of the anchor is constrained by rigid elements on either side of the joint. Designing the attachment so that its response is governed by a deformable link in the load path to the anchor is encouraged. This approach provides ductility and overstrength in the connection while protecting the anchor from overload. Ductile bolts should only be relied on as the primary ductile mechanism of a system if the bolts are designed to have adequate gauge length (using the unbonded strained length of the bolt) to accommodate the anticipated nonlinear displacements of the system at the design earthquake. Guidance for determining the gauge length can be found in Part 3 of the 2009 NEHRP provisions.

Anchors used to support towers, masts, and equipment are often provided with double nuts for leveling during installation. Where base-plate grout is specified at anchors with double nuts, it should not be relied on to carry loads because it can shrink and crack or be omitted altogether. The design should include the corresponding tension, compression, shear, and flexure loads.

Postinstalled anchors in concrete and masonry should be qualified for seismic loading through appropriate testing. The requisite tests for expansion and undercut anchors in concrete are given in ACI 355.2-07, *Qualification of Post-Installed Mechanical Anchors in Concrete and Commentary* (2007). Testing and assessment procedures based on the ACI standard that address expansion, undercut, screw, and adhesive anchors are incorporated in ICC-ES acceptance criteria. AC193, *Acceptance Criteria for Mechanical Anchors in Concrete Elements* (2012c), and AC308, *Acceptance Criteria for Post-Installed Adhesive Anchors in Concrete Elements* (2012d), refer to ACI 355.4-11, *Qualification of Post-Installed Adhesive Anchors in Concrete and Commentary* (2011c). These criteria, which include specific provisions for screw anchors and adhesive anchors, also reference ACI qualification standards for anchors. For postinstalled anchors in masonry, seismic prequalification procedures are contained in ICC-ES AC01, *Acceptance Criteria for Expansion Anchors in Masonry Elements* (2012b), AC58, *Acceptance Criteria for Adhesive Anchors in Masonry Elements* (2012a), and AC106, *Acceptance Criteria for Predrilled Fasteners (Screw Anchors) in Masonry* (2012e).

Other references to adhesives (such as in Section 13.5.7.2) apply not to adhesive anchors but to steel plates and other structural elements bonded or glued to the surface of another structural component with adhesive; such connections are generally nonductile.

C13.4.3 Installation Conditions. Prying forces on anchors, which result from a lack of rotational stiffness in the connected part, can be critical for anchor design and must be considered explicitly.

For anchorage configurations that do not provide a direct mechanism to transfer compression loads (for example, a base plate that does not bear directly on a slab or deck but is supported on a threaded rod), the design for overturning must reflect the actual stiffness of base plates, equipment, housing, and other elements in the load path when computing the location of the compression centroid and the distribution of uplift loads to the anchors.

C13.4.4 Multiple Attachments. Although the standard does not prohibit the use of single anchor connections, it is good practice to use at least two anchors in any load-carrying connection whose failure might lead to collapse, partial collapse, or disruption of a critical load path.

C13.4.5 Power-Actuated Fasteners. Restrictions on the use of power-actuated fasteners are based on observations of failures of sprinkler pipe runs in the 1994 Northridge earthquake. Although it is unclear from the record to what degree the failures occurred because of poor installation, product deficiency, overload, or consequential damage, the capacity of power-actuated fasteners in concrete often varies more than that of drilled postinstalled anchors. The shallow embedment, small diameter, and friction mechanism of these fasteners make them particularly susceptible to the effects of concrete cracking. The suitability of power-actuated fasteners to resist tension in concrete should be demonstrated by simulated seismic testing in cracked concrete.

Where properly installed in steel, power-actuated fasteners typically exhibit reliable cyclic performance. Nevertheless, they

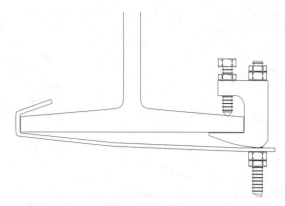

FIGURE C13.4-1 C-Type Beam Clamp Equipped with a Restraining Strap

should not be used singly to support suspended elements. Where used to attach cladding and metal decking, subassembly testing may be used to establish design capacities because the interaction among the decking, the subframe, and the fastener can only be estimated crudely by currently available analysis methods.

The exception permits the use of power-actuated fasteners for specific light-duty applications with upper limits on the load that can be resisted in these cases. All fasteners must have adequate capacity for the calculated loads, including prying forces.

The exception allows for the continued use of power-actuated fasteners in concrete for the vertical support of suspended acoustical tile or lay-in panel ceilings and for other light distributed systems such as small-diameter conduit held to the concrete surface with C-clips. Experience indicates that these applications have performed satisfactorily because of the high degree of redundancy and light loading. Other than ceilings, hung systems should not be included in this exception because of the potential for bending in the fasteners.

The exception for power-actuated fasteners in steel provides a conservative limit on loading. Currently, no accepted procedure exists for the qualification of power-actuated fasteners to resist earthquake loads.

C13.4.6 Friction Clips. The term *friction clip* is defined in Section 11.2 in a general way to encompass C-type beam clamps, as well as cold-formed metal channel (strut) connections. Friction clips are suitable to resist seismic forces provided that they are properly designed and installed, but under no circumstances should they be relied on to resist sustained gravity loads. C-type clamps must be provided with restraining straps, as shown in Fig. C13.4-1.

C13.5 ARCHITECTURAL COMPONENTS

For structures in Risk Categories I through III, the requirements of Section 13.5 are intended to reduce property damage and life-safety hazards posed by architectural components and caused by loss of stability or integrity. When subjected to seismic motion, components may pose a direct falling hazard to building occupants or to people outside the building (as in the case of parapets, exterior cladding, and glazing). Failure or displacement of interior components (such as partitions and ceiling systems in exits and stairwells) may block egress.

For structures in Risk Category IV, the potential disruption of essential function caused by component failure must also be considered.

Architectural component failures in earthquakes can be caused by deficient design or construction of the component, interrelationship with another component that fails, interaction with the structure, or inadequate attachment or anchorage. For architectural components, attachment and anchorage are typically the most critical concerns related to their seismic performance. Concerns regarding loss of function are most often associated with mechanical and electrical components. Architectural damage, unless severe, can be accommodated temporarily. Severe architectural damage is often accompanied by significant structural damage.

C13.5.1 General. Suspended architectural components are not required to satisfy the force and displacement requirements of Chapter 13, where prescriptive requirements are met. The requirements were relaxed in the 2005 edition of the standard to better reflect the consequences of the expected behavior. For example, impact of a suspended architectural ornament with a sheet metal duct may only dent the duct without causing a credible danger (assuming that the ornament remains intact). The reference to Section 13.2.3 allows the designer to consider such consequences in establishing the design approach.

Nonstructural components supported by chains or otherwise suspended from the structure are exempt from lateral bracing requirements, provided that they are designed not to inflict damage to themselves or any other component when subject to seismic motion. However, for the 2005 edition, it was determined that clarifications were needed on the type of nonstructural components allowed by these exceptions and the acceptable consequences of interaction between components. In ASCE 7-02, certain nonstructural components that could represent a fire hazard after an earthquake were exempted from meeting the Section 9.6.1 requirements. For example, gas-fired space heaters clearly pose a fire hazard after an earthquake, but were permitted to be exempted from the ASCE 7-02 Section 9.6.1 requirements. The fire hazard after the seismic event must be given the same level of consideration as the structural failure hazard when considering components to be covered by this exception. In addition, the ASCE 7-02 language was sometimes overly restrictive because it did not distinguish between credible seismic interactions and incidental interactions. In ASCE 7-02, if a suspended lighting fixture could hit and dent a sheet metal duct, it would have to be braced, although no credible danger is created by the impact. The new reference in Section 13.2.3 of ASCE 7-05 allowed the designer to consider whether the failures of the component and/or the adjacent components are likely to occur if contact is made. These provisions were carried into ASCE 7-10.

C13.5.2 Forces and Displacements. Partitions and interior and exterior glazing must accommodate story drift without failure that will cause a life-safety hazard. Design judgment must be used to assess potential life-safety hazards and the likelihood of life-threatening damage. Special detailing to accommodate drift for typical gypsum board or demountable partitions is unlikely to be cost-effective, and damage to these components poses a low hazard to life safety. Damage in these partitions occurs at low drift levels but is inexpensive to repair.

If they must remain intact after strong ground motion, nonstructural fire-resistant enclosures and fire-rated partitions require special detailing that provides isolation from the adjacent or enclosing structure for deformation equivalent to the calculated drift (relative displacement). In-plane differential movement between structure and wall is permitted. Provision must be made for out-of-plane restraint. These requirements are particularly important in steel or concrete moment-frame structures, which experience larger drifts. The problem is less likely to be encountered in stiff structures, such as those with shear walls.

Differential vertical movement between horizontal cantilevers in adjacent stories (such as cantilevered floor slabs) has occurred in past earthquakes. The possibility of such effects should be considered in the design of exterior walls.

C13.5.3 Exterior Nonstructural Wall Elements and Connections. Nonbearing wall panels that are attached to and enclose the structure must be designed to resist seismic (inertial) forces, wind forces, and gravity forces and to accommodate movements of the structure resulting from lateral forces and temperature change. The connections must allow wall panel movements caused by thermal and moisture changes and must be designed to prevent the loss of load-carrying capacity in the event of significant yielding. Where wind loads govern, common practice is to design connectors and panels to allow for not less than two times the story drift caused by wind loads determined, using a return period appropriate to the site location.

Design to accommodate seismic relative displacements often presents a greater challenge than design for strength. Story drifts can amount to 2 in. (50 mm) or more. Separations between adjacent panels are intended to limit contact and resulting panel misalignment or damage under all but extreme building response. Section 13.5.3, item 1, calls for a minimum separation of 1/2 in. (13 mm). For practical joint detailing and acceptable appearance, separations typically are limited to about 3/4 in. (19 mm). Manufacturing and construction tolerances for both wall elements and the supporting structure must be considered in establishing design joint dimensions and connection details.

Cladding elements, which are often stiff in-plane, must be isolated so that they do not restrain and are not loaded by drift of the supporting structure. Slotted connections can provide isolation, but connections with long rods that flex achieve the desired behavior without requiring precise installation. Such rods must be designed to resist tension and compression in addition to induced flexural stresses and brittle, low-cycle fatigue failure.

Full-story wall panels are usually rigidly attached to and move with the floor structure nearest the panel bottom and isolated at the upper attachments. Panels also can be vertically supported at the top connections with isolation connections at the bottom. An advantage of this configuration is that failure of an isolation connection is less likely to result in complete detachment of the panel because it tends to rotate into the structure rather than away from it.

To minimize the effects of thermal movements and shrinkage on architectural cladding panels, connection systems are generally detailed to be statically determinate. Because the resulting support systems often lack redundancy, exacerbating the consequences of a single connection failure, fasteners must be designed for amplified forces and connecting members must be ductile. The intent is to keep inelastic behavior in the connecting members while the more brittle fasteners remain essentially elastic. To achieve this intent, the tabulated a_p, R_p, and Ω_0 values produce fastener design forces that are about three times those for the connecting members.

Limited deformability curtain walls, such as aluminum systems, are generally light and can undergo large deformations without separating from the structure. However, care must be taken in design of these elements so that low deformability components (as defined in Section 11.2) that may be part of the system, such as glazing panels, are detailed to accommodate the expected deformations without failure.

In Table 13.5-1, veneers are classified as either limited or low-deformability elements. Veneers with limited deformability, such as vinyl siding, pose little risk. Veneers with low deformability, such as brick and ceramic tile, are highly sensitive to the performance of the supporting substrate. Significant distortion of the substrate results in veneer damage, possibly including separation from the structure. The resulting risk depends on the size and weight of fragments likely to be dislodged and on the height from which the fragments would fall. Detachment of large portions of the veneer can pose a significant risk to life. Such damage can be reduced by isolating veneer from displacements of the supporting structure. For structures with flexible lateral force-resisting systems, such as moment frames and buckling-restrained braced frames, approaches used to design nonbearing wall panels to accommodate story drift should be applied to veneers.

The limits on length to diameter ratios are needed to ensure proper connection performance. Recent full-scale building shake table tests conducted at University of California, San Diego, demonstrated that sliding connections perform well when the rod is short. Longer rods in sliding connections bind if there is significant bending and rotation in the rod, which may lead to a brittle failure. For rods that accommodate drift by flexure, longer rods reduce inelastic bending demands and provide better performance. Since anchor rods used in sliding and bending may undergo inelastic action, the use of mild steel improves ductility.

Threaded rods subjected to bending have natural notches (the threads) and are therefore a concern for fatigue. In high-seismic applications, the response may induce a high bending demand and low-cycle fatigue. Cold-worked threaded rod offers significantly reduced ductility unless annealed. Rods meeting the requirements of ASTM F1554, Grade 36, in their as-fabricated condition (i.e., after threading) provide the desired level of performance. ASTM 1554 rods that fulfill the requirements of Supplement 1 for Grade 55 Bars and Anchor Bolts are also acceptable. Other grades that may also be acceptable include ASTM A36, A307, A572, Grade 50, and A588. Other connection configurations and materials may be used, provided that they are approved in accordance with ASCE 7-16 Section 1.3.1.3 and are designed to accommodate the story drift without brittle failure.

The reference to D_p has been changed to D_{pI} to reflect consideration of the earthquake Importance Factor on drift demands. Connections should include a means for accommodating erection tolerance so that the required connection capacity is maintained.

C13.5.4 Glass. Glass is commonly secured to the window system framing by a glazing pocket built into the framing. This is commonly referred to as a *mechanically captured* or *dry-glazed* window system. Glass can also be secured to the window system framing with a structural silicone sealant. This is commonly referred to as a *wet-glazed* window system. Imposed loads are transferred from the glass to the window system framing through the adhesive bond of the structural silicone sealant. ASTM C1401 *Standard Guide for Structural Sealant Glazing* (2014b) provides guidance and reference standards for manufacture, testing, design and installation of structural silicone sealant. This standard addresses glazing sloped to a maximum of 15° from vertical. For glazing slopes exceeding 15°, additional general building code requirements pertaining to sloped glazing and skylights apply.

C13.5.5 Out-of-Plane Bending. The effects of out-of-plane application of seismic forces (defined in Section 13.3.1) on nonstructural walls, including the resulting deformations, must be considered. Where weak or brittle materials are used, conventional deflection limits are expressed as a proportion of the span. The intent is to preclude out-of-plane failure of heavy materials (such as brick or block) or applied finishes (such as stone or tile).

C13.5.6 Suspended Ceilings. Suspended ceiling systems are fabricated using a wide range of building materials with differing characteristics. Some systems (such as gypsum board, screwed or nailed to suspended members) are fairly homogeneous and should be designed as light-frame diaphragm assemblies, using the forces of Section 13.3 and the applicable material-specific design provisions of Chapter 14. Others are composed of discrete elements laid into a suspension system and are the subject of this section.

Seismic performance of ceiling systems with lay-in or acoustical panels depends on support of the grid and individual panels at walls and expansion joints, integrity of the grid and panel assembly, interaction with other systems (such as fire sprinklers), and support for other nonstructural components (such as light fixtures and HVACR systems). Observed performance problems include dislodgement of tiles because of impact with walls and water damage (sometimes leading to loss of occupancy) because of interaction with fire sprinklers.

Suspended lath and plaster ceilings are not exempted from the requirements of this section because of their more significant mass and the greater potential for harm associated with their failure. However, the prescriptive seismic provisions of Section 13.5.6.2 and ASTM E580 for acoustical tile and lay-in panel ceilings, including the use of compression posts, are not directly applicable to these systems primarily because of their behavior as a continuous diaphragm and greater mass. As such, they require more attention to design and detailing, in particular for the attachment of the hanger wires to the structure and main carriers, the attachment of the cross-furring channels to main carriers, and the attachment of lath to cross-furring channels. Attention should also be given to the attachment of light fixtures and diffusers to the ceiling structure. Bracing should consider both horizontal and vertical movement of the ceiling, as well as discontinuities and offsets. The seismic design and detailing of lath and plaster ceilings should use rational engineering methods to transfer seismic design ceiling forces to the building structural elements.

The performance of ceiling systems is affected by the placement of seismic bracing and the layout of light fixtures and other supported loads. Dynamic testing has demonstrated that splayed wires, even with vertical compression struts, may not adequately limit lateral motion of the ceiling system caused by straightening of the end loops. Construction problems include slack installation or omission of bracing wires caused by obstructions. Other testing has shown that unbraced systems may perform well where the system can accommodate the expected displacements, by providing both sufficient clearance at penetrations and wide closure members, which are now required by the standard.

With reference to the exceptions in Section 13.5.6,

- The first exemption is based on the presumption that lateral support is accomplished by the surrounding walls for areas equal to or less than 144 ft^2 (13.4 m^2) (e.g., a 12-ft by 12-ft (3.7-m by 3.7-m) room). The 144 ft^2 (13.4 m^2) limit corresponds historically to an assumed connection strength of 180 lb (4.5 N) and forces associated with requirements for suspended ceilings that first appeared in the 1976 Uniform Building Code.
- The second exemption assumes that planar, horizontal drywall ceilings behave as diaphragms (i.e., develop in-plane strength). This assumption is supported by the performance of drywall ceilings in past earthquakes.

C13.5.6.1 Seismic Forces. Where the weight of the ceiling system is distributed nonuniformly, that condition should be considered in the design because the typical T-bar ceiling grid has limited ability to redistribute lateral loads.

Table C13.5-1 Summary of Requirements for Acoustical Tile or Lay-in Panel Ceilings

Item	Seismic Design Category C	Seismic Design Categories D, E, and F
	Less Than or Equal to 144 ft^2	
NA	No requirements. (§1.4)	No requirements. (§1.4)
	Greater than 144 ft^2 but less than or equal to 1,000 ft^2	
Duty Rating	Only Intermediate or Heavy Duty Load Rated grid as defined by ASTM C635 may be used for commercial ceilings. (ASTM C635 sections 4.1.3.1, 4.1.3.2, & 4.1.3.3)	Heavy Duty Load Rating as defined in ASTM C635 is required. (§5.1.1)
Grid Connections	Minimum main tee connection and cross tee intersection strength of 60 lb. (§4.1.2)	Minimum main tee connection and cross tee intersection strength of 180 lb. (§5.1.2)
Vertical Suspension Wires	Vertical hanger wires must be a minimum of 12 gauge. (§4.3.1)	Vertical hanger wire must be a minimum of 12 gauge. (§5.2.7.1)
	Vertical hanger wires maximum 4 ft on center. (§4.3.1)	Vertical hanger wires maximum 4 ft on center. (§5.2.7.1)
	Vertical hanger wires must be sharply bent and wrapped with three turns in 3 in. or less. (§4.3.2)	Vertical hanger wires must be sharply bent and wrapped with three turns in 3 in. or less. (§5.2.7.2)
	All vertical hanger wires may not be more than 1 in 6 out of plumb without having additional wires counter-splayed. (§4.3.3)	All vertical hanger wires may not be more than 1 in 6 out of plumb without having additional wires counter-splayed. (§5.2.7.3)
	Any connection device from the vertical hanger wire to the structure above must sustain a minimum load of 90 lb. (§4.3.2)	Any connection device from the vertical hanger wire to the structure above must sustain a minimum load of 90 lb. (§5.2.7.2)
	Wires may not attach to or bend around interfering equipment without the use of trapezes. (§4.3.4)	Wires may not attach to or bend around interfering equipment without the use of trapezes. (§5.2.7.4)
Lateral Bracing	Lateral bracing is not permitted. Ceiling is intended to "float" relative to balance of structure. Tee connections may be insufficient to maintain integrity if braces were included. NOTE 1)	Not required under 1,000 ft^2. For ceiling areas under 1,000 ft^2, perimeter and tee connections are presumed to be sufficiently strong to maintain integrity whether bracing is installed or not. (§5.2.8.1)
Perimeter	Perimeter closure (molding) width must be a minimum of 7/8 in. (§4.2.2)	Perimeter closure (molding) width must be a minimum of 2 in. (§5.2.2)
	Perimeter closures with a support ledge of less than 7/8-in. shall be supported by perimeter vertical hanger wires not more than 8 in. from the wall. (§4.2.3)	Two adjacent sides must be connected to the wall or perimeter closure. (§5.2.3)
	A minimum clearance of 3/8 in. must be maintained on all four sides. (§4.2.4)	A minimum clearance of 3/4 in. must be maintained on the other two adjacent sides. (§5.2.3)
	Permanent attachment of grid ends is not permitted (§4.2.6)	Perimeter tees must be supported by vertical hanger wires not more than 8 in. from the wall. (§5.2.6)
	Perimeter tee ends must be prevented from spreading. (§4.2.5)	Perimeter tee ends must be prevented from spreading. (§5.2.4)
Light Fixtures	Light fixtures must be positively attached to the grid by at least two connections, each capable of supporting the weight of the lighting fixture. (§4.4.1 and NEC)	Light fixtures must be positively attached to the grid by at least two connections, each capable of supporting the weight of the lighting fixture. (NEC, §5.3.1)
	Surface-mounted light fixtures shall be positively clamped to the grid. (§4.4.2)	Surface-mounted light fixtures shall be positively clamped to the grid. (§5.3.2)
	Clamping devices for surface-mounted light fixtures shall have safety wires to the ceiling hanger or to the structure above. (§4.4.2)	Clamping devices for surface-mounted light fixtures shall have safety wires to the ceiling wire or to the structure above. (§5.3.2)
	Light fixtures and attachments weighing 10 lb or less require one number 12 gauge (minimum) hanger wire connected to the housing (e.g., canister light fixture). This wire may be slack. (§4.4.3)	When cross tees with a load-carrying capacity of less than 16 lb/ft are used, supplementary hanger wires are required. (§5.3.3)
	Light fixtures that weigh greater than 10 but less than or equal to 56 lb require two number 12 gauge (minimum) hanger wires connected to the housing. These wires may be slack. (§4.4.4)	Light fixtures and attachments weighing 10 lb or less require one 12-gauge minimum hanger wire connected to the housing and connected to the structure above. This wire may be slack. (§5.3.4)
	Light fixtures that weigh more than 56 lb require independent support from the structure. (§4.4.5)	Light fixtures that weigh greater than 10 but less than or equal to 56 lb require two number 12 gauge minimum hanger wires attached to the fixture housing and connected to the structure above. These wires may be slack. (§5.3.5)
	Pendant-hung light fixtures shall be supported by a minimum 9-gauge wire or other approved alternate. (§4.4.6)	Light fixtures that weigh more than 56 lb require independent support from the structure by approved hangers. (§5.3.6)
	Rigid conduit is not permitted for the attachment of fixtures. (§4.4.7)	Pendant-hung light fixtures shall be supported by a minimum 9-gauge wire or other approved support. (§5.3.7)
		Rigid conduit is not permitted for the attachment of fixtures. (§5.3.8)

Table C13.5-1 *(Continued)* **Summary of Requirements for Acoustical Tile or Lay-in Panel Ceilings**

Item	Seismic Design Category C	Seismic Design Categories D, E, and F
Mechanical Services	Flexibly mounted mechanical services weighing less than or equal to 20 lb must be positively attached to main runners or cross runners with the same load-carrying capacity as the main runners. (§4.5.1) Flexibly mounted mechanical services weighing more than 20 lb but less than or equal to 56 lb must be positively attached to main runners or cross runners with the same load-carrying capacity as the main runners and require two 12-gauge (minimum) hanger wires. These wires may be slack. (§4.5.2) Flexibly mounted mechanical services greater than 56 lb require direct support from the structure. (§4.5.3)	Flexibly mounted mechanical services weighing less than or equal to 20 lb must be positively attached to main runners or cross runners with the same load-carrying capacity as the main runners. (§5.4.1) Flexibly mounted mechanical services weighing more than 20 lb but less than or equal to 56 lb must be positively attached to main runners or cross runners with the same load-carrying capacity as the main runners and require two 12-gauge (minimum) hanger wires. These wires may be slack. (§5.4.2) Flexibly mounted mechanical services greater than 56 lb require direct support from the structure. (§5.4.3)
Supplemental Requirements	All ceiling penetrations must have a minimum of 3/8 in. clearance on all sides. (§4.2.4)	Direct concealed systems must have stabilizer bars or mechanically connected cross tees a maximum of 60 in. on center with stabilization within 24 in. of the perimeter. (§5.2.5) Bracing is required for ceiling plane elevation changes. (§5.2.8.6) Cable trays and electrical conduits shall be supported independently of the ceiling. (§5.2.8.7) 2,500 ft^2 All ceiling penetrations and independently supported fixtures or services must have closures that allow for a 1-in. movement. (§5.2.8.5) An integral ceiling sprinkler system may be designed by the licensed design professional to eliminate the required spacing of penetrations. (§5.2.8.8) A licensed design professional must review the interaction of nonessential ceiling components with essential ceiling components to prevent the failure of the essential components. (§5.7.1)
Partitions	The ceiling may not provide lateral support to partitions. (§4.6.1) Partitions attached to the ceiling must use flexible connections to avoid transferring force to the ceiling. (§4.6.1)	Partition attached to the ceiling and all partitions greater than feet in height shall be laterally braced to the building structure. This bracing must be independent of the ceiling. (§5.5.1)
Exceptions	The ceiling weight must be less than 2.5 lb=ft^2, otherwise the prescribed construction for Seismic Design Categories D, E, and F must be used. (§4.1.1)	None.

Greater than 1,000 ft^2 but less than or equal to 2,500 ft^2

Lateral Bracing	No additional requirements.	Lateral force bracing (4, 12 gauge splay wires) is required within 2 in. of main tee/cross tee intersection and splayed 90 deg apart in the plan view, at maximum 45-deg angle from the horizontal and located 12 ft on center in both directions, starting 6 ft from walls. (§5.2.8.1 & §5.2.8.2) Lateral force bracing must be spaced a minimum of 6 in. from unbraced horizontal piping or ductwork. (§5.2.8.3) Lateral force bracing connection strength must be a minimum of 250 lb. (§5.2.8.3) Rigid bracing designed to limit deflection at the point of attachment to less than 0.25 in. may be used in place of splay wires. Unless rigid bracing is used or calculations have shown that lateral deflection is less than 1/4 in., sprinkler heads and other penetrations shall have a minimum of 1-in. clear space in all directions. (§5.2.8.5)

Greater than 2,500 ft^2

Special Considerations	No additional requirements.	Seismic separation joints with a minimum or 3/4-in. axial movement, bulkhead, or full-height partitions with the usual 2-in. closure and other requirements. (§5.2.9.1) Areas defined by seismic separation joints, bulkheads, or full-height partitions must have a ratio of long to short dimensions of less than or equal to 4. (§5.2.9.1)

Notes: There are no requirements for suspended ceilings located in structures assigned to Seismic Design Categories A and B. Unless otherwise noted, all section references in parentheses (§) refer to ASTM E580 (2014).

C13.5.6.2 Industry Standard Construction for Acoustical Tile or Lay-In Panel Ceilings. The key to good seismic performance is sufficiently wide closure angles at the perimeter to accommodate relative ceiling motion and adequate clearance at penetrating components (such as columns and piping) to avoid concentrating restraining loads on the ceiling system.

Table C13.5-1 provides an overview of the combined requirements of ASCE/SEI 7 and ASTM E580 (2014a). Careful review of both documents is required to determine the actual requirements.

C13.5.6.2.1 Seismic Design Category C. The prescribed method for SDC C is a floating ceiling. The design assumes a small displacement of the building structure caused by the earthquake at the ceiling and isolates the ceiling from the perimeter. The vertical hanger wires are not capable of transmitting significant movement or horizontal force into the ceiling system, and therefore the ceiling does not experience significant force or displacement as long as the perimeter gap is not exceeded. All penetrations and services must be isolated from the building structure for this construction method to be effective. If this isolation is impractical or undesirable, the prescribed construction for SDCs D, E, and F may be used.

C13.5.6.2.2 Seismic Design Categories D through F. The industry standard construction addressed in this section relies on ceiling contact with the perimeter wall for restraint.

Typical splay wire lateral bracing allows for some movement before it effectively restrains the ceiling. The intent of the 2-in. (50-mm) perimeter closure wall angle is to permit back-and-forth motion of the ceiling during an earthquake without loss of support, and the width of the closure angle is important to good performance. This standard has been experimentally verified by large-scale testing conducted by ANCO Engineers, Inc., in 1983.

Extensive shake table testing using the protocol contained in ICC-ES AC156 by major manufacturers of suspended ceilings has been used to justify the use of perimeter clips designed to accommodate the same degree of movement as the closure angle while supporting the tee ends. These tests are conducted on 16-ft by 16-ft (4.9-m by 4.9-m) ceiling installations. Testing on larger ceiling systems reported by Rahmanishamsi et al. (2014) and Soroushian et al. (2012, 2014) indicates that the use of approved perimeter clips may lead to damage to the grid members and seismic clips, crushing of wall angles, and deformation of grid latches at moderate ground motion levels if the grid member loses contact with the horizontal leg of the closure angle or channel. A requirement has been added to screw the clips to the closure angle or channel to prevent this type of damage.

The requirement for a 1-in. (25-mm) clearance around sprinkler drops found in Section 13.5.6.2.2 (e) of ASCE/SEI 7-05 is maintained and is contained in ASTM E580.

This seismic separation joint is intended to break the ceiling into isolated areas, preventing large-scale force transfer across the ceiling. The new requirement to accommodate 3/4-in. (19-mm) axial movement specifies the performance requirement for the separation joint.

The requirement for seismic separation joints to limit ceiling areas to 2,500 ft^2 (232.3 m^2) is intended to prevent overload of the connections to the perimeter angle. Limiting the ratio of long to short dimensions to 4:1 prevents dividing the ceiling into long and narrow sections, which could defeat the purpose of the separation.

C13.5.6.3 Integral Construction. Ceiling systems that use integral construction are constructed of modular pre-engineered components that integrate lights, ventilation components, fire sprinklers, and seismic bracing into a complete system. They may include aluminum, steel, and PVC components and may be designed using integral construction of ceiling and wall. They often use rigid grid and bracing systems, which provide lateral support for all the ceiling components, including sprinkler drops. This bracing reduces the potential for adverse interactions among components and eliminates the need to provide clearances for differential movement.

C13.5.7 Access Floors

C13.5.7.1 General. In past earthquakes and in cyclic load tests, some typical raised access floor systems behaved in a brittle manner and exhibited little reserve capacity beyond initial yielding or failure of critical connections. Testing shows that unrestrained individual floor panels may pop out of the supporting grid unless they are mechanically fastened to supporting pedestals or stringers. This fault may be a concern, particularly in egress pathways.

For systems with floor stringers, it is accepted practice to calculate the seismic force, F_p, for the entire access floor system within a partitioned space and then distribute the total force to the individual braces or pedestals. For stringerless systems, the seismic load path should be established explicitly.

Overturning effects subject individual pedestals to vertical loads well in excess of the weight, W_p, used in determining the seismic force, F_p. It is unconservative to use the design vertical load simultaneously with the design seismic force for design of anchor bolts, pedestal bending, and pedestal welds to base plates. "Slip-on" heads that are not mechanically fastened to the pedestal shaft and thus cannot transfer tension are likely unable to transfer to the pedestal the overturning moments generated by equipment attached to adjacent floor panels.

To preclude brittle failure, each element in the seismic load path must have energy-absorbing capacity. Buckling failure modes should be prevented. Lower seismic force demands are allowed for special access floors that are designed to preclude brittle and buckling failure modes.

C13.5.7.2 Special Access Floors. An access floor can be a "special access floor" if the registered design professional opts to comply with the requirements of Section 13.5.7.2. Special access floors include construction features that improve the performance and reliability of the floor system under seismic loading. The provisions focus on providing an engineered load path for seismic shear and overturning forces. Special access floors are designed for smaller lateral forces, and their use is encouraged at facilities with higher nonstructural performance objectives.

C13.5.8 Partitions. Partitions subject to these requirements must have independent lateral support bracing from the top of the partition to the building structure or to a substructure attached to the building structure. Some partitions are designed to span vertically from the floor to a suspended ceiling system. The ceiling system must be designed to provide lateral support for the top of the partition. An exception to this condition is provided to exempt bracing of light (gypsum board) partitions where the load does not exceed the minimum partition lateral load. Experience has shown that partitions subjected to the minimum load can be braced to the ceiling without failure.

C13.5.9 Glass in Glazed Curtain Walls, Glazed Storefronts, and Glazed Partitions. The performance of glass in earthquakes falls into one of four categories:

1. The glass remains unbroken in its frame or anchorage.

2. The glass cracks but remains in its frame or anchorage while continuing to provide a weather barrier and to be otherwise serviceable.
3. The glass shatters but remains in its frame or anchorage in a precarious condition, likely to fall out at any time.
4. The glass falls out of its frame or anchorage, either in shards or as whole panels.

Categories 1 and 2 satisfy both Immediate Occupancy and Life Safety Performance Objectives. Although the glass is cracked in Category 2, immediate replacement is not required. Categories 3 and 4 cannot provide for immediate occupancy, and their provision of life safety depends on the post-breakage characteristics of the glass and the height from which it can fall. Tempered glass shatters into multiple, pebble-size fragments that fall from the frame or anchorage in clusters. These broken glass clusters are relatively harmless to humans when they fall from limited heights, but they could be harmful when they fall from greater heights.

C13.5.9.1 General. Eq. (13.5-2) is derived from Sheet Glass Association of Japan (1982) and is similar to an equation in Bouwkamp and Meehan (1960) that permits calculation of the story drift required to cause glass-to-frame contact in a given rectangular window frame. Both calculations are based on the principle that a rectangular window frame (specifically, one that is anchored mechanically to adjacent stories of a structure) becomes a parallelogram as a result of story drift, and that glass-to-frame contact occurs when the length of the shorter diagonal of the parallelogram is equal to the diagonal of the glass panel itself. The value $\Delta_{fallout}$ represents the displacement capacity of the system, and D_p represents the displacement demand.

The 1.25 factor in the requirements described above reflects uncertainties associated with calculated inelastic seismic displacements of building structures. Wright (1989) states that post-elastic deformations, calculated using the structural analysis process may well underestimate the actual building deformation by up to 30%. It would therefore be reasonable to require the curtain wall glazing system to withstand 1.25 times the computed maximum interstory displacement to verify adequate performance.

The reason for the second exception to Eq. (13.5-2) is that the tempered glass, if shattered, would not produce an overhead falling hazard to adjacent pedestrians, although some pieces of glass may fall out of the frame.

C13.5.9.2 Seismic Drift Limits for Glass Components. As an alternative to the prescriptive approach of Section 13.5.9.1, the deformation capacity of glazed curtain wall systems may be established by test.

C13.5.10 Egress Stairs and Ramps. In the Christchurch earthquake of February 22, 2011, several buildings using precast concrete stairs provided with a sliding joint at one end experienced stair collapse (Canterbury Earthquakes Royal Commission 2012). In one notable case, the 18-story Forsyth Barr office building, the structure was otherwise largely undamaged. In all cases, the primary cause of collapse was loss of vertical bearing at the end connection due to building drift that exceeded the support detail capacity. These stairs, in general, were intended to serve as egress routes, and occupants were trapped in some of these buildings following the earthquake. In U.S. practice, precast stairs (Fig. C13.5-1) are less common than steel-framed stairs (Fig. C13.5-2), which are generally considered to be inherently flexible. But in shake table tests conducted at the University of California, San Diego, as part of the Network for Earthquake Engineering Simulation (NEES) project, "Full-Scale Structural and Nonstructural Building System Performance during Earthquakes

FIGURE C13.5-1 Precast Stair
Source: Courtesy of Tindall Corp.

FIGURE C13.5-2 Steel-Framed Exit Stair
Source: Courtesy of Tara Hutchinson.

and Post-Earthquake Fire," connections of the commercial metal stair included in the test structure were shown to be brittle and susceptible to damage. Considering the critical nature of egress for life safety, specific attention to the ability of egress stairs to accept building drift demands is warranted. Effective sliding joints in typical steel stairs are complex to design and construct. Ductile connections, capable of accepting the drift without loss of vertical load-carrying capacity are often preferred. In such cases, sufficient ductility must be provided in these connections to accommodate multiple cycles at anticipated maximum drift levels. If drift is to be accommodated with full sliding connections lacking a fail-safe stop, additional length of bearing is required to prevent collapse where displacements exceed design levels. Where stair systems are rigidly attached to the structure, they must be included in the structure model, and the resultant forces must be accommodated, with overstrength, in the stair design.

These requirements do not apply to egress stair systems and ramps that are integral with the building structure since it is assumed that the seismic resistance of these systems is addressed in the overall building design. Examples include stairs and ramps comprising monolithic concrete construction, light-frame wood and cold-formed metal stair systems in multiunit residential construction, and integrally constructed masonry stairs.

C13.6 MECHANICAL AND ELECTRICAL COMPONENTS

These requirements, focused on design of supports and attachments, are intended to reduce the hazard to life posed by loss of component structural stability or integrity. The requirements increase the reliability of component operation but do not address functionality directly. For critical components where operability is vital, Section 13.2.2 provides methods for seismically qualifying the component.

Traditionally, mechanical components (such as tanks and heat exchangers) without rotating or reciprocating components are directly anchored to the structure. Mechanical and electrical equipment components with rotating or reciprocating elements are often isolated from the structure by vibration isolators (using rubber acting in shear, springs, or air cushions). Heavy mechanical equipment (such as large boilers) may not be restrained at all, and electrical equipment other than generators, which are normally isolated to dampen vibrations, usually is rigidly anchored (for example, switch gear and motor control centers).

Two distinct levels of earthquake safety are considered in the design of mechanical and electrical components. At the usual safety level, failure of the mechanical or electrical component itself because of seismic effects poses no significant hazard. In this case, design of the supports and attachments to the structure is required to avoid a life-safety hazard. At the higher safety level, the component must continue to function acceptably after the design earthquake. Such components are defined as designated seismic systems in Section 11.2 and may be required to meet the special certification requirements of Section 13.2.2.

Not all equipment or parts of equipment need to be designed for seismic forces. Where I_p is specified to be 1.0, damage to, or even failure of, a piece or part of a component does not violate these requirements as long as a life-safety hazard is not created. The restraint or containment of a falling, breaking, or toppling component (or its parts) by means of bumpers, braces, guys, wedges, shims, tethers, or gapped restraints to satisfy these requirements often is acceptable, although the component itself may suffer damage.

Judgment is required to fulfill the intent of these requirements; the key consideration is the threat to life safety. For example, a nonessential air handler package unit that is less than 4 ft (1.2 m) tall bolted to a mechanical room floor is not a threat to life as long as it is prevented from significant displacement by having adequate anchorage. In this case, seismic design of the air handler itself is unnecessary. However, a 10-ft (3.0-m) tall tank on 6-ft (1.8-m) long angles used as legs, mounted on a roof near a building exit does pose a hazard. The intent of these requirements is that the supports and attachments (tank legs, connections between the roof and the legs, and connections between the legs and the tank), and possibly even the tank itself be designed to resist seismic forces. Alternatively, restraint of the tank by guys or bracing could be acceptable.

It is not the intent of the standard to require the seismic design of shafts, buckets, cranks, pistons, plungers, impellers, rotors, stators, bearings, switches, gears, nonpressure retaining casings and castings, or similar items. Where the potential for a hazard to life exists, the design effort should focus on equipment supports, including base plates, anchorages, support lugs, legs, feet, saddles, skirts, hangers, braces, and ties.

Many mechanical and electrical components consist of complex assemblies of parts that are manufactured in an industrial process that produces similar or identical items. Such equipment may include manufacturers' catalog items and often are designed by empirical (trial-and-error) means for functional and transportation loadings. A characteristic of such equipment is that it may be inherently rugged. The term "rugged" refers to an ampleness of construction that provides such equipment with the ability to survive strong motions without significant loss of function. By examining such equipment, an experienced design professional usually should be able to confirm such ruggedness. The results of an assessment of equipment ruggedness may be used in determining an appropriate method and extent of seismic design or qualification effort.

The revisions to Table 13.6-1 in ASCE/SEI 07-10 were the result of work done in recent years to better understand the performance of mechanical and electrical components and their attachment to the structure. The primary concepts of flexible and rigid equipment and ductile and rugged behavior are drawn from SEAOC (1999), Commentary Section C107.1.7. Material on HVACR is based on ASHRAE (2000). Other material on industrial piping, boilers, and pressure vessels is based on the American Society of Mechanical Engineers codes and standards publications (ASME 2007, 2010a, 2010b).

C13.6.1 General. The exception allowing unbraced suspended components has been clarified, addressing concerns about the type of nonstructural components allowed by these exceptions, as well as the acceptable consequences of interaction between components. In previous editions of the standard, certain nonstructural components that could represent a fire hazard after an earthquake were exempt from lateral bracing requirements. In the revised exception, reference to Section 13.2.3 addresses such concerns while distinguishing between credible seismic interactions and incidental interactions.

The seismic demand requirements are based on component structural attributes of flexibility (or rigidity) and ruggedness. Table 13.6-1 provides seismic coefficients based on judgments of the component flexibility, expressed in the a_p term, and ruggedness, expressed in the R_p term. It may also be necessary to consider the flexibility and ductility of the attachment system that provides seismic restraint.

Entries for components and systems in Table 13.6-1 are grouped and described to improve clarity of application. Components are divided into three broad groups, within which they are further classified depending on the type of construction or expected seismic behavior. For example, mechanical components include "air-side" components (such as fans and air handlers) that experience dynamic amplification but are light and deformable; "wet-side" components that generally contain liquids (such as boilers and chillers) that are more rigid and somewhat ductile; and rugged components (such as engines, turbines, and pumps) that are of massive construction because of demanding operating loads and that generally perform well in earthquakes, if adequately anchored.

A distinction is made between components isolated using neoprene and those that are spring isolated. Spring-isolated components are assigned a lower R_p value because they tend to have less effective damping. Internally isolated components are classified explicitly to avoid confusion.

C13.6.2 Mechanical Components and C13.6.3 Electrical Components. Most mechanical and electrical equipment is inherently rugged and, where properly attached to the structure, has performed well in past earthquakes. Because the operational and transportation loads for which the equipment is designed typically are larger than those caused by earthquakes, these requirements focus primarily on equipment anchorage and attachments. However, designated seismic systems, which are required to function after an earthquake or which must maintain containment of flammable or hazardous materials, must themselves be designed for seismic forces or be qualified for seismic loading in accordance with Section 13.2.2.

The likelihood of post-earthquake operability can be increased where the following measures are taken:

1. Internal assemblies, subassemblies, and electrical contacts are attached sufficiently to prevent their being subjected to differential movement or impact with other internal assemblies or the equipment enclosure.
2. Operators, motors, generators, and other such components that are functionally attached to mechanical equipment by means of an operating shaft or mechanism are structurally connected or commonly supported with sufficient rigidity such that binding of the operating shaft is avoided.
3. Any ceramic or other nonductile components in the seismic load path are specifically evaluated.
4. Adjacent electrical cabinets are bolted together and cabinet lineups are prevented from impacting adjacent structural members.

Components that could be damaged, or could damage other components, and are fastened to multiple locations of a structure, must be designed to accommodate seismic relative displacements. Such components include bus ducts, cable trays, conduits, elevator guide rails, and piping systems. As discussed in Section C13.3.2.1, special design consideration is required where full story drift demands are concentrated in a fraction of the story height.

The values of a_p and R_p for air coolers (commonly known as fin fans) with integral support legs in Table 13.6-1 are taken from *Guidelines for Seismic Evaluation and Design of Petrochemical Facilities* (ASCE 2011). The values listed for "fans" in Table 13.6-1 ($a_p = 2.5$ and $R_p = 6$) are not intended for fin fans with integral support legs. (They do apply where fin fans are not supported on integral support legs.) As discussed in ASCE (2011), fin fans with integral support legs have not performed well in seismic events, such as the February 27, 2010, Chile earthquake.

Typically, fin fans are supported on pipe racks (Fig. C13.6-1). Where the fin fan is supported on legs, this configuration generally creates a condition where a relatively rigid mass is supported on flexible legs on top of a pipe rack and can result in significantly higher seismic force demands. The support legs should be braced in both directions. Knee braces should be avoided. Vertical bracing should intersect columns at panel points with beams. Where geometrically practical, chevron bracing may be used. Whenever possible, it is recommended that the fin fan should be designed without vendor-supplied integral legs and should be supported directly on the pipe rack structural steel. In such cases, the values of a_p and R_p for fans apply.

Regardless of whether the fin fan vendor or the engineering contractor provides the supporting steel, the structural steel directly supporting the air coolers should be designed to the same level of seismic detailing required of the pipe rack structural steel.

Mechanical components with similar construction details used in fin fans (such as air-cooled heat exchangers, condensing units, dry coolers, and remote radiators) are grouped with fin fans because similar behavior is assumed.

C13.6.4 Component Supports. The intent of this section is to require seismic design of all mechanical and electrical component supports to prevent sliding, falling, toppling, or other movement that could imperil life. Component supports are differentiated here from component attachments to emphasize that the supports themselves, as enumerated in the text, require seismic design even if they are fabricated by the mechanical or electrical component manufacturer. This need exists regardless of whether the mechanical or electrical component itself is designed for seismic loads.

C13.6.4.1 Design Basis. Standard supports are those developed in accordance with a reference document (Section 13.1.7). Where standard supports are not used, the seismic design forces and displacement demands of Chapter 13 are used with applicable material-specific design procedures of Chapter 14.

C13.6.4.2 Design for Relative Displacement. For some items, such as piping, seismic relative displacements between support points are of more significance than inertial forces. Components made of high-deformability materials such as steel or copper can accommodate relative displacements inelastically, provided that the connections also provide high deformability. Threaded and soldered connections exhibit poor ductility under inelastic displacements, even for ductile materials. Components made of less ductile materials can accommodate relative displacement effects only if appropriate flexibility or flexible connections are provided.

Detailing distribution systems that connect separate structures with bends and elbows makes them less prone to damage and less likely to fracture and fall, provided that the supports can accommodate the imposed loads.

C13.6.4.3 Support Attachment to Component. As used in this section, "integral" relates to the manufacturing process, not the location of installation. For example, both the legs of a cooling tower and the attachment of the legs to the body of the cooling tower must be designed, even if the legs are provided by the manufacturer and installed at the plant. Also, if the cooling tower has an $I_p = 1.5$, the design must address not only the attachments (e.g., welds and bolts) of the legs to the component but also local stresses imposed on the body of the cooling tower by the support attachments.

FIGURE C13.6-1 Fin Fan Elevated on Integral Supports

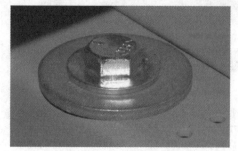

Failure of sheet metal base anchored with standard washer

Anchorage equipped with Belleville washer

FIGURE C13.6-2 Equipment Anchorage with Belleville Washers

Source: Courtesy of Philip Caldwell.

C13.6.4.5 Additional Requirements. As reflected in this section of the standard and in footnote b to Table 13.6-1, vibration-isolated equipment with snubbers is subject to amplified loads as a result of dynamic impact.

Most sheet metal connection points for seismic anchorage do not exhibit the same mechanical properties as bolted connections with structural elements. The use of Belleville washers improves the seismic performance of connections to equipment enclosures fabricated from sheet metal 7 gauge (0.18 in. (5 mm)) or thinner by distributing the stress over a larger surface area of the sheet metal connection interface, allowing for bolted connections to be torqued to recommended values for proper preload while reducing the tendency for local sheet metal tearing or bending failures or loosening of the bolted connection (Fig. C13.6-2). The intrinsic spring loading capacity of the Belleville washer assists with long-term preload retention to maintain integrity of the seismic anchorage.

Manufacturers test or design their equipment to handle seismic loads at the equipment "hard points" or anchor locations. The results of this design qualification effort are typically reflected in installation instructions provided by the manufacturer. It is imperative that the manufacturer's installation instructions be followed. Where such guidance does not exist, the registered design professional should design appropriate reinforcement.

C13.6.5 Distribution Systems: Conduit, Cable Tray, and Raceways. The term *raceway* is defined in several standards with somewhat varying language. As used here, it is intended to describe all electrical distribution systems including conduit, cable trays, and open and closed raceways. Experience indicates that a size limit of 2.5 in. (64 mm) can be established for the provision of flexible connections to accommodate seismic relative displacements that might occur between pieces of connected equipment because smaller conduit normally possesses the required flexibility to accommodate such displacements. See additional commentary pertaining to exemption of trapeze-supported systems in Section C13.1.4.

C13.6.6 Distribution Systems: Duct Systems. Experience in past earthquakes has shown that HVACR duct systems are rugged and perform well in strong ground shaking. Bracing in accordance with ANSI/SMACNA 001 (2000) has been effective in limiting damage to duct systems. Typical failures have affected only system function, and major damage or collapse has been uncommon. Therefore, industry standard practices should prove adequate for most installations. Expected earthquake damage is limited to opening of duct joints and tears in ducts. Connection details that are prone to brittle failures, especially hanger rods subject to large amplitude cycles of bending stress, should be avoided. See additional commentary in Section C13.1.4.

The amplification factor for duct systems has been increased from 1.0 to 2.5 because even braced duct systems are relatively flexible. The R_p values also have been increased so that the resulting seismic design forces are consistent with those determined previously.

Duct systems that carry hazardous materials or must remain operational during and after an earthquake are assigned a value of $I_p = 1.5$, and they require a detailed engineering analysis addressing leak tightness.

Lighter in-line components may be designed to resist the forces from Section 13.3 as part of the overall duct system design, whereby the duct attached to the in-line component is explicitly designed for the forces generated by the component. Where in-line components are more massive, the component must be supported and braced independently of the ductwork to avoid failure of the connections.

The requirements for flexible connections of unbraced piping to in-line components such as reheat coils applies regardless of the component weight.

C13.6.7 Distribution Systems: Piping and Tubing Systems. Because of the typical redundancy of piping system supports, documented cases of total collapse of piping systems in earthquakes are rare; however, pipe leakage resulting from excessive displacement or overstress often results in significant consequential damage and in some cases loss of facility operability. Loss of fluid containment (leakage) normally occurs at discontinuities such as threads, grooves, bolted connectors, geometric discontinuities, or locations where incipient cracks exist, such as at the toe or root of a weld or braze. Numerous building and industrial national standards and guidelines address a wide variety of piping systems materials and applications. Construction in accordance with the national standards referenced in these provisions is usually effective in limiting damage to piping systems and avoiding loss of fluid containment under earthquake conditions.

ASHRAE (2000) and MSS (2001) are derived in large part from the predecessors of SMACNA (2008). These documents may be appropriate references for use in the seismic design of piping systems. Because the SMACNA standard does not refer to pipe stresses in the determination of hanger and brace spacing, however, a supplementary check of pipe stresses may be necessary when this document is used. ASME piping rules as given in the ASME BPVC and ASME B31 parts B31.1, B31.3, B31.5, B31.9, and B31.12 are normally used for high-pressure, high-temperature piping but can also conservatively be applied to other lower pressure, lower temperature piping systems. Code-compliant seismic design manuals prepared specifically for proprietary systems may also be appropriate references.

Table 13.6-1 entries for piping previously listed the amplification factor related to the response of piping systems as rigid ($a_p = 1.0$) and values for component response modification factors lower than in the current table. However, it was realized that most piping systems are flexible and that the amplification factor values should reflect this fact; thus, a_p was increased to 2.5 and the R_p values were adjusted accordingly such that a_p/R_p remains roughly consistent with earlier provisions.

Although seismic design in accordance with Section 13.6.8 generally ensures that effective seismic forces do not fail piping, seismic displacements may be underestimated such that impact with nearby structural, mechanical, or electrical components could occur. In marginal cases, it may be advisable to protect the pipe with wrapper plates where impacts could occur, including at gapped supports. Insulation may in some cases also serve to protect the pipe from impact damage. Piping systems are typically designed for pressure containment, and piping designed with a factor of safety of three or more against pressure failure (rupture) may be inherently robust enough to survive impact with nearby structures, equipment, and other piping, particularly if the piping is insulated. Piping that has less than standard weight wall thickness may require the evaluation of the effects of impact locally on the pipe wall and may necessitate means to protect the pipe wall.

It is usually preferable for piping to be detailed to accommodate seismic relative displacements between the first seismic support upstream or downstream from connections and other seismically supported components or headers. This accommodation is preferably achieved by means of pipe flexibility or, where pipe flexibility is not possible, flexible supports. Piping not otherwise detailed to accommodate such seismic relative displacements must be provided with connections that have sufficient flexibility in the connecting element or in the component or header to avoid failure of the piping. The option to use a flexible connecting element may be less desirable because of the need for greater maintenance efforts to ensure continued proper function of the flexible element.

Grooved couplings, ball joints, resilient gasket compression fittings, other articulating-type connections, bellows expansion joints, and flexible metal hose are used in many piping systems and can serve to increase the rotational and lateral deflection design capacity of the piping connections.

Grooved couplings are classified as either rigid or flexible. Flexible grooved couplings demonstrate limited free rotational capacity. The free rotational capacity is the maximum articulating angle where the connection behaves essentially as a pinned joint with limited or negligible stiffness. The remaining rotational capacity of the connection is associated with conventional joint behavior, and design force demands in the connection are determined by traditional means.

Rigid couplings are typically used for high-pressure applications and usually are assumed to be stiffer than the pipe. Alternatively, rigid coupling may exhibit bilinear rotational stiffness with the initial rotational stiffness affected by installation.

Coupling flexibilities vary significantly between manufacturers, particularly for rigid couplings. Manufacturer's data may be available. Industrywide procedures for the determination of coupling flexibility are not currently available; however, some guidance for couplings may be found in the provisions for fire sprinkler piping, where grooved couplings are classified as either rigid or flexible on the basis of specific requirements on angular movement. In Section 3.5.4 of NFPA (2007), flexible couplings are defined as follows:

A listed coupling or fitting that allows axial displacement, rotation, and at least 1 degree of angular movement of the pipe without inducing harm on the pipe. For pipe diameters of 8 in. (203.2 mm) and larger, the angular movement shall be permitted to be less than 1 degree but not less than 0.5 degrees.

Couplings determined to be flexible on this basis are listed either with FM Global (2007) or UL (2004).

Piping component testing suggests that the ductility capacity of carbon steel threaded and flexible grooved piping component joints ranges between 1.4 and 3.0, implying an effective stress intensification of approximately 2.5. These types of connections have been classified as having limited deformability, and piping systems with these connections have R_p values lower than piping with welded or brazed joints.

The allowable stresses for piping constructed with ductile materials assumed to be materials with high deformability, and not designed in accordance with an applicable standard or recognized design basis, are based on values consistent with industrial piping and structural steel standards for comparable piping materials.

The allowable stresses for piping constructed with low-deformability materials, and not designed in accordance with an applicable standard or recognized design basis, are derived from values consistent with ASME standards for comparable piping materials.

For typical piping materials, pipe stresses may not be the governing parameter in determining the hanger and other support spacing. Other considerations, such as the capacity of the hanger and other support connections to the structure, limits on the lateral displacements between hangers and other supports to avoid impacts, the need to limit pipe sag between hangers to avoid the pooling of condensing gases, and the loads on connected equipment, may govern the design. Nevertheless, seismic span tables, based on limiting stresses and displacements in the pipe, can be a useful adjunct for establishing seismic support locations.

Piping systems' service loads of pressure and temperature also need to be considered in conjunction with seismic inertia loads. The potential for low ambient and lower than ambient operating temperatures should be considered in the designation of the piping system materials as having high or low deformability. High deformability may often be assumed for steels, particularly ASME listed materials operating at high temperatures, copper and copper alloys, and aluminum. Low deformability should be assumed for any piping material that exhibits brittle behavior, such as glass, ceramics, and many plastics.

Piping should be designed to accommodate relative displacements between the first rigid piping support and connections to equipment or piping headers often assumed to be anchors. Barring such design, the equipment or header connection could be designed to have sufficient flexibility to avoid failure. The specification of such flexible connections should consider the necessity of connection maintenance.

Where appropriate, a walkdown of the finally installed piping system by an experienced design professional familiar with seismic design is recommended, particularly for piping greater than 6-in. (152.4-mm) nominal pipe size, high-pressure piping, piping operating at higher than ambient temperatures, and piping containing hazardous materials. The need for a walkdown may also be related to the scope, function, and complexity of the piping system, as well as the expected performance of the facility. In addition to providing a review of seismic restraint location, orientation, and attachment to the structure, the walkdown verifies that the required separation exists between the piping and nearby structures, equipment, and other piping in the as-built condition.

C13.6.7.1 ASME Pressure Piping Systems. In Table 13.6-1, the increased R_p values listed for ASME B31-compliant piping systems are intended to reflect the more rigorous design, construction, and quality control requirements, as well as the intensified stresses associated with ASME B31 designs.

Materials meeting ASME toughness requirements may be considered high-deformability materials.

C13.6.7.2 Fire Protection Sprinkler Piping Systems. The lateral design procedures of NFPA (2007) have been revised for consistency with the ASCE/SEI 7 design approach while retaining traditional sprinkler system design concepts. Using conservative upper bound values of the various design parameters, a single lateral force coefficient, C_p, was developed. It is a function of the mapped short-period response parameter S_s. Stresses in the pipe and connections are controlled by limiting the maximum reaction at bracing points as a function of pipe diameter.

Other components of fire protection systems, e.g., pumps and control panels, are subject to the general requirements of ASCE/SEI 7.

Experience has shown that interaction of other nonstructural components and sprinkler drops and sprigs is a significant source of damage and can result in serious consequential damage as well as compromise the performance of the fire protection system. Clearance for between sprinkler drops and sprigs and other nonstructural components needs to be addressed beyond NFPA 13. The minimum clearance value provided is based on judgment observations in past earthquakes. It not the intent of this committee to require that sprinkler systems be field modified to accommodate these installed clearances if supports or equipment are installed after the sprinkler system is installed (i.e., the burden should not necessarily be on the sprinkler contractor to make the field modifications). It is the intent of this committee that the installation of permanently attached equipment, distribution systems, supports and fire sprinkler systems be coordinated such that the minimum clearance is maintained after their installation. As Building Information Systems become more widely used and nonstructural components and systems are detailed in the design phase of the project, maintaining these clearances should become easier to ensure by design.

C13.6.7.3 Exceptions. The conditions under which the force requirements of Section 13.3 may be waived are based on observed performance in past earthquakes. The limits on the maximum hanger or trapeze drop (hanger rod length) must be met by all the hangers or trapezes supporting the piping system. See additional commentary in Section C13.1.4.

C13.6.9 Utility and Service Lines. For essential facilities (Risk Category IV), auxiliary on-site mechanical and electrical utility sources are recommended.

Where utility lines pass through the interface of adjacent, independent structures, they must be detailed to accommodate differential displacement computed in accordance with Section 13.3.2 and including the C_d factor of Section 12.2.1.

As specified in Section 13.1.3, nonessential piping whose failure could damage essential utilities in the event of pipe rupture may be considered designated seismic systems.

C13.6.10 Boilers and Pressure Vessels. Experience in past earthquakes has shown that boilers and pressure vessels are rugged and perform well in strong ground motion. Construction in accordance with current requirements of the ASME *Boiler and Pressure Vessel Code* (ASME BPVC) has been shown to be effective in limiting damage to and avoiding loss of fluid containment in boilers and pressure vessels under earthquake conditions. It is, therefore, the intent of the standard that nationally recognized codes be used to design boilers and pressure vessels provided that the seismic force and displacement demands are equal to or exceed those outlined in Section 13.3. Where nationally recognized codes do not yet incorporate force and displacement requirements comparable to the requirements of Section 13.3, it is nonetheless the intent to use the design acceptance criteria and construction practices of those codes.

C13.6.11 Elevator and Escalator Design Requirements. The ASME *Safety Code for Elevators and Escalators* (ASME A17.1 has adopted many requirements to improve the seismic response of elevators; however, they do not apply to some regions covered by this chapter. These changes are to extend force requirements for elevators to be consistent with the standard.

C13.6.11.3 Seismic Controls for Elevators. ASME A17.1 Section 8.4.10.1.2, specifies the requirements for the location and sensitivity of seismic switches to achieve the following goals: (a) safe shutdown in the event of an earthquake severe enough to impair elevator operations, (b) rapid and safe reactivation of elevators after an earthquake, and (c) avoidance of unintended elevator shutdowns. This level of safety is achieved by requiring the switches to be in or near the elevator equipment room, by using switches located on or near building columns that respond to vertical accelerations that would result from P and S waves, and by setting the sensitivity of the switches at a level that avoids false shutdowns because of nonseismic sources of vibration. The trigger levels for switches with horizontal sensitivity (for cases where the switch cannot be located near a column) are based on the experience with California hospitals in the Northridge earthquake of 1994. Elevators in which the seismic switch and counterweight derail device have triggered should not be put back into service without a complete inspection. However, in the case where the loss of use of the elevator creates a life-safety hazard, an attempt to put the elevator back into service may be attempted. Operating the elevator before inspection may cause severe damage to the elevator or its components.

The building owner should have detailed written procedures in place defining for the elevator operator and/or maintenance personnel which elevators in the facility are necessary from a post-earthquake, life-safety perspective. It is highly recommended that these procedures be in place, with appropriate personnel training, before an event occurs that is strong enough to trip the seismic switch.

C13.6.11.4 Retainer Plates. The use of retainer plates is a low-cost provision to improve the seismic response of elevators.

C13.6.12 Rooftop Solar Panels. Rooftop solar panels without positive attachment to the roof structure are limited to low-profile panels with a low height-to-depth ratio that respond by sliding on the roof surface without overturning. The amount of roof slope is limited because studies show that panels on sloped surfaces tend to displace in the downslope direction when subjected to seismic shaking, and the displacement increases with greater roof slope.

Displacement-based design of panels includes verifying that the panel remains safe if displaced. It needs to be verified that there is roof capacity to support the weight of the displaced panel and that wiring to the panel can accommodate the design panel displacement without damage.

Eq. (13.6-1) conservatively assumes a minimum coefficient of friction between the solar panel and the roof of 0.4. In cold-weather regions, the effects on the friction coefficient should be considered for Seismic Design Categories D, E, and F.

Structural interconnection between portions of a panel must be of adequate design strength, in tension or compression, and stiffness in order to account for the potential that frictional resistance to sliding will be different under some portions of the panel as a result of varying normal force and actual instantaneous values of friction coefficient for a given roof surface material.

The requirement for unattached panel to be bounded by a curb or parapet is usually satisfied by a curb at the roof edge. In lieu of being bounded by curbs or parapets at roof edges and offsets, the panel may be set back a larger distance from the edge.

Analytical and experimental studies of the seismic response of unattached solar panels are reported by Schellenberg et al. (2012) and Maffei et al. (2013).

Shake table testing and nonlinear time history analysis may also be used to predict panel displacements; however, for unattached panels, it is necessary to use input motions appropriate for predicting sliding displacement, which can be affected by content in the low-frequency range. See SEAOC (2012) for guidance on the performance of such testing and analysis.

C13.6.13 Other Mechanical and Electrical Components. The material properties set forth in item 2 of this section are similar to those allowed in ASME BPVC and reflect the high factors of safety necessary for seismic, service, and environmental loads.

REFERENCES

American Concrete Institute (ACI). (2007). "Qualification of post-installed mechanical anchors in concrete and commentary." *ACI 355.2*, ACI, Farmington Hills, MI.

ACI. (2014). "Building code requirements for structural concrete and commentary." *ACI 318*, ACI, Farmington Hills, MI.

American National Standards Institute/Sheet Metal and Air-Conditioning Contractors National Association (ANSI/SMACNA). (2000). "Seismic restraint manual: Guidelines for mechanical systems." *ANSI/SMACNA 001*, Chantilly, VA.

ASCE. (2011). *Guidelines for seismic evaluation and design of petrochemical facilities*, 2nd Ed. ASCE, Reston, VA.

American Society of Heating, Refrigerating, and Air-Conditioning Engineers (ASHRAE). (2000). *Practical guide to seismic restraint, RP-812*. ASHRAE, Atlanta, GA.

American Society of Mechanical Engineers (ASME). (2007). "Safety code for elevators and escalators." *A17.1*, ASME, New York.

ASME. (2010a). "Boiler and pressure vessel code." *ASME*, New York.

ASME. (2010b). "Standard for the seismic design and retrofit of aboveground piping systems." *ASME*, New York.

ANCO Engineers, Inc. (1983). "Seismic hazard assessment of non-structural components—Phase I." *Final Report for the National Science Foundation from ANCO Engineers, Inc.*, Culver City, CA, September.

ASTM International. (2004). "Standard specification for the manufacture, performance, and testing of metal suspension systems for accoustical tile and lay-in panel ceiling." *ASTM C635*, West Conshohocken, PA.

ASTM. (2014a). "Standard practice for installation of ceiling suspension systems for acoustical tile and lay-in panels for areas subject to earthquake ground motion." *ASTM E580/E580M-14*, West Conshohocken, PA.

ASTM (2014b). "Standard guide for structural sealant glazing." *ASTM C1401*, West Conshohocken, PA.

Bachman, R. E., and Dowty, S. M. (2008). "Nonstructural component or nonbuilding structure?" *Bldg. Safety J.* (April–May).

Bachman, R. E., Drake, R. M., and Richter, P. J. (1993). "1994 Update to 1991 NEHRP provisions for architectural, mechanical, and electrical components and systems." *Letter report to the National Center for Earthquake Engineering Research*, Feb. 22, 1993.

Bouwkamp, J. G., and Meehan, J. F. (1960). "Drift limitations imposed by glass." Proc. 2nd World Conference on Earthquake Engineering, Tokyo, 1763–1778.

Canterbury Earthquakes Royal Commission. (2012). "The performance of Christchurch CBD buildings." *Final Report*, 2, 88–106, Canterbury, Wellington, UK.

Dowell, R. K., and Johnson, T. P. (2013). "Evaluation of seismic overstrength factors for anchorage into concrete via dynamic shaking table tests." *Report No. SERP – 13/09*, San Diego State University, CA.

FM Global. (2007). Approved standard for pipe couplings and fittings for aboveground fire protection systems. ⟨http://www.fmglobal.com/assets/pdf/fmapprovals/1920.pdf⟩.

International Code Council (ICC). (1976). *Uniform Building Code*, Whittier, CA.

Institute of Electrical and Electronics Engineers, Inc. (IEEE). (2005). "IEEE recommended practices for seismic design of substations" *IEEE 693-2005*, Piscataway, NJ.

International Code Council Evaluation Service (ICC-ES). (2010). "Seismic qualification by shake-table testing of nonstructural components and systems." *ICC-ES AC156*, Whittier, CA.

ICC-ES. (2012a). "Acceptance criteria for adhesive anchors in masonry elements." *ICC-ES AC58*, Whittier, CA.

ICC-ES. (2012b). "Acceptance criteria for expansion anchors in masonry elements." *ICC-ES AC01*, Whittier, CA.

ICC-ES. (2012c). "Acceptance criteria for mechanical anchors in concrete elements." *ICC-ES AC193*, Whittier, CA.

ICC-ES. (2012d). "Acceptance criteria for post-installed adhesive anchors in concrete elements." *ICC-ES AC308*, Whittier, CA.

ICC-ES. (2012e). "Acceptance criteria for predrilled fasteners (screw anchors) in masonry." *ICC-ES AC106*, Whittier, CA.

Kehoe, B., and Hachem, M. (2003). "Procedures for estimating floor accelerations." ATC 29-2, *Proceedings of the Seminar on Seismic Design, Performance, and Retrofit of Nonstructural Components in Critical Facilities*, Newport Beach, CA, 361–374.

Maffei, J., Fathali, S., Telleen, K., Ward, R., and Schellenberg, A. (2013). "Seismic design of ballasted solar arrays on low-slope roofs." *J. Struct. Eng.*, doi: 10.1061/(ASCE)ST.1943-541X.0000865.

Manufacturers Standardization Society of the Valve and Fitting Industry (MSS). (2001). "Bracing for piping systems seismic–wind–dynamic design, selection, application." *MSS SP-127*, MSS, Vienna, VA.

National Earthquake Hazards Reduction Program (NEHRP). (2009). NEHRP recommended provisions for seismic regulations for new buildings and other structures, NEHRP, Washington, DC.

National Fire Protection Association (NFPA). (2007). "Standard for the installation of sprinkler systems." *NFPA 13*, NFPA, Quincy, MA.

NFPA. (2011). "National electric code." *NFPA 70*, NFPA, Quincy, MA.

Network for Earthquake Engineering Simulation (NEES). (2011). "Full-scale structural and nonstructural building system performance during earthquake and post-earthquake fire." ⟨http://nees.ucsd.edu/projects/2011-five-story/⟩

Rahmanishamsi, E., Soroushian, S., and Maragakis, E. (2014). "Seismic response of ceiling/piping/partition systems in NEESR-GC system-level experiments," Proc. ASCE Structures Congress, Boston.

Schellenberg, A., Maffei, J., Miller, K., Williams, M., Ward, R., and Dent, M. (2012). "Shake-table testing of unattached rooftop solar arrays: Interim report, Subtask 4.1." *SunLink/Rutherford & Chekene Report for the California Solar Initiative* ⟨www.gosolarcalifornia.org/csi⟩, Sacramento, CA.

Sheet Glass Association of Japan. (1982). *Earthquake safety design of windows*, SGAJ, Tokyo.

Sheet Metal and Air-Conditioning Contractors National Association (SMACNA). (2008). *Seismic restraint manual: Guidelines for mechanical systems*, 3rd Ed., SMACNA, Chantilly, VA.

Soroushian, S., et al. (2014). "A comparative study of sub-system and system level experiments of suspension ceiling systems," Proc. 10th U.S. National Conference on Earthquake Engineering, Anchorage, AK.

Soroushian, S., Reinhorn, A., Rahmanishamsi, E., Ryu, K., and Maragakis, M. (2012). "Seismic Response of Ceiling/Sprinkler Piping Nonstructural Systems in NEES TIPS/NEES Nonstructural/NIED Collaborative Tests on a Full Scale 5-Story Building," Proc. ASCE Structures Congress, Chicago.

Soulages, J. R., and Weir, R. (2011). "Cyclic testing of pipe trapezes with rigid hanger assemblies," Proc. 80th Structural Engineers Association of California Annual Convention, Las Vegas, NV.

Structural Engineers Association of California (SEAOC). (1999). *Recommended lateral force requirements and commentary*, SEAOC, Sacramento, CA.

SEAOC. (2012). "Structural seismic requirements and commentary for rooftop solar photovoltaic arrays." *Report SEAOC-PV1-2012*, SEAOC, August.

Underwriter Laboratories (UL). (2004). "Rubber gasketed fittings for fire-protection service." *UL 213*, Northbrook, IL.

U.S. Department of Defense (DOD). (2007). "Seismic design for buildings." *UFC 3-310-04*, DOD, Unified Facilities Criteria, Washington, DC.

U.S. Army. (1986). "Seismic design guidelines for essential buildings." *TM 5-809-1*, Joint Departments of the Army, Navy, and Air Force., Washington, DC.

Wright, P. D. (1989). "The development of a procedure and rig for testing the racking resistance of curtain wall glazing." BRANZ Study Report 17, Building Research Association of New Zealand (BRANZ), Porirua, New Zealand.

OTHER REFERENCES (NOT CITED)

ACI. (2011a). "Building code requirements and specification for masonry structures and related commentaries." *ACI 530/530.1*, ACI, Farmington Hills, MI.

ACI. (2011b). "Building code requirements for structural concrete and commentary." *ACI 318*, ACI, Farmington Hills, MI.

ACI. (2011c). "Qualification of post-installed adhesive anchors in concrete and commentary." *ACI 355.4*, ACI, Farmington Hills, MI.

ASTM International. (2007). "Standard specification for the manufacture, performance, and testing of metal suspension systems for acoustical tile and lay-in panel ceilings." *ASTM C635/C635M*, West Conshohocken, PA.

ASTM. (2013a). "Standard specification for the manufacture, performance, and testing of metal suspension systems for acoustical tile and lay-in panel ceilings." *ASTM C635/C635M-13a*, West Conshohocken, Penn.

ASTM. (2013b). "Standard practice for installation of metal ceiling suspension systems for acoustical tile and lay-in panels." *ASTM C636/C636M-13*, West Conshohocken, PA.

Bachman, R. E, and Drake, R. M. (1996). "A study to empirically validate the component response modification factors in the 1994 NEHRP Provisions design force equations for architectural, mechanical, and electrical components." Letter report to the National Center for Earthquake Engineering Research, July.

Bachman, R. E., Drake, R. M., and Richter, P. J. (1993). "1994 update to 1991 NEHRP Provisions for architectural, mechanical, and electrical components and systems." Letter report to the National Center for Earthquake Engineering Research, February 22.

Behr, R. A., and Belarbi, A. (1996). "Seismic test methods for architectural glazing systems." *Earthq. Spectra*, 12(1), 129–143.

Behr, R. A., Belarbi, A., and Brown, A. T. (1995). "Seismic performance of architectural glass in a storefront wall system." *Earthq. Spectra*, 11(3), 367–391.

Drake, R. M., and Bachman, R. E. (1994). "1994 NEHRP provisions for architectural, mechanical, and electrical components." Proc. 5th United States National Conference on Earthquake Engineering, Chicago.

Drake, R. M., and Bachman, R. E. (1995). "Interpretation of instrumented building seismic data and implications for building codes." Proc. 1995 SEAOC Annual Convention, Squaw Creek, CA.

Drake, R. M., and Bachman, R. E. (1996). "NEHRP provisions for 1994 for nonstructural components." *J. Arch. Engrg.*, 2(1), 26–31.

Federal Emergency Management Agency (FEMA), FEMA E-74 Reducing the risks of nonstructural earthquake damage—A practical guide, 4th Ed., FEMA, Dec. 2012, 6-144–6-153.

Fleischmann, R. B., Restrepo, J. I., and Pampanin, S. (2014). "Damage Evaluations of Precast Concrete Structures in the 2010–2011 Canterbury Earthquake Sequence," *EERI Earthq. Spectra*, 30(1), 277–306.

Gates, W. E., and McGavin, G. (1998). "Lessons learned from the 1994 Northridge earthquake on the vulnerability of nonstructural systems." Proc. seminar on seismic design, retrofit, and performance of nonstructural components, ATC-29-1, Applied Technology Council, Redwood City, CA, 93–101.

Haroun, M. A., and Housner, G. W. (1981). "Seismic design of liquid storage tanks." *J. Tech. Councils of ASCE*, 107(1), 191–207.

Higgins, C. (2009) "Prefabricated steel stair performance under combined seismic and gravity loads," *J. Struct. Eng.*, doi: 10.1061/(ASCE)0733-9445 (2009)135:2(122), 122–129.

Pantelides, C. P., Truman, K. Z., Behr, R. A., and Belarbi, A. (1996). "Development of a loading history for seismic testing of architectural glass in a shop-front wall system." *Engrg. Struct.*, 18(12), 917–935.

Pantoli, E., Chen, M., Hutchinson, T., Underwood, G., and Hildebrand, M. (2013). "Shake table testing of a full-scale five-story building: seismic performance of precast concrete cladding panels," *4th ECCOMAS Thematic Conference on Computational Methods in Structural Dynamics and Earthquake Engineering (COMPDYN 2013)*, Kos Island, Greece, June 12–14.

Trautner, C., Hutchinson, T., Grosser, P. (2014). "Cyclic behavior of structural base plate connections with ductile fastening failure: component test results," 10th U.S. *National Conference on Earthquake Engineering*, Anchorage, AK.

CHAPTER C14
MATERIAL-SPECIFIC SEISMIC DESIGN AND DETAILING REQUIREMENTS

Because seismic loading is expected to cause nonlinear behavior in structures, seismic design criteria require not only provisions to govern loading but also provisions to define the required configurations, connections, and detailing to produce material and system behavior consistent with the design assumptions. Thus, although ASCE/SEI 7-10 is primarily a loading standard, compliance with Chapter 14, which covers material-specific seismic design and detailing, is required. In general, Chapter 14 adopts material design and detailing standards developed by material standards organizations. These material standards organizations maintain complete commentaries covering their standards, and such material is not duplicated here.

C14.0 SCOPE

The scoping statement in this section clarifies that foundation elements are subject to all of the structural design requirements of the standard.

C14.1 STEEL

C14.1.1 Reference Documents. This section lists a series of structural standards published by the American Institute of Steel Construction (AISC), the American Iron and Steel Institute (AISI), the American Society of Civil Engineers (ASCE/SEI), the Steel Deck Institute (SDI), and the Steel Joist Institute (SJI), which are to be applied in the seismic design of steel members and connections in conjunction with the requirements of ASCE/SEI 7. The AISC references are available free of charge in electronic format at www.aisc.org; the AISI references are available on www.steel.org; the SDI references are available as a free download at www.sdi.org; and the SJI references are available as a free download at www.steeljoist.org.

C14.1.2 Structural Steel

C14.1.2.1 General. This section adopts AISC 360 (2016) by direct reference. The specification applies to the design of the structural steel system or systems with structural steel acting compositely with reinforced concrete. In particular, the document sets forth criteria for the design, fabrication, and erection of structural steel buildings and other structures, where other structures are defined as structures designed, fabricated, and erected in a manner similar to buildings, with buildinglike vertical and lateral load-resisting elements. The document includes extensive commentary.

C14.1.2.2 Seismic Requirements for Structural Steel Structures

C14.1.2.2.1 Seismic Design Categories B and C. For the lower Seismic Design Categories (SDCs) B and C, a range of options are available in the design of a structural steel lateral force-resisting system. The first option is to design the structure to meet the design and detailing requirements in AISC 341 (2016) for structures assigned to higher SDCs, with the corresponding seismic design parameters (R, Ω_0, and C_d). The second option, presented in the exception, is to use an R factor of 3 (resulting in an increased base shear), an Ω_0 of 3, and a C_d value of 3 but without the specific seismic design and detailing required in AISC 341 (2016). The basic concept underlying this option is that design for a higher base shear force results in essentially elastic response that compensates for the limited ductility of the members and connections. The resulting performance is considered comparable to that of more ductile systems.

C14.1.2.2.2 Seismic Design Categories D through F. For the higher SDCs, the engineer must follow the seismic design provisions of AISC 341 (2016) using the seismic design parameters specified for the chosen structural system, except as permitted in Table 15.4-1. For systems other than those identified in Table 15.4-1, it is not considered appropriate to design structures without specific design and detailing for seismic response in these high SDCs.

C14.1.3 Cold-Formed Steel

C14.1.3.1 General. This section adopts two standards by direct reference: ANSI/AISI S100, *North American Specification for the Design of Cold-Formed Steel Structural Members* (2016), and ASCE/SEI 8, *Specification for the Design of Cold Formed Stainless Steel Structural Members* (2002).

Both of the adopted reference documents have specific limits of applicability. ANSI/AISI S100 (2016) (Section A1.1) applies to the design of structural members that are cold-formed to shape from carbon or low-alloy steel sheet, strip, plate, or bar not more than 1 in. (25 mm) thick. ASCE/SEI 8 (2002) (Section 1.1.1) governs the design of structural members that are cold-formed to shape from annealed and cold-rolled sheet, strip, plate, or flat bar stainless steels. Both documents focus on load-carrying members in buildings; however, allowances are made for applications in nonbuilding structures, if dynamic effects are considered appropriately.

Within each document, there are requirements related to general provisions for the applicable types of steel; design of elements, members, structural assemblies, connections, and joints; and mandatory testing. In addition, ANSI/AISI S100 contains a chapter on the design of cold-formed steel structural members and connections undergoing cyclic loading. Both standards contain extensive commentaries.

C14.1.3.2 Seismic Requirements for Cold-Formed Steel Structures. This section adopts three standards by direct reference—AISI S100 (2016), ASCE/SEI 8 (2002), and AISI S400 (2015). Cold-formed steel and stainless steel members that are part of a seismic force-resisting system listed in Table 12.2-1 must be detailed in accordance with the appropriate base standard: AISI S100 or ASCE 8.

The section also adopts a reference to AISI S400, which includes additional design provisions for a specific cold-formed steel seismic force-resisting system entitled "cold-formed steel—special bolted moment frame" or CFS-SBMF. Sato and Uang (2007) have shown that this system experiences inelastic deformation at the bolted connections because of slip and bearing during significant seismic events. To develop the designated mechanism, requirements based on capacity design principles are provided for the design of the beams, columns, and associated connections. The document has specific requirements for the application of quality assurance and quality control procedures.

C14.1.4 Cold-Formed Steel Light-Frame Construction

C14.1.4.1 General. This subsection of cold-formed steel relates to light-frame construction, which is defined as a method of construction where the structural assemblies are formed primarily by a system of repetitive wood or cold-formed steel framing members or subassemblies of these members (Section 11.2 of this standard). It adopts Section I4 of AISI S100 (2016), which directs the user to an additional suite of AISI standards, including ANSI/AISI S240 and ANSI/AISI S400.

In addition, all of these documents include commentaries to aid users in the correct application of their requirements.

C14.1.4.2 Seismic Requirements for Cold-Formed Steel Light-Frame Construction. Cold-formed steel structural members and connections in seismic force-resisting systems and diaphragms must be designed in accordance with the additional provisions of ANSI/AISI S400 in seismic design categories (SDC) D, E, or F, or wherever the seismic response modification coefficient, R, used to determine the seismic design forces is taken other than 3. In particular, this requirement includes all entries from Table 12.2-1 of this standard for "light-frame (cold-formed steel) walls sheathed with wood structural panels ... or steel sheets," "light-frame walls with shear panels of all other materials" (e.g., gypsum board and fiberboard panels), and "light-frame wall systems using flat strap bracing."

C14.1.4.3 Prescriptive Cold-Formed Steel Light-Frame Construction. This section adopts ANSI/AISI S230, *Standard for Cold-Formed Steel Framing—Prescriptive Method for One and Two Family Dwellings*, which applies to the construction of detached one- and two-family dwellings, townhouses, and other attached single-family dwellings not more than two stories in height using repetitive in-line framing practices (Section A1). This document includes a commentary to aid the user in the correct application of its requirements.

C14.1.5 Cold-Formed Steel Deck Diaphragms. This section adopts the applicable standards for the general design of cold-formed steel deck diaphragms and steel roof, noncomposite floor, and composite floor deck. The SDI standards also reference ANSI/AISI S100 for materials and determination of cold-formed steel cross section strength and specify additional requirements specific to steel deck design and installation.

Additionally, design of cold-formed steel deck diaphragms is to be based on ANSI/AISI S310. All fastener design values (welds, screws, power-actuated fasteners, and button punches) for attaching deck sheet to deck sheet or for attaching the deck to the building framing members must be per ANSI/AISI S310 or specific testing prescribed in ANSI/AISI 310. All cold-formed steel deck diaphragm and fastener design properties not specifically included in ANSI/AISI S310 must be approved for use by the authorities in whose jurisdiction the construction project occurs. Deck diaphragm in-plane design forces (seismic, wind, or gravity) must be determined per ASCE 7, Section 12.10.1. Cold-formed steel deck manufacturer test reports prepared in accordance with this provision can be used where adopted and approved by the authority having jurisdiction for the building project. The *Diaphragm Design Manual* produced by the Steel Deck Institute (2015) is also a reference for design values.

Cold-formed steel deck is assumed to have a corrugated profile consisting of alternating up and down flutes that are manufactured in various widths and heights. Use of flat sheet metal as the overall floor or roof diaphragm is permissible where designed by engineering principles, but it is beyond the scope of this section. Flat or bent sheet metal may be used as closure pieces for small gaps or penetrations or for shear transfer over short distances in the deck diaphragm where diaphragm design forces are considered.

Cold-formed steel deck diaphragm analysis must include design of chord members at the perimeter of the diaphragm and around interior openings in the diaphragm. Chord members may be steel beams attached to the underside of the steel deck designed for a combination of axial loads and bending moments caused by acting gravity and lateral loads.

Where diaphragm design loads exceed the bare steel deck diaphragm design capacity, then either horizontal steel trusses or a structurally designed concrete topping slab placed over the deck must be provided to distribute lateral forces. Where horizontal steel trusses are used, the cold-formed steel deck must be designed to transfer diaphragm forces to the steel trusses. Where a structural concrete topping over the deck is used as the diaphragm, the diaphragm chord members at the perimeter of the diaphragm and edges of interior openings must be either (a) designed flexural reinforcing steel placed in the structural concrete topping or (b) steel beams located under the deck with connectors (that provide a positive connection) as required to transfer design shear forces between the concrete topping and steel beams.

C14.1.7 Steel Cables. These provisions reference ASCE 19, *Structural Applications of Steel Cables for Buildings*, for the determination of the design strength of steel cables.

C14.1.8 Additional Detailing Requirements for Steel Piles in Seismic Design Categories D through F. Steel piles used in higher SDCs are expected to yield just under the pile cap or foundation because of combined bending and axial load. Design and detailing requirements of AISC 341 for H-piles are intended to produce stable plastic hinge formation in the piles. Because piles can be subjected to tension caused by overturning moment, mechanical means to transfer such tension must be designed for the required tension force, but not less than 10% of the pile compression capacity.

C14.2 CONCRETE

The section adopts by reference ACI 318 for structural concrete design and construction. In addition, modifications to ACI 318-14 are made that are needed to coordinate the provisions of that material design standard with the provisions of ASCE 7. Work is ongoing to better coordinate the provisions of the two documents (ACI 318 and ASCE 7) such that the provisions in Section 14.2 will be progressively reduced in future editions of ASCE 7.

C14.2.2.1 Definitions. Two definitions included here describe wall types for which definitions currently do not exist in ACI 318. These definitions are essential to the proper interpretation of the R and C_d factors for each wall type specified in Table 12.2-1.

A definition for *connector* has been added, which does not currently exist in ACI 318-14. Section 12.11 provides an alternative to the current diaphragm design procedure of Section 12.10. The alternative procedure is made mandatory for precast concrete diaphragms in structures assigned to SDC C, D, E, or F. The definition of *connector* is essential because the three design options (BDO, EDO, and RDO) are closely related to the connector classification, and the diaphragm design force reduction factor, R_s, depends on the design option.

The definition for *connection* in ACI 318-14 has also been supplemented, as it applies to this protocol.

C14.2.2.2 ACI 318, Section 10.7.6. ACI 318-14, Section 10.7.6.1.6, prescribes details of transverse reinforcement around anchor bolts in the top of a column or pedestal. This modification prescribes additional details for transverse reinforcement around such anchor bolts in structures assigned to SDCs C through F.

C14.2.2.3 Scope. This provision describes how the ACI 318-14 provisions should be interpreted for consistency with the ASCE 7 provisions.

C14.2.2.4 Intermediate Precast Structural Walls. Section 18.5 of ACI 318-14 imposes requirements on precast walls for moderate seismic risk applications. Ductile behavior is to be ensured by yielding of the steel elements or reinforcement between panels or between panels and foundations. This provision requires the designer to determine the deformation in the connection corresponding to the earthquake design displacement and then to check from experimental data that the connection type used can accommodate that deformation without significant strength degradation.

Several steel element connections have been tested under simulated seismic loading, and the adequacy of their load-deformation characteristics and strain capacity have been demonstrated (Schultz and Magana 1996). One such connection was used in the five-story building test that was part of the Precast Seismic Structural Systems (PRESSS) Phase 3 research. The connection was used to provide damping and energy dissipation, and it demonstrated a very large strain capacity (Nakaki et al. 2001). Since then, several other steel element connections have been developed that can achieve similar results (Banks and Stanton 2005 and Nakaki et al. 2005). In view of these results, it is appropriate to allow yielding in steel elements that have been shown experimentally to have adequate strain capacity to maintain at least 80% of their yield force through the full design displacement of the structure.

C14.2.2.6 Foundations. The intention is that there should be no conflicts between the provisions of ACI 318-14, Section 18.13, and ASCE 7, Sections 12.1.5, 12.13, and 14.2. However, the additional detailing requirements for concrete piles of Section 14.2.3 can result in conflicts with ACI 318-14 provisions if the pile is not fully embedded in the soil.

C14.2.2.7 Detailed Plain Concrete Shear Walls. Design requirements for plain masonry walls have existed for many years, and the corresponding type of concrete construction is the plain concrete wall. To allow the use of such walls as the lateral force-resisting system in SDCs A and B, this provision requires such walls to contain at least the minimal reinforcement specified in ACI 318-14, Section 14.6.2.2.

C14.2.3 Additional Detailing Requirements for Concrete Piles. Chapter 20 of PCI (2004) provides detailed information on the structural design of piles and on pile-to-cap connections for precast prestressed concrete piles. ACI 318-14 does not contain provisions governing the design and installation of portions of concrete piles, drilled piers, and caissons embedded in ground except for SDC D, E, and F structures.

C14.2.3.1.2 Reinforcement for Uncased Concrete Piles (SDC C). The transverse reinforcing requirements in the potential plastic hinge zones of uncased concrete piles in SDC C are a selective composite of two ACI 318-14 requirements. In the potential plastic hinge region of an intermediate moment-resisting concrete frame column, the transverse reinforcement spacing is restricted to the least of (1) eight times the diameter of the smallest longitudinal bar, (2) 24 times the diameter of the tie bar, (3) one-half the smallest cross-sectional dimension of the column, and (4) 12 in. (304.8 mm). Outside of the potential plastic hinge region of a special moment-resisting frame column, the transverse reinforcement spacing is restricted to the smaller of six times the diameter of the longitudinal column bars and 6 in. (152.4 mm).

C14.2.3.1.5 Reinforcement for Precast Nonprestressed Piles (SDC C). Transverse reinforcement requirements inside and outside of the plastic hinge zone of precast nonprestressed piles are clarified. The transverse reinforcement requirement in the potential plastic hinge zone is a composite of two ACI 318-14 requirements (see Section C14.2.3.1.2). Outside of the potential plastic hinge region, the transverse reinforcement spacing is restricted to 16 times the longitudinal bar diameter. This restriction should permit the longitudinal bars to reach compression yield before buckling. The maximum 8-in. (203.2-mm) tie spacing comes from current building code provisions for precast concrete piles.

C14.2.3.1.6 Reinforcement for Precast Prestressed Piles (SDC C). The transverse and longitudinal reinforcing requirements given in ACI 318-14, Chapter 21, were never intended for slender precast prestressed concrete elements and result in unbuildable piles. These requirements are based on PCI Committee on Prestressed Concrete Piling (1993).

Eq. (14.2-1), originally from ACI 318-14, has always been intended to be a lower bound spiral reinforcement ratio for larger diameter columns. It is independent of the member section properties and can therefore be applied to large- or small-diameter piles. For cast-in-place concrete piles and precast prestressed concrete piles, the spiral reinforcing ratios resulting from this formula are considered to be sufficient to provide moderate ductility capacities (Fanous et al. 2007).

Full confinement per Eq. (14.2-1) is required for the upper 20 ft (6.1 m) of the pile length where curvatures are large. The amount is relaxed by 50% outside of that length in view of lower curvatures and in consideration of confinement provided by the soil.

C14.2.3.2.3 Reinforcement for Uncased Concrete Piles (SDC D through F). The reinforcement requirements for uncased concrete piles are taken from current building code requirements and are intended to provide ductility in the potential plastic hinge zones (Fanous et al. 2007).

C14.2.3.2.5 Reinforcement for Precast Nonprestressed Piles (SDC D through F). The transverse reinforcement requirements for precast nonprestressed concrete piles are taken from the IBC (ICC 2012) requirements and should be adequate to provide ductility in the potential plastic hinge zones (Fanous et al. 2007).

C14.2.3.2.6 Reinforcement for Precast Prestressed Piles (SDC D through F). The reduced amounts of transverse reinforcement specified in this provision compared with those required for special moment frame columns in ACI 318-14 are justified by the results of the study by Fanous et al. (2007). The last paragraph provides minimum transverse reinforcement outside of the zone of prescribed ductile reinforcing.

C14.2.4 Additional Design and Detailing Requirements for Precast Concrete Diaphragms. Section 12.10.3 introduces an alternative procedure for the calculation of diaphragm design forces of Sections 12.10.1 and 12.10.2 and is made mandatory for precast concrete diaphragms in structures assigned to SDC C, D, E, or F. The diaphragm design force reduction factors, R_s, in Table 12.10-1 for precast concrete diaphragms are specifically tied to design and detailing requirements so that the ductility and overstrength necessary for expected diaphragm performance are achieved. Section 14.2.4 is based on the Diaphragm Seismic Design Methodology (DSDM), the product of a multiple-university research project termed the DSDM Project (Charles Pankow Foundation 2014), and gives detailing requirements for diaphragms constructed of precast concrete units in SDC C, D, E, or F consistent with the R_s factors. These detailing requirements are in addition to those of ACI 318, as modified by Section 14.2. The derivation of diaphragm design force reduction factors is described in Commentary Section C12.10.3.5.

Section C12.10.3.5 relates the global ductility required by the three design options defined in Section 11.2 to the local ductility of connectors measured at the maximum considered earthquake (MCE) level. The jointed nature of precast systems results in the load paths and deformations being largely determined by the connections across the joints. The connections may consist of either reinforced concrete topping slabs or discrete mechanical connectors. Since the diaphragm strains are concentrated at the joints, the connectors or the reinforcing in the topping slab must accommodate some strain demand.

C14.2.4.1 Diaphragm Seismic Demand Levels. Fig. 14.2-1 is used to determine diaphragm seismic demand level as a function of the diaphragm span and the diaphragm aspect ratio.

The diaphragm span defined in Section 14.2.4.1.1 is illustrated in Fig. C14.2-1. Most precast diaphragms contain precast units running in only one direction, and typically the maximum span is oriented perpendicular to the joints between the primary precast floor units. The connector or reinforcement deformability classifications and resulting R_s factors are calibrated relative to joint openings between the precast floor units and are thus based on the more typical orientation.

The diaphragm aspect ratio (AR) defined in Section 14.2.4.1.2 is also illustrated in Fig. C14.2-1.

The following lists provide details of seismic demand level classifications, determined in accordance with Fig. 14.2-1:

Low Seismic Demand Level

1. Diaphragms in structures assigned to SDC C.
2. Diaphragms in structures assigned to SDC D, E, or F with diaphragm span ≤75 ft (22.86 m), number of stories ≤3, and diaphragm aspect ratio <2.5.

Moderate Seismic Demand Level

1. Diaphragms in structures assigned to SDC D, E, or F with diaphragm span ≤75 ft (22.86 m) and number of stories >3 but ≤6.
2. Diaphragms in structures assigned to SDC D, E, or F with diaphragm span >75 ft (22.86 m) but ≤190 ft (57.91 m) and number of stories ≤2.
3. Diaphragms in structures assigned to SDC D, E, or F with diaphragm span >75 ft (22.86 m) but ≤140 ft (42.67 m) and number of stories >2 but ≤4.

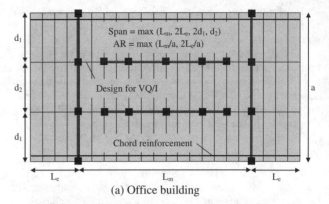

(a) Office building

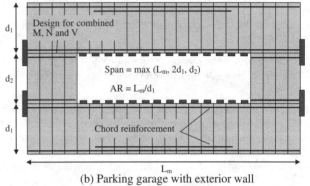

(b) Parking garage with exterior wall

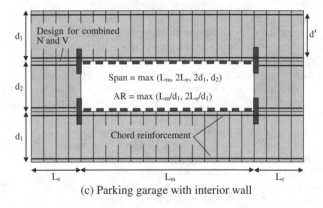

(c) Parking garage with interior wall

FIGURE C14.2-1 Diaphragm Dimensions

4. Diaphragms in structures assigned to SDC D, E, or F with diaphragm span ≤75 ft (22.86 m), number of stories ≤3, and diaphragm aspect ratio ≥2.5.
5. Diaphragms in structures assigned to SDC D, E, or F, categorized below as high seismic demand level, with diaphragm aspect ratio <1.5.

High Seismic Demand Level

1. Diaphragms in structures assigned to SDC D, E, or F with diaphragm span >190 ft (57.91 m).
2. Diaphragms in structures assigned to SDC D, E, or F with diaphragm span >140 ft (42.67 m)) and number of stories >2.
3. Diaphragms in structures assigned to SDC D, E, or F with diaphragm span >75 ft (22.86 m) and number of stories >4.
4. Diaphragms in structures assigned to SDC D, E, or F with number of stories >6.

Diaphragm Shear Overstrength Factor. The diaphragm shear overstrength factor, Ω_v, is applied to diaphragm shear reinforcement/connectors. The purpose of this factor is to keep the diaphragm shear response elastic while the diaphragm develops inelastic flexural action, as is anticipated for the basic design objective (BDO) in the MCE, and for the reduced design objective (RDO) for both the design earthquake and the MCE. No inelastic diaphragm response is anticipated for the elastic design objective (EDO).

The value of diaphragm shear overstrength factor is $\Omega_v = 1.4R_s$. The values of the diaphragm design force reduction factor, R_s, are 0.7, 1.0, and 1.4 for the EDO, BDO, and RDO, respectively. This value translates into diaphragm shear overstrength factors Ω_v of 1.0, 1.4, and 2.0 (rounded to one decimal place) for the EDO, BDO, and RDO, respectively.

The diaphragm shear overstrength factor, Ω_v, is applied to the diaphragm design forces and thus is a measure relative to the flexural strength of the diaphragm. As implied by the above-listed Ω_v values, the level of overstrength required relative to the diaphragm flexural strength varies with the design option. The RDO requires a higher overstrength than the BDO because of the larger anticipated inelastic action. For the EDO, no overstrength is required since the diaphragm design force itself targets elastic behavior in the MCE. It is noted that the absolute shear strength required in the design procedure is constant, regardless of design option, since the parameter R_s in the overstrength factor is canceled out by the R_s in the denominator of the diaphragm design force expression.

The Ω_v values represent upper bound constant values (for each diaphragm design objective) of parametric expressions developed for the required shear overstrength on the basis of detailed parametric studies performed using nonlinear dynamic time history analysis (NTHA) of analytical models of precast structures developed and calibrated on the basis of extensive large-scale physical testing. These precast structures were subjected to spectrum compatible ground motions scaled to the MCE in order to determine the required shear overstrength factors.

Precast diaphragms can be designed and detailed for ductile flexural response. However, to achieve the desired mechanism, potentially nonductile shear limit states have to be precluded. In order to prevent these shear failures, elastic shear response is targeted in the design procedure for both flexure-controlled and shear-controlled systems. Thus, the shear overstrength factor, Ω_v, is applied in diaphragm shear design.

The shear amplification factor values were obtained by bounding the maximum shear force V_{max} occurring in NTHA of the diaphragm at the critical shear joint as the diaphragm developed a flexural mechanism (in other regions of the floor) at MCE-level hazard and scaling it by the design shear, V_u. Accordingly:

- Ω_E, the diaphragm shear amplification factor for the EDO, is taken as unity ($\Omega_E = 1.0 \approx 1.4R_s$, where $R_s = 0.7$ for EDO) since elastic diaphragm response is expected in the MCE for EDO.
- Ω_B, the diaphragm shear amplification factor for the BDO, is taken as an upper bound on the V_{max}/V_u ratio for the BDO design under MCE-level hazard.
- Ω_R, the diaphragm shear amplification factor for the RDO, is taken as an upper bound on the V_{max}/V_u ratio for the RDO design under MCE-level hazard.

Fig. C14.2-2 shows a scatter plot of the V_{max}/V_u ratios from NTHA for different numbers of stories (n) and diaphragm aspect ratios (ARs) at the maximum considered earthquake. The data represent the mean of the maximum responses from five ground motions. The expression provided for Ω_v, $\Omega_v = 1.4R_s$, is plotted as a horizontal dashed line on each plot, indicating that the expression provides a constant upper bound for the anticipated required elastic shear forces for all design cases.

C14.2.4.2 Diaphragm Design Options. The intent of the design procedure is to provide the diaphragm with the proper combination of strength and deformation capacity in order to survive anticipated seismic events. Three different design options are provided to the designer to accomplish this objective, ranging from a fully elastic diaphragm design under the MCE to designs that permit significant inelastic deformations in the diaphragm under the design earthquake. The motivation for this approach is the recognition that, under certain conditions, a precast diaphragm designed to remain fully elastic up to the MCE may not be economical or reliable. Under other conditions, however, a diaphragm designed to remain elastic up to the MCE will be satisfactory and may be most desirable.

The methodology allows the designer three options related to deformation capacity:

1. An elastic design option (EDO), where the diaphragm is designed to the highest force levels, is calibrated to keep the diaphragm elastic not only for the design earthquake but also in an MCE. In exchange for the higher design force, this option permits the designer to detail the diaphragm with the low deformability element (LDE) connector or reinforcement that need not meet any specific deformation capacity requirements (tension deformation capacity less than 0.3 in. (7.6 mm)). This option is limited in its use through the introduction of diaphragm seismic demand levels, which are based on building height, diaphragm geometry, and seismic hazard level. The use of the EDO is not permitted if the diaphragm seismic demand level is high.

2. A basic design objective (BDO) is one in which the diaphragm is designed to a force level calibrated to keep the diaphragm elastic in the design earthquake but not necessarily in the MCE. The design force level is lower than that required for the EDO, but this option requires moderate deformability element (MDE) connectors or

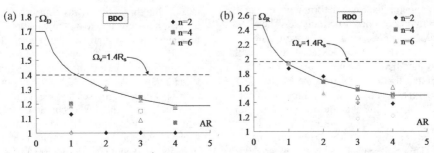

FIGURE C14.2-2 Diaphragm Shear Amplification Factor Results from NTHA at MCE: (a) BDO; (b) RDO

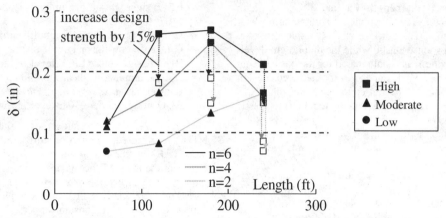

FIGURE C14.2-3 Diaphragm Maximum Joint Opening in NTHA for Basic Design Objective Designs under the MCE

reinforcement or better to provide an inelastic deformation capacity sufficient to survive the anticipated deformation demands in an MCE. This option and the RDO require the use of a diaphragm shear overstrength factor, Ω_v, to ensure that a nonductile shear failure does not occur before the connectors or reinforcement reaches its intended inelastic deformation. Note that *inelastic deformation* is associated with joint *opening* caused by diaphragm flexure, not joint sliding deformation caused by shear.

3. A reduced design option (RDO) is one in which the diaphragm is designed for the lowest design force level.

Because the design force level is lower than in the BDO, some yielding in the diaphragm is anticipated in the design earthquake. The force levels have been calibrated to keep diaphragm inelastic deformation demands in an MCE within the allowable deformation capacity for the high deformability element (HDE), the highest classification of precast diaphragm connector or reinforcement (see Section 14.2.4.3).

Each design option can be used with its associated seismic demand level or a lower seismic demand level. A 15% diaphragm force increase penalty is applied when a diaphragm design option is used for a seismic demand level that is one higher than its associated seismic demand level. A design option cannot be used for a seismic demand level two higher than the associated seismic demand level, i.e., the elastic design option cannot be used for the high seismic demand level.

The BDO has two performance targets: (1) elastic diaphragm response in the design earthquake, and (2) diaphragm connector/reinforcement deformation demands (i.e., joint opening) in the MCE within the allowable deformation capacity of connector/reinforcement in the moderate deformability element (MDE) category, δ_a^{MD}. The diaphragm design force levels for the BDO are aligned to the former requirement. Thus, the attainment of the second performance target hinges on the selection of the value for δ_a^{MD} relative to the diaphragm inelastic deformation demands anticipated for the maximum considered earthquake. These anticipated deformation demands were established through nonlinear dynamic time history analysis (NTHA) of precast structures with diaphragms designed to the BDO force levels and subjected to spectrum compatible ground motions scaled to the MCE.

It should be recognized that practical considerations also exist in the selection of δ_a^{MD}. The allowable deformation of high deformability elements (HDEs), δ_a^{HD}, (as required for the RDO) was established based on the best performing existing precast diaphragm connectors. This performance resulted in an HDE allowable deformation capacity $\delta_a^{HD} = 0.4$ in. ($\delta_a^{HD} = 10.2$ mm). (Note that the allowable value is 2/3 of the qualification value, thus HDEs are required to have a demonstrated deformation capacity of 0.6 in. (15.2 mm) in qualification testing, as was achieved by the best performing existing connectors). Given that low deformability elements (LDEs) do not have a deformation requirement, the MDE allowable deformation value should reside somewhere near half the HDE value, or $\delta_a^{MD} = 0.2$ in. ($\delta_a^{MD} = 5.1$ mm).

The NTHA results for the MCE are shown in Fig. C14.2-3. These results show that $\delta_a^{MD} = 0.2$ in . ($\delta_a^{MD} = 5.1$ mm) was an appropriate and viable choice for the MDEs used in the BDO, provided that the diaphragms were in the moderate seismic demand level (solid triangles in Fig. C14.2-3) or in the low seismic demand level (solid circles in Fig. C14.2-3). However, this value did not produce satisfactory designs for diaphragms in the high seismic demand level (solid squares in Fig. C14.2-3), and thus some measure is required to bring the design procedure in conformance.

A choice exists in how to modify the design procedure to resolve this nonconformance to the design target: (a) The allowable deformation ranges for the diaphragm connectors/reinforcement could be modified (i.e., a more stringent qualification deformation requirement for MDE, leading to an increase in δ_a^{MD}); (b) the diaphragm force levels could be increased across the board (i.e., change the design earthquake performance target for elastic diaphragm response from the diaphragm yield point itself to a lower value within the diaphragm elastic range); or (c) create a special requirement for the nonconforming diaphragm case (i.e., increase the diaphragm forces only for nonconforming cases). The first choice did not align well with the typical deformation capacities of existing connectors and would not produce evenly sized deformation ranges for the LDE, MDE, and HDE classifications. The second choice not only produces overly conservative designs for many cases, but it also blurs the clean BDO performance target of elastic diaphragm response in the design earthquake. For these reasons, the third choice was considered the most desirable.

Thus, rather than increase the value of δ_a^{MD} to accommodate the diaphragms in the high seismic demand level, it was decided to keep $\delta_a^{MD} = 0.2$ in. ($\delta_a^{MD} = 5.1$ mm) and create a special requirement for conformance in the case of diaphragms in the high seismic demand level. As each design option was developed with an associated diaphragm seismic demand level in mind, and the nonconformance did not occur at the associated level, i.e., the moderate seismic demand level, but instead at the high seismic demand level, the special requirement can be considered a measure for using a diaphragm design option with a more demanding seismic demand level.

The special requirement is an increase in the design force for the nonconforming case. The magnitude of the design force increase is 15%. The manner in which this value was established is also shown in Fig. C14.2-3. As mentioned previously, the solid squares indicate the maximum diaphragm connector/reinforcement deformation (joint opening demand) for the BDO for high diaphragm seismic demand levels and indicate demands greater than $\delta_a^{MD} = 0.2$ in. (5.1 mm). The open squares indicate the maximum diaphragm connector/reinforcement deformation for these same cases with the 15% increase in diaphragm force. This design force increase is seen to bring the deformation demand within the allowable limit. The same design force increase is enforced in Section 14.2.4.2.1 for use of the EDO with the moderate seismic demand level, though this provision was not based on any quantitative analytical results.

C14.2.4.3 Diaphragm Connector or Joint Reinforcement Deformability. The precast diaphragm seismic design methodology (DSDM) uses an approach that requires knowledge of the diaphragm connector or reinforcement stiffness, deformation capacity, and strength to effectively and efficiently design the diaphragm system for seismic forces. To meet this need, it is critical that the connector or reinforcement properties be determined in a repeatable, reproducible, and consistent manner so that existing and new connections can be used effectively in the diaphragm system. The qualification protocol provides an experimental approach for the determination of connector or reinforcement properties.

Precast concrete diaphragms deform mostly by the strains that occur at the joints between the precast concrete units. The requirements for reinforcement or connector deformability come from the need for the connections to accommodate these strains at the joints. A connection is an assembly of connectors, including the linking parts, welds, and anchorage to concrete. Mechanical connectors are identified as the primary parts that make the connection, but the deformation capacity identified with the connector represents the performance of the entire link across the joint. Qualification of the deformation capacity of the connector, then, is dependent on the details of the entire load path across the joint. The use in design of a connector qualified by testing is only valid when the design incorporates the complete connector detailing, as tested.

The diaphragm reinforcement classifications are high deformability elements (HDEs), moderate deformability elements (MDEs), and low deformability elements (LDEs). The threshold values of tension deformation capacity for each connector or reinforcement class were selected by considering the range of the ultimate (cyclic tension opening) deformations exhibited by the various precast diaphragm connectors examined in the DSDM experimental program (Naito et al. 2006, 2007). Based on these results, a threshold deformation of 0.6 in. (15.2 mm) was selected for HDE connector or reinforcement and 0.3 in. (7.6 mm) for MDE connector or reinforcement. There is no deformation requirement for LDE reinforcement.

A factor of safety of 1.5 was introduced into the design procedure by establishing the allowable maximum joint opening value at 2/3 of the connector's reliable and maximum joint opening deformation capacity. The 2/3 factor leads to maximum allowable deformations of 0.4 in. (10.2 mm) and 0.2 in. (5.1 mm) for the high deformability element (HDE) and the moderate deformability element (MDE), respectively. No deformation capacity requirement is needed for the low deformability element (LDE), since this classification of connector or reinforcement is used with designs that result in fully elastic diaphragm response up to the MCE. The allowable maximum joint openings were used as targets in the analytical parametric studies to calibrate the design factors.

A few further comments are given about the connector or reinforcement classification:

1. The diaphragm connector or reinforcement classification is based on inelastic deformation associated with joint opening caused by diaphragm flexure, not joint sliding deformation caused by shear.
2. The diaphragm connector or reinforcement classification applies to the chord reinforcement and shear reinforcement. Other reinforcement (collector/anchorages, secondary connections to spandrels, and similar items) may have different requirements or characteristics.
3. In meeting the required maximum deformation capacity using the testing protocols in the qualification procedure, the required cumulative inelastic deformation capacity is also met.

C14.2.4.3.5 Deformed Bar Reinforcement. Deformed bar reinforcement can be considered to be high deformability elements (HDEs), provided that certain conditions are met.

C14.2.4.3.6 Special Inspection. The purpose of this requirement is to verify that the detailing required in HDEs is properly executed through inspection personnel who are qualified to inspect these elements. Qualifications of inspectors should be acceptable to the jurisdiction enforcing the general building code.

C14.2.4.4 Precast Concrete Diaphragm Connector and Joint Reinforcement Qualification Procedure. This section provides a qualification procedure using experimental methods to assess the in-plane strength, stiffness, and deformation capacity of precast concrete diaphragm connectors and reinforcement. The methodology was developed as part of the DSDM research program specifically for diaphragm flange-to-flange connections and is intended to provide the required connector or reinforcement properties and classification for use in the seismic design procedure.

C14.2.4.4.1 Test Modules. Test modules are fabricated and tested to evaluate the performance of a precast concrete connection. Fig. C14.2-4 illustrates an example test module. It is required that multiple tests be conducted to assess repeatability and consistency. The test module should represent the geometry and thickness of the precast concrete components that will be connected. All connectors and reinforcement should be installed and welded in accordance with the manufacturer's published installation instructions. The results or the data generated are limited to connections built to the specified requirements.

Reduced scale connectors with appropriate reductions in maximum aggregate size following laws of similitude can be

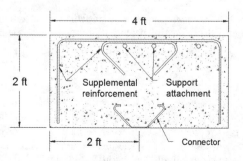

FIGURE C14.2-4 Test Module

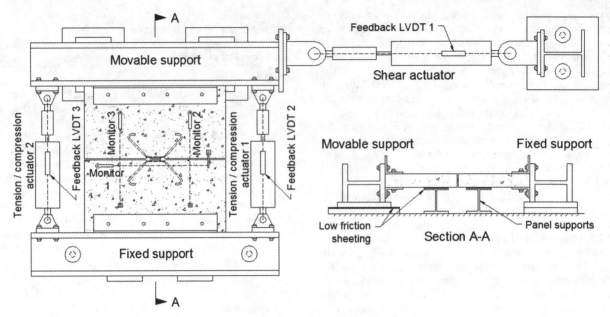

FIGURE C14.2-5 Possible Test Setup

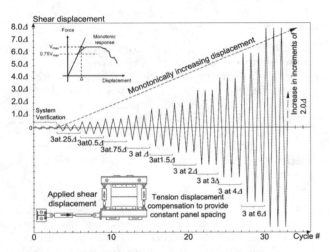

FIGURE C14.2-6 Shear Loading Protocol

used as research tools to gain knowledge but are not to be used for connector qualification.

C14.2.4.4.3 Test Configuration. A possible setup is illustrated in Fig. C14.2-5. Three independently controlled actuators are used, two providing axial displacement and one providing shear displacement to the connection.

C14.2.4.4.4 Instrumentation. Use of actuator transducers is not recommended because of potential slip in the test fixture.

C14.2.4.4.5 Loading Protocols. Figs. C14.2-6 and C14.2-7 illustrate the shear and tension/compression loading protocols for use in testing.

C14.2.4.4.6 Measurement Indices, Test Observations, and Acquisition of Data. Quantitative data should be recorded from each test, such that interpretation can be made of the performance of the test module. For in-plane tests, the axial and shear force and deformations should be recorded. Photographs should be taken to illustrate the condition of the test module at the initiation and completion of testing as well as at points through the testing history. Ideally, photos should be taken at the end of each group of cycles. Test history photos taken at points of interest, such as cracking, yielding, and peak load, and post test photos are adequate for most evaluations.

The backbone curve is adopted to represent a simple approximation of the load-deformation response of the connection. The points are defined in terms of the resistances P_a, P_1, P_b, P_2, P_{2a}, and P_3, and the displacements Δ_a, Δ_1, Δ_b, Δ_2, Δ_{2a}, and Δ_3, respectively.

As depicted in Fig. 14.2-3, the Type 1 curve is representative of ductile behavior where there is an elastic range (Point 0 to Point 1 on the curve) and an inelastic range (Point 1 to Point 3 on the curve), followed by loss of force-resisting capacity. The Type 2 curve is representative of ductile behavior where there is an elastic range (Point 0 to Point 1) and an inelastic range (Point 1 to Point 2 on the curve), followed by substantial loss of force-resisting capacity. Some connections may exhibit a small peak strength with limited ductility. For these cases, the Alternate Type 2 curve is recommended. The Type 3 curve is representative of a brittle or nonductile behavior where there is an elastic range (Point 0 to Point 1) followed by loss of strength. Deformation-controlled elements conform to Type 1 or Type 2, but not Type 2 Alternate, response with $\Delta_2 \geq 2\Delta_1$. All other responses are classified as force-controlled. An example of test data is included in Ren and Naito (2013).

C14.2.4.4.7 Response Properties. The reliable and stable maximum deformation capacity is defined for design code purposes as the connector deformation at peak load, Point 2 on the backbone curve, obtained in testing following the loading protocols defined here. All analytical calibrations were performed for a reliable and stable maximum deformation capacity corresponding to a deformation where the strength reduces to 80% of P_2, which is similar to the beam–column connection deformation capacity definition for steel structures in AISC 341. Thus, an added degree of conservatism is provided in the definition proposed for the design code.

Deformation Category. The category ranges were determined from finite element analysis of a database of diaphragm systems under a range of seismic demands. Alternate deformation limits

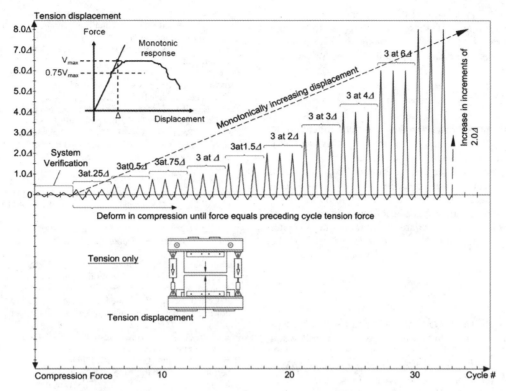

FIGURE C14.2-7 Tension/Compression Loading Protocol

can be used if supporting data are provided. It should be noted that the connector or joint reinforcement classification is based solely on *tension* deformation capacity (as stated in Section 14.2.4.3), whereas the qualification procedure applies equally to, and requires, both tension and shear tests. In other words, while both tension and shear characterization are required to determine the needed strengths, the connector classification is based solely on the tension testing.

Tensile Strength. The design factors for flexural strength are calibrated to the yield point of the chord connectors, not to their peak strength. For instance, for the EDO, elastic response of the diaphragm under the MCE is being targeted, so this response is aligned to the yield strength, not the peak strength. For consistency, the BDO and RDO factors are also calibrated to this same level, i.e., yield. So the nominal strength of the connectors is based on P_1, not P_2. Using P_2 creates a situation where yield should be anticipated in the diaphragm for the EDO, and larger inelastic deformations for the BDO and RDO.

Shear Strength. The intention is for the diaphragm system to remain elastic under shear demands. Consequently, the inelastic shear force capacity of connections is not considered. Because of the existence of low stiffness connections, limits are placed on the allowable deformation at which the force P_1 can be determined.

C14.2.4.4.8 Test Report. The minimum information that must be included in a test report is spelled out.

C14.3 COMPOSITE STEEL AND CONCRETE STRUCTURES

This section provides guidance on the design of composite and hybrid steel–concrete structures. Composite structures are defined as those incorporating structural elements made of steel and concrete portions connected integrally throughout the structural element by mechanical connectors, bonds, or both. Hybrid structures are defined as consisting of steel and concrete structural elements connected together at discrete points. Composite and hybrid structural systems mimic many of the existing steel (moment and braced frame) and reinforced concrete (moment frame and wall) configurations but are given their own design coefficients and factors in Table 12.2-1. Their design is based on ductility and energy dissipation concepts comparable to those used in conventional steel and reinforced concrete structures, but it requires special attention to the interaction of the two materials because it affects the stiffness, strength, and inelastic behavior of the members, connections, and systems.

C14.3.1 Reference Documents. Seismic design for composite structures assigned to SDCs D, E, or F is governed primarily by AISC 341. Composite design provisions in ANSI/AISC 341 are less prescriptive than those for structural steel and provide flexibility for designers to use analytical tools and results of research in their practice. Composite structures assigned to SDC A, B, or C may be designed according to principles outlined in AISC 360 and ACI 318. ANSI/AISC 360 and ACI 318 provide little guidance on connection design; therefore, designers are encouraged to review ANSI/AISC 341 for guidance on the design of joint areas. Differences between older AISC and ACI provisions for cross-sectional strength for composite beam–columns have been minimized by changes in the latest AISC 360, and AISC 360 refers to ACI 318 for much of the design of reinforced concrete components of composite structures. However, there is not uniform agreement between the provisions in ACI 318 and AISC 360 regarding detailing, limits on material strengths, stability, and strength for composite beam–columns. The composite design provisions in ANSI/AISC 360 are considered to be current.

C14.3.4 Metal-Cased Concrete Piles. Design of metal-cased concrete piles, which are analogous to circular concrete filled tubes, is governed by Sections 14.2.3.1.3 and 14.2.3.2.4 of this standard. The intent of these provisions is to require metal-cased concrete piles to have confinement and protection against long-term deterioration comparable to that for uncased concrete piles.

C14.4 MASONRY

This section adopts by reference and then makes modifications to TMS 402 and TMS 602. In past editions of this standard, modifications to the TMS referenced standards were also made. During the development of the 2016 edition of TMS standards, each of these modifications was considered by the TMS 402/602 committee. Some were incorporated directly into the TMS standards. These modifications have accordingly been removed from the modifications in this standard. Work is ongoing to better coordinate the provisions of the two documents so that the provisions in Section 14.4 are significantly reduced or eliminated in future editions.

C14.5 WOOD

C14.5.1 Reference Documents. Two national consensus standards are adopted for seismic design of engineered wood structures: the *National Design Specification* (AWC NDS-15), and the *Special Design Provisions for Wind and Seismic* (AWC SDPWS-15). Both of these standards are presented in dual allowable stress design (ASD) and load and resistance factor design (LRFD) formats. Both standards reference a number of secondary standards for related items such as wood materials and fasteners. AWC SDPWS addresses general principles and specific detailing requirements for shear wall and diaphragm design and provides tabulated nominal unit shear capacities for shear wall and diaphragm sheathing and fastening. The balance of member and connection design is to be in accordance with the AWC NDS.

REFERENCES

American Concrete Institute (ACI). (2014). "Building code requirements for structural concrete and commentary," *ACI 318*, Farmington Hills, MI.
American Institute of Steel Construction (AISC). (2016a). "Seismic provisions for structural steel buildings." *ANSI/AISC 341*, Chicago.
AISC. (2016b). "Specification for structural steel buildings." *ANSI/AISC 360*, Chicago.
American Iron and Steel Institute (AISI). (2015). "North American standard for cold-formed steel structural framing." *ANSI/AISI S240*, Washington, DC.
AISI. (2015). "Standard for cold-formed steel framing—Prescriptive method for one and two family dwellings." *ANSI/AISI S230*, Washington, DC.
AISI. (2015). "North American standard for seismic design of cold-formed steel structural systems" *ANSI/AISI S400*, Washington, DC.
AISI. (2016). "North American specification for the design of cold-formed steel structural members." *ANSI/AISI S100*, Washington, DC.
AISI. (2016). "North American standard for the design of profiled steel diaphragm panels" *ANSI/AISI S310*, Washington, DC.
ASCE. (2002). "Specification for the design of cold-formed stainless steel structural members." *ASCE/SEI 8-02*. Reston, VA.
ASCE. (2010). "Structural applications of steel cables for buildings." *ASCE 19-10*, Reston, VA.
American Wood Council (AWC). (2008). "Special design provisions for wind and seismic (wind and seismic)." *ANSI/AWC SDPWS-2012*, Leesburg, VA.
AWC. (2012). "National design specification (NDS) for wood construction with commentary." *ANSI/AWC NDS-2012*, Leesburg, VA.
Banks, G., and Stanton, J. (2005). "Panel-to-panel connections for hollow-core shear walls subjected to seismic loading." *Proc., 2005 PCI Convention*, Precast/Prestressed Concrete Institute, Chicago.
Fanous, A., Sritharan, S., Suleiman, M., and Arulmoli, A. (2007). *Minimum spiral reinforcement requirements and lateral displacement limits for prestressed concrete piles in high seismic regions*. Department of Civil, Construction and Environmental Engineering, Iowa State University, ISU-ERI Ames Report, Ames, IA.
International Code Council (ICC). (2012). "International building code." Country Club Hills, IL.
Masonry Standards Joint Committee (MSJC). (2005a). "Building code requirements for masonry structures." *ACI 530-05/ASCE/SEI 5-05/TMS 402-5*, Farmington Hills, MI.
MSJC. (2005b). "Specification for masonry structures." *ACI 530.1-05/ASCE/SEI 6-05/TMS 602-05*, Farmington Hills, MI.
Naito, C., Peter, W., Cao, L. (2006). "Development of a seismic design methodology for precast diaphragms—Phase 1 Summary Report," *ATLSS Report No. 06-03*, January, ATLSS Center, Lehigh University, Bethlehem, PA.
Naito, C., Ren, R., Jones, C., Cullent, T. (2007). "Development of a seismic design methodology for precast diaphragms—Connector performance, Phase 1B," *ATLSS Report No. 07-04*, June, ATLSS Center, Lehigh University, Bethlehem, PA.
Nakaki, S., Stanton, J. F., and Sritharan, S. (2001). "The PRESSS Five-Story Precast Concrete Test Building, University of California, San Diego, La Jolla, California," *PCI J.*, 46(5), 20–26.
Nakaki, S., Becker, R., Oliva, M. G., and Paxson, D. (2005). "New connections for precast wall systems in high seismic regions." *Proc., 2005 PCI Convention*, Precast/Prestressed Concrete Institute, Chicago.
PCI Committee on Prestressed Concrete Piling. (1993). "Recommended practice for design, manufacture and installation of prestressed concrete piling." *PCI J.*, 38(2), 14–41.
Precast/Prestressed Concrete Institute (PCI). (2004). "Precast/prestressed concrete piles." *Bridge design manual*, BM-20-04. PCI, Chicago.
Ren, R., and Naito, C. J. (2013). "Precast concrete diaphragm connector performance database." *J. Struct. Eng.*, 139(1), 15–27.
Sato, A., and Uang, C.-M. (2007). "Development of a seismic design procedure for cold-formed steel bolted frames," *Report No. SSRP-07/16*, University of California, San Diego.
Schultz, A. E., and Magana, R. A. (1996). "Seismic behavior of connections in precast concrete walls." *Proc., Mete A. Sozen Symposium, SP-162*, American Concrete Institute, Farmington Hills, MI, 273–311.
Steel Deck Institute (SDI). (2015). *Diaphragm design manual*, 4th Ed., DDMO4, Glenshaw, PA.
The Masonry Society (TMS). (2016). "Building code requirements and specification for masonry structures." *TMS 402-16*, Longmont, CO.
The Masonry Society (TMS). (2016). "Specification for masonry structures." *TMS 602-16*, Longmont, CO.

OTHER REFERENCES (NOT CITED)

American Institute of Timber Construction. (2005). *Timber construction manual*, 5th Ed. Wiley & Sons, New York.
APA–The Engineered Wood Association. (1994). "Northridge California earthquake." *T94-5*, Tacoma, WA.
APA–The Engineered Wood Association. (2004). "Diaphragms and shear walls design/construction guide." *L350*, Tacoma, WA.
Applied Technology Council. (1981). "Guidelines for the design of horizontal wood diaphragms." *ATC-7*, Redwood City, CA.
Bora, C., Oliva, M. G., Nakaki, S. D., and Becker, R. (2007). "Development of a precast concrete shear-wall system requiring special code acceptance." *PCI J.*, 52(1), 122–135.
Breyer, D., Fridley, K., Jr., Pollack, D., and Cobeen, K. (2006). *Design of wood structures ASD/LRFD*, 6th Ed., McGraw-Hill, New York.
Canadian Wood Council. (1995). *Wood reference handbook*. Canadian Wood Council, Ottawa.
CWC. (2005). *Wood design manual*. Canadian Wood Council, Ottawa.
Charles Pankow Foundation. (2014). "Seismic design methodology document for precast concrete diaphragms." CPF, Vancouver, WA, January 23.
Cobeen, K. (2004). "Recent developments in the seismic design and construction of woodframe buildings." *Earthquake engineering from engineering seismology to performance-based engineering*, Y. Bozorgia, and V. Bertero, eds., CRC Press, Boca Raton, FL.
Consortium of Universities for Research in Earthquake Engineering (CUREE). (2004). *Recommendations for earthquake resistance in the design and construction of woodframe buildings, CUREE W-30*. CUREE, Richmond, CA.

Dolan, J. D. (2003). "Wood structures." *Earthquake engineering handbook*, W-F. Chen, and C. Scawthorn, eds., CRC Press, Boca Raton, FL.

Earthquake Engineering Research Institute (EERI). (1996). "Northridge earthquake reconnaissance report." *Earthq. Spectra*, Chapter 6, Suppl. C to Vol. 11, EERI, Oakland, CA.

Faherty, K. F., and Williamson, T. G. (1989). *Wood engineering and construction handbook*. McGraw-Hill, New York.

Federal Emergency Management Agency (FEMA). (2003). "NEHRP recommended provisions for seismic regulations for new buildings and other structures." *FEMA 450*, FEMA, Building Seismic Safety Council, Washington, DC.

FEMA. (2005). *Coastal construction manual*, 3rd Ed., FEMA 55. FEMA, Washington, D.C.

Forest Products Laboratory. (1986). *Wood: Engineering design concepts*. Materials Education Council, Pennsylvania State University, University Park, PA.

Goetz, K. H., Hoor, D., Moehler, K., and Natterer, J. (1989). *Timber design and construction source book: A comprehensive guide to methods and practice*. McGraw-Hill, New York.

Hoyle, R. J., and Woeste, F. E. (1989). *Wood technology in the design of structures*. Iowa State University Press, Ames.

International Code Council (ICC). (2006). "ICC standard on the design and construction of log structures," *Third Draft*. Country Club Hills, IL.

Ishizuka, T., and Hawkins, N. M. (1987). "Effect of bond deterioration on the seismic response of reinforced and partially prestressed concrete ductile moment resistant frames." *Report SM 87-2*. Department of Civil Engineering, University of Washington, Seattle.

Karacabeyli, E., and Popovsky, M. (2003). "Design for earthquake resistance." *Timber engineering*, H. Larsen, and S. Thelandersson, eds., John Wiley & Sons, New York.

Keenan, F. J. (1986). *Limit states design of wood structures*. Morrison Hershfield, North York, Ontario.

Lee, N. H., Kim, K. S., Bang, C. J., and Park, K. R. (2007). "Tensile-headed anchors with large diameter and deep embedment in concrete." *ACI Struct. J.*, 104(4), 479–486.

Lee, N. H., Park, K. R., and Suh, Y. P. (2010). "Shear behavior of headed anchors with large diameters and deep embedments." *ACI Struct. J.*, 107(2), 146–156.

Masonry Standards Joint Committee (MSJC). (1999). "Building code requirements for masonry structures." *ACI 530-99/ASCE 5-99/TMS 402-99*, "Specification for masonry structures"; *ACI 530.1-99/ASCE 6-99/TMS 602-99*, "Commentary on building code requirements for masonry structures, commentary on specification for masonry structures," known as *MSJC standards (code and specification)*, The Masonry Society, Boulder, CO, American Concrete Institute, Farmington Hills, MI, ASCE, Reston, VA.

Nakaki, S. D., Stanton, J. F., and Sritharan, S. (1999). "An overview of the PRESSS five-story precast test building." *PCI J.*, 44(2), 26–39.

Park, R., and Thompson, K. J. (1977). "Cyclic load tests on prestressed and partially prestressed beam-column joints." *PCI J.*, 22(5), 84–110.

Sherwood, G. E., and Stroh, R. C. (1989). "Wood-frame house construction." *Agricultural handbook 73*, U.S. Government Printing Office, Washington, DC.

Somayaji, Shan. (1992). *Structural wood design*. West Publishing Co., St. Paul, MN.

Stalnaker, J. J., and Harris, E. C. (1996). *Structural design in wood*, 2nd Ed., McGraw-Hill, New York.

Structural Engineers Association of California (SEAOC). (1999). *Recommended lateral force requirements and commentary*. SEAOC, Sacramento, CA.

Structural Engineers Association of Northern California (SEAONC). (2005). *Guidelines for seismic evaluation and rehabilitation of tilt-up buildings and other rigid wall/flexible diaphragm structures*. SEAONC, Sacramento, CA.

U.S. Department of Agriculture, National Oceanic and Atmospheric Administration. (1971). San Fernando, California, Earthquake of February 9, 1971. NOAA, Washington, DC.

U.S. Department of the Army, Navy, and Air Force. (1992). "Seismic design for buildings." *TM5-809-10* (Tri-Services Manual). U.S. Government Printing Office, Washington, DC.

CHAPTER C15
SEISMIC DESIGN REQUIREMENTS FOR NONBUILDING STRUCTURES

C15.1 GENERAL

C15.1.1 Nonbuilding Structures. Building codes traditionally have been perceived as minimum standards for the design of nonbuilding structures, and building code compliance of these structures is required by building officials in many jurisdictions. However, requirements in the industry reference documents are often at odds with building code requirements. In some cases, the industry documents need to be altered, whereas in other cases, the building codes need to be modified. Registered design professionals are not always aware of the numerous accepted documents within an industry and may not know whether the accepted documents are adequate. One of the intents of Chapter 15 of the standard is to bridge the gap between building codes and existing industry reference documents.

Differences between the ASCE/SEI 7 design approaches for buildings and industry document requirements for steel multilegged water towers (Fig. C15.1-1) are representative of this inconsistency. Historically, such towers have performed well when properly designed in accordance with American Water Works Association (AWWA) standards and industry practices. Those standards and practices differ from the ASCE/SEI 7 treatment of buildings in that tension-only rods are allowed, upset rods are preloaded at the time of installation, and connection forces are not amplified.

Chapter 15 also provides an appropriate link so that the industry reference documents can be used with the seismic ground motions established in the standard. Some nonbuilding structures are similar to buildings and can be designed using sections of the standard directly, whereas other nonbuilding structures require special analysis unique to the particular type of nonbuilding structure.

Building structures, vehicular bridges, electrical transmission towers, hydraulic structures (e.g., dams), buried utility lines and their appurtenances, and nuclear reactors are excluded from the scope of the nonbuilding structure requirements, although industrial buildings are permitted per Chapter 11 to use the provisions in Chapter 15 for nonbuilding structures with structural systems similar to buildings, provided that specific conditions are met. The excluded structures are covered by other well-established design criteria (e.g., electrical transmission towers and vehicular bridges), are not under the jurisdiction of local building officials (e.g., nuclear reactors and dams), or require technical considerations beyond the scope of the standard (e.g., buried utility lines and their appurtenances).

C15.1.2 Design. Nonbuilding structures and building structures have much in common with respect to design intent and expected performance, but there are also important differences. Chapter 15 relies on other portions of the standard where possible and provides special notes where necessary.

There are two types of nonbuilding structures: those with structural systems similar to buildings and those with structural systems not similar to buildings. Specific requirements for these two cases appear in Sections 15.5 and 15.6.

C15.1.3 Structural Analysis Procedure Selection. Nonbuilding structures that are similar to buildings are subject to the same analysis procedure limitations as building structures. Nonbuilding structures that are not similar to buildings are subject to those limitations and are subject to procedure limitations prescribed in applicable specific reference documents.

For many nonbuilding structures supporting flexible system components, such as pipe racks (Fig. C15.1-2), the supported piping and platforms generally are not regarded as rigid enough to redistribute seismic forces to the supporting frames.

For nonbuilding structures supporting very stiff (i.e., rigid) system components, such as steam turbine generators (STGs) and heat recovery steam generators (HRSGs) (Fig. C15.1-3), the supported equipment, ductwork, and other components (depending on how they are attached to the structure) may be rigid enough to redistribute seismic forces to the supporting frames. Torsional effects may need to be considered in such situations.

FIGURE C15.1-1 Steel multilegged water tower

Source: Courtesy of CB&I LLC; reproduced with permission.

FIGURE C15.1-2 Steel pipe rack

Source: Courtesy of CB&I LLC; reproduced with permission.

FIGURE C15.1-3 Heat recovery steam generators

Source: Courtesy of CB&I LLC; reproduced with permission.

Section 12.6 presents seismic analysis procedures for building structures based on the Seismic Design Category (SDC); the fundamental period, T; and the presence of certain horizontal or vertical irregularities in the structural system. Where the fundamental period is greater than or equal to $3.5\ T_s$ (where $T_s = S_{D1}/S_{DS}$), the use of the equivalent lateral force procedure is not permitted in SDCs D, E, and F. This requirement is based on the fact that, unlike the dominance of the first mode response in case of buildings with lower first mode period, higher vibration modes do contribute more significantly in situations when the first mode period is larger than $3.5\ T_s$. For buildings that exhibit classic flexural deformation patterns (such as slender shear-wall or braced-frame systems), the second mode frequency is at least 3.5 times the first mode frequency, so where the fundamental period exceeds $3.5\ T_s$, the higher modes have larger contributions to the total response because they occur near the peak of the design response spectrum.

It follows that dynamic analysis (modal response spectrum analysis or response history analysis) may be necessary to properly evaluate buildinglike nonbuilding structures if the first mode period is larger than $3.5\ T_s$ and the equivalent lateral force analysis is sufficient for nonbuilding structures that respond as single-degree-of-freedom systems.

The recommendations for nonbuilding structures provided in the following are intended to supplement the designer's judgment and experience. The designer is given considerable latitude in selecting a suitable analysis method for nonbuilding structures.

Buildinglike Nonbuilding Structures. Table 12.6-1 is used in selecting analysis methods for buildinglike nonbuilding structures, but, as illustrated in the following three conditions, the relevance of key behavior must be considered carefully:

1. Irregularities: Table 12.6-1 requires dynamic analysis for SDC D, E, and F structures that have certain horizontal or vertical irregularities. Some of these building irregularities (defined in Section 12.3.2) are relevant to nonbuilding structures. The weak- and soft-story vertical irregularities (Types 1a, 1b, 5a, and 5b of Table 12.3-2) are pertinent to the behavior of buildinglike nonbuilding structures. Other vertical and horizontal irregularities may or may not be relevant, as described below.
   a. Horizontal irregularities: Horizontal irregularities of Types 1a and 1b affect the choice of analysis method, but these irregularities apply only where diaphragms are rigid or semirigid, and some buildinglike nonbuilding structures have either no diaphragms or flexible diaphragms.
   b. Vertical irregularities: Vertical irregularity Type 2 is relevant where the various levels actually support significant loads. Where a buildinglike nonbuilding structure supports significant mass at a single level while other levels support small masses associated with stair landings, access platforms, and so forth, dynamic response is dominated by the first mode, so the equivalent lateral force procedure may be applied. Vertical irregularity Type 3 addresses large differences in the horizontal dimension of the seismic force-resisting system in adjacent stories because the resulting stiffness distribution can produce a fundamental mode shape unlike that assumed in the development of the equivalent lateral force procedure. Because the concern relates to stiffness distribution, the horizontal dimension of the seismic force-resisting system, not of the overall structure, is important.
2. Arrangement of supported masses: Even where a nonbuilding structure has buildinglike appearance, it may not behave like a building, depending on how masses are attached. For example, the response of nonbuilding structures with suspended vessels and boilers cannot be determined reliably using the equivalent lateral force procedure because of the pendulum modes associated with the significant mass of the suspended components. The resulting pendulum modes, while potentially reducing story shears and base shear, may require large clearances to allow pendulum motion of the supported components and may produce excessive demands on attached piping. Dynamic analysis is highly recommended in such cases, with consideration for appropriate impact forces in the absence of adequate clearances.
3. Relative rigidity of beams: Even where a classic building model may seem appropriate, the equivalent lateral force

procedure may underpredict the total response if the beams are flexible relative to the columns (of moment frames) or the braces (of braced frames). This underprediction occurs because higher modes associated with beam flexure may contribute more significantly to the total response (even if the first mode response is at a period less than $3.5\ T_s$). This situation of flexible beams can be especially pronounced for nonbuilding structures because the "normal" floors common to buildings may be absent. Therefore, the dynamic analysis procedures are suggested for buildinglike nonbuilding structures with flexible beams.

Nonbuilding Structures Not Similar to Buildings. The (static) equivalent lateral force procedure is based on classic building dynamic behavior, which differs from the behavior of many nonbuilding structures not similar to buildings. As discussed below, several issues should be considered for selecting either an appropriate method of dynamic analysis or a suitable distribution of lateral forces for static analysis.

1. Structural geometry: The dynamic response of nonbuilding structures with a fixed base and a relatively uniform distribution of mass and stiffness, such as bottom-supported vertical vessels, stacks, and chimneys, can be represented adequately by a cantilever (shear building) model. For these structures, the equivalent lateral force procedure provided in the standard is suitable. This procedure treats the dynamic response as being dominated by the first mode. In such cases, it is necessary to identify the first mode shape (using, for instance, the Rayleigh–Ritz method or other classical methods from the literature) for distribution of the dynamic forces. For some structures, such as tanks with low height-to-diameter ratios storing granular solids, it is conservative to assume a uniform distribution of forces. Dynamic analysis is recommended for structures that have neither a uniform distribution of mass and stiffness nor an easily determined first mode shape.
2. Number of lateral supports: Cantilever models are obviously unsuitable for structures with multiple supports. Fig. C15.1-4 shows a nonbuilding braced frame structure that provides nonuniform horizontal support to a piece of equipment. In such cases, the analysis should include coupled model effects. For such structures, an application of the equivalent lateral force method could be used,

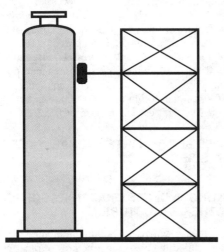

FIGURE C15.1-4 Multiple lateral supports

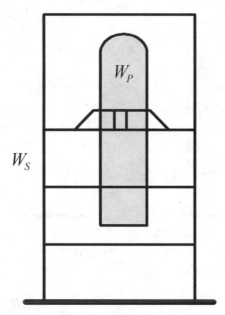

FIGURE C15.1-5 Unusual support of dead weight

depending on the number and locations of the supports. For example, most beam-type configurations lend themselves to application of the equivalent lateral force method.
3. Method of supporting dead weight: Certain nonbuilding structures (such as power boilers) are supported from the top. They may be idealized as pendulums with uniform mass distribution. In contrast, a suspended platform may be idealized as a classic pendulum with concentrated mass. In either case, these types of nonbuilding structures can be analyzed adequately using the equivalent lateral force method by calculating the appropriate frequency and mode shape. Fig. C15.1-5 shows a nonbuilding structure containing lug-supported equipment with W_P greater than $0.25(W_S + W_P)$. In such cases, the analysis should include a coupled system with the mass of the equipment and the local flexibility of the supports considered in the model. Where the support is located near the nonbuilding structure's vertical location of the center of mass, a dynamic analysis is recommended.
4. Mass irregularities: Just as in the case of buildinglike nonbuilding structures, the presence of significantly uneven mass distribution is a situation where the equivalent lateral force method is not likely to provide a very accurate and perhaps unconservative force distribution. The dynamic analysis methods are recommended in such situations. Fig. C15.1-6 illustrates two such situations. In part (a), a mass irregularity exists if W_1 is greater than $1.5W_2$ or less than $0.67W_2$. In part (b), a mass irregularity exists if W_3 is greater than either $1.5W_2$ or $1.5W_4$.
5. Torsional irregularities: Structures in which the fundamental mode of response is torsional or in which modes with significant mass participation exhibit a prominent torsional component may also have inertial force distributions that are significantly different from those predicted by the equivalent lateral force method. In such cases, dynamic analyses should be considered. Fig. C15.1-7 illustrates one such case where a vertical vessel is attached to a secondary vessel with W_2 greater than about $0.25(W_1 + W_2)$.
6. Stiffness and strength irregularities: Just as for buildinglike nonbuilding structures, abrupt changes in the distribution

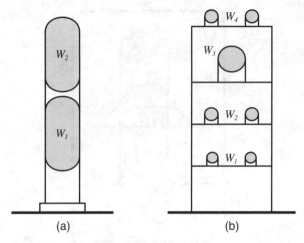

FIGURE C15.1-6 Mass irregularities

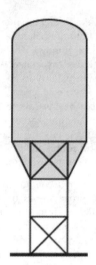

FIGURE C15.1-8 Soft-story irregularity

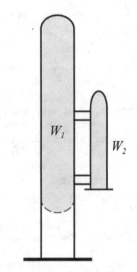

FIGURE C15.1-7 Torsional irregularity

of stiffness or strength in a nonbuilding structure not similar to buildings can result in substantially different inertial forces from those indicated by the equivalent lateral force method. Fig. C15.1-8 represents one such case. For structures that have such configurations, consideration should be given to the use of dynamic analysis procedures. Even where dynamic analysis is required, the standard does not define in any detail the degree of modeling; an adequate model may have a few dynamic degrees of freedom or tens of thousands of dynamic degrees of freedom. The important point is that the model captures the significant dynamic response features so that the resulting lateral force distribution is valid for design. The designer is responsible for determining whether dynamic analysis is warranted and, if so, the degree of detail required to address adequately the seismic performance.

7. Coupled response: Where the weight of the supported structure is large compared with the weight of the supporting structure, the combined response can be affected significantly by the flexibility of the supported nonbuilding structure. In that case, dynamic analysis of the coupled system is recommended. Examples of such structures are shown in Fig. C15.1-9. Part (a) shows a flexible nonbuilding structure with W_p greater than $0.25(W_s + W_p)$, supported by a relatively flexible structure; the flexibility of the supports and attachments should be considered. Part (b) shows flexible equipment connected by a large-diameter, thick-walled pipe and supported by a flexible structure; the structures should be modeled as a coupled system including the pipe.

Distributed mass cantilever structures have over several cycles of ASCE 7 had their R values reduced and/or special detailing requirements added to improve their performance. The exceptions to the modal scaling rules of Section 12.9 listed in Section 15.1.3 for distributed mass cantilever structures recognize this improvement in performance.

C15.1.4 Nonbuilding Structures Sensitive to Vertical Ground Motions. Traditionally, ASCE 7 did not provide guidance to address designing for a separate vertical ground motion. Historically, this omission has not been a problem for buildings because there is inherent strength in the vertical direction because of the margin that is developed when the dead load and live load are applied. However, this is not necessarily the case for nonbuilding structures. Many nonbuilding structures are sensitive to vertical motions and do not have the benefit of the inherent strength that exists in buildings. Examples of some structures are liquid and granular storage tanks or vessels, suspended structures (such as boilers), and nonbuilding structures incorporating horizontal cantilevers. Such structures are required to incorporate Section 11.9 into the design of the structure in lieu of applying the traditional vertical ground motion of $0.2S_{DS}$.

C15.2 THIS SECTION INTENTIONALLY LEFT BLANK; SEE SECTION C15.8

C15.3 NONBUILDING STRUCTURES SUPPORTED BY OTHER STRUCTURES

There are instances where nonbuilding structures not similar to buildings are supported by other structures or other nonbuilding structures. This section specifies how the seismic design loads for

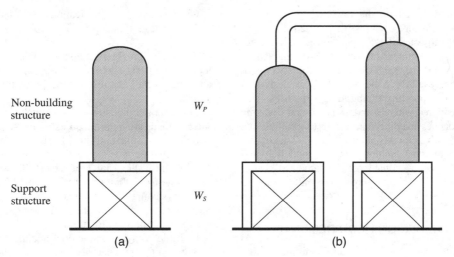

FIGURE C15.1-9 Couple system

such structures are to be determined and the detailing requirements that are to be satisfied in the design.

C15.3.1 Less Than 25% Combined Weight Condition. In many instances, the weight of the supported nonbuilding structure is relatively small compared with the weight of the supporting structure, such that the supported nonbuilding structure has a relatively small effect on the overall nonlinear earthquake response of the primary structure during design-level ground motions. It is permitted to treat such structures as nonstructural components and to use the requirements of Chapter 13 for their design. The ratio of secondary component weight to total weight of 25% at which this treatment is permitted is based on judgment and was introduced into code provisions in the 1988 *Uniform Building Code* by the SEAOC Seismology Committee. Analytical studies, typically based on linear elastic primary and secondary structures, indicate that the ratio should be lower, but the SEAOC Seismology Committee judged that the 25% ratio is appropriate where primary and secondary structures exhibit nonlinear behavior that tends to lessen the effects of resonance and interaction. In cases where a nonbuilding structure (or nonstructural component) is supported by another structure, it may be appropriate to analyze in a single model. In such cases, it is intended that seismic design loads and detailing requirements be determined following the procedures of Section 15.3.2. Where there are multiple large nonbuilding structures, such as vessels supported on a primary nonbuilding structure, and the weight of an individual supported nonbuilding structure does not exceed the 25% limit but the combined weight of the supported nonbuilding structures does, it is recommended that the combined analysis and design approach of Section 15.3.2 be used. It is also suggested that dynamic analysis be performed in such cases because the equivalent lateral force procedure may not capture some important response effects in some members of the supporting structure.

Where the weight of the supported nonbuilding structure does not exceed the 25% limit and a combined analysis is performed, the following procedure should be used to determine the F_p force of the supported nonbuilding structure based on Eq. (13.3-4):

1. A modal analysis should be performed in accordance with Section 12.9. The base shear of the combined structure and nonbuilding structure should be taken as no less than 85% of the equivalent lateral force procedure base shear.
2. For a component supported at level i, the acceleration at that level should be taken as a_i, the total shear just below level i divided by the seismic weight at and above level i.
3. The elastic value of the component shear force coefficient should next be determined as the shear force from the modal analysis at the point of attachment of the component to the structure divided by the weight of the component. This value is preliminarily taken as $a_i a_p$. Because a_p cannot be taken as less than 1.0, the value of a_p is taken as $a_i a_p / a_i$, except that the final value a_p need not be taken as greater than 2.5 and should not be taken as less than 1.0. The final value of $a_i a_p$ should be the final value of a_i determined in step 2 multiplied by the final value of a_p determined earlier in this step.
4. The resulting value of $a_i a_p$ should be used in Eq. (13.3-4); the resulting value of F_p is subject to the maximum and minimum values of Eqs. (13.3-2) and (13.3-3), respectively.

C15.3.2 Greater Than or Equal to 25% Combined Weight Condition. Where the weight of the supported structure is relatively large compared with the weight of the supporting structure, the overall response can be affected significantly. The standard sets forth two analysis approaches, depending on the rigidity of the nonbuilding structure. The determination of what is deemed rigid or flexible is based on the same criteria used for nonstructural components.

Where the supported nonbuilding structure is rigid, it is acceptable to treat the supporting structure as a nonbuilding structure similar to a building and to determine its design loads and detailing using the requirements of Section 15.5. The design of the rigid nonbuilding structure and its anchorage is determined using the requirements of Chapter 13 with the amplification factor, a_p, taken as 1.0. However, this condition is relatively rare because the flexibility of any directly supporting members in the primary structure, such as floor beams, must be considered in determining the period of the component.

In the usual case, where the supported nonbuilding structure is flexible, a combined model of the supporting structure and the supported nonbuilding structure is used. The design loads and detailing are determined based on the lower R value of the supported nonbuilding structure or supporting structure.

Although not specifically mentioned in Section 15.3.2, another approach is permitted. A nonlinear response history analysis of the combined system can be performed in accordance with Section 16.2, and the results can be used for the design of both the supported and supporting nonbuilding structures. This option should be considered where standard static and dynamic elastic analysis approaches may be inadequate to evaluate the earthquake response (such as for suspended boilers). This option should be used with extreme caution because modeling and interpretation of results require considerable judgment. Because of this sensitivity, Section 16.2 requires independent design review.

C15.4 STRUCTURAL DESIGN REQUIREMENTS

This section specifies the basic coefficients and minimum design forces to be used to determine seismic design loads for nonbuilding structures. It also specifies height limits and restrictions. As with building structures, it presumes that the first step in establishing the design forces is to determine the design base shear for the structure.

There are two types of nonbuilding structures: those with structural systems similar to buildings and those with structural systems not similar to buildings. Specific requirements for these two cases appear in Sections 15.5 and 15.6.

Table 15.4-1 contains the response modification coefficient (R) for nonbuilding structures similar to buildings. Table 15.4-2 contains the response modification coefficient for nonbuilding structures not similar to buildings. Every response modification coefficient has associated design and detailing requirements to ensure the required ductility associated with that response modification coefficient value (e.g., AISC 341). Some structures, such as pipe racks, do not resemble a traditional building in that they do not house people or have such things as walls and bathrooms. These structures have lateral force-resisting systems composed of braced frames and moment frames similar to a traditional building. Therefore, pipe racks are considered nonbuilding structures similar to buildings. The response modification coefficient for a pipe rack should be taken from Table 15.4-1 for the appropriate lateral force-resisting system used, and the braced frames and/or moment frames used must meet all of the design and detailing requirements associated with the R value selected (see Section 15.5.2, Pipe Racks).

Most major power distribution facility (power island) structures, such as HRSG support structures, steam turbine pedestals, coal boiler support structures, pipe racks, air inlet structures, and duct support structures, also resist lateral forces predominantly by use of buildinglike framing systems such as moment frames, braced frames, or cantilever column systems. Therefore, their response modification coefficient should be selected from Table 15.4-1, and they must meet all the design and detailing requirements associated with the response modification coefficient selected.

Many nonbuilding structures, such as flat-bottom tanks, silos, and stacks, do not use braced frames or moment frames similar to those found in buildings to resist seismic loads. Therefore, they have their own unique response modification coefficient, which can be found in Table 15.4-2.

For nonbuilding structures with lateral systems composed predominantly of buildinglike framing systems, such as moment frames, braced frames, or cantilever column systems, it would be inappropriate to extrapolate the descriptions in Table 15.4-2, resulting in inappropriately high response modification coefficients and the elimination of detailing requirements.

Once a response modification coefficient is selected from the tables, Section 15.4.1 provides additional guidance.

C15.4.1 Design Basis. Separate tables provided in this section identify the basic coefficients, associated detailing requirements, and height limits and restrictions for the two types of nonbuilding structures.

For nonbuilding structures similar to buildings, the design seismic loads are determined using the same procedures used for buildings as specified in Chapter 12, with two exceptions: fundamental periods are determined in accordance with Section 15.4.4, and Table 15.4-1 provides additional options for structural systems. Although only Section 12.8 (the equivalent lateral force procedure) is specifically mentioned in Section 15.4.1, Section 15.1.3 provides the analysis procedures that are permitted for nonbuilding structures.

In Table 15.4-1, seismic coefficients, system restrictions, and height limits are specified for a few nonbuilding structures similar to buildings. The values of R, Ω_0, and C_d; the detailing requirement references; and the structural system height limits are the same as those in Table 12.2-1 for the same systems, except for ordinary moment frames. In Chapter 12, increased height limits for ordinary moment frame structural systems apply to metal building systems, whereas in Chapter 15 they apply to pipe racks with end plate bolted moment connections. The seismic performance of pipe racks was judged to be similar to that of metal building structures with end plate bolted moment connections, so the height limits were made the same as those specified in previous editions.

Table 15.4-1 also provides lower R values with less restrictive height limits in SDCs D, E, and F based on good performance in past earthquakes. For some options, no seismic detailing is required if very low values of R (and corresponding high seismic design forces) are used. The concept of extending this approach to other structural systems is the subject of future research using the methodology developed in FEMA P-695 (FEMA 2009).

For nonbuilding structures not similar to buildings, the seismic design loads are determined as in Chapter 12 with three exceptions: the fundamental periods are determined in accordance with Section 15.4.4, the minima are those specified in Section 15.4.1(2), and the seismic coefficients are those specified in Table 15.4-2.

Some entries in Table 15.4-2 may seem to be conflicting or confusing. For example, the first major entry is for elevated tanks, vessels, bins, or hoppers. A subset of this entry is for tanks on braced or unbraced legs. This subentry is intended for structures where the supporting columns are integral with the shell (such as an elevated water tank). Tension-only bracing is allowed for such a structure. Where the tank or vessel is supported by buildinglike frames, the frames are to be designed in accordance with all of the restrictions normally applied to building frames. Section 15.3 provides provisions for nonbuilding structures supported by buildinglike frames. Beginning with the 2005 edition of ASCE 7, Table 15.4-2 contained an entry for "Tanks or vessels supported on structural towers similar to buildings." Under certain circumstances, text provided with this table entry conflicted with the requirements of Section 15.3. If the weight of the nonbuilding structure is relatively small compared to the weight of the structure (less than 25% of the weight of the structure) or the nonbuilding structure is rigid, the supported nonbuilding structure can be treated as a nonstructural component and the values of the supporting structure seismic coefficients can be taken from Table 15.4-1. Under these circumstances, the deleted entry was correct. However, if the weight of the supported nonbuilding structure is not small and the nonbuilding structure is flexible (which is generally the case especially when you consider the vertical and rocking flexibility of supporting floor beams), the seismic coefficients are determined as the most conservative.

The accidental torsion requirements of Section 12.8.4.2 were formulated primarily for use in building structures. The primary factors that contribute to accidental torsion are lateral force-resisting systems that are located primarily near the center of the structure rather than the perimeter, disproportionate concentration of inelastic demands in system components, the effects of nonstructural elements, uncertainties in defining the structure's stiffness characteristics, and spatial variation (and rotational components of ground motions) of horizontal input motions applied to long structures. Inherently torsionally resistant systems as defined in Section 15.4.1, Item 5, with R values less than or equal to 3.5 are not expected to have inelastic demands of a level that would require additional consideration of accidental torsion. Additionally, nonbuilding structures rarely contain significant nonstructural elements that are not accounted for explicitly in the design of these structures and typically have very well-known mass and stiffness characteristics. Nonbuilding structures also rarely, if ever, have their lateral force-resisting systems located at the center of the structure in plan rather than at the perimeter. The requirement that the calculated center of rigidity at each diaphragm is greater than 5% of the plan dimension of the diaphragm in each direction from the calculated center of mass of the diaphragm prevents configurations of lateral force-resisting elements that are inherently susceptible to the effects of torsion from being exempted from the effects of accidental torsion. Spatial variations of ground motions should be considered in the design of structures of considerable length. If there are significant variations between full and empty weights of the structure, the inherent torsion caused by these variations should be considered in the design of the structure. If there is a nonuniform distribution of mass in silos or bins storing bulk materials because of multiple filling or discharge points, multiple hoppers, nonuniform funnel flow, bulk material behavior, or other operational considerations, the inherent torsion caused by these conditions should be considered in the design of the silo or bin.

C15.4.1.1 Importance Factor. The Importance Factor for a nonbuilding structure is based on the risk category defined in Chapter 1 of the standard or the building code being used in conjunction with the standard. In some cases, reference standards provide a higher Importance Factor, in which case the higher Importance Factor is used.

If the Importance Factor is taken as 1.0 based on a hazard and operability (HAZOP) analysis performed in accordance with Chapter 1, the third paragraph of Section 1.5.3 requires careful consideration; worst-case scenarios (instantaneous release of a vessel or piping system) must be considered. HAZOP risk analysis consultants often do not make such assumptions, so the design professional should review the HAZOP analysis with the HAZOP consultant to confirm that such assumptions have been made to validate adjustment of the Importance Factor. Clients may not be aware that HAZOP consultants do not normally consider the worst-case scenario of instantaneous release but tend to focus on other, more hypothetical, limited-release scenarios, such as those associated with a 2-in.^{2} (1,290 mm^2) hole in a tank or vessel.

C15.4.2 Rigid Nonbuilding Structures. The definition of rigid (having a natural period of less than 0.06 s) was selected judgmentally. Below that period, the energy content of seismic ground motion is generally believed to be very low, and therefore the building response is not likely to be excessively amplified.

Also, it is unlikely that any building will have a first mode period as low as 0.06 s, and it is even unusual for a second mode period to be that low. Thus, the likelihood of either resonant behavior or excessive amplification becomes quite small for equipment that has periods below 0.06 s.

The analysis to determine the period of the nonbuilding structure should include the flexibility of the soil subgrade.

C15.4.3 Loads. As for buildings, the seismic weight must include the range of design operating weight of permanent equipment.

C15.4.4 Fundamental Period. A significant difference between building structures and nonbuilding structures is that the approximate period formulas and limits of Section 12.8.2.1 may not be used for nonbuilding structures. In lieu of calculating a specific period for a nonbuilding structure for determining seismic lateral forces, it is of course conservative to assume a period of $T = T_s$, which results in the largest lateral design forces. Computing the fundamental period is not considered a significant burden because most commonly used computer analysis programs can perform the required calculations.

C15.4.7 Drift, Deflection, and Structure Separation. Nonbuilding structure drifts, deflection, and structure separation are calculated using strength design factored load combinations in order to be compatible with the seismic load definition and the definition of the C_d factors. This philosophy is consistent with that of drift, deflections, and structure separation for buildings defined in Chapter 12.

C15.4.8 Site-Specific Response Spectra. Where site-specific response spectra are required, they should be developed in accordance with Chapter 21 of the standard. If determined for other recurrence intervals, Section 21.1 applies, but Sections 21.2 through 21.4 apply only to risk-targeted maximum considered earthquake (MCE$_R$) determinations. Where other recurrence intervals are used, it should be demonstrated that the requirements of Chapter 15 also are satisfied.

C15.4.9 Anchors in Concrete or Masonry. Many nonbuilding structures rely on the ductile behavior of anchor bolts to justify the response modification factor, R, assigned to the structure. Nonbuilding structures typically rely more heavily on anchorage to provide system ductility. The additional requirements of Section 15.4.9 provide additional anchorage strength and ductility to support the response modification factors assigned to these systems. The addition of Section 15.4.9 provides a consistent treatment of anchorage for nonbuilding structures.

C15.4.9.4 ASTM F1554 Anchors. ASTM F1554 contains a requirement that is not consistent with the anchor requirements found in Chapter 15. Section 6.4 of ASTM F1554 allows the anchor supplier to substitute weldable Grade 55 anchors for Grade 36 anchors without the approval of the registered design professional. Because many nonbuilding structures rely on the ability of the anchors to stretch to justify the response modification factor, R, assigned to the structure, a higher yield anchor cannot be allowed to be substituted for a lower yield anchor without the approval of the registered design professional. Except where anchors are specified and are designed as ductile steel anchors in accordance with ACI 318, Section 17.2.3.4.3(a), or where the design must meet the requirements of Section 15.7.5 or Section 15.7.11.7b, this provision does not prohibit ductility from being provided by another element of the structure. In that case, the ASTM F1554 anchors would be designed for the corresponding forces.

C15.4.10 Requirements for Nonbuilding Structure Foundations on Liquefiable Sites. Section 12.13.9 allows shallow foundation to be built on liquefiable soils with a number of restrictions. Many nonbuilding structures are sensitive to large foundation settlements. This sensitivity is caused by restraint imposed by interconnecting piping and equipment and the buckling sensitivity of shell structures. Therefore, in order to build these structures on shallow foundations on liquefiable soils, it must be demonstrated that the foundation, nonbuilding structure not similar to buildings, and connecting systems can be designed for the soil strength loss, the anticipated settlements from lateral spreading, and total and differential settlements induced by MCE_G earthquake ground motions.

C15.5 NONBUILDING STRUCTURES SIMILAR TO BUILDINGS

C15.5.1 General. Although certain nonbuilding structures exhibit behavior similar to that of building structures, their functions and occupancies are different. Section 15.5 of the standard addresses the differences.

C15.5.2 Pipe Racks. Freestanding pipe racks supported at or below grade with framing systems that are similar to building systems are designed in accordance with Section 12.8 or 12.9 and Section 15.4. Single-column pipe racks that resist lateral loads should be designed as inverted pendulums.

Based on good performance in past earthquakes, Table 15.4-1 sets forth the option of lower R values and less restrictive height limits for structural systems commonly used in pipe racks. The R value versus height limit tradeoff recognizes that the size of some nonbuilding structures is determined by factors other than traditional loadings and results in structures that are much stronger than required for seismic loadings. Therefore, the ductility demand is generally much lower than that for a corresponding building. The intent is to obtain the same structural performance at the increased heights. This option proves to be economical in most situations because of the relative cost of materials and construction labor. The lower R values and increased height limits of Table 15.4-1 apply to nonbuilding structures similar to buildings; they cannot be applied to building structures. Table C15.5-1 illustrates the R values and height limits for a 70-ft (21.3-m) high steel ordinary moment frame (OMF) pipe rack.

C15.5.3.1 Steel Storage Racks. The two approaches to the design of steel storage racks set forth by the standard are intended to produce comparable results. The specific revisions to the Rack Manufacturers Institute (RMI) specification cited in earlier editions of this standard and the detailed requirements of the ANSI/RMI MH 16.1 specification reflect the recommendations of FEMA 460 (FEMA 2005).

Although the ANSI/RMI MH 16.1 specification reflects the recommendations of FEMA 460 (FEMA 2005), the anchorage provisions of the ANSI/RMI MH 16.1 specification are not in conformance with ASCE/SEI 7. Therefore, specific anchorage requirements were added in Sections 15.5.3.1.1 and 15.5.3.1.2.

These recommendations address the concern that storage racks in warehouse-type retail stores may pose a greater seismic risk to the general public than exists in low-occupancy warehouses or more conventional retail environments. Under normal conditions, retail stores have a far higher occupant load than an ordinary warehouse of a comparable size. Failure of a storage rack system in a retail environment is much more likely to cause personal injury than would a similar failure in a storage warehouse. To provide an appropriate level of additional safety in areas open to the public, an Importance Factor of 1.50 is specified. Storage rack contents, though beyond the scope of the standard, may pose a potentially serious threat to life should they fall from the shelves in an earthquake. It is recommended that restraints be provided, as shown in Fig. C15.5-1, to prevent the contents of rack shelving open to the general public from falling during strong ground shaking.

C15.5.3.2 Steel Cantilevered Storage Racks. The two approaches to the design of steel cantilevered storage racks set forth by the standard are intended to produce comparable results. The specific development of a new RMI standard to include the detailed requirements of the new ANSI/RMI MH 16.3 (2016) specification, reflect the unique characteristics of this structural storage system, along with the recommendations of FEMA 460, *Seismic Considerations for Steel Storage Racks Located in Areas Accessible to the Public.*

The values of R, C_d, and Ω_0 added to Table 15.4-1 for Steel Cantilever Storage Racks were taken directly from Table 2.7.2.2.3 (1) of ANSI/RMI MH 16.3.

The anchorage provisions of the ANSI/RMI MH 16.3 specification are not in conformance with ASCE/SEI 7. Therefore, specific anchorage requirements were added in Section 15.5.3.2.1.

These recommendations address the concern that steel cantilevered storage racks in warehouse-type retail stores may pose a greater seismic risk to the general public than exists in low-occupancy warehouses or more conventional retail

Table C15.5-1 *R* Value Selection Example for Steel OMF Pipe Racks

Seismic Design Category	R	ASCE/SEI 7-10 Table	System	Seismic Detailing Requirements
C	3.5	12.2-1 or 15.4-1	Steel ordinary moment frame (OMF)	AISC 341
C	3	12.2-1	Structural steel systems not specifically detailed for seismic resistance	None
D or E	2.5	15.4-1	Steel OMF with permitted height increase	AISC 341
D, E, or F	1	15.4-1	Steel OMF with unlimited height	AISC 341

FIGURE C15.5-1 Merchandise restrained by netting

Source: FEMA 460 Seismic Considerations for Steel Storage Racks.

FIGURE C15.5-2 Examples of Structural Towers That are Not Integral with the Supported Tank

Source: (left) Courtesy of Chevron; reproduced with permission. (right) Courtesy of CB&I LLC; reproduced with permission.

environments. Under normal conditions, retail stores have a far higher occupant load than an ordinary warehouse of a comparable size. Failure of a steel cantilevered storage rack system in a retail environment is much more likely to cause personal injury than would a similar failure in a storage warehouse. To provide an appropriate level of additional safety in areas open to the public, an Importance Factor of 1.50 is specified. Steel cantilevered storage rack contents, though beyond the scope of the standard, may pose a potentially serious threat to life should they fall from the shelves in an earthquake. It is recommended that restraints be provided, as shown in Figure C15.5-1, to prevent the contents of rack shelving open to the general public from falling during strong ground shaking.

All systems in ANSI/MH16.3, Table 2.7.2.2.3(1) are ordinary systems. For all systems in SDC B and C, the values in ANSI/MH16.3 (2016), Table 2.7.2.2.3(1) for R, Ω_0, and C_d correspond to the values shown in Table 12.2-1 for Steel Systems Not Specifically Detailed for Seismic Resistance, Excluding Cantilever Column Systems. No seismic detailing is required. For hot-rolled steel systems in SDC D, E, and F, the values in ANSI/MH16.3, Table 2.7.2.2.3(1) for R, Ω_0, and C_d correspond to the values shown in Table 15.4-1 for ordinary systems with permitted height increase except that no height limits apply. The hot-rolled steel systems are detailed to AISC 341. For cold-formed steel systems in SDC D, E, and F, the values in ANSI/MH16.3 (2016), Table 2.7.2.2.3(1) for R, Ω_0, and C_d correspond to the values shown in Table 15.4-1 for ordinary systems with unlimited height. Seismic detailing is not required for the cold-formed steel systems.

C15.5.4 Electrical Power-Generating Facilities. Electrical power plants closely resemble building structures, and their performance in seismic events has been good. For reasons of mechanical performance, lateral drift of the structure must be limited. The lateral bracing system of choice has been the concentrically braced frame. In the past, the height limits on braced frames in particular have been an encumbrance to the design of large power-generating facilities. Based on acceptable past performance, Table 15.4-1 permits the use of ordinary concentrically braced frames with both lower R values and less restrictive height limits. This option is particularly effective for boiler buildings, which generally are 300 ft (91.4 m) or more high. A peculiarity of large boiler buildings is the general practice of suspending the boiler from the roof structures; this practice results in an unusual mass distribution, as discussed in Section C15.1.3.

FIGURE C15.5-3 Examples of Structural Towers That are Integral with the Supported Tank

Source: Courtesy of CB&I LLC; reproduced with permission.

C15.5.5 Structural Towers for Tanks and Vessels. The requirements of this section apply to structural towers that are not integral with the supported tank. Elevated water tanks designed in accordance with AWWA D100 are not subject to Section 15.5.5. A structural tower supporting a tank or vessel is considered integral with the supported tank or vessel where the tank or vessel shell acts as a part of the seismic force-resisting system of the supporting tower.

Examples of structural towers that are not integral with the supported tank are shown in Fig. C15.5-2. Examples of structural towers that are integral with the supported tank are shown in Fig. C15.5-3. Examples of structural towers that are integral with the supported tank include column-supported elevated water tanks designed to AWWA D100 and column-supported liquid and gas spheres designed to ASME BVPC, Section VIII.

C15.5.6 Piers and Wharves. Current industry practice recognizes the distinct differences between the two categories of piers and wharves described in the standard. Piers and wharves with public occupancy, described in Section 15.5.6.2, are commonly treated as the "foundation" for buildings or buildinglike structures; design is performed using the standard, likely under the jurisdiction of the local building official. Piers and wharves without occupancy by the general public are often treated differently and are outside the scope of the standard; in many cases, these structures do not fall under the jurisdiction of building officials, and design is performed using other industry-accepted approaches.

Design decisions associated with these structures often reflect economic considerations by both owners and local, regional, or state jurisdictional entities with interest in commercial development. Where building officials have jurisdiction but lack experience analyzing pier and wharf structures, reliance on other industry-accepted design approaches is common.

Where occupancy by the general public is not a consideration, seismic design of structures at major ports and marine terminals often uses a performance-based approach, with criteria and methods that are very different from those used for buildings, as provided in the standard. Design approaches most commonly used are generally consistent with the practices and criteria described in the following documents: *Seismic Design Guidelines for Port Structures* (2001); Ferritto et al. (1999); Priestley et al. (1996); Werner (1998); *Marine Oil Terminal Engineering and Maintenance Standards* (2005).

These alternative approaches have been developed over a period of many years by working groups within the industry, and they reflect the historical experience and performance characteristics of these structures, which are very different from those of building structures.

The main emphasis of the performance-based design approach is to provide criteria and methods that depend on the economic importance of a facility. Adherence to the performance criteria in the documents listed does not seek to provide uniform margins of collapse for all structures; their application is expected to provide at least as much inherent life safety as for buildings designed using the standard. The reasons for the higher inherent level of life safety for these structures include the following:

1. These structures have relatively infrequent occupancy, with few working personnel and very low density of personnel. Most of these structures consist primarily of open area, with no enclosed structures that can collapse onto personnel. Small control buildings on marine oil terminals or similar secondary structures are commonly designed in accordance with the local building code.
2. These pier or wharf structures typically are constructed of reinforced concrete, prestressed concrete, or steel and are highly redundant because of the large number of piles supporting a single wharf deck unit. Tests done at the University of California at San Diego for the Port of Los Angeles have shown that high ductilities (10 or more) can be achieved in the design of these structures using practices currently used in California ports.
3. Container cranes, loading arms, and other major structures or equipment on piers or wharves are specifically designed not to collapse in an earthquake. Typically, additional piles and structural members are incorporated into the wharf or pier specifically to support such items.
4. Experience has shown that seismic "failure" of wharf structures in zones of strong seismicity is indicated not by collapse but by economically irreparable deformations of the piles. The wharf deck generally remains level or slightly tilting but shifts out of position. Earthquake loading on properly maintained marine structures has never induced complete failure that could endanger life safety.
5. The performance-based criteria of the listed documents address reparability of the structure. These criteria are much more stringent than collapse prevention criteria and create a greater margin for life safety.

Lateral load design of these structures in low, or even moderate, seismic regions often is governed by other marine conditions.

C15.6 GENERAL REQUIREMENTS FOR NONBUILDING STRUCTURES NOT SIMILAR TO BUILDINGS

Nonbuilding structures not similar to buildings exhibit behavior markedly different from that of building structures. Most of these types of structures have reference documents that address their unique structural performance and behavior. The ground motion in the standard requires appropriate translation to allow use with industry standards.

C15.6.1 Earth-Retaining Structures. Section C11.8.3 presents commonly used approaches for the design of nonyielding walls and yielding walls for bending, overturning, and sliding, taking into account the varying soil types, importance, and site seismicity.

C15.6.2 Chimneys and Stacks

C15.6.2.1 General. The design of stacks and chimneys to resist natural hazards generally is governed by wind design considerations. The exceptions to this general rule involve locations with high seismicity, stacks and chimneys with large elevated masses, and stacks and chimneys with unusual geometries. It is prudent to evaluate the effect of seismic loads in all but those areas with the lowest seismicity. Although not specifically required, it is recommended that the special seismic details required elsewhere in the standard be considered for application to stacks and chimneys.

C15.6.2.2 Concrete Chimneys and Stacks. Concrete chimneys typically possess low ductility, and their performance is especially critical in the regions around large (breach) openings because of reductions in strength and loss of confinement for vertical reinforcement in the jamb regions around the openings. Earthquake-induced chimney failures have occurred in recent history (in Turkey in 1999) and have been attributed to strength and detailing problems (Kilic and Sozen 2003). Therefore, the R value of 3 traditionally used in ASCE/SEI 7-05 for concrete stacks and chimneys was reduced to 2, and detailing requirements for breach openings were added in the 2010 edition of this standard.

C15.6.2.3 Steel Chimneys and Stacks. Guyed steel stacks and chimneys generally are lightweight. As a result, the design loads caused by natural hazards generally are governed by wind. On occasion, large flares or other elevated masses located near the top may require in-depth seismic analysis. Although it does not specifically address seismic loading, Chapter 6 of Troitsky (1990) provides a methodology appropriate for resolution of the seismic forces defined in the standard in addition to the requirements found in ASME STS-1.

C15.6.4 Special Hydraulic Structures. The most common special hydraulic structures are baffle walls and weirs that are used in water treatment and wastewater treatment plants. Because there are openings in the walls, during normal operations the fluid levels are equal on each side of the wall, exerting no net horizontal force. Sloshing during a seismic event can exert large forces on the wall, as illustrated in Fig. C15.6-1. The walls can fail unless they are designed properly to resist the dynamic fluid forces.

C15.6.5 Secondary Containment Systems. This section reflects the judgment that designing all impoundment dikes for the MCE_R ground motion when full and sizing all impoundment dikes for the sloshing liquid height is too conservative. Designing an impoundment dike as full for the MCE_R assumes failure of the

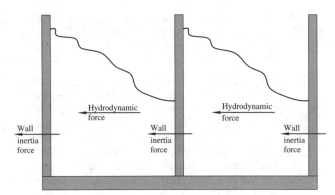

FIGURE C15.6-1 Wall forces

primary containment and occurrence of a significant aftershock. Such significant aftershocks (of the same magnitude as the MCE_R ground motion) are rare and do not occur in all locations. Although explicit design for aftershocks is not a requirement of the standard, secondary containment must be designed full for an aftershock to protect the general public. The use of two-thirds of the MCE_R ground motion as the magnitude of the design aftershock is supported by Bath's law, according to which the maximum expected aftershock magnitude may be estimated to be 1.2 scale units below the main shock magnitude.

The risk assessment and risk management plan described in Section 1.5.2 are used to determine where the secondary containment must be designed full for the MCE_R. The decision to design secondary containment for this more severe condition should be based on the likelihood of a significant aftershock occurring at the particular site, considering the risk posed to the general public by the release of hazardous material from the secondary containment.

Secondary containment systems must be designed to contain the sloshing liquid height where the release of liquid would place the general public at risk by exposing them to hazardous materials, by scouring of foundations of adjacent structures, or by causing other damage to adjacent structures.

C15.6.5.1 Freeboard. Eq. (15.6-1) was revised in ASCE 7-10 to return to the more exact theoretical formulation for sloshing liquid height instead of the rounded value introduced in ASCE/SEI 7-05. The rounded value in part accounted for maximum direction of response effects. Because the ground motion definition in ASCE/SEI 7-10 was changed and the maximum direction of response is now directly accounted for, it is no longer necessary to account for these effects by rounding up the theoretical sloshing liquid height factor in Eq. (15.6-1).

C15.6.6 Telecommunication Towers. Telecommunication towers support small masses, and their design generally is governed by wind forces. Although telecommunication towers have a history of experiencing seismic events without failure or significant damage, seismic design in accordance with the standard is required.

Typically bracing elements bolt directly (without gusset plates) to the tower legs, which consist of pipes or bent plates in a triangular plan configuration.

C15.6.7 Steel Tubular Support Structures for Onshore Wind Turbine Generator Systems. The most common support structures for large onshore wind turbine generator systems are steel tubular towers. Recommendations for the design of these structures can be found in ASCE/AWEA (2011). ASCE/AWEA (2011) applies to wind turbines that have a rotor-swept area

FIGURE C15.6-2 Typical Steel Tubular Support Structure for Onshore Wind Turbine Generator Systems

Source: Courtesy of GE Power; reproduced with permission.

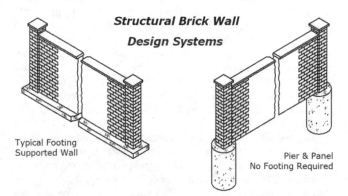

FIGURE C15.6-3 Typical Cantilever Wall Systems Falling under the Requirements of Section 15.6.8.

Source: Courtesy of J.G. Soules; reproduced with permission.

greater than 2,153 ft^2 (200 m^2). These recommendations are to be used in conjunction with seismic lateral forces determined in accordance with Section 15.4. A typical steel tubular support structure for an onshore wind turbine generator system is shown in Fig. C15.6-2.

C15.6.8 Ground-Supported Cantilever Walls or Fences. Ground-supported cantilever walls and fences constructed from masonry, concrete, timber, or a combination of materials, including steel, are common. Such walls are often used as sound barrier walls or to limit access to residential subdivisions. Ground-supported cantilever walls and fences include walls supported by a footing and pier and panel/pilaster and panel wall systems (Fig. C15.6-3) as long as these systems are not supported laterally in the out-of-plane direction above grade. An example of a masonry ground-supported cantilever wall is shown in Fig. C15.6-4. Many improperly designed ground-supported cantilever walls and fences constructed from masonry or concrete have experienced problems and have failed during seismic events

FIGURE C15.6-4 Typical Masonry Ground-Supported Cantilever Wall

as evidenced in Section 6.3.9.1 of FEMA E-74 (2012), *Reducing the Risks of Nonstructural Earthquake Damage—A Practical Guide*.

The provisions for ground-supported cantilever walls and fences more than 6 ft (1.83 m) high were contained in prior issues of the Uniform Building Code, including the 1997 Uniform Building Code (ICBO 1997). When the International Building Code was developed, the provisions were inadvertently dropped and were not incorporated in ASCE 7. Walls of all heights should be properly designed. The 6-ft (1.83-m) height has been retained from the 1997 Uniform Building Code as the minimum height at which these provisions apply because walls less than 6 ft (1.83 m) high are not deemed to present as significant a risk to life safety.

The seismic design parameters chosen for this system are based on those given in Table 15.4-2 for "all other self-supporting structures, tanks, or vessels not covered above or by reference standards that are not similar to buildings" except that all height limits were changed to no limit (NL), considering that the structure is a cantilever wall. Cantilever walls covered by these provisions can be of any material or combination of materials; therefore, a relatively low value of R was chosen to account for these material combinations. Additionally, pilasters incorporated in many of these wall systems are essentially ordinary cantilever columns. Ordinary cantilever columns in ASCE 7 tend to have low R values irrespective of the material used.

A decision was made by the ASCE 7 Seismic Subcommittee that a ground-supported freestanding wall or fence was a nonbuilding structure not similar to a building and should fall under the provisions of Chapter 15 instead of Chapter 13.

C15.7 TANKS AND VESSELS

C15.7.1 General. Methods for seismic design of tanks, currently adopted by a number of reference documents, have evolved from earlier analytical work by Jacobsen (1949), Housner (1963), Velestos (1974), Haroun and Housner (1981), and others. The procedures used to design flat-bottom storage tanks and liquid containers are based on the work of Housner (U.S. Department of Energy, 1963) and Wozniak and Mitchell (1978). The reference documents for tanks and vessels have specific requirements to safeguard against catastrophic failure of the primary structure based on observed behavior in seismic events since the 1930s. Other methods of analysis, using flexible shell models, have been proposed but at present are beyond the scope of the standard.

The industry-accepted design methods use three basic steps:

1. Dynamic modeling of the structure and its contents. When a liquid-filled tank is subjected to ground acceleration, the lower portion of the contained liquid, identified as the impulsive component of mass, W_i, acts as if it were a solid mass rigidly attached to the tank wall. As this mass accelerates, it exerts a horizontal force, P_i, on the wall; this force is directly proportional to the maximum acceleration of the tank base. This force is superimposed on the inertia force of the accelerating wall itself, P_s. Under the influence of the same ground acceleration, the upper portion of the contained liquid responds as if it were a solid mass flexibly attached to the tank wall. This portion, which oscillates at its own natural frequency, is identified as the convective component, W_c, and exerts a horizontal force, P_c, on the wall. The convective component oscillations are characterized by sloshing whereby the liquid surface rises above the static level on one side of the tank and drops below that level on the other side.
2. Determination of the period of vibration, T_i, of the tank structure and the impulsive component and determination of the natural period of oscillation (sloshing), T_c, of the convective component.
3. Selection of the design response spectrum. The response spectrum may be site specific, or it may be constructed on the basis of seismic coefficients given in national codes and standards. Once the design response spectrum is constructed, the spectral accelerations corresponding to T_i and T_c are obtained and are used to calculate the dynamic forces P_i, P_s, and P_c.

Detailed guidelines for the seismic design of circular tanks, incorporating these concepts to varying degrees, have been the province of at least four industry reference documents: AWWA D100 for welded steel tanks (since 1964); API 650 for petroleum storage tanks; AWWA D110 for prestressed, wire-wrapped tanks (since 1986); and AWWA D115 for prestressed concrete tanks stressed with tendons (since 1995). In addition, API 650 and API 620 contain provisions for petroleum, petrochemical, and cryogenic storage tanks. The detail and rigor of analysis prescribed in these documents have evolved from a semistatic approach in the early editions to a more rigorous approach at present, reflecting the need to include the dynamic properties of these structures.

The requirements in Section 15.7 are intended to link the latest procedures for determining design-level seismic loads with the allowable stress design procedures based on the methods in the standard. These requirements, which in many cases identify specific substitutions to be made in the design equations of the reference documents, will assist users of the standard in making consistent interpretations.

ACI has published ACI 350.3-06 (2006), *Seismic Design of Liquid-Containing Concrete Structures*. This document, which addresses all types of concrete tanks (prestressed and nonprestressed, circular, and rectilinear), has provisions that are unfortunately not consistent with the seismic criteria of ASCE/SEI 7. However, the document, when combined with the modifications required in Section 15.7.7.3, serves as both a practical "how-to" loading reference and a guide to supplement application of ACI 318, Chapter 18.

C15.7.2 Design Basis. In the case of the seismic design of nonbuilding structures, standardization requires adjustments to industry reference documents to minimize existing inconsistencies among them, while recognizing that structures designed and built over the years in accordance with these documents have performed well in earthquakes of varying severity. Of the inconsistencies among reference documents, the ones most important to seismic design relate to the base shear equation. The traditional base shear takes the following form:

$$V = \frac{ZIS}{R_w} CW \qquad (C15.7\text{-}1)$$

An examination of those terms as used in the different references reveals the following:

1. Z, S: The seismic zone coefficient, Z, has been rather consistent among all the documents because it usually has been obtained from the seismic zone designations and maps in the model building codes. However, the soil profile coefficient, S, does vary from one document to another. In some documents, these two terms are combined.
2. I: The Importance Factor, I, has varied from one document to another, but this variation is unavoidable and understandable because of the multitude of uses and degrees of importance of tanks and vessels.
3. C: The coefficient C represents the dynamic amplification factor that defines the shape of the design response spectrum for any given ground acceleration. Because C is primarily a function of the frequency of vibration, inconsistencies in its derivation from one document to another stem from at least two sources: differences in the equations for the determination of the natural frequency of vibration, and differences in the equation for the coefficient itself. (For example, for the shell/impulsive liquid component of lateral force, the steel tank documents use a constant design spectral acceleration [constant C] that is independent of the "impulsive" period, T.) In addition, the value of C varies depending on the damping ratio assumed for the vibrating structure (usually between 2% and 7% of critical).
4. Where a site-specific response spectrum is available, calculation of the coefficient C is not necessary except in the case of the convective component (coefficient C_c), which is assumed to oscillate with 0.5% of critical damping and whose period of oscillation is usually long (greater than 2.5 s). Because site-specific spectra are usually constructed for high damping values (3% to 7% of critical) and because the site-specific spectral profile may not be well-defined in the long-period range, an equation for C_c applicable to a 0.5% damping ratio is necessary to calculate the convective component of the seismic force.
5. R_w: The response modification factor, R_w, is perhaps the most difficult to quantify, for a number of reasons. Although R_w is a compound coefficient that is supposed to reflect the ductility, energy-dissipating capacity, and redundancy of the structure, it is also influenced by serviceability considerations, particularly in the case of liquid-containing structures.

In the standard, the base shear equation for most structures has been reduced to $V = C_s W$, where the seismic response coefficient, C_s, replaces the product ZSC/R_w. C_s is determined from the design spectral response acceleration parameters S_{DS} and S_{D1} (at short periods and at a period of 1, respectively), which in turn are obtained from the mapped MCE_R spectral accelerations S_s and S_1. As in the case of the prevailing industry reference documents, where a site-specific response spectrum is available, C_s is replaced by the actual values of that spectrum.

The standard contains several bridging equations, each designed to allow proper application of the design criteria of a particular reference document in the context of the standard. These bridging equations associated with particular types of liquid-containing structures and the corresponding reference documents are discussed below. Calculation of the periods of vibration of the impulsive and convective components is in accordance with the reference documents, and the detailed resistance and allowable stresses for structural elements of each industry structure are unchanged, except where new information has led to additional requirements.

It is expected that the bridging equations of Section 15.7.7.3 will be eliminated as the relevant reference documents are updated to conform to the standard. The bridging equations previously provided for AWWA D100 and API 650 already have been eliminated as a result of updates of these documents.

Tanks and vessels are sensitive to vertical ground motions. Traditionally, the approach has been to apply a vertical seismic coefficient equal to $0.2S_{DS}$ to the design. This design approach came from the process used to design buildings and may underestimate the vertical response of the tank and its contents. For noncylindrical tanks, the increase in the hydrostatic pressure caused by vertical excitation has taken the form of $0.4S_{av}$, where S_{av} is determined in accordance with Section 15.7.2 and Section 11.9. This pressure is combined directly with the hydrodynamic loads induced from lateral ground motions. The result is equal to 100% horizontal plus 40% vertical. The response of cylindrical tanks to vertical motions is well known and documented in various papers. Unless otherwise specified in a reference document, the vertical period T_v, may be determined by

$$T_v = 2\pi \sqrt{\frac{\gamma_L R H_L^2}{gtE}} \qquad (C15.7\text{-}2)$$

where

γ_L = Unit weight of stored liquid;
R = Tank radius to the inside of the wall;
H_L = liquid height inside the tank;
g = acceleration caused by gravity in consistent units;
t = average shell thickness; and
E = modulus of elasticity of shell.

Eq. (C15.7-2) comes from ACI 350.3 (2006) and is based on a rigid response of the liquid to vertical ground motions. Additional documents, such as Section 7.7.1 of ASCE's *Guidelines for the Seismic Design of Oil and Gas Pipeline Systems* (1984) provide solutions to determine the response of a flexible tank to vertical ground motions. The response of the structure itself is set equal to 0.4 times the peak of the vertical response spectra. Using the peak of the vertical response spectra recognizes the vertical stiffness of the tank walls. This load is combined directly with loads produced from lateral ground motions. The result is equal to 100% horizontal plus 40% vertical. It should also be noted that R has been added to Eq. (15.7-1). R is included in the calculation of hoop stress because the response of the tank shell caused by the added hoop tension from vertical ground motions is no different than the response of the tank shell caused by the added hoop tension from horizontal ground motions. ACI 350.3 has used this philosophy for many years.

C15.7.3 Strength and Ductility. As is the case for building structures, ductility and redundancy in the lateral support systems for tanks and vessels are desirable and necessary for good seismic performance. Tanks and vessels are not highly redundant structural systems, and therefore ductile materials and well-designed connection details are needed to increase the capacity of the vessel to absorb more energy without failure. The critical performance of many tanks and vessels is governed by shell stability requirements rather than by yielding of the structural elements. For example, contrary to building structures, ductile stretching of anchor bolts is a desirable energy absorption component where tanks and vessels are anchored. The performance of cross-braced towers is highly dependent on the ability of the horizontal compression struts and connection details to develop fully the tension yielding in the rods. In such cases, it is also important to preclude both premature failure in the threaded portion of the connection and failure of the connection of the rod to the column before yielding of the rod.

The changes made to Section 15.7.3(a) are intended to ensure that anchors and anchor attachments are designed such that the anchor yields (stretches) before the anchor attachment to the structure fails. The changes also clarify that the anchor rod embedment requirements are to be based on the requirements of Section 15.7.5 and not Section 15.7.3(a).

C15.7.4 Flexibility of Piping Attachments. Poor performance of piping connections (tank leakage and damage) caused by seismic deformations is a primary weakness observed in seismic events. Although commonly used piping connections can impart mechanical loads to the tank shell, proper design in seismic areas results in only negligible mechanical loads on tank connections subject to the displacements shown in Table 15.7-1. API 650 treats the values shown in Table 15.7-1 as allowable stress-based values and therefore requires that these values be multiplied by 1.4 where strength-based capacity values are required for design.

The displacements shown in Table 15.7-1 are based on movements observed during past seismic events. The vertical tank movements listed are caused by stretch of the mechanical anchors or steel tendons (in the case of a concrete tank) for mechanically anchored tanks or the deflection caused by bending of the bottom of self-anchored tanks. The horizontal movements listed are caused by the deformation of the tank at the base.

In addition, interconnected equipment, walkways, and bridging between multiple tanks must be designed to resist the loads and accommodate the displacements imposed by seismic forces. Unless connected tanks and vessels are founded on a common rigid foundation, the calculated differential movements must be assumed to be out of phase.

C15.7.5 Anchorage. Many steel tanks can be designed without anchors by using annular plate detailing in accordance with reference documents. Where tanks must be anchored because of overturning potential, proper anchorage design provides both a shell attachment and an embedment detail that allows the bolt to yield without tearing the shell or pulling the bolt out of the foundation. Properly designed anchored tanks have greater reserve strength to resist seismic overload than do unanchored tanks.

To ensure that the bolt yields (stretches) before failure of the anchor embedment, the anchor embedment must be designed in accordance with ACI 318, Eq. (17.4.1.2), and must be provided with a minimum gauge length of eight bolt diameters. Gauge length is the length of the bolt that is allowed to stretch. It may include part of the embedment length into the concrete that is not bonded to the bolt. A representation of gauge length is shown in Fig. C15.7-1.

It is also important that the bolt not be significantly oversized to ensure that the bolt stretches. The prohibition on using the load combinations with overstrength of Section 12.4.3 is intended to accomplish this goal.

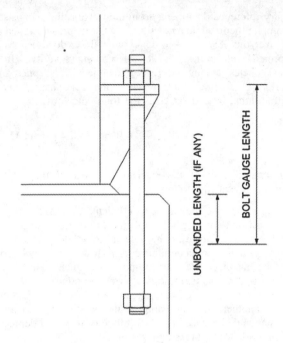

FIGURE C15.7-1 Bolt gauge length

Where anchor bolts and attachments are misaligned such that the anchor nut or washer does not bear evenly on the attachment, additional bending stresses in threaded areas may cause premature failure before anchor yielding.

C15.7.6 Ground-Supported Storage Tanks for Liquids

C15.7.6.1 General. The response of ground storage tanks to earthquakes is well documented by Housner (1963), Wozniak and Mitchell (1978), Velestos (1974), and others. Unlike building structures, the structural response of these tanks is influenced strongly by the fluid–structure interaction. Fluid–structure interaction forces are categorized as sloshing (convective) and rigid (impulsive) forces. The proportion of these forces depends on the geometry (height-to-diameter ratio) of the tank. API 650, API 620, AWWA D100-11, AWWA D110, AWWA D115, and ACI 350.3 provide the data necessary to determine the relative masses and moments for each of these contributions.

The standard requires that these structures be designed in accordance with the prevailing reference documents, except that the height of the sloshing wave, δ_s, must be calculated using Eq. (15.7-13). API 650 and AWWA D100-11 include this requirement in their latest editions.

Eqs. (15.7-10) and (15.7-11) provide the spectral acceleration of the sloshing liquid for the constant-velocity and constant-displacement regions of the response spectrum, respectively. The 1.5 factor in these equations is an adjustment for 0.5% damping. An exception in the use of Eq. (15.7-11) was added for the 2010 edition of this standard. The mapped values of T_L were judged to be unnecessarily conservative by the ASCE 7 Seismic Subcommittee in light of actual site-specific studies carried out since the introduction of the T_L requirements of ASCE/SEI 7-05. These studies indicate that the mapped values of T_L appear to be very conservative based on observations during recent large earthquakes, especially the 2010 M_w 8.8 Chilean earthquake, where the large amplifications at very long periods (6–10 s) were not evident either in the ground motion records or in the behavior of long-period structures (particularly sloshing in tanks). Because a revision of the T_L maps is a time-consuming task that was not possible during the 2010 update cycle, an exception was added to allow the use of site-specific values that are less than the mapped values with a floor of 4 s or one-half the mapped value of T_L. The exception was added under Section 15.7.6 because, for nonbuilding structures, the overly conservative values for T_L are primarily an issue for tanks and vessels. Discussion of the site-specific procedures can be found in the Commentary for Chapter 22.

Small-diameter tanks and vessels are more susceptible to overturning and vertical buckling. As a general rule, a greater ratio of H/D produces lower resistance to vertical buckling. Where H/D is greater than 2, overturning approaches "rigid mass" behavior (the sloshing mass is small). Large-diameter tanks may be governed by additional hydrodynamic hoop stresses in the middle regions of the shell.

The impulsive period (the natural period of the tank components and the impulsive component of the liquid) is typically in the 0.25–0.6 s range. Many methods are available for calculating the impulsive period. The Veletsos flexible-shell method is commonly used by many tank designers. For example, see Veletsos (1974) and Malhotra et al. (2000).

C15.7.6.1.1 Distribution of Hydrodynamic and Inertia Forces. Most of the reference documents for tanks define reaction loads at the base of the shell–foundation interface, without indicating the distribution of loads on the shell as a function of height. ACI 350.3 specifies the vertical and horizontal distribution of such loads.

The overturning moment at the base of the shell in the industry reference documents is only the portion of the moment that is transferred to the shell. The total overturning moment also includes the variation in bottom pressure, which is an important consideration for design of pile caps, slabs, or other support elements that must resist the total overturning moment. Wozniak and Mitchell (1978) and U.S. Department of Energy TID-7024 (1963) provide additional information.

C15.7.6.1.2 Sloshing. In past earthquakes, sloshing contents in ground storage tanks have caused both leakage and noncatastrophic damage to the roof and internal components. Even this limited damage and the associated costs and inconvenience can be significantly mitigated where the following items are considered:

1. Effective masses and hydrodynamic forces in the container;
2. Impulsive and pressure loads at
   a. The sloshing zone (that is, the upper shell and edge of the roof system);
   b. The internal supports (such as roof support columns and tray supports); and
   c. The internal equipment (such as distribution rings, access tubes, pump wells, and risers); and
3. Freeboard (which depends on the sloshing wave height).

When no freeboard is required, a minimum freeboard of $0.7\delta_s$ is recommended for economic considerations. Freeboard is always required for tanks assigned to Risk Category IV.

Tanks and vessels storing biologically or environmentally benign materials typically do not require freeboard to protect the public health and safety. However, providing freeboard in areas of frequent seismic occurrence for vessels normally operated at or near top capacity may lessen damage (and the cost of subsequent repairs) to the roof and upper container. The exception to the minimum required freeboard per Table 15.7-3 for open-top tanks was added because it is rare for damage to occur that would impair the functionality of the facility when water or municipal wastewater overtops an open-top tank, provided that measures have been taken to intercept and properly handle the resulting overflow.

The sloshing liquid height specified in Section 15.7.6.1.2 is based on the design earthquake defined in the standard. For economic reasons, freeboard for tanks assigned to Risk Category I, II, or III may be calculated using a fixed value of T_L equal to 4 s (as indicated in Section 15.7.6.1.2, c) but using the appropriate Importance Factor taken from Table 1.5-2. Because of life-safety and operational functionality concerns, freeboard for tanks assigned to Risk Category IV must be based on the mapped value of T_L. Because use of the mapped value of T_L results in the theoretical maximum value of freeboard, the calculation of freeboard in the case of Risk Category IV tanks is based on an Importance Factor equal to 1.0 (as indicated in Section 15.7.6.1.2 b).

If the freeboard provided is less than the computed sloshing height, δ_s, the sloshing liquid impinges on the roof in the vicinity of the roof-to-wall joint, subjecting it to a hydrodynamic force. This force may be approximated by considering the sloshing wave as a hypothetical static liquid column that has a height, δ_s. The pressure exerted at any point along the roof at a distance y_s above the at-rest surface of the stored liquid may be assumed to be equal to the hydrostatic pressure exerted by the hypothetical liquid column at a distance $\delta_s - y_s$ from the top of that column. A better approximation of the pressure exerted on the roof is found in Malhotra (2005, 2006).

Another effect of a less-than-full freeboard is that the restricted convective (sloshing) mass "converts" into an impulsive mass, thus increasing the impulsive forces. This effect should be taken into account in the tank design. A method for converting the restricted convective mass into an impulsive mass is found in Malhotra (2005, 2006). It is recommended that sufficient freeboard to accommodate the full sloshing height be provided wherever possible.

Eq. (15.7-13) was revised to use the theoretical formulation for sloshing wave height instead of the rounded value introduced in ASCE/SEI 7-05. The rounded value of Eq. (15.6-1) increased the required freeboard by approximately 19%, thereby significantly increasing the cost of both secondary containment and large-diameter, ground-supported storage tanks. See Section C15.6.5.1 for additional commentary on freeboard.

C15.7.6.1.4 Internal Elements. Wozniak and Mitchell (1978) provide a recognized analysis method for determining the lateral loads on internal components caused by sloshing liquid.

C15.7.6.1.5 Sliding Resistance. Historically, steel ground-supported tanks full of product have not slid off foundations. A few unanchored, empty tanks or bulk storage tanks without steel bottoms have moved laterally during earthquake ground shaking. In most cases, these tanks may be returned to their proper locations. Resistance to sliding is obtained from the frictional resistance between the steel bottom and the sand cushion on which bottoms are placed. Because tank bottoms usually are crowned upward toward the tank center and are constructed of overlapping, fillet-welded, individual steel plates (resulting in a rough bottom), it is reasonably conservative to take the ultimate coefficient of friction on concrete as 0.70 (AISC 1986), and therefore a value of tan 30° ($=0.577$) for sand is used in design. The value of 30° represents the internal angle of friction of sand and is conservatively used in design. The vertical weight of the tank and contents, as reduced by the component of vertical acceleration, provides the net vertical load. An orthogonal combination of vertical and horizontal seismic forces, following the procedure in Section 12.5.3, may be used. In recent years, a significant issue has been the prevention of subsurface pollution caused by tank bottom corrosion and leakage. To prevent this problem, liners are often used with the tank foundation. When some of these liners are used, sliding of the tank and/or foundation caused by the seismic base shear may be an issue. If the liner is completely contained within a concrete ring-wall foundation, the liner's surface is not the critical plane to check for sliding. If the liner is placed within an earthen foundation or is placed above or completely below a concrete foundation, it is imperative that sliding be evaluated. It is recommended that the sliding resistance factor of safety be at least 1.5.

C15.7.6.1.6 Local Shear Transfer. The transfer of seismic shear from the roof to the shell and from the shell to the base is accomplished by a combination of membrane shear and radial shear in the wall of the tank. For steel tanks, the radial (out-of-plane) seismic shear is very small and usually is neglected; thus, the shear is assumed to be resisted totally by membrane (in-plane) shear. For concrete walls and shells, which have a greater radial shear stiffness, the shear transfer may be shared. The ACI 350.3-06 (2006) commentary provides further discussion.

C15.7.6.1.7 Pressure Stability. Internal pressure may increase the critical buckling capacity of a shell. Provision to include pressure stability in determining the buckling resistance of the shell for overturning loads is included in AWWA D100-11 (2011). Recent testing on conical and cylindrical shells with internal pressure yielded a design methodology for resisting permanent loads in addition to temporary wind and seismic loads (Miller et al. 1997).

C15.7.6.1.8 Shell Support. Anchored steel tanks should be shimmed and grouted to provide proper support for the shell and to reduce impact on the anchor bolts under reversible loads. The high bearing pressures on the toe of the tank shell may cause inelastic deformations in compressible material (such as fiberboard), creating a gap between the anchor and the attachment. As the load reverses, the bolt is no longer snug and an impact of the attachment on the anchor can occur. Grout is a structural element and should be installed and inspected as an important part of the vertical and lateral force-resisting system.

C15.7.6.1.9 Repair, Alteration, or Reconstruction. During their service life, storage tanks are frequently repaired, modified, or relocated. Repairs often are related to corrosion, improper operation, or overload from wind or seismic events. Modifications are made for changes in service, updates to safety equipment for changing regulations, or installation of additional process piping connections. It is imperative that these repairs and modifications be designed and implemented properly to maintain the structural integrity of the tank or vessel for seismic loads and the design operating loads.

The petroleum steel tank industry has developed specific guidelines in API 653 that are statutory requirements in some states. It is recommended that the provisions of API 653 also be applied to other liquid storage tanks (e.g., water, wastewater, and chemical) as it relates to repairs, modifications, or relocation that affect the pressure boundary or lateral force-resisting system of the tank or vessel.

C15.7.7 Water Storage and Water Treatment Tanks and Vessels. The AWWA design requirements for ground-supported steel water storage structures use allowable stress design procedures that conform to the requirements of the standard.

C15.7.7.1 Welded Steel. AWWA D100 refers to ASCE 7-05 and repeats the ASCE 7-05 seismic design ground motion maps within the body of the document. A requirement is added in this section to point the user to the ground motions in the current version of ASCE 7. The clause in AWWA D100, Section 13.5.4.4, "unless otherwise specified" in the context of the determination of seismic freeboard can result in seismic freeboard below that required by ASCE 7 and is therefore disallowed.

C15.7.7.2 Bolted Steel. A clarification on the ground motions to use in design is added and restrictions are added on the use of Type 6 tanks in AWWA D103 (2009). AWWA D103 refers to ASCE 7-05 and repeats the ASCE 7-05 ground motion maps within the body of the document. Therefore, a clarifying statement is added to point the user to the seismic design ground motions in the current version of ASCE 7. A Type 6 tank is a concrete-bottom bolted steel shell tank with an embedded steel base setting ring. Type 6 tanks are considered to be mechanically anchored. There are no requirements for the anchorage design or bottom design (other than ACI 318) in AWWA D103. For the tank to be considered mechanically anchored, the tank bottom cannot uplift. In this case, the tank bottom is the foundation. If the bottom/foundation uplifts, the tank is now a self-anchored tank and the additional shell compression that develops must be taken into account in the design. That is why J in equation 14–32 of AWWA D103 (2009) is limited to 0.785.

C15.7.7.3 Reinforced and Prestressed Concrete. A review of ACI 350.3 (2006), *Seismic Design of Liquid-Containing Concrete Structures and Commentary*, revealed that this document is not in general agreement with the seismic provisions of ASCE/SEI 7-10.

This section was clarified to note that the Importance Factor, I, and the response modification factor, R, are to be specified by ASCE/SEI 7 and not the reference document. The descriptions used in ACI 350.3 to determine the applicable values of the Importance Factor and response modification factor do not match those used in ASCE/SEI 7.

It was noted that the ground motions for determining the convective (sloshing) seismic forces specified in ACI 350.3 were not the same and are actually lower than those specified by ASCE/SEI 7. ACI 350.3 essentially redefines the long-period transition period, T_L. This alternate transition period allows large-diameter tanks to have significantly lower convective forces and lower seismic freeboard than those permitted by the provisions of ASCE/SEI 7. Therefore, Section 15.7.7.3 was revised to require that the convective acceleration be determined according to the procedure found in Section 15.7.6.1.

C15.7.8 Petrochemical and Industrial Tanks and Vessels Storing Liquids

C15.7.8.1 Welded Steel. The American Petroleum Institute (API) uses an allowable stress design procedure that conforms to the requirements of the standard.

The most common damage to tanks observed during past earthquakes includes the following:

1. Buckling of the tank shell near the base because of excessive axial membrane forces. This buckling damage is usually evident as "elephant foot" buckles a short distance above the base or as diamond-shaped buckles in the lower ring. Buckling of the upper ring also has been observed.
2. Damage to the roof caused by impingement on the underside of the roof of sloshing liquid with insufficient freeboard.
3. Failure of piping or other attachments that are overly restrained.
4. Foundation failures.

Other than the above damage, the seismic performance of floating roofs during earthquakes has generally been good, with damage usually confined to the rim seals, gauge poles, and ladders. However, floating roofs have sunk in some earthquakes because of lack of adequate freeboard or the proper buoyancy and strength required by API 650. Similarly, the performance of open-top tanks with top wind girder stiffeners designed per API 650 has been generally good.

C15.7.8.2 Bolted Steel. Bolted steel tanks are often used for temporary functions. Where use is temporary, it may be acceptable to the jurisdictional authority to design bolted steel tanks for no seismic loads or for reduced seismic loads based on a reduced return period. For such reduced loads based on reduced exposure time, the owner should include a signed removal contract with the fixed removal date as part of the submittal to the authority having jurisdiction.

C15.7.9 Ground-Supported Storage Tanks for Granular Materials

C15.7.9.1 General. The response of a ground-supported storage tank storing granular materials to a seismic event is highly dependent on its height-to-diameter (H/D) ratio and the characteristics of the stored product. The effects of intergranular friction are described in more detail in Section C15.7.9.3.1 (increased lateral pressure), C15.7.9.3.2 (effective mass), and C15.7.9.3.3 (effective density).

Long-term increases in shell hoop tension because of temperature changes after the product has been compacted also must be included in the analysis of the shell; Anderson (1966) provides a suitable method.

C15.7.9.2 Lateral Force Determination. Seismic forces acting on ground-supported liquid storage tanks are divided between impulsive and convective (sloshing) components. However, in a ground-supported storage tank for granular materials, all seismic forces are of the impulsive type and relate to the period of the storage tank itself. Because of the relatively short period of a tank shell, the response is normally in the constant acceleration region of the response spectrum, which relates to S_{DS}. Therefore, the seismic base shear is calculated as follows:

$$V = \frac{S_{DS}}{\left(\frac{R}{I}\right)} W_{\text{effective}} \qquad (C15.7-3)$$

where V, S_{DS}, I, and R have been previously defined, and $W_{\text{effective}}$ is the gross weight of the stored product multiplied by an effective mass factor and an effective density factor, as described in Sections C15.7.9.3.2 and C15.7.9.3.3, plus the dead weight of the tank. Unless substantiated by testing, it is recommended that the product of the effective mass factor and the effective density factor be taken as no less than 0.5 because of the limited test data and the highly variable properties of the stored product.

C15.7.9.3 Force Distribution to Shell and Foundation

C15.7.9.3.1 Increased Lateral Pressure. In a ground-supported tank storing granular materials, increased lateral pressures develop as a result of rigid body forces that are proportional to ground acceleration. Information concerning design for such pressure is scarce. Trahair et al. (1983) describe both a simple, conservative method and a difficult, analytical method using failure wedges based on the Mononobe–Okabe modifications of the classical Coulomb method.

C15.7.9.3.2 Effective Mass. For ground-supported tanks storing granular materials, much of the lateral seismic load can be transferred directly into the foundation, via intergranular shear, before it can reach the tank shell. The effective mass that loads the tank shell is highly dependent on the H/D ratio of the tank and the characteristics of the stored product. Quantitative information concerning this effect is scarce, but Trahair et al. (1983) describe a simple, conservative method to determine the effective mass. That method presents reductions in effective mass, which may be significant, for H/D ratios less than 2. This effect is absent for elevated tanks.

C15.7.9.3.3 Effective Density. Granular material stored in tanks (both ground-supported and elevated) does not behave as a solid mass. Energy loss through intergranular movement and grain-to-grain friction in the stored material effectively reduces the mass subject to horizontal acceleration. This effect may be quantified by an effective density factor less than 1.0.

Based on Chandrasekaran and Jain (1968) and on shake table tests reported in Chandrasekaran et al. (1968), ACI 313 (1997) recommends an effective density factor of not less than 0.8 for most granular materials. According to Chandrasekaran and Jain (1968), an effective density factor of 0.9 is more appropriate for materials with high moduli of elasticity, such as aggregates and metal ores.

C15.7.9.3.4 Lateral Sliding. Most ground-supported steel storage tanks for granular materials rest on a base ring and do not have a steel bottom. To resist seismic base shear, a partial bottom or annular plate is used in combination with anchor bolts or a curb angle. An annular plate can be used alone to resist the seismic base shear through friction between the plate and the foundation, in which case the friction limits of Section 15.7.6.1.5 apply. The curb angle detail serves to keep the base of the shell round while allowing it to move and flex under seismic load. Various base details are shown in Fig. 13 of Kaups and Lieb (1985).

C15.7.9.3.5 Combined Anchorage Systems. This section is intended to apply to combined anchorage systems that share loads based on their relative stiffnesses, and not to systems where sliding is resisted completely by one system (such as a steel annular plate) and overturning is resisted completely by another system (such as anchor bolts).

C15.7.10 Elevated Tanks and Vessels for Liquids and Granular Materials

C15.7.10.1 General. The three basic lateral load-resisting systems for elevated water tanks are defined by their support structure:

1. Multilegged braced steel tanks (trussed towers, as shown in Fig. C15.1-1);
2. Small-diameter, single-pedestal steel tank (cantilever column, as shown in Fig. C15.7-2); and
3. Large-diameter, single-pedestal tanks of steel or concrete construction (load-bearing shear walls, as shown in Fig. C15.7-3).

Unbraced multilegged tanks are uncommon. These types of tanks differ in their behavior, redundancy, and resistance to overload. Multilegged and small-diameter pedestal tanks have longer fundamental periods (typically greater than 2 s) than the shear wall type tanks (typically less than 2 s). The lateral load failure mechanisms usually are brace failure for multilegged tanks, compression buckling for small-diameter steel tanks, compression or shear buckling for large-diameter steel tanks, and shear failure for large-diameter concrete tanks. Connection, welding, and reinforcement details require careful attention to mobilize the full strength of these structures. To provide a greater margin of safety, R factors used with elevated tanks typically are less than those for other comparable lateral load-resisting systems.

FIGURE C15.7-2 Small-diameter, single-pedestal steel tank

Source: Courtesy of CB&I LLC; reproduced with permission.

(a) Steel   (b) Concrete

FIGURE C15.7-3 Large-diameter, single-pedestal tank

Source: Courtesy of CB&I LLC; reproduced with permission.

C15.7.10.4 Transfer of Lateral Forces into Support Tower. The vertical loads and shears transferred at the base of a tank or vessel supported by grillage or beams typically vary around the base because of the relative stiffness of the supports, settlements, and variations in construction. Such variations must be considered in the design for vertical and horizontal loads.

C15.7.10.5 Evaluation of Structures Sensitive to Buckling Failure. Nonbuilding structures that are designed with limited structural redundancy for lateral loads may be susceptible to total failure when loaded beyond the design loads. This phenomenon is particularly true for shell-type structures that exhibit unstable postbuckling behavior, such as tanks and vessels supported on shell skirts or pedestals. Evaluation for this critical condition ensures stability of the nonbuilding structure for governing design loads.

The design spectral response acceleration, S_a, used in this evaluation includes site factors. The I/R coefficient is taken as 1.0 for this critical check. The structural capacity of the shell is taken as the critical buckling strength (that is, the factor of safety is 1.0). Vertical and orthogonal combinations need not be considered for this evaluation because the probability of peak values occurring simultaneously is very low.

The intent of Section 15.7.10.5 and Table 15.4-2 is that skirt-supported vessels must be checked for seismic loads based on $I_e/R = 1.0$ if the structure falls in Risk Category IV or if an R value of 3.0 is used in the design of the vessel. For the purposes of this section, a skirt is a thin-walled steel cylinder or cone used to support the vessel in compression. Skirt-supported vessels fail in buckling, which is not a ductile failure mode. Therefore, a more conservative design approach is required. The $I_e/R = 1.0$ check typically governs the design of the skirt over using loads determined with an R factor of 3 in a moderate to high area of seismic activity. The only benefit of using an R factor of 3 in this case is in the design of the foundation. The foundation is not required to be designed for the $I_e/R = 1.0$ load. Section 15.7.10.5, item b, states that resistance of the structure shall be defined as the critical buckling resistance of the element for the $I_e/R = 1.0$ load. This stipulation means that the support skirt can be designed based on critical buckling (factor of safety of 1.0). The critical buckling strength of a skirt can be determined using a number of published sources. The two most common methods for determining the critical buckling strength of a skirt are the ASME BVPC (2007), Section VIII, Division 2, 2008 Addenda, Paragraph 4.4, using a factor of safety of 1.0 and AWWA D100-05 (2006a), Section 13.4.3.4. To use these methods, the radius, length, and thickness of the skirt; modulus of elasticity of the steel; and yield strength of the steel are required. These methods take into account both local buckling and slenderness effects of the skirt. Under no circumstance should the theoretical buckling strength of a cylinder, found in many engineering mechanics texts, be used to determine the critical buckling strength of the skirt. The theoretical value, based on a perfect cylinder, does not take into account imperfections built into real skirts. The theoretical buckling value is several times greater than the actual value measured in tests. The buckling values found in the suggested references above are based on actual tests.

Examples of applying the ASME BVPC (2007), Section VIII, Division 2, 2008 Addenda, Paragraph 4.4, and AWWA D100-05 (2006a), Section 13.4.3.4, buckling rules are shown in Figs. C15.7-4 and C15.7-5.

C15.7.10.7 Concrete Pedestal (Composite) Tanks. A composite elevated water storage tank is composed of a welded steel tank for watertight containment, a single-pedestal concrete support structure, a foundation, and accessories. The lateral load-resisting system is a load-bearing concrete shear wall. ACI 371R (1998), referenced in previous editions of ASCE 7, has been replaced with AWWA D107 (2010). Because AWWA D107-10 is based on the seismic design ground motions from ASCE 7-05, a requirement was added in Section 15.7.10.7 to require the use of the seismic design ground motions from Section 11.4.

C15.7.11 Boilers and Pressure Vessels. The support system for boilers and pressure vessels must be designed for the seismic forces and displacements presented in the standard. Such design must include consideration of the support, the attachment of the support to the vessel (even if "integral"), and the body of the vessel itself, which is subject to local stresses imposed by the support connection.

C15.7.12 Liquid and Gas Spheres. The commentary in Section C15.7.11 also applies to liquid and gas spheres.

C15.7.13 Refrigerated Gas Liquid Storage Tanks and Vessels. Even though some refrigerated storage tanks and

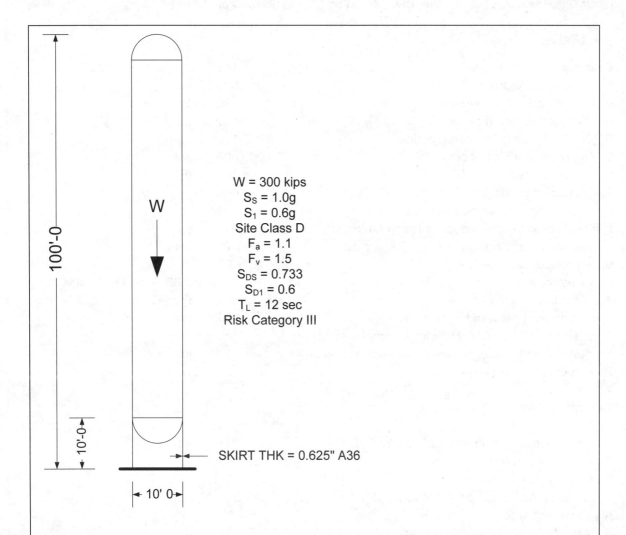

Vertical Vessel

Vessel Period

$$T = \frac{7.78}{10^6}\left(\frac{H}{D}\right)^2\sqrt{\frac{12wD}{t}} = \frac{7.78}{10^6}\left(\frac{100}{10}\right)^2\sqrt{\frac{12\left(\frac{300,000}{100}\right)10}{0.625}} = 0.591 \text{ seconds}$$

Determine C_s per ASCE/SEI 7-10 Section 12.8.1.1 with I/R = 1.0

$T_i \leq T_s \Rightarrow C_s = S_{DS} = 0.733$
Base Shear, V, = 0.733(300) = 219.9 kips

Per ASCE/SEI 7-10 Section 12.8.3 for T = 0.591 seconds, k = 1.045
Centroid for a distributed mass cantilever structure = [(k+1) / (k+2)]H = 67.16 ft
Overturning Moment = 219.9(67.16) = 14768 ft-kips

Determine Stresses at Base of Skirt

Axial Stress = P/A = 300000 / (π(10)12(0.625)) = 1274 psi
Bending Stress = M/S = 14768(1000)12 / [(0.625)π(10(12))²/4] = 25072 psi

FIGURE C15.7-4 Example Problem Using ASME BPVC (2007), Section VIII, Division 2, 2008 Addenda, Paragraph 4.4

continues

Input Data for ASME Section VIII, Div. 2 Buckling Checks (Paragraph 4.4)

Input Values

Course =	Skirt	
t = thickness of vessel section =	0.625	in.
H_T = top elevation of course =	120	in.
H_B = bottom elevation of course =	0	in.
D_o = outer diameter of vessel section =	120	in.
E_y = material modulus of elasticity =	29,000,000	psi
S_y = material yield strength =	36,000	psi
P_{ext} = external pressure =	0.000	psi
f_a = axial comp membrane stress from axial load =	1,274	psi
f_b = axial comp membrane stress from bending =	25,072	psi
V = net section shear force =	219,900	lbs
V_{phi} = applied shear force angle =	90	deg.
C_m = coefficient =	1	0.85, 1, or $0.6 - 0.4(M_1/M_2)$
K_u = effective length factor =	2.1	Free-Fixed
L_u = maximum unbraced length =	1,200	in.
L = design length vessel section for external pressure =	120	in.
L = design length vessel section for axial compression =	120	in.
FS = Input Factor of Safety =	1.0	

Calculated Values

R_o = outer radius of shell section =	60	in.
R = radius to center of shell =	59.6875	in.
R_m = vessel mean radius =	59.6875	in.
r = radius of gyration of cyl = $(1/4)(D_o^2 + D_i^2)^{0.5}$ =	42.2	in.
A = Cross sectional area of cylinder =	234.4	in^2
f_q = axial compressive membrane stress = $P_{ext}\pi D_i^2/4A$ =	0.0	psi

4.4.12 Combined Loadings and Allowable Compressive Stresses (continued)

b) Axial Compressive Stress Acting Alone - The allowable axial compressive membrane stress of a cylinder subject to an axial compressive load acting alone, F_{xa}, is computed by following equations.

1) For lambda$_c$ ≤ 0.15 (Local Buckling)

F_{xa} = min [F_{xa1}, F_{xa2}]		(4.4.61)
F_{xa1} = S_y / FS	for D_0/t ≤ 135	(4.4.62)
F_{xa1} = 466 S_y / [(331 + D_0/t) FS]	for 135 < D_0/t < 600	(4.4.63)
F_{xa1} = 0.5 S_y / FS	for 135 < D_0/t < 600	(4.4.64)
F_{xa2} = F_{xe} / FS		(4.4.65)

where: F_{xe} = C_x E_y t/D_0 (4.4.66)

C_x = min[409 c /[389 + D_0/t], 0.9]	for D_0/t < 1247	(4.4.67)
C_x = 0.25 c	for 1247 ≤ D_0/t ≤ 2000	(4.4.68)
c = 2.64	for M_x ≤ 1.5	(4.4.69)
c = 3.13 / $M_x^{0.42}$	for 1.5 < M_x < 15	(4.4.70)
c = 1.0	for M_x ≥ 15	(4.4.71)

FIGURE C15.7-4 (*Continued*) Example Problem Using ASME BPVC (2007), Section VIII, Division 2, 2008 Addenda, Paragraph 4.4

continues

4.4.12 Combined Loadings and Allowable Compressive Stresses (continued)

$$M_x = L/(R_o t)^{1/2} \quad (4.4.124)$$

where L is the design length of a vessel section between lines of support

$D_o/t = 192.00 \quad 135 < D_o/t < 600$
$M_x = L/(R_o t)^{1/2} = 19.60 > 15 \quad\quad c = 1.0000$

$D_o/t < 1247 \quad\quad C_x = \min[409 c / [389 + D_o/t], 0.9] = 0.7039587 \quad (4.4.67)$

$$F_{xe} = C_x E_y t / D_o = 106{,}327 \text{ psi} \quad (4.4.66)$$

Calculate F_{xa1} | Calculate F_{xa2}
$F_{ic} = 466 * S_y / (331 + D_o/t) = 32{,}076 \text{ psi}$ | $F_{ic} = 106{,}327 \text{ psi}$
Use Input FS = 1.0 | Use Input FS = 1.0
$F_{xa1} = 466 * S_y / [(331 + D_o/t) FS] = 32{,}076 \text{ psi}$ | $F_{xa2} = F_{xe}/FS = 106{,}327 \text{ psi} \quad (4.4.65)$

Calculate F_{xa}

$$F_{xa} = \min[F_{xa1}, F_{xa2}] = 32{,}076 \text{ psi} \quad (4.4.61)$$

$\text{lambda}_c = (K)(L_u) / [(\pi)(r_g)][(F_{xa})(FS)/E]^{0.5} = 0.6321 \quad 0.15 < \text{lambda}_c < 1.147$

2) For $\text{lambda}_c > 0.15$ and $K_u L_u / r_g < 200$ (Column Buckling)

$F_{ca} = F_{xa}(1 - 0.74(\text{lambda}_c - 0.15))^{0.3} \quad 0.15 < \text{lambda}_c < 1.147 \quad (4.4.72)$
$F_{ca} = 0.88 F_{xa} / (\text{lambda}_c)^2 \quad\quad \text{lambda}_c \geq 1.147 \quad (4.4.73)$

$KL_u/r_g = 59.7 < 200$
Lambda$_c > 0.15$ and $KL_u/r_g < 200$ therefore:

$$F_{ca} = F_{xa}(1 - 0.74(\text{lambda}_c - 0.15))^{0.3} = 28{,}100 \text{ psi}$$

c) Compressive Bending Stress - The allowable axial compressive membrane stress of a cylindrical shell subject to a bending moment acting across the full circular cross section F_{ba}, is computed using the following equations.

$F_{ba} = F_{xa}$ | for $135 \leq D_o/t \leq 2000$ | (4.4.74)
$F_{ba} = 466 S_y / [(331 + D_o/t) FS]$ | for $100 < D_o/t < 135$ | (4.4.75)
$F_{ba} = 1.081 S_y / FS$ | for $D_o/t < 100$ and $y > 0.11$ | (4.4.76)
$F_{ba} = (1.4 - 2.9 y) S_y / FS$ | for $D_o/t < 100$ and $y < 0.11$ | (4.4.77)
where: $y = S_y D_o / E_y t$ | | (4.4.78)

$D_o/t = 192$
$y = S_y D_o / E_y t = 0.2383 \quad\quad F_{ic} = F_{xa}$
$\quad\quad\quad\quad\quad\quad\quad\quad\quad\quad\quad\quad F_{ic} = 32{,}076 \text{ psi}$

$D_o/t = 192 > 135$ (see Sect. 3.1.1) \quad Use Input FS = 1.0

$$F_{ba} = F_{xa} = 32{,}076 \text{ psi}$$

FIGURE C15.7-4 (*Continued*) Example Problem Using ASME BPVC (2007), Section VIII, Division 2, 2008 Addenda, Paragraph 4.4

continues

d) Shear Stress - The allowable shear stress of a cylindrical shell, F_{va}, is computed using the following equations.

$$F_{va} = n_v F_{ve} / FS \qquad (4.4.79)$$
$$\text{where: } F_{ve} = a_v C_v E t / D_o \qquad (4.4.80)$$

$C_v = 4.454$	for $M_x < 1.5$	(4.4.81)
$C_v = (9.64 / M_x^2)(1 + 0.0239 M_x^3)^{1/2}$	for $1.5 < M_x < 26$	(4.4.82)
$C_v = 1.492 / (M_x)^{1/2}$	for $26 < M_x < 4.347 D_o / t$	(4.4.83)
$C_v = 0.716 (t / D_o)^{1/2}$	for $M_x > 4.347 D_o / t$	(4.4.84)
$a_v = 0.8$	for $D_o / t < 500$	(4.4.85)
$a_v = 1.389 - 0.218 \log_{10}(D_o / t)$	for $D_o / t > 500$	(4.4.86)
$n_v = 1.0$	for $F_{ve} / S_y < 0.48$	(4.4.87)
$n_v = 0.43 S_y / F_{ve} + 0.1$	for $0.48 < F_{ve} / S_y < 1.7$	(4.4.88)
$n_v = 0.6 S_y / F_{ve}$	for $F_{ve} / S_y > 1.7$	(4.4.89)

$D_o / t = 192$
$M_x = L / (R_o t)^{0.5} = 19.596$
$\qquad 1.5 < M_x < 26 \qquad C_v = (9.64/M_x^2)(1+0.0239 M_x^3)^{0.5} = 0.3376$
$D_o / t < 500 \qquad a_v = \quad 0.8000$
$\qquad F_{ve} = a_v C_v E t / D_o = 40{,}793 \text{ psi}$
$\qquad F_{ve} / S_y = 1.13 \quad 0.48 < F_{ve} / S_y < 1.7$
$n_v = 0.43 S_y / F_{ve} + 0.1 = 0.4795$
$F_{ic} = 19{,}559 \text{ psi}$

$$\text{Use Input FS} = 1.000$$

$$F_{va} = n_v F_{ve} / FS = 19{,}559 \text{ psi}$$

4.4.12 Combined Loadings and Allowable Compressive Stresses (continued)

Axial Compressive Stress, Compressive Bending Stress, and Shear - the allowable compressive stress for the combination of uniform axial compression, axial compression due to bending, and shear in the absence of hoop compression.

Let $K_s = 1 - (f_v / F_{va})^2$ \qquad (4.4.105)

For $0.15 < \text{lambda}_c < 1.2$
$\qquad \text{lambda}_c = \quad 0.63 \quad \text{(Section 3.2)} \qquad 0.15 < \text{lambda}_c < 1.2 \quad \text{OK}$

$f_a /(K_s F_{ca}) + (8/9)(\text{delta}) f_b /(K_s F_{ba}) < 1.0 \qquad f_a /(K_s F_{ca}) > 0.2$	(4.4.112)
$f_a /(2 K_s F_{ca}) + (\text{delta}) f_b /(K_s F_{ba}) < 1.0 \qquad f_a /(K_s F_{ca}) < 0.2$	(4.4.113)

$$K_s = 1 - (f_v / F_{va})^2 = \quad 0.9977$$
$$F_e = (\pi)^2 E / [K_s L_u / r]^2 = 80{,}287 \text{ psi}$$
$$\text{delta} = C_m / [1 - f_a FS / F_e] = 1.0161$$

$$f_a /(K_s F_{ca}) = 0.045442959 < 0.2$$

$$f_a /(2 K_s F_{ca}) + (\text{delta}) f_b /(K_s F_{ba}) = 0.82 \; < 1.0 \qquad \text{OK!}$$

FIGURE C15.7-4 (*Continued*) Example Problem Using ASME BPVC (2007), Section VIII, Division 2, 2008 Addenda, Paragraph 4.4

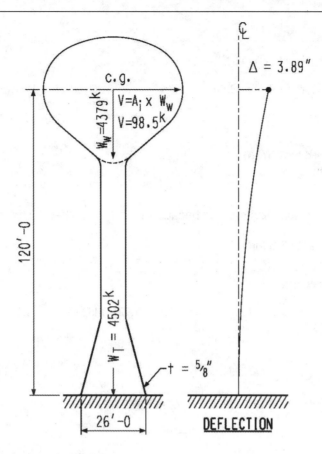

Single Pedestal Elevated Water Tank

Seismic Information

$S_s = 0.162$, $S_1 = 0.077$

Site Class C, $F_a = 1.2$, $F_v = 1.7$

$S_{DS} = 0.130$, $S_{D1} = 0.087$, $T_L = 12$ s

Risk Category IV

$T_s = 0.674$ s

Tank Information

Structure Period $T_i = 3.88$ s

Class 2 Material: A36 ($F_y = 36$ kip/in.^{2})

Skirt angle (from vertical) = 15 deg

Weight of tank and water, $W_w = 4{,}379$ kip

Weight of tank, tower, and water, $W_T = 4{,}502$ kip

$KL/r = 50$

Determine S_{ai} per AWWA D100-05, Section 13.2.7.2

$T_s < T_i \leq T_L \Rightarrow S_{ai} = S_{D1}/T_i = 0.087/3.88 = 0.0225$

FIGURE C15.7-5 Example Problem Using AWWA D100-05, Section 13.4.3.4

continues

Determine Critical Buckling Acceleration ($I_i/1.4\, R_i = 1$)

Per Section 13.4.3.4, $A_i = S_{ai}$ for critical buckling check (A_i in AWWA D100-05 is the same as C_s in ASCE/SEI 7-10)

$A_i = 0.0225$

Lateral Displacement Caused by S_{ai} (P-Δ)

The final deflected position of the water centroid is an iterative process and must account for the additional moment applied to the structure because of the P-Δ effect. The deflection from the critical buckling deflection is equal to 3.89 in.

Check Skirt at Base of Tower

Seismic overturning moment at base of tower without P-Δ = 11,928 ft-kip (includes mass of tower).

Seismic overturning moment at base of tower with P-Δ = 11,928 ft-kip + 4,379 kip × 3.89 in./12 in. per ft

= 13,348 ft-kip

Area of skirt = $\pi(26 \times 12)(0.625) = 612.6$ in.^{2}

Section modulus of skirt = $\pi(26 \times 12)^2/4 \times 0.625 = 47,784$ in.^{3}

Skirt stress caused by axial load = $4,502(1,000)/(612.6 \times \cos 15) = 7,608$ lb/in.^{2}

Skirt stress caused by moment = $13,348(12)(1000)/(47,784 \times \cos 15) = 3,470$ lb/in.^{2}

Determine Critical Buckling Stress

$R = 13 \times 12/\cos 15 = 161.5$ in.

$t/R = 0.625/161.5 = 0.0039$

For Class 2 material, $KL/r = 50$, and $t/R = 0.0039$, determine allowable axial compressive stress, F_a, from Table 13 of AWWA D100-05.

$F_a = 9,882$ lb/in.^{2}

Per AWWA D100-05, Section 13.4.3.4,

Critical buckling stress = $2F_a = 19,764$ lb/in.^{2}

For Class 2 material and $t/R = 0.0039$, determine allowable bending compressive stress, F_b, from Table 11 of AWWA D100-05.

$F_b = F_L = 10,380$ lb/in.^{2}

Per AWWA D100-05, Section 13.4.3.4,

Critical bending stress = $2F_b = 20.760$ lb/in.^{2}

Check unity per AWWA D100-05, Section 3.3.1:

$7,608/19,764 + 3,470/20,760 = 0.552 \leq 1.0$ OK

FIGURE C15.7-5 (*Continued*) Example Problem Using AWWA D100-05, Section 13.4.3.4

vessels, such as those storing liquefied natural gas, are required to be designed for ground motions and performance goals in excess of those found in the standard, all such structures must also meet the requirements of this standard as a minimum. All welded steel refrigerated storage tanks and vessels must be designed in accordance with the requirements of the standard and the requirements of API 620.

C15.7.14 Horizontal, Saddle-Supported Vessels for Liquid or Vapor Storage. Past practice has been to assume that a horizontal, saddle-supported vessel (including its contents) behaves as a rigid structure (with natural period, T, less than 0.06 s). For this situation, seismic forces would be determined using the requirements of Section 15.4.2. For large horizontal, saddle-supported vessels (length-to-diameter ratio of 6 or more), this assumption can be unconservative, so Section 15.7.14.3 requires that the natural period be determined assuming the vessel to be a simply supported beam.

C15.8 CONSENSUS STANDARDS AND OTHER REFERENCED DOCUMENTS

Chapter 15 of this standard makes extensive use of reference documents in the design of nonbuilding structures for seismic forces; see Chapter 23. The documents referenced in Chapter 15 are industry documents commonly used to design specific types of nonbuilding structures. The vast majority of these reference documents contain seismic provisions that are based on the seismic ground motions of the 1997 Uniform Building Code (ICBO 1997) or earlier editions of the UBC. To use these reference documents, Chapter 15 modifies the seismic force provisions of these reference documents through the use of "bridging equations." The standard only modifies industry documents that specify seismic demand and capacity. The bridging equations are intended to be used directly with the other provisions of the specific reference documents. Unlike the other provisions of the standard, if the reference documents are written in terms of allowable stress design, then the bridging equations are shown in allowable stress design format. In addition, the detailing requirements referenced in Tables 15.4-1 and Table 15.4-2 are expected to be followed, as well as the general requirements found in Section 15.4.1. The usage of reference documents in conjunction with the requirements of Section 15.4.1 are summarized in Table C15.8-1.

Currently, only four reference documents have been revised to meet the seismic requirements of the standard. AWWA D100, API 620, API 650, and ANSI/RMI MH 16.1 have been adopted by reference in the standard without modification, except that height limits are imposed on "elevated tanks on symmetrically braced legs (not similar to buildings)" in AWWA D100, and the anchorage requirements of Section 15.4.9 are imposed on steel storage racks in ANSI/RMI MH 16.1. Three of these reference documents apply to welded steel liquid storage tanks.

REFERENCES

American Concrete Institute (ACI). (1997). "Standard practice for the design and construction of concrete silos and stacking tubes for storing granular materials," *ACI 313*. Farmington Hills, MI.
ACI. (1998). "Guide to the analysis, design, and construction of concrete-pedestal water towers," *ACI 371R*. Farmington Hills, MI.
ACI. (2006). "Seismic design of liquid-containing concrete structures," *ACI 350.3-06*. Farmington Hills, MI.
ACI. (2014), "Building code requirements for structural concrete and commentary," *ACI 318*. Farmington Hills, MI.
American Institute of Steel Construction (AISC). (1986). *Load and resistance factor design specification for structural steel buildings*, Chicago.
AISC. (2016). "Seismic provisions for structural steel buildings," *AISC 341*, Chicago.
Anderson, P. F. (1966). "Temperature stresses in steel grain storage tanks." *Civil Eng.* 36(1), 74.
ASCE. (1984). "Guidelines for the seismic design of oil and gas pipeline systems." *ASCE*, Reston, VA.
ASCE/American Wind Energy Association (ASCE/AWEA). (2011). "Recommended practice for large land-based wind turbine support structures." *RP2011*, ASCE, Reston, VA.
ASCE/SEI. (1995). Minimum design loads for buildings and other structures, 7-05, including Supplement 2, ASCE, Reston, VA.
ASCE/SEI. (2010). Minimum design loads for buildings and other structures, 7-10, including Supplement 2, ASCE, Reston, VA.
American Society of Mechanical Engineers (ASME). (2007). Boiler and pressure vessel code (BPVC), New York.
ASME. (2011). Steel stacks, *ASME STS-1*, New York.
ASTM International. (ASTM). (2015). "Standard specification for anchor bolts, steel, 36, 55, and 105-ksi yield strength," *ASTM F1554*, West Conshohocken, PA.
American Water Works Association (AWWA). (2005). "Welded steel tanks for water storage," AWWA D100, Denver.
AWWA. (2006a). "Welded steel tanks for water storage," *AWWA D100-05*, American Water Works Association, Denver.
AWWA. (2006b). "Tendon-prestressed concrete water tanks," *American Water Works Association D115*, Denver.
AWWA. (2011). "Welded steel tanks for water storage," *American Water Works Association D100*, Denver.
AWWA. (2009). "Factory-coated bolted steel tanks for water storage," *American Water Works Association D103*, Denver.
AWWA. (2010). "Composite elevated tanks for water storage," *American Water Works Association D107*, Denver.
AWWA. (2013). "Wire- and strandwound circular prestressed concrete water tanks," *AWWA D110*, Denver.
Chandrasekaran, A. R., and Jain, P. C. (1968). "Effective live load of storage materials under dynamic conditions." *Ind. Concrete J.* 42(9), 364–365.
Chandrasekaran, A. R., Saini, S. S., and Jhamb, I. C. (1968). "Live load effects on dynamic behavior of structures." *J. Inst. Eng.*, 48, 850–859.
Federal Emergency Management Agency (FEMA). (2005). "Seismic considerations for steel storage racks located in areas accessible to the public,

Table C15.8-1 Usage of Reference Documents in Conjunction with Section 15.4.1

Subject	Requirement
R, Ω_0, and C_d values, detailing requirements, and height limits	Use values and limits in Tables 12.2-1, 15.4-1, or 15.4-2 as appropriate. Values from the reference document are not to be used.
Minimum base shear	Use the appropriate value from Eq. (15.4-1) or (15.4-2) for nonbuilding structures not similar to buildings. For structures containing liquids, gases, and granular solids supported at the base, the minimum seismic force cannot be less than that required by the reference document.
Importance Factor	Use the value from Section 15.4.1.1 based on Risk Category. Importance Factors from the reference document are not to be used unless they are greater than those provided in the standard.
Vertical distribution of lateral load	Use requirements of Section 12.8.3 or Section 12.9 or the applicable reference document.
Seismic provisions of reference documents	The seismic force provisions of reference documents may be used only if they have the same basis as Section 11.4 and the resulting values for total lateral force and total overturning moment are no less than 80% of the values obtained from the standard.
Load combinations	Load combinations specified in Section 2.3 (LRFD) or Section 15 (includes ASD load combinations of Section 2.4) must be used.

"*FEMA 460*. Building Seismic Safety Council, National Institute of Building Sciences, Washington, DC.

FEMA. (2009). "Quantification of Building Seismic Performance Factors," FEMA P695, Applied Technology Council, Redwood City, CA.

FEMA. (2012). Reducing the risks of nonstructural earthquake damage—A practical guide. FEMA E-74, Applied Technology Council, Redwood City, CA.

Ferritto, J., Dickenson, S., Priestley, N., Werner, S., Taylor, C., and Burke, D., et al. (1999). *Seismic criteria for California marine oil terminals*, Vols. 1 and 2, Technical Report TR-2103-SHR, Naval Facilities Engineering Service Center, Port Hueneme, CA.

International Conference of Building Officials (ICBO). (1997). *Uniform building code*, Whittier, CA.

Kaups, T., and Lieb, J. M. (1985). *A practical guide for the design of quality bulk storage bins and silos*, Chicago Bridge & Iron, Plainfield, IL.

Kilic, S., and Sozen, M. (2003). "Evaluation of effect of August 17, 1999, Marmara earthquake on two tall reinforced concrete chimneys." *ACI Struct. J.*, 100(3), 357–364.

Malhotra, P. K. (2005). "Sloshing loads in liquid-storage tanks with insufficient freeboard." *Earthq. Spectra*, 21(4), 1185–1192.

Malhotra, P. K. (2006). "Earthquake induced sloshing in cone and dome roof tanks with insufficient freeboard." *J. Struct. Eng. Intl., IABSE*, 16(3), 222–225.

Malhotra, P. K., Wenk, T., and Wieland, M. (2000). "Simple procedure for seismic analysis of liquid-storage tanks." *J. Struct. Eng. Intl, IABSE*, 10(3), 197–201.

Marine oil terminal engineering and maintenance standards. (2005). Title 24, Part 2, California Building Code, Chapter 31F.

Miller, C. D., Meier, S. W., and Czaska, W. J. (1997). "Effects of internal pressure on axial compressive strength of cylinders and cones." Paper presented at the Structural Stability Research Council Annual Technical Meeting, June.

Priestley, M. J. N., Siebel, F., and Calvi, G. M. (1996). *Seismic design and retrofit of bridges*, New York.

Rack Manufacturers Institute (RMI). (2012). "Specification for the design, testing, and utilization of industrial steel storage racks," *ANSI/RMI MH 16.1*, Charlotte, NC.

RMI. (2016). "Specification for the design, testing, and utilization of industrial steel cantilevered storage racks," *ANSI/RMI MH 16.3*, Charlotte, NC.

Seismic design guidelines for port structures. (2001). Working Group No. 34 of the Maritime Navigation Commission

Trahair, M. S., Abel, A., Ansourian, P., Irvine, H. M., and Rotter, J. M. (1983). *Structural design of steel bins for bulk solids*. Australian Institute of Steel Construction, Sydney.

Troitsky, M. S. (1990). *Tubular steel structures—Theory and design*. The James F. Lincoln Arc Welding Foundation, Mentor, OH.

U.S. Department of Energy. (1963). "Nuclear reactors and earthquakes." *TID-7024*, U.S. Atomic Energy Commission, Washington, DC.

Veletsos, A. S. (1974). "Seismic effects in flexible liquid-storage tanks." In *Proc. 5th World Conference on Earthquake Engineering*, Rome, 630–639.

Werner, S. D., ed. (1998). *Seismic guidelines for ports, Monograph No. 12*, ASCE, Reston, VA.

Wozniak, R. S., and Mitchell, W. W. (1978). "Basis of seismic design provisions for welded steel oil storage tanks." Presented at the Session on Advances in Storage Tank Design, American Petroleum Institute, Refining, 43rd Midyear Meeting, Toronto, May 9.

OTHER REFERENCES (NOT CITED)

American Concrete Institute (ACI). (2008), "Code requirements for reinforced concrete chimneys and commentary," *ACI 307*. Farmington Hills, MI.

American Petroleum Institute (API). (2014a). "Design and construction of large, welded, low pressure storage tanks, 12th Ed. Addendum 1," *API 620*, Washington, DC.

API. (2014b). "Welded steel tanks for oil storage, 12th Ed., Addendum 1," *API 650*, Washington, DC.

API. (2014c). "Tank, inspection, repair, alteration, and reconstruction, 5th Ed.," *API 653*, Washington, DC.

API. (2014d). "Specification for bolted tanks for storage of production liquids, 16th Ed.," *API 12B*, Washington, DC.

ASCE. (1997). *Design of secondary containment in petrochemical facilities*, Reston, VA.

ASCE. (2011). *Guidelines for seismic evaluation and design of petrochemical facilities*, 2nd Ed., Reston, VA.

Drake, R. M., and Walter, R. J. (2010). "Design of Structural Steel Pipe Racks," *AISC Eng. J.*, 4th Quarter, 241–251.

International Conference of Building Officials (ICBO). (1988). *Uniform building code*, Whittier, CA.

National Fire Protection Association (NFPA). (2013). "Standard for the production, storage, and handling of liquefied natural gas (LNG)," *NFPA 59A*, Quincy, MA.

Soules, J. G., "The seismic provisions of the 2006 IBC – Nonbuilding structure criteria," *Proc. 8th National Conference on Earthquake Engineering*, San Francisco, April 18, 2006.

CHAPTER C16
NONLINEAR RESPONSE HISTORY ANALYSIS

C16.1 GENERAL REQUIREMENTS

C16.1.1 Scope. Response history analysis is a form of dynamic analysis in which response of the structure to a suite of ground motions is evaluated through numerical integration of the equations of motions. In nonlinear response history analysis, the structure's stiffness matrix is modified throughout the analysis to account for the changes in element stiffness associated with hysteretic behavior and P-delta effects. When nonlinear response history analysis is performed, the R, C_d, and Ω_0 coefficients considered in linear procedures are not applied because the nonlinear analysis directly accounts for the effects represented by these coefficients.

Nonlinear response history analysis is permitted to be performed as part of the design of any structure and is specifically required to be performed for the design of certain structures incorporating seismic isolation or energy dissipation systems. Nonlinear response history analysis is also frequently used for the design of structures that use alternative structural systems or do not fully comply with the prescriptive requirements of the standard in one or more ways. Before this edition, ASCE 7 specified that nonlinear response history analyses be performed using ground motions scaled to the design earthquake level and that design acceptance checks be performed to ensure that mean element actions do not exceed two-thirds of the deformations at which loss of gravity-load-carrying capacity would occur. In this edition of ASCE 7, a complete reformulation of these requirements was undertaken to require analysis at the Risk-Targeted Maximum Considered Earthquake (MCE_R) level and also to be more consistent with the target reliabilities indicated in Section 1.3.1.3.

The target collapse reliabilities given in Table 1.3-2 are defined such that, when a building is subjected to MCE_R ground motion, not greater than a 10% probability of collapse exists for Risk Category I and II structures. For Risk Category III and IV structures, these maximum collapse probabilities are reduced to 5% and 2.5%, respectively.

There are additional performance expectations for Risk Category III and IV structures that go beyond the collapse safety performance goals (e.g., limited damage and postearthquake functionality for lower ground motion levels). These enhanced performance goals are addressed in this chapter by enforcing an $I_e > 1.0$ in the linear design step (which is consistent with the approach taken in the other design methods of Chapter 12) and also by considering I_e in acceptance checks specified in Section 16.4.

It is conceptually desirable to create a Chapter 16 response history analysis (RHA) design process that explicitly evaluates the collapse probability and ensures that the performance goal is fulfilled. However, explicit evaluation of collapse safety is a difficult task requiring (a) a structural model that is able to directly simulate the collapse behavior, (b) use of numerous nonlinear response history analyses, and (c) proper treatment of many types of uncertainties. This process is excessively complex and lengthy for practical use in design. Therefore, Chapter 16 maintains the simpler approach of *implicitly* demonstrating adequate performance through a prescribed set of analysis rules and acceptance criteria. Even so, this implicit approach does not preclude the use of more advanced procedures that explicitly demonstrate that a design fulfills the collapse safety goals. Such more advanced procedures are permitted by Section 1.3.1.3 of this standard. An example of an advanced explicit procedure is the building-specific collapse assessment methodology in Appendix F of FEMA P-695 (FEMA 2009b).

C16.1.2. Linear Analysis. As a precondition to performing nonlinear response history analysis, a linear analysis in accordance with the requirements of Chapter 12 is required. Any of the linear procedures allowed in Chapter 12 may be used. The purpose of this requirement is to ensure that structures designed using nonlinear response history analyses meet the minimum strength and other criteria of Chapter 12, with a few exceptions. In particular, when performing the Chapter 12 evaluations it is permitted to take the value of Ω_0 as 1.0 because it is felt that values of demand obtained from the nonlinear procedure is a more accurate representation of the maximum forces that will be delivered to critical elements, considering structural overstrength, than does the application of the judgmentally derived factors specified in Chapter 12. Similarly, it is permitted to use a value of 1.0 for the redundancy factor, ρ, because it is felt that the inherent nonlinear evaluation of response to MCE_R shaking required by this chapter provides improved reliability relative to the linear procedures of Chapter 12. For Risk Category I, II, and III structures, it is permitted to neglect the evaluation of story drift when using the linear procedure because it is felt that the drift evaluation performed using the nonlinear procedure provides a more accurate assessment of the structure's tolerance to earthquake-induced drift. However, linear drift evaluation is required for Risk Category IV structures because it is felt that this level of drift control is important to attaining the enhanced performance desired for such structures.

As with other simplifications permitted in the linear analysis required under this section, it is also permitted to use a value of 1.0 for the torsional amplification, A_x, when performing a nonlinear analysis if accidental torsion is explicitly modeled in the nonlinear analysis. Although this does simplify the linear analysis somewhat, designers should be aware that the resulting structure may be more susceptible to torsional instability when performing the nonlinear analysis. Therefore, some designers may find it expedient to use a value of A_x consistent with the linear procedures as a means of providing a higher likelihood that the nonlinear analysis will result in acceptable outcomes.

C16.1.3 Vertical Response Analysis. Most structures are not sensitive to the effects of response to vertical ground shaking, and there is little evidence of the failure of structures in earthquakes resulting from vertical response. However, some nonbuilding structures and building structures with long spans, cantilevers, prestressed construction, or vertical discontinuities in their gravity-load-resisting systems can experience significant vertical earthquake response that can cause failures. The linear procedures of Chapter 12 account for these effects in an approximate manner through use of the $0.2S_{DS}D$ term in the load combinations. When nonlinear response history analysis is performed for structures with sensitivity to vertical response, direct simulation of this response is more appropriate than use of the approximate linear procedures. However, in order to properly capture vertical response to earthquake shaking, it is necessary to accurately model the stiffness and distribution of mass in the vertical load system, including the flexibility of columns and horizontal framing. This effort can considerably increase the complexity of analytical models. Rather than requiring this extra effort in all cases where vertical response can be significant, this chapter continues to rely on the approximate approach embedded in Chapter 12 for most cases. However, where the vertical load path is discontinuous and where vertical response analysis is required by Chapter 15, Chapter 16 does require explicit modeling and analysis of vertical response. Since in many cases the elements sensitive to vertical earthquake response are not part of the seismic force-resisting system, it is often possible to decouple the vertical and lateral response analyses, using separate models for each.

Appropriate accounting for the effects of vertical response to ground shaking requires that horizontal framing systems, including floor and roof systems, be modeled with distributed masses and sufficient vertical degrees of freedom to capture their out-of-plane dynamic characteristics. This increased fidelity in modeling of the structure's vertical response charactersitics will significantly increase the size and complexity of models. As a result, the chapter requires direct simulation of vertical response only for certain structures sensitive to those effects and relies on the procedures of Chapter 12 to safeguard the vertical response of other structures.

C16.1.4 Documentation. By its nature, most calculations performed using nonlinear response history analysis are contained within the input and output of computer software used to perform the analysis. This section requires documentation, beyond the computer input and output, of the basic assumptions, approaches, and conclusions so that thoughtful review may be performed by others including peer reviewers and the authority having jurisdiction. This section requires submittal and review of some of these data before the analyses are performed in order to ensure that the engineer performing the analysis/design and the reviewers are in agreement before substantive work is performed.

C16.2 GROUND MOTIONS

C16.2.1 Target Response Spectrum. The target response spectrum used for nonlinear dynamic analysis is the maximum direction MCE_R spectrum determined in accordance with Chapter 11 or Chapter 21. Typical spectra determined in accordance with those procedures are derived from uniform hazard spectra (UHSs) and modified to provide a uniform risk spectrum (URS), or alternatively, a deterministic MCE spectrum. UHSs have been used as the target spectra in design practice since the 1980s. The UHS is created for a given hazard level by enveloping the results of seismic hazard analysis for each period (for a given probability of exceedance). Accordingly, it is generally a conservative target spectrum if used for ground motion selection and scaling, especially for large and rare ground motions, unless the structure exhibits only elastic first-mode response. This inherent conservatism comes from the fact that the spectral values at each period are not likely to all occur in a single ground motion. This limitation of the UHS has been noted for many years (e.g., Bommer et al. 2000; Naeim and Lew 1995; Reiter 1990). The same conservatism exists for the URS and deterministic MCE spectra that serves as the basis for Method 1.

Method 2 uses the conditional mean spectrum (CMS), an alternative to the URS that can be used as a target for ground motion selection in nonlinear response history analysis (e.g., Baker and Cornell 2006; Baker 2011; Al Atik and Abrahamson 2010).

To address the conservatism inherent in analyses using URSs as a target for ground motion selection and scaling, the CMS instead conditions the spectrum calculation on a spectral acceleration at a single period and then computes the mean (or distribution of) spectral acceleration values at other periods. This conditional calculation ensures that the resulting spectrum is reasonably likely to occur and that ground motions selected to match the spectrum have an appropriate spectral shape consistent with naturally occurring ground motions at the site of interest. The calculation is no more difficult than the calculation of a URS and is arguably more appropriate for use as a ground motion selection target in risk assessment applications. The spectrum calculation requires disaggregation information, making it a site-specific calculation that cannot be generalized to other sites. It is also period-specific, in that the conditional response spectrum is conditioned on a spectral acceleration value at a specified period. The shape of the conditional spectrum also changes as the spectral amplitude changes (even when the site and period are fixed). Fig. C16.2-1 provides examples of CMSs for an example site in Palo Alto, California, anchored at four different candidate periods. The UHS for this example site is also provided for comparison.

As previously discussed, the URS is a conservative target spectrum for ground motion selection, and the use of CMS target spectra is more appropriate for representing anticipated MCE_R ground motions at a specified period. A basic CMS-type approach was used in the analytical procedures of the FEMA P-695 (FEMA 2009b) project, the results of which provided the initial basis for establishing the 10% probability of collapse goal shown in Table 1.3-2. Therefore, the use of CMS target spectra in the

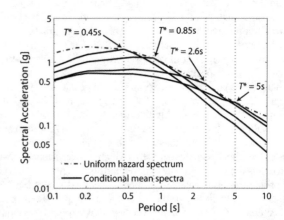

FIGURE C16.2-1 Example Conditional Mean Spectra for a Palo Alto Site Anchored for 2% in 50-Year Motion at $T = 0.45s$, 0.85s, 2.6s, and 5s

Source: NIST 2011

Chapter 16 RHA design procedure is also internally consistent with how the collapse probability goals of Table 1.3-2 were developed.

The URS (or deterministic MCE) target spectrum is retained in Section 16.2.1.1 (as a simpler and more conservative option) as the specified target spectrum, and the CMS is permitted as an alternate in Section 16.2.1.2. Whereas CMS appropriately captures the earthquake energy and structural response at a particular period resulting from a particular scenario earthquake, it is not capable of capturing the MCE_R level response associated with other scenarios that are component to the MCE_R spectrum. Therefore, when using CMS, it may be necessary to use several conditioning periods and associated targets to develop conditional mean spectra in order to fully capture the structure's response to different earthquake scenarios. The recommended procedure includes the following steps for creating the site-specific scenario response spectra.

1. Select those periods that correspond to periods of vibration that significantly contribute to the building's inelastic dynamic response. This selection includes a period near the fundamental period of the building, or perhaps a slightly extended period to account for inelastic period lengthening (e.g., $1.5T_1$). In buildings where the fundamental response periods in each of two orthogonal axes is significantly different, a conditioning period associated with each direction is needed. It also likely requires periods near the translational second-mode periods. When selecting these significant periods of response, the elastic periods of response should be considered (according to the level of mass participation for each of these periods), and the amount of first-mode period elongation caused by inelastic response effects should also be considered.
2. For each period selected above, create a scenario spectrum that matches or exceeds the MCE_R value at that period. When developing the scenario spectrum, (a) perform site-specific disaggregation to identify earthquake events likely to result in MCE_R ground shaking, and then (b) develop the scenario spectrum to capture one or more spectral shapes for dominant magnitude and distance combinations revealed by the disaggregation.
3. Enforce that the envelope of the scenario spectra not be less than 75% of the MCE_R spectrum (from Method I) for any period within the period range of interest (as defined in Section 16.2.3.1).

After the target spectra are created, each target response spectrum is then used in the remainder of the response history analyses process and the building must be shown to meet the acceptance criteria for each of the scenarios.

The primary purpose of the 75% floor value is to provide a basis for determining how many target spectra are needed for analysis. For small period ranges, fewer targets are needed, and more target spectra are needed for buildings where a wider range of periods are important to the structural response (e.g., taller buildings). When creating the target spectra, some spectral values can also be artificially increased to meet the requirements of this 75% floor. A secondary reason for the 75% floor is to enforce a reasonable lower bound. The specific 75% threshold value was determined using several examples; the intention is that this 75% floor requirement will be fulfilled through the use of two target spectra in most cases. From the perspective of collapse risk, the requirement of being within 75% of the MCE_R at all periods may introduce some conservatism, but the requirement adds robustness to the procedure by ensuring that the structure is subjected to ground motions with near-MCE_R-level intensities at all potentially relevant periods. Additionally, this requirement ensures that demands unrelated to collapse safety, such as higher mode-sensitive force demands, can be reasonably determined from the procedure.

C16.2.2 Ground Motion Selection. Before this edition of ASCE 7, Chapter 16 required a minimum of three ground motions for nonlinear response history analysis. If three ground motions were used, the procedures required evaluation of structural adequacy using the maximum results obtained from any of the ground motions. If seven or more motions were used, mean results could be used for evaluation. Neither three nor seven motions are sufficient to accurately characterize either mean response or the record-to-record variability in response. In the 2016 edition of the standard, the minimum number of motions was increased to 11. The requirement for this larger number of motions was not based on detailed statistical analyses, but rather was judgmentally selected to balance the competing objectives of more reliable estimates of mean structural responses (through use of more motions) against computational effort (reduced by using fewer motions). An advantage of using this larger number of motions is that if unacceptable response is found for more than one of the 11 motions, this does indicate a significant probability that the structure will fail to meet the 10% target collapse reliability for Risk Category I and II structures of Section 1.3.1.3. This advantage is considered in the development of acceptance criteria discussed in Section C16.4.

All real ground motions include three orthogonal components. For most structures, it is only necessary to consider response to horizontal components of ground shaking. However, consideration of vertical components is necessary for structures defined as sensitive to vertical earthquake effects.

Section 11.4.1 defines near-fault sites as sites located within 9.3 mi (15 km) of the surface projection of faults capable of producing earthquakes of magnitude 7.0 or greater and within 6.2 mi (10 km) of the surface projection of faults capable of producing earthquakes of magnitude 6.0 or greater, where the faults must meet minimum annual slip rate criteria. Such near-fault sites have a reasonable probability of experiencing ground motions strongly influenced by rupture directivity effects. These effects can include pulse-type ground motions (e.g., Shahi et al. 2011) observable in velocity histories and polarization of ground motions such that the maximum direction of response tends to be in the direction normal to the fault strike. The issue of pulse-type ground motions affects the manner by which individual ground motions are selected for the site and applied to the structure.

Selection of Ground Motions for Sites That Are Not Near-Fault. The traditional approach has been to select (and/or simulate) ground motions that have magnitudes, fault distances, source mechanisms, and site soil conditions that are roughly similar to those likely to cause the ground motion intensity level of interest (e.g., Stewart et al. 2002) and not to consider the spectral shape in the ground motion selection. In many cases, the response spectrum is the property of a ground motion most correlated with the structural response (Bozorgnia et al. 2009) and should be considered when selecting ground motions. When spectral shape is considered in the ground motion selection, the allowable range of magnitudes, distances, and site conditions can be relaxed so that a sufficient number of ground motions with appropriate spectral shapes are available.

The selection of recorded motions typically occurs in two steps, as explained in the following illustration. Step 1 involves preselecting the ground motion records in the database (e.g., Anchenta et al. 2015) that have reasonable source mechanisms, magnitude, site soil conditions, range of usable frequencies, and site-to-source distance. In completing this preselection, it is permissible to use relatively liberal ranges because Step 2 can involve selecting motions that provide good matches to a target spectrum of interest (and matching to a target spectrum tends to implicitly account for many of the above issues). Step 2 in the selection process is to select the final set of motions from those preselected in Step 1.

In the first step, the following criteria should be used to filter out ground motions that should not be considered as candidates in the final selection process:

- **Source Mechanism:** Ground motions from differing tectonic regimes (e.g., subduction versus active crustal regions) often have substantially differing spectral shapes and durations, so recordings from appropriate tectonic regimes should be used whenever possible.
- **Magnitude:** Earthquake magnitude is related to the duration of ground shaking, so using ground motions from earthquakes with appropriate magnitudes should already have approximately the appropriate durations. Earthquake magnitude is also related to the shape of the resulting ground motion's response spectrum, though spectral shape is considered explicitly in Step 2 of the process, and so this is not a critical factor when identifying ground motions from appropriate magnitude earthquakes.
- **Site Soil Conditions:** Site soil conditions (Site Class) exert a large influence on ground motions but are already reflected in the spectral shape used in Step 2. For Step 1, reasonable limits on site soil conditions should be imposed but should not be too restrictive as to unnecessarily limit the number of candidate motions.
- **Usable Frequency of the Ground Motion:** Only processed ground motion records should be considered for RHA. Processed motions have a usable frequency range; in active regions, the most critical parameter is the lowest usable frequency. It is important to verify that the usable frequencies of the record (after filtering) accommodate the range of frequencies important to the building response; this frequency (or period) range is discussed in this next section on scaling.
- **Period/Frequency Sampling:** Ground motion recordings are discretized representations of continuous functions. The sampling rate for the recorded data can vary from as little as 0.001 seconds to as much as 0.02 seconds depending on the recording instrument and processing. If the sampling rate is too coarse, important characteristics of the motion, particularly in the high-frequency range, can be lost. On the other hand, the finer the sampling rate, the longer the analysis will take. Particularly for structures with significant response at periods less than 0.1 second, caution should be used to ensure that the sampling rate is sufficiently fine to capture the motion's important characteristics. As a general guideline, discretization should include at least 100 points per decade of significant response. Thus, for a structure with significant response at a period of 0.1 second, time steps should not be greater than 0.001 second.
- **Site-to-Source Distance:** The distance is a lower priority parameter to consider when selecting ground motions. Studies investigating this property have all found that response history analyses performed using ground motions from different site-to-source distances but otherwise equivalent properties produce practically equivalent demands on structures.

Once the preselection process has been completed, Step 2 is undertaken to select the final set of ground motions according to the following criteria:

- **Spectral Shape:** The shape of the response spectrum is a primary consideration when selecting ground motions.
- **Scale Factor:** It is also traditional to select motions such that the necessary scale factor is limited; an allowable scale factor limit of approximately 0.25 to 4 is not uncommon.
- **Maximum Motions from a Single Event:** Many also think it important to limit the number of motions from a single seismic event, such that the ground motion set is not unduly influenced by the single event. This criterion is deemed less important than limiting the scale factor, but imposing a limit of only three or four motions from a single event would not be unreasonable for most cases.

Further discussion of ground motion selection is available in NIST GCR 11-917-15 (NIST 2011), *Selecting and Scaling Earthquake Ground Motions for Performing Response-History Analyses*.

Near-fault sites have a probability of experiencing pulse-type ground motions. This probability is not unity, so only a certain fraction of selected ground motions should exhibit pulselike characteristics, while the remainder can be nonpulse records selected according to the standard process described above. The probability of experiencing pulselike characteristics is dependent principally on (1) distance of site from fault; (2) fault type (e.g., strike slip or reverse); and (3) location of hypocenter relative to site, such that rupture occurs toward or away from the site.

Criteria (1) and (2) are available from conventional disaggregation of probabilistic seismic hazard analysis. Criterion (3) can be computed as well in principal but is not generally provided in a conventional hazard analysis. However, for the long ground motion return periods associated with MCE_R spectra, it is conservative and reasonable to assume that the fault rupture is toward the site for the purposes of evaluating pulse probabilities. Empirical relations for evaluating pulse probabilities in consideration of these criteria are given in NIST GCR 11-917-15 (2011) and in Shahi et al. (2011).

Once the pulse probability is identified, the proper percentage of pulselike records should be enforced in the ground motion selection. For example, if the pulse probability is 30% and 11 records are to be used, then 3 or 4 records in the set should exhibit pulselike characteristics in at least one of the horizontal components. The PEER Ground Motion Database can be used to identify records with pulse-type characteristics. The other criteria described in the previous section should also be considered to identify pulselike records that are appropriate for a given target spectrum and set of disaggregation results.

C16.2.3 Ground Motion Modification. Two procedures for modifying ground motions for compatibility with the target spectrum are available: amplitude scaling and spectral matching. Amplitude scaling consists of applying a single scaling factor to the entire ground motion record such that the variation of earthquake energy with structural period found in the original record is preserved. Amplitude scaling preserves record-to-record variability; however, individual ground motions that are amplitude scaled can significantly exceed the response input of the target spectrum at some periods, which can tend to overstate the importance of higher mode response in some structures. In spectral matching techniques, shaking amplitudes are modified by differing amounts at differing periods, and in some cases additional wavelets of energy are added to or subtracted from the motions, such that the response spectrum of the modified motion closely resembles the target spectrum. Some spectral matching techniques are incapable of preserving important characteristics of velocity pulses in motions and should not be used for near-fault sites where these effects are important. Spectral matching does not generally preserve the record-to-record response variability observed when evaluating a structure for unmodified motions, but it can capture the mean response well, particularly if nonlinear response is moderate.

Vertical response spectra of earthquake records are typically significantly different than the horizontal spectra. Therefore, regardless of whether amplitude scaling or spectral matching is used, separate scaling of horizontal and vertical effects is required.

C16.2.3.1 Period Range for Scaling or Matching. The period range for scaling of ground motions is selected such that the ground motions accurately represent the MCE_R hazard at the structure's fundamental response periods, periods somewhat longer than this to account for period lengthening effects associated with nonlinear response and shorter periods associated with a higher mode response. Before the 2016 edition of the standard, ground motions were required to be scaled between periods of $0.2T$ and $1.5T$. The lower bound was selected to capture higher mode response, and the upper bound, period elongation effects. In the 2016 edition, nonlinear response history analyses are performed at the MCE_R ground motion level. Greater inelastic response is anticipated at this level as compared to the design spectrum, so the upper bound period has accordingly been raised from $1.5T$ to $2.0T$, where T is redefined as the *maximum* fundamental period of the building (i.e., the maximum of the fundamental periods in both translational directions and the fundamental torsional period). This increase in the upper bound period is also based on recent research, which has shown that the $1.5T$ limit is too low for assessing ductile frame buildings subjected to MCE_R motions (Haselton and Baker 2006).

For the lower bound period, the $0.2T$ requirement is now supplemented with an additional requirement that the lower bound also should capture the periods needed for 90% mass participation in both directions of the building. This change is made to ensure that when used for tall buildings and other long-period structures, the ground motions are appropriate to capture response in higher modes that have significant response.

In many cases, the substructure is included in the structural model, and this inclusion substantially affects the mass participation characteristics of the system. Unless the foundation system is being explicitly designed using the results of the response history analyses, the above 90% modal mass requirement pertains only to the superstructure behavior; the period range does not need to include the very short periods associated with the subgrade behavior.

C16.2.3.2 Amplitude Scaling. This procedure is similar to those found in earlier editions of the standard, but with the following changes:

1. Scaling is based directly on the maximum direction spectrum, rather than the square root of the sum of the squares spectrum. This change was made for consistency with the MCE_R ground motion now being explicitly defined as a maximum direction motion.
2. The approach of enforcing that the average spectrum "does not fall below" the target spectrum is replaced with requirements that (a) the average spectrum "matches the target spectrum" and (b) the average spectrum does not fall below 90% of the target spectrum for any period within the period range of interest. This change was made to remove the conservatism associated with the average spectrum being required to *exceed* the target spectrum at *every* period within the period range.

The scaling procedure requires that a maximum direction response spectrum be constructed for each ground motion. For some ground motion databases, this response spectrum definition is already precomputed and publicly available (e.g., for the Ancheta 2012). The procedure basically entails computing the maximum acceleration response to each ground motion pair for a series of simple structures that have a single mass. This procedure is repeated for structures of different periods, allowing construction of the spectrum. A number of software tools can automatically compute this spectrum for a given time–history pair.

Fig. C16.2-2 shows an example of the scaling process for an example site and structure. This figure shows how the average of the maximum direction spectra meets the target spectrum (a) and

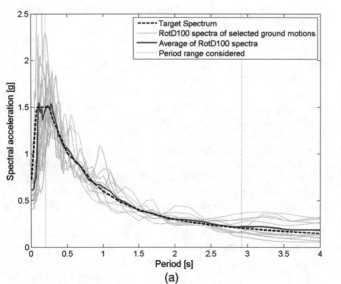
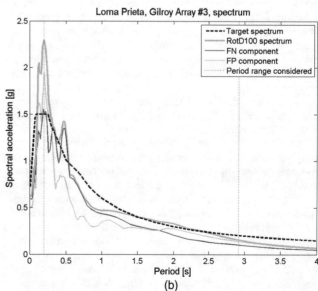

FIGURE C16.2-2 Ground Motion Scaling for an Example Site and Structure, Showing (a) the Ground Motion Spectra for All 11 Motions and (b) an Example for the Loma Prieta, Gilroy Array #3 Motion

shows more detail for a single Loma Prieta motion in the scaled ground motion set (b).

C16.2.3.3 Spectral Matching. Spectral matching of ground motions is defined as the modification of a real recorded earthquake ground motion in some manner such that its response spectrum matches a desired target spectrum across a period range of interest. There are several spectral matching procedures in use, as described in the NIST GCR 11-917-15 report (NIST 2011). The recommendations in this report should be followed regarding appropriate spectral matching techniques to be applied.

This section requires that when spectral matching is applied, the average of the maximum direction spectra of the matched motions must exceed the target spectrum over the period range of interest; this is intentionally a more stringent requirement, as compared to the requirement for scaled unmatched motions, because the spectral matching removes variability in the ground motion spectra and also has the potential to predict lower mean response (e.g., Luco and Bazzurro 2007; Grant and Diaferia 2012).

The specific technique used to perform spectral matching is not prescribed. It is possible to match both components of motion to a single target spectrum or to match the individual components to different spectra, as long as the average maximum direction spectra for the matched records meets the specified criteria.

Spectral matching is not allowed for near-fault sites, unless the pulse characteristics of the ground motions are retained after the matching process has been completed. This is based on the concern that, when common spectral matching methods are used, the pulse characteristics of the motions may not be appropriately retained.

C16.2.4 Application of Ground Motions to the Structural Model. This section explains the guidelines for ground motion application for both non-near-fault and near-fault sites.

Sites That Are Not Near-Fault. In this standard, the maximum direction spectral acceleration is used to describe the ground motion intensity. This spectral acceleration definition causes a perceived directional dependence to the ground motion. However, the direction in which the maximum spectral acceleration occurs is random at distances beyond 5 km (3.1 mi) from the fault (Huang et al. 2008), does not necessarily align with a principal direction of the building, and is variable from period to period. Accordingly, for the analysis to result in an unbiased prediction of structural response, the ground motions should be applied to the structure in a random orientation to avoid causing a biased prediction of structural response. True random orientation is difficult to achieve. Instead, the standard specifies that the average of the spectra applied in each direction should be similar to each other, such that unintentional bias in the application of motion, with one building axis experiencing greater demand than the other, is avoided.

Near-Fault Sites. Some recorded ground motions obtained from instruments located near zones of fault rupture have exhibited motion of significantly different character in one direction than the other. When this effect, known as directionality, occurs, it is common for the component of motion perpendicular to the fault to be stronger than that parallel to the fault and also for the fault-normal component to exhibit large velocity pulses. Sites located close to faults and that can experience motion having these characteristics are termed near-fault in this standard. For such sites, the fault-normal and fault-parallel components of recorded ground motions should be maintained and applied to the corresponding orientations of the structure.

It is important to note that not all near-fault records exhibit these characteristics and also that when records do have these characteristics the direction of maximum motion is not always aligned perpendicular to the fault strike. If appropriate selection of records is performed, some of the records used in the analysis should have these characteristics and some not. For those records that do exhibit directionality, the direction of strong shaking is generally aligned at varying azimuths, as occurred in the original recordings. It is also important to note that because ground motions have considerable variability in their characteristics, it is specifically not intended that buildings be designed weaker in the fault-parallel direction than in the fault-normal direction.

C16.3 MODELING AND ANALYSIS

C16.3.1 Modeling. Nonlinear response history analysis offers several advantages over linear response history analysis, including the ability to model a wide variety of nonlinear material behaviors, geometric nonlinearities (including P-delta and large displacement effects), gap opening and contact behavior, and nonlinear viscous damping, and to identify the likely spatial and temporal distributions of inelasticity. Nonlinear response history analysis has several disadvantages, including increased effort to develop the analytical model, increased time to perform the analysis (which is often complicated by difficulties in obtaining converged solutions), sensitivity of computed response to system parameters, large amounts of analysis results to evaluate, and the inapplicability of superposition to combine live, dead, and seismic load effects.

While computation of collapse probability is not necessary, it is important to note that mathematical models used in the analysis should have the capability to determine if collapse occurs when the structure is subjected to MCE_R level ground motions. The ability to predict collapse is important because the global acceptance criteria in Section 16.4.1.1 allow collapse (or unacceptable response) to occur for only one of the 11 ground motions for Risk Category I and II buildings and allows no such responses for Risk Category III and IV buildings. Development of models with the ability to predict collapse requires attributes such as cyclic loss of strength and stiffness, low cycle fatigue failure, and geometric nonlinearity.

Although analytical models used to perform linear analysis in accordance with Chapter 12 typically do not include representation of elements other than those that compose the intended lateral-force-resisting system, the gravity-load-carrying system and some nonstructural components can add significant stiffness and strength. Because the goal of nonlinear response history analysis is to accurately predict the building's probable performance, it is important to include such elements in the analytical model and also to verify that the behavior of these elements will be acceptable. This inclusion may mean that contribution of stiffness and strength from elements considered as nonparticipating elements in other portions of this standard should be included in the response history analysis model. Since structures designed using nonlinear response history analysis must also be evaluated using linear analyses, this analysis ensures that the strength of the intended seismic force-resisting system is not reduced relative to that of structures designed using only the linear procedures.

Expected material properties are used in the analysis model, attempting to characterize the expected performance as closely as possible. It is suggested that expected properties be selected considering actual test data for the proposed elements. Where test data are not readily available, the designer may consider estimates as found in ASCE 41 and the PEER TBI Guidelines (Bozorgnia et al. 2009). Guidance on important considerations in modeling may also be found in *Nonlinear Structural Analysis for Seismic Design*, NIST GCR 10-917-5 (NIST 2010).

Two-dimensional structural models may be useful for initial studies and for checking some specific issues in a structure; however, the final structural model used to confirm the structural performance should be three-dimensional.

For certain structures, the response under both horizontal and vertical ground motions should be considered. NIST GCR 11-917-15 (NIST 2011) provides some guidance to designers considering the application of vertical ground motions. To properly capture the nonlinear dynamic response of structures where vertical dynamic response may have a significant influence on structural performance, it is necessary to include vertical mass in the mathematical model. Typically the vertical mass must be distributed across the floor and roof plates to properly capture vertical response modes. Additional degrees of freedom (e.g., nodes at quarter points along the span of a beam) need to be added to capture this effect, or horizontal elements need to be modeled with consistent mass. Numerical convergence problems caused by large oscillatory vertical accelerations have been noted (NIST 2012) where base rotations caused by wall cracking in fiber wall models are the primary source of vertical excitation. See also the Commentary on Chapter 22.

Consideration of the additional vertical load of $(0.2S_{DS}) * D$, per Section 12.4.2, is inappropriate for response history analysis. Response history analyses are desired to reflect actual building response to the largest extent possible. Applying an artificial vertical load to the analysis model before application of a ground motion results in an offset in the yield point of elements carrying gravity load because of the initial artificial stress. Similarly, applying an artificial vertical load to the model at the conclusion of a response history analysis is not indicative of actual building response. If vertical ground motions are expected to significantly affect response, application of vertical shaking to the analysis model is recommended. It should be noted that vertical response often occurs at higher frequencies than lateral response, and hence, a finer analysis time-step might be required when vertical motions are included.

For structures composed of planar seismic force-resisting elements connected by floor and roof diaphragms, the diaphragms should be modeled as semirigid in plane, particularly where the vertical elements of the seismic force-resisting system are of different types (such as moment frames and walls). Biaxial bending and axial force interaction should be considered for corner columns, nonrectangular walls, and other similar elements.

Nonlinear response history analysis is load path dependent, with the results depending on combined gravity and lateral load effects. The MCE shaking and design gravity load combinations required in ASCE 7 have a low probability of occurring simultaneously. Therefore, the gravity load should instead be a realistic estimate of the expected loading on a typical day in the life of the structure. In this chapter, two gravity load cases are used. One includes an expected live loading characterizing probable live loading at the time of the Maximum Considered Earthquake shaking, and the other, no live load. The case without live load is required to be considered only for those structures where live load constitutes an appreciable amount of the total gravity loading. In those cases, structural response modes can be significantly different, depending on whether the live load is present. The dead load used in this analysis should be determined in a manner consistent with the determination of seismic mass. When used, the live load is reduced from the nominal design live load to reflect both the low probability of the full design live load occurring simultaneously throughout the building and the low probability that the design live load and Maximum Considered Earthquake shaking will occur simultaneously.

The reduced live load values, of $0.8L_0$ for live loads that exceed 100 lb/ft^2 (4.79 kN/m^2) and $0.4L_0$ for all other live loads, were simply taken as the maximum reduction allowable in Sections 4.7.2 and 4.7.3.

Gravity loads are to be applied to the nonlinear model first and then ground shaking simulations applied. The initial application of gravity load is critical to the analysis, so member stresses and displacements caused by ground shaking are appropriately added to the initially stressed and displaced structure.

C16.3.3 P-Delta Effects. P-delta effects should be realistically included, regardless of the value of the elastic story stability coefficient $\theta = P\Delta I_e/(Vh)$. The elastic story stability coefficient is not a reliable indicator of the importance of P-delta during large inelastic deformations. This problem is especially important for dynamic analyses with large inelastic deformations because significant ratcheting can occur. During these types of analyses, when the global stiffness starts to deteriorate and the tangent stiffness of story shear to story drift approaches zero or becomes negative, P-delta effects can cause significant ratcheting (which is a precursor to dynamic instability) of the displacement response in one direction. The full reversal of drifts is no longer observed, and the structural integrity is compromised. To ascertain the full effect of P-delta effects for a given system, a comparison of static pushover curves from a P-delta model and non-P-delta model can be compared.

When including P-delta effects, it is important to capture not only the second-order behavior associated with lateral displacements but also with global torsion about the vertical axis of the system. Additionally, the gravity load used in modeling P-delta effects must include 100% of the gravity load in the structure. For these reasons, the use of a single "leaning column," where much of a structure's vertical weight is lumped at a single vertical coordinate, is discouraged, and instead, the structure's vertical load should be distributed throughout the structure in a realistic manner, either through direct modeling of the gravity system or by appropriately distributed "leaning columns."

In some structures, in addition to considering P-delta effects associated with global structural deformation, it is also important to consider local P-delta effects associated with the local deformation of members. This is particularly important for slender elements subject to buckling.

C16.3.4 Torsion. Inherent torsion is actual torsion caused by differences in the location of the center of mass and center of rigidity throughout the height of the structure. Accidental torsion effects per Section 12.8.4.2 are artificial effects that attempt to account for actual variations in load and material strengths during building operation that differ from modeling assumptions. Some examples of this difference would be nonuniformity of the actual mass in the building, unaccounted for openings in the diaphragm, torsional foundation input motion caused by the ground motion being out of phase at various points along the base, the lateral stiffness of the gravity framing, variation in material strength and stiffness caused by typical construction tolerances, and incidental stiffness contribution by the nonstructural elements.

When the provision for accidental torsion was first introduced, it was to address buildings that have no inherent torsion but are sensitive to torsional excitation. Common examples of this type of configuration are cruciform core or I-shaped core buildings. In reality, many things can cause such a building to exhibit some torsional response. None of the aforementioned items are typically included in the analysis model; therefore, the accidental torsion approach was introduced to ensure that the structure has some minimum level of resistance to incidental twisting under seismic excitation.

The accidental torsion also serves as an additional check to provide more confidence in the torsional stability of the structure. During the initial proportioning of the structure using linear analysis (per Section 16.1.1), accidental torsion is required to be enforced in accordance with Section 12.8.4.2. When there is no inherent torsion in the building, accidental torsion is a crucial step in the design process because this artificial offset in the center of mass is a simple way to force a minimum level of twisting to occur in the building. The accidental torsion step (i.e., the required 5% force offsets) is also important when checking for plan irregularities in symmetric and possibly torsionally flexible buildings. Where there is already inherent torsion in the building, additional accidental torsion is not generally a crucial requirement (though still required, in accordance with Section 12.8.4.2) because the building model will naturally twist during analysis, and no additional artificial torsion is required for this twisting to occur. However, for buildings exhibiting either torsional or extreme torsional irregularities, inclusion of accidental torsion in the nonlinear analysis is required by this standard to assist in identification of potential nonlinear torsional instability.

C16.3.5 Damping. Viscous damping can be represented by combined mass and stiffness (Rayleigh) damping. To ensure that the viscous damping does not exceed the target level in the primary response modes, the damping is typically set at the target level for two periods, one above the fundamental period and one below the highest mode frequency of significance. For very tall buildings, the second and even third modes can have significant contributions to response; in this case, the lower multiple on T_1 may need to be reduced to avoid excessive damping in these modes.

Viscous damping may alternatively be represented by modal damping, which allows for the explicit specification of the target damping in each mode.

Various studies have shown that the system damping may vary with time as the structure yields, and in some cases, damping well above the target levels can temporarily exist. Zareian and Medina (2010) provide recommendations for implementation of damping in such a way that the level of viscous damping remains relatively constant throughout the response.

The level of structural damping caused by component-level hysteresis can vary significantly based on the degree of inelastic action. Typically, hysteretic damping provides a damping contribution less than or equal to 2.5% of critical.

Damping and/or energy dissipation caused by supplemental damping and energy dissipation elements should be explicitly accounted for with component-level models and not included in the overall viscous damping term.

C16.3.6 Explicit Foundation Modeling. The PEER TBI guidelines (Bozorgnia et al. 2009) and NIST GCR 12-917-21 (NIST 2012) both recommend inclusion of subterranean building levels in the mathematical model of the structure. The modeling of the surrounding soil has several possible levels of sophistication, two of which are depicted below in (b) and (c) of Fig. C16.3-1, which are considered most practical for current practice. For an MCE_R-level assessment, which is the basis for the Chapter 16 RHA procedure, the rigid bathtub model is preferred by PEER TBI (Bozorgnia et al. 2009) and NIST (2012) (Fig. C16.3-1c). This model includes soil springs and dashpots, and identical horizontal ground motions are input at each level of the basement. Such a modeling approach, where the soil is modeled in the form of springs and/or dashpots (or similar methods) placed around the foundation, is encouraged but is not required. When spring and dashpot elements are included in the structural model, horizontal input ground motions are applied to the ends of the horizontal soil elements rather than being applied to the foundation directly. A simpler but less accurate model is to exclude the soil springs and dashpots from the numerical model and apply the horizontal ground motions at the bottom level of the basement (Fig. C16.3-1b), which is fixed at the base. Either the fixed-base (Fig. C16.3-1b) or bathtub (Fig. C16.3-1c) approaches are allowed, but the bathtub approach is encouraged because it is more accurate.

For the input motions, the PEER TBI (Bozorgnia et al. 2009) guidelines allow the use of either the free-field motion, which is the motion defined in Section 16.2.2, or a foundation input motion modified for kinematic interaction effects. Guidelines for modeling kinematic interaction are contained in NIST (2012).

More sophisticated procedures for soil–structure interaction modeling, including the effects of multisupport excitation, can also be applied in RHA. Such analyses should follow the guidelines presented in NIST (2012).

Approximate procedures for the evaluation of foundation springs are provided in Chapter 19 of this standard.

C16.4.1 Global Acceptance Criteria

C16.4.1.1 Unacceptable Response. This section summarizes the criteria for determining unacceptable response and how the criteria were developed. It must be made clear that these unacceptable response acceptance criteria are not the primary acceptance criteria that ensure adequate collapse safety of the building; the primary acceptance criteria are the story drift criteria and the element-level criteria discussed later in Section C16.4. The

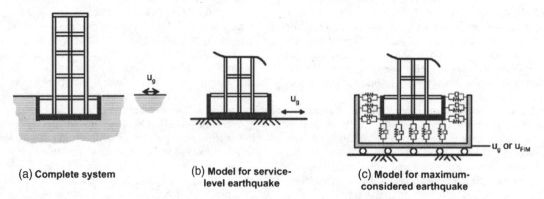

FIGURE C16.3-1 Illustration of the Method of Inputting Ground Motions into the Base of the Structural Model
Source: NIST 2011.

unacceptable response acceptance criteria were developed to be a secondary protection to supplement the primary criteria. Unacceptable responses result in instabilities and loss of gravity load support. Consequently, if it can be shown that after a deformation controlled element reaches its (collapse prevention) limit, the model is able to redistribute demands to other elements, this would not constitute unacceptable response. The acceptance criteria were intentionally structured in this manner because there is high variability in unacceptable response (as described in this section) and the other primary acceptance criteria are much more stable and reliable (because they are based on mean values of 11 motions rather than the extreme response of 11 motions).

When performing nonlinear analysis for a limited suite of ground motions, the observance of a single unacceptable response (or, conversely, the observance of no unacceptable responses) is statistically insignificant. That is, it is reasonably probable that no collapses will be observed in a small suite of analyses, even if the structure has a greater than 10% chance of collapse at MCE_R shaking levels. It is also possible that a structure with less than a 10% chance of collapse at MCE_R shaking levels will still produce an unacceptable response for one ground motion in a small suite. In order for statistics on the number of unacceptable responses in a suite of analyses to produce meaningful indication of collapse probability, a very large suite of analyses must be performed. Furthermore, the observance or nonobservance of an unacceptable response depends heavily on how the ground motions were selected and scaled (or spectrally matched) to meet the target spectrum.

Since the observance or nonobservance of an unacceptable response is not statistically meaningful, the standard does not rely heavily on the prohibition of unacceptable responses in the attempt to "prove" adequate collapse safety. The many other acceptance criteria of Section 16.4 are relied upon to implicitly ensure adequate collapse safety of the building. If one desired to expand the unacceptable response acceptance criteria to provide true meaningful collapse safety information about the building, a more complex statistical inference approach would need to be used. This is discussed further below.

The statistical insignificance of unacceptable response in a small suite of analyses leaves a large open question about how to interpret the meaning of such responses when they occur. Even though occurrence of a single unacceptable response is statistically meaningless, the occurrence of many unacceptable responses (e.g., 5 of 11) does indicate that the collapse probability is significantly in excess of 10%. Additionally, a conscientious structural designer is concerned about such occurrence, and the occurrences of unacceptable responses may provide the designer with some insight into possible vulnerabilities in the structural design.

Some engineers presume that the acceptance criteria related to *average* response effectively disallow any unacceptable responses (because you cannot average in an infinite response), while others presume that *average* can also be interpreted as *median*, which could allow almost half of the ground motions to cause unacceptable response.

The statistics presented below are provided to help better interpret the meaning of observance of a collapse or other type of unacceptable response in a suite of analyses. These simple statistics are based on predicting the occurrence of collapse (or other unacceptable response) using a binomial distribution, based on the following assumptions:

- The building's collapse probability is exactly 10% at the MCE_R level.
- Collapse probability is lognormally distributed and has a dispersion (lognormal standard deviation) of 0.6. This

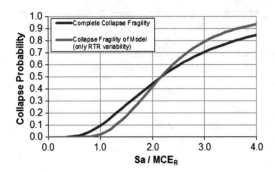

FIGURE C16.4-1 Collapse Fragilities for a Building with $P[C|MCE_R] = 10\%$ and $\beta_{COL,RTR} = 0.40$

value includes all sources of uncertainty and variability (e.g., record-to-record variability, modeling uncertainty). The value of 0.6 is the same value used in creating the risk-consistent hazard maps for ASCE 7-10 (FEMA 2009a) and is consistent with the values used in FEMA P-695 (FEMA 2009b).

- The record-to-record variability ranges from 0.25 to 0.40. This is the variability in the collapse capacity that would be expected from the analytical model. This value is highly dependent on the details of the ground motion selection and scaling; values of 0.35 to 0.45 are expected for motions that are not fit tightly to the target spectrum, and values of 0.2 to 0.3 are expected for spectrally matched motions (FEMA 2009b).

Fig. C16.4-1 shows collapse fragility curves for a hypothetical building that has a 10% collapse probability conditioned on MCE_R motion ($P[C|MCE_R] = 10\%$) with an assumed record-to-record collapse uncertainty of 0.40 and a total collapse uncertainty of 0.60. The figure shows that the median collapse capacity must be a factor of 2.16 above the MCE_R ground motion level, that the probability of collapse is 10% at the MCE_R when the full variability is included (as required), but that the probability of collapse is only 2.7% at the MCE_R when only the record-to-record variability is included. This 2.7% collapse probability is what would be expected from the structural model that is used in the RHA assessment procedure.

Table C16.4-1 shows the probability of observing n collapses in a suite of 11 ground motions for a structure that has different values of $P[C|MCE_R]$.

Table C16.4-1 shows that for a building meeting the $P[C|MCE_R] = 10\%$ performance goal, there is a 74% chance of

Table C16.4-1 Likelihood of Observing Collapses in 11 Analyses, Given Various MCE_R Collapse Probabilities and a Record-to-Record Uncertainty of 0.4

Number of Collapses	Likelihood for Various $P[C\|MCE_R]$ Values				
	0.05	0.10	0.15	0.20	0.30
0 of 11	0.93	**0.74**	0.51	0.30	0.07
1 of 11	0.07	**0.23**	0.36	0.38	0.21
2 of 11	0	**0.03**	0.11	0.22	0.29
3 of 11	0	0	0.02	0.08	0.24
4 of 11	0	0	0	0.02	0.13
5 of 11	0	0	0	0	0.05

Table C16.4-2 Likelihood of Observing Collapses in 11 Analyses, Given Various MCE$_R$ Collapse Probabilities and a Record-to-Record Uncertainty of 0.25

Number of Collapses	Likelihood for Various P[C|MCE$_R$] Values				
	0.05	0.10	0.15	0.20	0.30
0 of 11	1.00	**0.99**	0.93	0.79	0.30
1 of 11	0	**0.01**	0.07	0.19	0.38
2 of 11	0	**0**	0	0.02	0.22
3 of 11	0	**0**	0	0	0.08
4 of 11	0	**0**	0	0	0.02
5 of 11	0	**0**	0	0	0

observing no collapses, a 23% chance of observing one collapse, a 3% chance of observing two collapses, and virtually no chance of observing more than two collapses. In comparison, for a building with P[C|MCE$_R$] = 20%, there is a 30% chance of observing no collapses, a 38% chance of observing one collapse, a 22% chance of observing two collapses, and a 10% chance of observing more than two collapses.

This table illustrates that

- Even if no collapses are observed in a set of 11 records, this does not, in any way, prove that the P[C|MCE$_R$] = 10% performance goal has been met. For example, even for a building with P[C|MCE$_R$] = 20%, there is still a 30% chance that no collapses will be observed in the analysis. Therefore, the other noncollapse acceptance criteria (e.g., criteria for drifts and element demands) must be relied upon to enforce the 10% collapse probability goal.
- If the P[C|MCE$_R$] = 10% performance goal is met, it is highly unlikely (only a 3% chance) that two collapses will be observed in the set of 11 records. Therefore, an acceptance criterion that prohibits two collapses is reasonable.

The collapse likelihoods show in Table C16.4-1 are based on a relatively large record-to-record variability value of 0.40. Table C16.4-2 illustrates similar statistics for the case when the record-to-record variability is suppressed in ground motion selection and scaling, such as occurs with spectral matching. This table shows that, for a building meeting the P[C|MCE$_R$] = 10% performance goal and with record-to-record variability taken as 0.25, the likelihood of observing a collapse response is very low. This is why no unacceptable responses are permitted in the suite of analyses when spectral matching is used.

For Risk Category I and II structures, if more than 11 ground motions are used for analysis, then additional unacceptable responses may be permissible. Two unacceptable responses would be permissible if 20 or more motions are used, and three unacceptable responses are permissible when 30 or more motions are used. For Risk Category III and IV structures, the collapse probability goals are 6% and 3%, respectively, at the MCE$_R$ level. When the above computations are redone using these lower collapse probability targets, this shows that the acceptance criteria should require that no motions of the 11 produce an unacceptable response for these categories.

Typically, mean building response values (story drifts, element deformations, and forces) are used in acceptance evaluations, where the "mean" is the simple statistical average for the response parameter of interest. When an unacceptable response occurs, it is not possible to compute a mean value of the building response values because one of the 11 response quantities is undefined. In this case, rather than the mean, the standard requires use of the counted median response multiplied by 1.2 but not less than the mean response from the remaining motions.

To compute the median value, the unacceptable response is assumed as larger than the other responses and then, assuming that 11 analyses were performed, the counted median value is taken to be the 6th largest value from the set of 11 responses. The 1.2 factor is based on a reasonable ratio of mean to median values for a lognormal distribution ($\beta = 0.4$ results in mean/median = 1.08, $\beta = 0.5$ results in mean/median = 1.13, $\beta = 0.6$ results in mean/median = 1.20, and $\beta = 0.7$ results in mean/median = 1.28).

The requirement to also check the mean of the remaining 10 response results is simply an added safeguard to ensure that the 1.2 × median value does not underpredict the mean response values that should be used when checking the acceptance criteria.

Although currently the purpose of this acceptance criterion is not to quantify the structure's collapse probability under MCE$_R$ ground motions, the acceptance criterion can be recast to do so in future provisions. The collapse probability can be inferred from analysis results and compared to the target value (e.g., 10% for structures in Risk Category I or II). In this alternate light, existing statistical inference theory can be used to determine the number of acceptable responses, and the number of ground motions required to conclude that the proposed design may have an acceptable collapse probability.

As was done in the previous section, analysis results can be thought of as following a binomial distribution. Based on this distribution, one could use the observed counts of collapsed and noncollapsed responses (indicated by unacceptable and acceptable responses) to estimate the collapse probability of the proposed design in a manner that accounts for the uncertainty in the estimated collapse probability. This uncertainty depends on the total number of ground motions. If few ground motions are used, there is a large uncertainty in the collapse probability. If many ground motions are used, there is a small uncertainty. For example, compare a set of 11 ground motions with 1 unacceptable response to a set of 110 ground motions with 10 unacceptable responses. Both sets have a most likely unacceptable response probability of 9.1%. The design with 1 unacceptable and 10 acceptable responses has only a 34% chance that its unacceptable response probability is 10% or less. The design with 10 unacceptable and 100 acceptable responses has a 56% chance that its unacceptable response probability is 10% or less.

In the current acceptance criterion, the choice to require 11 ground motions follows from the need to have confidence in the average values of the resulting element-level and story-level responses (Section C16.2.3.1). These element-level and story-level responses are then used to *implicitly* demonstrate adequate collapse safety. If future provisions seek to *explicitly* ensure that the proposed design has an acceptable collapse probability, then this unacceptable response acceptance criterion should be revised using statistical inference theory to establish the number of required ground motions and the maximum number of unacceptable responses, as well as the element- and story-level response limits.

C16.4.1.2 Story Drift. The limit on mean story drift was developed to be consistent with the linear design procedures of this standard. To this end, the basic Table 12.12-1 story drift limits are the following:

- Increased by a factor of 1.5, to reflect the analysis being completed at the MCE$_R$ ground motion level rather than at 2/3 of the MCE$_R$ level, and
- Increased by another factor of 1.25, to reflect an average ratio of R/C_d.

These two above increases are the basis for the requirement that the mean story drift be limited to 1.9 (which was rounded to 2.0) of the standard Table 12.12-1 limits.

The masonry-specific drift limits of Table 12.12-1 are not enforced in this section because the component-level acceptance criteria of Section 16.4.2 are expected to result in equivalent performance (i.e., a masonry building designed in accordance with Chapter 16 is expected to have similar performance to a masonry building designed using linear analysis methods and the more stringent drift limits of Table 12.12-1).

The standard does not require checks on residual drift. Residual drifts are an indicator of incipient dynamic instability, and a prudent engineer checks for this instability. Limiting residual drifts is an important consideration for postearthquake operability and for limiting financial losses, but such performance goals are not included in the scope of the ASCE 7 standard. For Risk Category I and II buildings, the ASCE 7 standard is primarily meant to ensure the protection of life safety. Additionally, residual drifts can be extremely difficult to predict reliably with available structural analysis tools.

C16.4.2 Element-Level Acceptance Criteria. The element-level acceptance criteria require classification of each element action as either force-controlled or deformation-controlled, similar to the procedures of ASCE 41. Note that this is done for each *element action*, rather than for each *element*. For example, for a single column element, the flexural behavior may be classified as a deformation-controlled action, whereas the axial behavior may be classified as a force-controlled action.

Deformation-controlled actions are those that have reliable inelastic deformation capacity. Force-controlled actions pertain to brittle modes where inelastic deformation capacity cannot be ensured. Based on how the acceptance criteria are structured, any element action that is modeled elastically must be classified as being force-controlled.

Some examples of force-controlled actions are

- Shear in reinforced concrete (other than diagonally reinforced coupling beams).
- Axial compression in columns.
- Punching shear in slab–column joints without shear reinforcing.
- Connections that are not explicitly designed for the strength of the connected component, such as some braces in braced frames.
- Displacement of elements resting on a supporting element without rigid connection (such as slide bearings).
- Axial forces in diaphragm collectors.

Some examples of deformation-controlled actions are

- Shear in diagonally reinforced coupling beams.
- Flexure in reinforced concrete columns and walls.
- Axial yielding in buckling restrained braces.
- Flexure in special moment frames.

Section 16.4.2 further requires categorization of component actions as critical, ordinary, or noncritical based on the consequence of their exceeding strength or deformation limits. Because of the differences in consequence, the acceptance criteria are developed differently for each of the above classifications of component actions. An element's criticality is judged based on the extent of collapse that may occur, given the element's failure, and also a judgment as to whether the effect of the element's failure on seismic resistance is substantive. An element's failure could be judged to have substantial effect on the structure's seismic resistance if analysis of a model of the building without the element present predicts unacceptable performance, while analysis with the element present does not.

Limits placed on response quantities are correlated to building performance and structural reliability. In order for compliance with these limits to meaningfully characterize overall performance and reliability, grouping of certain component actions for design purposes may be appropriate. For example, while symmetrical design forces may be obtained for symmetrical structures using equivalent lateral force and modal response spectrum analysis procedures, there is no guarantee that component actions in response history analysis of symmetrical models will be the same—or even similar—for identical components arranged symmetrically. Engineering judgment should be applied to the design to maintain symmetry by using the greater demands (that is, the demands on the more heavily loaded component determined using the appropriate factor on its mean demand) for the design of both components. For this purpose, using the mean demands of the pair of components would not be appropriate because this method would reduce the demand used for design of the more heavily loaded component.

Though this point is perhaps trivial in the case of true symmetry, it is also a concern in nonsymmetrical structures. For these buildings, it may be appropriate to group structural components that are highly similar either in geometric placement or purpose. The demands determined using the suite mean (the mean response over all ground motions within a suite) may be very different for individual components within this grouping. This is a result both of the averaging process and the limited explicit consideration of ground motion to structure orientation in the provisions. Although the analysis may indicate that only a portion of the grouped components do not meet the provisions, the engineer ought to consider whether such nonconformance should also suggest redesign in other similar elements. Thus response history analysis places a higher burden on the judgment of the engineer to determine the appropriate methods for extracting meaningful response quantities for design purposes.

C16.4.2.1 Force-Controlled Actions. The acceptance criteria for force-controlled actions follow the framework established by the PEER TBI guidelines (Bozorgnia et al. 2009), shown in Eq. (C16.4-1):

$$\lambda F_u \leq \phi F_{n,e} \qquad (C16.4\text{-}1)$$

where λ is a calibration parameter, F_u is the mean demand for the response parameter of interest, ϕ is the strength reduction factor from a material standard, and $F_{n,e}$ is the nominal strength computed from a material standard considering expected material properties.

To determine appropriate values of λ, we begin with the collapse probability goals of Table 1.3-2 (for Risk Categories I and II) for MCE_R motions. These collapse probability goals include a 10% chance of a total or partial structural collapse and a 25% chance of a failure that could result in endangerment of individual lives. For the assessment of collapse, we then make the somewhat conservative assumption that the failure of a single critical force-controlled component would result in a total or partial structural collapse of the building.

Focusing first on the goal of a 10% chance of a total or partial structural collapse, we assume that the component force demand and component capacity both follow a lognormal distribution and that the estimate of $F_{n,e}$ represents the true expected strength of the component. We then calibrate the λ value required to achieve

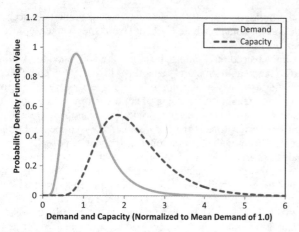

FIGURE C16.4-2 Illustration of Component Capacity and Demand Lognormal Distributions (Normalized to a Mean Capacity of 1.0); the Mean Component Capacity Is Calibrated to Achieve $P[C|MCE_R] = 10\%$

the 10% collapse probability goal. This value is depicted in Fig. C16.4-2, which shows the lognormal distributions of component capacity and component demand.

The calibration process is highly dependent on the uncertainties in component demand and capacity. Table C16.4-3a shows typical uncertainties in force *demand* for analyses at the MCE_R ground motion level for both the general case and the case where the response parameter is limited by a well-defined yield mechanism. Table C16.4-3b shows typical uncertainty values for the component *capacity*. The values are based on reference materials, as well as the collective experience and professional judgment of the development team.

In the calibration process, the λ and ϕ values both directly affect the required component strength. Therefore, the calibration is completed to determine the required value of λ/ϕ needed to fulfill the 10% collapse safety objective. This calibration is done by assuming a value of λ/ϕ, convolving the lognormal distributions of demand and capacity and iteratively determining the capacity required to meet the 10% collapse safety objective by adjusting λ/ϕ.

Table C16.4-4 reports the final λ/ϕ values that come from such integration.

It should be clearly stated that this approach of calibrating the λ/ϕ ratio means that the final acceptance criterion is independent of the ϕ value specified by a material standard. If it is desirable for the acceptance criteria to be partially dependent on the value of ϕ, then the uncertainty factors of Table C16.4-3b would need to be made dependent on the ϕ value in some manner.

Since the Table C16.4-4 values are similar, for simplicity the acceptance criterion is based on $\lambda/\phi = 2.0$ for all cases, and a separate case for the existence of a well-defined mechanism is not included. Additionally, the strength term is defined slightly differently. For Risk Categories III and IV, this full calculation was redone using the lower collapse probability goals of 6% and 3%, respectively, and it was found that scaling the force demands by I_e sufficiently achieves these lower collapse probability goals.

This statistical calculation was then repeated for the goal of 25% chance of a failure that could result in endangerment of individual lives. This resulted in a required ratio of 1.5 for such force-controlled failure modes; deemed as "ordinary."

Force-controlled actions are deemed noncritical if the failure does not result in structural collapse or any meaningful endangerment to individual lives; this occurs in situations where gravity forces can reliably redistribute to an alternate load path and no failure will ensue. For noncritical force-controlled components, the acceptance criteria allow the use of $\lambda = 1.0$.

Where an industry standard does not define expected strength, expected (or mean) strength, F_e, is computed as follows. First, a standard strength-prediction equation is used from a material standard, using a strength reduction factor, ϕ, of 1.0; the expected material properties are also used in place of nominal material properties. In some cases, this estimate of strength ($F_{n,e}$) may still be conservative in comparison with the mean expected strength shown by experimental tests (F_e) caused by inherent conservatism in the strength equations adopted by the materials standards. If such conservatism exists, the $F_{n,e}$ value may be multiplied by a "component reserve strength factor" greater than 1.0 to produce the estimate of the mean expected strength (F_e). This process is illustrated in Fig. C16.4-3, which shows the $F_e/F_{n,e}$ ratios for the shear strengths from test data of reinforced concrete shear walls (Wallace et al. 2013). This figure shows that the ratio of $F_e/F_{n,e}$

Table C16.4-3a Assumed Variability and Uncertainty Values for Component Force Demand

Demand Dispersion (β_D)		
General	Well-Defined Mechanism	Variabilities and Uncertainties in the Force Demand
0.40	0.20	Record-to-record variability (for MCE_R ground motions)
0.20	0.20	Uncertainty from estimating force demands using structural model
0.13	0.06	Variability from estimating force demands from mean of only 11 ground motions
0.46	**0.29**	$\beta_{D-\text{Total}}$

Table C16.4-3b Assumed Variability and Uncertainty Values for Component Force Capacity

Capacity Dispersion (β_C)		
General	Well-Defined Mechanism	Variabilities and Uncertainties in the Final As-Built Capacity of the Component
0.30	0.30	Typical variability in strength equation for $F_{n,e}$ (from available data)
0.10	0.10	Typical uncertainty in strength equation for $F_{n,e}$ (extrapolation beyond available data)
0.20	0.20	Uncertainty in as-built strength because of construction quality and possible errors
0.37	**0.37**	$\beta_{C-\text{Total}}$

Table C16.4-4 Required Ratios of λ/ϕ to Achieve the 10% Collapse Probability Objective

Dispersion	Required Ratios of λ/ϕ
General	2.1
Well-Defined Mechanism	1.9

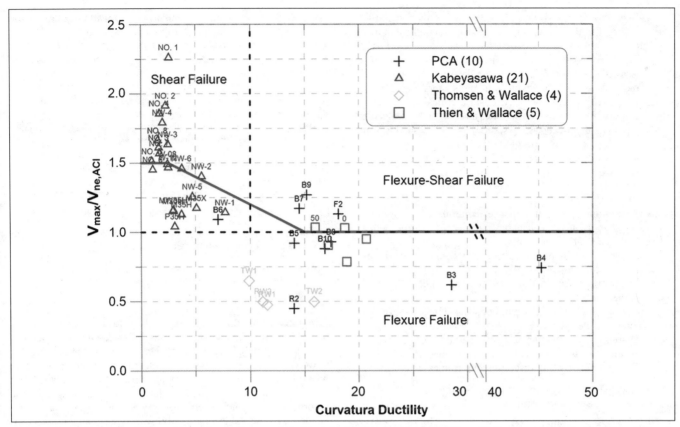

FIGURE C16.4-3 Expected Shear Strengths (in Terms of $F_e/F_{n,e}$) for Reinforced Concrete Shear Walls When Subjected to Various Levels of Flexural Ductility

Source: Courtesy of John Wallace.

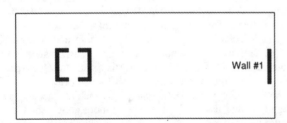

FIGURE C16.4-4 Plan View of Sample Building Showing Arrangement of Concrete Shear Walls

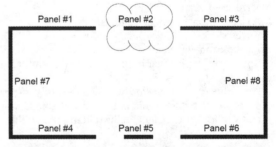

FIGURE C16.4-5 Plan View of Sample Building Showing Components of a Reinforced Concrete Core Shear Wall

depends on the flexural ductility of the shear wall, demonstrating that $F_e = 1.0\ F_{n,e}$ is appropriate for the shear strength in the zone of high flexural damage and $F_e = 1.5\ F_{n,e}$ may be appropriate in zones with no flexural damage.

For purposes of comparison, Eq. (C16.4-1) is comparable to the PEER TBI acceptance criteria (Bozorgnia et al. 2009) for the case that $\phi = 0.75$ and $F_e = 1.0\ F_{n,e}$.

The exception allows for use of the capacity design philosophy for force-controlled components that are "protected" by inelastic fuses, such that the force delivered to the force-controlled component is limited by the strength of the inelastic fuse.

The following are some examples of force-controlled actions that are deemed to be critical actions:

- Steel Moment Frames (SMF):
  - Axial compression forces in columns caused by combined gravity and overturning forces
  - Combined axial force, bending moments, and shear in column splices
  - Tension in column base connections (unless modeled inelastically, in which case it would be a deformation-controlled component)
- Steel Braced Frames (BRBF - Buckling Restrained Braced Frame, SCBF - Special Concentrically Braced Frames):
  - Axial compression forces in columns caused by combined gravity and overturning forces
  - Combined axial force, bending moments, and shear in column splices
  - Tension in brace and beam connections
  - Column base connections (unless modeled inelastically)
- Concrete Moment Frames:
  - Axial compression forces in columns caused by combined gravity and overturning forces
  - Shear force in columns and beams
- Concrete or Masonry Shear Walls:

- Shear in concrete shear wall, in cases when there is limited ability for the shear force to transfer to adjacent wall panels. For cases of isolated shear walls (i.e., wall #1 in Fig. C16.4-4), the shear force in this isolated wall is deemed as a critical action. In contrast, the shear force in a one-wall pier that is in a group of wall piers (e.g., panel #2 of Fig. C16.4-5) need not be deemed a critical action (especially when determining whether an analysis is deemed to represent an unacceptable response). For this case of a group of wall piers, it may be appropriate to consider the sum of the wall shears to be the critical action (e.g., the sum of wall shears in panels #1, #2, and #3 of Fig C16.4-5).
- Axial (plus flexural) compression in concrete shear wall (for most cases)
- Axial compression in outrigger columns
- Axial (plus flexural) tension in outrigger column splices
- Other Types of Components:
  - Shear forces in piles and pile cap connections (unless modeled inelastically)
  - Shear forces in shallow foundations (unless modeled inelastically)
  - Punching shear in slabs without shear reinforcing (unless modeled inelastically)
  - Diaphragms that transfer a substantial amount of force (from more than one story)
  - Elements supporting discontinuous frames and walls

The following are some examples of force-controlled actions that are deemed to be ordinary actions:

- Steel Moment Frames (SMF):
  - Shear force in beams and columns
  - Column base connections (unless modeled inelastically)
  - Welded or bolted joints (as distinct from the inelastic action of the overall connection) between moment frame beams and columns
- Steel Braced Frames (BRBF, SCBF):
  - Axial tension forces in columns caused by overturning forces (unless modeled inelastically)
- Concrete Moment Frames:
  - Splices in longitudinal beam and column reinforcement
- Concrete or Masonry Shear Walls:
  - An ordinary classification would only apply in special cases where failure would not cause widespread collapse and would cause minimal reduction in the building seismic resistance.
- Other Types of Components:
  - Axial forces in diaphragm collectors (unless modeled inelastically)
  - Shear and chord forces in diaphragms (unless modeled inelastically)
  - Pile axial forces

Table C16.4-5a Assumed Variability and Uncertainty Values for Component Deformation Demand

Demand Dispersion (β_D)	Variabilities and Uncertainties in the Deformation Demand
0.40	Record-to-record variability (for MCE_R ground motions)
0.20	Uncertainty from estimating deformation demands using structural model
0.13	Variability from estimating deformation demands from mean of only 11 ground motions
0.46	$\beta_{D-Total}$

Table C16.4-5b Assumed Variability and Uncertainty Values for Component Deformation Capacity

Capacity Dispersion (β_C)	Variabilities and Uncertainties in the Final As-Built Deformation Capacity of the Component
0.60	Typical variability in prediction equation for deformation capacity (from available data)
0.20	Typical uncertainty in prediction equation for deformation capacity (extrapolation beyond data)
0.20	Uncertainty in as-built deformation capacity because of construction quality and errors
0.66	$\beta_{C-Total}$

The following are some examples of force-controlled actions that could be deemed noncritical actions:

- Any component where the failure would not result in either collapse or substantive loss of the seismic resistance of the structure.

C16.4.2.2 Deformation-Controlled Actions. While substantive data exist to indicate the capacity of force-controlled actions, there are relatively few laboratory data to indicate the deformation at which a deformation-controlled element action reaches a level where loss of vertical load-carrying capacity occurs. There are a number of reasons for this, including the following: (1) the deformation at which such loss occurs can be very large and beyond the practical testing capability of typical laboratory equipment; (2) many researchers have tested such components with the aim of quantifying useful capacity for elements of a seismic force-resisting system and have terminated testing after substantial degradation in strength has occurred, even though actual failure has not yet been experienced; and (3) testing of gravity-load-bearing elements to failure can be dangerous and destructive of test equipment. Therefore, lacking a comprehensive database on the reliable collapse capacity of different deformation-controlled element actions, the standard defaults to acceptance criteria contained in ASCE 41. However, the standard does present alternative criteria, which directly use the expected deformation at which loss of vertical load-carrying capability occurs, in the recognition that use of such values is more consistent with the collapse goals of Section C1.3.1 and also in the hopes that data on the deformation capacity of elements will eventually be available for use.

To determine appropriate inelastic deformation limits for this alternative procedure, a process similar to that used for force-controlled actions is used. Table C16.4-5a shows the assumed uncertainties in deformation demand for structural analyses for MCE_R ground motions. Table C16.4-5b similarly shows assumed uncertainties in the component deformation capacity at the point that loss in vertical load-carrying capacity occurs. These β_C values are larger than the comparable values for force-controlled components because the uncertainty is quite large when trying to quantify the deformation at which loss of vertical load-carrying capability occurs.

The results of integration show that the mean deformation capacity must be a factor of 3.2 larger than the mean deformation demand in order to meet the 10% collapse safety objective (for total or partial structural collapse) for MCE_R ground motions. Using the inverse of this value leads to a requirement that the mean deformation demand be limited to less than 0.3 of the mean deformation capacity.

This 0.3 limit is quite conservative and assumes that immediate collapse results when the deformation capacity is exceeded in a single component. Such immediate collapse may occur in some uncommon cases where no alternative load path exists; however, in most cases, there is at least one alternative load path and the gravity loads will redistribute and delay the occurrence of vertical collapse. Note that the use of a 0.3 ratio in the acceptance criterion assumes that there is a 100% probability of building collapse when the deformation capacity is exceeded in a single component; the use of a 0.5 ratio instead implies a 40% probability of building collapse when the deformation capacity is exceeded in a single component. These are the acceptance criteria used for critical deformation-controlled actions.

This statistical calculation was then redone for the goal of a 25% chance of a failure that would result in endangerment of individual lives. The results of integration show that the mean deformation capacity must be a factor of 2.0 larger than the mean deformation demand in order to meet the 25% goal for MCE_R ground motions; using the inverse of this value, this leads to a requirement that the mean deformation demand be limited to less than 0.5 of the mean deformation capacity.

For noncritical deformation-controlled actions, by definition, the failure of such a component would not result in any collapse and also would not result in substantive loss in the seismic strength of the structure. Accordingly, for such a case, the inelastic deformation is not limited by the Section 16.4.2.2 acceptance criterion (because there is no meaningful consequence of failure for such component), but the inelastic deformation of such component is still limited by the unacceptable response criterion of Section 16.4.1.1 (i.e., the component must be adequately modeled up to the deformation levels that the component experiences in the structural simulation).

The following are some examples of deformation-controlled actions that are deemed to be critical actions:

- Steel Moment Frames (SMF)
  - Hinge rotations in beams and columns leading to significant strength/stiffness degradation
  - Deformations of nonductile gravity beam-to-column connections
- Steel Braced Frames (BRBF, SCBF)
  - Axial deformations (tension/compression) in braces
  - Hinge rotations in beams and columns leading to significant strength/stiffness degradation
  - Deformations of nonductile gravity beam-to-column connections
- Concrete Moment Frames:
  - Hinge rotations in beams and columns leading to significant strength/stiffness degradation
  - Deformations of nonductile slab–column connections in reinforced concrete gravity systems
- Concrete Shear Walls:
  - Tensile strains in longitudinal wall reinforcement
  - Compression strains in longitudinal wall reinforcement and concrete
  - Flexural hinging or shear yielding of coupling beams
  - Deformations of nonductile slab–column or slab–wall connections in reinforced concrete gravity systems
- Other Types of Components:
  - Soil uplift and bearing deformations in shallow foundations (when modeled inelastically)
  - Tensile pullout deformations or compression bearing deformations of pile foundations (when modeled inelastically)

The following are some examples of deformation-controlled actions that are deemed to be ordinary actions:

- Steel Moment Frames (SMF):
  - Deformations of ductile gravity beam-to-column connections
- Steel Braced Frames (BRBF, SCBF, or nonconforming braced frames):
  - Deformations of ductile gravity beam-to-column connections
- Concrete Moment Frames:
  - Deformations of ductile slab–column connections in reinforced concrete gravity systems
- Concrete Shear Walls:
  - Deformations of ductile slab–column or slab–wall connections in reinforced concrete gravity systems

The following are some examples of deformation-controlled actions that could be deemed noncritical actions:

- Deformations in a coupling beam in a shear wall system, in the case that the failure of the coupling beam neither results in any collapse nor substantive loss to seismic resistance.

C16.4.2.3 Elements of the Gravity Force-Resisting System. The basic deformation-compatibility requirement of ASCE 7-10, Section 12.12.5 is imposed for gravity-system components, which are not part of the established seismic force-resisting system, using the deformation demands predicted from response history analysis under MCE_R-level ground motions, as opposed to evaluation under linear analysis.

If an analyst wanted to further investigate the performance of the gravity system (which is not required), the most direct and complete approach (but also the most time-consuming) would be to directly model the gravity system components as part of the structural model and then impose the same acceptance criteria used for the components of the seismic force-resisting system. An alternative approach (which is more common) would be to model the gravity system in a simplified manner and verify that the earthquake-imposed force demands do not control over the other load combinations and/or to verify that the mean gravity system deformations do not exceed the deformation limits for deformation-controlled components.

REFERENCES

Al Atik, L., and Abrahamson, N. (2010). "An improved method for nonstationary spectral matching." *Earthq. Spectra* 26(3), 601–617.

Ancheta, T. D., Darragh, R. B., Stewart, J. P., Seyhan, E., Silva, W. J., Chious, B. S. J., et al. (2012). "PEER NGA-West 2 database." Pacific Earthquake Engineering Research Center, Berkeley, CA.

Baker, J. W. (2011). "Conditional mean spectrum: Tool for ground motion selection." *J. Struct. Eng.*, 137(3), 322–331. http://dx.doi.org/10.1061/(ASCE)ST.1943-541X.0000215.

Baker, J. W., and Cornell, C. A. (2006). "Correlation of response spectral values for multi-component ground motions." *Bull. Seismol. Soc. Am.* 96(1), 215–227.

Bommer, J. J., Scott, S. G., and Sarma, S. K. (2000). "Hazard-consistent earthquake scenarios." *Soil Dyn. Earthq. Eng.* 19(4), 219–231.

Bozorgnia, Y., Crouse, C. B., Hamburger, R. O., Klemencic, R., Krawinkler, H., Malley, J. O., et al. (2009). *Guidelines for performance-based seismic design of tall buildings*, Pacific Earthquake Engineering Research Center, Berkeley, CA.

Federal Emergency Management Agency (FEMA). (2009a). *NEHRP recommended provisions for seismic regulation for buildings and other structures*, FEMA P-753, FEMA, Washington, DC.

FEMA. (2009b). *Quantification of building seismic performance factors*, FEMA P-695, FEMA, Washington, DC.

Grant, D. N., and Diaferia, R. (2012). "Assessing adequacy of spectrum-matched ground motions for response history analysis," *Earthq. Eng. Struct. Dyn.* 42(9), 1265–1280. http://dx.doi.org/10.1002/eqe.2270.

Haroun, M. A., and Housner, G. W. (1981). "Seismic design of liquid storage tanks." *J. Techn. Coun.*. ASCE, New York 107(1), 191–207.

Housner, G. W. (1963). "The dynamic behavior of water tanks." *Bull. Seismol. Soc. Am.* 53(2), 381–387.

Huang, Y. N., Whittaker, A. S., and Luco, N. (2008). "Maximum spectral demands in the near fault region," *Earthq. Spectra*, 24, 319–341.

Jacobsen, L. S. (1949). "Impulsive hydrodynamics of fluid inside a cylindrical tank and of fluid surrounding a cylindrical pier." *Bull. Seismol. Soc. Am.* 39(3), 189–203.

Luco, N., and Bazzurro, P. (2007). "Does amplitude scaling of ground motion records result in biased nonlinear structural drift responses?" *Earthq. Eng. Struct. Dyn.* 36, 1813–1835.

Naeim, F., and Lew, M. (1995). On the use of design spectrum compatible time histories. *Earthq. Spectra* 11(1), 111–127.

National Institute of Standards and Technology (NIST). (2011). *Selecting and scaling earthquake ground motions for performing response-history analyses.* GCR 11-917-15. NEHRP Consultants Joint Venture, NIST, Gaithersburg, MD.

NIST. (2012). *Soil-structure interaction for building structures*, GCR 12-917-21. NEHRP Consultants Joint Venture, NIST, Gaithersburg, MD.

NIST. (2010). *Nonlinear structural analysis for seismic design.* NIST GCR 10-917-5, NIST, Gaithersburg, MD.

NIST (National Institute of Standards, and Technology). (2011). *Selecting and scaling earthquake ground motions for performing response-history analyses*, NIST GCR 11-917-15, NIST, Gaithersburg, MD.

Reiter, L. (1990). *Earthquake hazard analysis: Issues and insights.* Columbia University Press, New York.

Schnabel, P. B., Lysmer, J., and Seed, H. B. (1972). "SHAKE—A computer program for earthquake response analysis of horizontally layered sites." *Report EERC 72-12*, Earthquake Engineering Research Center, University of California, Berkeley, CA.

Shahi, S. K., Ling, T., Baker, J. W., and Jayaram, N. (2011). *New ground motion selection procedures and selected motions for the PEER transportation research program*, Pacific Earthquake Engineering Research Center, Berkeley, CA.

Stewart, J. P., Chiou, S. J., Bray, J. D., Graves, R. W., Sommerville, P. G., and Abrahamson, N. A. (2002). "Ground motion evaluation procedures for performance-based design." *J. Soil Dyn. Earthq. Eng.*, 22, 9–12.

Wallace, J. W., Segura, C., and Tran, T. (2013). "Shear design of structural walls," Proc., 10th International Conference on Urban Earthquake Engineering, Tokyo Institute of Technology, Tokyo.

Zareian, F., and Medina, R. (2010). "A practical method for proper modeling of structural damping in inelastic plane structural systems," *J. Comput. Struct.* 88(1–2), 45–53.

OTHER REFERENCES (NOT CITED)

ASCE. (2014). "Seismic evaluation and retrofit of existing buildings." ASCE/SEI Standard 41-13, ASCE, Reston, VA.

Haselton, C. B., and Deierlein, G. G. (2007). *Assessing seismic collapse safety of modern reinforced concrete frame buildings*, PEER Report 2007/08, Pacific Earthquake Engineering Research Center, University of California, Berkeley, CA.

CHAPTER C17
SEISMIC DESIGN REQUIREMENTS FOR SEISMICALLY ISOLATED STRUCTURES

C17.1 GENERAL

Seismic isolation, also referred to as base isolation because of its common use at the base of building structures, is a design method used to substantially decouple the response of a structure from potentially damaging horizontal components of earthquake motions. This decoupling can result in response that is significantly reduced from that of a conventional, fixed-base building.

The significant damage to buildings and infrastructure following large earthquakes over the last three decades has led to the rapid growth of seismic isolation technology and the development of specific guidelines for the design and construction of seismically isolated buildings and bridges in the United States, as well as standardized testing procedures of isolation devices.

Design requirements for seismically isolated building structures were first codified in the United States as an appendix to the 1991 Uniform Building Code, based on "General Requirements for the Design and Construction of Seismic-Isolated Structures" developed by the State Seismology Committee of the Structural Engineers Association of California. In the intervening years, those provisions have developed along two parallel tracks into the design requirements in Chapter 17 of the ASCE/SEI 7 standard and the rehabilitation requirements in Section 9.2 of ASCE/SEI 41 (2007), *Seismic Rehabilitation of Existing Buildings*. The design and analysis methods of both standards are similar, but ASCE/SEI 41 allows more relaxed design requirements for the superstructure of rehabilitated buildings. The basic concepts and design principles of seismic isolation of highway bridge structures were developed in parallel and first codified in the United States in the 1990 AASHTO provisions *Guide Specifications for Seismic Isolation Design*. The subsequent version of this code (AASHTO 1999) provides a systematic approach to determining bounding limits for analysis and design of isolator mechanical properties.

The present edition of the ASCE/SEI 7, Chapter 17, provisions contains significant modifications with respect to superseded versions, intended to facilitate the design and implementation process of seismic isolation, thus promoting the expanded use of the technology. Rather than addressing a specific method of seismic isolation, the standard provides general design requirements applicable to a wide range of seismic isolation systems. Because the design requirements are general, testing of isolation-system hardware is required to confirm the engineering parameters used in the design and to verify the overall adequacy of the isolation system. Use of isolation systems whose adequacy is not proved by testing is prohibited. In general, acceptable systems (a) maintain horizontal and vertical stability when subjected to design displacements, (b) have an inherent restoring force defined as increasing resistance with increasing displacement, (c) do not degrade significantly under repeated cyclic load, and (d) have quantifiable engineering parameters (such as force-deflection characteristics and damping).

The lateral force-displacement behavior of isolation systems can be classified into four categories, as shown in Fig. C17.1-1, where each idealized curve has the same design displacement, D_D.

A linear isolation system (Curve A) has an effective period that is constant and independent of the displacement demand, where the force generated in the superstructure is directly proportional to the displacement of the isolation system.

A hardening isolation system (Curve B) has a low initial lateral stiffness (or equivalently a long effective period) followed by a relatively high second stiffness (or a shorter effective period) at higher displacement demands. Where displacements exceed the design displacement, the superstructure is subjected to increased force demands, while the isolation system is subject to reduced displacements, compared to an equivalent linear system with equal design displacement, as shown in Fig. C17.1-1.

A softening isolation system (Curve C) has a relatively high initial stiffness (short effective period) followed by a relatively low second stiffness (longer effective period) at higher displacements. Where displacements exceed the design displacement, the superstructure is subjected to reduced force demands, while the isolation system is subject to increased displacement demand than for a comparable linear system.

The response of a purely sliding isolation system without lateral restoring force capabilities (Curve D) is governed by friction forces developed at the sliding interface. With increasing

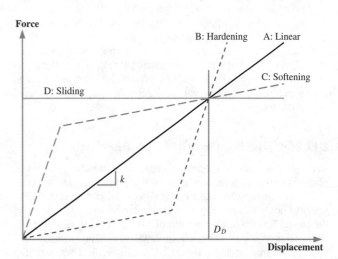

FIGURE C17.1-1 Idealized Force-Deflection Relationships for Isolation Systems (Stiffness Effects of Sacrificial Wind-Restraint Systems Not Shown for Clarity)

displacements, the effective period lengthens while loads on the superstructure remain constant. For such systems, the total displacement caused by repeated earthquake cycles is highly dependent on the characteristics of the ground motion and may exceed the design displacement, D_D. Since these systems do not have increasing resistance with increasing displacement, which helps to recenter the structure and prevent collapse, the procedures of the standard cannot be applied, and use of the system is prohibited.

Chapter 17 establishes isolator design displacements, shear forces for structural design, and other specific requirements for seismically isolated structures based on MCE_R only. All other design requirements, including loads (other than seismic), load combinations, allowable forces and stresses, and horizontal shear distribution, are the same as those for conventional, fixed-base structures. The main changes incorporated in this edition of the provisions include the following:

- Modified calculation procedure for the elastic design base shear forces from the design earthquake (DE) event to the MCE_R event using a consistent set of upper and lower bound stiffness properties and displacements. This modification simplifies the design and analysis process by focusing only on the MCE_R event.
- Relaxed permissible limits and criteria for the use of the equivalent lateral force (ELF) procedure. This modification minimizes the need to perform complex and computationally expensive nonlinear time history analyses to design the superstructure and isolation system on many base-isolated structures.
- Enhanced definitions of design properties of the isolation system.
- Use of nominal properties in the design process of typical isolation bearings specified by the manufacturers based on prior prototype testing.
- These nominal properties are adjusted using the newly incorporated AASHTO (1999) lambda factor concept to account for response uncertainties and obtain upper and lower bound properties of the isolation system for the design process.
- New method for the vertical distribution of lateral forces associated with the ELF method of design.
- Simplified approach for incorporating a 5% accidental mass eccentricity in nonlinear time history analyses.
- Reduction in the required number of peer reviewers on a seismic isolation project from the current three to five to a minimum of one peer reviewer. Also, peer reviewers are not required to attend the prototype tests.
- Calculation procedure to estimate permanent residual displacements that may occur in seismic isolation applications with relatively long period high yield/friction levels, and small yield displacements under a wide range of earthquake intensity.

C17.2 GENERAL DESIGN REQUIREMENTS

In an ideal seismic isolation application, the lateral displacement of the structure is primarily accommodated through large lateral displacement or deformation of the isolation system rather than internal deformation of the superstructure above. Accordingly, the lateral force-resisting system of the superstructure above the isolation system is designed to have sufficient stiffness and strength to prevent large, inelastic displacements. Therefore, the standard contains criteria that limit the inelastic response of the superstructure. Although damage control is not an explicit objective of the standard, design to limit inelastic response of the structural system directly reduces the level of damage that would otherwise occur during an earthquake. In general, isolated structures designed in accordance with the standard are expected to

1. resist minor and moderate levels of earthquake ground motion without damage to structural elements, nonstructural components, or building contents, and
2. resist major levels of earthquake ground motion without failure of the isolation system, significant damage to structural elements, extensive damage to nonstructural components, or major disruption to facility function.

Isolated structures are expected to perform considerably better than fixed-based structures during moderate and major earthquakes. Table C17.2-1 compares the expected performance of isolated and fixed-based structures designed in accordance with the standard. Actual performance of an isolated structure should be determined by performing nonlinear time history analyses and computing interstory drifts and floor acceleration demands for an array of ground motions. Those results can be used to compute postearthquake repair costs of the structure using the FEMA P-58 performance-based earthquake engineering (PBEE) methodology (FEMA 2012) and/or large-scale simulations of direct and indirect costs using HAZUS software (FEMA 1999). Evaluation of seismic performance enhancement using seismic isolation should include its impact on floor accelerations, as well as interstory drifts, because these elements are key engineering demand parameters affecting damage in mechanical, electrical, and plumbing (MEP) equipment, ceilings and partitions, and building contents.

Loss of function or discontinued building operation is not included in Table C17.2-1. For certain fixed-based facilities, loss of function would not be expected unless there is significant structural and nonstructural damage that causes closure or restricted access to the building. In other cases, a facility with only limited or no structural damage would not be functional as a result of damage to vital nonstructural components or contents. Seismic isolation, designed in accordance with these provisions, would be expected to mitigate structural and nonstructural damage and to protect the facility against loss of function. The postearthquake repair time required to rehabilitate the structure can also be determined through a FEMA P-58 PBEE evaluation.

Observed structural or nonstructural damage in fixed-based buildings caused by moderate and large earthquakes around the world have typically been associated with high-intensity lateral ground motion excitation rather than vertical acceleration. Gravity design procedures for typical structures result in structural sections and dimensions with relatively high safety factors for

Table C17.2-1 Performance Expected for Minor, Moderate, and Major Earthquakes

	Earthquake Ground Motion Level[a]		
Performance Measure	Minor	Moderate	Major
Life safety: Loss of life or serious injury is not expected	F, I	F, I	F, I
Structural damage: Significant structural damage is not expected	F, I	F, I	I
Nonstructural damage: Significant nonstructural or content damage is not expected	F, I	I	I

[a]F indicates fixed base; I indicates isolated.

seismic resistance. Therefore, current code provisions for fixed-based (or isolated) buildings only require use of a vertical earthquake component, E_v, obtained from static analysis procedures per Sections 12.2.4.6 and 12.2.7.1, defined as $0.2 S_{DS} D$ under the design earthquake, where D is the tributary dead load rather than explicit incorporation of vertical ground motions in the design analysis process. For seismic isolation, it should be noted that the term $0.2 S_{DS}$ is replaced with $0.2 S_{MS}$.

However, similar to fixed-based buildings, consideration of horizontal ground motion excitation alone may underestimate the acceleration response of floors and other building components. Portions of fixed-based and isolated structures may be especially sensitive to adverse structural response amplification induced by vertical ground motions including long spans, vertical discontinuities, or large cantilever elements. Certain nonstructural components, such as acoustic tile suspended ceiling systems, are also particularly vulnerable to the combination of vertical and horizontal ground motion effects. These building subassemblies or components may warrant additional vertical considerations. In addition, isolators with relatively low tributary gravity load and isolators located below columns that form part of the lateral force-resisting system can potentially have net uplift or tensile displacements caused by combined large vertical ground motion accelerations and global overturning. This uplift or bearing tension may induce high impact forces on the substructure, jeopardize the stability of the bearings, or result in bearing rupture.

Base-isolated structures located near certain fault characteristics that produce large vertical accelerations (e.g., hanging wall in reverse and reverse/oblique faults) are also more vulnerable and therefore may also require consideration of vertical ground motion excitation.

Vertical ground acceleration may affect the behavior of axial-load dependent isolation systems in the horizontal direction caused by potential coupling between horizontal and vertical response of the building structure.

Building response parameters that are expected to be affected by vertical excitation are vertical floor spectra and axial load demand on isolation bearings and columns, as discussed in Section C17.2.4.6. Isolated buildings with significant horizontal–vertical coupling are also expected to impart additional horizontal accelerations to the building at the frequencies of coupled modes that match the vertical motions.

If it is elected to investigate the effect of vertical ground motion acceleration on building response, one of the following analysis methods is suggested:

- Response spectrum analysis using horizontal and vertical spectrum (upward and downward).
- Response spectrum analysis using a vertical spectrum, combined with horizontal response spectrum analysis results using orthogonal combinations corresponding to the 100%–30%–30% rule.
- Three-dimensional response history analysis following the recommendations of Section C17.3.3 with explicit inclusion of vertical ground motion acceleration records.
- Horizontal response history analysis following the provisions of Section 17.3.3 considering the two limiting initial gravity load conditions defined per Section 17.2.7.1. Note that this analysis affects the effective characteristics of axial load-dependent isolators with resulting changes in base shear and displacement demands.

The structural model in these analyses should be capable of capturing the effects of vertical response and vertical mass participation, and should include the modeling recommendations in Section C17.6.2.

C17.2.4 Isolation System

C17.2.4.1 Environmental Conditions. Environmental conditions that may adversely affect isolation system performance must be investigated thoroughly. Specific requirements for environmental considerations on isolators are included in the new Section 17.2.8. Unlike conventional materials whose properties do not vary substantially with time, the materials used in seismic isolators are typically subject to significant aging effects over the life span of a building structure. Because the testing protocol of Section 17.8 does not account for the effects of aging, contamination, scragging (temporary degradation of mechanical properties with repeated cycling), temperature, velocity effects, and wear, the designer must account for these effects by explicit analysis. The approach to accommodate these effects, introduced in the AASHTO specifications (AASHTO 1999), is to use property modification factors as specified in Section 17.2.8.4.

C17.2.4.2 Wind Forces. Lateral displacement over the depth of the isolation region resulting from wind loads must be limited to a value similar to that required for other stories of the superstructure.

C17.2.4.3 Fire Resistance. Where fire may adversely affect the lateral performance of the isolation system, the system must be protected to maintain the gravity-load resistance and stability required for the other elements of the superstructure supported by the isolation system.

C17.2.4.4 Lateral Restoring Force. The restoring force requirement is intended to limit residual displacements in the isolation system resulting from any earthquake event so that the isolated structure will adequately withstand aftershocks and future earthquakes. The potential for residual displacements is addressed in Section C17.2.6.

C17.2.4.5 Displacement Restraint. The use of a displacement restraint to limit displacements beyond the design displacement is discouraged. Where a displacement restraint system is used, explicit nonlinear response history analysis of the isolated structure for the MCE_R level is required using the provisions of Chapter 16 to account for the effects of engaging the displacement restraint.

C17.2.4.6 Vertical-Load Stability. The vertical loads used to assess the stability of a given isolator should be calculated using bounding values of dead load, live load, and the peak earthquake demand at the MCE_R level. Because earthquake loads are reversible in nature, peak earthquake load should be combined with bounding values of dead and live load in a manner that produces both the maximum downward force and the maximum upward force on any isolator. Stability of each isolator should be verified for these two extreme values of vertical load at peak MCE_R displacement of the isolation system. In addition, all elements of the isolation system require testing or equivalent measures that demonstrate their stability for the MCE_R ground motion levels. This stability can be demonstrated by performing a nonlinear static analysis for an MCE_R response displacement of the entire structural system, including the isolation system, and showing that lateral and vertical stability are maintained. Alternatively, this stability can be demonstrated by performing a nonlinear dynamic analysis for the MCE_R motions using the same inelastic reductions as for the design earthquake (DE) and acceptable capacities except that member and connection strengths can be taken as their nominal strengths with resistance factors, ϕ, taken as 1.0.

Vertical ground motion excitation affects bounding axial loads on isolation bearings and vertical stability design checks. The E component of load combination 5 of Section 2.3.2 should consider the maximum of E_v per code or the dynamic amplification from analysis when significant vertical acceleration is anticipated per Section C17.2.

C17.2.4.7 Overturning. The intent of this requirement is to prevent both global structural overturning and overstress of elements caused by localized uplift. Isolator uplift is acceptable as long as the isolation system does not disengage from its horizontal-resisting connection details. The connection details used in certain isolation systems do not develop tension resistance, a condition which should be accounted for in the analysis and design. Where the tension capacity of an isolator is used to resist uplift forces, design and testing in accordance with Sections 17.2.4.6 and 17.8.2.5 must be performed to demonstrate the adequacy of the system to resist tension forces at the total maximum displacement.

C17.2.4.8 Inspection and Replacement. Although most isolation systems do not require replacement following an earthquake event, access for inspection, repair, and replacement must be provided. In some cases (Section 17.2.6), recentering may be required. The isolation system should be inspected periodically as well as following significant earthquake events, and any damaged elements should be repaired or replaced.

C17.2.4.9 Quality Control. A testing and inspection program is necessary for both fabrication and installation of the isolator units. Because of the rapidly evolving technological advances of seismic isolation, reference to specific standards for testing and inspection is difficult for some systems, while reference for some systems is possible (e.g., elastomeric bearings should follow ASTM D4014 requirements (ASTM 2012). Similar standards are yet to be developed for other isolation systems. Special inspection procedures and load testing to verify manufacturing quality should therefore be developed for each project. The requirements may vary depending on the type of isolation system used. Specific requirements for quality control testing are now given in Section 17.8.5.

C17.2.5 Structural System

C17.2.5.2 Minimum Building Separations. A minimum separation between the isolated structure and other structures or rigid obstructions is required to allow unrestricted horizontal translation of the superstructure in all directions during an earthquake event. The separation dimension should be determined based on the total design displacement of the isolation system, the maximum lateral displacement of the superstructure above the isolation, and the lateral deformation of the adjacent structures.

C17.2.5.4 Steel Ordinary Concentrically Braced Frames. Section 17.5.4.2 of this standard implies that only seismic force-resisting systems permitted for fixed-based building applications are permitted to be used in seismic isolation applications. Table 12.2-1 limits the height of steel ordinary concentrically braced frames (OCBFs) in fixed-based multistory buildings assigned Seismic Design Categories D and E to 35 ft (10.7 m) and does not permit them in buildings assigned to Seismic Design Category F. Section 17.2.5.4 permits them to be used for seismic isolation applications to heights of 160 ft (48.8 m) in buildings assigned to Seismic Design Categories D, E, and F, provided that certain additional requirements are satisfied. The additional design requirements that must be satisfied include that the building must remain elastic at the design earthquake level (i.e., $R_I = 1.0$), that the moat clearance displacement, D_{TM}, be increased by 20%, and that the braced frame be designed to satisfy Section F1.7 of AISC 341. It should be noted that currently permitted OCBFs in seismically isolated buildings assigned to Seismic Design Categories D and E also need to satisfy Section F1.7 of AISC 341.

Seismic isolation has the benefit of absorbing most of the displacement of earthquake ground motions, allowing the seismic force-resisting system to remain essentially elastic. Restrictions in Chapter 17 on the seismic force-resisting system limit the inelastic reduction factor to a value of 2 or less to ensure essentially elastic behavior. A steel OCBF provides the benefit of providing a stiff superstructure with reduced drift demands on drift-sensitive nonstructural components while providing significant cost savings as compared to special systems. Steel OCBFs have been used in the United States for numerous (perhaps most) new seismically isolated essential facility buildings since the seismic isolation was first introduced in the 1980s. Some of these buildings have had heights as high as 130 ft (39.6 m). The 160-ft (48.8-m) height limit was permitted for seismic isolation with OCBFs in high seismic zones when seismic isolation was first introduced in the building code as an appendix to the UBC in 1991. When height limits were restricted for fixed-based OCBFs in the 2000 NEHRP Recommended Provisions, it was not recognized the effect the restriction could have on the design of seismically isolated buildings. The Section 17.2.5.4 change rectifies that oversight. It is the judgment of this committee that height limits should be increased to the 160-ft (48.8-m) level, provided that the additional conditions are met.

The AISC Seismic Committee (Task Committee-9) studied the concept of steel OCBFs in building applications to heights of 160 ft (48.8 m) in high seismic areas. They decided that additional detailing requirements are required, which are found in Section F1.7 of AISC 341.

There has been some concern that steel ordinary concentrically braced frames may have an unacceptable collapse hazard if ground motions greater than MCE$_R$ cause the isolation system to impact the surrounding moat wall. While there has not been a full FEMA P-695 (FEMA 2009) study of ordinary steel concentrically braced frame systems, a recent conservative study of one structure using OCBFs with $R_I = 1$ on isolation systems performed by Armin Masroor at SUNY Buffalo (Masroor and Mosqueda 2015) indicates that an acceptable risk of collapse (10% risk of collapse given MCE ground motions) is achieved if a 15–20% larger isolator displacement is provided. The study does not include the backup capacity of gravity connections or the influence of concrete-filled metal deck floor systems on the collapse capacity. Even though there is no requirement to consider ground motions beyond the risk-targeted maximum considered earthquake ground motion in design, it was the judgment of this committee to provide additional conservatism by requiring 20% in moat clearance. It is possible that further P-695 studies will demonstrate that the additional 1.2 factor of displacement capacity may not be needed.

C17.2.5.5 Isolation System Connections. This section addresses the connections of the structural elements that join isolators together. The isolators, joining elements, and connections comprise the isolation system. The joining elements are typically located immediately above the isolators; however, there are many ways to provide this framing, and this section is not meant to exclude other types of systems. It is important to note that the elements and the connections of the isolation system are designed for V_b level forces, while elements immediately above the isolation system are designed for V_s level forces.

Although ductility detailing for the connections in the isolation system is not required, and these elements are designed to remain elastic with V_b level forces using $R = 1.0$, in some cases it may still be prudent to incorporate ductility detailing in these connections (where possible) to protect against unforeseen loading. This incorporation has been accomplished in the past by providing connection details similar to those used for a seismic force-resisting system of Table 12.2-1, with connection moment and shear strengths beyond the code minimum requirements. Ways of accomplishing this include factoring up the design forces for these connections, or providing connections with moment and shear strengths capable of developing the expected plastic moment strength of the beam, similar to AISC 341 or ACI 318 requirements for ordinary moment frames (OMFs).

C17.2.6 Elements of Structures and Nonstructural Components.
To accommodate the differential horizontal and vertical movement between the isolated building and the ground, flexible utility connections are required. In addition, stiff elements crossing the isolation interface (such as stairs, elevator shafts, and walls) must be detailed to accommodate the total maximum displacement without compromising life safety provisions.

The effectiveness and performance of different isolation devices in building structures under a wide range of ground motion excitations have been assessed through numerous experimental and analytical studies (Kelly et al. 1980, Kelly and Hodder 1981, Kelly and Chaloub 1990; Zayas et al. 1987; Constantinou et al. 1999; Warn and Whittaker 2006; Buckle et al. 2002; Kelly and Konstantinidis 2011). The experimental programs included in these studies have typically consisted of reduced-scale test specimens, constructed with relatively high precision under laboratory conditions. These studies initially focused on elastomeric bearing devices, although in recent years the attention has shifted to the single- and multiconcave friction pendulum bearings. The latter system provides the option for longer isolated periods.

Recent full-scale shake table tests (Ryan et al. 2012) and analytical studies (Katsaras 2008) have shown that the isolation systems included in these studies with a combination of longer periods, relatively high yield/friction levels and small yield displacements will result in postearthquake residual displacements. In these studies, residual displacements ranging from 2 to 6 in. (50 to 150 mm) were measured and computed for isolated building structures with a period of 4 seconds or greater and a yield level in the range of 8 to 15% of the structure's weight. This permanent offset may affect the serviceability of the structure and possibly jeopardize the functionality of elements crossing the isolation plane (such as fire protection and weatherproofing elements, egress/entrance details, elevators, and joints of primary piping systems). Since it may not be possible to recenter some isolation systems, isolated structures with such characteristics should be detailed to accommodate these permanent offsets.

The Katsaras report (2008) provides recommendations for estimating the permanent residual displacement in any isolation system based on an extensive analytical and parametric study. The residual displacements measured in full-scale tests (Ryan et al. 2012) are reasonably predicted by this procedure, which

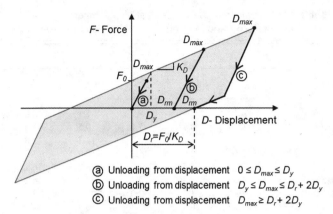

FIGURE C17.2-1 Definitions of Static Residual Displacement D_{rm} for a Bilinear Hysteretic System

uses an idealized bilinear isolation system, shown in Fig. C17.2-1. The three variables that affect the residual displacement are the isolated period (based on the second slope stiffness K_D), the yield/friction level (F_0), and the yield displacement (D_y).

The procedure for estimating the permanent residual displacement, D_{rd} (see Eq. C17.2-1) is a function of the system yield displacement D_y, the static residual displacement, $D_r = F_0/K_p$, and D_{rm}, which is a function of D_m, the maximum earthquake displacement shown in Table C17.2-2. For most applications, D_{rm} is typically equal to D_r.

$$D_{rd} = \frac{0.87 D_{rm}}{\left(1 + 4.3 \frac{D_{rm}}{D_r}\right)\left(1 + 31.7 \frac{D_y}{D_r}\right)} \quad (C17.2\text{-}1)$$

Thus, there is a simple two-step process to estimate the permanent residual displacement, D_{rd}:

- Calculate the static residual displacement, D_r, based on the isolated period (using the second slope stiffness, K_D) and the yield/friction levels. Table C17.2-3 provides values of D_r for a range of periods from 2.5 to 20 seconds and a range of yield/friction levels from 0.03 W to 0.15 W.
- Using the value of D_r calculated for the isolation system and the yield displacement, D_y, of the system, the permanent residual displacement, D_{rd}, can be calculated from Eq. (C17.2-1), and Tables C17.2-4 and C17.2-5 provide the residual displacements for earthquake displacements (D_m) of 10 in. and 20 in. (250 mm to 500 mm), respectively.

The cells with bold type in Tables C17.2-4 and C17.2-5 correspond to permanent residual displacements exceeding 2.0 in. (50 mm). Note that for yield displacements of approximately 2.0 in. (50 mm), residual displacements will not occur for most isolation systems.

Table C17.2-2 Values of Static Residual Displacement, D_{rm}

Range of Maximum Displacement, D_{max}	Static Residual Displacement, D_{rm}
$0 \leq D_{max} \leq D_y$	0
$D_y \leq D_{max} < D_r + 2D_y$	$D_r(D_{max} - D_y)/(D_r + D_y)$
$D_r + 2D_y \leq D_{max}$	D_r

Table C17.2-3 Values of Static Residual Displacement, D_r (in.), for Various Isolated Periods, T (s), and Yield/Friction Levels, F_0

	F_0				
T (s)	0.03	0.06	0.09	0.12	0.15
2.5	1.8	3.6	5.3	7.1	8.9
2.8	2.4	4.7	7.1	9.5	11.9
3.5	3.6	7.1	10.7	14.2	17.8
4.0	4.7	9.5	14.2	19.0	23.7
5.0	7.2	14.5	21.7	28.9	36.1
5.6	9.2	18.5	27.7	37.0	46.2
6.0	10.7	21.3	32.0	42.7	53.3
7.0	14.2	28.4	42.7	56.9	71.1
8.0	18.7	37.4	56.2	74.9	93.6
9.0	23.7	47.4	71.1	94.8	118.5
20.1	118.5	237.0	355.5	474.0	592.5

Note: 1 in. = 25 mm.

Table C17.2-4 Permanent Residual Displacement, D_{rd}, for a Maximum Earthquake Displacement, D_m, of 10 in. (250 mm)

	D_y (in.)							
D_r (in.)	0.005	0.01	0.02	0.20	0.39	0.59	0.98	1.97
4.0	0.63	0.60	0.56	0.25	0.16	0.11	0.07	0.04
7.9	1.28	1.25	1.21	0.73	0.50	0.39	0.26	0.14
11.9	1.86	1.84	1.79	1.22	0.90	0.71	0.50	0.27
15.8	**2.32**	**2.30**	**2.25**	1.67	1.29	1.04	0.75	0.43
19.8	**2.72**	**2.70**	**2.66**	**2.07**	1.65	1.37	1.01	0.59
23.7	**3.08**	**3.06**	**3.02**	**2.43**	1.99	1.68	1.27	0.76
27.7	**3.39**	**3.37**	**3.34**	**2.75**	**2.30**	1.97	1.51	0.92
31.6	**3.68**	**3.66**	**3.62**	**3.05**	**2.59**	**2.24**	1.75	1.09
35.6	**3.93**	**3.91**	**3.87**	**3.32**	**2.85**	**2.49**	1.97	1.25
39.5	**4.16**	**4.14**	**4.11**	**3.56**	**3.09**	**2.73**	**2.19**	1.41

Note: 1 in. = 25 mm.
Bold values designate D_{rd} values of 2 inches or more.

Table C17.2-5 Permanent Residual Displacements, D_{rd}, for a Maximum Earthquake Displacement, D_m, of 20 in. (500 mm)

	D_y (in.)							
D_r (in.)	0.005	0.01	0.02	0.20	0.39	0.59	0.98	1.97
4.0	0.63	0.60	0.56	0.25	0.16	0.11	0.07	0.04
7.9	1.28	1.25	1.21	0.73	0.50	0.39	0.26	0.15
11.9	1.93	1.90	1.85	1.28	0.95	0.76	0.54	0.31
15.8	**2.58**	**2.55**	**2.50**	1.86	1.45	1.19	0.87	0.52
19.8	**3.23**	**3.20**	**3.15**	**2.47**	1.98	1.65	1.24	0.75
23.7	**3.75**	**3.72**	**3.67**	**2.97**	**2.45**	**2.08**	1.59	0.99
27.7	**4.22**	**4.20**	**4.15**	**3.45**	**2.90**	**2.50**	1.95	1.24
31.6	**4.67**	**4.64**	**4.60**	**3.90**	**3.33**	**2.90**	**2.30**	1.50
35.6	**5.08**	**5.06**	**5.02**	**4.32**	**3.74**	**3.30**	**2.65**	1.76
39.5	**5.47**	**5.45**	**5.41**	**4.72**	**4.13**	**3.67**	**2.99**	**2.02**

Note: 1 in. = 25 mm.
Bold values designate D_{rd} values of 2 inches or more.

C17.2.8 Isolation System Properties. This section defines and combines sources of variability in isolation system mechanical properties measured by prototype testing, permitted by manufacturing specification tolerances, and occurring over the life span of the structure because of aging and environmental effects. Upper bound and lower bound values of isolation system component behavior (e.g., for use in response history analysis procedures) and maximum and minimum values of isolation system effective stiffness and damping based on these bounding properties (e.g., for use in equivalent lateral force procedures) are established in this section. Values of property modification factors vary by product and cannot be specified generically in the provisions. Typical "default" values for the more commonly used systems are provided below. The designer and peer reviewer are responsible for determining appropriate values of these factors on a project-specific and product-specific basis.

This section also refines the concept of bounding (upper bound and lower bound) values of isolation system component behavior by

1. Explicitly including variability caused by manufacturing tolerances, aging, and environmental effects. ASCE/SEI 7-10 only addressed variability associated with prototype testing and
2. Simplifying design by basing bounding measures of amplitude-dependent behavior on only MCE_R ground motions. ASCE/SEI 7-10 used both design earthquake (DE) and MCE_R ground motions.

The new section also refines the concept of maximum and minimum effective stiffness and damping of the isolation system by use of revised formulas that

1. Define effective properties of the isolation system on bounding values of component behavior (i.e., same two refinements, described above) and
2. Eliminates the intentional conservatism of ASCE/SEI 7-10 that defines minimum effective damping in terms of maximum effective stiffness.

C17.2.8.2 Isolator Unit Nominal Properties. Isolator manufacturers typically supply nominal design properties that are reasonably accurate and can be confirmed by prototype tests in the design and construction phases. These nominal properties should be based on past prototype tests as defined in Section 17.8.2; see Fig. C17.2-2.

C17.2.8.3 Bounding Properties of Isolation System Components. The methodology for establishing lower and upper bound values for isolator basic mechanical properties based on property modification factors was first presented in Constantinou et al. (1999). It has since then been revised in Constantinou et al. (2007) based on the latest knowledge of

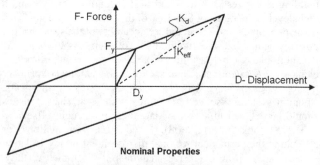

FIGURE C17.2-2 Example of the Nominal Properties of a Bilinear Force Deflection System

lifetime behavior of isolators. The methodology presented uses property modification factors to adjust isolator nominal properties based on considerations of natural variability in properties, effects of heating during cyclic motion, and the effects of aging, contamination, ambient temperature and duration of exposure to that temperature, and history of loading. The nominal mechanical properties should be based on prototype (or representative) testing on isolators not previously tested, at normal temperature and under dynamic loading.

The methodology also modifies the property modification factors to account for the unlikely situation of having several events of low probability of occurrence occur at the same time (i.e., maximum earthquake, aging, and low temperature) by use of property adjustment factors that are dependent on the significance of the structure analyzed (values range from 0.66 for a typical structure to 1.0 for a critical structure). This standard presumes that the property adjustment factor is 0.75. However, the registered design professional may opt to use the value of 1.0 based on the significance of the structure (e.g., health-care facilities or emergency operation centers) or based on the number of extreme events considered in the establishment of the property modification factor. For example, if only aging is considered, then a property adjustment factor of unity is appropriate.

Examples of application in the analysis and design of bridges may be found in Constantinou et al. (2011). These examples may serve as guidance in the application of the methodology in this standard. Constantinou et al. (2011) also presents procedures for estimating the nominal properties of lead-rubber and friction pendulum isolators, again based on the assumption that prototype test data are not available. Data used in the estimation of the range of properties were based on available test data, all of which were selected to heighten heating effects. Such data would be appropriate for cases of high-velocity motion and large lead core size or high friction values.

Recommended values for the specification tolerance on the average properties of all isolators of a given size isolator are typically in the $\pm 10\%$ to $\pm 15\%$ range. For a $\pm 10\%$ specification tolerance, the corresponding lambda factors would be $\lambda_{(spec, max)} = 1.10$ and $\lambda_{(spec, min)} = 0.90$. Variations in individual isolator properties are typically greater than the tolerance on the average properties of all isolators of a given size as presented in Section 17.2.8.4. It is recommended that the isolator manufacturer be consulted when establishing these tolerance values.

Section 17.2.8.4 requires the isolation system to be designed with consideration given to environmental conditions, including aging effects, creep, fatigue, and operating temperatures. The individual aging and environmental factors are multiplied together and then the portion of the lambda factor differing from unity is reduced by 0.75 based on the assumption that not all of the maximum values will occur simultaneously. As part of the design process, it is important to recognize that there will be additional variations in the nominal properties because of manufacturing. The next section specifies the property modification factors corresponding to the manufacturing process or default values if manufacturer-specific data are not available. These factors are combined with the property modification factors (Section 17.2.8.4) to determine the maximum and minimum properties of the isolators (Section 17.2.8.5) for use in the design and analysis process.

The lambda-test values $\lambda_{test, max}$ and $\lambda_{test, min}$ are determined from prototype testing and shall bound the variability and degradation in properties caused by speed of motion, heating effects, and scragging from Item 2 of Section 17.8.2.2. The registered design professional (RDP) shall specify whether this testing is performed quasi-statically, as in Item 2(a), or dynamically, as in Item 2(b). When testing is performed quasi-statically, the dynamic effects shall be accounted for in analysis and design using appropriate adjustment of the lambda-test values.

Item 3 of the testing requirements of Section 17.8.2.2 is important for property determination since it is common to Item 2. Using this testing, the lambda-test values $\lambda_{test, max}$ and $\lambda_{test, min}$ may be determined by three fully reversed cycles of dynamic (at the effective period T_M) loading at the maximum displacement $1.0 D_M$ on full-scale specimens. This test regime incorporates the effects of high-speed motion. The upper and lower bound values of K_d shall also envelop the $0.67 D_M$ and $1.0 D_M$ tests of Item 2 of Section 17.8.2.2. Therefore, the lambda-test values bound the effects of heating and scragging. As defined by Section 17.2.8.2, the nominal property of interest is defined as the average among the three cycles of loading. $\lambda_{test, max}$ shall be determined as the ratio of the first cycle property to the nominal property value. $\lambda_{test, min}$ shall be determined as the ratio of the property value at a representative cycle, determined by the RDP, to the nominal property value. The number of cycles shall be representative of the accepted performance of the isolation system for the local seismic hazard conditions, with the default cycle being the third cycle. A critique and guidance are provided in McVitty and Constantinou (2015).

C17.2.8.4 Property Modification Factors. The lambda factors are used to establish maximum and minimum mathematical models for analysis, the simplest form of which is the linear static procedure used to assess the minimum required design base shear and system displacements. More complex mathematical models account for various property variation effects explicitly (e.g., velocity, axial load, bilateral displacement, and instantaneous temperature). In this case, the cumulative effect of the lambda factors reduces (the combined lambda factor is closer to 1.0). However, some effects, such as specification tolerance and aging, are likely to always remain since they cannot be accounted for in mathematical models. Default lambda factors are provided in Table C17.2-6 as isolators from unknown manufacturers that do not have qualification test data. Default lambda factors are provided in Table C17.2-7 for most common types of isolators fabricated by quality manufacturers. Note that this table does not have any values of property modification factors for the actual stiffness (K_d) of sliding isolators. It is presumed that sliding isolators, whether flat or spherical, are produced with sufficiently high accuracy that their actual stiffness characteristics are known. The RDP may assign values of property modification factors different than unity for the actual stiffness of sliding bearings on the basis of data obtained in the prototype testing or on the basis of lack of experience with unknown manufacturers. Also note that this table provides values of property modification factors to approximately account for uncertainties in the materials and manufacturing methods used. These values presume lack of test data or incomplete test data and unknown manufacturers. For example, the values in Table C17.2-6 for sliding bearings presume unknown materials for the sliding interfaces so that there is considerable uncertainty in the friction coefficient values. Also, the data presume that elastomers used in elastomeric bearings have significant scragging and aging. Moreover, for lead-rubber bearings, the data in the table presume that there is considerable uncertainty in the starting value (before any hysteretic heating effects) of the effective yield strength of lead.

Accordingly, there is a considerable range in the upper and lower values of the property modification factors. Yet, these values should be used with caution since low-quality fabricators

Table C17.2-6 Default Upper and Lower Bound Multipliers for Unknown Manufacturers

Variable	Unlubricated Interfaces, μ or Q_d	Lubricated (Liquid) Interfaces, μ or Q_d	Plain Low Damping Elastomeric, K	Lead Rubber Bearing (LRB), K_d	Lead Rubber Bearing (LRB), Q_d	High-Damping Rubber (HDR), K_d	High-Damping Rubber (HDR), Q_d
Example: Aging and Environmental Factors							
Aging, λ_a	1.3	1.8	1.3	1.3	1	1.4	1.3
Contamination, λ_c	1.2	1.4	1	1	1	1	1
Example Upper Bound, $\lambda_{(ae, max)}$	1.56	2.52	1.3	1.3	1	1.4	1.3
Example Lower Bound, $\lambda_{(ae, min)}$	1.	1	1	1	1	1	1
Example: Testing Factors							
All cyclic effects, Upper	1.3	1.3	1.3	1.3	1.6	1.5	1.3
All cyclic effects, Lower	0.7	0.7	0.9	0.9	0.9	0.9	0.9
Example Upper Bound, $\lambda_{(test, max)}$	1.3	1.3	1.3	1.3	1.6	1.5	1.3
Example Lower Bound, $\lambda_{(test, min)}$	0.7	0.7	0.9	0.9	0.9	0.9	0.9
$\lambda_{(PM, max)} = (1 + (0.75 * (\lambda_{(ae, max)} - 1))) * \lambda_{(test, max)}$	1.85	2.78	1.59	1.59	1.6	1.95	1.59
$\lambda_{(PM, min)} = (1 - (0.75 * (1 - \lambda_{(ae, min)})) * \lambda_{(test, min)}$	0.7	0.7	0.9	0.9	0.9	0.9	0.9
Lambda factor for Spec. Tolerance, $\lambda_{(spec, max)}$	1.15	1.15	1.15	1.15	1.15	1.15	1.15
Lambda factor for Spec. Tolerance, $\lambda_{(spec, min)}$	0.85	0.85	0.85	0.85	0.85	0.85	0.85
Upper Bound Design Property Multiplier	2.12	3.2	1.83	1.83	1.84	2.24	1.83
Lower Bound Design Property Multiplier	0.6	0.6	0.77	0.77	0.77	0.77	0.77
Default Upper Bound Design Property Multiplier	2.1	3.2	1.8	1.8	1.8	2.2	1.8
Default Lower Bound Design Property Multiplier	0.6	0.6	0.8	0.8	0.8	0.8	0.8

Note: λ_{PM} is the lambda value for testing and environmental effects.

Table C17.2-7 Default Upper and Lower Bound Multipliers for Quality Manufacturers

Variable	Unlubricated PTFE, μ	Lubricated PTFE, μ	Rolling/ Sliding, K2	Plain Elastomerics, K	Lead rubber bearing (LRB), K2	Lead rubber bearing (LRB), Q_d	High-Damping Rubber (HDR), Q_d	High-Damping Rubber (HDR), K_d
Example: Aging and Environmental Factors								
Aging, λ_a	1.10	1.50	1.00	1.10	1.10	1.00	1.20	1.20
Contamination, $\lambda+$	1.10	1.10	1.00	1.00	1.00	1.00	1.00	1.00
Example Upper Bound, $\lambda_{(ae, max)}$	1.21	1.65	1.00	1.10	1.10	1.00	1.20	1.20
Example Lower Bound, $\lambda_{(ae, min)}$	1.00	1.00	1.00	1.00	1.00	1.00	1.00	1.00
Example: Testing Factors								
All cyclic effects, Upper	1.20	1.30	1.00	1.03	1.03	1.30	1.50	1.30
All cyclic effects, Lower	0.95	0.95	1.00	0.98	0.98	0.95	0.95	0.95
Example Upper Bound, $\lambda_{(test, max)}$	1.20	1.30	1.00	1.03	1.03	1.30	1.50	1.30
Example Lower Bound, $\lambda_{(test, min)}$	0.95	0.95	1.00	0.98	0.98	0.95	0.95	0.95
$\lambda_{(PM, max)} = (1 + (0.75 * (\lambda_{(ae, max)} - 1))) * \lambda_{(test, max)}$	1.39	1.93	1.00	1.11	1.11	1.30	1.73	1.50
$\lambda_{(PM, min)} = (1 - (0.75 * (1 - \lambda_{(ae, min)})) * \lambda_{(test, min)}$	0.95	0.95	1.00	0.98	0.98	0.95	0.95	0.95
Lambda factor for Spec. Tolerance, $\lambda_{(spec, max)}$	1.15	1.15	1.00	1.15	1.15	1.15	1.15	1.15
Lambda factor for Spec. Tolerance, $\lambda_{(spec, min)}$	0.85	0.85	1.00	0.85	0.85	0.85	0.85	0.85
Upper Bound Design Property Multiplier	1.60	2.22	1.00	1.27	1.27	1.50	1.98	1.72
Lower Bound Design Property Multiplier	0.81	0.81	1.00	0.83	0.83	0.81	0.81	0.81
Default Upper Bound Design Property Multiplier	1.6	2.25	1	1.3	1.3	1.5	2	1.7
Default Lower Bound Design Property Multiplier	0.8	0.8	1	0.8	0.8	0.8	0.8	0.8

Note: λ_{PM} is the lambda value for testing and environmental effects.

could use materials and vulcanization and manufacturing processes that result in even greater property variations. The preferred approach for establishing property modification factors is through rigorous qualification testing of materials and manufacturing methods by a quality manufacturer, and dynamic prototype testing of full-size specimens, and by quality control testing at project-specific loads and displacements. These test data on similar-sized isolators take precedence over the default values.

For elastomeric isolators, lambda factors and prototype tests may need to address axial–shear interaction, bilateral deformation, load history including first cycle effects and the effects of scragging of virgin elastomeric isolators, ambient temperature, other environmental loads, and aging effects over the design life of the isolator.

For sliding isolators, lambda factors and prototype tests may need to address contact pressure, rate of loading or sliding velocity, bilateral deformation, ambient temperature, contamination, other environmental loads, and aging effects over the design life of the isolator.

Rate of loading or velocity effects are best accounted for by dynamic prototype testing of full-scale isolators. Property modification factors for accounting for these effects may be used in lieu of dynamic testing.

Generally, ambient temperature effects can be ignored for most isolation systems if they are in conditioned space where the expected temperature varies between 30°F and 100°F.

The following comments are provided in the approach to be followed for the determination of the bounding values of mechanical properties of isolators:

1. Heating effects (hysteretic or frictional) may be accounted for on the basis of a rational theory (e.g., Kalpakidis and Constantinou 2008, 2009; Kalpakidis et al. 2010) so that only the effects of uncertainty in the nominal values of the properties, aging, scragging, and contamination need to be considered. This is true for lead-rubber bearings where lead of high purity and of known thermomechanical properties is used. For sliding bearings, the composition of the sliding interface affects the relation of friction to temperature and therefore cannot be predicted by theory alone. Moreover, heating generated during high-speed motion may affect the bond strength of liners. Given that there are numerous sliding interfaces (and that they are typically proprietary), that heating effects in sliding bearings are directly dependent on pressure and velocity, and that size is important in the heating effects (Constantinou et al. 2007), full-scale dynamic prototype and production testing are very important for sliding bearings.

2. Heating effects are important for sliding bearings and the lead core in lead-rubber bearings. They are not important and need not be considered for elastomeric bearings of either low or high damping. The reason for this is described in Constantinou et al. (2007), where it has been shown, based on theory and experimental evidence, that the rise in temperature of elastomeric bearings during cyclic motion (about one degree centigrade per cycle) is too small to significantly affect their mechanical properties. Prototype and production testing of full-size specimens at the expected loads and displacements should be sufficient to detect poor material quality and poor material bonding in plain elastomeric bearings, even if done quasi-statically.

3. Scragging and recovery to the virgin rubber properties (see Constantinou et al. 2007 for details) are dependent on the rubber compound, size of the isolator, the vulcanization process, and the experience of the manufacturer. Also, it has been observed that scragging effects are more pronounced for rubber of low shear modulus and that the damping capacity of the rubber has a small effect. It has also been observed that some manufacturers are capable of producing low-modulus rubber without significant scragging effects, whereas others cannot. It is therefore recommended that the manufacturer should present data on the behavior of the rubber under virgin conditions (not previously tested and immediately after vulcanization) so that scragging property modification factors can be determined. This factor is defined as the ratio of the effective stiffness in the first cycle to the effectiveness stiffness in the third cycle, typically obtained at a representative rubber shear strain (e.g., 100%). It has been observed that this factor can be as high as, or can exceed, a value of 2.0 for shear-modulus rubber less than or equal to 0.45 MPa (65 psi). Also, it has been observed that some manufacturers can produce rubber with a shear modulus of 0.45 MPa (65 psi) and a scragging factor of approximately 1.2 or less. Accordingly, it is preferred to establish this factor by testing for each project or to use materials qualified in past projects.

4. Aging in elastomeric bearings has in general small effects (typically increases in stiffness and strength of the order of 10% to 30% over the lifetime of the structure), provided that scragging is also minor. It is believed that scragging is mostly the result of incomplete vulcanization, which is thus associated with aging as chemical processes in the rubber continue over time. Inexperienced manufacturers may produce low shear modulus elastomers by incomplete vulcanization, which should result in significant aging.

5. Aging in sliding bearings depends on the composition of the sliding interface. There are important concerns with bimetallic interfaces (Constantinou et al. 2007), even in the absence of corrosion, so that they should be penalized by large aging property modification factors or simply not used. Also, lubricated interfaces warrant higher aging and contamination property modification factors. The designer can refer to Constantinou et al. (2007) for detailed values of the factor depending on the conditions of operation and the environment of exposure. Note that lubrication is meant to be *liquid* lubrication typically applied either directly at the interface or within dimples. Solid lubrication in the form of graphite or similar materials that are integrated in the fabric of liners and used in contact with stainless steel for the sliding interface does not have the problems experienced by liquid lubrication.

C17.2.8.5 Upper Bound and Lower Bound Force-Deflection Behavior of Isolation System Components. An upper and lower bound representation of each type of isolation system component shall be developed using the lambda factors developed in Section 17.2.8.4. An example of a bilinear force deflection loop is shown in Fig. C17.2-2. In C17.2-3, the upper and lower bound lambda factors are applied to the nominal properties of the yield/friction level and the second or bilinear slope of the lateral force-displacement curve to determine the

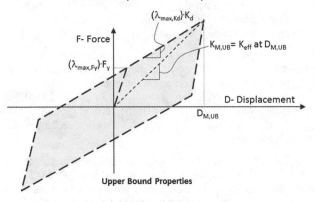
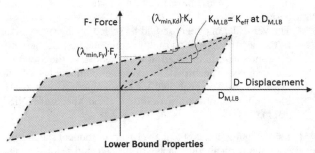

FIGURE C17.2-3 Example of the Upper and Lower Bound Properties of a Bilinear Force Deflection System

upper and lower bound representation of an isolation system component. The nomenclature shown in Fig. C17.2-3 is important to note. The effective stiffness and effective damping are calculated for both the upper and lower bound properties at the corresponding D_M. The maximum and minimum effective stiffness and effective damping are then developed from these upper and lower bound lateral force-displacement relationships in Section 17.2.8.6.

C17.3 SEISMIC GROUND MOTION CRITERIA

C17.3.1 Site-Specific Seismic Hazard. This new section consolidates existing site-specific hazard requirements from other sections.

C17.3.3 MCE_R Ground Motion Records. The MCE_R spectrum is constructed from the S_{MS}, S_{M1} parameters of Section 11.4.5, or 11.4.6, or 11.4.7.

When vertical excitation is included in isolated building response history analysis or response spectrum analysis, it is recommended that the vertical design spectra be computed by one of the following methods:

1. 2009 NEHRP Provisions (FEMA 2009) in new Chapter 23, equivalent to Annex A of Chapter 15, where the term S_{DS} is replaced with S_{MS}. The vertical spectrum is computed based on near-fault or far-fault conditions through the parameter S_s (short-period horizontal spectral acceleration for the site), as well as soil conditions (site classification).
2. Site-specific seismic hazard analysis using ground motion prediction equations for vertical shaking.
3. Multiplying the ordinates of the target spectrum corresponding to horizontal shaking by empirically based vertical-to-horizontal ratios that may be dependent on vertical period, site class, and proximity to fault.
4. Other approaches discussed in NIST GCR 11-917-15 (NIST 2011) consisting of a vertical conditional spectrum or conditional mean spectrum, envelope scaling, and mean spectral matching, or others.

Where response history analysis procedures are used, MCE_R ground motions should consist of not less than seven pairs of appropriate horizontal acceleration components.

Where vertical excitation is included in isolated building response history analysis, scaling of the vertical ground motion component may follow one of the following recommended procedures:

- The vertical motions are spectrally matched to the design vertical spectrum using a vertical period range of $0.2T_v$ to $1.5T_v$, where T_v is the building's primary vertical period of vibration. A wider period range may be considered because of uncertainty in the estimation of the primary vertical period of the building.
- The vertical component should be scaled by the same factor as the horizontal ground motion component(s). If the vertical component is included in the response of the structure, the response spectra of the vertical components of the records should be evaluated for reasonableness by comparing their spectra with a design vertical spectrum (NIST 2011).

If achieving a spectral fit to the vertical component spectrum is desirable, the vertical components of the selected records can be scaled by different factors than those used for horizontal components. Amplitude scaling of vertical components to a target vertical spectrum can be used using a least square error fit to a vertical period range of $0.2T_v$ to $1.5T_v$, where T_v is the building's primary vertical period of vibration. A wider period range may be considered in this case because of uncertainty in the estimation of the primary vertical period of the building.

C17.4 ANALYSIS PROCEDURE SELECTION

Three different analysis procedures are available for determining design-level seismic loads: the equivalent lateral force (ELF) procedure, the response spectrum procedure, and the response history procedure. For the ELF procedure, simple equations computing the lateral force demand at each level of the building structure (similar to those for conventional, fixed-base structures) are used to determine peak lateral displacement and design forces as a function of spectral acceleration and isolated-structure period and damping. The provisions of this section permit increased use of the ELF procedure, recognizing that the ELF procedure is adequate for isolated structures whose response is dominated by a single translational mode of vibration and whose superstructure is designed to remain essentially elastic (limited ductility demand and inelastic deformations) even for MCE_R level ground motions. The ELF procedure is now permitted for the design of isolated structures at all sites (except Site Class F) as long as the superstructure is regular (as defined in new Section 17.2.2), has a fixed-base period (T) that is well separated from the isolated period (T_{min}), and the isolation system meets certain "response predictability" criteria with which typical and commonly used isolation systems comply.

The design requirements for the structural system are based on the forces and drifts obtained from the MCE_R earthquake using a consistent set of upper and lower bound isolation system properties, as discussed in Section C17.5. The isolation system—including all connections, supporting structural elements, and the "gap"—is required to be designed (and tested) for 100% of MCE_R demand. Structural elements above the isolation system are now designed to remain essentially elastic for the MCE_R earthquake. A similar fixed-base structure would be designed for design earthquake loads ($2/3MCE_R$) reduced by a factor of 6 to 8 rather than the MCE_R demand reduced by a factor of up to 2 for a base-isolated structure.

C17.5 EQUIVALENT LATERAL FORCE PROCEDURE

The lateral displacements given in this section approximate peak earthquake displacements of a single-degree-of-freedom, linear-elastic system of period, T, and effective damping, β. Eqs. (17.5-1) and (17.5-3) of ASCE 7-10 provided the peak displacement in the isolation system at the center of mass for both the DE and MCE_R earthquakes, respectively. In these prior equations, as well as the current equation, the spectral acceleration terms at the isolated period are based on the premise that the longer period portion of the response spectra decayed as $1/T$. This is a conservative assumption and is the same as that required for design of a conventional, fixed-base structure of period T_M. A damping factor B, is used to decrease (or increase) the computed displacement demand where the effective damping coefficient of the isolation system is greater (or smaller) than 5% of critical damping. A comparison of values obtained from Eq. (17.5-1) and those obtained from nonlinear time history analyses are given in Kircher et al. (1988) and Constantinou et al. (1993).

The ELF formulas in this new edition compute minimum lateral displacements and forces required for isolation system design based only on MCE_R level demands, rather than on a combination of design earthquake and MCE_R levels, as in earlier editions of the provisions.

The calculations are performed separately for upper bound and lower bound isolation system properties, and the governing case shall be considered for design. Upper bound properties typically, but not always, result in a lower maximum displacement (D_M), higher damping (β_M), and higher lateral forces (V_b, V_{st}, V_s, and k).

Section 17.2.8 relates bounding values of effective period, stiffness, and damping of the isolation system to upper bound and lower bound lateral force-displacement behavior of the isolators.

C17.5.3 Minimum Lateral Displacements Required for Design

C17.5.3.1 Maximum Displacement. The provisions of this section reflect the MCE_R-only basis for design and define maximum MCE_R displacement in terms of MCE_R response spectral acceleration, S_{M1}, at the appropriate T.

In addition, and of equal significance, the maximum displacement (D_M) and the damping modification factor (B_M) are determined separately for upper bound and lower bound isolation system properties. In earlier provisions, the maximum displacement (D_M) was defined only in terms of the damping associated with lower bound displacement, and this damping was combined with the upper bound stiffness to determine the design forces. This change is theoretically more correct, but it removes a significant conservatism in the ELF design of the superstructure. This reduction in superstructure design conservatism is offset by the change from design earthquake to MCE_R ground motions as the basis for superstructure design forces.

C17.5.3.2 Effective Period at the Maximum Displacement. The provisions of this section are revised to reflect the MCE_R-only basis for design and associated changes in terminology (although maintaining the concept of effective period). The effective period T_M is also determined separately for the upper and lower bound isolation properties.

C17.5.3.3 Total Maximum Displacement. The provisions of this section are revised to reflect the MCE_R-only basis for design and associated changes in terminology. Additionally, the formula for calculating total (translational and torsional) maximum MCE_R displacement has been revised to include a term and corresponding equations that reward isolation systems configured to resist torsion.

The isolation system for a seismically isolated structure should be configured to minimize eccentricity between the center of mass of the superstructure and the center of rigidity of the isolation system, thus reducing the effects of torsion on the displacement of isolation elements. For conventional structures, allowance must be made for accidental eccentricity in both horizontal directions. Fig. C17.5-1 illustrates the terminology used in the standard. Eq. (17.5-3) provides a simplified formula for estimating the response caused by torsion in lieu of a more refined analysis. The additional component of displacement caused by torsion increases the design displacement at the corner of a structure by about 15% (for one perfectly square in plan) to about 30% (for one long and rectangular in plan) if the eccentricity is 5% of the maximum plan dimension. These calculated torsional displacements correspond to structures with an isolation system whose stiffness is uniformly distributed in plan. Isolation systems that have stiffness concentrated toward the perimeter of the structure, or certain sliding systems that minimize the effects of mass eccentricity, result in smaller torsional displacements.

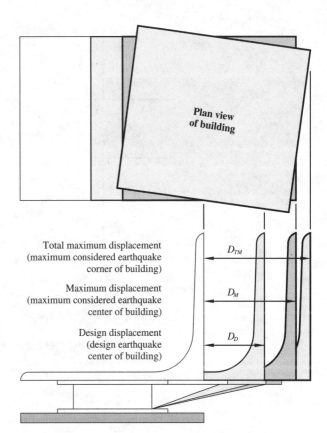

FIGURE C17.5-1 Displacement Terminology

The standard permits values of D_{TM} as small as $1.15D_M$, with proper justification.

C17.5.4 Minimum Lateral Forces Required for Design. Fig. C17.5-2 illustrates the terminology for elements at, below, and above the isolation system. Eq. (17.5-5) specifies the peak elastic seismic shear for design of all structural elements at or below the isolation system (without reduction for ductile response). Eq. (17.5-7) specifies the peak elastic seismic shear for design of structural elements above the isolation system. For structures that have appreciable inelastic-deformation capability, this equation includes an effective reduction factor ($R_I = 3R/8$ not exceeding 2). This factor ensures essentially elastic behavior of the superstructure above the isolators.

These provisions include two significant philosophic changes in the method of calculating the elastic base shear for the structure. In ASCE 7-10 and earlier versions of the provisions, the elastic design base shear forces were determined from the design earthquake (DE) using a mixture of the upper bound effective stiffness and the maximum displacement obtained using the lower bound properties of the isolation system, as shown schematically in Fig. C17.5-3. This was known to be conservative. The elastic design base shear is now calculated from the MCE_R event with a consistent set of upper and lower bound stiffness properties, as shown in Eq. (17.5-5) and Fig. C17.5-3.

A comparison of the old elastic design base shears for a range of isolation system design parameters and lambda factors using the ASCE 7-10 provisions and those using these new provisions is shown in Table C17.5-1. This comparison assumes that the DE is 2/3 the MCE_R and the longer period portion of both spectra decay as S_1/T. Table C17.5-1 shows a comparison between elastic design base shear calculated using the ASCE/SEI 7-10

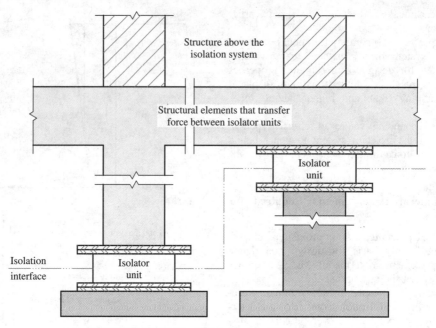

FIGURE C17.5-2 Isolation System Terminology

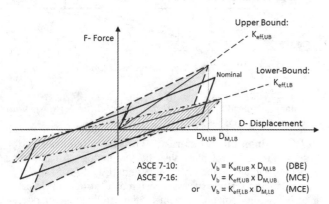

FIGURE C17.5-3 Nominal, Upper Bound, and Lower Bound Bilinear Hysteretic Properties of Typical Isolator Bearing

and 7-16 editions for a range of yield levels, second slopes, and bounding property multipliers.

The dark gray cells in Table C17.5-1 indicate that the new elastic design base shears are more than 10% higher than the old provisions; the light gray cells indicate that the new elastic base shears are 0 to 10% higher than the old provisions; and the white cells indicate that the new elastic base shears are less than the old provisions.

C17.5.4.1 Isolation System and Structural Elements below the Base Level. The provisions of this section are revised to reflect the MCE_R-only basis for design and associated changes in terminology. A new paragraph was added to this section to clarify that unreduced lateral loads should be used to determine overturning forces on the isolation system.

C17.5.4.2 Structural Elements above the Base Level. The provisions of this section are revised to reflect the MCE_R-only basis for design and associated changes in terminology, including the new concept of the "base level" as the first floor immediately above the isolation system.

An exception has been added to allow values of R_I to exceed the current limit of 2.0, provided that the pushover strength of the superstructure at the MCE_R drift or $0.015h_{sx}$ story drift exceeds (by 10%) the maximum MCE_R force at the isolation interface (V_b). This exception directly addresses required strength and associated limits on inelastic displacement for MCE_R demands. The pushover method is addressed in ASCE 41 (2007).

A new formula (Eq. (17.5-7)) now defines lateral force on elements above the base level in terms of reduced seismic weight (seismic weight excluding the base level), and the effective damping of the isolation system, based on recent work (York and Ryan 2008). In this formulation, it is assumed that the base level is located immediately (within 3.0 ft (0.9m) of top of isolator) above the isolation interface. When the base level is not located immediately above the isolation interface (e.g., there is no floor slab just above the isolators), the full (unreduced) seismic weight of the structure above the isolation interface is used in Eq. (17.5-7) to conservatively define lateral forces on elements above the base level.

C17.5.4.3 Limits on V_s. The provisions of this section are revised to reflect the MCE_R-only basis for design and associated changes in terminology.

In Section 17.5.4.3, the limits given on V_s are revised to clarify that the force required to fully activate the isolation system should be based on either the upper bound force-deflection properties of the isolation system or 1.5 times nominal properties, whichever is greater. Other limits include (a) the yield/friction level to fully activate the isolation system and (b) the ultimate capacity of a sacrificial wind-restraint system that is intended to fail and release the superstructure during significant lateral load.

These limits are needed so that the superstructure does not yield prematurely before the isolation system has been activated and significantly displaced.

C17.5.5 Vertical Distribution of Force. The provisions of this section are revised to incorporate a more accurate distribution of shear over height considering the period of the superstructure and the effective damping of the isolation system. The specified

Table C17.5-1 Comparison of Elastic Design Base Shears between ASCE 7-10 and 7-16

	Upper Bound Multipliers			K_d	Yield Level	Lower Bound Multipliers			K_d	Yield Level
$MCE_R\ S_1 = 1.5$				1.15	1.6				0.85	0.85
$T2$ (s)	2.00	2.00	3.00	3.00	4.00	4.00	5.00	5	6	6
Yield Level	0.05	0.10	0.05	0.10	0.05	0.10	0.05	0.1	0.05	0.1
New, V_b/W	0.80	0.66	0.47	0.42	0.33	0.33	0.26	0.28	0.21	0.26
ASCE 7-16/ASCE 7-10	1.14	1.02	1.08	0.91	1.02	0.84	0.96	0.83	0.91	0.82
				1.0	1.6				1.0	0.85
New, V_b/W	0.77	0.71	0.52	0.42	0.35	0.31	0.26	0.27	0.21	0.25
ASCE 7-16/ASCE 7-10	1.32	1.25	1.39	1.01	1.25	0.88	1.24	1.02	1.16	1.12
$MCE_R\ S_1 = 1.0$				1.15	1.6				0.85	0.85
$T2$ (s)	2.00	2.00	3.00	3.00	4.00	4.00	5.00	5	6	6
Yield Level	0.05	0.10	0.05	0.10	0.05	0.10	0.05	0.1	0.05	0.1
New, V_b/W	0.47	0.43	0.29	0.30	0.21	0.23	0.17	0.23	0.15	0.21
ASCE 7-16/ASCE 7-10	1.08	0.91	0.99	0.83	0.91	0.65	0.84	0.76	0.84	0.71
				1.35	1.5				0.85	0.85
New, V_b/W	0.54	0.47	0.33	0.32	0.24	0.29	0.19	0.22	0.16	0.20
ASCE 7-16/ASCE 7-10	1.12	0.99	1.05	0.90	0.99	0.92	0.94	0.82	0.90	0.81
				1.3	1.3				0.85	0.85
New, V_b/W	0.55	0.47	0.33	0.31	0.24	0.24	0.18	0.20	0.15	0.18
ASCE 7-16/ASCE 7-10	1.22	1.10	1.16	1.01	1.10	0.94	1.05	0.91	1.01	0.89

Note: Dark gray cells indicate that the new elastic design base shears are more than 10% higher than the old provisions; light gray cells indicate 0–10% higher than old provisions.

method for vertical distribution of forces calculates the force at the base level immediately above the base isolation plane, then distributes the remainder of the base shear among the levels above. That is, the mass of the "base slab" above the isolators is not included in the vertical distribution of forces.

The proposed revision to the vertical force distribution is based on recent analytical studies (York and Ryan 2008 in collaboration with Structural Engineers Association of Northern California's Protective Systems Subcommittee PSSC). Linear theory of base isolation predicts that base shear is uniformly distributed over the height of the building, while the equivalent lateral force procedure of ASCE 7-10 prescribes a distribution of lateral forces that increase linearly with increasing height. The uniform distribution is consistent with the first mode shape of an isolated building, and the linear distribution is consistent with the first mode shape of a fixed-base building. However, a linear distribution may be overly conservative for an isolated building structure, especially for one- or two-story buildings with heavy base mass relative to the roof.

The principle established in the York and Ryan (2008) study was to develop two independent equations: one to predict the superstructure base shear V_{st} relative to the base shear across the isolators V_b, and a second to distribute V_{st} over the height of the building. Considering a reduction in V_{st} relative to V_b allowed for the often significant inertial forces at the base level, which can be amplified because of disproportionate mass at the base level, to be accounted for in design. The study also assumed that the superstructure base shear was distributed over the height using a 'k' distribution (i.e., lateral force $\infty w_x h_x^k$ where w_x is the weight and h_x the height to level x), where $k = 0$ is a uniform distribution and $k = 1$ is a linear distribution. In the study, representative base-isolated multistory single-bay frame models were developed, and response history analysis was performed with a suite of 20 motions scaled to a target spectrum corresponding to the effective isolation system parameters. Regression analysis was performed to develop a best fit (relative to median results from response history analysis) of the superstructure to base shear ratio and k factor as a function of system parameters. The equations recommended in York and Ryan (2008) provided the best "goodness of fit" among several considered, with R^2 values exceeding 0.95. Note that Eqs. (17.5-8) and (17.5-11) in the code change are the same as Eqs. (15) and (17) in York and Ryan (2008), with one modification: the coefficient for k in Eq. (17.5-11) has been modified to reflect that the reference plane for determining height should be taken as the plane of isolation, which is below the isolated base slab.

It is difficult to confirm in advance whether the upper bound or lower bound isolation system response will govern the design of the isolation system and structure. It is possible, and even likely, that the distribution corresponding to upper bound isolation system properties will govern the design of one portion of the structure, and the lower bound distribution will govern another. For example, lower bound isolation system response may produce a higher displacement, D_M, a lower damping, β_M, but also a higher base shear, V_b. This difference could result in a vertical force distribution that governs for the lower stories of the building. The corresponding upper bound case, with lower displacement, D_M, but higher damping, β_M, might govern design of the upper part of the structure, even though the base shear, V_b, is lower.

The proposal to adopt the approach in York and Ryan (2008) is part of an overall revamp that will permit the equivalent static force method to be extended to a wider class of buildings. In York and Ryan (2008), the current method was shown to be quite conservative for systems with low to medium levels of damping combined with stiff superstructures but unconservative for highly damped systems or systems with relatively flexible superstructures.

The proposal has undergone a high level of scrutiny by the code committee. First, regression analysis was performed using the original York and Ryan (2008) response history data set to fit several alternative distributions suggested by code committee members that were intuitively more appealing. In all cases, the equations recommended in York and Ryan (2008) were shown to best fit the data. Second, a few code committee members appropriately attempted to validate the equations using independently generated response history analysis data sets. Much discussion ensued following the discovery that the equations were unconservative for a class of one- and two-story buildings with long isolation periods and high levels of effective damping in the isolation system. This was most noticeable for one- and two-story buildings, i.e., with relatively low W_{st}/W ratios, predominantly single-mode fixed-base response, and where T_{fb} aligned with the period based on the initial stiffness of the isolation system, T_{k1}. The York and Ryan (2008) data set was confirmed to contain similar cases to those generated independently, and the unconservatism was rationalized as a natural outcome of the regression approach. In an attempt to remove the unconservatism, equations were fit to the 84th percentile (median $+1\sigma$) vertical force distributions based on the original York and Ryan (2008) data set. However, the resulting distributions were unacceptably conservative and thus rejected.

The York and Ryan (2008) data set was subsequently expanded to broaden the range of fixed-base periods for low-rise structures and to provide additional confirmation of the independent data set. In addition, isolation system hysteresis loop shape was identified as the most significant factor in the degree of higher mode participation, resulting in increased V_{st}/V_b ratio and k factor. The provisions now identify this variable as needing a more conservative k factor.

When computing the vertical force distribution using the equivalent linear force procedure, the provisions now divide isolation systems into two broad categories according to the shape of the hysteresis loop. Systems that have an abrupt transition between preyield and postyield response (or preslip and postslip for friction systems) are described as "strongly bilinear" and have been found to typically have higher superstructure accelerations and forces. Systems with a gradual or multistage transition between pre- and postyield response are described as "weakly bilinear" and were observed to have relatively lower superstructure accelerations and forces, at least for systems that fall within the historically adopted range of system strength/friction values (nominal isolation system force at zero displacement, $F_o = 0.03 \times W$ to $0.07 \times W$).

This limitation is acceptable because isolation systems with strength levels that fall significantly outside the upper end of this range are likely to have upper bound properties that do not meet the limitations of Section 17.4.1, unless the postyield stiffness or hazard level is high. Care should also be taken when using the equations to assess the performance of isolation systems at lower hazard levels because the equivalent damping can increase beyond the range of applicability of the original work.

Additional description of the two hysteresis loop types are provided in Table C17.5-2. An example of a theoretical loop for each system type is shown in Fig. C17.5-4.

Capturing this acceleration and force increase in the equivalent linear force procedure requires an increase in the V_{st}/V_b ratio (Eq. (17.5-7)) and the vertical force distribution k factor (Eq. (17.5-11)). Consequently, the provisions require a different exponent to be used in Eq. (17.5-7) for a system that exhibits "strongly bilinear" behavior. Similar differences were observed in the k factor (Eq. (17.5-11)), but these findings were judged to be insufficiently well developed to include in the provisions at this time, and the more conservative value for "strongly bilinear" systems was adopted for both system types.

The exception in Section 17.5.5 is a tool to address the issue identified in the one- and two-story buildings on a project-specific basis and to simplify the design of seismically isolated structures by eliminating the need to perform time-consuming and complex response history analysis of complete 3D building models each time the design is changed. At the beginning of the project, a response history analysis of a simplified building model (e.g., a stick model on isolators) is used to establish a custom inertia force distribution for the project. The analysis of the 3D building model can then be accomplished using simple static analysis techniques.

The limitations on use of the equivalent linear force procedure (Section 17.4.1) and on the response spectrum analysis procedure (Section 17.4.2.1) provide some additional limits. Item 7a in Section 17.4.1 requires a minimum restoring force, which effectively limits postyield stiffness to $K_d > F_o/D_M$ and also limits effective damping to 32% for a bilinear system.

Items 2 and 3 in Section 17.4.1 limit the effective period, $T_M \leq 4.5$ s and effective damping, $\beta_M \leq 30\%$ explicitly.

C17.5.6 Drift Limits. Drift limits are divided by C_d/R for fixed-base structures since displacements calculated for lateral loads reduced by R are multiplied by C_d before checking drift. The C_d term is used throughout the standard for fixed-base structures to approximate the ratio of actual earthquake response to response calculated for reduced forces. Generally, C_d is 1/2 to 4/5 the value of R. For isolated structures, the R_I factor is used both to reduce lateral loads and to increase displacements (calculated for reduced lateral loads) before checking drift. Equivalency would be obtained if the drift

Table C17.5-2 Comparison of "Strongly Bilinear" and "Weakly Bilinear" Isolation Systems

System Type and Equation Term[a]	Pre- to Postyield Transition Characteristics	Cyclic Behavior Below Bilinear Yield/Slip Deformation	Example of Hysteresis Loop Shape	Example Systems[b]
"Strongly bilinear" $(1-3.5\beta_M)$	Abrupt transition from preyield or preslip to postyield or postslip	Essentially linear elastic, with little energy dissipation	Fig. C17.5-4a	• Flat sliding isolators with rigid backing • Single-concave FPS • Double-concave FPS with same friction coefficients top and bottom
"Weakly bilinear" $(1-2.5\beta_M)$	Smooth or multistage transition from preyield or preslip to postyield or postslip	Exhibits energy dissipation caused by yielding or initial low-level friction stage slip	Fig. C17.5-4b	• Elastomeric and viscous dampers • Triple-concave FPS • High-damping rubber • Lead-rubber • Elastomeric-backed sliders

[a]Equation term refers to the exponent in Eq. (17.5-11).
[b]FPS is friction pendulum system.

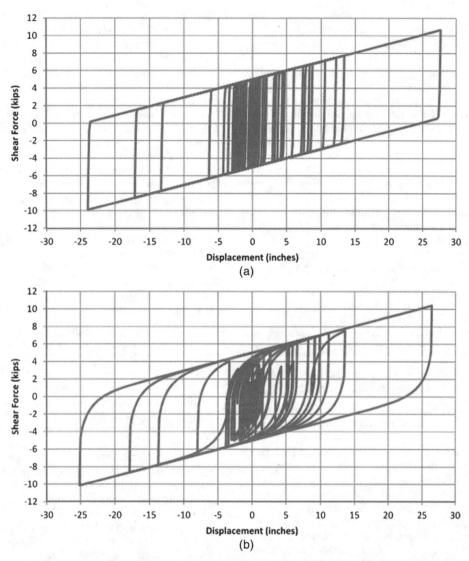

FIGURE C17.5-4 Example Isolation System Example Loops

limits for both fixed-base and isolated structures were based on their respective R factors. It may be noted that the drift limits for isolated structures generally are more conservative than those for conventional, fixed-base structures, even where fixed-base structures are assigned to Risk Category IV. The maximum story drift permitted for design of isolated structures is constant for all risk categories.

C17.6 DYNAMIC ANALYSIS PROCEDURES

This section specifies the requirements and limits for dynamic procedures.

A more detailed or refined study can be performed in accordance with the analysis procedures described in this section, compatible with the minimum requirements of Section 17.5. Reasons for performing a more refined study include

1. The importance of the building.
2. The need to analyze possible structure-isolation system interaction where the fixed-base period of the building is greater than one-third of the isolated period.
3. The need to explicitly model the deformational characteristics of the lateral force-resisting system where the structure above the isolation system is irregular.

4. The desirability of using site-specific ground motion data, especially for very soft or liquefiable soils (Site Class F) or for structures located where S_1 is greater than 0.60.
5. The desirability of explicitly modeling the deformational characteristics of the isolation system. This point is especially important for systems that have damping characteristics that are amplitude-dependent, rather than velocity-dependent, because it is difficult to determine an appropriate value of equivalent viscous damping for these systems.

Where response history analysis is used as the basis for design, the design displacement of the isolation system and design forces in elements of the structure above are computed from the average of seven pairs of ground motion, each selected and scaled in accordance with Section 17.3.2.

The provisions permit a 10% reduction of V_b below the isolation system and 20% reduction of V_b for the structure above the isolators if the structure is of regular configuration. The displacement reduction should not be greater than 20% if a dynamic analysis is performed.

In order to avoid the need to perform a large number of nonlinear response history analyses that include the suites of ground motions, the upper and lower bound isolator properties, and five or more locations of the center of mass, this provision allows the center-of-mass analysis results to be scaled and used to account for the effects of mass eccentricity in different building quadrants.

The following is a recommended method of developing appropriate amplification factors for deformations and forces for use with center-of-mass nonlinear response history analyses (NRHAs) which account for the effects of accidental torsion. The use of other rationally developed amplification factors is permitted.

The most critical directions for shifting the calculated center of mass are such that the accidental eccentricity adds to the inherent eccentricity in each orthogonal direction at each level. For each of these two eccentric mass positions, and with lower bound isolator properties, the suite of NRHA analyses should be run and the results processed in accordance with Section 17.6.3.4. The analysis cases are defined in Table C17.6-1.

The results from Cases IIa and IIb are then compared in turn to those from Case I. The following amplification factors (ratio of Case IIa or IIb response to Case I response) are computed:

1. The amplification of isolator displacement at the plan location with the largest isolator displacement;
2. The amplification of story drift in the structure at the plan location with the highest drift, enveloped over all stories; and
3. The amplification of frame-line shear forces at each story for the frame subjected to the maximum drift.

The larger of the two resulting scalars on isolator displacement should be used as the displacement amplification factor; the larger of the two resulting scalars on drift should be used as the deformation amplification factor; and the larger of the two resulting scalars on force should be used as the force amplification factor. Once the amplification factors are established, the effects of accidental eccentricity should be considered as follows.

The nonlinear response history analysis procedure should be carried out for the inherent mass eccentricity case only, considering both upper and lower bound isolator properties. For each isolator property variation, response quantities should be computed in accordance with Section 17.6.3.4. All resulting isolator displacements should be increased by the displacement amplification factor, all resulting deformation response quantities should be increased by the deformation amplification factor, and all resulting force quantities should be increased by the force amplification before being used for evaluation or design.

The procedure for scaling of dynamic analysis results to the ELF-based minima described in Section 17.6.4.3 is slightly different for response spectrum versus response history analysis. The reason for this difference is that it is necessary to create a consistent basis of comparison between the dynamic response quantities and the ELF-based minima (which are based on the maximum direction). When response spectrum analysis is performed, the isolator displacement, base shear, and story shear at any level used for comparison with the ELF-based minima already correspond to a single, maximum direction of excitation.

Thus, the vector sum of the 100%/30% directional combination rule (as described in Section 17.6.3.3) need not be used. Note, however, that while the 100%/30% directional combination rule is not required in scaling response spectrum analysis results to the ELF-based minima of Section 17.6.4.3, the 100%/30% directional combination rule is still required for design of the superstructure by response spectrum analysis, per Section 17.6.3.3. When nonlinear response history analysis is performed, the isolator displacement and base shear for each ground motion is calculated as the maximum of the vector sum of the two orthogonal components (of displacement or base shear) at each time step. The average of the maxima over all ground motions of these displacement and base shear vector-sum values is then used for comparison with the ELF-based minimum displacement and base shear per Section 17.6.4.3.

C17.6.2 Modeling. Capturing the vertical response of a building structure with a high degree of confidence may be a challenging task. Nonetheless, when the effects of vertical shaking are to be included in the analysis and/or design process of an isolated building structure, the following modeling recommendations are provided:

1. Vertical mass: All beams, columns, shear walls, and slabs should be included in the model, and the vertical mass should be distributed appropriately across the footprint of each floor.
2. Foundation properties: A range of soil properties and foundation damping should be considered in the analysis procedure since horizontal and vertical ground motion excitation can significantly affect building response.
3. Soil–foundation–structure interaction effects: Foundation damping, embedment, and base slab averaging may alter the vertical motions imparted on the structure as compared to the free-field motions.
4. Degrees of freedom: Additional degrees of freedom (e.g., nodes along the span of a beam or slab) will need to be added to the model to capture vertical effects.
5. Reduced time step: Since vertical ground motion excitation and building response often occur at higher frequencies than lateral excitation and response, a finer analysis time step might be required when vertical motions are included.

C17.6.3.4 Response History Analysis Procedure. For sites identified as near-fault, each pair of horizontal ground motion components shall be rotated to the fault-normal and fault-parallel directions of the causative faults and applied to the building in such orientation.

For all other sites, each pair of horizontal ground motion components shall be applied to the building at orthogonal orientations such that the mean of the component response spectra for the records applied in each direction is approximately equal (±10%) to the mean of the component response spectra of all records applied for the period range specified in Section 17.3.3. Peer review would be the judge of "approximately equal."

C17.7 DESIGN REVIEW

The provisions allow for a single peer reviewer to evaluate the isolation system design. The reviewer should be a registered design professional (RDP), and if the engineer of record (EOR) is required to be a structural engineer (SE), the owner may consider ensuring that there is one SE on the peer review team. On more significant structures, it is likely that the design review panel may

Table C17.6-1 Analysis Cases for Establishing Amplification Factors

Case	Isolator Properties	Accidental Eccentricity
I	Lower bound	No
IIa	Lower bound	Yes, X direction
IIb	Lower bound	Yes, Y direction

include two or three individuals, but for many isolated structures, a single, well-qualified peer reviewer is sufficient. If a manufacturer with unknown experience in the United States is selected as the supplier, the building owner may require the design reviewer to attend prototype tests.

The standard requires peer review to be performed by registered design professionals who are independent of the design team and other project contractors. The reviewer or review panel should include individuals with special expertise in one or more aspects of the design, analysis, and implementation of seismic isolation systems.

The peer reviewer or review panel should be identified before the development of design criteria (including site-specific ground-shaking criteria) and isolation system design options. Furthermore, the review panel should have full access to all pertinent information and the cooperation of the general design team and regulatory agencies involved in the project.

C17.8 TESTING

The design displacements and forces determined using the standard assume that the deformational characteristics of the isolation system have been defined previously by comprehensive testing. If comprehensive test data are not available for a system, major design alterations in the structure may be necessary after the tests are complete. This change would result from variations in the isolation system properties assumed for design and those obtained by test. Therefore, it is advisable that prototype tests of systems be conducted during the early phases of design if sufficient prototype test data are not available from a given manufacturer.

The design displacements and forces determined using the standard are based on the assumption that the deformational characteristics of the isolation system have been defined previously by comprehensive qualification and prototype testing. Variations in isolator properties are addressed by the use of property variation factors that account for expected variation in isolator and isolation system properties from the assumed nominal values. In practice, past prototype test data are very likely to have been used to develop the estimated nominal values and associated lambda factors used in the design process, as described in Section 17.2.8.4.

When prototype testing is performed in accordance with Section 17.8.2, it serves to validate and check the assumed nominal properties and property variation factors used in the design. Where project-specific prototype testing is not performed, it is possible to perform a subset of the checks described below on the isolator unit and isolation system test properties using data from the quality control test program, described in Section 17.8.5.

C17.8.2.2 Sequence and Cycles. Section 17.2.8.4 describes the method by which minimum and maximum isolator properties for design and analysis are established using property variation or lambda (λ) factors to account for effects such as specification tolerance, cyclic degradation, and aging. The structural analysis is therefore performed twice, and the resulting demands are enveloped for design. For force-based design parameters and procedures, this requirement is relatively straightforward, as typically one case or the other governs, primarily, but not always, the upper bound. However, for components dependent on both force and deformation, e.g., the isolators, there exist two sets of axial load and displacement values for each required test. Lower bound properties typically result in larger displacements and smaller axial loads, whereas upper bound properties typically result in smaller displacements and larger axial loads. To avoid requiring that a complete set of duplicate tests be performed for the lower and upper bound conditions, Section 17.8.2.2 requires the results to be enveloped, combining the larger axial demands from one case with the larger displacements from the other. Strictly, these demands and displacement do not occur simultaneously, but the enveloping process is conservative.

The enveloping process typically results in test axial loads that correspond to the maximum properties and displacements that correspond to minimum properties. Hence, the test results determined using the enveloped demands may not directly relate to the design properties or analysis results determined for maximum and minimum properties separately. However, since the test demands envelop the performance range for the project, the registered design professional is able to use them to determine appropriate properties for both linear and nonlinear analysis using the same philosophy as provided here.

Two alternate testing protocols are included in Section 17.8.2.2. The traditional three-cycle tests are preserved in Item 2(a) for consistency with past provisions. These tests can be performed dynamically but have often been performed at slow speed consistent with the capability of manufacturers' testing equipment. The alternate test sequence provided in Item 2(b) is more suited to full-scale dynamic cyclic testing.

The Item (3) test displacement has been changed from D_D to D_M, reflecting the focus of the provisions on only the MCE_R event. Since this test is common to both test sequences 2(a) and 2(b), it becomes important for property determination. This is the only test required to be repeated at different axial loads when isolators are also axial load-carrying elements, which is typically the case. This change was made to counter the criticism that the total test sequence of past provisions represented the equivalent energy input of many MCE_R events back to back and that prototype test programs could not be completed in a reasonable time if any provision for isolator cooling and recovery was included.

The current test program is therefore more reflective of code-minimum required testing. The RDP and/or the isolator manufacturer may wish to perform additional testing to more accurately characterize the isolator for a wider range of axial loads and displacements than is provided here. For example, this might include performing the Item 2(b) dynamic test at additional axial loads once the code-required sequence is complete.

Heat effects for some systems may become significant, and misleading, if insufficient cooling time is included between adjacent tests. As a consequence, in test sequence 4 only five cycles of continuous dynamic testing are required as this is a limit of most test equipment. The first-cycle or scragging effects observed in some isolators may recover with time, so back-to-back testing may result in an underestimation of these effects. Refer to Constantinou et al. (2007) and Kalpakidis and Constantinou (2008) for additional information. The impact of this behavior may be mitigated by basing cyclic lambda factors on tests performed relatively early in the sequence before these effects become significant.

C17.8.2.3 Dynamic Testing. Section 17.8.2.3 clarifies when dynamic testing is required. Many common isolator types exhibit velocity dependence, however, this testing can be expensive and can only be performed by a limited number of test facilities. The intent is not that dynamic testing of isolators be performed for every project. Sufficient dynamic test data must be available to characterize the cyclic performance of the isolator, in particular the change in isolator properties during the test, i.e., with respect to the test average value. Dynamic testing must therefore be used to establish the $\lambda_{(test, min)}$ and $\lambda_{(test, max)}$ values used in Section 17.2.8.4, since these values are typically underestimated from slow-speed test data. If project prototype or production testing is to be performed at slow speeds, this testing would also be used to establish factors that account for the effect of velocity and heating on the test average values of k_{eff}, k_d, and E_{loop}. These factors can either be thought of as a separate set of velocity-correction factors to be applied on test average values, or they can be incorporated into the $\lambda_{(test, min)}$ and $\lambda_{(test, max)}$ values themselves.

It may also be possible to modify the isolator mathematical model, for example, to capture some or all of the isolator velocity dependence, e.g., the change in yield level of the lead core in a lead rubber bearing (LRB).

If project-specific prototype testing is undertaken, it may be necessary to adjust the test sequence in recognition of the capacity limitations of the test equipment, and this notion is now explicitly recognized in Section 17.8.2.2. For example, tests that simultaneously combine maximum velocity and maximum displacement may exceed the capacity of the test equipment and may also not be reflective of earthquake shaking characteristics. A more detailed examination of analysis results may be required to determine the maximum expected velocity corresponding to the various test deformation levels and to establish appropriate values for tests.

Refer to Constantinou et al. (2007) for additional information.

C17.8.2.4 Units Dependent on Bilateral Load. All types of isolators have bilateral load dependence to some degree. The mathematical models used in the structural analysis may include some or all of the bilateral load characteristics for the particular isolator type under consideration. If not, it may be necessary to examine prototype test data to establish the impact on the isolator force-deformation response as a result of the expected bilateral loading demands. A bounding approach using lambda (λ) factors is one method of addressing bilateral load effects that cannot be readily incorporated in the isolator mathematical model.

Bilateral isolator testing is complex, and only a few test facilities are capable of performing these tests. Project-specific bilateral load testing has not typically been performed for isolation projects completed to date. In lieu of performing project-specific testing, less restrictive similarity requirements may be considered by the registered design professional compared to those required for test data submitted to satisfy similarity for Sections 17.8.2.2 and 17.8.2.5. Refer to Constantinou et al. (2007) for additional information.

C17.8.2.5 Maximum and Minimum Vertical Load. The exception to Section 17.8.2.5 permits that the tests may be performed twice, once with demands resulting from upper bound properties and once with lower bound properties. This option may be preferable for these isolator tests performed at D_{TM} since the isolator will be closer to its ultimate capacity.

C17.8.2.7 Testing Similar Units. Section 17.8.2.7 now provides specific limits related to the acceptability of data from testing of similar isolators. A wider range of acceptability is permitted for dynamic test data.

1. The submitted test data should demonstrate the manufacturers' ability to successfully produce isolators that are comparable in size to the project prototypes, for the relevant dimensional parameters, and to test them under force and displacement demands equal to or comparable to those required for the project.
2. It is preferred that the submitted test data necessary to satisfy the registered design professional and design review be for as few different isolator types and test programs as possible. Nonetheless, it may be necessary to consider data for isolator A to satisfy one aspect of the required project prototype test program, and data from isolator B for another.
3. For more complex types of testing, it may be necessary to accept a wider variation of isolator dimension or test demands than for tests that more fundamentally establish the isolator nominal operating characteristics, e.g., the testing required to characterize the isolator for loading rate dependence (Section 17.8.2.3) and bilateral load dependence (Section 17.8.2.4).
4. The registered design professional is not expected to examine quality control procedures in detail to determine whether the proposed isolators were manufactured using sufficiently similar methods and materials. Rather, it is the responsibility of the manufacturer to document the specific differences, if any, preferably via traceable quality control documentation and to substantiate that any variations are not significant.
5. In some cases, the manufacturer may not wish to divulge proprietary information regarding methods of isolator fabrication, materials, or quality control procedures. These concerns may or may not be alleviated by confidentiality agreements or other means to limit the distribution and publication of sensitive material. Regardless, the final acceptability of the test information of similar units is at the sole discretion of the registered design professional and the design review, and not the manufacturer.
6. Similarity can be especially problematic in a competitive bid situation, when successful selection may hinge on the success of one supplier in eliminating the need to fabricate and test project-specific prototype isolators. This requirement can be addressed by determining acceptability of similarity data before bid or by including more detailed similarity acceptance provisions in the bid documentation than have been provided herein.

Refer to Constantinou et al. (2007) and Shenton (1996) for additional information.

C17.8.3 Determination of Force-Deflection Characteristics. The method of determining the isolator effective stiffness and effective damping ratio is specified in Eqs. (17.8-1) and (17.8-2). Explicit direction is provided for establishment of effective stiffness and effective damping ratio for each cycle of test. A procedure is also provided for fitting a bilinear loop to a given test cycle, or to an average test loop to determine the postyield stiffness, k_d. This process can be performed several different ways; however, the fitted bilinear loop should also match effective stiffness and energy dissipated per cycle from the test. Once k_d is established, the other properties of the bilinear loop (e.g., f_y, f_o) all follow from the bilinear model.

Depending on the isolator type and the degree of sophistication of the isolator hysteresis loop adopted in the analysis, additional parameters may also be calculated, such as different friction coefficients, tangent stiffness values, or trilinear loop properties.

These parameters are used to develop a mathematical model of the isolator test hysteresis that replicates, as near as possible, the observed test response for a given test cycle. The model should result in a very close match to the effective stiffness and effective damping ratio and should result in a good visual fit to the hysteresis loop with respect to the additional parameters. The mathematic loop model must, at a minimum, match the effective stiffness and loop area from the test within the degree of variation adopted within the $\lambda_{(spec,min)}$ to $\lambda_{(spec,max)}$ range.

Data from the first cycle (or half cycle) of testing is not usually representative of full-cycle behavior and is typically discarded by manufacturers during data processing. An additional cycle (or half cycle) is added at the end to provide the required number of test cycles from which data can be extracted. However, the first cycle of a test is often important when establishing upper bound isolator properties and should be included when determining the $\lambda_{(test,min)}$ and $\lambda_{(test,max)}$ factors. The form of the test loop, however, is different to that of a full-scale loop, particularly for multistage isolator systems such as the double- or triple-concave friction pendulum system. This form may require different hysteresis parameters to be considered than the ones described by the bilinear model in Fig. 17.8-1. The provisions permit the use of different methods for fitting the loop, such as a straight-line fit of k_d directly to the hysteresis curve extending to D_M and then determining k_1 to match E_{loop}, or an alternate is defining D_y and F_y by visual fit and then determining k_d to match E_{loop}.

The effective stiffness and effective damping ratio are required in linear static and linear response spectrum analysis. However, even if a nonlinear response history analysis is performed, these parameters are still required to check the required minimum lateral displacements and lateral forces of Sections 17.5.3 and 17.5.4, respectively.

C17.8.4 Test Specimen Adequacy. For each isolator type, the effective stiffness and effective damping ratio for a given test axial load, test displacement, and cycle of test are determined in accordance with Section 17.8.3. For the dynamic test sequence in Item 2(a) in Section 17.8.2.2, there are two cycles at each increment of test displacement; for the traditional slow-speed sequence, there are three.

However, as part of a seismic isolation system, the axial load on a given isolator varies during a single complete cycle of loading. The required range of variation is assumed to be defined by the test load combinations required in Section 17.2.4.6, and the appropriate properties for analysis are assumed to be the average of the properties at the three axial loads. The test performed for Item (3) in Section 17.8.2.2 is critical to this evaluation since it is the three-cycle test performed at all three axial loads common to both the dynamic and slow-speed sequence.

In addition, since all isolators must undergo the same total horizontal cyclic loading as part of the same system, it is therefore assumed to be appropriate to assemble the total seismic isolation system properties using the following sequence:

1. Average the test results for a given isolator and cycle of loading across the three test axial loads. Also compute corresponding test lambda factors for each isolator type.
2. Sum up the total isolation system properties for each cycle of loading according to the number of isolators of each type.
3. Determine the maximum and minimum values of total system effective stiffness over the required three cycles of testing and the corresponding values of the effective damping ratio. Also compute the test lambda factors for the overall isolation system.

Two sets of test lambda factors emerge from this process, those applicable to individual isolators determined in (1) and those applicable to the overall isolation system properties determined in (3). In general, the test lambda factors for individual isolator tests are similar to those for each isolator type, which are similar to that for the overall isolation system. If this is the case, it may be more convenient to simplify the lambda factors assumed during design to reflect reasonable envelope values to be applied to all isolator types.

However, if the test lambda factors that emerge from project-specific prototype testing differ significantly from those assumed during design, it may be helpful to build up the system properties as described above, since the unexpectedly high test lambda factors for one isolator type may be offset by test lambda factors for another isolator type that were lower than the assumed values. In this circumstance, the prototype test results may be considered acceptable, provided that the torsional behavior of the system is not significantly affected and that the isolator connection and adjacent members can accommodate any resulting increase in local force demands.

Also, note that a subset of the isolation system properties can be determined from quality assessment and quality control (production) testing. This testing is typically performed at an axial load corresponding to the average $D + 0.5L$ axial load for the isolator type and to a displacement equal to $2/3(D_M)$. Keep in mind that isolator properties with target nominal three-cycle values estimated to match the average test value across three axial loads may not exactly match the values from production testing at the average dead load.

This result is most commonly observed with effective stiffness and effective damping ratio values for friction-based isolators since the average of the three test axial loads required in Section 17.8.2.2 does not exactly match that present in the isolator during the lateral analysis (the seismic weight, typically $1.0 \times$ Dead Load). In this case, some additional adjustment of properties may be required. Once the test effective stiffness and effective damping ratio of the isolation system have been established, these are compared to the values assumed for design in Section 17.2.8.4, defined by the nominal values and the values of $\lambda_{(test,max)}$ and $\lambda_{(test,min)}$.

In practice, instead of performing prototype tests for direct use in analysis, it may be simpler to use prototype test data or data from acceptable past testing of similar units (see Section 17.8.2.7) to establish isolator property dependence relationships for such things as axial load or velocity. If relationships are established for applicable hysteresis-loop parameters, such as yield force, friction ratio, initial stiffness, and postyield stiffness, these can be used to generate the required isolator unit and isolation system effective stiffness and effective damping ratios for the project over the required operating range.

C17.8.5 Production Tests. The number of production isolation units to be tested in combined compression and shear is 100%. Both quasi-static and dynamic tests are acceptable for all types of isolators. If a quasi-static test is used, it must have been performed as a part of the prototype tests. The registered design professional (RDP) is responsible for defining in the project specifications the scope of the manufacturing quality control test program. The RDP decides on the acceptable range of variations in the measured properties of the production isolation units. All (100%) of the isolators of a given type and size are tested in combined compression and shear, and the allowable variation of the mean should be within the specified tolerance of Section 17.2.8.4 (typically $\pm 10\%$ or $\pm 15\%$). Individual isolators may be permitted a wider variation ($\pm 15\%$ or $\pm 20\%$) from the nominal design properties. For example, the mean of the characteristic strength, Q, for all tested isolators might be permitted to vary no more than $\pm 10\%$ from the specified value of Q, but the characteristic strength for any individual isolation unit might be permitted to vary no more than $\pm 15\%$ from the specified value of Q.

Another commonly specified allowable range of deviation from specified properties is ±15% for the mean value of all tested isolation units, and ±20% for any single isolation unit.

The combined compression and shear testing of the isolators reveals the most relevant characteristics of the completed isolation unit and permits the RDP to verify that the production isolation units provide load-deflection behavior that is consistent with the structural design assumptions. Although vertical load-deflection tests have sometimes been specified in quality control testing programs, these test data are typically of little value. Consideration should be given to the overall cost and schedule effects of performing multiple types of quality control tests, and only those tests that are directly relevant to verifying the design properties of the isolation units should be specified.

Where project-specific prototype testing in accordance with Section 17.8.2 is not performed, the production test program should evaluate the performance of each isolator unit type for the property variation effects from Section 17.2.8.4.

REFERENCES

American Association of State Highway and Transportation Officials (AASHTO). (1990). *Guide specifications for seismic isolation design*. AASHTO, Washington, DC.

AASHTO. (1999). *Guide specifications for seismic isolation design*. American Association of State Highway and Transportation Officials, Washington, DC.

ANSI/American Institute of Steel Construction (AISC). "Seismic provisions for structural steel buildings." *ANSI/AISC 341*, Chicago.

ASCE. (2007). "Seismic rehabilitation of existing buildings." *ASCE/SEI 41-06*, ASCE, Reston, VA.

ASTM International. (2012). "Standard specification for plain and steel-laminated elastomeric bearings for bridges." *D4014*. ASTM International, West Conshohocken, PA.

Buckle, I. G., Nagarajaiah, S, Ferrel, K. (2002). "Stability of elastomeric isolation bearings: Experimental study." *ASCE J. Struct. Eng.* 128, 3–11.

Constantinou, M. C., Kalpakidis, I., Filiatrault, A., and Ecker Lay, R. A. (2011). "LRFD-based analysis and design procedures for bridge bearings and seismic isolators." *Report No. MCEER-11-0004*, Multidisciplinary Center for Earthquake Engineering Research, Buffalo, NY.

Constantinou, M. C., Tsopelas, P., Kasalanati, A., and Wolff, E. D. (1999). "Property modification factors for seismic isolation bearings." *MCEER-99-0012*, Multidisciplinary Center for Earthquake Engineering Research, Buffalo, NY.

Constantinou, M. C., Whittaker, A. S., Kalpakidis, Y., Fenz, D. M., and Warn, G. P. (2007). "Performance of seismic isolation hardware under service and seismic loading." *MCEER-07-0012*, Multidisciplinary Center for Earthquake Engineering Research, Buffalo, NY.

Constantinou, M. C., Winters, C. W., and Theodossiou, D. (1993). "Evaluation of SEAOC and UBC analysis procedures. Part 2: Flexible superstructure." *Proc., Seminar on Seismic Isolation, Passive Energy Dissipation and Active Control*, ATC Report 17-1. Applied Technology Council, Redwood City, CA.

Federal Emergency Management Agency (FEMA). (1999). "HAZUS software." *Federal Emergency Management Agency*, Washington, DC.

FEMA. (2003). *NEHRP recommended seismic provisions for new buildings and other structures*, Federal Emergency Management Agency, Washington, DC.

FEMA. (2009a). "Quantification of building seismic performance factors." *P-695*. Federal Emergency Management Agency, Washington, DC.

FEMA. (2009b). *NEHRP recommended seismic provisions for new buildings and other structures*, Federal Emergency Management Agency, Washington, DC.

FEMA. (2012). "Seismic performance assessment of buildings." *P-58*. Federal Emergency Management Agency, Washington, DC.

International Council of Building Officials (ICBO). (1991). *Uniform Building Code*, Whither, CA.

Kalpakidis, I. V., and Constantinou, M. C. (2008). "Effects of heating and load history on the behavior of lead-rubber bearings," *MCEER-08-0027*, Multi-disciplinary Center for Earthquake Engineering Research, Buffalo, NY.

Kalpakidis, I. V., and Constantinou, M. C. (2009). "Effects of heating on the behavior of lead-rubber bearings. I: Theory." *J. Struct. Eng.*, 135(12), 1440–1449.

Kalpakidis, I. V., Constantinou, M. C., and Whittaker, A. S. (2010). "Modeling strength degradation in lead-rubber bearings under earthquake shaking," *Earthq. Eng. Struct. Dyn.* 39(13), 1533–1549.

Katsaras, A. (2008). "Evaluation of current code requirements for displacement restoring capability of seismic isolation systems and proposals for revisions." Project No. GOCE-CT-2003-505488, LessLoss Project cofounded by European Commission with 6th Framework.

Kelly, J. M., and Chaloub, M. S. (1990). "Earthquake simulator testing of a combined sliding bearing and rubber bearing isolation system." Report No. UCB/EERC-87/04, University of California, Berkeley.

Kelly, J. M., and Hodder, S. B. (1981). "Experimental study of lead and elastomeric dampers for base isolation systems." Report No. UCB/EERC-81/16, University of California, Berkeley.

Kelly, J. M., and Konstantinidis, D. A. (2011). *History of multi-layered rubber bearings*. John Wiley and Sons, New York.

Kelly, J. M., Skinner, M. S., Beucke, K. E. (1980). "Experimental testing of an energy absorbing seismic isolation system." Report No. UCB/EERC-80/35, University of California, Berkeley.

Kircher, C. A., Lashkari, B., Mayes, R. L., and Kelly, T. E. (1988). "Evaluation of nonlinear response in seismically isolated buildings." *Proc., Symposium on Seismic, Shock and Vibration Isolation*, ASME Pressure Vessels and Piping Conference, New York.

Masroor, A., and Mosqueda, G. (2015). "Assessing the Collapse Probability of Base-Isolated Buildings Considering Pounding to Moat Walls Using the FEMA P695 Methodology." *Earthq. Spectra* 31(4), 2069–2086.

McVitty, W., and Constantinou, M.C. (2015). "Property Modifications factors for Seismic Isolators: Design guidance for buildings." *MCEER Report No. 000-2015*.

National Institute of Standards and Technology (NIST). (2011). *Selecting and scaling earthquake ground motions for performing response-history analyses*, GCR 11-917-15, National Institute of Standards and Technology, Gaithersburg, MD.

Ryan, K. L., Coria, C. B., Dao, N. D., (2012). "Large scale earthquake simulation for hybrid lead rubber isolation system designed with consideration for nuclear seismicity." U.S. Nuclear Regulatory Commission CCEER 13-09.

Shenton, H. W., III., (1996). Guidelines for pre-qualification, prototype, and quality control testing of seismic isolation systems, NISTIR 5800.

York, K., and Ryan, K. (2008). "Distribution of lateral forces in base-isolated buildings considering isolation system nonlinearity." *J. Earthq. Eng.*, 12, 1185–1204.

Zayas, V., Low, S., and Mahin, S. (1987). "The FPS earthquake resisting system." Report No. UCB/EERC-87-01; University of California, Berkeley.

OTHER REFERENCES (NOT CITED)

Applied Technology Council. (ATC). (1982). "An investigation of the correlation between earthquake ground motion and building performance." *ATC Report 10*. ATC, Redwood City, CA.

Lashkari, B., and Kircher, C. A. (1993). "Evaluation of SEAOC and UBC analysis procedures. Part 1: Stiff superstructure." *Proc., Seminar on seismic isolation, passive energy dissipation and active control*. Applied Technology Council, Redwood City, CA.

Warn, G. P., and Whittaker, A. W. (2006). "Performance estimates in seismically isolated bridge structures." *Eng. Struct.*, 26, 1261–1278.

Warn, G. P., and Whittaker, A. S. (2004). "Performance estimates in seismically isolated bridge structures." *Eng. Struct.* 26, 1261–1278.

CHAPTER C18
SEISMIC DESIGN REQUIREMENTS FOR STRUCTURES WITH DAMPING SYSTEMS

C18.1 GENERAL

The requirements of this chapter apply to all types of damping systems, including both displacement-dependent damping devices of hysteretic or friction systems and velocity-dependent damping devices of viscous or viscoelastic systems (Soong and Dargush 1997, Constantinou et al. 1998, Hanson and Soong 2001). Compliance with these requirements is intended to produce performance comparable to that for a structure with a conventional seismic force-resisting system, but the same methods can be used to achieve higher performance.

The damping system (DS) is defined separately from the seismic force-resisting system (SFRS), although the two systems may have common elements. As illustrated in Fig. C18.1-1, the DS may be external or internal to the structure and may have no shared elements, some shared elements, or all elements in common with the SFRS. Elements common to the DS and the SFRS must be designed for a combination of the loads of the two systems. When the DS and SFRS have no common elements, the damper forces must be collected and transferred to members of the SFRS.

C18.2 GENERAL DESIGN REQUIREMENTS

C18.2.1 System Requirements. Structures with a DS must have an SFRS that provides a complete load path. The SFRS must comply with all of the height, Seismic Design Category, and redundancy limitations and with the detailing requirements specified in this standard for the specific SFRS. The SFRS without the damping system (as if damping devices were disconnected) must be designed to have not less than 75% of the strength required for structures without a DS that have that type of SFRS (and not less than 100% if the structure is horizontally or vertically irregular). The damping systems, however, may be used to meet the drift limits (whether the structure is regular or irregular). Having the SFRS designed for a minimum of 75% of the strength required for structures without a DS provides safety in the event of damping system malfunction and produces a composite system with sufficient stiffness and strength to have controlled lateral displacement response.

The analysis and design of the SFRS under the base shear, V_{min}, from Eqs. (18.2-1) or (18.2-2) or, if the exception applies, under the unreduced base shear, V, should be based on a model of the SFRS that excludes the damping system.

C18.2.1.2 Damping System. The DS must be designed for the actual (unreduced) MCE_R forces (such as peak force occurring in damping devices) and deflections. For certain elements of the DS (such as the connections or the members into which the damping devices frame), other than damping devices, limited yielding is permitted provided that such behavior does not affect damping system function or exceed the amount permitted for elements of conventional structures by the standard.

Furthermore, force-controlled actions in elements of the DS must consider seismic forces that are 1.2 times the computed average MCE_R response. Note that this increase is applied for each *element action*, rather than for each *element*. Force-controlled actions are associated with brittle failure modes where inelastic deformation capacity cannot be ensured. The 20% increase in seismic force for these actions is required to safeguard against undesirable behavior.

C18.2.2 Seismic Ground Motion Criteria. It is likely that many projects incorporating a supplemental damping system simply use design earthquake (DE) and MCE_R spectra based on the mapped values referenced in Chapter 11. Site-specific spectra are always permitted and must be used for structures on Site Class F.

When nonlinear response history analysis is used, ground motions are selected, scaled or matched and applied in accordance with the procedures of Chapter 16, with the exception that a minimum of 7 rather than 11 ground motions are required. The use of 7 motions is consistent with current practice for design of code-compliant structures, and 7 is considered an adequate number to estimate the mean response for a given hazard level. No other provisions of Chapter 16 apply to structures incorporating supplemental damping systems.

C18.2.3 Procedure Selection. The nonlinear response history procedure for structures incorporating supplemental damping systems is the preferred procedure, and Chapter 18 is structured accordingly. This method, consistent with the majority of current practice, provides the most realistic predictions of the seismic response of the combined SFRS and DS. If the nonlinear response history procedure is adopted, the relevant sections of Chapter 18 are 18.1 through 18.6.

However, via the exception, response spectrum (RS) and equivalent lateral force (ELF) analysis methods can be used for design of structures with damping systems that meet certain configuration and other limiting criteria (for example, at least two damping devices at each story configured to resist torsion). In such cases, additional nonlinear response history analysis is used to confirm peak responses when the structure is located at a site with S_1 greater than or equal to 0.6. The analysis methods of damped structures are based on nonlinear static "pushover" characterization of the structure and calculation of peak response using effective (secant) stiffness and effective damping properties of the first (pushover) mode in the direction of interest. These concepts are used in Chapter 17 to characterize the force-deflection properties of isolation systems, modified to incorporate explicitly the effects of ductility demand (post-yield response) and higher mode response of structures with dampers. Similar to conventional structures, damped structures generally

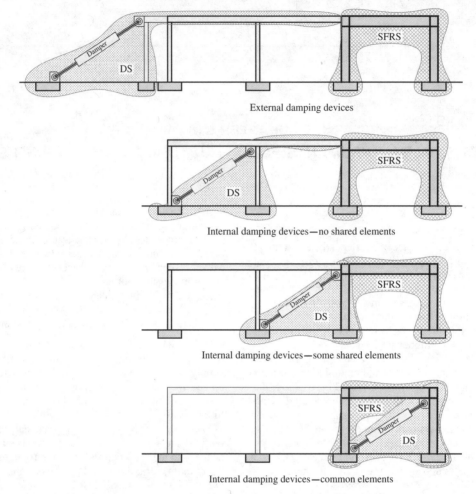

FIGURE C18.1-1 Damping System (DS) and Seismic Force-Resisting System (SFRS) Configurations

yield during strong ground shaking, and their performance can be influenced strongly by response of higher modes.

The RS and ELF procedures presented in Chapter 18 have several simplifications and limits, outlined as follows:

1. A multiple-degree-of-freedom (MDOF) structure with a damping system can be transformed into equivalent single-degree-of-freedom (SDOF) systems using modal decomposition procedures. This procedure assumes that the collapse mechanism for the structure is an SDOF mechanism so that the drift distribution over height can be estimated reasonably using either the first mode shape or another profile, such as an inverted triangle. Such procedures do not strictly apply to either yielding buildings or buildings that are nonproportionally damped.
2. The response of an inelastic SDOF system can be estimated using equivalent linear properties and a 5% damped response spectrum. Spectra for damping greater than 5% can be established using damping coefficients, and velocity-dependent forces can be established either by using the pseudovelocity and modal information or by applying correction factors to the pseudovelocity.
3. The nonlinear response of the structure can be represented by a bilinear hysteretic relationship with zero postelastic stiffness (elastoplastic behavior).
4. The yield strength of the structure can be estimated either by performing simple plastic analysis or by using the specified minimum seismic base shear and values of R, Ω_0, and C_d.
5. Higher modes need to be considered in the equivalent lateral force procedure to capture their effects on velocity-dependent forces. This requirement is reflected in the residual mode procedure.

FEMA 440 (2005) presents a review of simplified procedures for the analysis of yielding structures. The combined effects of the simplifications mentioned above are reported by Ramirez et al. (2001) and Pavlou and Constantinou (2004) based on studies of three-story and six-story buildings with damping systems designed by the procedures of the standard. The RS and ELF procedures of the standard are found to provide conservative predictions of drift and predictions of damper forces and member actions that are of acceptable accuracy when compared to results of nonlinear dynamic response history analysis. When designed in accordance with the standard, structures with damping systems are expected to have structural performance at least as good as that of structures without damping systems. Pavlou and Constantinou (2006) report that structures with damping systems designed in accordance with the standard provide the benefit of reduced secondary system response, although this benefit is restricted to systems with added viscous damping.

If either the RS or ELF procedures are adopted, the relevant sections of Chapter 18 are Sections 18.1, 18.2, 18.5, 18.6, and 18.7.

C18.2.4.1 Device Design. Damping devices may operate on a variety of principles and may use materials that affect their short-term and long-term performance. This commentary provides guidance on the behavior of some of these devices in order to justify the language in the standard and in order to assist the engineer in deciding on the upper and lower bound values of mechanical properties of the devices for use in analysis and design.

Damping devices that have found applications or have potential for application may be classified as follows:

1. Fluid viscous dampers (or oil dampers) that operate on the principle of orificing of fluid, typically some form of oil (Constantinou et al. 2007). These devices are typically highly engineered and precision made so that their properties are known within a narrow range. That is, when the devices are tested, their properties show small variability. One issue is heating that may have significant effects (Makris et al. 1998), which can be alleviated or eliminated by using accumulators or by using materials with varying thermal expansion properties so that the orifice size is automatically adjusted with varying temperature.

    However, their long-term behavior may be affected by a variety of potential problems:

    a. Devices using accumulators include valves that may fail over time depending on the quality of construction and history of operation. It is not possible to know if and when a valve may fail.
    b. Fluid is maintained in the device by seals between the body and the moving piston of the device, which may leak either as a result of wear caused by excessive cumulative travel or poor construction. For buildings, excessive cumulative travel is rarely an issue. When seals leak, the output of the device reduces, depending on the reduction of internal pressure of the device. It is recommended that potential leakage of oil not be considered in establishing lower bound values of property modification factors (as it is not possible to know) but rather a periodic inspection and maintenance program recommended by the manufacturer be used to detect problems and make corrections.
    c. Orifices may be very small in diameter and therefore may result in clogging when impure oil is used or the oil is contaminated by particles of rubber used in the sealing of fluid in poorly constructed devices or by metal particles resulting from internal corrosion or because of oil cavitation when poor-quality materials are used. Typically, rubber should not be used in sealing and parts should be threaded rather than welded or connected by posttensioning. Larger diameter orifices should be preferred.

2. Viscoelastic fluid or solid devices. These devices operate on the principle of shearing of highly viscous fluids or viscoelastic solids. These viscous fluids and viscoelastic solids have a strong dependence of properties on frequency and temperature. These effects should be assessed by qualification testing. Their long-term behavior is determined by the behavior of the fluid or solid used, both of which are expected to harden with time. The engineer should ask the supplier for data on the aging of the material based on observations in real time. Information based on accelerated aging is not useful and should not be used (Constantinou et al. 2007).

3. Metallic yielding devices. Yielding steel devices are typically manufactured of steel with yield properties that are known within a narrow range. Nevertheless, the range of values of the yield strength can be determined with simple material tests. Also, testing some of the devices should be used to verify the information obtained in coupon testing. Aging is of least concern because corrosion may only slightly reduce the section geometrical properties. An inspection and maintenance program should eliminate the concern for aging.

4. Friction devices. Friction devices operate on the principle of preloaded sliding interfaces. There are two issues with such devices:

    a. The preload may reduce over time because of creep in sliding interface materials or the preloading arrangement, or wear in the sliding interface when there is substantial service-load related motion or after high-speed seismic motion. It is not possible to know what the preload may be within the lifetime of the structure, but the loss may be minimized when high-strength bolts are used and high-strength/low-wear materials are used for the sliding interface.
    b. The friction coefficient at the sliding interface may substantially change over time. The engineer is directed to Constantinou et al. (2007) for a presentation on the nature of friction and the short-term and long-term behavior of some sliding interfaces. In general, reliable and predictable results in the long-term friction may be obtained when the sliding interface consists of a highly polished metal (typically stainless steel) in contact with a nonmetallic softer material that is loaded to high pressure under confined conditions so that creep is completed in a short time. However, such interfaces also result in low friction (and thus are typically used in sliding isolation bearings). The engineer is referred to Chapter 17 and the related commentary for such cases. Desirable high friction (from a performance standpoint) may be obtained by use of metal to metal sliding interfaces. However, some of these interfaces are absolutely unreliable because they promote severe additional corrosion and they should never be used (British Standards Institution 1983). Other bimetallic interfaces have the tendency to form solid solutions or intermetallic compounds with one another when in contact without motion. This tendency leads to cold welding (very high adhesion or very high friction). Such materials are identified by compatibility charts (Rabinowicz 1995). The original Rabinowicz charts categorized pairs of metals as incompatible (low adhesion) to compatible and identical (high adhesion). Based on that characterization, identical metals and most bimetallic interfaces should be excluded from consideration in sliding interfaces. Excluding interfaces that include lead (too soft), molybdenum, silver, and gold (too expensive), only interfaces of tin–chromium, cadmium–aluminum, and copper–chromium are likely to have low adhesion. Of these, the tin–chromium interface has problems of additional corrosion (British Standards Institution 1983) and should not be used. Accordingly, only bimetallic interfaces of cadmium–aluminum and copper–chromium may be useful. The materials in these interfaces have similar hardness so that creep-related effects are expected to be important, leading to increased true area of contact and increased friction force over time (Constantinou et al. 2007). This increase leads to the conclusion that all bimetallic

interfaces result in significant changes in friction force over time that are not possible to predict, and therefore these types of interfaces should not be used.

5. **Lead extrusion devices.** These devices operate on the principle of extruding lead through an orifice. The behavior of the device is dependent on the rate of loading and temperature, and its force output reduces with increasing cycling because of heating effects. These effects can be quantified by testing so that the nominal properties and property modification factors can be established. Leakage of lead during the lifetime of the device is possible during operation and provided that the seals fail, although the effects cannot be expected to be significant. Leakage is preventable by use of proven sealing technologies and by qualification testing to verify. (Skinner et al. 1993).

The registered design professional (RDP) must define the ambient temperature and the design temperature range. The ambient temperature is defined as the normal in-service temperature of the damping device. For devices installed in interior spaces, this temperature may be taken as 70°F, and the design temperature range could come from the project mechanical engineer. For devices installed exposed to exterior temperature variation, the ambient temperature may be taken as the annual average temperature at the site, and the design temperature range may be taken as the annual minimum and maximum temperatures. Since the design temperature range is implicitly tied to MCE_R analysis through λ factors for temperature, the use of maximum and minimum temperatures over the design life of the structure are considered too severe.

C18.2.4.4 Nominal Design Properties. Device manufacturers typically supply nominal design properties that are reasonably accurate based on previous prototype test programs. The nominal properties can be confirmed by project-specific prototype tests during either the design or construction phases of the project.

C18.2.4.5 Maximum and Minimum Damper Properties

Specification Tolerance on Nominal Design Properties. As part of the design process, it is important to recognize that there are variations in the production damper properties from the nominal properties. This difference is caused by manufacturing variation. Recommended values for the specification tolerance on the average properties of all devices of a given type and size are typically in the $\pm 10\%$ to $\pm 15\%$ range. For a $\pm 10\%$ specification tolerance, the corresponding λ factors would be $\lambda_{(spec,max)} = 1.1$ and $\lambda_{(spec,min)} = 0.9$. Variations for individual device properties may be greater than the tolerance on the average properties of all devices of a given type and size. It is recommended that the device manufacturer be consulted when establishing these tolerance values.

Property Variation (λ) Factors and Maximum and Minimum Damper Properties. Section 18.2.4.5 requires the devices to be analyzed and designed with consideration given to environmental conditions, including the effects of aging, creep, fatigue, and operating temperatures. The individual aging and environmental factors are multiplied together, and then the portion of the resulting λ factor (λ_{ae}) differing from unity is reduced by 0.75 based on the assumption that not all of the maximum/minimum aging and environmental values occur simultaneously.

Results of prototype tests may also indicate the need to address device behavior whereby tested properties differ from the nominal design properties because of test-related effects. Such behavior may include velocity effects, first cycle effects, and any other testing effects that cause behavior different from the nominal

design properties. This behavior is addressed through a testing λ factor (λ_{test}), which is a multiple of all the individual testing effects.

The specification (λ_{spec}), environmental (λ_{ae}) and testing (λ_{test}) factors are used to establish maximum (λ_{max}) and minimum (λ_{min}) damper properties for each device type and size for use in mathematical models of the damped structure in accordance with Eqs. (18.2-3a) and (18.2-3b). These factors are typically applied to whatever parameters govern the mathematical representation of the device.

It should be noted that more sophisticated mathematical models account for various property variation effects directly (e.g., velocity or temperature). When such models are used, the cumulative effect of the λ factors reduce (become closer to 1.0) since some of the typical behaviors contributing to λ_{max} and λ_{min} are already included explicitly in the model. Some effects, such as specification tolerance and aging, will likely always remain since they cannot be accounted for in mathematical models.

Example

Data from prototype testing, as defined in Section 18.6.1, are used to illustrate the λ factors and the maximum and minimum values to be used in analysis and design. The fluid viscous damper under consideration has the following nominal force-velocity constitutive relationship, with kips and inch units:

$$F = C \operatorname{sgn}(V) |V|^\alpha = 128 \operatorname{sgn}(v) |V|^{0.38}$$

The solid line in Fig. C18.2-1 depicts the nominal force-displacement relationship.

Prototype tests of damper corresponding to the following conditions were conducted:

- Force-velocity characteristic tests, all conducted at ambient temperature of 70°F.
  - 10 full cycles performed at various amplitudes.
- Temperature tests, three fully reversible cycles conducted at various velocities at the following temperatures:
  - 40°F
  - 70°F
  - 100°F

The data from prototype tests for each cycle (maximum and negative) are shown as data points in Fig. C18.2-1.

Also shown in the figure are the variations from nominal in the force-velocity relationships for this damper. The relationships are

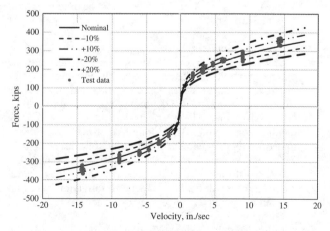

FIGURE C18.2-1 Force-Velocity Relationship for a Nonlinear Viscous Damper

obtained by changing the damper constant (C) value. No variation is considered for the velocity exponent, α. The following diagrams are shown:

- A pair of lines corresponding to damper nominal constitutive relationship computed with the *C* value increased or decreased by 10%. These lines account for the λ_{test} factors as defined in Section 18.2.5.4: $\lambda_{(test,max)} = 1.1$, $\lambda_{(test,min)} = 0.9$.

  For these particular devices, the variation in properties caused by aging and environmental factors is taken as ±5% ($\lambda_{(ae,max)} = 1.05$, $\lambda_{(ae,min)} = 0.95$), and the specification tolerance is set at ±5% ($\lambda_{(spec,max)} = 1.05$, $\lambda_{(spec,min)} = 0.95$). These values should be developed in conjunction with the device manufacturer based on their history of production damper test data and experience with aging and other environmental effects. Using these values in Eqs. (18.2-3a) and (18.2-3b) results in $\lambda_{max} = 1.20$ and $\lambda_{min} = 0.82$. These values satisfy the minimum variation requirements of Section 18.2.4.5. They are rounded to $\lambda_{max} = 1.2$ and $\lambda_{min} = 0.8$.

- A pair of lines corresponding to the cumulative maximum and minimum λ values (accounting for testing, specification tolerance, and other factors listed in Section 18.2.4.5) computed with the nominal *C* value increased or decreased by 20%.

For this example, analysis with minimum and maximum damper properties should be conducted by using 80% and 120% of the nominal value for *C*, respectively. The analysis with maximum damper properties typically produces larger damper forces for use in the design of members and connections, whereas the analysis with minimum damper properties typically produces less total energy dissipation and hence larger drifts.

C18.2.4.6 Damping System Redundancy. This provision is intended to discourage the use of damping systems with low redundancy in any story. At least four damping devices should be provided in each principal direction, with at least two devices in each direction on each side of the center of stiffness to control torsional response. In cases where there is low damping system redundancy by this definition, all damping devices in all stories must be capable of sustaining increased displacements (with associated forces) and increased velocities (with associated displacements and forces) relative to a system with adequate redundancy. The penalty is 130%.

C18.3 NONLINEAR RESPONSE HISTORY PROCEDURE

Those elements of the SFRS and the DS that respond essentially elastically at MCE_R (based on a limit of 1.5 times the expected strength calculated using $\phi = 1$) are permitted to be modeled elastically. Modeling parameters and acceptance criteria provided in ASCE 41, with a performance objective defined in Table 2.2, as modified in this chapter, are deemed satisfactory to meet the requirements of this section.

The hardware of all damping devices (for example, the cylinder of a piston-type device) and the connections between the damping devices and the remainder of the structure must remain elastic at MCE_R (see Section 18.2.1.2). The nonlinear behavior of all other elements of both the SFRS and the DS must be modeled based on test data, which must not be extrapolated beyond the tested deformations. Strength and stiffness degradation must be included if such behavior is indicated. However, the damping system must not become nonlinear to such an extent that its function is impaired.

Table C18.3-1 Analysis Cases for Establishing Amplification Factors

Case	Damper Properties	Accidental Eccentricity
I	Minimum	No
IIa	Minimum	Yes, *X* direction
IIb	Minimum	Yes, *Y* direction

Nonlinear response history analysis (NRHA) is performed at both the design earthquake (DE) and the MCE_R levels. Accidental eccentricity is included at MCE_R but need not be included at the DE level, since the SFRS design checks from Section 18.2.1.1 include accidental eccentricity. However, the results from the NRHA at DE using a model of the combined SFRS and DS must be used to recheck all elements of the SFRS, since the checks of Section 18.2.1.1 are conducted using a representation of the structure excluding the damping system. This requirement is defined in Section 18.4.1. The damping system is designed and evaluated based on the results of the MCE_R analyses, as defined in Section 18.4.2.

For sites classified as near-fault, individual pairs of horizontal ground motion components must be applied to the model to reflect the fault-normal and fault-parallel directions. For all other sites, each pair of horizontal ground motion components should be applied to the building at orthogonal orientations such that the mean of the component response spectra for the records applied in each direction is approximately equal (close to ±10%) to the mean of the component response spectra of all records applied for the period range specified in Section 18.2.2.2. The design reviewer would be the judge of what constitutes "approximately equal."

C18.3.2 Accidental Mass Eccentricity. In order to avoid the need to perform a large number of nonlinear response history analyses that include the suites of ground motions, the upper and lower bound damper properties, and five or more locations of the center of mass, the exception in this provision allows the center-of-mass analysis results to be scaled and used to account for the effects of mass eccentricity in different building quadrants.

The following is one suggested method of developing appropriate amplification factors for deformations and forces for use with center-of-mass NRHAs to account for the effects of accidental eccentricity. The use of other rationally developed amplification factors is permitted and encouraged given that the artificial shift of the center of mass changes the dynamic characteristics of the analyzed structure and may lead to the paradox of reduced torsional response with increasing accidental eccentricity (Basu et al. 2014).

The most critical directions for moving the calculated center of mass are such that the accidental eccentricity adds to the inherent eccentricity in each orthogonal direction at each level. For each of these two eccentric mass positions, and with minimum damper properties, the suite of NRHAs should be run and the results processed in accordance with Section 18.3.3. The analysis cases are defined in Table C18.3-1.

The results from Cases IIa and IIb are then compared in turn to those from Case I. The following amplification factors (ratio of Case IIa or IIb response to Case I response) are computed:

(a) The amplification for story drift in the structure at the plan location with the highest drift, enveloped over all stories;
(b) The amplification for frame-line shear forces at each story for the frame subjected to the maximum drift.

The larger of the two resulting scalars on drift should be used as the deformation amplifier, and the larger of the two resulting scalars on force should be used as the force amplifier. Once the amplification factors are established, the effects of accidental eccentricity should be considered as follows.

The NRHA procedure should be run for the inherent mass eccentricity case only, considering both maximum and minimum damper properties. For each damper property variation, response quantities should be computed in accordance with Section 18.3.3. All resulting deformation response quantities should be increased by the deformation amplifier, and all resulting force quantities should be increased by the force amplifier before being used for evaluation or design.

C18.4 SEISMIC LOAD CONDITIONS AND ACCEPTANCE CRITERIA FOR NONLINEAR RESPONSE HISTORY PROCEDURE

C18.4.1 Seismic Force-Resisting System. All elements of the SFRS are checked under two conditions. First, the SFRS (excluding the damping system) is checked under the minimum base shear requirements of Section 18.2.1.1. Second, the demands from the NRHA at DE (with a model of the combined SFRS and DS) must be used to recheck all elements of the SFRS.

There are three limiting values for the analytically computed drift ratios at MCE_R. Table 12.12-1 lists the allowable drifts for structures. These limiting drift ratios are checked against drift ratio demands computed from the code procedure. Since the code design is an implied DE intensity, the drift ratios in the table are also intended to be used at analysis conducted at this level.

1. 3% limit: For most common structures, the DE allowable drift ratio (Δ_a/h) is 2%. Because for most cases, the ratio of MCE_R to DE intensity is 1.5, then the allowable drift ratio at MCE_R becomes 3% (1.5 × 2%).
2. 1.9 factor: When NRHA analysis is used, the code (Section 16.2.4.3 of ASCE 7-10) allows the DE drift ratios computed from analysis to be limited to 125% of the DE drift ratio limits of Table 12.12-1. Therefore, the MCE_R drift ratios are limited to 1.9 (approximately equal to 1.5 × 1.25) of limits of Table 12.12-1.
3. $1.5R/C_d$ factor: The deflections δ_x of Eq. (12.8-15) are computed by amplifying the deflections computed from analysis by the deflection amplification factor (C_d). The elastic deflections used in Chapter 12 themselves are computed at DE intensity using elastic analysis with forces that are reduced by the response modification factor, R. Thus, for the purpose of comparing drift ratios computed from NRHA with Table 12.12-1, the entries of the table need to be modified by the R/C_d factor for comparison at DE level. Therefore, the allowable drift ratios at MCE_R correspond to $1.5R/C_d$ of entries of the table.

Example: Five-Story Steel Special Moment Frames in Risk Category I or II

- Allowable drift ratio from Table 12.12-1 = 2%.
- Allowable drift ratio for structures with dampers using NRHA then would be the smallest of
  - 3%,
  - 1.9 × 2% = 3.8%, and
  - 1.5 × (8/5.5) × 2% = 4.4%.
- 3% controls. Thus, all computed drift ratios from NRHA should be 3% or less at MCE_R.

C18.5 DESIGN REVIEW

The independent design review of many structures incorporating supplemental damping may be performed adequately by one registered and appropriately experienced design professional. However, for projects involving significant or critical structures, it is recommended that a design review panel consisting of two or three registered and appropriately experienced design professionals be used.

C18.6 TESTING

C18.6.1.2 Sequence and Cycles of Testing. The use of $1/(1.5T_1)$ as the testing frequency is based on a softening of the combined SFRS and DS associated with a system ductility of approximately 2. Test 2 (d) in Section 18.6.1.2 ensures that the prototype damper is tested at the maximum force from analysis.

It should be noted that velocity-dependent devices (for example, those devices characterized by $F = C v^\alpha$) are not intended to be characterized as frequency-dependent under item 4 of this section.

C18.6.1.3 Testing Similar Devices. In order for existing prototype test data to be used to satisfy the requirement of Section 18.6.1, the conditions of this provision must be satisfied. It is imperative that identical manufacturing and quality control procedures be used for the preexisting prototype and the project-specific production damping devices. The precise interpretations of "similar dimensional characteristics, internal construction, and static and dynamic internal pressures" and "similar maximum strokes and forces" are left to the RDP and the design review team. However, variations in these characteristics of the preexisting prototype device beyond approximately ±20% from the corresponding project-specific values should be cause for concern.

C18.6.1.4 Determination of Force-Velocity-Displacement Characteristics. When determining nominal properties (item 2) for damping devices whose first-cycle test properties differ significantly from the average properties of the first three cycles, an extra cycle may be added to the test, and the nominal properties may be determined from the average value using data from the second through fourth cycles. In this case, the effect of first-cycle properties must be addressed explicitly and included in the λ_{max} factor. It should be noted that if the property variation methodology of Sections 18.2.4.4 and 18.2.4.5 is applied consistently, the maximum and minimum design properties (Eqs. (18.2-4a) and (18.2-4b)) will be identical, regardless of whether the nominal properties are taken from the average of cycles 1 through 3 or cycles 2 through 4.

C18.6.2 Production Tests. The registered design professional is responsible for defining in the project specifications the scope of the production damper test program, including the allowable variation in the average measured properties of the production damping devices. The registered design professional must decide on the acceptable variation of damper properties on a project-by-project basis. This range must agree with the specification tolerance from Section 18.2.4.5. The standard requires that all production devices of a given type and size be tested.

Individual devices may be permitted a wider variation (typically ±15% or ±20%) from the nominal design properties. For example, in a device characterized by $F = C v^\alpha$, the mean of the force at a specified velocity for all tested devices might be permitted to vary no more than ±10% from the specified value of force, but the force at a specified velocity for any individual device might be permitted to vary no more than ±15% from the specified force.

The production dynamic cyclic test is identical (except for three versus five cycles) to one of the prototype tests of Section 18.6.1.2, so that direct comparison of production and prototype damper properties is possible.

The exception is intended to cover those devices that would undergo yielding or be otherwise damaged under the production test regime. The intent is that piston-type devices be 100% production tested, since their properties cannot be shown to meet the requirements of the project specifications without testing. For other types of damping devices, whose properties can be demonstrated to be in compliance with the project specifications by other means (for example, via material testing and a manufacturing quality control program), the dynamic cyclic testing of 100% of the devices is not required. However, in this case, the RDP must establish an alternative production test program to ensure the quality of the production devices. Such a program would typically focus on such things as manufacturing quality control procedures (identical between prototype and production devices), material testing of samples from a production run, welding procedures, and dimensional control. At least one production device must be tested at 0.67 times the MCE_R stroke at a frequency equal to $1/(1.5T_1)$, unless the complete project-specific prototype test program has been performed on an identical device. If such a test results in inelastic behavior in the device, or the device is otherwise damaged, that device cannot be used for construction.

C18.7 ALTERNATE PROCEDURES AND CORRESPONDING ACCEPTANCE CRITERIA

This section applies only to those cases where either the RS or the ELF procedure is adopted.

C18.7.1 Response-Spectrum Procedure and C18.7.2 Equivalent Lateral Force Procedure

Effective Damping. In the standard, the reduced response of a structure with a damping system is characterized by the damping coefficient, B, based on the effective damping, β, of the mode of interest. This approach is the same as that used for isolated structures. Like isolation, effective damping of the fundamental mode of a damped structure is based on the nonlinear force-deflection properties of the structure. For use with linear analysis methods, nonlinear properties of the structure are inferred from the overstrength factor, Ω_0, and other terms.

Fig. C18.7-1 illustrates reduction in design earthquake response of the fundamental mode caused by increased effective damping (represented by coefficient, B_{1D}). The capacity curve is a plot of the nonlinear behavior of the fundamental mode in spectral acceleration-displacement coordinates. The reduction caused by damping is applied at the effective period of the fundamental mode of vibration (based on the secant stiffness).

In general, effective damping is a combination of three components:

1. Inherent Damping (β_I)—Inherent damping of the structure at or just below yield, excluding added viscous damping (typically assumed to be 2–5% of critical for structural systems without dampers).
2. Hysteretic Damping (β_H)—Postyield hysteretic damping of the seismic force-resisting system and elements of the damping system at the amplitude of interest (taken as 0% of critical at or below yield).
3. Added Viscous Damping (β_V)—The viscous component of the damping system (taken as 0% for hysteretic or friction-based damping systems).

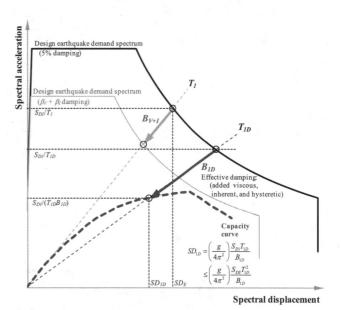

FIGURE C18.7-1 Effective Damping Reduction of Design Demand

Both hysteretic damping and added viscous damping are amplitude-dependent, and the relative contributions to total effective damping change with the amount of postyield response of the structure. For example, adding dampers to a structure decreases postyield displacement of the structure and, hence, decreases the amount of hysteretic damping provided by the seismic force-resisting system. If the displacements are reduced to the point of yield, the hysteretic component of effective damping is zero and the effective damping is equal to inherent damping plus added viscous damping. If there is no damping system (as in a conventional structure), effective damping simply equals inherent damping.

Linear Analysis Methods. The section specifies design earthquake displacements, velocities, and forces in terms of design earthquake spectral acceleration and modal properties. For equivalent lateral force (ELF) analysis, response is defined by two modes: the fundamental mode and the residual mode. The residual mode is used to approximate the combined effects of higher modes. Although typically of secondary importance to story drift, higher modes can be a significant contributor to story velocity and, hence, are important for design of velocity-dependent damping devices. For response spectrum analysis, higher modes are explicitly evaluated.

For both the ELF and the response spectrum analysis procedures, response in the fundamental mode in the direction of interest is based on assumed nonlinear (pushover) properties of the structure. Nonlinear (pushover) properties, expressed in terms of base shear and roof displacement, are related to building capacity, expressed in terms of spectral coordinates, using mass participation and other fundamental-mode factors shown in Fig. C18.7-2. The conversion concepts and factors shown in Fig. C18.7-2 are the same as those defined in Chapter 9 of ASCE/SEI 41 (2014), which addresses seismic rehabilitation of a structure with damping devices.

Where using linear analysis methods, the shape of the fundamental-mode pushover curve is not known, so an idealized elastoplastic shape is assumed, as shown in Fig. C18.7-3. The idealized pushover curve is intended to share a common point with the actual pushover curve at the design earthquake displacement, D_{1D}. The idealized curve permits definition of the global

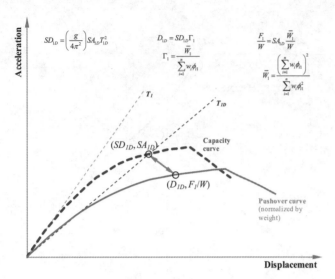

FIGURE C18.7-2 Pushover and Capacity Curves

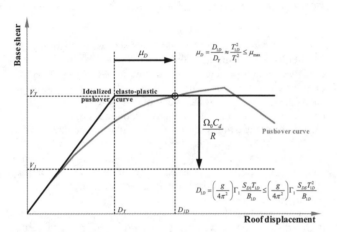

FIGURE C18.7-3 Pushover and Capacity Curves

ductility demand caused by the design earthquake, μ_D, as the ratio of design displacement, D_{1D}, to yield displacement, D_Y. This ductility factor is used to calculate various design factors; it must not exceed the ductility capacity of the seismic force-resisting system, μ_{max}, which is calculated using factors for conventional structural response. Design examples using linear analysis methods have been developed and found to compare well with the results of nonlinear time history analysis (Ramirez et al. 2001).

Elements of the damping system are designed for fundamental-mode design earthquake forces corresponding to a base shear value of V_Y (except that damping devices are designed and prototypes are tested for maximum considered earthquake response). Elements of the seismic force-resisting system are designed for reduced fundamental-mode base shear, V_1, where force reduction is based on system overstrength (represented by Ω_0), multiplied by C_d/R for elastic analysis (where actual pushover strength is not known). Reduction using the ratio C_d/R is necessary because the standard provides values of C_d that are less than those for R. Where the two parameters have equal value and the structure is 5% damped under elastic conditions, no adjustment is necessary. Because the analysis methodology is based on calculating the actual story drifts and damping device displacements (rather than the displacements

Table C18.7-1 Values of Damping Coefficient, B

Effective Damping, β (%)	Table 17.5-1 of ASCE/SEI 7 (2010), AASHTO (2010), CBC (2013), seismically isolated structures)	Table 18.6-1 of ASCE/SEI 7 (2010) (structures with damping systems)	FEMA 440 (2005)	Eurocode 8 (2005)
2	0.8	0.8	0.8	0.8
5	1.0	1.0	1.0	1.0
10	1.2	1.2	1.2	1.2
20	1.5	1.5	1.5	1.6
30	1.7	1.8	1.8	1.9
40	1.9	2.1	2.1	2.1
50	2.0	2.4	2.4	2.3

calculated for elastic conditions at the reduced base shear and then multiplied by C_d), an adjustment is needed. Because actual story drifts are calculated, the allowable story drift limits of Table 12.12-1 are multiplied by R/C_d before use.

C18.7.3 Damped Response Modification

C18.7.3.1 Damping Coefficient. Values of the damping coefficient, B, in Table 18.7-1 for design of damped structures are the same as those in Table 17.5-1 for isolated structures at damping levels up to 20% but extend to higher damping levels based on results presented in Ramirez et al. (2001). Table C18.7-1 compares values of the damping coefficient as found in the standard and various resource documents and codes. FEMA 440 (2005) and Eurocode 8 (2005) present equations for the damping coefficient, B, whereas the other documents present values of B in tabular format.

The equation in FEMA 440 is

$$B = \frac{4}{5.6 - \ln(100\beta)}$$

The equation in Eurocode 8 (2005) is

$$B = \sqrt{\frac{0.05 + \beta}{0.10}}$$

C18.7.3.2 Effective Damping. The effective damping is calculated assuming that the structural system exhibits perfectly bilinear hysteretic behavior characterized by the effective ductility demand, μ, as described in Ramirez et al. (2001). Effective damping is adjusted using the hysteresis loop adjustment factor, q_H, which is the actual area of the hysteresis loop divided by the area of the assumed perfectly bilinear hysteretic loop. In general, values of this factor are less than unity. In Ramirez et al. (2001), expressions for this factor (which they call Quality Factor) are too complex to serve as a simple rule. Eq. (18.7-49) provides a simple estimate of this factor. The equation predicts correctly the trend in the constant acceleration domain of the response spectrum, and it is believed to be conservative for flexible structures.

C18.7.4 Seismic Load Conditions and Acceptance Criteria for RSA and ELF Procedures

C18.7.4.5 Seismic Load Conditions and Combination of Modal Responses. Seismic design forces in elements of the damping system are calculated at three distinct stages: maximum displacement, maximum velocity, and maximum acceleration.

All three stages need to be checked for structures with velocity-dependent damping systems. For displacement-dependent damping systems, the first and third stages are identical, whereas the second stage is inconsequential.

Force coefficients C_{mFD} and C_{mFV} are used to combine the effects of forces calculated at the stages of maximum displacement and maximum velocity to obtain the forces at maximum acceleration. The coefficients are presented in tabular form based on analytic expressions presented in Ramirez et al. (2001) and account for nonlinear viscous behavior and inelastic structural system behavior.

REFERENCES

American Association of State Highway and Transportation Officials (AASHTO). (2010). *Guide specifications for seismic isolation design*, Washington, DC.

ASCE/SEI. (2014), "Seismic evaluation and retrofit of existing buildings." *ASCE/SEI 41-13*, Reston, VA.

Basu, D., Constantinou, M. C., and Whittaker, A. S. (2014). "An equivalent accidental eccentricity to account for the effects of torsional ground motion on structures." *Eng. Struct.*, 69, 1–11.

British Standards Institution. (1983). *Commentary on corrosion at bimetallic contacts and its alleviation*, PD6484:1979, London.

California Buildings Standards Commission (CBC). (2013). *California building code*. Sacramento, CA.

Constantinou, M. C., Soong, T. T., and Dargush, G. F. (1998). *Passive energy dissipation systems for structural design and retrofit, Monograph 1*, Multidisciplinary Center for Earthquake Engineering Research, University of Buffalo, State University of New York, Buffalo.

Constantinou, M. C., Whittaker, A. S., Kalpakidis, Y., Fenz, D. M., and Warn, G. P. (2007). "Performance of seismic isolation hardware under service and seismic loading." *Report No. MCEER-07-0012*, Multidisciplinary Center for Earthquake Engineering Research, Buffalo, NY.

European Committee for Standardization (Eurocode 8). (2005). "Design of structures for earthquake resistance. Part 2: Bridges." *Eurocode 8*, European Committee for Standardization.

Federal Emergency Management Agency (FEMA). (2005). "Improvement of nonlinear static seismic analysis procedures." *FEMA 440*, FEMA, Washington, DC.

Hanson, R. D., and Soong, T. T. (2001). "Seismic design with supplemental energy dissipation devices." *MNO-8*, Earthquake Engineering Research Institute, Oakland, CA.

Makris, N., Roussos, Y., Whittaker, A. S., and Kelly, J. M. (1998). "Viscous heating of fluid dampers. I: Large-amplitude motions." *J. Eng. Mech.*, 124(11), 1217–1223.

Pavlou, E., and Constantinou, M. C. (2004). "Response of elastic and inelastic structures with damping systems to near-field and soft-soil ground motions." *Engrg. Struct.* 26, 1217–1230.

Pavlou, E., and Constantinou, M. C. (2006). "Response of nonstructural components in structures with damping systems." *J. Struct. Engrg.*, 132(7), 1108–1117.

Rabinowicz, E. (1995). *Friction and wear of materials*, John Wiley & Sons, New York.

Ramirez, O. M., Constantinou, M. C., Kircher, C. A., Whittaker, A., Johnson, M., Gomez, J. D., et al. (2001). "Development and evaluation of simplified procedures of analysis and design for structures with passive energy dissipation systems." *Technical Report MCEER-00-0010*, Revision 1, Multidisciplinary Center for Earthquake Engineering Research, University of Buffalo, State University of New York, Buffalo.

Skinner, R. I., Robinson, W. H., and McVerry, G. H. (1993). *An introduction to seismic isolation*, John Wiley & Sons, Chichester, UK.

Soong, T. T., and Dargush, G. F. (1997). *Passive energy dissipation systems in structural engineering*, John Wiley & Sons, London.

OTHER REFERENCES (NOT CITED)

Miyamoto, H. K., Gilani, A. S. J., Wada, A., and Ariyaratana, C. (2011). "Identifying the collapse hazard of steel special moment-frame buildings with viscous dampers using the FEMA P-695 methodology." *Earthq. Spectra*, 27(4), 1147–1168.

Newmark, N. M., and Hall, W. J. (1969). "Seismic design criteria for nuclear reactor facilities." *Proc., 4th World Conference in Earthquake Engineering*, Santiago, Chile.

Ramirez, O. M., Constantinou, M. C., Gomez, J., Whittaker, A. S., and Chrysostomou, C. Z. (2002a). "Evaluation of simplified methods of analysis of yielding structures with damping systems." *Earthq. Spectra*, 18(3), 501–530.

Ramirez, O. M., Constantinou, M. C., Whittaker, A. S., Kircher, C. A., and Chrysostomou, C. Z. (2002b). "Elastic and inelastic seismic response of buildings with damping systems." *Earthq. Spectra*, 18(3), 531–547.

Ramirez, O. M., Constantinou, M. C., Whittaker, A. S., Kircher, C. A., Johnson, M. W., and Chrysostomou, C. Z. (2003). "Validation of 2000 NEHRP provisions equivalent lateral force and modal analysis procedures for buildings with damping systems." *Earthq. Spectra*, 19(4), 981–999.

Structural Engineers Association of California (SEAOC). (2013). *2012 IBC SEAOC structural/seismic design manual Volume 5: Examples for seismically isolated buildings and buildings with supplemental damping*, Sacramento, CA.

Whittaker, A. S., Constantinou, M. C., Ramirez, O. M., Johnson, M. W., and Chrysostomou, C. Z. (2003). "Equivalent lateral force and modal analysis procedures of the 2000 NEHRP provisions for buildings with damping systems." *Earthq. Spectra*, 19(4), 959–980.

CHAPTER C19
SOIL–STRUCTURE INTERACTION FOR SEISMIC DESIGN

C19.1 GENERAL

In an earthquake, the shaking is transmitted up through the structure from the geologic media underlying and surrounding the foundation. The response of a structure to earthquake shaking is affected by interactions among three linked systems: the structure, the foundation, and the geologic media underlying and surrounding the foundation. The analysis procedures in Chapters 12 and 15 idealize the response of the structure by applying forces to the structure, which is typically assumed to have a fixed base at the foundation–soil interface. In some cases, the flexibility of the foundation elements and underlying soils is included in the analysis model. The forces that are applied to the structure are devised based on parameters representing free-field ground motions. The term "free-field" refers to motions not affected by structural vibrations or the foundation characteristics of the specific structure and represents the condition for which the design spectrum is derived using the procedures given in Chapter 11 and Chapter 21. In most cases, however, the motions at the foundation that are imparted to the structure are different from the free-field motions. This difference is caused by the effects of the interaction of the structure and the geologic media. A seismic soil–structure interaction (SSI) analysis evaluates the collective response of these systems to a specified free-field ground motion.

SSI effects are absent for the theoretical condition of rigid geologic media, which is typical of analytical models of structures. Accordingly, SSI effects reflect the differences between the actual response of the structure and the response for the theoretical, rigid base condition. Visualized within this context, two following SSI effects can significantly affect the response of structures:

1. **Foundation Deformations.** Flexural, axial, and shear deformations of foundation elements occur as a result of loads applied by the superstructure and the supporting geologic media. Additionally, the underlying geologic media deforms because of loads from the foundations. Such deformations represent the seismic demand for which foundation components should be designed. These deformations can also significantly affect the overall system behavior, especially with respect to damping.
2. **Inertial SSI Effects.** Inertia developed in a vibrating structure gives rise to base shear, moment, and torsional excitation, and these loads in turn cause displacements and rotations of the foundation relative to the free-field displacement. These relative displacements and rotations are only possible because of flexibility in the soil–foundation system, which can significantly contribute to the overall structural flexibility in some cases. Moreover, the relative foundation free-field motions give rise to energy dissipation via radiation damping (i.e., damping associated with wave propagation into the ground away from the foundation, which acts as the wave source) and hysteretic soil damping, and this energy dissipation can significantly affect the overall damping of the soil–foundation–structure system. Because these effects are rooted in the structural inertia, they are referred to as inertial interaction effects.
3. **Kinematic SSI Effects.** Kinematic SSI results from the presence of foundation elements on or in soil that are much stiffer than the surrounding soil. This difference in stiffness causes foundation motions to deviate from free-field motion as a result of base slab averaging and embedment effects.

Chapter 19 addresses both types of SSI effects. Procedures for calculating kinematic and inertial SSI effects were taken from recommendations in NIST GCR 12-917-21 (NIST 2012). Further discussion of SSI effects can be found in this NIST document and some of the references cited therein.

Substantial revisions have been made to Chapter 19 in this edition of ASCE 7. They include (1) the introduction of formulas for the stiffness and damping of rectangular foundations, (2) revisions to the formulas for the reduction of base shear caused by SSI, (3) reformulation of the effective damping ratio of the SSI system, (4) introduction of an effective period lengthening ratio, which appears in the formula for the effective damping ratio of the SSI system, and which depends on the expected structural ductility demand, and (5) the introduction of kinematic SSI provisions. Most of these revisions come from the NIST GCR 12-917-21 (NIST 2012) report on SSI. However, the basic model of the inertial SSI system has remained the same since SSI provisions were first introduced in the ATC 3-06 report (ATC 1978).

The first effect, foundation deformation, is addressed by explicitly requiring the design professional to incorporate the deformation characteristics of the foundation into their analysis model. Including foundation deformations is essential for understanding soil–structure interaction (SSI). Therefore, the flexibility of the foundation must be modeled to capture translational and rotational movement of the structure at the soil–foundation interface.

For the linear procedures, this requirement to model the flexibility of the foundation and soil means that springs should be placed in the model to approximate the effective linear stiffness of the deformations of the underlying geologic media and the foundation elements. This could be done by placing isolated spring elements under the columns and walls, by explicitly modeling the foundation elements and geologic media in the mathematical model, or some combination of the two. For the response history procedure, this would mean that in addition to the stiffness of the subsurface media and foundation elements, the nonlinear parameters of those materials would be incorporated into the analytical model. Because of the uncertainty in estimating the stiffness and deformation capacity of geologic media, upper and lower bound estimates of the properties should be used and the condition that produces the more conservative change in response parameters from a fixed-base structure must be used.

Inertial interaction effects are addressed through the consideration of foundation damping. Inertial interaction in structures tends to be important for stiff structural systems such as shear walls and braced frames, particularly where the foundation soil is relatively soft. The provisions provide a method for estimating radiation damping and soil hysteretic damping.

The two main kinematic interaction effects are included in these provisions: base slab averaging and embedment effects. The kinematic interaction effects cause the motion input into the structure to be different from the free-field motions. The provisions provide a means by which a free-field, site-specific response spectrum can be modified to account for these kinematic interaction effects to produce a foundation-input spectrum.

Site classes A and B are excluded from Chapter 19 because the dynamic interaction between structures and rock is minimal based on theory. Furthermore, there are no empirical data to indicate otherwise.

Section 19.1.1 prohibits using the cap of S_s included in Section 12.8.1.3 because of the belief that structures meeting the requirements of that section have performed satisfactorily in past earthquakes, partially because of SSI effects. Taking advantage of that predetermined cap on S_s and then subsequently reducing the base shear caused by SSI effects may therefore amount to double-counting the SSI effects.

C19.2 SSI ADJUSTED STRUCTURAL DEMANDS

When the equivalent lateral force procedure is used, the equivalent lateral force is computed using the period of the flexible base structure and is modified for the SSI system damping. For the modal analysis procedure, a response spectrum, which has been modified for the SSI system damping and then divided by (R/I_e), is input into the mathematical model. The lower bound limit on the design base shear based on the equivalent lateral force procedures per Section 12.9.1.4 still applies, but the equivalent lateral force base shear modified to account for SSI effects replaces the base shear for the fixed-base case.

For both the equivalent lateral force and response spectrum procedures, the total reduction caused by SSI effects is limited to a percentage of the base shear determined in accordance with Section 12.8.1, which varies based on the R factor. This limitation on potential reductions caused by SSI reflects the limited understanding of how the effects of SSI interact with the R factor. All of the SSI effects presented herein are based on theoretical linear elastic models of the structure and geologic media. That is why reductions of 30% are permitted for $R = 3$ or less. It is felt that those systems exhibit limited inelastic response and therefore, a larger reduction in the design force caused by SSI should be permitted. For higher R factor systems, where significant damping caused by structural yielding is expected, the contribution of foundation damping is assumed to have little effect on the reduction of the response. Some reduction is permitted because of (1) an assumed period lengthening resulting from the incorporation of base flexibility, (2) potential reduction in mass participation in the fundamental mode because two additional degrees of freedom are present caused by translation and rotation of the base, and (3) limited foundation damping interacting with the structural damping.

Reductions to the response spectrum caused by the SSI system damping and kinematic SSI effects are for the elastic 5% damped response spectrum typically provided to characterize free-field motion. In addition, studies have indicated that there is a fair amount of uncertainty in the amount of kinematic SSI when measured reductions between the free-field motion and the foundation input motion are compared with the theoretical models (Stewart 2000).

Reductions for kinematic SSI effects are not permitted for the equivalent lateral force and modal response spectrum procedures. The equations for predicting the kinematic SSI effects are based on modifications to the linear elastic response spectrum. Studies have not been performed to verify if they are similarly valid for inelastic response spectra, on which the R factor procedures are based. Additionally, the amount of the reduction for kinematic SSI effects is dependent on the period of the structure, with the greatest modifications occurring in the short period range. Because the fundamental periods of most structures lengthen as they yield, what would potentially be a significant reduction at the initial elastic period may become a smaller reduction as the structure yields. Without an understanding of how the period may lengthen in the equivalent lateral force or modal response spectrum procedures, there is a potential for a user to overestimate the reduction in the response parameters caused by kinematic SSI effects. Thus, their use is not permitted.

All types of SSI effects are permitted to be considered in a response history analysis per Chapter 16. If SSI effects are considered, the site-specific response spectrum should be used as the target to which the acceleration histories are scaled. The requirement to use a site-specific response spectrum was placed in the provisions because of the belief that it provided a more realistic definition of the earthquake shaking than is provided by the design response spectrum and MCE_R response spectrum in accordance with Sections 11.4.6 and 11.4.7. A more realistic spectrum was required for proper consideration of SSI effects, particularly kinematic SSI effects. The design response spectrum and MCE_R response spectrum, in accordance with Sections 11.4.6 and 11.4.7, use predetermined factors to modify the probabilistic or deterministic response spectrum for the soil conditions. These factors are sufficient for most design situations. However, if SSI effects are to be considered and the response spectrum modified accordingly, then more accurate representations of how the underlying geologic media alter the spectral ordinates should be included before the spectrum is modified because of the SSI effects.

A site-specific response spectrum that includes the effects of SSI can be developed with explicit consideration of SSI effects by modifying the spectrum developed for free-field motions through the use of the provisions in Sections 19.3 and 19.4. If the foundation damping is not specifically modeled in the analytical model of the structure, the input response spectrum can include the effects of foundation damping. Typically, the base slab averaging effect is not explicitly modeled in the development of a site-specific response spectrum and the provisions in Section 19.4.1 are used to modify the free-field, site-specific response spectrum to obtain the foundation input spectrum. Embedment effects can be modeled directly by developing the site-specific spectrum at the foundation base level, as opposed to the ground surface. Alternatively, the site-specific spectrum for the free field can be developed at the ground level and the provisions of Section 19.4.2 can be used to adjust it to the depth corresponding to the base of foundation.

The limitations on the reductions from the site-specific, free-field spectrum to the foundation input spectrum are based on several factors. The first is the scatter between measured ratios of foundation input motion to free-field motion versus the ratios from theoretical models (Stewart 2000). The second is the inherent variability of the properties of the underlying geologic media over the footprint of the structure. Whereas there is a requirement to bound the flexibility of the soil and foundation springs, there are no corresponding bounding requirements applied to the geologic media parameters used to compute the foundation damping and kinematic SSI. The last factor is the aforementioned lack of research into the interaction between SSI effects and yielding structures. Some studies have shown that there are reductions for most cases of SSI when coupled with an R factor based approach (Jarernprasert et al. 2013).

A limitation was placed on the maximum reduction for an SSI modified site-specific response spectrum with respect to the response spectrum developed based on the USGS ground-motion parameters and the site coefficients. This limitation is caused by similar concerns expressed in Section C21.3 regarding the site-specific hazard studies generating unreasonably low response spectra. There is a similar concern that combining SSI effects with site-specific ground motions could significantly reduce the seismic demand from that based on the USGS ground-motion parameters and the site coefficients. However, it was recognized that these modifications are real and the limit could be relaxed, but not eliminated, if there were (1) adequate peer review of the site-specific seismic hazard analysis and the methods used to determine the reductions attributable to SSI effects and (2) approval of the jurisdictional authority.

Peer review would include, but not be limited, to the following:

1. Development of the site-specific response spectrum used to scale the ground motions;
2. Determination of foundation stiffness and damping, including the properties of the underlying subsurface media used in the determination;
3. Confirmation that the base slab and first slab above the base are sufficiently rigid to allow base slab averaging to occur, including verification that the base slab is detailed to act as a diaphragm; and
4. Assumptions used in the development of the soil and radiation damping ratios.

The SSI effects can be used in a response history analysis per Chapter 16. Two options for the modeling of the SSI are as follows:

1. Create a nonlinear finite element (FE) model of the structure, foundation, and geologic media. The mesh for the geologic media should extend to an appropriate depth and horizontal distance away from the foundation with transmitting boundaries along the sides to absorb outgoing seismic waves generated by the foundation. The motion should be input at the base of the FE model and should propagate upward as shear waves. The free-field response spectrum can be reduced for kinematic SSI only per the provisions in Section 19.4, but embedment effects would not be allowed in the reduction because the waves propagating up from the depth of the foundation to the surface would automatically include kinematic effects of embedment.
2. Create a nonlinear finite element model of the structure and foundation, with springs and dashpots attached to the perimeter walls and base of the foundation to account for the soil–foundation interaction. Guidance on the development of dashpots can be found in NIST GCR 12-917-21 (NIST 2012). The free-field response spectrum can be reduced for kinematic SSI per Section 19.4, but embedment effects may or may not be allowed in the reduction depending on whether or not (i) the motion is allowed to vary with depth along the embedded portion of the foundation, and (ii) the free-field motion used as input motion is defined at the ground surface or at the bottom of the basement. The dashpots would account for the radiation and hysteretic damping of the geologic media, either per Section 19.3 or more detailed formulations.

C19.3 FOUNDATION DAMPING

The procedures in Section 19.3 are used to estimate an SSI system damping ratio, β_0, based on the underlying geologic media and interaction of the structure and its foundation with this geologic media. There are two main components that contribute to foundation damping: soil hysteretic damping and radiation damping. The provisions in this section provide simplified ways to approximate these effects. However, they are complex phenomena and there are considerably more detailed methods to predict their effects on structures. The majority of the provisions in this section are based on material in NIST GCR 12-917-21 (NIST 2012). Detailed explanations of the background of these provisions, supplemental references, and more sophisticated methods for predicting radiation damping can be found in that report. However, those references do not provide the derivation of the effective period lengthening ratio, $(\tilde{T}/T)_{\text{eff}}$ given by Eq. (19.3-2). This ratio appears in the equation for β_0 (Eq. 19.3-1), and it is derived from the total displacement of the mass of the SSI oscillator model resulting from a horizontal force applied to the mass. A component of this displacement is the displacement of the mass relative to its base, and it is equal to the ductility demand, μ, times the elastic displacement of the mass relative to the base. The other components of the total displacement arise from displacement of the translational foundation spring (K_y or K_r) and the translation resulting from the rotational foundation spring (K_{xx} or K_{rr}). The period lengthening ratio, $(\tilde{T}/T)$ appearing in Eq. (19.3-2) is derived in the same manner assuming that $\mu = 1$.

Radiation damping refers to energy dissipation from wave propagation away from the vibrating foundation. As the ground shaking is transmitted into the structure's foundation, the structure itself begins to translate and rock. The motion of the foundation relative to the free-field motion creates waves in the geologic media, which can act to counter the waves being transmitted through the geologic media caused by the earthquake shaking. The interference is dependent on the stiffness of the geologic media and the structure, the size of the foundation, type of underlying geologic media, and period of the structure. The equations for radiation damping in Section 19.3.3 were taken from NIST GCR 12-917-21 (NIST 2012); details of the derivation are found in Givens (2013).

In Section 19.3.3, the equations for K_y and K_{xx}, for rectangular foundations, and the associated damping ratios, β_y and β_{xx}, come from Pais and Kausel (1988) and are listed in Table 2-2a and Table 2-3a in the NIST report. The corresponding static stiffness equations for circular foundations in Section 19.3.4 were taken from Veletsos and Verbic (1973); the other equations appearing in Section 19.3.4 were adapted from equations in the NIST report. The foundation stiffness and damping equations in these two sections apply to surface foundations. The reasons for excluding embedment effects are explained in the third paragraph from the end of this subsection.

Soil hysteretic damping occurs because of shearing within the soil and at the soil–foundation interface. Values of the equivalent viscous damping ratio, β_s, to model the hysteretic damping can be obtained from site response analysis or Table 19.3-3.

Foundation damping effects, modeled by β_f, tend to be important for stiff structural systems such as shear walls and

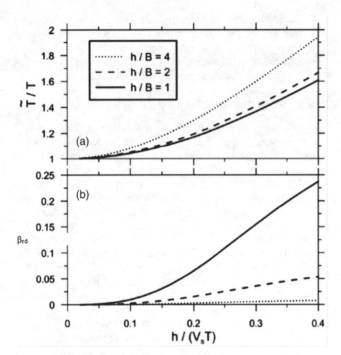

FIGURE C19.3-1 Example of Radiation Damping
Source: NIST 2012.

braced frames, particularly where they are supported on relatively soft soil sites, such as Site Classes D and E. This effect is determined by taking the ratio of the fundamental period of the structure, including the flexibility of the foundation and underlying subsurface media (flexible-base model) and the fundamental period of the structure assuming infinite rigidity of the foundation and underlying subsurface media (fixed-base model). Analytically, this ratio can be determined by computing the period of the structure with the foundation/soil springs in the model and then replacing those springs with rigid support.

Fig. C19.3-1 illustrates the effect of the period ratio, $\tilde{T}/T$, on the radiation damping, β_r, which typically accounts for most of the foundation damping. $\tilde{T}/T$ is the ratio of the fundamental period of the SSI system to the period of the fixed-base structure. The figure shows that for structures with larger height, h, to foundation half-width, B, aspect ratios, the effects of foundation damping become less. In this figure, the aspect ratio of the foundation is assumed to be square.

These inertial interaction effects are influenced considerably by the shear modulus of the underlying subgrade, specifically the modulus that coincides with the seismic shaking being considered. As noted in the standard, shear modulus G can be evaluated from small-strain shear wave velocity as $G = (G/G_o)G_o = (G/G_o)\gamma v_{so}^2/g$ (all terms defined in the standard). Shear wave velocity, v_{so}, should be evaluated as the average small-strain shear wave velocity within the effective depth of influence below the foundation. The effective depth should be taken as half the lesser dimension of the foundation, which in the provisions is defined as B. Methods for measuring v_{so} (preferred) or estimating it from other soil properties are summarized elsewhere (e.g., Kramer 1996).

The radiation damping procedure is conservative and underestimates the foundation damping for shaking in the long direction where the foundation aspect ratios exceed 2:1 but could be potentially unconservative where wall and frame elements are close enough so that waves emanating from distinct foundation components destructively interfere with each other across the period range of interest. That is why the limit of spacing of the vertical lateral force-resisting elements is imposed on the use of these provisions.

For structures supported on footings, the formulas for radiation damping can generally be used with B and L calculated using the footprint dimensions of the entire structure, provided that the footings are interconnected with grade beams and/or a sufficiently rigid slab on grade. An exception can occur for structures with both shear walls and frames, for which the rotation of the foundation beneath the wall may be independent of that for the foundation beneath the column (this type is referred to as weak rotational coupling). In such cases, B and L are often best calculated using the dimensions of the wall footing. Very stiff foundations like structural mats, which provide strong rotational coupling, are best described using B and L values that reflect the full foundation dimension. Regardless of the degree of rotational coupling, B and L should be calculated using the full foundation dimension if foundation elements are interconnected or continuous. Further discussion can be found in FEMA 440 (FEMA 2005) and NIST GCR 12-917-21 (NIST 2012).

The radiation damping provisions conservatively exclude the effects of embedment. Embedment typically increases the amount of radiation damping if the basement or below-grade foundation stays in contact with the soil on all sides. Because there is typically some gapping between the soil and the sides of the basement or foundation, these embedment effects may be less than the models predict. There are some additional issues with the procedures for embedded foundations. For the case where the embedment is significant but the soils along the sides are much more flexible than the bearing soils, a high impedance contrast between the first two layers is recognized as a potential problem regardless of the embedment. The NIST GCR 12-917-21 (NIST 2012) report therefore recommends ignoring the additional contributions caused by embedment but still using the soil properties derived below the embedded base.

The equations in Sections 19.3.3 and 19.3.4 are for shallow foundations. This is not to say that radiation damping does not occur with deep (pile or caisson) foundation systems, but the phenomenon is more complex. Soil layering and group effects are important, and there are the issues of the possible contributions of the bottom structural slab and pile caps. Because the provisions are based on the impedance produced by a rigid plate in soil, these items cannot be easily taken into account. Therefore, more detailed modeling of the soil and the embedded foundations is required to determine the foundation impedances. The provisions permit such modeling but do not provide specific guidance for it. Guidance can be found for example in NIST GCR 12-917-21 (NIST 2012) and its references.

Soil hysteretic damping occurs as seismic waves propagate through the subsurface media and reach the base of the structure, and it can have an effect on the overall system damping when the soil strains are high. Table 19.3-3 in the provisions was derived based on relationships found in EPRI (1993) and Vucetic and Dobry (1991) that relate the ratio between G/G_0 to cyclic shear strain in the soil, and then to soil damping. The values in the table are based on conservative assumptions about overburden pressures on granular soils and plasticity index of clayey soils. This simplified approach does not preclude the geotechnical engineer from providing more detailed estimates of soil damping. However, the cap on reductions in the seismic demand are typically reached at around an additional 5% hysteretic damping ratio (10% total damping ratio), and further reductions would require peer review.

C19.4 KINEMATIC SSI EFFECTS

Kinematic SSI effects are broadly defined as the difference between the ground motion measured in a free-field condition and the motion which would be measured at the structure's foundation, assuming that it and the structure were massless (i.e., inertial SSI was absent). The differences between free-field and foundation input motions are caused by the characteristics of the structure foundation, exclusive of the soil and radiation damping effects in the preceding section. There are two main types of kinematic interaction effects: base slab averaging and embedment. The provisions provide simplified methods for capturing these effects. The basis for the provisions and additional background material can be found in FEMA 440 (FEMA 2005) and NIST GCR 12-917-21 (NIST 2012).

FEMA 440 (FEMA 2005) specifically recommends against applying these provisions to very soft soil sites such as E and F. These provisions allow kinematic SSI for Site Class E but retain the prohibition for Site Class F. That is not to say that kinematic interaction effects are not present at Site Class F sites, but that these specific provisions should not be used; rather, more detailed site-specific assessments are permitted to be used to determine the possible modifications at those sites.

In addition to the prescriptive methods contained in the standard, there are also provisions that allow for direct computation of the transfer function of the free-field motion to a foundation input motion caused by base slab averaging or embedment. Guidance on how to develop these transfer functions can be found in NIST GCR 12-917-21 (NIST 2012) and the references contained therein.

C19.4.1 Base Slab Averaging. Base slab averaging refers to the filtering of high-frequency portions of the ground shaking caused by the incongruence of motion over the base. For this filtering to occur, the base of the structure must be rigid or semirigid with respect to the vertical lateral force-resisting elements and the underlying soil. If the motions are out of phase from one end of the foundation to the other and the foundation is sufficiently rigid, then the motion on the foundation would be different from the ground motion at either end. The ground motions at any point under the structure are not in phase with ground motions at other points along the base of the structure. That incongruence leads to interference over the base of the structure, which translates into the motions imparted to the foundation, which are different from the ground motions. Typically, this phenomenon results in a filtering out of short-period motions, which is why the reduction effect is much more pronounced in structures with short fundamental periods, as illustrated in Fig. C19.4-1.

Fig. C19.4-1 illustrates the increase in reduction as the base area parameter, b_e, increases. This parameter is computed as the square root of the foundation area. Therefore, for larger foundations, base slab averaging effects are more significant.

For base slab averaging effects to occur, foundation components must be interconnected with grade beams or a concrete slab that is sufficiently stiff to permit the base to move as a unit and allow this filtering effect to occur. That is why requirements are placed on the rigidity of the foundation diaphragm relative to the vertical lateral force-resisting elements at the first story. Additionally, requirements are placed on the floor diaphragm or roof diaphragm, in the case of a one-story structure needing to be stiff in order for this filtering of ground motion to occur. FEMA 440 (FEMA 2005) indicates that there is a lack of data regarding this effect when either the base slab is not interconnected or the floor diaphragms are flexible. It is postulated that reductions between the ground motion and the foundation input motion may still occur. Because cases like this have not been studied in FEMA 440 (FEMA 2005) and NIST GCR

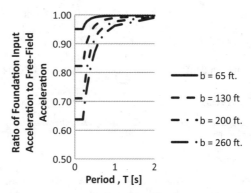

FIGURE C19.4-1 Example of Base Slab Averaging Response Spectra Ratios

12-917-21 (NIST 2012) explicitly, the requirements for foundation connectivity and stiff or rigid diaphragms above the foundation have been incorporated into the provisions.

The underlying models have only been studied up to an effective base size of 260 ft (79.2 m), which is why that limitation has been placed on Eq. (19.4-4). FEMA 440 (FEMA 2005) postulates that this effect is likely to still occur for larger base areas, but there has not been sufficient study to compare the underlying equations to data at larger effective base sizes.

Also, because the reduction can become quite significant and because studies of these phenomena have indicated variability between the theoretically predicted modifications and actual measured modifications (Stewart et al. 1999, Stewart 2000), a 0.75 factor is applied to the equations that are found in NIST GCR 12-917-21 (NIST 2012) to provide an upper bound estimate of the reduction factors with respect to the theoretical models. This is why the equations differ from those found in FEMA 440 (FEMA 2005).

Lastly, the method has not been rigorously studied for structures on piles (NIST 2012); however, it is considered reasonable to extend the application to pile-supported structures in which the pile caps are in contact with the soil and are laterally connected to one another. Another justification is that some of the empirical data for kinematic SSI come from pile-supported structures.

C19.4.2 Embedment. The kinematic interaction effects caused by embedment occur because the seismic motions vary with depth below the ground surface. It is common for these effects to be directly considered in a site-specific response spectrum by generating response spectra and acceleration histories at the embedded base of the structure instead of the ground surface. If that is not done, then these effects can be accounted for using the provisions in this section. However, these provisions should not be used if the response spectrum has already been developed at the embedded base of the structure. The embedment effect model was largely based on studies of structures with basements. The provisions can also be applied to structures with embedded foundations without basements where the foundation is laterally connected at the plane taken as the embedment depth. However, the provisions are not applicable to embedded individual spread footings.

As with base slab averaging, the reduction can become quite significant, and studies of these phenomena have indicated variability between the theoretically predicted modifications and actual measured modifications (Stewart et al. 1999). Again, a 0.75 factor is applied to the equations found in NIST GCR 12-917-21 (NIST 2012) to provide a slightly conservative estimate of the reductions with respect to the theoretical models. This is why the equations differ from those found in FEMA 440 (FEMA

2005) and NIST GCR 12-917-21 (NIST 2012). Additionally, the underlying models upon which the provisions are based have only been validated in NIST GCR 12-917-21 (NIST 2012) up to an effective embedment depth of approximately 20 ft (6.096 m), which is why a depth limitation has been placed on Eq. (19.2-4).

REFERENCES

Applied Technology Council (ATC). (1978). *Tentative Provisions for the Development of Seismic Regulations for Buildings*, ATC-3-06, Redwood City, CA.

Electrical Power Research Institute (EPRI). (1993). *Guidelines for determining design basis ground motions*, EPRI TR-102293, EPRI, Palo Alto, CA.

Federal Emergency Management Agency (FEMA). (2005). *Improvement of nonlinear static seismic analysis procedures*, FEMA 440. FEMA, Washington, DC.

Givens, M. J. (2013). *Dynamic soil-structure interaction of instrumented buildings and test structures*, Ph.D. Thesis, University of California, Los Angeles, CA.

Jarernprasert, S., Bazan-Zurita, E., and Bielak, J. (2013). "Seismic soil-structure interaction response of inelastic structures." *Soil Dyn. Earthq. Eng.*, 47, 132–143.

Kramer, S. L. (1996). *Geotechnical earthquake engineering*, Prentice Hall, Upper Saddle River, NJ.

National Institute of Standards and Technology (NIST). (2012). *Soil-structure interaction for building structures*, NIST GCR 12-917-21. NIST, Gaithersburg, MD.

Pais, A., and Kausel, E. (1988). "Approximate formulas for the dynamic stiffness of rigid foundations." *Soil Dyn. Earthq. Eng.*, 7(4), 213–227.

Stewart, J. P. (2000). "Variations between foundation-level and free-field earthquake ground motions," *Earthq. Spectra*, 16(2), 511–532.

Stewart, J. P., Seed, R. B., and Fenves, G. L. (1999). "Seismic soil-structure interaction in buildings. II: Empirical findings." *J. Geotech. Geoenvir. Engrg.*, 125(1), 38–48.

Veletsos, A. S., and Verbic, B. (1973). "Vibration of viscoelastic foundations." *Earthq. Engrg. Struct. Dyn.* 2(1), 87–105.

Vucetic, M., and R., Dobry (1991). "Effect of soil plasticity on cyclic response." *J. Geotech. Eng.*, 117, 89–107.

CHAPTER C20
SITE CLASSIFICATION PROCEDURE FOR SEISMIC DESIGN

C20.1 SITE CLASSIFICATION

Site classification procedures are given in Chapter 20 for the purpose of classifying the site and determining site coefficients and site-adjusted risk-targeted maximum considered earthquake ground motions in accordance with Section 11.4.3. Site classification procedures are also used to define the site conditions for which site-specific site response analyses are required to obtain site ground motions in accordance with Section 11.4.7 and Chapter 21.

C20.3 SITE CLASS DEFINITIONS

C20.3.1 Site Class F. Site Class F conditions are conditions for which the site coefficients F_a and F_v in Tables 11.4-1 and 11.4-2 may not be applicable for site response analyses required by Section 11.4.7; they are defined in this section. For three of the categories of Site Class F soils—Category 1 liquefiable soils, Category 3 very high plasticity clays, and Category 4 very thick soft/medium stiff clays—exceptions to the requirement to conduct site response analyses are given, provided that certain conditions and requirements are satisfied. These exceptions are discussed below.

Category 1. For liquefiable soils in Category 1, an exception to conducting site response analyses was developed by Technical Subcommittee 3, Foundations and Geotechnical Considerations, of the BSSC Provisions Update Committee and was first published in the 2000 NEHRP Provisions (FEMA 2001). The exception is made for short-period structures, defined for purposes of the exception as having fundamental periods of vibration equal to or less than 0.5 s. For such structures, it is permissible to determine site coefficients F_a and F_v from Tables 11.4-1 and 11.4-2 assuming that liquefaction does not occur because ground motion data obtained in liquefied soil areas during earthquakes indicate that short-period ground motions are generally reduced in amplitude because of liquefaction, whereas long-period ground motions may be amplified by liquefaction. Note, however, that this exception does not affect the requirement in Section 11.8 to assess liquefaction potential as a geologic hazard and develop hazard mitigation measures if required.

Categories 3 and 4. For very high plasticity clays in Category 3 and very thick soft/medium stiff clays in Category 4, exceptions to conducting site response analysis were evaluated and developed by an ad hoc geotechnical task committee formed to support the BSSC Simplified Seismic Project in 2011 and 2012. Exceptions for Categories 3 and 4 were limited to sites of expected low ground motions, i.e., Seismic Design Category B (SDC B) as defined in Tables 11.6-1 and 11.6-2.

For Category 3 very high plasticity clays, the task committee evaluated published research and conducted supplemental analyses to evaluate effects of increasing soil plasticity on soil amplification. From these evaluations, exceptions were developed for scaling the F_a and F_v site coefficients upward from the values given in Tables 11.4-1 and 11.4-2 for Site Class D or E by factors that are a function of soil plasticity as quantified by soil plasticity index (PI).

For Category 4 very thick soft/medium stiff clays, the task committee reviewed analyses conducted for a 1992 Site Response Workshop (Dobry et al. 2000) (at which recommendations were developed for site classifications and site coefficients that have been in the ASCE 7 standard through ASCE 7-10) and conducted supplemental analyses to evaluate effects of clay thickness on soil amplification. These analyses indicated that maximum amplifications should be insensitive to soil thickness and that site coefficients F_a and F_v for Site Class E should be adequate or conservative in most cases for soft/medium stiff clays thicker than 120 ft (36.576 m). A supplemental analysis indicated that soft/medium stiff clay thicknesses greater than 120 ft (36.576 m) would be uncommon for sites in the United States because of the higher overburden pressures at these depths that would generally result in clays being in a category of "stiff" rather than "soft to medium stiff." Clays meeting the Category 4 definition of Site Class F have undrained shear strengths, s_u, less than 1,000 psf (47.9 kN/m^2), whereas stiff clays have higher s_u values.

Sections C20.3.2 through C20.3.5. These sections and Table 20.3-1 provide definitions for Site Classes A through E. Except for the additional definitions for Site Class E in Section 20.3.2, the site classes are defined fundamentally in terms of the average small-strain shear wave velocity in the top 100 ft (30 m) of the soil or rock profile. If shear wave velocities are available for the site, they should be used to classify the site. However, recognizing that in many cases shear wave velocities are not available for the site, alternative definitions of the site classes also are included. These definitions are based on geotechnical parameters: standard penetration resistance for cohesionless soils and rock, and standard penetration resistance and undrained shear strength for cohesive soils. The alternative definitions are intended to be conservative because the correlation between site coefficients and these geotechnical parameters is more uncertain than the correlation with shear wave velocity. That is, values of F_a and F_v tend to be smaller if the site class is based on shear wave velocity rather than on the geotechnical parameters. Also, the site class definitions should not be interpreted as implying any specific numerical correlation between shear wave velocity and standard penetration resistance or undrained shear strength.

Although the site class definitions in Sections 20.3.2 through 20.3.5 are straightforward, there are aspects of these assessments that may require additional judgment and interpretation. Highly variable subsurface conditions beneath a building footprint could result in overly conservative or unconservative site classification. Isolated soft soil layers within an otherwise firm soil site may not affect the overall site response if the predominant soil conditions do not include such strata. Conversely, site response studies have shown that continuous, thin, soft clay strata may affect the site amplification.

The site class should reflect the soil conditions that affect the ground motion input to the structure or a significant portion of the structure. For structures that receive substantial ground motion input from shallow soils (for example, structures with shallow spread footings, with laterally flexible piles, or with basements where substantial ground motion input to the structure may come through the sidewalls), it is reasonable to classify the site on the basis of the top 100 ft (30 m) of soils below the ground surface. Conversely, for structures with basements supported on firm soils or rock below soft soils, it may be reasonable to classify the site on the basis of the soils or rock below the mat, if it can be justified that the soft soils contribute very little to the response of the structure.

Buildings on sites with sloping bedrock or highly variable soil deposits across the building area require careful study because the input motion may vary across the building (for example, if a portion of the building is on rock and the rest is over weak soils). Site-specific studies including two- or three-dimensional modeling may be used in such cases to evaluate the subsurface conditions and site and superstructure response. Other conditions that may warrant site-specific evaluation include the presence of low shear wave velocity soils below a depth of 100 ft (30 m), location of the site in a sedimentary basin, or subsurface or topographic conditions with strong two- and three-dimensional site response effects. Individuals with appropriate expertise in seismic ground motions should participate in evaluations of the need for and nature of such site-specific studies.

C20.4 DEFINITIONS OF SITE CLASS PARAMETERS

Section 20.4 provides formulas for defining site classes in accordance with definitions in Section 20.3 and Table 20.3-1. Eq. (20.4-1) is for determining the effective average small-strain shear wave velocity, $\bar{v}_s$, to a depth of 100 ft (30 m) at a site. This equation defines $\bar{v}_s$ as 100 ft (30 m) divided by the sum of the times for a shear wave to travel through each layer within the upper 100 ft (30 m), where travel time for each layer is calculated as the layer thickness divided by the small-strain shear wave velocity for the layer. It is important that this method of averaging be used because it may result in a significantly lower effective average shear wave velocity than the velocity that would be obtained by directly averaging the velocities of the individual layers.

For example, consider a soil profile that has four 25-ft (7.62-m) thick layers with shear wave velocities of 500 ft/s (152.4 m/s), 1,000 ft/s (304.8 m/s), 1,500 ft/s (457.2 m/s), and 2,000 ft/s (609.2 m/s). The arithmetic average of the shear wave velocities is 1,250 ft/s (381.0 m/s) (corresponding to Site Class C), but Eq. (20.4-1) produces a value of 960 ft/s (292.6 m/s) (corresponding to Site Class D). The Eq. (20.4-1) value is appropriate because the four layers are being represented by one layer with the same wave passage time.

Eq. (20.4-2) is for classifying the site using the average standard penetration resistance blow count, $\bar{N}$, for cohesionless soils, cohesive soils, and rock in the upper 100 ft (30 m). A method of averaging analogous to the method of Eq. (20.4-1) for shear wave velocity is used. The maximum value of N that may be used for any depth of measurement in soil or rock is 100 blows/ft (305 blows/m). For the common situation where rock is encountered, the standard penetration resistance, N, for rock layers is taken as 100.

Eqs. (20.4-3) and (20.4-4) are for classifying the site using the standard penetration resistance of cohesionless soil layers, $\bar{N}_{ch}$, and the undrained shear strength of cohesive soil layers, $\bar{s}_u$, within the top 100 ft (30 m). These equations are provided as an alternative to using Eq. (20.4-2), for which N values in all geologic materials in the top 100 ft (30 m) are used. Where using Eqs. (20.4-3) and (20.4-4), only the respective thicknesses of cohesionless soils and cohesive soils within the top 100 ft (30 m) are used.

REFERENCES

Dobry, R., Borcherdt, R. D., Crouse, C. B., Idriss, I. M., Joyner, W. B., Martin, G. R., et al. (2000). "New site coefficients and site classification system used in recent building seismic code provisions." *Earthq. Spectra*, 16(1), 41–67.

Federal Emergency Management Agency, (FEMA). (2001). *NEHRP recommended provisions for seismic regulations for new buildings and other structures*. FEMA 368. Building Seismic Safety Council for FEMA, Washington, DC.

CHAPTER C21
SITE-SPECIFIC GROUND MOTION PROCEDURES FOR SEISMIC DESIGN

C21.0 GENERAL

Site-specific procedures for computing earthquake ground motions include dynamic site response analyses and probabilistic and deterministic seismic hazard analyses (PSHA and DSHA), which may include dynamic site response analysis as part of the calculation. Use of site-specific procedures may be required in lieu of the general procedure in Sections 11.4.2 through 11.4.7; Section C11.4.8 in 7-16 explains the conditions under which the use of these procedures is required. Such studies must be comprehensive and must incorporate current scientific interpretations. Because there is typically more than one scientifically credible alternative for models and parameter values used to characterize seismic sources and ground motions, it is important to formally incorporate these uncertainties in a site-specific analysis. For example, uncertainties may exist in seismic source location, extent, and geometry; maximum earthquake magnitude; earthquake recurrence rate; ground motion attenuation; local site conditions, including soil layering and dynamic soil properties; and possible two- or three-dimensional wave-propagation effects. The use of peer review for a site-specific ground motion analysis is encouraged.

Site-specific ground motion analysis can consist of one of the following approaches: (a) PSHA and possibly DSHA if the site is near an active fault, (b) PSHA/DSHA followed by dynamic site response analysis, and (c) dynamic site response analysis only. The first approach is used to compute ground motions for bedrock or stiff soil conditions (not softer than Site Class D). In this approach, if the site consists of stiff soil overlying bedrock, for example, the analyst has the option of either (a) computing the bedrock motion from the PSHA/DSHA and then using the site coefficient (F_a and F_v) tables in Section 11.4.3 to adjust for the stiff soil overburden or (b) computing the response spectrum at the ground surface directly from the PSHA/DSHA. The latter requires the use of attenuation equations for computing stiff soil-site response spectra (instead of bedrock response spectra).

The second approach is used where softer soils overlie the bedrock or stiff soils. The third approach assumes that a site-specific PSHA/DSHA is not necessary but that a dynamic site response analysis should or must be performed. This analysis requires the definition of an outcrop ground motion, which can be based on the 5% damped response spectrum computed from the PSHA/DSHA or obtained from the general procedure in Section 11.4. A representative set of acceleration time histories is selected and scaled to be compatible with this outcrop spectrum. Dynamic site response analyses using these acceleration histories as input are used to compute motions at the ground surface. The response spectra of these surface motions are used to define a maximum considered earthquake (MCE) ground motion response spectrum.

The approaches described in the aforementioned have advantages and disadvantages. In many cases, user preference governs the selection, but geotechnical conditions at the site may dictate the use of one approach over the other. If bedrock is at a depth much greater than the extent of the site geotechnical investigations, the direct approach of computing the ground surface motion in the PSHA/DSHA may be more reasonable. On the other hand, if bedrock is shallow and a large impedance contrast exists between it and the overlying soil (i.e., density times shear wave velocity of bedrock is much greater than that of the soil), the two-step approach might be more appropriate.

Use of peak ground acceleration as the anchor for a generalized site-dependent response spectrum is discouraged because sufficiently robust ground motion attenuation relations are available for computing response spectra in western U.S. and eastern U.S. tectonic environments.

C21.1 SITE RESPONSE ANALYSIS

C21.1.1 Base Ground Motions. Ground motion acceleration histories that are representative of horizontal rock motions at the site are required as input to the soil model. Where a site-specific ground motion hazard analysis is not performed, the MCE response spectrum for Site Class B (rock) is defined using the general procedure described in Section 11.4.1. If the model is terminated in material of Site Class A, C, or D, the input MCE response spectrum is adjusted in accordance with Section 11.4.3. The U.S. Geological Survey (USGS) national seismic hazard mapping project website (earthquake.usgs.gov/hazards/products/conterminous) includes hazard deaggregation options that can be used to evaluate the predominant types of earthquake sources, magnitudes, and distances contributing to the probabilistic ground motion hazard. Sources of recorded acceleration time histories include the databases of the Consortium of Organizations for Strong Motion Observation Systems (COSMOS) Virtual Data Center website (www.cosmos-eq.org), the Pacific Earthquake Engineering Research (PEER) Center Strong Motion Database website (peer.berkeley.edu/products/strong_ground_motion_db.html), and the U.S. National Center for Engineering Strong Motion Data (NCESMD) website (http://www.strongmotioncenter.org). Ground motion acceleration histories at these sites generally were recorded at the ground surface and hence apply for an outcropping condition and should be specified as such in the input to the site response analysis code (Kwok et al. 2007 have additional details).

C21.1.2 Site Condition Modeling. Modeling criteria are established by site-specific geotechnical investigations that should include (a) borings with sampling; (b) standard penetration tests (SPTs), cone penetrometer tests (CPTs), and/or other subsurface investigative techniques; and (c) laboratory testing to establish the soil types, properties, and layering. The depth to rock or stiff soil material should be established from these investigations. Investigation should extend to bedrock or, for very deep soil profiles, to material in which the model is terminated. Although it is preferable to measure shear wave velocities in all soil layers, it is also possible to estimate shear wave velocities based on measurements available for similar soils in the local area or through correlations with soil types and properties. A number of such correlations are summarized by Kramer (1996).

Typically, a one-dimensional soil column extending from the ground surface to bedrock is adequate to capture first-order site response characteristics. For very deep soils, the model of the soil columns may extend to very stiff or very dense soils at depth in the column. Two- or three-dimensional models should be considered for critical projects when two- or three-dimensional wave propagation effects may be significant (for example, sloping ground sites). The soil layers in a one-dimensional model are characterized by their total unit weights and shear wave velocities from which low-strain (maximum) shear moduli may be obtained and by relationships defining the nonlinear shear stress–strain behavior of the soils. The required relationships for analysis are often in the form of curves that describe the variation of soil shear modulus with shear strain (modulus reduction curves) and by curves that describe the variation of soil damping with shear strain (damping curves). In a two- or three-dimensional model, compression wave velocities or moduli or Poisson ratios also are required. In an analysis to estimate the effects of liquefaction on soil site response, the nonlinear soil model also must incorporate the buildup of soil pore water pressures and the consequent reductions of soil stiffness and strength. Typically, modulus reduction curves and damping curves are selected on the basis of published relationships for similar soils (for example, Vucetic and Dobry 1991, Electric Power Research Institute 1993, Darendeli 2001, Menq 2003, and Zhang et al. 2005). Site-specific laboratory dynamic tests on soil samples to establish nonlinear soil characteristics can be considered where published relationships are judged to be inadequate for the types of soils present at the site. Shear and compression wave velocities and associated maximum moduli should be selected based on field tests to determine these parameters or, if such tests are not possible, on published relationships and experience for similar soils in the local area. The uncertainty in the selected maximum shear moduli, modulus reduction and damping curves, and other soil properties should be estimated (Darendeli 2001, Zhang et al. 2008). Consideration of the ranges of stiffness prescribed in Section 12.13.3 (increasing and decreasing by 50%) is recommended.

C21.1.3 Site Response Analysis and Computed Results. Analytical methods may be equivalently linear or nonlinear. Frequently used computer programs for one-dimensional analysis include the equivalent linear program SHAKE (Schnabel et al. 1972, Idriss and Sun 1992) and the nonlinear programs FLAC (Itasca 2005); DESRA-2 (Lee and Finn 1978); MARDES (Chang et al. 1991); SUMDES (Li et al. 1992); D-MOD_2 (Matasovic 2006); DEEPSOIL (Hashash and Park 2001); TESS (Pyke 2000); and OpenSees (Ragheb 1994, Parra 1996, and Yang 2000). If the soil response induces large strains in the soil (such as for high acceleration levels and soft soils), nonlinear programs may be preferable to equivalent linear programs. For analysis of liquefaction effects on site response, computer programs that incorporate pore water pressure development (effective stress analyses) should be used (for example, FLAC, DESRA-2, SUMDES, D-MOD_2, TESS, DEEPSOIL, and OpenSees). Response spectra of output motions at the ground surface are calculated as the ratios of response spectra of ground surface motions to input outcropping rock motions. Typically, an average of the response spectral ratio curves is obtained and multiplied by the input MCE response spectrum to obtain the MCE ground surface response spectrum. Alternatively, the results of site response analyses can be used as part of the PSHA using procedures described by Goulet et al. (2007) and programmed for use in OpenSHA (www.opensha.org; Field et al. 2005). Sensitivity analyses to evaluate effects of soil-property uncertainties should be conducted and considered in developing the final MCE response spectrum.

C21.2 RISK-TARGETED MAXIMUM CONSIDERED EARTHQUAKE (MCE_R) GROUND MOTION HAZARD ANALYSIS

Site-specific risk-targeted maximum considered earthquake (MCE_R) ground motions are based on separate calculations of site-specific probabilistic and site-specific deterministic ground motions.

Both the probabilistic and deterministic ground motions are defined in terms of 5% damped spectral response in the maximum direction of horizontal response. The maximum direction in the horizontal plane is considered the appropriate ground motion intensity parameter for seismic design using the equivalent lateral force (ELF) procedure of Section 12.8 with the primary intent of avoiding collapse of the structural system.

Most ground motion relations are defined in terms of average (geometric mean) horizontal response. Maximum response in the horizontal plane is greater than average response by an amount that varies with period. Maximum response may be reasonably estimated by factoring average response by period-dependent factors, such as 1.1 at short periods and 1.3 at a period of 1.0 s (Huang et al. 2008). The maximum direction was adopted as the ground motion intensity parameter for use in seismic design in lieu of explicit consideration of directional effects.

C21.2.1 Probabilistic (MCE_R) Ground Motions. Probabilistic seismic hazard analysis (PSHA) methods and subsequent computations of risk-targeted probabilistic ground motions based on the output of PSHA are sufficient to define MCE_R ground motion at all locations except those near highly active faults. Descriptions of current PSHA methods can be found in McGuire (2004). The primary output of PSHA methods is a so-called hazard curve, which provides mean annual frequencies of exceeding various user-specified ground motion amplitudes. Risk-targeted probabilistic ground motions are derived from hazard curves using one (or both for comparison purposes) of the methods described in the following two subsections.

C21.2.1.1 Method 1. The simpler but more approximate method of computing a risk-targeted probabilistic ground motion for each spectral period in a response spectrum is to first interpolate from a site-specific hazard curve the ground motion for a mean annual frequency corresponding to 2% probability of exceedance in 50 years (namely 1/2,475 per year). Then this "uniform-hazard" ground motion is factored by a so-called risk coefficient for the site location that is based on those mapped in Figs. 22-18 and 22-19. Via the method explained in the next subsection, the mapped risk coefficients have been computed from the USGS hazard curves for Site Class B and spectral periods of 0.2 and 1.0 s.

C21.2.1.2 Method 2. The direct method of computing risk-targeted probabilistic ground motions uses the entire site-specific hazard curve that results from PSHA. The computation is detailed in Luco et al. (2007). Summarizing, the hazard curve is combined with a collapse fragility (or probability distribution of the ground motion amplitude that causes collapse) that depends on the risk-targeted probabilistic ground motion itself. The combination quantifies the risk of collapse. Iteratively, the risk-targeted probabilistic ground motion is modified until combination of the corresponding collapse fragility with the hazard curve results in a risk of collapse of 1% in 50 years. This target is based on the average collapse risk across the western United States that is expected to result from design for the probabilistic MCE ground motions in ASCE 7.

C21.2.2 Deterministic (MCE$_R$) Ground Motions. Deterministic ground motions are to be based on characteristic earthquakes on all known active faults in a region. The magnitude of a characteristic earthquake on a given fault should be a best estimate of the maximum magnitude capable for that fault but not less than the largest magnitude that has occurred historically on the fault. The maximum magnitude should be estimated considering all seismic-geologic evidence for the fault, including fault length and paleoseismic observations. For faults characterized as having more than a single segment, the potential for rupture of multiple segments in a single earthquake should be considered in assessing the characteristic maximum magnitude for the fault.

For consistency, the same attenuation equations and ground motion variability used in the PSHA should be used in the deterministic seismic hazard analysis (DSHA). Adjustments for directivity and/or directional effects should also be made, when appropriate. In some cases, ground motion simulation methods may be appropriate for the estimation of long-period motions at sites in deep sedimentary basins or from great ($M \geq 8$) or giant ($M \geq 9$) earthquakes, for which recorded ground motion data are lacking.

Values of the site coefficients (F_a and F_v) for setting the deterministic (MCE$_R$) ground motion floor are introduced to incorporate both site amplification and spectrum shape adjustment as described in the research study "Investigation of an Identified Short-Coming in the Seismic Design Procedures of ASCE 7-16 and Development of Recommended Improvements for ASCE 7-16" (Kircher 2015). This study found that the shapes of the response spectra of ground motions were not accurately represented by the shape of the design response spectrum of Figure 11.4-1 for the following site conditions and ground motion intensities: (1) Site Class D where values of $S_1 \geq 0.2$; and (2) Site Class E where values of $S_S \geq 1.0$ and/or $S_1 \geq 0.2$. An adjustment of the corresponding values of F_a and F_v was required to account for this difference in spectrum shape, which was causing the design response spectrum to underestimate long-period motions. Two options were considered to address this shortcoming. For the first option, the subject study developed values of new "spectrum shape adjustment" factors (C_a and C_v) that could be used with site factors (F_a and F_v) to develop appropriate values of design ground motions (S_{DS} and S_{D1}). The second option, ultimately adopted by ASCE 7-16, circumvents the need for these new factors by requiring site-specific analysis for Site Class D site conditions where values of $S_1 \geq 0.2$, and for Site Class E site conditions where values of $S_S \geq 1.0$ and/or $S_1 \geq 0.2$ (i.e., new requirements of Section 11.4.8 of ASCE 7-16). The spectrum shape adjustment factors developed by the subject study for Option 1 provide the basis for the values of site coefficients (F_a and F_v) proposed for Section 21.2.2 and Section 21.3 that incorporate both site amplification and adjustment for spectrum shape. Specifically, the proposed value of $F_v = 2.5$ for Site Class D is based on the product of 1.7 (Site Class D amplification at $S_1 = 0.6$, without spectrum shape adjustment) and 1.5 (spectrum shape adjustment factor); the proposed value of $F_v = 4.0$ is based on the product of 2.0 (Site Class E amplification at $S_1 = 0.6$ without spectrum shape adjustment) and 2.0 (spectrum shape adjustment factor), where values of spectrum shape adjustment are taken from Section 6.2.2 (Table 11.4-4) of the subject study. The proposed value of $F_a = 1.0$ is based on the product of 0.8 (Site Class E amplification at $S_S = 1.5$ without spectrum shape adjustment) and 1.25 (spectrum shape adjustment factor), where the value of the spectrum shape adjustment is taken from Section 6.2.2 (Table 11.4-3) of the subject study. Site amplification adjusted for spectrum shape effects is approximately independent of ground motion intensity and, for simplicity, the proposed values of site factors adjusted for spectrum shape are assumed to be valid for all ground motion intensities.

C21.2.3 Site-Specific MCE$_R$. Because of the deterministic lower limit on the MCE$_R$ spectrum (Fig. 21.2-1), the site-specific MCE$_R$ ground motion is equal to the corresponding risk-targeted probabilistic ground motion wherever it is less than the deterministic limit (e.g., 1.5g and 0.6g for 0.2 and 1.0 s, respectively, and Site Class B). Where the probabilistic ground motions are greater than the lower limits, the deterministic ground motions sometimes govern, but only if they are less than their probabilistic counterparts. On the MCE$_R$ ground motion maps in ASCE/SEI 7-10, the deterministic ground motions govern mainly near major faults in California (like the San Andreas) and Nevada. The deterministic ground motions that govern are as small as 40% of their probabilistic counterparts.

C21.3 DESIGN RESPONSE SPECTRUM

Eighty percent of the design response spectrum determined in accordance with Section 11.4.6 was established as the lower limit to prevent the possibility of site-specific studies generating unreasonably low ground motions from potential misapplication of site-specific procedures or misinterpretation or mistakes in the quantification of the basic inputs to these procedures. Even if site-specific studies were correctly performed and resulted in ground motion response spectra less than the 80% lower limit, the uncertainty in the seismic potential and ground motion attenuation across the United States was recognized in setting this limit. Under these circumstances, the allowance of up to a 20% reduction in the design response spectrum based on site-specific studies was considered reasonable.

As described in Section 21.2.2, values of the site coefficients (F_a and F_v) for setting the deterministic (MCE$_R$) ground motion floor are introduced to incorporate both site amplification and spectrum shape adjustment.

Although the 80% lower limit is reasonable for sites not classified as Site Class F, an exception has been introduced at the end of this section to permit a site class other than E to be used in establishing this limit when a site is classified as F. This revision eliminates the possibility of an overly conservative design spectrum on sites that would normally be classified as Site Class C or D.

C21.4 DESIGN ACCELERATION PARAMETERS

The S_{DS} criteria of Section 21.4 are based on the premise that the value of the parameter S_{DS} should be taken as 90% of peak value of site-specific response spectral acceleration regardless of the period (greater than or equal to 0.2 s) at which the peak value of response spectral acceleration occurs. Consideration of periods beyond 0.2 s recognizes that site-specific studies (e.g., softer site conditions) can produce response spectra with ordinates at periods greater than 0.2 s that are significantly greater than those at 0.2 s. Periods less than 0.2 s are excluded for consistency with the 0.2-s period definition of the short-period ground motion parameter, S_s, and recognizing that certain sites, such as Central and Eastern United States (CEUS) sites, could have peak response at very short periods that would be inappropriate for defining the value of the parameter S_{DS}. The upper bound limit of 5 s precludes unnecessary checking of response at periods that cannot govern the peak value of site-specific response spectral acceleration. Ninety percent (rather than 100%) of the peak value of site-specific response spectral acceleration is considered appropriate for defining the parameter S_{DS} (and the domain of constant acceleration) since most short-period structures have a design period that is not at or near the period of peak response spectral acceleration. Away from the period of peak response, response spectral accelerations are less, and the domain of constant acceleration is adequately described by 90% of the peak value. For those short-period structures with a design period at or near the period of peak response spectral acceleration, anticipated yielding of the structure during MCE_R ground motions effectively lengthens the period and shifts dynamic response to longer periods at which spectral demand is always less than that at the peak of the spectrum.

The S_{D1} criteria of Section 21.4 are based on the premise that the value of the parameter S_{D1} should be taken as 100% of the peak value of site-specific response spectral acceleration for a period range, $1 s \leq T \leq 2 s$, for stiffer sites $v_{s,30}$ ft/s > 1,200 ft/s ($v_{s,30}$ m/s > 366 m/s) similar to the previous requirements of Section 21.4 of ASCE 7-10 and for a period range, $1 s \leq T \leq 5 s$, for softer sites $v_{s,30}$ ft/s ≤ 1,200 ft/s ($v_{s,30}$ m/s > 366 m/s), which are expected to have peak values of response spectral velocity at periods greater than 2 s. The criteria use the maximum value of the product, TS_a, over the period range of interest to effectively identify the period at which the peak value of response spectral velocity occurs. Consideration of periods beyond 1 s accounts for the possibility that the assumed $1/T$ proportionality for the constant velocity portion of the design response spectrum begins at periods greater than 1 s or is actually $1/T^n$ (where $n < 1$). Periods less than 1 s are excluded for consistency with the definition of the 1-s ground motion parameter, S_1. Peak velocity response is expected to occur at periods less than or equal to 5 s, and periods beyond 5 s are excluded by the criteria to avoid potential misuse of very long period ground motions that may not be reliable. One hundred percent (rather than a reduced percentage) of the peak value of site-specific response spectral acceleration at the period of peak velocity response is considered appropriate for defining the value of the parameter S_{D1} since response spectral accelerations can be approximately proportional to the assumed $1/T$ shape of the domain of constant velocity for design periods of interest.

C21.5 MAXIMUM CONSIDERED EARTHQUAKE GEOMETRIC MEAN (MCE$_G$) PEAK GROUND ACCELERATION

Site-specific requirements for determination of peak ground acceleration (PGA) are provided in a new Section 21.5 that is parallel to the procedures for developing site-specific response spectra in Section 21.2. The site-specific MCE peak ground acceleration, PGA_M, is taken as the lesser of the probabilistic geometric mean peak ground acceleration of Section 21.5.1 and the deterministic geometric mean peak ground acceleration of Section 21.5.2. Similar to the provisions for site-specific spectra, a deterministic lower limit is prescribed for PGA_M with the intent to limit application of deterministic ground motions to the site regions containing active faults where probabilistic ground motions are unreasonably high. However, the deterministic lower limit for PGA_M (in g) is set at a lower value, $0.5\ F_{PGA}$, than the value set for the zero-period response spectral acceleration, $0.6\ F_a$. The rationale for the value of the lower deterministic limit for spectra is based on the desire to limit minimum spectral values, for structural design purposes, to the values given by the 1997 Uniform Building Code (UBC) for Zone 4 (multiplied by a factor of 1.5 to adjust to the MCE level). This rationale is not applicable to PGA_M for geotechnical applications, and therefore a lower value of $0.5\ F_{PGA}$ was selected. Section 21.5.3 of ASCE 7-10 states that the site-specific MCE peak ground acceleration cannot be less than 80% of PGA_M derived from the PGA maps. The 80% limit is a long-standing base for site-specific analyses in recognition of the uncertainties and limitations associated with the various components of a site-specific evaluation.

REFERENCES

Chang, C.-Y., Mok, C. M., Power, M. S., and Tang, Y. K. (1991). "Analysis of ground response at lotung large-scale soil-structure interaction experiment site." *Report NP-7306-SL*. EPRI, Palo Alto, Calif.

Darendeli, M. (2001). "Development of a new family of normalized modulus reduction and material damping curves." Ph.D. Dissertation, Department of Civil Engineering, University of Texas, Austin.

Electric Power Research Institute (EPRI). (1993). "Guidelines for determining design basis ground motions." *Report EPRI TR-102293*. EPRI, Palo Alto, CA.

Field, E. H., Gupta, N., Gupta, V., Blanpied, M., Maechling, P., and Jordan, T. H. (2005). "Hazard calculations for the WGCEP-2002 forecast using OpenSHA and distributed object technologies." *Seismol. Res. Lett.*, 76, 161–167.

Goulet, C. A., Stewart, J. P., Bazzurro, P., and Field, E. H. (2007). "Integration of site-specific ground response analysis results into probabilistic seismic hazard analyses." Paper 1486, *Proc., 4th Intl. Conference on Earthquake Geotechnical Engineering*, Thessaloniki, Greece, CRC Press, Boca Raton, FL.

Hashash, Y. M. A., and Park, D. (2001). "Non-linear one-dimensional seismic ground motion propagation in the Mississippi embayment." *Engrg. Geol.*, 62(1-3), 185–206.

Huang, Y.-N., Whittaker, A. S., and Luco, N. (2008). "Maximum spectral demands in the near-fault region." *Earthq. Spectra*, 24(1), 319–341.

Idriss, I. M., and Sun, J. I. (1992). *User's manual for SHAKE91*. Center for Geotechnical Modeling, Department of Civil and Environmental Engineering, University of California, Davis.

Itasca Consulting Group. (2005). *FLAC, Fast Langrangian Analysis of Continua*, v. 5.0. Itasca Consulting Group, Minneapolis, MN.

Kircher, C. A. (2015). Investigation of an identified short-coming in the seismic design procedures of ASCE 7-16 and development of recommended improvements for ASCE 7-16, prepared for Building Seismic Safety Council, National Institute of Building Sciences, Washington, DC, prepared by Kircher & Associates, Consulting Engineers, Palo Alto, CA, March 15, 2015. https://c.ymcdn.com/sites/www.nibs.org/resource/resmgr/BSSC2/Seismic_Factor_Study.pdf

Kramer, S. L. (1996). *Geotechnical earthquake engineering*, Prentice-Hall, Englewood Cliffs, NJ.

Kwok, A. O. L., Stewart, J. P., Hashash, Y. M. A., Matasovic, N., Pyke, R., Wang, Z., et al. (2007). "Use of exact solutions of wave propagation problems to guide implementation of nonlinear seismic ground response analysis procedures." *J. Geotech. Geoenviron. Engrg.* 133(11), 1385–1398.

Lee, M. K. W., and Finn, W. D. L. (1978). *DESRA-2, Dynamic effective stress response analysis of soil deposits with energy transmitting boundary including assessment of liquefaction potential*, Soil Mechanics Series 36, Department of Civil Engineering, University of British Columbia, Vancouver.

Li, X. S., Wang, Z. L., and Shen, C. K. (1992). *SUMDES, a nonlinear procedure for response analysis of horizontally-layered sites subjected to multi-directional earthquake loading*. Department of Civil Engineering, University of California, Davis.

Luco, N., Ellingwood, B. R., Hamburger, R. O., Hooper, J. D., Kimball, J. K., and Kircher, C. A. (2007). "Risk-targeted versus current seismic design maps for the conterminous United States." *Proc. SEAOC 76th Annual Convention*. Structural Engineers Association of California, Sacramento, CA.

Matasovic, N. (2006). "D-MOD_2—A computer program for seismic response analysis of horizontally layered soil deposits, earthfill dams, and solid waste landfills." *User's manual*, GeoMotions, LLC, Lacey, WA, 20 (plus Appendices).

McGuire, R. K. (2004). *Seismic hazard and risk analysis*, Monograph, MNO-10. Earthquake Engineering Research Institute, Oakland, CA, 221.

Menq, F. (2003). "Dynamic properties of sandy and gravely soils." Ph.D. Dissertation, Department of Civil Engineering, University of Texas, Austin.

Parra, E. (1996). "Numerical modeling of liquefaction and lateral ground deformation including cyclic mobility and dilation response in soil systems." Ph.D. Dissertation, Department of Civil Engineering, Rensselaer Polytechnic Institute, Troy, NY.

Pyke, R. M. (2000). "TESS: A computer program for nonlinear ground response analyses." TAGA Engineering Systems & Software, Lafayette, CA.

Ragheb, A. M. (1994). "Numerical analysis of seismically induced deformations in saturated granular soil strata." Ph.D. Dissertation, Department of Civil Engineering, Rensselaer Polytechnic Institute, Troy, NY.

Schnabel, P. B., Lysmer, J., and Seed, H. B. (1972). *SHAKE: A Computer program for earthquake response analysis of horizontally layered sites*, Report EERC 72-12, Earthquake Engineering Research Center, University of California, Berkeley.

Vucetic, M., and Dobry, R. (1991). "Effect of soil plasticity on cyclic response." *J. Geotech. Engrg.*, 117(1), 89–107.

Yang, Z. (2000). "Numerical modeling of earthquake site response including dilation and liquefaction." Ph.D. Dissertation, Department of Civil Engineering and Engineering Mech., Columbia University, New York.

Zhang, J., Andrus, R. D., and Juang, C. H. (2005). "Normalized shear modulus and material damping ratio relationships." *J. Geotech. Geoenviron. Engrg.*, 131(4), 453–464.

Zhang, J., Andrus, R. D., and Juang, C. H. (2008). "Model uncertainty in normalized shear modulus and damping relationships." *J. Geotech. Geoenviron. Engrg.*, 134(1), 24–36.

OTHER REFERENCES (NOT CITED)

Abrahamson, N. A. (2000). "Effects of rupture directivity on probabilistic seismic hazard analysis." *Proc. 6th Intl. Conference on Seismic Zonation*, Earthquake Engineering Research Institute, Oakland, CA.

Somerville, P. G., Smith, N. F., Graves, R. W., and Abrahamson, N. A. (1997). "Modification of empirical strong ground motion attenuation relations to include the amplitude and duration effects of rupture directivity." *Seismol. Res. Lett.*, 68, 199–222.

CHAPTER C22
SEISMIC GROUND MOTION, LONG-PERIOD TRANSITION, AND RISK COEFFICIENT MAPS

Like the NEHRP Provisions (2009) and ASCE/SEI 7-10, this standard continues to use risk-targeted maximum considered earthquake (MCE_R) ground motion contour maps of 0.2-s and 1.0-s spectral response accelerations (Figs. 22-1 through 22-8), maximum considered earthquake geometric mean (MCE_G) peak ground acceleration maps (Figs. 22-9 through 22-13), and mapped risk coefficients at 0.2 s and 1 s (Figs. 22-18 and 22-19). However, the basis for these mapped values for the conterminous United States has been updated by the U.S. Geological Survey (USGS), as described below.

Furthermore, consistent with the site-specific procedures of Section 21.2.1.2 of ASCE/SEI 7-10 and this standard, but unlike the MCE_R ground motion maps of ASCE/SEI 7-10 (and unlike the site-specific procedures of the 2009 *NEHRP Recommended Provisions for Seismic Regulations for New Buildings and Other Structures*), the logarithmic standard deviation of the collapse fragility used in determining the mapped MCE_R values for the conterminous United States has also been updated (from 0.8 to 0.6). Although this standard also continues to use mapped long-period transition periods (Figs. 22-14 through 22-17), these mapped values have not been updated. Even for the conterminous United States, no significant changes are expected to the deaggregation computations that underlie the mapped long-period transition periods in ASCE/SEI 7-10.

The MCE_R ground motion, MCE_G peak ground acceleration, and risk coefficient maps incorporate the latest seismic hazard models developed by the USGS for the U.S. National Seismic Hazard Maps, including the latest seismic, geologic, and geodetic information on earthquake rates and associated ground shaking.

For the conterminous United States, the latest USGS model is documented in Petersen et al. (2013, 2014). This 2014 model supersedes versions released in 1996, 2002, and 2008. The most significant changes for the 2014 model fall into four categories, as follows:

1. For Central and Eastern United States (CEUS) sources:
   - Developed a moment magnitude-based earthquake catalog through 2012, replacing the 2008 m_b-based catalog;
   - Updated earthquake catalog completeness estimates, catalog of statistical parameters, treatment of nontectonic seismicity, and treatment of magnitude uncertainty;
   - Updated the distribution for maximum magnitude (M_{max}) for background earthquakes based on a new analysis of global earthquakes in stable continental regions;
   - Updated the zonation for maximum magnitude, keeping the two-zone model that distinguishes craton and margin zones used in previous maps, and added a new four-zone model based on the Central and Eastern U.S. Seismic Source Characterization for Nuclear Facilities Project (CEUS-SSCn 2012) delineating the craton, Paleozoic margin, Mesozoic margin, and Gulf Coast;
   - Updated the smoothing algorithms for background seismicity, keeping the previous fixed-length Gaussian smoothing model, and adding a nearest-neighbor-type adaptive smoothing model;
   - Updated the New Madrid source model, including fault geometry, recurrence rates of large earthquakes, and alternative magnitudes from M6.6–M8.0 (keeping the highest weight about M7.5);
   - Adapted seismic sources such as Charleston, Wabash, Charlevoix, Commerce lineament, East Rift Margin, Mariana based on the CEUS-SSCn (2012) model; and
   - Updated the treatment of earthquakes that are potentially induced by underground fluid injection.

2. For Intermountain West and Pacific Northwest crustal sources:
   - Considered recommendations from the Basin and Range Province Earthquake Working Group on magnitude-frequency distributions for fault sources, smoothing parameters, comparison of historical and modeled seismicity rates, treatment of magnitude uncertainty, assessment of maximum magnitude, modeling of antithetic fault pairs, slip rate uncertainties, and dip uncertainty for normal faults (Lund 2012);
   - Updated the earthquake catalog and treatment of magnitude uncertainty in rate calculations;
   - Incorporated dips for normal faults of 35°, 50°, and 65° but applied the fault earthquake rate using only the 50° dip to the three alternatives;
   - Updated fault parameters for faults in Utah based on new data sets and models supplied by the Utah Geological Survey and the Working Group on Utah Earthquake Probabilities;
   - Introduced new combined geologic and geodetic inversion models for assessing fault slip rates on fault sources;
   - Implemented new models for Cascadia earthquake-rupture geometries and rates based on onshore (paleo-tsunami) and offshore (turbidite) studies;
   - Updated the model for deep (intraslab) earthquakes along the coasts of Oregon and Washington, including a new depth distribution for intraslab earthquakes;
   - Allowed for an M_{max} up to M8.0 for crustal and intraslab earthquakes; and
   - Added the Tacoma fault source and updated the South Whidbey Island fault source in Washington.

3. For California sources:
   - The USGS worked in cooperation with the Southern California Earthquake Center and the state of California to develop a new seismic source model based on the Uniform California Earthquake Rupture Forecast, Version 3 (WGCEP 2013) and new earthquake forecasts for

California, which include many more multisegment ruptures than in previous editions of the maps. These models were developed over the past several years and involved a major update of the methodology for calculating earthquake recurrence.
4. For ground motion models (or "attenuation relations"):
   - Included new earthquake ground motion models for active shallow crustal earthquakes (NGA-West 2) and subduction zone-related interface and intraslab earthquakes;
   - Adjusted the additional epistemic uncertainty model to account for regional variability and data availability;
   - Updated ground motion prediction equation weights using a new residual analysis based on the Next Generation Attenuation (NGA)-East ground motion database, reevaluated model weights in light of a preliminary Electric Power Research Institute (EPRI 2013) ground motion study, and included newly published ground motion prediction equations for stable continental regions;
   - Incorporated new and evaluated older ground motion models: five equations for the Western United States (WUS), nine for the CEUS, and four for the subduction interface and intraslab earthquakes; and
   - Increased the maximum distance from 200 km (124.3 mi) to 300 km (186.4 mi) when calculating ground motion from WUS crustal sources.

The 2014 updated National Seismic Hazard Maps differ from the 2008 maps in a variety of ways. The new ground motions vary locally depending on complicated changes in the underlying models. In the CEUS, the new earthquake catalog, completeness models, smoothing algorithms, magnitude uncertainty adjustments, and fault models increase the hazard in some places, and the new ground motion model-weighting scheme generally lowers the ground motions. The resulting maps for the CEUS can differ by ±20% compared to the 2008 maps because of interactions between the various parts of models summarized in the bullets above. In the Intermountain West region, the combined geologic and geodetic inversion models increase the hazard along the Wasatch fault and central Nevada region, but the new NGA-West2 ground motions tend to lower the hazard on the hanging walls of normal faults with respect to the 2008 maps. These counteracting effects can result in complicated patterns of changes. In the Pacific Northwest, the new Cascadia source model causes the hazard to increase by up to 40% in the southern Cascadia subduction zone because of the addition of possible M8 and greater earthquakes, but the model causes the hazard to decrease slightly along the northern Cascadia subduction zone because of reduced earthquake rates relative to the 2008 USGS hazard model. Subduction ground motions from the new models fall off faster with distance than motions in previous models, but they also tend to be higher near fault ruptures. In California, the new UCERF3 model (WGCEP 2013) accounts for earthquakes that rupture multiple faults, yielding larger magnitudes than applied in the previous model, but with smaller recurrence rates. However, they also include new ground motion models for strike-slip earthquakes, new slip rates from combined geodetic-geologic inversions, new faults, and an adaptive smoothing seismicity model that can locally increase the hazard compared to the previous model. At a specific site, it is important to examine all model changes, documented in Petersen et al. (2013) and Petersen (2014), to determine why the ground motions may have increased or decreased.

The combined effects of the updates to the USGS hazard model and to the collapse-fragility logarithmic standard deviation (or "beta value") on the MCE_R and MCE_G ground motion maps are demonstrated in Tables C22-2 through C22-4 for the same 34 locations considered in the 2009 and 2015 *NEHRP Provisions* listed in Table C22-1. In the tables, the MCE_R (S_S and S_1) and MCE_G (PGA) ground motions from the proposed maps are compared with those from ASCE/SEI 7-10 (and their equivalents from the 1997 Uniform Building Code). Furthermore, the updated site coefficients of this standard and those of ASCE/SEI 7-10 are applied to the corresponding mapped values to provide examples of the *design* spectral response accelerations (S_{DS} and S_{D1}) and *site-adjusted* peak ground accelerations (PGA_M) for an undetermined site class (Site Class D in ASCE/SEI 7-10, the worst case of Site Classes C and D in this standard). Lastly, the seismic design categories (SDCs) corresponding to the design spectral response accelerations are also compared in the tables.

It is important to bear in mind that the updated S_{DS}, S_{D1}, and PGA_M values in the ASCE/SEI 7-16 columns of the tables and the updated SDCs include the approved changes to the site coefficients, an up to 20% increase for an undetermined site class. Nevertheless, from Tables C22-2 and C22-3, it is apparent that the more severe SDC from S_{DS} and S_{D1} does not change for all but two of the 34 locations. The exceptions are the locations in Los Angeles and San Mateo, where the SDC based on S_{D1} (and S_1) alone decreases from E to D. For the Las Vegas location, the SDC based on S_{DS} alone increases from C to D, but the SDC based on S_{D1} was already D in ASCE/SEI 7-10. Close examination of the S_S map reveals a few additional areas where the SDC based on S_{DS} alone increases relative to ASCE/SEI 7-10, most notably:

- In southeastern New Hampshire, central Virginia, and at the border between Tennessee and North Carolina, S_S increases because of (i) inclusion of a widely used adaptive (as opposed to fixed-length) algorithm for smoothing historical seismicity rates, which increases hazard in areas of clustered historical seismicity, and (ii) changes to historical earthquake magnitudes and their rates based on the Central and Eastern U.S. Seismic Source Characterization for Nuclear Facilities project (CEUS-SSCn 2012), funded by the U.S. Department of Energy, the Electric Power Research Institute, and the U.S. Nuclear Regulatory Commission.
- In southwestern Oklahoma, S_S increases because of inclusion of a much broader range of potential earthquake magnitudes and rates for the Meers fault, based on CEUS-SSCn (2012).

At the aforementioned 34 locations, the ASCE/SEI 7-16 PGA values of Table C22-4, which have only been affected by the updates to the USGS National Seismic Hazard Maps (not the updated beta value), are within ±20% of the respective ASCE/SEI 7-10 values, with the exceptions below. Recall that a 20% decrease is the most allowed when a site-specific hazard analysis is performed in accordance with Chapter 21.

- For the San Diego location, the increase in the PGA value (and S_S value discussed below) is due to a combination of the addition of more offshore faults, the consideration of geodetic (GPS) data, the inclusion of lower magnitude earthquakes that can contribute significantly to PGA values (and 0.2-s spectral response accelerations), and the increases in the ground motions for large-magnitude earthquakes near strike-slip faults from the updated NGA-West 2 attenuation relations.
- For Vallejo, the increase in the PGA value is primarily caused by lengthening of the West Napa fault based on

Table C22-1 Latitudes and Longitudes for which MCE$_R$ and MCE$_G$ Ground Motions from ASCE/SEI 7-16 and ASCE/SEI 7-10 are Compared in Tables C22-2 through C22-4

Region	City and Location of Site			County or Metropolitan Statistical Area	
	Name	Latitude	Longitude	Name	Population
Southern California	Los Angeles	34.05	−118.25	Los Angeles	9,948,081
	Century City	34.05	−118.40		
	Northridge	34.20	−118.55		
	Long Beach	33.80	−118.20		
	Irvine	33.65	−117.80	Orange	3,002,048
	Riverside	33.95	−117.40	Riverside	2,026,603
	San Bernardino	34.10	−117.30	San Bernardino	1,999,332
	San Luis Obispo	35.30	−120.65	San Luis Obispo	257,005
	San Diego	32.70	−117.15	San Diego	2,941,454
	Santa Barbara	34.45	−119.70	Santa Barbara	400,335
	Ventura	34.30	−119.30	Ventura	799,720
	Total Population—S. California		22,349,098	Population—8 Counties	21,374,788
Northern California	Oakland	37.80	−122.25	Alameda	1,502,759
	Concord	37.95	−122.00	Contra Costa	955,810
	Monterey	36.60	−121.90	Monterey	421,333
	Sacramento	38.60	−121.50	Sacramento	1,233,449
	San Francisco	37.75	−122.40	San Francisco	776,733
	San Mateo	37.55	−122.30	San Mateo	741,444
	San Jose	37.35	−121.90	Santa Clara	1,802,328
	Santa Cruz	36.95	−122 05	Santa Cruz	275,359
	Vallejo	38.10	−122.25	Solano	423,473
	Santa Rosa	38.45	−122.70	Sonoma	489,290
	Total Population—N. California		14,108,451	Population—10 Counties	8,621,978
Pacific Northwest	Seattle	47.60	−122.30	King, WA	1,826,732
	Tacoma	47.25	−122.45	Pierce, WA	766,878
	Everett	48.00	−122.20	Snohomish, WA	669.887
	Portland	45.50	−122.65	Portland Metro, OR (3)	1,523,690
	Total Population—OR and WA		10,096,556	Population—6 Counties	4,787,187
Other WUS	Salt Lake City	40.75	−111.90	Salt Lake, UT	978,701
	Boise	43.60	−116.20	Ada/Canyon, ID (2)	532,337
	Reno	39.55	−119.80	Washoe, NV	396,428
	Las Vegas	36.20	−115.15	Clarke, NV	1,777,539
	Total Population—ID/UT/NV		6,512,057	Population—5 Counties	3,685,005
CEUS	St. Louis	38.60	−90.20	St. Louis MSA (16)	2,786,728
	Memphis	35.15	−90.05	Memphis MSA (8)	1,269,108
	Charleston	32.80	−79.95	Charleston MSA (3)	603,178
	Chicago	41.85	−87.65	Chicago MSA (7)	9,505,748
	New York	40.75	−74.00	New York MSA (23)	18,747,320
	Total Population—MO/TN/SC/IL/NY		48,340,918	Population—57 Counties	32,912,082

Note: The 34 locations come from the 2009 and 2015 *NEHRP Provisions*. It is important to note that these locations are each just one of many in the named cities, and their ground motions may be significantly different than those at other locations in the cities.

the Statewide Community Fault Model (see WGCEP 2013).
- For Reno, the increase in the PGA value is primarily caused by the NGA-West 2 attenuation relations. Note that the NGA-West 2 attenuation relations are based on double the strong-motion data used for the NGA-West 1 relations.
- For Las Vegas, the increase in the PGA value (and S_S and S_1 values discussed below) is primarily caused by an increase in the estimated rate of earthquakes on the Eglington fault, based on recent studies and a recommendation from the state geologist of Nevada.
- For Memphis, the increase in the PGA value is caused by consideration of an alternative model for the New Madrid Seismic Zone based on the aforementioned CEUS-SSCn reference (2012).
- For Charleston, the increase in the PGA value (and S_S value discussed below) is caused by reevaluation of the data from the Charleston earthquakes and consequent revision of the Charleston seismic source model (CEUS-SSCn 2012). The USGS adopted this revised model based on its own analysis, as well as the recommendations of its steering committee and participants of a regional workshop.

The ASCE/SEI 7-16 S_S values of Table C22-2 have been affected by the update to the ASCE/SEI 7-10 beta value (ground motion changes of up to approximately 10%), in addition to the updated USGS National Seismic Hazard Maps. The only locations where the S_S values have changed by more than 20% with respect to ASCE/SEI 7-10 are San Diego and Santa Barbara, California; Las Vegas, Nevada; and Charleston, South Carolina. Please see the explanation below of the changes in the USGS

Table C22-2 A Comparison of the Short-Period Design Spectral Response Accelerations (S_{DS} Values) from this Standard and ASCE/SEI 7-10 and Their Equivalents from the 1997 Uniform Building Code for the 34 Locations Considered in the 2009 and 2015 *Provisions*

Region	Location Name	1997 UBC		ASCE/SEI 7-10			ASCE/SEI 7-16		
		Zone	$2.5 \cdot C_a$	S_S (g)	S_{DS} (g)*	SDC_S**	S_S (g)	S_{DS} (g)***	SDC_S**
Southern California	Los Angeles	4	1.10	2.40	1.60	D	1.97	1.58	D
	Century City	4 (NF)	1.32	2.16	1.44	D	2.11	1.69	D
	Northridge	4	1.10	1.69	1.13	D	1.74	1.39	D
	Long Beach	4 (NF)	1.43	1.64	1.10	D	1.68	1.35	D
	Irvine	4	1.10	1.55	1.03	D	1.25	1.00	D
	Riverside	4	1.10	1.50	1.00	D	1.50	1.20	D
	San Bernardino	4 (NF)	1.32	2.37	1.58	D	2.33	1.86	D
	San Luis Obispo	4	1.10	1.12	0.78	D	1.09	0.87	D
	San Diego	4 (NF)	1.43	1.25	0.84	D	1.58	1.26	D
	Santa Barbara	4 (NF)	1.43	2.83	1.89	D	2.12	1.70	D
	Ventura	4 (NF)	1.43	2.38	1.59	D	2.02	1.62	D
	Weighted Mean		**1.25**	**1.83**	**1.22**		**1.75**	**1.40**	
Northern California	Oakland	4 (NF)	1.43	1.86	1.24	D	1.88	1.51	D
	Concord	4	1.10	2.08	1.38	D	2.22	1.78	D
	Monterey	4	1.10	1.53	1.02	D	1.33	1.06	D
	Sacramento	3	0.90	0.67	0.57	D	0.57	0.51	D
	San Francisco	4	1.10	1.50	1.00	D	1.50	1.20	D
	San Mateo	4 (NF)	1.28	1.85	1.23	D	1.80	1.44	D
	San Jose	4	1.10	1.50	1.00	D	1.50	1.20	D
	Santa Cruz	4	1.10	1.52	1.01	D	1.59	1.27	D
	Vallejo	4 (NF)	1.19	1.50	1.00	D	1.50	1.20	D
	Santa Rosa	4 (NF)	1.65	2.51	1.67	D	2.41	1.93	D
	Weighted Mean		**1.18**	**1.60**	**1.08**		**1.59**	**1.28**	
Pacific Northwest	Seattle			1.36	0.91	D	1.40	1.12	D
	Tacoma			1.30	0.86	D	1.36	1.08	D
	Everett			1.27	0.85	D	1.20	0.96	D
	Portland			0.98	0.72	D	0.89	0.71	D
	Weighted Mean			**1.22**	**0.83**		**1.20**	**0.96**	
Other WUS	Salt Lake City			1.54	1.03	D	1.54	1.24	D
	Boise			0.31	0.32	B	0.31	0.32	B
	Reno			1.50	1.00	D	1.47	1.17	D
	Las Vegas			0.49	0.46	C	0.65	0.55	D
	Weighted Mean			**0.85**	**0.65**		**0.92**	**0.77**	
CEUS	St. Louis			0.44	0.42	C	0.46	0.44	C
	Memphis			1.01	0.74	D	1.02	0.82	D
	Charleston			1.15	0.80	D	1.42	1.13	D
	Chicago			0.13	0.14	A	0.12	0.13	A
	New York			0.28	0.29	B	0.29	0.30	B
	Weighted Mean			**0.30**	**0.29**		**0.30**	**0.30**	

Note: It is important to bear in mind that the design spectral response accelerations (S_{DS} values) and the seismic design categories in the table include the effects of the updated site coefficients of this standard.
*The ASCE/SEI 7-10 S_{DS} values are calculated using the ASCE/SEI 7-10 F_a site coefficients, for an undetermined site class (assigned Site Class D in ASCE/SEI 7-10).
**The ASCE/SEI 7-16 S_{DS} values are calculated using the updated ASCE/SEI 7-16 F_a site coefficients, also for an undetermined site class (assigned the worst case of Site Classes C and D in ASCE/SEI 7-16).
***The SDC_S categories corresponding to the S_{DS} values (and Risk Category I/II/III) are assigned using Table 11.6-1 and the SDC E definition (of ASCE/SEI 7-10 and 7-16) alone.

National Seismic Hazard Maps at the Santa Barbara location and the explanations stated previously for the San Diego, Las Vegas, and Charleston locations. For the other locations explained previously—Vallejo, California; Reno, Nevada; and Memphis, Tennessee—the S_S values have changed by at most 2%.

- For the Santa Barbara location, the decrease in the S_1 value is a combination of the decrease in ground motions over reverse faults from the NGA-West 2 attenuation relations and the fact that more multifault earthquakes have been allowed in UCERF3 (WGCEP 2013), relative to the hazard model underlying the ASCE/SEI 7-10 ground motion maps.

This, in effect, lowers the rate of earthquakes and, hence, lowers the probabilistic ground motions.

The ASCE/SEI 7-16 S_1 values of Table C22-3 have also been affected by the update to the ASCE/SEI 7-10 beta value (changes of up to approximately 10%) and the updated USGS National Seismic Hazard Maps. The only locations where the S_1 values have changed by more than 20% with respect to ASCE/SEI 7-10 are Irvine, Santa Barbara, and Las Vegas. Please see the explanation below of the changes in the USGS National Seismic Hazard Maps at the Irvine, California, location and the previous explanations for the Santa Barbara and Las Vegas

Table C22-3 A Comparison of the 1.0-s MCE_R Design Spectral Response Accelerations (S_{D1} values) from this Standard and ASCE/SEI 7-10 and Their Equivalents from the 1997 Uniform Building Code for the 34 Locations Considered in the 2009 and 2015 *Provisions*

Region	Location Name	1997 UBC Zone	2.5*C_a	ASCE/SEI 7-10 S_1 (g)	S_{D1} (g)*	SDC_1***	ASCE/SEI 7-16 S_1 (g)	S_{D1} (g)**	SDC_1***
Southern California	Los Angeles	4 (NF)	0.72	0.84	0.84	E	0.70	0.79	D
	Century City	4 (NF)	0.93	0.80	0.80	E	0.75	0.85	E
	Northridge	4	0.64	0.60	0.60	D	0.60	0.68	D
	Long Beach	4 (NF)	1.02	0.62	0.62	D	0.61	0.69	D
	Irvine	4	0.64	0.57	0.57	D	0.45	0.55	D
	Riverside	4	0.64	0.60	0.60	D	0.58	0.67	D
	San Bernardino	4 (NF)	0.93	1.08	1.08	E	0.93	1.06	E
	San Luis Obispo	4 (NF)	0.77	0.43	0.45	D	0.40	0.51	D
	San Diego	4 (NF)	1.02	0.48	0.49	D	0.53	0.62	D
	Santa Barbara	4 (NF)	1.02	0.99	0.99	E	0.77	0.88	E
	Ventura	4 (NF)	1.02	0.90	0.90	E	0.76	0.86	E
	Weighted Mean		**0.83**	**0.70**	**0.70**		**0.63**	**0.73**	
Northern California	Oakland	4 (NF)	1.04	0.75	0.75	D	0.72	0.81	D
	Concord	4 (NF)	0.77	0.73	0.73	D	0.67	0.76	D
	Monterey	4 (NF)	0.77	0.56	0.56	D	0.50	0.60	D
	Sacramento	3	0.54	0.29	0.35	D	0.25	0.35	D
	San Francisco	4 (NF)	0.74	0.64	0.64	D	0.60	0.68	D
	San Mateo	4 (NF)	0.95	0.86	0.86	E	0.74	0.83	D
	San Jose	4 (NF)	0.69	0.60	0.60	D	0.60	0.68	D
	Santa Cruz	4 (NF)	0.72	0.60	0.60	D	0.60	0.68	D
	Vallejo	4 (NF)	0.87	0.60	0.60	D	0.60	0.68	D
	Santa Rosa	4 (NF)	1.28	1.04	1.04	E	0.94	1.06	E
	Weighted Mean		**0.81**	**0.65**	**0.65**		**0.61**	**0.70**	
Pacific Northwest	Seattle			0.53	0.53	D	0.49	0.59	D
	Tacoma			0.51	0.51	D	0.47	0.57	D
	Everett			0.48	0.49	D	0.43	0.53	D
	Portland			0.42	0.44	D	0.39	0.50	D
	Weighted Mean			**0.48**	**0.49**		**0.45**	**0.55**	
Other WUS	Salt Lake City			0.56	0.56	D	0.55	0.65	D
	Boise			0.11	0.17	C	0.11	0.17	C
	Reno			0.52	0.52	D	0.52	0.61	D
	Las Vegas			0.17	0.24	D	0.21	0.30	D
	Weighted Mean			**0.30**	**0.34**		**0.32**	**0.41**	
CEUS	St. Louis			0.17	0.24	D	0.16	0.25	D
	Memphis			0.35	0.40	D	0.35	0.45	D
	Charleston			0.37	0.41	D	0.41	0.52	D
	Chicago			0.06	0.10	B	0.06	0.10	B
	New York			0.07	0.11	B	0.06	0.10	B
	Weighted Mean			**0.09**	**0.14**		**0.09**	**0.13**	

Note: It is important to bear in mind that the design spectral response accelerations (S_{D1} values) and the seismic design categories in the table include the effects of the updated site coefficients of this standard.
*The ASCE/SEI 7-10 S_{D1} values are calculated using the ASCE/SEI 7-10 F_v site coefficients, for an undetermined site class (assigned Site Class D in ASCE/SEI 7-10).
**The ASCE/SEI 7-16 S_{D1} values are calculated using the updated ASCE/SEI 7-16 F_v site coefficients, also for an undetermined site class (assigned the worst case of Site Classes C and D in ASCE/SEI 7-16).
***The SDC_1 categories corresponding to the S_{D1} values (and Risk Category I/II/III) are assigned using Table 11.6-2 and the SDC E definition (of ASCE/SEI 7-10 and 7-16) alone.

locations. For the other locations explained previously—San Diego, Vallejo, Reno, Memphis, and Charleston—the S_1 values have changed by at most 13%.

- For the Irvine location, the decrease in the S_1 values is caused primarily by a decrease in ground motions over reverse faults from the NGA-West 2 attenuation relations, and secondarily to the allowance for more multifault earthquakes in UCERF 3 (WGCEP 2013) that is described in the preceding bullet.

In summary, with the updates to the USGS National Seismic Hazard maps for the conterminous United States and the updated logarithmic standard deviation of the collapse fragilities, (i) the Seismic Design Categories for 32 of the 34 Table C22-1 locations do not change with respect to ASCE/SEI 7-10; (ii) the PGA values change by −15% to +17% for 28 of the 34 locations; (iii) the S_S values change by −19% to +7 for 30 of the locations; and (iv) the S_1 values change by −17% to +13% for 31 of the locations.

Like previous versions of the USGS national seismic hazard model, the 2014 model purposefully excludes swarms of earthquakes that may be causally related to industrial fluid processes, such as hydrocarbon production or wastewater disposal. The excluded swarms are identified in Figure 15 of Petersen (2014). Whereas an average of 21 earthquakes per year of magnitude greater than 3 occurred from 1967 to 2000 in the CEUS, more

Table C22-4 A Comparison of the MCE$_G$ Peak Ground Accelerations (PGA Values) from this Standard and ASCE/SEI 7-10 for the 34 Locations Considered in the 2009 and 2015 *Provisions*

Region	Location Name	ASCE/SEI 7-10 PGA (g)	ASCE/SEI 7-10 PGA$_M$ (g)*	ASCE/SEI 7-16 PGA (g)	ASCE/SEI 7-16 PGA$_M$ (g)**
Southern California	Los Angeles	0.91	0.91	0.84	1.01
	Century City	0.81	0.81	0.91	1.09
	Northridge	0.62	0.62	0.71	0.86
	Long Beach	0.64	0.64	0.74	0.89
	Irvine	0.60	0.60	0.53	0.63
	Riverside	0.50	0.50	0.50	0.60
	San Bernardino	0.91	0.91	0.98	1.18
	San Luis Obispo	0.44	0.47	0.48	0.58
	San Diego	0.57	0.57	0.72	0.86
	Santa Barbara	1.09	1.09	0.93	1.11
	Ventura	0.91	0.91	0.88	1.06
	Weighted Mean	**0.70**	**0.70**	**0.74**	**0.89**
Northern California	Oakland	0.72	0.72	0.79	0.95
	Concord	0.79	0.79	0.90	1.07
	Monterey	0.59	0.59	0.58	0.69
	Sacramento	0.23	0.31	0.24	0.32
	San Francisco	0.57	0.57	0.58	0.70
	San Mateo	0.73	0.73	0.78	0.93
	San Jose	0.50	0.50	0.57	0.69
	Santa Cruz	0.59	0.59	0.67	0.81
	Vallejo	0.51	0.51	0.62	0.74
	Santa Rosa	0.97	0.97	1.02	1.22
	Weighted Mean	**0.59**	**0.60**	**0.65**	**0.78**
Pacific Northwest	Seattle	0.56	0.56	0.60	0.72
	Tacoma	0.50	0.50	0.50	0.60
	Everett	0.52	0.52	0.51	0.62
	Portland	0.42	0.46	0.40	0.48
	Weighted Mean	**0.50**	**0.51**	**0.51**	**0.61**
Other WUS	Salt Lake City	0.67	0.67	0.70	0.84
	Boise	0.12	0.19	0.14	0.21
	Reno	0.50	0.50	0.62	0.74
	Las Vegas	0.20	0.28	0.28	0.37
	Weighted Mean	**0.34**	**0.39**	**0.41**	**0.51**
CEUS	St. Louis	0.23	0.31	0.27	0.36
	Memphis	0.50	0.50	0.61	0.73
	Charleston	0.75	0.75	0.93	1.12
	Chicago	0.07	0.11	0.06	0.09
	New York	0.17	0.25	0.18	0.26
	Weighted Mean	**0.17**	**0.23**	**0.18**	**0.25**

Note: It is important to bear in mind that the site-adjusted peak ground accelerations (PGA$_M$ values) in the table include the effects of the updated site coefficients of this standard.
*The ASCE/SEI 7-10 PGA$_M$ values are calculated using the ASCE/SEI 7-10 F_{PGA} site coefficients, for an undetermined site class (assigned Site Class D in ASCE/SEI 7-10).
**The ASCE/SEI 7-16 PGA$_M$ values are calculated using the updated ASCE/SEI 7-16 F_{PGA} site coefficients, also for an undetermined site class (assigned the worst case of Site Classes C and D in ASCE/SEI 7-16).

than 300 such earthquakes have occurred from 2010 through 2012. Thus, in the areas of the excluded swarms, the seismic hazard might be higher than that estimated by the 2014 USGS model; on the other hand, it could decrease significantly in the coming years with changes in the fluid processes. Treatment of the potentially induced earthquake swarms in hazard modeling is a topic of active research.

In 2012, the USGS developed seismic hazard models for Guam and the Northern Mariana Islands (Guam/NMI) and for American Samoa using the same type of seismic hazard analysis that underlies the 2008 model for the conterminous United States. The hazard models for the islands are documented in Mueller et al. (2012) and Petersen et al. (2012), respectively. In comparing the MCE$_R$ ground motion maps derived from these USGS hazard models to the geographically constant values stipulated for Guam and American Samoa (Tutuila) in the 2010 and previous editions of ASCE/SEI 7, it is important to bear in mind that the latter were not computed via seismic hazard modeling. According to the commentary of the NEHRP Provisions (1997), the geographically constant values were merely conversions, via rough approximations, from values on the NEHRP Provisions (1994) maps that had been in use for nearly 20 years. As such, they did not take into account the 1993 Guam earthquake that was the largest ever recorded in the region and caused considerable damage, the 2009 earthquake near American Samoa that caused a tsunami, nor the 2008 "Next Generation Attenuation (NGA)" and another 2006 empirical ground motion prediction equation that have now been used for both Guam/NMI and American Samoa. This and other such information is directly used in the seismic hazard modeling that is the basis for the MCE$_R$ ground motion, MCE$_G$ peak ground acceleration, and risk coefficient maps for Guam/NMI and American Samoa in this standard.

RISK-TARGETED MAXIMUM CONSIDERED EARTHQUAKE (MCE$_R$) GROUND MOTION MAPS

As introduced in the NEHRP Provisions (2009) and ASCE/SEI 7-10, the MCE$_R$ ground motion maps are derived from underlying USGS seismic hazard models in a manner that is significantly different from that of the mapped values of MCE ground motions in previous editions of the NEHRP Provisions and ASCE/SEI 7. These differences include use of (1) probabilistic ground motions that are risk-targeted, rather than uniform hazard, (2) deterministic ground motions that are based on the 84th percentile (approximately 1.8 times median), rather than 1.5 times median response spectral acceleration for sites near active faults, and (3) ground motion intensity that is based on maximum, rather than the average (geometrical mean), response spectra acceleration in the horizontal plane.

The MCE$_R$ ground motion maps have been prepared in accordance with the site-specific procedures of Section 21.2. More specifically, they represent the lesser of probabilistic ground motions defined in Section 21.2.1 and deterministic ground motions defined in Section 21.2.2, in accordance with Section 21.2.3. The preparation of the probabilistic and deterministic ground motions is described below.

The probabilistic ground motions have been calculated using Method 2 of Section 21.2.1 and the latest USGS hazard curves (of mean annual frequency of exceedance versus ground motion level) computed in accordance with Section 21.2 at gridded locations covering the United States and its territories. The USGS hazard curves are first converted from geometric-mean ground motions (output by the ground motion attenuation relations available to the USGS) to ground motions in the maximum direction of horizontal spectral response acceleration, with one exception. The USGS hazard curves for Hawaii, without conversion, are deemed to represent the maximum-response ground motions because of the attenuation relations applied there. For the other regions, the conversions were done by applying the factors specified in the site-specific procedures (Section 21.2) of ASCE/SEI 7-10 and this standard, namely 1.1 at 0.2 s and 1.3 at 1.0 s. The collapse fragilities used in calculating the probabilistic ground motions have a logarithmic standard deviation (or "beta value") of 0.6, as specified in ASCE/SEI 7-10 and this standard (Section 21.2.1), for the conterminous United States, Guam and the Northern Mariana Islands, and American Samoa. For the other regions (Hawaii, Puerto Rico and the U.S. Virgin Islands, and Alaska), where the latest USGS hazard curves predate the change of the logarithmic standard deviation from the NEHRP Provisions (2009) to ASCE/SEI 7-10, the beta value is 0.8. Please see Luco et al. (2007) for more information on the development of risk-targeted probabilistic ground motions.

The deterministic ground motions have been calculated using the "characteristic earthquakes on all known active faults" (quoted from Section 21.2.2) that the USGS uses in computing the probabilistic hazard curves. The largest characteristic magnitude considered by the USGS on each fault, excluding any lower weighted magnitudes from the USGS logic tree for epistemic uncertainty, is used for the deterministic ground motions. The active faults considered for the deterministic ground motions are those that have evidence of slip during Holocene time (the past 12,000 years, approximately), plus those with reported geologic rates of slip larger than 0.0004 in./year (0.1 mm/year). This slip rate can result in a magnitude 7 earthquake, which on average corresponds to 3.94 ft (1.2 m) of slip (Wells and Coppersmith 1994), over a 12,000-year time period; 0.0004 in./year (0.1 mm/year) also is the slip rate assigned by the Working Group on California Earthquake Probabilities (WGCEP 2013) to faults that, with the information available, could only be categorized as having a slip rate less than 0.0008 in./year (0.2 mm/year). At a user-input location, the fault (among hundreds) and corresponding magnitude that govern its deterministic ground motion is output by the USGS web tool briefly described in a section of this commentary below. For all the deterministic faults and magnitudes, the USGS has computed median (50th percentile), geometric-mean ground motions. To convert to maximum-response ground motions, the same scale factors described in the preceding paragraph for probabilistic ground motions are applied. To approximately convert to 84th percentile ground motions, the maximum-response ground motions are multiplied by 1.8.

MAXIMUM CONSIDERED EARTHQUAKE GEOMETRIC MEAN (MCE$_G$) PGA MAPS

Like the NEHRP Provisions (2009) and ASCE/SEI 7-10, but not previous editions, this standard includes contour maps of maximum considered earthquake geometric mean (MCE$_G$) peak ground acceleration, PGA, Figs. 22-9 through 22-13, for use in geotechnical investigations (Section 11.8.3). In contrast to the MCE$_R$ ground motion maps, the maps of MCE$_G$ PGA are defined in terms of geometric mean (rather than maximum direction) intensity and a 2% in 50-year hazard level (rather than a 1% in 50-year risk). Like the MCE$_R$ ground motion maps, the maps of MCE$_G$ PGA are governed near major active faults by deterministic values defined as 84th-percentile ground motions. The MCE$_G$ PGA maps have been prepared in accordance with the site-specific procedures of Section 21.5 of ASCE/SEI 7-10 and this standard.

LONG-PERIOD TRANSITION MAPS

The maps of the long-period transition period, T_L (Figs. 22-14 through 22-17), were introduced in ASCE/SEI 7-05. They were prepared by the USGS in response to Building Seismic Safety Council recommendations and were subsequently included in the NEHRP Provisions (2003). See Section C11.4.6 for a discussion of the technical basis of these maps. The value of T_L obtained from these maps is used in Eq. (11.4-7) to determine values of S_a for periods greater than T_L.

The exception in Section 15.7.6.1, regarding the calculation of S_{ac}, the convective response spectral acceleration for tank response, is intended to provide the user the option of computing this acceleration with three different types of site-specific procedures: (a) the procedures in Chapter 21, provided that they cover the natural period band containing T_c, the fundamental convective period of the tank-fluid system; (b) ground motion simulation methods using seismological models; and (c) analysis of representative accelerogram data. Elaboration of these procedures is provided below.

With regard to the first procedure, attenuation equations have been developed for the western United States (Next Generation Attenuation, e.g., Power et al. 2008) and for the central and eastern United States (e.g., Somerville et al. 2001) that cover the period band, 0 to 10 s. Thus, for $T_c \leq 10$ s, the fundamental convective period range for nearly all storage tanks, these attenuation equations can be used in the same probabilistic seismic hazard analysis (PSHA) and deterministic seismic hazard analysis (DSHA) procedures described in Chapter 21 to compute $S_a(T_c)$. The 1.5 factor in Eq. (15.7-11), which converts a 5% damped spectral acceleration to a 0.5% damped value, could then be applied to obtain S_{ac}. Alternatively, this factor could be established by statistical analysis of 0.5% damped and 5% damped response spectra of accelerograms representative of the ground motion expected at the site.

In some regions of the United States, such as the Pacific Northwest and southern Alaska, where subduction-zone earthquakes dominate the ground motion hazard, attenuation equations for these events only extend to periods between 3 and 5 s, depending on the equation. Thus, for tanks with T_c greater than these periods, other site-specific methods are required.

The second site-specific method to obtain S_a at long periods is simulation through the use of seismological models of fault rupture and wave propagation (e.g., Graves and Pitarka 2004, Hartzell and Heaton 1983, Hartzell et al. 1999, Liu et al. 2006, and Zeng et al. 1994). These models could range from simple seismic source-theory and wave-propagation models, which currently form the basis for many of the attenuation equations used in the central and eastern United States, for example, to more complex numerical models that incorporate finite fault rupture for scenario earthquakes and seismic wave propagation through 2D or 3D models of the regional geology, which may include basins. These models are particularly attractive for computing long-period ground motions from great earthquakes ($M_w \geq \sim 8$) because ground motion data are limited for these events. Furthermore, the models are more accurate for predicting longer period ground motions because (a) seismographic recordings may be used to calibrate these models and (b) the general nature of the 2D or 3D regional geology is typically fairly well resolved at these periods and can be much simpler than would be required for accurate prediction of shorter period motions.

A third site-specific method is the analysis of the response spectra of representative accelerograms that have accurately recorded long-period motions to periods greater than T_c. As T_c increases, the number of qualified records decreases. However, as digital accelerographs continue to replace analog accelerographs, more recordings with accurate long-period motions are becoming available. Nevertheless, a number of analog and digital recordings of large and great earthquakes are available that have accurate long-period motions to 8 s and beyond. Subsets of these records, representative of the earthquake(s) controlling the ground motion hazard at a site, can be selected. The 0.5% damped response spectra of the records can be scaled using seismic source theory to adjust them to the magnitude and distance of the controlling earthquake. The levels of the scaled response spectra at periods around T_c can be used to determine S_{ac}. If the subset of representative records is limited, then this method should be used in conjunction with the aforementioned simulation methods.

RISK COEFFICIENT MAPS

Like those in the NEHRP Provisions (2009) and ASCE/SEI 7-10 (where they were introduced), the risk coefficient maps in this standard (Figs. 22-18 and 22-19) provide factors, C_{RS} and C_{R1}, that are used in the site-specific procedures of Chapter 21 (Section 21.2.1.1, Method 1). These factors are implicit in the MCE_R ground motion maps.

The mapped risk coefficients are the ratios of (i) risk-targeted probabilistic ground motions (for 1% in 50 years collapse risk) derived from the USGS probabilistic seismic hazard curves, as described in the MCE_R ground motion maps section above, to (ii) corresponding uniform-hazard (2% in 50 years ground motion exceedance probability) ground motions that are simply interpolated from the USGS hazard curves. Note that these ratios (risk coefficients) are invariant to maximum-response scale factors that are applied to both the numerator and denominator.

GROUND MOTION WEB TOOL

The USGS has developed a companion web tool that calculates location-specific spectral values based on latitude and longitude. The calculated values are based on the gridded values used to prepare the maps. The spectral values can be adjusted for site class effects within the program using the site classification procedure in Section 20.1 and the site coefficients in Sections 11.4 and 11.8. The companion tool may be accessed at the USGS Earthquake Hazards Program website or through other hazards mapping tools. The tool should be used to establish spectral values for design because the maps found in this chapter are too small to provide accurate spectral values for many sites.

UNIFORM HAZARD AND DETERMINISTIC GROUND MOTION MAPS

Implicit in the MCE_R ground motion, MCE_G PGA, and risk coefficient maps provided are uniform hazard (2% in 50 years ground motion exceedance probability) and deterministic (84th percentile) ground motions. The NEHRP *Provisions* (2009) provided maps of such uniform hazard and deterministic ground motions, but *ASCE/SEI 7-10* and this standard do not. Instead, uniform hazard and deterministic ground motion maps consistent with this chapter are provided. Furthermore, values from these maps can be obtained via the ground motion software tool previously described.

It is important to note that the provided uniform hazard ground motion maps are for maximum direction of horizontal spectral response acceleration. As such, they are different than the maps of geometric mean spectral response acceleration provided elsewhere on the USGS Earthquake Hazards Program website. The provided deterministic ground motion maps are also for the maximum direction, but no geometric mean counterparts are provided. The USGS prepares the deterministic ground motion maps solely for the purposes of this standard, following the definition of deterministic ground motions in Section 21.2.2 (with the 84th percentile approximated as 1.8 times the median).

REFERENCES

CEUS-SSCn. (2012). Central and Eastern United States seismic source characterization for nuclear facilities: Electric Power Research Institute, U.S. Department of Energy and U.S. Nuclear Regulatory Commission. EPRI, Palo Alto, CA.

Electric Power Research Institute (EPRI). (2013). "EPRI (2004, 2006) Ground-Motion Model (GMM) Review Project," *EPRI Technical Report*, Product ID 3002000717, http://www.epri.com

Graves, R. W., and Pitarka, A. (2004). "Broadband time history simulation using a hybrid approach." Paper 1098, *Proc., 13th World Conference on Earthquake Engineering*, Vancouver.

Hartzell, S., and Heaton, T. (1983). "Inversion of strong ground motion and teleseismic waveform data for the fault rupture history of the 1979 Imperial Valley, California, earthquake." *Bull. Seismol. Soc. of Am.*, 73, 1553–1583.

Hartzell, S., Harmsen, S., Frankel, A., and Larsen, S. (1999). "Calculation of broadband time histories of ground motion: Comparison of methods and validation using strong ground motion from the 1994 Northridge earthquake." *Bull. Seismol. Soc. of Am.*, 89, 1484–1504.

Liu, P., Archuleta, R. J., and Hartzell, S. H. (2006). "Prediction of broadband ground-motion time histories: Hybrid low/high-frequency method with correlated random source parameters." *Bull. Seismol. Soc. of Am.*, 96, 2118–2130.

Luco, N., Ellingwood, B. R., Hamburger, R. O., Hooper, J. D., Kimball, J. K., and Kircher, C. A. (2007). "Risk-targeted vs. current seismic design maps for the conterminous United States." *Proc., Structural Engineers Association of California 76th Annual Convention*. SEAOC, Sacramento, CA.

Lund, W. R., ed. (2012). "Basin and range province earthquake working group II—Recommendations to the U.S. Geological Survey national seismic hazard mapping program for the 2014 update of the national seismic hazard maps: Utah Geological Survey." *USGS Open File Report 591*, Utah Geological Survey, Salt Lake City.

Mueller, C. S., Haller, K. M., Luco, N., Petersen, M. D., and Frankel, A. D. (2012). "Seismic hazard assessment for Guam and the Northern Mariana Islands," *USGS Open File Report 2012-1015*. USGS, Golden, CO.

National Earthquake Hazards Reduction Program (NEHRP). (1994). "Recommended provisions for seismic regulations for new buildings," *FEMA 222*. Building Seismic Safety Council, National Institute of Building Sciences, Washington, DC.

NEHRP. (1997). "Recommended provisions for seismic regulations for new buildings and other structures," *FEMA 302*. Building Seismic Safety Council, National Institute of Building Sciences, Washington, DC.

NEHRP. (2003). "Recommended provisions for seismic regulations for new buildings and other structures," *FEMA 450*. Building Seismic Safety Council, National Institute of Building Sciences, Washington, DC.

NEHRP. (2009). "Recommended provisions for seismic regulations for new buildings and other structures," *FEMA 750*. Building Seismic Safety Council, National Institute of Building Sciences, Washington, DC.

NEHRP. (2015). "Recommended provisions for seismic regulations for new buildings and other structures," *FEMA 1050*. Building Seismic Safety Council, National Institute of Building Sciences, Washington, DC.

Petersen, M.D. (2014). "Documentation for the 2014 update of the United States national seismic hazard maps," *USGS Open File Report 2014-1091*. USGS, Reston, VA.

Petersen, M. D., Moschetti, M. P., Powers, P. M., Mueller, C. S., Haller, K. M., Frankel, A. D., et al. (2013). "Documentation for the 2014 update of the United States National seismic hazard maps." USGS Administrative Report prepared for the Building Seismic Safety Council, Washington, DC.

Petersen, M. D., S. C., Harmsen, K. S., Rukstales, C. S., Mueller, D. E., McNamara, N., Luco, and M., Walling (2012). "Seismic Hazard of American Samoa and Neighboring South Pacific Islands: Data, Methods, Parameters, and Results," *USGS Open File Report 2008–1087*. USGS, Golden, CO.

Power, M., Chiou, B., Abrahamson, N., Bozorgnia, Y., Shantz, T., and Roblee, C. (2008). "An overview of the NGA project." *Earthquake Spectra Special Issue on the Next Generation of Ground Motion Attenuation (NGA) Project*. Earthquake Engineering Research Institute, Oakland, CA.

Somerville, P. G., Collins, N., Abrahamson, N., Graves, R., and Saikia, C. (2001). "Earthquake source scaling and ground motion attenuation relations for the Central and Eastern United States." *Final Report to the USGS under Contract 99HQGR0098*. USGS, Reston, VA.

Wells, D. L., and Coppersmith, K. J. (1994). "New empirical relationships among magnitude, rupture length, rupture width, rupture area, and surface displacement." *Bull. Seismol. Soc. of America* 84, 974–1002.

Working Group on California Earthquake Probabilities (WGCEP). (2013). "Uniform California earthquake rupture forecast, v. 3 (UCERF3)–The time-independent model." *USGS Open File Report 2013-1165*.

Zeng, Y., Anderson, J. G., and Yu, G. (1994). "A composite source model for computing synthetic strong ground motions." *Geophys. Res. Lett.*, 21, 725–728.

CHAPTER C23
SEISMIC DESIGN REFERENCE DOCUMENTS

There is no Commentary for Chapter 23.

CHAPTER C24
RESERVED FOR FUTURE COMMENTARY

CHAPTER C25
RESERVED FOR FUTURE COMMENTARY

CHAPTER C26
WIND LOADS: GENERAL REQUIREMENTS

C26.1 PROCEDURES

Chapter 26 is the first of six chapters devoted to the wind load provisions. It provides the basic wind design parameters that are applicable to the various wind load determination methodologies contained in Chapters 27 through 31. Specific items covered in Chapter 26 include definitions, basic wind speed, exposure categories, internal pressures, elevation effects, enclosure classification, gust effects, and topographic factors.

C26.1.1 Scope. The procedures specified in this standard provide wind pressures and forces for the design of the main wind force resisting system (MWFRS) and of components and cladding (C&C) of buildings and other structures. The procedures involve the determination of wind directionality and velocity pressure, the selection or determination of an appropriate gust-effect factor, and the selection of appropriate pressure or force coefficients. The procedures account for the level of structural reliability required, the effects of differing wind exposures, the speed-up effects of certain topographic features such as hills and escarpments, and the size and geometry of the building or other structure under consideration. The procedures differentiate between rigid and flexible buildings and other structures, and the results generally envelope the most critical load conditions for the design of MWFRS as well as C&C.

The pressure and force coefficients provided in Chapters 27, 28, 29, and 30 have been assembled from the latest boundary-layer wind tunnel and full-scale tests and from previously available literature. Because the boundary-layer wind tunnel results were obtained for specific types of building, such as low- or high-rise buildings and buildings that have specific types of structural framing systems, the designer is cautioned against indiscriminate interchange of values among the figures and tables.

C26.1.2 Permitted Procedures. The wind load provisions provide several procedures (as illustrated in Fig. 26.1-1) from which the designer can choose.
For MWFRS:

1. Directional Procedure for Buildings of All Heights (Chapter 27).
2. Envelope Procedure for Low-Rise Buildings (Chapter 28).
3. Directional Procedure for Building Appurtenances and Other Structures (Chapter 29).
4. Wind Tunnel Procedure for All Buildings and Other Structures (Chapter 31).

For C&C:

1. Analytical Procedure for Buildings and Building Appurtenances (Chapter 30).
2. Wind Tunnel Procedure for All Buildings and Other Structures (Chapter 31).

A "simplified method" for which the designer can select wind pressures directly from a table without any calculation, where the building meets all the requirements for application of the method, is provided for designing buildings using the Directional Procedure (Chapter 27, Part 2), the Envelope Procedure (Chapter 28, Part 2), and the Analytical Procedure for Components and Cladding (Chapter 30).

Limitations. The provisions given under Section 26.1.2 apply to the majority of site locations and buildings and other structures, but for some projects, these provisions may be inadequate. Examples of site locations and buildings and other structures (or portions thereof) that may require other approved standards, special studies using applicable recognized literature pertaining to wind effects, or using the Wind Tunnel Procedure of Chapter 31, include

1. Site locations that have channeling effects or wakes from upwind obstructions. Channeling effects can be caused by topographic features (e.g., a mountain gorge) or buildings (e.g., a neighboring tall building or a cluster of tall buildings). Wakes can be caused by hills, buildings, or other structures.
2. Buildings with unusual or irregular geometric shape, including barrel vaults, arched roofs, and other buildings whose shape (in plan or vertical cross section) differs significantly from the shapes in Figs. 27.3-1, 27.3-2, 27.3-3, 27.3-7, 28.3-1, and 30.3-1 through 30.3-7. Unusual or irregular geometric shapes include buildings with multiple setbacks, curved facades, or irregular plans resulting from significant indentations or projections, openings through the building, or multitower buildings connected by bridges.
3. Buildings or other structures with response characteristics that result in substantial vortex-induced and/or torsional dynamic effects, or dynamic effects resulting from aeroelastic instabilities such as flutter or galloping. Such dynamic effects are difficult to anticipate, being dependent on many factors, but should be considered when any one or more of the following apply:

- The height of the building or other structure is more than 400 ft (122 m).
- The height of the building or other structure is greater than 4 times its minimum effective width B_{min}, as defined below.
- The lowest natural frequency of the building or other structure is less than $n_1 = 0.25$ Hz.
- The reduced velocity $\bar{V}_{\bar{z}}/(n_1 B_{min}) > 5$, where $\bar{z} = 0.6h$ and $\bar{V}_{\bar{z}}$ is the mean hourly velocity in ft/s (m/s) at height $\bar{z}$.

The minimum effective width B_{min} is defined as the minimum value of $\sum h_i B_i / \sum h_i$ considering all wind directions. The summations are performed over the height of the building or other structure for each wind direction, where h_i is the height above grade of level i and B_i is the width at level i normal to the wind direction.

4. Bridges, cranes, electrical transmission lines, guyed masts, highway signs and lighting structures, telecommunication towers, and flagpoles.

When undertaking detailed studies of the dynamic response to wind forces, the fundamental frequencies of the building or other structure in each direction under consideration should be established using the structural properties and deformational characteristics of the resisting elements in a properly substantiated analysis, and not using approximate equations based on height.

Shielding. Because of the lack of reliable analytical procedures for predicting the effects of shielding provided by buildings and other structures or by topographic features, reductions in velocity pressure caused by shielding are not permitted under the provisions of this chapter. However, this does not preclude the determination of shielding effects and the corresponding reductions in velocity pressure by means of the Wind Tunnel Procedure in Chapter 31.

C26.2 DEFINITIONS

Several important definitions given in the standard are discussed in the following text. These terms are used throughout the standard and are provided to clarify application of the standard provisions.

BUILDING, ENCLOSED; BUILDING, OPEN; BUILDING, PARTIALLY ENCLOSED; BUILDING, PARTIALLY OPEN: These definitions relate to the proper selection of internal pressure coefficients, (GC_{pi}). "Enclosed," "open," and "partially enclosed" buildings are specifically defined. All other buildings are considered to be "partially open" by definition, although there may be large openings in two or more walls. An example of this would be a parking garage through which the wind can easily pass but which does not meet the definition for either an open or a partially enclosed building. The internal pressure coefficient for such a building would be ±0.18, and the internal pressures would act on the solid areas of the walls and roof. The standard also specifies that a building that meets both the "open" and "partially enclosed" definitions should be considered "open."

BUILDING OR OTHER STRUCTURE, FLEXIBLE: A building or other structure is considered "flexible" if it contains a significant dynamic resonant response. Resonant response depends on the gust structure contained in the approaching wind, on wind loading pressures generated by the wind flow about the building, and on the dynamic properties of the building or structure. Gust energy in the wind is smaller at frequencies above about 1 Hz. Therefore, the resonant response of most buildings and structures with lowest natural frequency above 1 Hz are sufficiently small that resonant response can often be ignored. The natural frequency of buildings or other structures greater than 60 ft (18.3 m) in height is determined in accordance with Sections 26.11.1 and 26.11.2. When buildings or other structures have a height exceeding 4 times the least horizontal dimension or when there is reason to believe that the natural frequency is less than 1 Hz (natural period greater than 1 s), the natural frequency of the structure should be investigated. Approximate equations for natural frequency or period for various building and structure types in addition to those given in Section 26.11.2 for buildings are contained in Commentary Section C26.11.

BUILDING OR OTHER STRUCTURE, REGULAR-SHAPED: Defining the limits of applicability of the various procedures within the standard requires a balance between the practical need to use the provisions past the range for which data have been obtained and restricting use of the provisions past the range of realistic application. Wind load provisions are based primarily on wind tunnel tests on shapes shown in Figs. 27.3-1, 27.3-2, 27.3-3, 27.3-7, 28.3-1, and 30.3-1 through 30.3-7. Extensive wind tunnel tests on actual structures under design show that relatively large changes from these shapes can, in many cases, have minor changes in wind load, while in other cases seemingly small changes can have relatively large effects, particularly on cladding pressures. Wind loads on complicated shapes are frequently smaller than those on the simpler shapes of Figs. 27.3-1, 27.3-2, 27.3-7, 28.3-1, and 30.3-1 through 30.3-7, and so wind loads determined from these provisions are expected to envelop most structure shapes. Buildings or other structures that are clearly unusual should be designed using the Wind Tunnel Procedure of Chapter 31.

BUILDING OR OTHER STRUCTURE, RIGID: The defining criterion for "rigid," in comparison to "flexible," is that the natural frequency is greater than or equal to 1 Hz. A general guidance is that most rigid buildings and structures have height-to-minimum-width less than 4. The provisions of Sections 26.11.1 and 26.11.2 provide methods for calculating natural frequency (period = 1/natural frequency), and Commentary Section C26.11 provides additional guidance.

COMPONENTS AND CLADDING (C&C): Components receive wind loads directly or from cladding and transfer the load to the MWFRS. Cladding receives wind loads directly. Examples of components include, but are not limited to, fasteners, purlins, girts, studs, sheathing, roof decking, certain trusses, and elements of trusses receiving wind loads from cladding. Examples of cladding include, but are not limited to, wall coverings, curtain walls, roof coverings, sheathing, roof decking, exterior windows, and doors. Components can be part of the MWFRS when they act as elements in shear walls or roof diaphragms, but they may also be loaded directly by wind as individual elements. The designer should use appropriate loads for design of components, which may require certain components to be designed for more than one type of wind loading; for example, long-span roof trusses should be designed for loads associated with MWFRS, and individual members of trusses should also be designed for C&C loads (Mehta and Marshall 1998).

DIAPHRAGM: This definition for diaphragm in wind load applications, for the case of untopped steel decks, differs somewhat from the definition used in Section 12.3 because diaphragms under wind loads are expected to remain essentially elastic.

EFFECTIVE WIND AREA, A: Effective wind area is the area of the building surface used to determine (GC_p). This area does not necessarily correspond to the area of the building surface contributing to the force being considered. Two cases arise. In the usual case, the effective wind area does correspond to the area tributary to the force component being considered. For example, for a cladding panel, the effective wind area may be equal to the total area of the panel. For a cladding fastener, the effective wind area is the area of cladding secured by a single fastener. A mullion may receive wind from several cladding panels. In this case, the effective wind area is the area associated with the wind load that is transferred to the mullion.

The second case arises where components such as roofing panels, wall studs, or roof trusses are spaced closely together. The area served by the component may become long and narrow. To better approximate the actual load distribution in such cases, the width of the effective wind area used to evaluate (GC_p) need not be taken as less than one-third the length of the area. This increase in effective wind area has the effect of reducing the average wind pressure acting on the component. Note, however, that this effective wind area should only be used in determining the (GC_p) in Figs. 30.3-1 through 30.3-6. The induced wind load should be applied over the actual area tributary to the component being considered.

For membrane roof systems, the effective wind area is the area of an insulation board (or deck panel if insulation is not used) if the boards are fully adhered (or the membrane is adhered directly to the deck). If the insulation boards or membrane are mechanically attached or partially adhered, the effective wind area is the area of the board or membrane secured by a single fastener or individual spot or row of adhesive.

For windows, doors, and other fenestration assemblies, the effective wind area for typical single-unit assemblies can be taken as the overall area of the assembly. For assemblies comprised of more than one unit mulled together or for more complex fenestration systems, it is recommended that the fenestration product manufacturer be consulted for guidance on the appropriate effective wind area to use when calculating the design wind pressure for product specification purposes.

The definition of effective wind area for rooftop solar panels and arrays is similar to that for components and cladding. As with C&C, the width of the effective wind area need not be less than one-third its length (which is typically equal to the span of the framing element being considered). The induced wind pressure is calculated per Fig. 29.4-4 using this effective wind area, and the wind pressure is then applied over the actual area tributary to the element.

Effective wind area is equal to the tributary area except in cases where the exception is invoked that the width of the effective wind area need not be less than one-third its length. In such cases, the effective wind area can be taken as larger than the tributary area.

Tributary area for a spanning structural member of a solar array depends on the span length of that member times the perpendicular distances to adjacent parallel members. For a support point or fastener, tributary area depends on the span of members framing into that support point.

Tributary area (and effective wind area) can depend on the characteristics of the solar array support system and the load path.

For a roof-bearing system that has different load paths for upward, downward, and lateral forces, the appropriate effective wind area for each direction of forces is used.

If the support system for the solar array has adequate strength, stiffness, and interconnectedness to span across a support or ballast point that is subject to yielding or uplift, the effective wind area can be correspondingly increased, provided that strengths are not governed by brittle failure and that the deformation of the array is evaluated and does not result in adverse performance. It should be noted that effective wind areas for uplift are usually much smaller than for lateral (drag) forces for ballasted arrays.

MAIN WIND FORCE RESISTING SYSTEM (MWFRS): The MWFRS can consist of a structural frame or an assemblage of structural elements that work together to transfer wind loads acting on the entire building or structure to the ground. Structural elements such as cross-bracing, shear walls, roof trusses, and roof diaphragms are part of the MWFRS when they assist in transferring overall loads (Mehta and Marshall 1998).

WIND-BORNE DEBRIS REGIONS: Wind-borne debris regions are defined to alert the designer to areas requiring consideration of missile impact design. These areas are located within hurricane-prone regions where there is a high risk of glazing failure caused by the impact of wind-borne debris.

C26.3 SYMBOLS

The following additional symbols and notation are used herein:

A_{ob} = average area of open ground surrounding each obstruction;
n = reference period, in years;
P_a = annual probability of wind speed exceeding a given magnitude [Eq. (C26.5-3)];
P_n = probability of exceeding design wind speed during n years [Eq. (C26.5-3)];
S_{ob} = average frontal area presented to the wind by each obstruction;
V_t = wind speed averaged over t seconds (see Fig. C26.5-1), in mi/h (m/s);
V_{3600} = mean wind speed averaged over 1 h (see Fig. C26.5-1), in mi/h (m/s); and
β = damping ratio (percentage of critical damping).

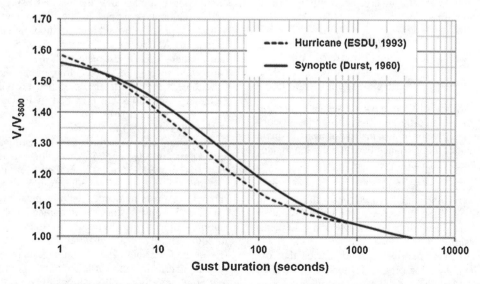

FIGURE C26.5-1 Maximum Speed Averaged over t (s) to Hourly Mean Speed

C26.4 GENERAL

C26.4.3 Wind Pressures Acting on Opposite Faces of Each Building Surface. Section 26.4.3 is included in the standard to ensure that internal and external pressures acting on a building surface are taken into account by determining a net pressure from the algebraic sum of those pressures. For additional information on the application of the net C&C wind pressure acting across a multilayered building envelope system, including air-permeable cladding, refer to Section C30.1.1.

C26.5 WIND HAZARD MAP

C26.5.1 Basic Wind Speed. All the wind speed maps in ASCE 7-16 have been updated, based on (i) a new analysis of nonhurricane wind data available through 2010, and (ii) improvements to the hurricane simulation model, which better account for the translation speed effects of fast-moving storms and the transition from hurricanes to extratropical storms in the northern latitudes (i.e., transition from warm core to cold core low-pressure systems). Separate wind speed maps are now provided for Risk Category III and Risk Category IV buildings and structures, recognizing the higher reliabilities required for essential facilities and facilities whose failure could pose a substantial hazard to the community. Location-specific basic wind speeds may be determined using http://windspeedatcouncil.org/windspeed. This website provides wind speeds to the nearest mile per hour based on a defined location using either latitude/longitude or an address. The website results use the same data used to develop the paper maps currently in the standard. Wind speeds are provided to the user for each of the Risk Categories, each of the serviceability periods, and a comparison speed for ASCE 7-10, ASCE 7-05 (3-s peak gust), and ASCE 7-93 (fastest mile).

In the 2016 edition, microzoned "effective" wind speed maps for Hawaii were added in accordance with the strength design return periods, including the effect of topography. The Hawaii effective wind speeds are algebraically formulated to include the macroscale and mesoscale terrain-normalized values of K_{zt} and K_d (Chock et al. 2005), i.e., $V_{\text{effective}}$ is the basic windspeed V multiplied by $\sqrt{(K_{zt} \times K_d/0.85)}$, so that the engineer is permitted to more conveniently use the standard values of K_{zt} of 1.0 and K_d as given in Table 26.6-1. Note that local site conditions of finer toposcale, such as ocean promontories and local escarpments, should still be examined. Spatial resolution scales for digital modeling, including terrain effects, are conventionally described in the recognized literature as follows:

Scale	Spatial Resolution
Toposcale	32–656 ft (10–200 m)
Mesoscale	656 ft–3.1 mi (200 m–5 km)
Macroscale	3.1 mi–311 mi (5–500 km)

The decision in ASCE 7-10 to move to separate strength design wind speed maps for different Risk Categories in conjunction with a wind load factor of 1.0, instead of using a single map in conjunction with an Importance Factor and a load factor of 1.6, relied on several considerations:

1. A strength-level design wind speed map brings the wind loading approach in line with that used for seismic loads, in that they both are aimed at achieving uniform risk rather than uniform hazard and eliminate the use of a load factor for strength design.
2. Having separate maps removes inconsistencies that occurred with the use of Importance Factors, which varied with location, and allows for the geographical description of zones affected by nonhurricane winds only and by both hurricane and non-hurricane winds as a function of mean recurrence interval (MRI).
3. Each map has the same MRI for design wind speeds in those two zones.
4. By providing the design wind speed directly, the maps more clearly inform owners and their consultants about the storm intensities for which the buildings and other structures are designed.

Selection of Return Periods. The methodology for selection of the return periods used in ASCE 7-10 (Vickery et al. 2010) has been modified for ASCE 7-16. In order to determine a return period for each Risk Category consistent with the target reliabilities in Table C1.3-1, the ASCE 7 Load Combinations Subcommittee conducted a reliability analysis that incorporated new data on the directionality factor. The nominal design value $K_d = 0.85$ was based on a relatively simple directional analysis conducted as part of the original ANSI A58/ASCE 7 load factor development. One of the underlying assumptions of the original analysis was that the wind directionality factor, $K_d = 0.85$, was unbiased because only limited data on the effects of wind directionality were available at the time. More recent research by Isyumov et al. (2013), simulating three building geometries at two different locations, indicates that the ASCE 7 nominal values of K_d are affected by a bias (defined as the ratio of the mean value, μK_d, to the nominal value, K_{dn}). The nominal value of K_d is conservative for both tropical and extratropical winds because the mean value is less than the nominal value. Additional reliability analyses were performed to examine the effect of K_d on the return period and associated reliability. The subcommittee found that the following return periods for each Risk Category are consistent with the target reliabilities in the first row of Table 1.3-1: Risk Category I: 300 years; Risk Category II: 700 years; Risk Category III: 1,700 years; and Risk Category IV: 3,000 years.

Wind Speed. The wind speed maps of Figs. 26.5-1 and 26.5-2 present basic wind speeds for the contiguous United States, Alaska, Hawaii, and other selected locations. The wind speeds correspond to 3-s gust speeds at 33 ft (10 m) above ground for Exposure Category C. Because the wind speeds of Figs. 26.5-1 and 26.5-2 reflect conditions at airports and similar open-country exposures, they do not account for the effects of significant topographic features such as those described in Section 26.8. In ASCE 7-16, wind speeds in nonhurricane-prone areas of the contiguous United States are mapped using contours to better reflect regional variations in the extreme wind climate. Point values are provided to aid interpolation, in a style similar to that used in the ASCE 7 seismic hazard maps. Summaries of the data and methods used to estimate both the nonhurricane and hurricane wind speeds are given below, along with a description of how these wind speeds are combined to make the final maps. Detailed descriptions are provided in Pintar et al. (2015) and Lombardo et al. (2016).

Nonhurricane Wind Speeds. The nonhurricane wind speeds for the contiguous United States were estimated from peak gust speed data collected at 575 meteorological stations. The data at each station were extracted from the meteorological records and classified by storm type, thunderstorm or nonthunderstorm after removal of gusts associated with tropical cyclones (i.e., hurricanes and tropical storms). Recorded peak gusts from each station were corrected as needed to standardize the observations to equivalent 3-s peak gusts at 10 m (33 ft) height over open (Exposure C) terrain. At each station, there were at least 15 years of data, and there were sufficient numbers of both thunderstorm and nonthunderstorm observations to account for their potential differences when estimating wind speeds with specified mean recurrence intervals. The estimation was performed in two stages. In the first stage, a peaks-over-thresholds (POT) model was fitted to the data from each station. The POT model used was the Poisson process model first described in Pickands (1971) and extended in Smith (1989) to allow the parameters of the Poisson process to be time dependent. This model allowed for differentiation between thunderstorm and nonthunderstorm winds. The Poisson process has a tail length parameter that may be set to zero, leading to Gumbel-like tails for the distribution of wind speeds. Such distributional tails were used in this work, consistent with past practice in wind engineering. The fitted POT models allowed for the estimation of wind speeds for any required mean recurrence interval at all stations. In the second stage, local regression (Cleveland and Devlin 1988) was used to interpolate wind speeds at all points of a fine regular grid covering the contiguous United States for all required mean recurrence intervals. This had the effect of spatially smoothing the noisy station estimates. The smoothed wind speed estimates provided the basis for creating the isotach maps.

Limited data were available on the Washington and Oregon coast. In this region, a special wind region was defined to permit local jurisdictions to select speeds based on local knowledge and analysis. Speeds in the Aleutian Islands and in the interior of Alaska were established from gust data. Insufficient data were available for a detailed coverage of the mountainous regions, so gust data in Alaska were not corrected for potential terrain influence. It is possible that wind speeds in parts of Alaska would be smaller if the topographic wind speed-up effect on recorded wind speeds were taken into account. In Alaska, the maps for each return period were determined by multiplying the 50-year MRI contours given in ASCE 7-10 Fig. CC-3 by a factor, F_{RA}, equal to

$$F_{RA} = 0.45 + 0.085 \ln(12T) \quad (C26.5\text{-}1)$$

where T is the return period in years (Peterka and Shahid 1998). The resulting contours were interpolated to the nearest 10 mi/h, except for the innermost and outermost contours, which were rounded to the nearest 5 mi/h.

Hurricane Wind Speeds. The hurricane wind speeds are based on the results of a Monte Carlo simulation model generally described in Applied Research Associates (2001), Vickery and Wadhera (2008a, b), and Vickery et al. (2009a, b, 2010). The hurricane simulation model used to develop the wind speeds in ASCE 7-16 included two updates to the model used for ASCE 7-10. A reduced translation speed effect for fast-moving storms (NRC 2011) was incorporated, and a simple extratropical transition model was also implemented, where the surface winds are reduced linearly by up to 10% over the latitude range 37 N to 45 N. This reduction approximates transitioning from a hurricane boundary layer to an extratropical storm boundary layer. The effects of the model revisions are to slightly reduce hurricane speeds in the northeast, extending from Maine to Virginia.

Combination of Nonhurricane and Hurricane Wind Speed Data. Nonhurricane wind speeds and hurricane wind speeds were estimated for return periods ranging from 10 years to 100,000 years. The nonhurricane and hurricane winds were then combined as statistically independent events using Eq. (C26.5-2), the same general approach that has been used in previous editions of ASCE 7:

$$P_a(v > V) = 1 - P_{NH}(v < V)P_H(v < V) \quad (C26.5\text{-}2)$$

where

$P_a(v > V)$ is the annual exceedance probability for the combined wind hazards,
$P_{NH}(v < V)$ is the annual nonexceedance probability for nonhurricane winds, and
$P_H(v < V)$ is the annual nonexceedance probability for hurricane winds.

The combined winds were interpolated to yield the combined wind hazard curves for MRIs associated with each of the wind speed maps. In cases where the hurricane contours are unchanged from ASCE 7-10, the shape files from these previous maps were used to ensure continuity between the maps.

Correlation of Basic Wind Speed Map with the Saffir-Simpson Hurricane Wind Scale. Hurricane intensities are reported by the National Hurricane Center (NHC) according to the Saffir-Simpson Hurricane Wind Scale (NHC 2015), shown in Table C26.5-1. This scale has found broad usage by hurricane forecasters and local and federal agencies responsible for short-range evacuation of residents during hurricane alerts, as well as long-range disaster planners and the news media. The scale contains five categories of hurricanes and distinguishes them based on wind speed intensity.

The wind speeds used in the Saffir-Simpson Hurricane Wind Scale are defined in terms of a sustained wind speed with a 1-min averaging time at 33 ft (10 m) over open water. The ASCE 7 standard by comparison uses a 3-s gust speed at 33 ft (10 m) above ground in Exposure C (defined as the basic wind speed, and shown in the wind speed map, Figs. 26.5-1 and 26.5-2). The sustained wind speed over water in Table C26.5-2 cannot be converted to a peak gust wind speed using the Durst curve of Fig. C26.5-1, which is only valid for wind blowing over open terrain (Exposure C). An approximate relationship between the wind speeds in ASCE 7 and the Saffir-Simpson scale, based on recent data which indicate that the sea surface roughness remains approximately constant for mean hourly speeds in excess of 67 mi/h (30 m/s), is shown in Table C26.5-2. The table provides the sustained wind speeds of the Saffir-Simpson Hurricane Wind Scale over water, equivalent-intensity gust wind speeds over water, and equivalent-intensity gust wind speeds over land. For a storm of a given intensity, Table C26.5-2 takes into consideration both the reduction in wind speed as the storm moves from over water to over land because of changes in surface roughness, and the change in the gust factor as the storm moves from over water to over land (Vickery et al 2009a; Simiu et al. 2007).

Table C26.5-3 shows the design wind speed from the ASCE 7 basic wind speed maps (Figs. 26.5-1 and 26.5-2) for various

Table C26.5-1 Saffir-Simpson Hurricane Wind Scale

Hurricane Category	Sustained Wind Speed[a] mph (m/s)	Types of Damage Due to Hurricane Winds
1	74–95 (33–42)	Very dangerous winds will produce some damage
2	96–110 (43–49)	Extremely dangerous winds will cause extensive damage
3	111–129 (50–57)	Devastating damage will occur
4	130–156 (58–69)	Catastrophic damage will occur
5	≥157 (70)	Highly catastrophic damage will occur

[a]1-minute average wind speed at 33 ft (10 m) above open water.

Table C26.5-2 Approximate Relationship between Wind Speeds in ASCE 7 and Saffir-Simpson Hurricane Wind Scale

Saffir-Simpson Hurricane Category	Sustained Wind Speed over Water[a]		Gust Wind Speed over Water[b]		Gust Wind Speed Over Land[c]	
	mph	m/s	mph	m/s	mph	m/s
1	74–95	33–42	90–116	40–51	81–105	36–47
2	96–110	43–49	117–134	52–59	106–121	48–54
3	111–129	50–57	135–157	60–70	122–142	55–63
4	130–156	58–69	158–190	71–84	143–172	64–76
5	>157	>70	>191	>85	>173	>77

[a] 1-min average wind speed at 33 ft (10 m) above open water.
[b] 3-s gust wind speed at 33 ft (10 m) above open water.
[c] 3-s gust wind speed at 33 ft (10 m) above open ground in Exposure Category C. This column has the same basis (averaging time, height, and exposure) as the basic wind speed from Figs. 26.5-1 and 26.5-2.

locations along the hurricane coastline from Maine to Texas, and for Hawaii, Puerto Rico, and the Virgin Islands. Tables C26.5-4 through C26.5-6 show the basic wind speeds for Risk Category II, III, and IV buildings and other structures in terms of the hurricane category equivalents on the Saffir-Simpson Hurricane Wind Scale. These wind speeds represent an approximate limit state; structures designed to withstand the wind loads specified in this standard, which are also appropriately constructed and maintained, should have a high probability of surviving hurricanes of the intensities shown in Tables C26.5-4 through C26.5-6 without serious structural damage from wind pressure alone.

Tables C26.5-2 through C26.5-6 are intended to help users of the standard better understand design wind speeds as used in this standard in relation to wind speeds reported by weather forecasters and the news media, who commonly use the Saffir-Simpson Hurricane Wind Scale. The Saffir-Simpson hurricane category equivalent Exposure C gust wind speed values given in Tables C26.5-2 through C26.5-6, which are associated with a given sustained wind speed, should be used as a guide only.

Table C26.5-3 Basic Wind Speeds at Selected Coastal Locations in Hurricane-Prone Areas

Location	Coordinates (decimal degrees)		Basic Wind Speeds (mph)		
	Latitude	Longitude	Risk Cat II (700-yr)	Risk Cat III (1,700-yr)	Risk Cat IV (3,000-yr)
Bar Harbor, Maine	44.3813	−68.1968	110	119	124
Hampton Beach, New Hampshire	42.9107	−70.8102	116	124	129
Boston, Massachusetts	42.3578	−71.0012	120	129	133
Hyannis, Massachusetts	41.6359	−70.2901	132	139	145
Newport, Rhode Island	41.453	−71.3058	131	139	144
New Haven, Connecticut	41.2803	−72.9327	121	129	134
Southhampton, New York	40.871	−72.3844	130	138	145
Manhattan, New York	40.7005	−74.0135	115	127	130
Atlantic City, New Jersey	39.3536	−74.4336	126	135	140
Rehoboth Beach, Delaware	38.7167	−75.0752	121	130	137
Ocean City, Maryland	38.3314	−75.0835	128	136	140
Virginia Beach, Virginia	36.8306	−75.9691	123	132	137
Wrightsville Beach, North Carolina	34.1973	−77.8014	146	156	160
Folly Beach, South Carolina	32.6496	−79.9512	149	158	166
Sea Island, Georgia	31.179	−81.3472	131	145	150
Jacksonville Beach, Florida	30.2836	−81.387	129	140	150
Melbourne Beach, Florida	28.0684	−80.5564	151	162	170
Miami Beach, Florida	25.7643	−80.1309	170	183	190
Key West, Florida	24.5477	−81.7843	180	200	200
Clearwater, Florida	27.9658	−82.8042	146	153	158
Panama City Beach, Florida	30.1558	−85.7744	135	146	149
Gulf Shores, Alabama	30.2486	−87.6808	160	172	177
Biloxi, Mississippi	30.3924	−88.8887	160	176	183
Slidell, Louisiana	30.2174	−89.824	142	152	158
Cameron, Louisiana	29.7761	−93.2921	144	154	158
Galveston, Texas	29.2663	−94.826	150	159	167
Port Aransas, Texas	27.8346	−97.0446	150	157	163
Hawaii	n/a	n/a	130	145	150
San Juan, Puerto Rico	18.4501	−66.0367	160	170	175
Virgin Islands	n/a	n/a	165	175	180

Notes:
1. All wind speeds in Table C26.5-3 are 3-s gust wind speeds at 33 ft (10 m) above ground for Exposure Category C.
2. The basic wind speed in hurricane–prone regions can vary significantly over a city or county; the values shown are for randomly selected points along the coast for communities listed in the tables. Wind speeds at other locations within those communities may be greater or less than the values shown in the table.
3. Conversion of mph to m/s: mph × 0.44704 = m/s.

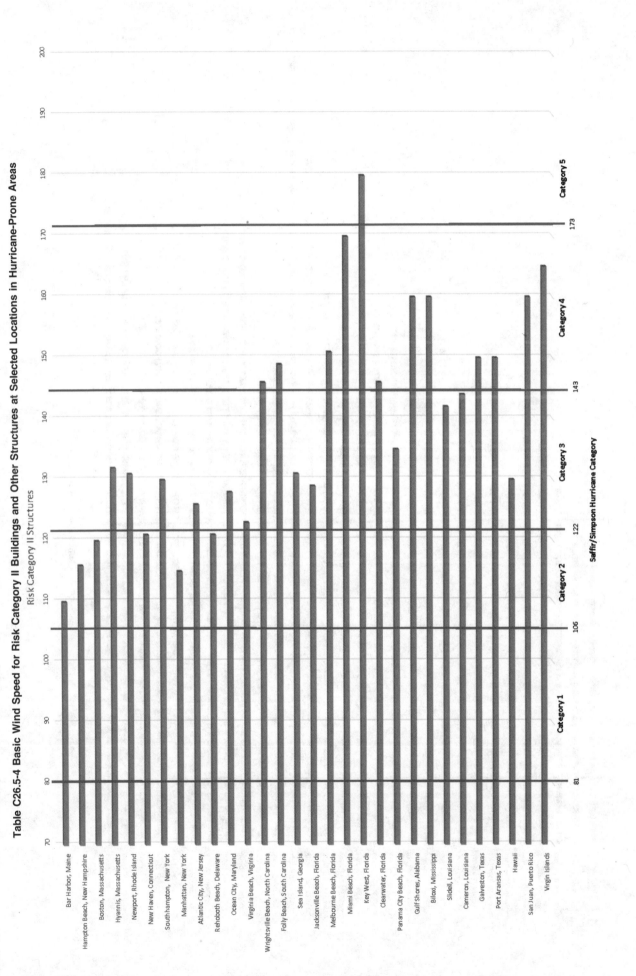

Table C26.5-4 Basic Wind Speed for Risk Category II Buildings and Other Structures at Selected Locations in Hurricane-Prone Areas

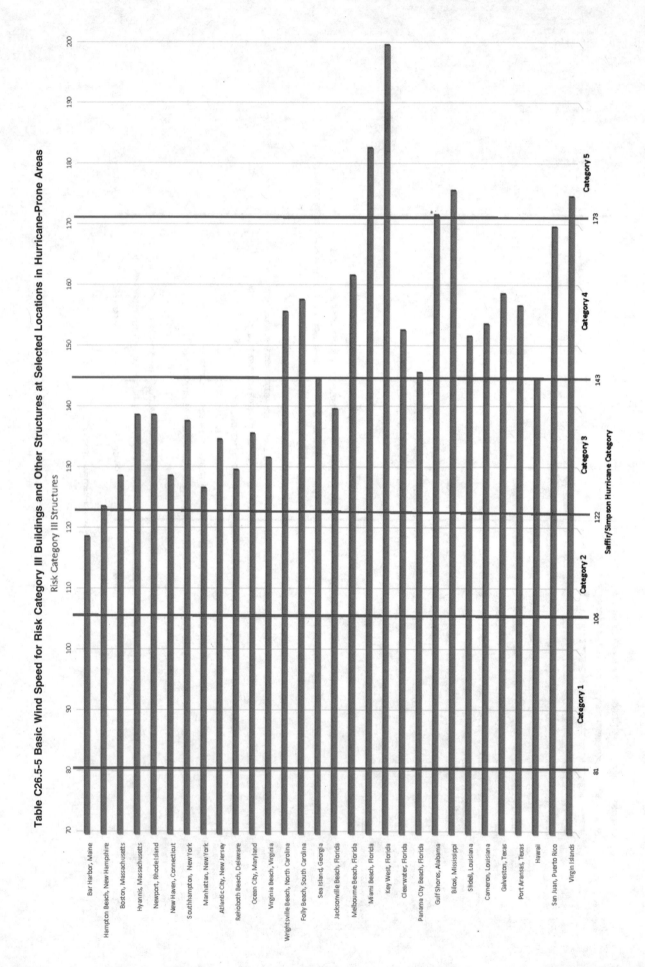

Table C26.5-5 Basic Wind Speed for Risk Category III Buildings and Other Structures at Selected Locations in Hurricane-Prone Areas

Table C26.5-6 Basic Wind Speed for Risk Category IV Buildings and Other Structures at Selected Locations in Hurricane-Prone Areas

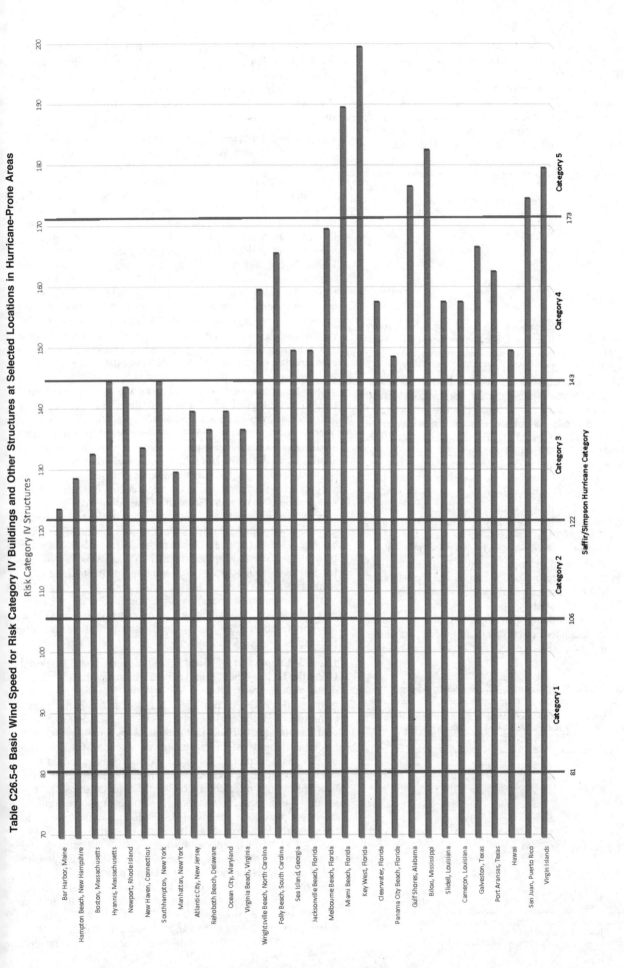

The gust wind speeds associated with a given sustained wind speed may vary with storm size and intensity, as suggested in Vickery et al. (2009a).

Wind Speeds for Serviceability Design. For serviceability applications such as drift and habitability, the Appendix C Commentary presents maps of peak gust wind speeds at 33 ft (10 m) above ground in Exposure C conditions for return periods of 10, 25, 50, and 100 years (Figs. CC.2-1 through CC.2-4).

The probability, P_n, that the wind speed associated with a certain annual probability, P_a, will be equaled or exceeded at least once during an exposure period of n years is given by

$$P_n = 1 - (1 - P_a)^n \quad \text{(C26.5-3)}$$

where

$$P_a = 1 - e^{(-1/\mathrm{MRI})} \quad \text{(C26.5-4)}$$

For MRIs of about 10 years or longer, P_a is very closely approximated by the reciprocal of the mean recurrence interval, i.e., $P_a \approx 1/MRI$.

As an example, if a wind speed is based upon $P_a = 0.02/\text{year}$ (50-year MRI), the probability that this speed will be equaled or exceeded (at least once) during a 25-year period is 0.40 (i.e., 40%), and the probability of being equaled or exceeded in a 50-year period is 64%. Similarly, if a wind speed is based upon $P_a = 0.00143$ (700-year MRI), the probability that this speed will be equaled or exceeded during a 25-year period is 3.5%, and the probability of being equaled or exceeded in a 50-year period is 6.9%.

Some products have been evaluated, and test methods have been developed based on design wind speeds that are consistent with the unfactored load effects typically used in allowable stress design (ASD). Table C26.5-7 provides a comparison of the strength design-based wind speeds used in the ASCE 7-10 and 7-16 basic wind speed maps and the ASCE 7-05 basic wind speeds used in these product evaluation reports and test methods. A column of values is also provided to allow comparison with ASCE 7-93 basic wind speeds.

C&C Rating for Building Envelope Products. Building envelope products that have been tested to air pressure standards (such as ASTM E330, CSA A123.21, or other standards that incorporate a safety factor) are typically rated for an allowable stress design wind pressure $(0.6W)$ rather than a strength design pressure $(1.0W)$ or wind speed. In order to properly select products tested and rated in this manner, the C&C pressures determined from Chapter 30 should be adjusted for the allowable stress design load factor of $0.6W$ in Section 2.4.1.

C26.5.2 Special Wind Regions. Although the wind speed maps of Figs. 26.5-1 and 26.5-2 are valid for most regions of the country, there are special regions in which wind speed anomalies are known to exist. Some of these special regions are noted in Figs. 26.5-1 and 26.5-2. In 2016, the special wind regions were restored to the areas originally designated in ANSI A58.1 through ASCE 7-93. The regions had inadvertently been changed in ASCE 7-95 because of an unintended graphical error. The special wind regions around the Great Lakes in the northeast United States and in the Puget Sound area near Seattle were omitted intentionally in ASCE 7-95 because of lack of meteorological data demonstrating the existence of wind speeds higher than could be explained by exposure. Winds blowing over mountain ranges or through gorges or river valleys in these special regions can develop speeds that are substantially higher than the values indicated on the map. When selecting basic wind speeds in these special regions, use of regional climatic data and consultation with a wind engineer or meteorologist is advised.

It is also possible that anomalies in wind speeds exist on a micrometeorological scale. For example, wind speed-up over hills and escarpments is addressed in Section 26.8. Wind speeds over complex terrain may be better determined by wind tunnel studies as described in Chapter 31. Adjustments of wind speeds should be made at the micrometeorological scale on the basis of wind engineering or meteorological advice and used in accordance with the provisions of Section 26.5.3 when such adjustments are warranted. Because of the complexity of mountainous terrain and valley gorges in Hawaii, there are topographic wind speed-up effects that cannot be addressed solely by Fig. 26.8-1 (Applied Research Associates 2001). In the Hawaii special wind region, research and analysis have established that there are special K_{zt} topographic effect adjustments (Chock et al. 2005).

The southernmost special wind region in California experiences Santa Ana winds (dry, mountain downslope winds). The appropriate boundaries of this region are difficult to quantify because of a lack of data.

C26.5.3 Estimation of Basic Wind Speeds from Regional Climatic Data. When using regional climatic data in accordance with the provisions of Section 26.5.3 and in lieu of the basic wind speeds given in Figs. 26.5-1 and 26.5-2, the user is cautioned that the gust factors, velocity pressure exposure coefficients, gust-effect factors, pressure coefficients, and force coefficients of this standard are intended for use with the 3-s gust speed at 33 ft (10 m) above ground in terrain with open exposure. It is necessary, therefore, that regional climatic data based on a different averaging time, for example, hourly mean, 10-min mean, fastest minute, or fastest mile, be adjusted to reflect peak gust speeds at 33 ft (10 m) above ground in terrain with open exposure. The results of statistical studies of wind-speed records, reported by Durst (1960) for extratropical winds and for hurricanes (Vickery et al. 2000), are given in Fig. C26.5-1, which defines the relation between wind speed averaged over time t in seconds, V_t, and the hourly wind speed, V_{3600}. The hurricane simulation model described in Section C26.5.1 uses the gust-factor curve from ESDU (1982, 1993), which has been shown to be valid for hurricane winds (Vickery and Skerlj 2005). Similar conclusions regarding hurricane gust

Table C26.5-7 Basic Wind Speeds: ASCE 7-93 to ASCE 7-16

ASCE 7-95 through ASCE 7-05 Basic Wind Speed (3-s gust in mi/h)	ASCE 7-10 and ASCE 7-16 Basic Wind Speed (3-s gust, mi/h)	ASCE 7-93 and Prior Editions Basic Wind Speed (fastest mile, mi/h)
85	108[a]	71
90	114[a]	76
100	126	85
105	133	90
110	139	95
120	152	104
130	164	114
140	177	123
145	183	128
150	190	133
170	215	152

[a]In ASCE 7-10 wind speed values of 108 mi/h and 114 mi/h were rounded to 110 mi/h and 115 mi/h, respectively.
Note: Conversion of mi/h to m/s: mi/h × 0.44704 = m/s.

factors were drawn by Jung and Masters (2013). The relation between wind speed averaging times for extratropical winds in any terrain exposure is given in Section 11.2.4.2 of Simiu (2011).

In using local data, it should be emphasized that sampling errors can lead to large uncertainties in specification of the wind speed. Sampling errors are the errors associated with the limited size of the climatological data samples (e.g., years of record of extreme speeds). It is possible to have a 20-mi/h (8.9-m/s) error in the estimated extreme wind speed at an individual station with a record length of 30 years. When short local records are used to estimate extreme wind speeds, care and conservatism should be exercised in their use.

If meteorological data are used to justify a wind speed lower than the basic wind speed from Figs. 26.5-1 and 26.5-2, an analysis of sampling error is required. This can be accomplished by showing that the difference between the estimated speed and the basic wind speed from Figs. 26.5-1 and 26.5-2 is at least two to three times the standard deviation of the sampling error (Simiu and Scanlan 1996).

C26.6 WIND DIRECTIONALITY

The wind load factor 1.3 in ASCE 7-95 included a "wind directionality factor" with a nominal value of 0.85 (Ellingwood 1981; Ellingwood et al. 1982). This factor accounts for two effects: (1) The reduced probability of maximum winds coming from any given direction, and (2) the reduced probability of the maximum pressure coefficient occurring for any given wind direction. The nominal wind directionality factor (denoted by K_d in the standard) is tabulated in Table 26.6-1 for different structure types. As new research becomes available, this factor can be directly modified. Nominal values for the factor were established from references in the literature and collective committee judgment. A value of 0.85 might be more appropriate if a triangular trussed frame is shrouded in a round cover. A value of 0.95 might be more appropriate for a round structure that has a nonaxisymmetrical lateral load resistance system.

C26.7 EXPOSURE

The descriptions of the surface roughness categories and exposure categories in Section 26.7 have been expressed as far as possible in easily understood verbal terms that are sufficiently precise for most practical applications. Upwind surface roughness conditions required for Exposures B and D are shown schematically in Figs. C26.7-1 and C26.7-2, respectively. Aerial photographs showing examples of Exposures B, C, and D are shown in Figs. C26.7-5 through C26.7-7. For cases where the designer wishes to make a more detailed assessment of the surface roughness category and exposure category, the following more-mathematical description is offered for guidance (Irwin 2006). The ground surface roughness is best measured in terms of a roughness length parameter called z_0. Each of the surface roughness categories B through D corresponds to a range of values of this parameter, as does the even-rougher category A used in older versions of the standard in heavily built-up urban areas but removed

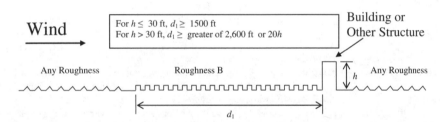

FIGURE C26.7-1 Upwind Surface Roughness Conditions Required for Exposure B

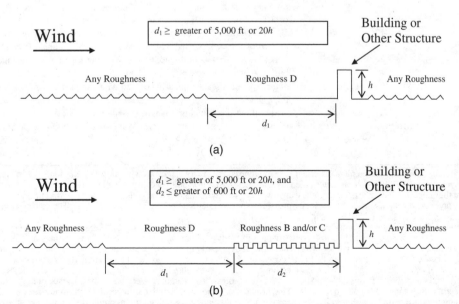

FIGURE C26.7-2 Upwind Surface Roughness Conditions Required for Exposure D, for the Cases with (a) Surface Roughness D Immediately Upwind of the Building, and (b) Surface Roughness B and/or C Immediately Upwind of the Building

Table C26.7-1 Range of z_0 by Exposure Category

Exposure Category	Lower Limit of z_0, ft (m)	Typical Value of z_0, ft (m)	Upper Limit of z_0, ft (m)	z_0 Inherent in Tabulated K_z Values in Table 26.10-1, ft (m)
A	$2.3\,(0.7) \leq z_0$	6.6 (2)	—	—
B	$0.5\,(0.15) \leq z_0$	1.0 (0.3)	$z_0 < 2.3\,(0.7)$	0.66 (0.20)
C	$0.033\,(0.01) \leq z_0$	0.066 (0.02)	$z_0 < 0.5\,(0.15)$	0.066 (0.02)
D	—	0.016 (0.005)	$z_0 < 0.033\,(0.01)$	0.016 (0.005)

in the recent editions. The range of z_0 in ft (m) for each terrain category is given in Table C26.7-1. Exposure A has been included in Table C26.7-1 as a reference that may be useful when using the Wind Tunnel Procedure. Further information on values of z_0 in different types of terrain can be found in Simiu and Scanlan (1996) and Table C26.7-2 based on Davenport et al. (2000) and Wieringa et al. (2001).

The roughness classifications in Table C26.7-2 are not intended to replace the use of exposure categories as required in the standard for structural design purposes. However, the terrain roughness classifications in Table C26.7-2 may be related to exposure categories by comparing z_0 values between Tables C26.7-1 and C26.7-2. For example, the z_0 values for Classes 3 and 4 in Table C26.7-2 fall within the range of z_0 values for Exposure C in Table C26.7-1. Similarly, the z_0 values for Classes 5 and 6 in Table C26.7-2 fall within the range of z_0 values for Exposure B in Table C26.7-1.

Research described in Powell et al. (2003), Donelan et al. (2004), and Vickery et al. (2009a) showed that the drag coefficient over the ocean in high winds in hurricanes does not continue to increase with increasing wind speed, as was previously believed (e.g., Powell 1980). These studies showed that the sea surface drag coefficient, and hence the aerodynamic roughness of the ocean, reached a maximum at mean wind speeds of about 67 mi/h (30 m/s). There is some evidence that the drag coefficient actually decreases (i.e., the sea surface becomes aerodynamically smoother) as the wind speed increases further (Powell et al. 2003) or as the hurricane radius decreases (Vickery et al. 2009a). The consequences of these studies are that the surface roughness over the ocean in a hurricane is consistent with that of Exposure D rather than Exposure C. Consequently, the use of Exposure D along the hurricane coastline is required.

For Exposure B, the tabulated values of K_z correspond to $z_0 = 0.66$ ft (0.2 m), which is below the typical value of 1 ft (0.3 m), whereas for Exposures C and D they correspond to the typical

Table C26.7-2 Davenport Classification of Effective Terrain Roughness

Class	z_0, ft (m)[a]	α[b]	z_g, ft (m)[b]	z_d (ft or m)[c]	Wind Flow and Landscape Description[d]
1	0.0007 (0.0002)	12.9	509 (155)	$z_d = 0$	**Sea:** Open sea or lake (irrespective of wave size), tidal flat, snow-covered flat plain, featureless desert, tarmac, and concrete, with a free fetch of several kilometers.
2	0.016 (0.005)	11.4	760 (232)	$z_d = 0$	**Smooth:** Featureless land surface without any noticeable obstacles and with negligible vegetation, e.g., beaches, pack ice without large ridges, marsh, and snow-covered or fallow open country.
3	0.1 (0.03)	9.0	952 (290)	$z_d = 0$	**Open:** Level country with low vegetation (e.g., grass) and isolated obstacles with separations of at least 50 obstacle heights, e.g., grazing land without windbreaks, heather, moor, and tundra, runway area of airports. Ice with ridges across-wind.
4	0.33 (0.10)	7.7	1,107 (337)	$z_d = 0$	**Roughly open:** Cultivated or natural area with low crops or plant covers, or moderately open country with occasional obstacles (e.g., low hedges, isolated low buildings, or trees) at relative horizontal distances of at least 20 obstacle heights.
5	0.82 (0.25)	6.8	1,241 (378)	$z_d = 0.2 z_H$	**Rough:** Cultivated or natural area with high crops or crops of varying height and scattered obstacles at relative distances of 12 to 15 obstacle heights for porous objects (e.g., shelterbelts) or 8 to 12 obstacle heights for low solid objects (e.g., buildings).
6	1.64 (0.5)	6.2	1,354 (413)	$z_d = 0.5 z_H$	**Very rough:** Intensely cultivated landscape with many rather large obstacle groups (large farms, clumps of forest) separated by open spaces of about 8 obstacle heights. Low, densely planted major vegetation like bushland, orchards, young forest. Also, area moderately covered by low buildings with interspaces of 3 to 7 building heights and no high trees.
7	3.3 (1.0)	5.7	1,476 (450)	$z_d = 0.7 z_H$	**Skimming:** Landscape regularly covered with similar-size large obstacles, with open spaces of the same order of magnitude as obstacle heights, e.g., mature regular forests, densely built-up area without much building height variation.
8	$\geq$matu ($\geq$mat)	5.2	1,610 (490)	Analysis by wind tunnel advised	**Chaotic:** City centers with mixture of low-rise and high-rise buildings, or large forests of irregular height with many clearings. (Analysis by wind tunnel advised.)

[a] The surface roughness length, z_0, represents the physical effect that roughness objects (obstacles to wind flow) on the Earth's surface have on the shape of the atmospheric boundary layer wind velocity profile as determined by the logarithmic law and used in the ESDU model.
[b] The power law uses α on the shape of the atmospheric boundary layer wz_g, representing the height at which geostrophic wind flow begins to occur, as the basis for determining the boundary layer wind velocity profile and velocity pressure exposure coefficients (see Section C26.10.1). The values provided in this table are based on the published z_0 values and use of Eqs. (C26.10-3) and (C26.10-4).
[c] The zero plane displacement height, z_d, is the elevation above ground that the base of the logarithmic law (and power law) wind profile must be elevated to accurately depict the boundary layer wind flow. Below z_d and less than some fraction of the typical height, z_H, of obstacles causing roughness, the near ground wind flow is characterized as a turbulent exchange with the boundary layer wind flow above resulting in significant shielding effects under uniform to moderately uniform roughness conditions (e.g., Classes 5 through 7 in this table). In this condition, the effective mean roof height, h_{eff}, may then be determined as $h-z_d$ (but not less than 15 ft or 4.6 m) for the purpose of determining MWFRS wind loads acting on a building structure located within such a roughness class. Appropriate values of z_d for a given site may vary widely, and those shown in this table should be used with professional judgment. Because of the presence of highly turbulent flow at elevations near or below z_d (except perhaps structures embedded in uniform Class 7 roughness), use of an effective mean roof height should not be applied for the determination of components and cladding wind loads. In Class 8 roughness where wind flow disruptions can be highly nonuniform, channeling effects and otherwise "chaotic" wind flow patterns can develop between and below the height of obstacles to wind flow. For this reason, a wind tunnel study is generally advised.
[d] Use of these wind flow and landscape descriptions should result in no greater than one roughness class error, corresponding to a maximum +/- of these wind q_h.

value of z_0. The reason for the difference in Exposure B is that this category of terrain, which is applicable to suburban areas, often contains open patches, such as highways, parking lots, and playing fields. These open patches cause local increases in the wind speeds at their edges. By using an exposure coefficient corresponding to a lower-than-typical value of z_0, some allowance is made for this. The alternative would be to introduce a number of exceptions to the use of Exposure B in suburban areas, which would add an undesirable level of complexity.

The value of z_0 for a particular terrain can be estimated from the typical dimensions of surface roughness elements and their spacing on the ground area using an empirical relationship, found by Lettau (1969), which is

$$z_0 = 0.5 H_{ob} \frac{S_{ob}}{A_{ob}} \qquad (C26.7\text{-}1)$$

where

H_{ob} = average height of the roughness in the upwind terrain;
S_{ob} = average vertical frontal area per obstruction presented to the wind; and
A_{ob} = average area of ground occupied by each obstruction, including the open area surrounding it.

Vertical frontal area is defined as the area of the projection of the obstruction onto a vertical plane normal to the wind direction. The area S_{ob} may be estimated by summing the approximate vertical frontal areas of all obstructions within a selected area of upwind fetch and dividing the sum by the number of obstructions in the area. The average height, H_{ob}, may be estimated in a similar way by averaging the individual heights rather than using the frontal areas. Likewise, A_{ob} may be estimated by dividing the size of the selected area of upwind fetch by the number of obstructions in it.

As an example, if the upwind fetch consists primarily of single-family homes with typical height H_{ob} = 20 ft (6 m), vertical frontal area (including some trees on each lot) of 1,000 ft² (100 m²), and ground area per home of 10,000 ft² (1,000 m²), then z_0 is calculated to be $z_0 = 0.5 \times 20 \times 1,000/10,000 = 1$ ft (0.3 m), which falls into Exposure Category B, according to Table C26.7-1.

Trees and bushes are porous and are deformed by strong winds, which reduce their effective frontal areas (ESDU 1993). For conifers and other evergreens, no more than 50% of their gross frontal area can be taken to be effective in obstructing the wind. For deciduous trees and bushes, no more than 15% of their gross frontal area can be taken to be effective in obstructing the wind. *Gross frontal area* is defined in this context as the projection onto a vertical plane (normal to the wind) of the area enclosed by the envelope of the tree or bush.

Ho (1992) estimated that the majority of buildings (perhaps as much as 60–80%) have an exposure category corresponding to Exposure B. While the relatively simple definition in the standard normally suffices for most practical applications, the designer is often in need of additional information, particularly with regard to the effect of large openings or clearings (e.g., large parking lots, freeways, or tree clearings) in the otherwise "normal" ground Surface Roughness B. The following is offered as guidance for these situations:

1. The simple definition of Exposure B given in the body of the standard, using the surface roughness category definition, is shown pictorially in Fig. C26.7-1. This definition applies for the Surface Roughness B condition prevailing 2,630 ft (800 m) upwind with insufficient "open patches," as defined in the following procedure to disqualify the use of Exposure B. This procedure on the net effect of these open patches applies where the prevailing exposure beyond 2,600 ft (792 m) is Exposure B.
2. An open area in the Surface Roughness B large enough to have a significant effect on the exposure category determination is defined as an "open patch." To be considered an "open patch," an open area meets the following:
    a. Open areas should be greater than the minimum areas given by Fig. C26.7-4. Interpolation shall be used between the reference distances of 500 ft, 1,500 ft, and 2,600 ft (152, 457, and 790 m, respectively) to determine the intermediate minimum open patch area criteria.
    b. The open area shall have minimum dimensions given by conditions i, ii, or iii below and have length to width ratios between 0.5 and 2.0.
        i. Within 500 ft (152 m) of the building or structure, an open area greater than or equal to approximately 164 ft (50 m) in length or width.
        ii. At 1,500 ft (457 m) upwind from the building or structure, an open area greater than or equal to approximately 328 ft (100 m) in length or width.
        iii. At 2,600 ft (790 m) upwind from the building or structure, an open area greater than or equal to approximately 500 ft (152 m) in length or width.
3. Open patches separated by less than the along-wind dimension of the larger patch shall be treated as equivalent to a single open patch with length equal to the sum of the individual patch along-wind dimensions and width determined to provide an area equal to the sum of the individual open patch areas.
4. A circular sector is an area defined by a limiting radius from the center and an arc, in this case, 45° per Section 26.7.4. If the proportion of open patch within any of the sectors defined by the three radii above is less than 25% of the sector area, the sector is considered to meet the requirements for Exposure Category B. Where the proportion of open patches within any 45° sector within any of the three radii of 500 ft (152 m), 1,500 ft (457 m), or the greater of 2,600 ft (790 m) or 20 times the height of the building or structure exceeds 25% of the sector area but is not greater than 50%, the values of K_z are taken as the average of the Exposure B and C values within 100 ft (31 m) height above grade. Above 100 ft (31 m), Exposure B values of K_z shall still apply. Where the proportion of open patches within any of the sectors defined by the three radii of the building or structure exceeds 50%, the values of K_z shall be based on Exposure C.
5. The procedure for evaluation of the net effect of open patches of Surface Roughness C or D on the use of Exposure Category B is shown pictorially in Figs. C26.7-3 and C26.7-4. Note that the plan location of any open patch may have a different effect for different wind directions.

This above procedure is a simplification derived from a boundary-layer model, and therefore more exact results for the velocity profile may be achieved through direct use of an accepted boundary-layer model that is capable of addressing the effects of open areas within a regime defined by the surface roughness parameters given in Table C26.7-2.

Aerial photographs, representative of each exposure type, are included in Figs. C26.7-5 to C26.7-7 to aid the user in establishing the proper exposure for a given site. Obviously, the proper assessment of the exposure is a matter of good engineering judgment. This fact is particularly true in light of the possibility that the exposure could change in one or more wind directions due to future demolition and/or development.

Diagrams

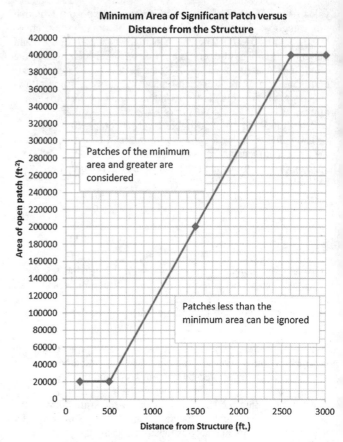

Notes

1. For each selected wind direction at which the wind loads are to be determined, the exposure of the building or structure shall be determined for the two upwind sectors extending 45° either side of the selected wind direction.
2. Consider open patches of sizes equal to or greater than the areas given in Figure C26.7-4 per Commentary Section C26.7.
3. Determine the proportion of open patches in any 45-degree sector within radii of 500 ft, 1,500 ft, or the greater of 2,600 ft or 20 times the height of the structure.
4. If the proportion of open patch within any of the three radii above is less than 25% of the sector area, the sector is considered to meet the requirements for Exposure B. Where the proportion within any of the three radii above exceeds 25% of the sector area but is not greater than 50%, the values of K_z are taken as the average of the Exposure B and C values within 100 ft height above grade. Above 100 ft, Exposure B values shall still apply. Where the proportion of open patches within any of the three radii of the structure exceeds 50%, the values of K_z shall be based on Exposure C.
5. Apply the exposure requirements of Section 26.7.4 once the directional exposures are determined for each sector. See Commentary Section C26.7.4.

FIGURE C26.7-3 Exposure B with Upwind Open Patches Sector Analysis

C26.7.4 Exposure Requirements. Section 26.5.1 of the standard requires that a structure be designed for winds from all directions. A rational procedure to determine directional wind loads is described here. Wind load for buildings using Section 27.3.1 and Figs. 27.3-1, 27.3-2, or 27.3-3 are determined for eight wind directions at 45° intervals, with four falling along primary building axes, as shown in Fig. C26.7-8. For each of the eight directions, upwind exposure is determined for each of two 45° sectors, one on each side of the wind direction axis. The sector with the exposure giving highest loads is used to define wind loads for that direction. For example, for winds from the north, the exposure from Sector 1 or 8, whichever gives the higher load, is used. For wind from the east, the exposure from Sector 2 or 3, whichever gives the highest load, is used. For wind coming from the northeast, the more exposed of Sectors 1 or 2 is used to determine full x and y loading individually, and then 75% of these loads are to be applied in each direction at the same time, according to the requirements of Section 27.3.5 and Fig. 27.3-8. The procedure defined in this section for determining wind loads in each design direction is not to be confused with the determination of the wind directionality factor, K_d. The K_d factor determined from Section 26.6 and Table 26.6-1 applies for all design wind directions. See Section C26.6.

C&C loads for all buildings and MWFRS loads for low-rise buildings are determined using the upwind exposure for the single surface roughness in one of the eight sectors of Fig. C26.7-8 that gives the highest wind loads.

FIGURE C26.7-4 Minimum Area of Individual Open Patches Affecting Qualification of Exposure B

C26.8 TOPOGRAPHIC EFFECTS

This section specifies when topographic effects need to be applied to a particular structure (Means et al. 1996). In an effort to exclude situations where little or no topographic effect exists, Condition 2 recognizes the fact that the topographic feature should protrude significantly above (by a factor of 2 or more) upwind terrain features before it becomes a factor. For example, if a significant upwind terrain feature has a height of 35 ft (10.7 m) above its base elevation and has a top elevation of 100 ft (30.5 m) above mean sea level, then the topographic feature (hill, ridge, or escarpment) must have at least the H specified and extend to elevation 170 ft (52 m) above mean sea level [100 ft + $(2 \times 35 \text{ ft})$] or $[31 \text{ m} + (2 \times 10 \text{ m})]$ in S.I. within the 2-mi (3.2-km) radius specified.

Condition 5 specifies a lower height, H, for consideration of topographic effects in Exposure C and D than for Exposure B (Means et al. 1996), and observation of actual wind damage has shown that the affected height, H, is less in Exposures C and D.

Buildings sited on the upper half of an isolated hill or escarpment may experience significantly higher wind speeds than buildings situated on level ground. The topographic feature (2D ridge or escarpment, or 3D axisymmetrical hill) is described

FIGURE C26.7-5(a) Exposure B: Suburban Residential Area with Mostly Single-Family Dwellings. Low-Rise Structures, Less Than 30 ft (9.1 m) High, in the Center of the Photograph Have Sites Designated as Exposure B with Surface Roughness Category B Terrain around the Site for a Distance Greater Than 1,500 ft (457 m) in Any Wind Direction

FIGURE C26.7-5(b) Exposure B: Urban Area with Numerous Closely Spaced Obstructions Having the Size of Single-Family Dwellings or Larger. For All Structures Shown, Terrain Representative of Surface Roughness Category B Extends More Than 20 Times the Height of the Structure or 2,600 ft (792 m), Whichever Is Greater, in the Upwind Direction

by two parameters, H and L_h. H is the height of the hill or difference in elevation between the crest and that of the upwind terrain. L_h is the distance upwind of the crest to where the ground elevation is equal to half the height of the hill. K_{zt} is determined from three multipliers, K_1, K_2, and K_3, which are obtained from Fig. 26.8-1. K_1 is related to the shape of the topographic feature; the maximum speed-up near the crest, K_2, accounts for the reduction in speed-up with distance upwind or downwind of the crest; and K_3 accounts for the reduction in speed-up with height above the local ground surface.

FIGURE C26.7-5(c) Exposure B: Structures in the Foreground Are Located in Exposure B. Structures in the Center Top of the Photograph Adjacent to the Clearing to the Left, Which Is Greater Than Approximately 656 ft (200 m) in Length, are Located in Exposure C When Wind Comes from the Left over the Clearing. (See Figure C26.7-4)

FIGURE C26.7-6(a) Exposure C: Flat Open Grassland with Scattered Obstructions Having Heights Generally Less Than 30 ft (9.1 m)

The multipliers listed in Fig. 26.8-1 are based on the assumption that the wind approaches the hill along the direction of maximum slope, causing the greatest speed-up near the crest. The average maximum upwind slope of the hill is approximately $H/2L_h$, and measurements have shown that hills with slopes of less than about 0.10 ($H/L_h < 0.20$) are unlikely to produce significant speed-up of the wind. For values of $H/L_h > 0.5$, the speed-up effect is assumed to be independent of slope. The

FIGURE C26.7-6(b) Exposure C: Open Terrain with Scattered Obstructions Having Heights Generally Less Than 30 ft (9.1 m). For Most Wind Directions, all One-Story Structures with a Mean Roof Height Less Than 30 ft (9.1 m) in the Photograph are Less Than 1,500 ft (457 m) or 10 Times the Height of the Structure, Whichever Is Greater, from an Open Field that Prevents the Use of Exposure B

FIGURE C26.7-7 Exposure D: A Building at the Shoreline (Excluding Shorelines in Hurricane-Prone Regions) with wind Flowing over Open Water for a Distance of at Least One Mile. Shorelines in Exposure D Include Inland Waterways, the Great Lakes, and Coastal Areas of California, Oregon, Washington, and Alaska

speed-up principally affects the mean wind speed rather than the amplitude of the turbulent fluctuations, and this fact has been accounted for in the values of K_1, K_2, and K_3 given in Fig. 26.8-1. Therefore, values of K_{zt} obtained from Fig. 26.8-1 are intended for use with velocity pressure exposure coefficients, K_h and K_z, which are based on gust speeds.

It is not the intent of Section 26.8 to address the general case of wind flow over hilly or complex terrain for which engineering

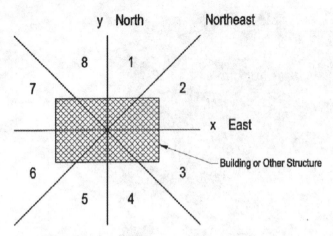

FIGURE C26.7-8 Determination of Wind Loads from Different Directions

judgment, expert advice, or the Wind Tunnel Procedure as described in Chapter 31 may be required. Background material on topographic speed-up effects may be found in the literature (Jackson and Hunt 1975; Lemelin et al. 1988; Walmsley et al. 1986).

The designer is cautioned that, at present, the standard contains no provision for vertical wind speed-up because of a topographic effect, even though this phenomenon is known to exist and can cause additional uplift on roofs. Additional research is required to quantify this effect before it can be incorporated into the standard.

C26.9 GROUND ELEVATION FACTOR

The ratio of air pressure and density at elevation z relative to the standard values at $z=0$, with constant temperature, is given by the barometric formula

$$p_z/p_0 = \rho_z/\rho_0 = e^{-gz/RT}$$

where

g = acceleration of gravity = 32.174 ft/s² (9.807 m/s²),
R = gas constant of air = 1,718 lb-ft/slug/°R (287 N-m/kg/K), and
T = absolute temperature = 518°R (288K).

With these values and elevation z = elevation of ground above sea level, the ratio is determined from the formulas given in Table 26.9-1 where $K_e = \rho_z/\rho_0$. For reference, a more complete version of Table 26.9-1 including air density values is provided in Table C26.9-1.

$K_e = 1.0$ is permitted in all cases. While this is somewhat unconservative for elevations below sea level, the committee believes it is reasonable to permit this since the effect is very small for all areas below sea level in the U.S. (0 to −300 ft in Death Valley, a maximum of 1 percent increase in air density), and is likely to be reduced even more due to higher average temperatures.

C26.10 VELOCITY PRESSURE

C26.10.1 Velocity Pressure Exposure Coefficient. The velocity pressure exposure coefficient K_z can be obtained using the equations:

$$K_z = 2.01 \left(\frac{z}{z_g} \right)^{2/\alpha} \quad \text{for } 15 \text{ ft} \leq z \leq z_g \quad \text{(C26.10-1)}$$

$$K_z = 2.01 \left(\frac{z}{z_g} \right)^{2/\alpha} \quad \text{for } 4.6 \text{ m} \leq z \leq z_g \quad \text{(C26.10-1.si)}$$

$$K_z = 2.01 \left(\frac{15}{z_g} \right)^{2/\alpha} \quad \text{for } z \leq 15 \text{ ft} \quad \text{(C26.10-2)}$$

$$K_{zu} = 2.01 \left(\frac{20}{Z_g} \right)^{2/\alpha} \quad \text{for } z \leq 4.6 \text{ m} \quad \text{(C26.10-2.si)}$$

in which values of α and z_g are given in Table 26.11-1. These equations are now given in Table 26.10-1 to aid the user.

The values of α given in Table 26.11-1 define gust profiles. The mean profiles implied by these α values are based on 4.5, 7.0, and 10 for Exposures B, C, and D, respectively. These have been in use for the underlying mean velocity provided for K_z since 1982. The $\bar{\alpha}$ values in Table 26.11-1 based on 4.0, 6.5, and 9.0 are used only to calculate the gust factor, G_f, for flexible structures in Section 26.11.5.

Changes were implemented in ASCE 7-98, including truncation of K_z values for Exposures A and B below heights of 100 ft (30.5 m) and 30 ft (9.1 m), respectively, applicable to Components and Cladding (C&C) and the Envelope Procedure. Exposure A was eliminated in the 2002 edition.

In the ASCE 7-05 standard, the K_z expressions were unchanged from ASCE 7-98. However, the possibility of interpolating between the standard exposures using a rational method was added in the ASCE 7-05 edition. One rational method is provided in the following text.

To a reasonable approximation, the empirical exponent α and gradient height z_g in the preceding expressions [Eqs. (C26.10-1) and (C26.10-2)] for exposure coefficient K_z may be related to the roughness length z_0 (where z_0 is defined in Section C26.7) by the relations

$$\alpha = c_1 z_0^{-0.133} \quad \text{(C26.10-3)}$$

and

$$z_g = c_2 z_0^{0.125} \quad \text{(C26.10-4)}$$

where

$c_1 = 5.65$ and $c_2 = 450$, when units of z_0 and z_g are m, and
$c_1 = 6.62$ and $c_2 = 1{,}273$, when units of z_0 and z_g are ft.

The preceding relationships are based on matching the ESDU boundary-layer model (ESDU 1982, 1993) empirically with the power law relationship in Eqs. (C26.10-1) and (C26.10-2), the

Table C26.9-1. Gound Elevation Factor including Air Density

Ground elevation, z_g		Air density, ρ		Ratio, K_e
ft	m	slug/ft³	kg/m³	ρ/ρ_0
−10,000	−305	0.000247	1.269	1.04
0	0	0.000238	1.224	1.00
1,000	305	0.000229	1.180	0.96
2,000	610	0.000221	1.138	0.93
3,000	914	0.000213	1.098	0.90
4,000	1,219	0.000206	1.059	0.86
5,000	1,524	0.000198	1.021	0.83
6,000	1,829	0.000191	0.985	0.80
7,000	2,134	0.000185	0.950	0.78
8,000	2,438	0.000178	0.916	0.75
9,000	2,743	0.000172	0.883	0.72
10,000	3,048	0.000166	0.852	0.70

ESDU model being applied at latitude 35° with a gradient wind of 168 mi/h (75 m/s). If z_0 has been determined for a particular upwind fetch, Eqs. (C26.10-1) through (C26.10-4) can be used to evaluate K_z. The correspondence between z_0 and the parameters α and z_g implied by these relationships does not align exactly with that described in the commentary to ASCE 7-95 and 7-98. However, the differences are relatively small and not of practical consequence. The ESDU boundary-layer model has also been used to derive the following simplified method (Irwin 2006) of evaluating K_z following a transition from one surface roughness to another. For more precise estimates, the reader is referred to the original ESDU model (Harris and Deaves 1981; ESDU 1990, 1993).

In uniform terrain, the wind travels a sufficient distance over the terrain for the planetary boundary layer to reach an equilibrium state. The exposure coefficient values in Table 26.11-1 are intended for this condition. Suppose that the site is a distance x miles downwind of a change in terrain. The equilibrium value of the exposure coefficient at height z for the terrain roughness downwind of the change will be denoted by K_{zd}, and the equilibrium value for the terrain roughness upwind of the change will be denoted by K_{zu}. The effect of the change in terrain roughness on the exposure coefficient at the site can be represented by adjusting K_{zd} by an increment ΔK, thus arriving at a corrected value K_z for the site:

$$K_z = K_{zd} + \Delta K \quad (C26.10\text{-}5)$$

In this expression, ΔK is calculated using

$$\Delta K = (K_{33,u} - K_{33,d}) \frac{K_{zd}}{K_{33,d}} F_{\Delta K}(x)$$

$$|\Delta K| \leq |K_{zu} - K_{zd}| \quad (C26.10\text{-}6)$$

where $K_{33,d}$ and $K_{33,u}$ are, respectively, the downwind and upwind equilibrium values of exposure coefficient at 33 ft (10 m) height, and the function $F_{\Delta K}(x)$ is given by

$$F_{\Delta K}(x) = \frac{\log_{10}(\frac{x_1}{x})}{\log_{10}(\frac{x_1}{x_0})} \quad (C26.10\text{-}7)$$

For $x_0 < x < x_1$:

$$F_{\Delta k}(x) = 1 \quad \text{for } x < x_0$$

$$F_{\Delta k}(x) = 0 \quad \text{for } x > x_1$$

In the preceding relationships,

$$x_0 = c_3 \times 10^{-(K_{33,d} - K_{33,u})^2 - 2.3} \quad (C26.10\text{-}8)$$

The constant $c_3 = 0.621$ mi (1.0 km). The length $x_1 = 6.21$ mi (10 km) for $K_{33,d} < K_{33,u}$ (wind going from smoother terrain upwind to rougher terrain downwind) or $x_1 = 62.1$ mi (100 km) for $K_{33,d} > K_{33,u}$ (wind going from rougher terrain upwind to smoother terrain downwind).

The above description is in terms of a single roughness change. The method can be extended to multiple roughness changes. The extension of the method is best described by an example. Fig. C26.10-1 shows wind with an initial profile characteristic of Exposure D encountering an expanse of B roughness, followed by a further expanse of D roughness and then some more B roughness again before it arrives at the building site. This situation is representative of wind from the sea flowing over an outer strip of land, then a coastal waterway, and then some suburban roughness before arriving at the building site. The above method for a single roughness change is first used to

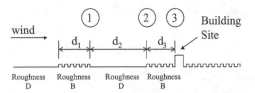

FIGURE C26.10-1 Multiple Roughness Changes Due to Coastal Waterway

compute the profile of K_z at Station 1 in Fig. C26.10-1. Call this profile $K_z^{(1)}$. The value of ΔK for the transition between Stations 1 and 2 is then determined using the equilibrium value of $K_{33,u}$ for the roughness immediately upwind of Station 1, i.e., as though the roughness upwind of Station 1 extended to infinity. This value of ΔK is then added to the equilibrium value $K_{zd}^{(2)}$ of the exposure coefficient for the roughness between Stations 1 and 2 to obtain the profile of K_z at Station 2, which we will call $K_z^{(2)}$. Note however, that the value of $K_z^{(2)}$ in this way cannot be any lower than $K_z^{(1)}$. The process is then repeated for the transition between Stations 2 and 3. Thus, ΔK for the transition from Station 2 to Station 3 is calculated using the value of $K_{33,u}$ for the equilibrium profile of the roughness immediately upwind of Station 2, and the value of $K_{33,d}$ for the equilibrium profile of the roughness downwind of Station 2. This value of ΔK is then added to $K_{zd}^{(2)}$ to obtain the profile $K_z^{(3)}$ at Station 3, with the limitation that the value of $K_z^{(3)}$ cannot be any higher than $K_z^{(2)}$.

Example 1: Single Roughness Change. Suppose that the building is 66 ft high, and its local surroundings are suburban with a roughness length $z_0 = 1$ ft. However, the site is 0.37 mi downwind of the edge of the suburbs, beyond which the open terrain is characteristic of open country with $z_0 = 0.066$ ft. From Eqs. (C26.10-1), (C26.10-3), and (C26.10-4), for the open terrain,

$$\alpha = c_1 z_0^{-0.133} = 6.62 \times 0.066^{-0.133} = 9.5$$

$$z_g = c_2 z_0^{0.125} = 1{,}273 \times 0.066^{0.125} = 906 \text{ ft}$$

Therefore, applying Eq. (C26.10-1) at 66 ft (20 m) and 33 ft (10 m) heights,

$$K_{zu} = 2.01 \left(\frac{66}{906}\right)^{2/9.5} = 1.16$$

and

$$K_{33,u} = 2.01 \left(\frac{33}{906}\right)^{2/9.5} = 1.00$$

Similarly, for the suburban terrain,

$$\alpha = c_1 z_0^{-0.133} = 6.62 \times 1.0^{-0.133} = 6.62$$

$$z_g = c_2 z_0^{0.125} = 1{,}273 \times 1.0^{0.125} = 1{,}273 \text{ ft}$$

Therefore,

$$K_{zd} = 2.01 \left(\frac{66}{1{,}273}\right)^{2/6.62} = 0.82$$

and

$$K_{33,d} = 2.01 \left(\frac{33}{1{,}273}\right)^{2/6.62} = 0.67$$

From Eq. (C26.10-8),

$$x_0 = c_3 \times 10^{-(K_{33,d}-K_{33,u})^2 - 2.3} = 0.621$$
$$\times 10^{-(0.62-1.0)^2 - 2.3} = 0.00241 \text{ mi}$$

From Eq. (C26.10-7),

$$F_{\Delta K}(x) = \frac{\log_{10}(6.21/0.36)}{\log_{10}(6.21/0.00241)} = 0.36$$

Therefore, from Eq. (C26.10-6),

$$\Delta K = (1.00 - 0.67)\frac{0.82}{0.67}0.36 = 0.15$$

Note that because $|\Delta K|$ is 0.15, which is less than the 0.38 value of $\Delta|K_{33,u} - K_{33,d}|$, 0.15 is retained. Finally, from Eq. (C26.10-5), the value of K_z is

$$K_z = K_{zd} + \Delta K = 0.82 + 0.15 = 0.97$$

Because the value 0.97 for K_z lies between the values 0.88 and 1.16, which would be derived from Table 26.11-1 for Exposures B and C, respectively, it is an acceptable interpolation. If it falls below the Exposure B value, then the Exposure B value of K_z is to be used. The value $K_z = 0.97$ may be compared with the value 1.16 that would be required by the simple 2,600-ft fetch length requirement of Section 26.7.3.

The most common case of a single roughness change where an interpolated value of K_z is needed is for the transition from Exposure C to Exposure B, as in the example just described. For this particular transition, using the typical values of z_0 of 0.066 ft and 1.0 ft, the preceding formulas can be simplified to

$$K_z = K_{zd}\left(1 + 0.146\log_{10}\left(\frac{6.21}{x}\right)\right)$$
$$K_{zB} \leq K_z \leq K_{zC} \quad \text{(C26.10-9)}$$

where x is in miles, and K_{zd} is computed using $\alpha = 6.62$. K_{zB} and K_{zC} are the exposure coefficients in the standard Exposures B and C, respectively. Fig. C26.10-2 illustrates the transition from terrain roughness C to terrain roughness B from this expression. Note that it is acceptable to use the typical z_0 rather than the lower limit for Exposure B in deriving this

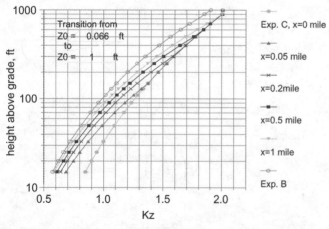

FIGURE C26.10-2 Transition from Terrain Roughness C to Terrain Roughness B, Eq. (26.10-9)

Table C26.10-1 Tabulated Exposure Coefficients

Transition from sea to Station 1	$K_{33,u}$	$K_{33,d}$	$K_{50,d}$	$F_{\Delta K}$	ΔK_{50}	$K_{50}^{(1)}$
	1.215	0.667	0.758	0.220	0.137	0.895
Transition from Station 1 to Station 2	$K_{33,u}$	$K_{33,d}$	$K_{50,d}$	$F_{\Delta K}$	ΔK_{50}	$K_{50}^{(2)}$
	0.667	1.215	1.301	0.324	−0.190	1.111
Transition from Station 2 to Station 3	$K_{33,u}$	$K_{33,d}$	$K_{50,d}$	$F_{\Delta K}$	ΔK_{50}	$K_{50}^{(3)}$
	1.215	0.667	0.758	0.498	0.310	1.067

Note: The equilibrium values of the exposure coefficients, $K_{33,u}$, $K_{33,d}$ and $K_{50,d}$ (downwind value of K_z at 50 ft), were calculated from Eq. (C26.10-1) using α and z_g values obtained from Eqs. (C26.10-3) and (C26.10-4) with the roughness values given. Then $F_{\Delta K}$ is calculated using Eqs. (C26.10-7) and (C26.10-8), and then the value of ΔK at 50-ft height, ΔK_{50}, is calculated from Eq. (C26.10-6). Finally, the exposure coefficient at 50 ft at Station i, $K_{50}^{(i)}$, is obtained from Eq. (C26.10-5).

formula because the rate of transition of the wind profiles is dependent on average roughness over significant distances, not local roughness anomalies. The potential effects of local roughness anomalies, such as parking lots and playing fields, are covered by using the standard Exposure B value of exposure coefficient, K_{zB}, as a lower limit to the calculated value of K_z.

Example 2: Multiple Roughness Change. Suppose we have a coastal waterway situation, as illustrated in Fig. C26.10-1, where the wind comes from open sea with roughness type D, for which we assume $z_0 = 0.01$ ft and passes over a strip of land 1 mi wide, which is covered in buildings that produce typical B type roughness, i.e., $z_0 = 1$ ft. It then passes over a 2-mi-wide strip of coastal waterway where the roughness is again characterized by the open water value $z_0 = 0.01$ ft. It then travels over 0.1 mi of roughness type B ($z_0 = 1$ ft) before arriving at the site, Station 3 in Fig. C26.10-1, where the exposure coefficient is required at the 50-ft height. The exposure coefficient at Station 3 at a 50-ft height is calculated as shown in Table C26.10-1.

The value of the exposure coefficient at 50 ft at Station 3 is seen from the table to be 1.067. This is above that for Exposure B, which would be 0.81, but well below that for Exposure D, which would be 1.27, and similar to that for Exposure C, which would be 1.09.

C26.10.2 Velocity Pressure. The basic wind speed is converted to velocity pressure q_z in lb/ft² (N/m²) at height z by the use of Eq. (26.10-1).

The constant in this equation reflects the mass density of air for the standard atmosphere, that is, temperature of 59°F (15°C) and sea level pressure of 29.92 in. of mercury (101.325 kPa), equal to 0.0765 lbm/ft³ or 0.002378 slug/ft³ or 0.002378 lb-s²/ft⁴ (1.225 kg/m³), and dimensions associated with wind speed in mi/h (m/s). The constant is obtained as follows:

Dynamic pressure from Bernoulli's law:

$$p = \tfrac{1}{2}\rho V^2$$

with V in mi/h:

$$p = \tfrac{1}{2}(0.002378 \text{ lb s}^2/\text{ft}^4)[V \text{ mi/h}(88 \text{ ft/s}/60 \text{ mi/h})^2]$$
$$= 0.00256 V^2 \text{ lb/ft}^2$$

with V in m/s:

$$p = \tfrac{1}{2}(1.225 \text{ kg/m}^3)(1 \text{ N}/1 \text{ kg m/s}^2)(V \text{ m/s})^2$$
$$= 0.613 V^2 \text{ N/m}^2 \quad \text{(si)}$$

Values of air density other than the standard-atmosphere values above may be adjusted using the factor K_e as described in Section C26.9.

Table C26.11-1. Gust-Effect Factor, Example

Item	Value	Source
Example Calculation—Gust-Effect Factors		
DEFAULT FACTOR FOR RIGID BUILDING (Requires $n_1 \geq 1$)		
G Gust-effect factor	0.85	26.9.1
ALTERNATE CALCULATED FACTOR FOR RIGID BUILDING (requires $n_1 \geq 1$)		
h Mean roof height	600 ft (183 m)	User spec
B Width normal to wind	100 ft (30 m)	User spec
D Depth parallel to wind	100 ft (30 m)	User spec
$\bar{z}$ Effective structure height	360 ft (110 m)	$0.6h$ (26.9.4)
Exposure category	B	
c Turbulence intensity at 10 m	0.3	Table 26.9-1
$I_{\bar{z}}$ Turbulence intensity at eff. height	0.201	(26.9-7)
l Turbulence length scale at 10 m	320 ft (98 m)	Table 26.9-1
$\bar{\epsilon}$ Power law exponent of turbulent length scale profile	1/3	Table 26.9-1
$L_{\bar{z}}$ Turbulence length scale at eff. height	710 ft (216 m)	(26.9-9)
Q^2 Background response (squared)	0.616	(26.9-8)
g_Q Background load peak factor	3.4	26.9.4
g_v Velocity peak factor	3.4	26.9.4
G Calculated gust-effect factor	0.818	(26.9-6)
ADDITIONAL CALCULATIONS FOR FLEXIBLE BUILDING (all n_1)		
V Basic wind speed	115 mph (51 m/s)	
n_1 Fundamental natural frequency in direction of wind	0.2 Hz	Analysis or rational approximation
β Damping ratio	0.01	Rational assignment
$\bar{\alpha}$ Power law exponent of mean wind speed profile	0.25	Table 26.9-1
$\bar{b}$ Gust factor 1/F at 10 m	0.45	Table 26.9-1
$\bar{V}_{\bar{z}}$ Mean wind speed at effective height	138 ft/s (42.1 m/s)	26.9-16
N_1 Reduced natural frequency	1.029	26.9-14
R_n Resonance response factor for n	0.129	26.9-13
Example Calculation—Gust-Effect Factors (*Continued*)		
η_h Vertical decay parameter	4.002	$26.9.5: 4.6n_1h/\bar{V}_{\bar{z}}$
η_B Cross-wind decay parameter	0.667	$26.9.5: 4.6n_1B/\bar{V}_{\bar{z}}$
η_L Along-wind decay parameter	2.233	$26.9.5: 15.4n_1L/\bar{V}_{\bar{z}}$
R_h Resonant factor for h	0.219	(26.9-15a)
R_B Resonant factor for B	0.671	(26.9-15a)
R_L Resonant factor for L	0.349	(26.9-15a)
R^2 Resonant response (squared)	1.313	(26.9-12)
g_R Resonant peak factor	3.787	(26.9-11)
G_f Gust-effect factor	1.173	(26.9-10)

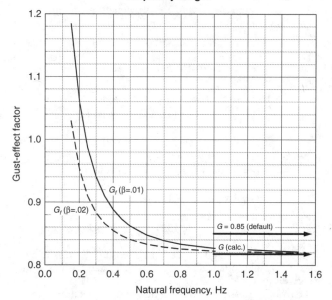

Gust-effect factor relations for example building in extended frequency range:

C26.11 GUST EFFECTS

This standard specifies a single, conservative, gust-effect factor of 0.85 for rigid buildings. As an option, the designer can incorporate specific features of the wind environment and building size to more accurately calculate an alternate but more accurate gust-effect factor that accounts for the decorrelation of wind gusts over the size of the structure. One such procedure is located in the body of the standard (Solari and Kareem 1998). Neither of these factors accounts for dynamic amplification caused by vibration of the structure, but they are considered acceptable for rigid structures as defined in the standard. The alternate calculated gust factor is 5–10% lower than the value of 0.85 permitted in the standard without calculation.

A third gust-effect factor, G_f, is provided for flexible buildings and structures that do not meet the requirements of rigid structures that the fundamental natural frequency, n_1, is greater than or equal to 1 Hz. This factor also accounts for the building size and gust size in the same manner as the alternate calculated factor for rigid buildings, but it also accounts for dynamic amplification caused by the design wind speed, the fundamental natural frequency of vibration, and the damping ratio.

Example: Calculation of the gust-effect factors for a subject building is demonstrated in Table C26.11-1. The frequency-dependent relationship among all factors is illustrated in the graph at the end of this table. The flexible factor, G_f, may be used for all cases but is required when $n_1 < 1$. This factor gradually approaches the alternate calculated factor for rigid cases, G, as the natural frequency exceeds 1, especially for higher levels of damping, but it always exceeds G. The difference is deemed negligible for n_1 greater than 1, so G, which is considerably simpler to calculate, is offered as an acceptable alternative. The default value of $G = 0.85$, which requires no calculation, is offered as an even more convenient alternative when $n_1 > 1$, if the greater conservatism is acceptable to the designer. In addition, the default value results in a large abrupt change in the gust-effect value for cases that have a natural frequency close to 1, which may be awkward for a designer to reconcile. A designer is free to use any other rational procedure in the approved literature, as stated in Section 26.11.5.

The gust-effect factors account for loading effects in the along-wind direction caused by wind turbulence–structure interaction. They do not include allowances for across-wind loading effects, vortex shedding, instability caused by galloping or flutter, or amplification of aerodynamic torsion caused by building vibration in a pure torsional mode. For structures susceptible to loading effects that are not accounted for in the gust-effect factor, information should be obtained from recognized literature (Kareem 1992, 1985; Gurley and Kareem 1993; Solari 1993a, b; Zhou et al. 2002; Chen and Kareem 2004; Bernardini et al. 2013a) or from wind tunnel tests.

Along-Wind Response. The maximum along-wind displacement response can be approximated by a static analysis of the structure under the action of loads multiplied by the appropriate gust-effect factor, as defined in the standard. Such displacements are based on the static elastic curve of the structure and are reasonably accurate when the resonant response is small compared to the mean and background responses. For highly flexible structures, where the response is dominated by resonance, more accurate values, including variation with height and dynamic responses such as acceleration, can be calculated as described in the following sections. These response components are needed for strength and serviceability limit states.

The maximum along-wind displacement as a function of height above the ground surface is given by

$$X_{max}(z) = \frac{\phi(z)\rho B h C_{fx} \hat{V}_{\bar{z}}^2}{2m_1(2\pi n_1)^2} K G_f \qquad (C26.11\text{-}1)$$

where

$\phi(z)$ = fundamental model shape, $\phi(z) = (z/h)^\xi$;
ξ = mode shape power-law exponent;
ρ = air density;
C_{fx} = mean along-wind force coefficient;
m_1 modal mass = $\int_0^h \mu(z)\phi^2(z)dz$;
n_1 = fundamental natural frequency;
$\mu(z)$ = mass per unit height; $K = (1.65)^{\hat{\alpha}}/(\hat{\alpha} + \xi + 1)$; and
$\hat{V}_{\bar{z}}$ = 3-s gust speed at height $\bar{z}$.

$$\hat{V}_{\bar{z}} = \hat{b}(z/33)^{\hat{\alpha}} V \left(\frac{88}{60}\right)$$

where V is the 3-s gust speed in Exposure C (mi/h) at the reference height (obtained from Figs. 26.5-1 and 26.5-2); $\hat{b}$ and $\hat{\alpha}$ are given in Table 26.11-1.

The root-mean-square (rms) along-wind acceleration $\sigma_{\ddot{x}}(z)$ as a function of height above the ground surface is given by

$$\sigma_{\ddot{x}}(z) = \frac{0.85\phi(z)\rho B h C_{fx} \bar{V}_{\bar{z}}^2}{m_1} I_{\bar{z}} K R \qquad (C26.11\text{-}2)$$

where $\bar{V}_{\bar{z}}$ is the mean hourly wind speed at height $\bar{z}$, ft/s.

$$\bar{V}_{\bar{z}} = \bar{b}\left(\frac{\bar{z}}{33}\right)^{\bar{\alpha}} V \left(\frac{88}{60}\right) \qquad (C26.11\text{-}3)$$

where $\bar{b}$ and $\bar{\alpha}$ are defined in Table 26.11-1.

The maximum along-wind acceleration as a function of height above the ground surface is given by

$$\ddot{X}_{max}(z) = g_{\ddot{x}}\sigma_{\ddot{x}}(z) \qquad (C26.11\text{-}4)$$

$$g_{\ddot{x}} = \sqrt{2\ln(n_1 T)} + \frac{0.5772}{\sqrt{2\ln(n_1 T)}} \qquad (C26.11\text{-}5)$$

where T = length of time over which the minimum acceleration is computed, usually taken to be 3,600 s to represent 1 h.

Example calculations of maximum along-wind displacement, rms along-wind acceleration, and maximum along-wind acceleration are given in Table C26.11-2.

Approximate Fundamental Frequency. To estimate the dynamic response of structures, knowledge of the fundamental frequency (lowest natural frequency) of the structure is essential. This value would also assist in determining whether the dynamic

Table C26.11-2. Example Calculation

Example Calculation – Along-wind Response
See Table C26.9-1 for additional items not shown

Item	Value	Source
V Basic wind speed	115 mph	Figure 26.5-1A
ρ Air density	0.0024 slug/ft^3	Site elevation near sea level
C_{pw} External pressure coefficient, windward wall	0.8	Figure 27.4-1
C_{pw} External pressure coefficient, leeward wall	−0.8	Figure 27.4-1
C_{fx} Along-wind force coefficient	1.3	
ξ Mode shape power law exponent	1	Analysis of structure
$\hat{\alpha}$ Peak velocity power law exponent	1/7	Table 26.9-1
$\hat{b}$ Velocity profile parameter	0.84	Table 26.9-1
$\hat{V}_{\bar{z}}$ 3-s gust velocity at height $\bar{z}$	199 ft/s	(C26.9-1a)
K Modal load parameter	0.501	
ρ_b Building density	12 lbm/ft^3, 0.3727 slug/ft^3	Building design
μ Building mass per unit height	3727 slug/ft	
m_1 Modal mass	745,400 slug	
MAXIMUM ALONG-WIND DISPLACEMENT AT BUILDING TOP		
$\phi(h)$ Mode shape at $z = h$	1.0	
$X_{max}(h)$	1.86 ft	(C26.9-1)
RMS ALONG-WIND ACCELERATION AT BUILDING TOP		
V Basic wind speed, 10-yr MRI	76 mph	Figure CC-1
$\bar{V}_{\bar{z}}$ Mean velocity at height $\bar{z}$	91.2 ft/s	(C26.9-3)
R Resonant response factor	0.755	(26.9-12)
$\sigma_{\ddot{x}}(h)$	0.135 ft/s^2	(C26.9-2)
MAXIMUM ALONG-WIND ACCELERATION AT BUILDING TOP		
T Time period for maximum	3600 s	Traditional
$g_{\ddot{x}}$ Peak factor	3.79	(C26.9-5)
$\ddot{X}_{max}(h)$	0.512 ft/s^2	(C26.9-4)

response estimates are necessary. Most computer codes used in the analysis of structures would provide estimates of the natural frequencies of the structure being analyzed. For the preliminary design stages, some empirical relationships for building period T_a ($T_a = 1/n_1$) are available in the earthquake-related chapters of this standard. However, these expressions are based on recommendations for earthquake design with inherent bias toward higher estimates of fundamental frequencies (Goel and Chopra 1997, 1998). For wind design applications, these values may be unconservative because an estimated frequency higher than the actual frequency would yield lower values of the gust-effect factor and thus a lower design wind pressure. However, Goel and Chopra (1997, 1998) also cite lower bound estimates of frequency that are more suited for use in wind applications and are now given in Section 26.11.2; graphs of these expressions are shown in Fig. C26.11-1.

Because these expressions are based on regular buildings, limitations based on height and slenderness are required. The effective length, L_{eff}, uses a height-weighted average of the along-wind length of the building for slenderness evaluation. The top portion of the building is most important; hence, the height-weighted average is appropriate. This method is an appropriate first-order equation for addressing buildings with setbacks. Explicit calculation of the gust-effect factor per the other methods given in Section 26.11 can still be performed.

Observations from wind tunnel testing of buildings where frequency is calculated using analysis software show that the

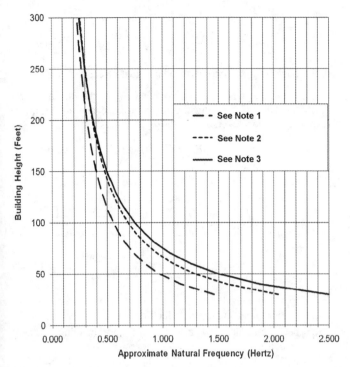

Notes

1. Eq. (26.11-2): $22.2/h^{0.8}$
2. Eq. (26.11-3): $43.5/h^{0.9}$
3. Eq. (26.11-4): $75/h$

FIGURE C26.11-1 Equations for Approximate Lower-bound Natural Frequency n_a versus Building Height

following expression for frequency can be used for steel and concrete buildings less than about 400 ft (122 m) in height:

$$n_1 = 100/H \text{(ft) average value} \quad (C26.11\text{-}6)$$

$$n_a = 75/H \text{(ft) lower bound value} \quad (C26.11\text{-}7)$$

Eq. (C26.11-7) for the lower bound value is provided in Section 26.11.3.

Based on full-scale measurements of buildings under the action of wind, the following expression has been proposed for wind applications (Zhou and Kareem 2001a; Zhou et al. 2002):

$$f_{n1} = 150/h \text{ (ft)} \quad (C26.11\text{-}8)$$

This frequency expression is based on older buildings and overestimates the frequency common in U.S. construction for smaller buildings less than 400 ft (122 m) in height, but it becomes more accurate for tall buildings greater than 400 ft (122 m) in height.

Studies in Japan involving a suite of buildings under low-amplitude excitations have led to the following expressions for natural frequencies of buildings (Sataka et al. 2003):

$$n_1 = 220/h \text{ (ft) (concrete buildings)} \quad (C26.11\text{-}9)$$

$$n_1 = 164/h \text{ (ft) (steel buildings)} \quad (C26.11\text{-}10)$$

These expressions result in higher frequency estimates than those obtained from the general expression given in Eqs. (C26.11-6) through (C26.11-8), particularly since the Japanese data set has limited observations for the more flexible buildings sensitive to wind effects, and Japanese construction tends to be stiffer.

For cantilevered masts or poles of uniform cross section (in which bending action dominates)

$$n_1 = (0.56/h^2)\sqrt{(EI/m)}$$

where EI is the bending stiffness of the section and m is the mass/unit height. This formula may be used for masts with a slight taper, using average value of EI and m (ECCS 1978).

An approximate formula for cantilevered, *tapered*, circular poles (ECCS 1978) is

$$n_1 \approx [\lambda/(2\pi h^2)]\sqrt{(EI/m)} \quad (C26.11\text{-}12)$$

where h is the height of the pole, and E, I, and m are calculated for the cross section at the base. λ depends on the wall thicknesses at the tip and base, e_t and e_b, and external diameter at the tip and base, d_t and d_b, according to the following formula:

$$\lambda = \left[1.9\exp\left(\frac{-4d_t}{d_b}\right)\right] + \left[\frac{6.65}{0.9 + (e_t/e_b)^{0.666}}\right] \quad (C26.11\text{-}13)$$

Eq. (C26.11-12) reduces to Eq. (C26.11-11) for uniform masts. For freestanding lattice towers (without added ancillaries, such as antennas or lighting frames) (Standards Australia 1994):

$$n_1 \approx 1{,}500 w_a/h^2 \quad (C26.11\text{-}14)$$

where w_a is the average width of the structure in meters and h is tower height. An alternative formula for lattice towers (with added ancillaries) (Wyatt 1984) is

$$n_1 = \left(\frac{L_N}{h}\right)^{2/3}\left(\frac{w_b}{h}\right)^{1/2} \quad (C26.11\text{-}15)$$

where w_b = tower base width and $L_N = 270$ m for square base towers, or 230 m for triangular base towers.

Structural Damping. Structural damping is a measure of energy dissipation in a vibrating structure that results in bringing the structure to a quiescent state. The damping is defined as the ratio of the energy dissipated in one oscillation cycle to the maximum amount of energy in the structure in that cycle. There are as many structural damping mechanisms as there are modes of converting mechanical energy into heat. The most important mechanisms are material damping and interfacial damping.

In engineering practice, the damping mechanism is often approximated as viscous damping because it leads to a linear equation of motion. This damping measure, in terms of the damping ratio, is usually assigned based on the construction material, for example, steel or concrete. The calculation of dynamic load effects requires damping ratio as an input. In wind applications, damping ratios of 1% and 2% are typically used in the United States for steel and concrete buildings at serviceability levels, respectively, while ISO (1997) suggests 1% and 1.5% for steel and concrete, respectively. Damping ratios for buildings under ultimate strength design conditions may be significantly higher, and 2.5% to 3% is commonly assumed. Damping values for steel support structures for signs, chimneys, and towers may be much lower than buildings and may fall in the range of 0.15–0.5%. Damping values of special structures like steel stacks can be as low as 0.2–0.6% and 0.3–1.0% for unlined and lined steel chimneys, respectively (ASME 1992; CICIND 1999). These values may provide some guidance for design. Damping levels used in wind load applications are smaller than the 5% damping ratios common in seismic applications because buildings subjected to wind loads respond essentially elastically, whereas buildings subjected to design-level earthquakes respond inelastically at higher damping levels.

Because the level of structural response in the strength and serviceability limit states is different, the damping values associated with these states may differ. Furthermore, because of the number of mechanisms responsible for damping, the limited full-scale data manifest a dependence on factors such as material, height, and type of structural system and foundation (Kijewski-Correa et al. 2013). The Committee on Damping of the Architectural Institute of Japan suggests different damping values for these states based on a large damping database described in Sataka et al. (2003).

Recently, the NatHaz group has developed an interactive database of full-scale experimentally determined modal damping ratios based on the database (Kareem et al. 2012). The database is publicly available at https://vortex-winds.org and is equipped with a query-based Web interface for the rapid identification of modal damping ratios that satisfy specific requirements, such as geometric form, structural system, construction material, foundation type, and building use. A recent publication offers a data-driven model of damping that has been validated with several full-scale studies (Spence and Kareem 2014).

In addition to structural damping, aerodynamic damping may be experienced by a structure oscillating in air. In general, the aerodynamic damping contribution is quite small compared to the structural damping, and it is positive in low to moderate wind speeds. Depending on the structural shape, at some wind velocities the aerodynamic damping may become negative, which can lead to unstable oscillations. In these cases, reference should be made to recognized literature or a wind tunnel study.

Alternate Procedure to Calculate Wind Loads. The concept of the gust-effect factor implies that the effect of gusts can be adequately accounted for by multiplying the mean wind load distribution with height by a single factor. This is an approximation. If a more accurate representation of gust effects is required, the alternative procedure in this section can be used. It takes account of the fact that the inertial forces created by the building's mass, as it moves under wind action, have a different distribution with height than the mean wind loads or the loads caused by the direct actions of gusts (Zhou and Kareem 2001a; Chen and Kareem 2004). The alternate formulation of the equivalent static load distribution uses the peak base bending moment and expresses it in terms of inertial forces at different building levels. A base bending moment, instead of the base shear as in earthquake engineering, is used for the wind loads because it is less sensitive to deviations from a linear mode shape. For a more detailed discussion on this wind-loading procedure, see Zhou and Kareem (2001a, 2001b) and Chen and Kareem (2004).

Alternate Procedure: Along-Wind Equivalent Static Wind Loading. The equivalent static wind loading for the mean, background, and resonant components is obtained using the procedure outlined in this section.

Mean wind load component, $\overline{P}_j$, is given by

$$\overline{P}_j = q_j \times C_p \times A_j \times \overline{G} \qquad \text{(C26.11-16)}$$

where

j = floor level;
q_j = velocity pressure at height z_j;
z_j = height of the jth floor above the ground level;
C_p = external pressure coefficient; and
$\overline{G} = 0.925 \times (1 + 1.7 g_v I_{\bar{z}})^{-1}$ = gust velocity factor.

Peak background wind load component, $\hat{P}_{Bj}$, at the jth floor level is given similarly by

$$\hat{P}_{Bj} = \overline{P}_j \times G_B/\overline{G} \qquad \text{(C26.11-17)}$$

where

$$G_B = 0.925 \times \left(\frac{1.7 I_{\bar{z}} \times g_Q Q}{1 + 1.7 g_v I_{\bar{z}}}\right)$$

is the background component of the gust-effect factor.

Peak resonant wind load component, $\hat{P}_{Rj}$, at the jth floor level is obtained by distributing the resonant base bending moment response to each level:

$$\hat{P}_{Rj} = C_{Mj} \hat{M}_R \qquad \text{(C26.11-18)}$$

$$C_{Mj} = \frac{w_j \phi_j}{\sum w_j \phi_j z_j} \qquad \text{(C26.11-19)}$$

$$\hat{M}_R = \overline{M} \cdot G_R/\overline{G} \qquad \text{(C26.11-20)}$$

$$\overline{M} = \sum_{j=1,n} \overline{P}_j \cdot z_j \qquad \text{(C26.11-21)}$$

where

C_{Mj} = vertical load distribution factor;
$\hat{M}_R$ = peak resonant component of the base bending moment response;
w_j = portion of the total gravity load of the building located or assigned to level j;
ϕ_j = first structural mode shape value at level j;
$\overline{M}$ = mean base bending produced by mean wind load;
$G_R = 0.925 \times [(1.7 I_{\bar{z}} \times g_R R)/(1 + 1.7 g_v I_{\bar{z}})]$ resonant component of the gust-effect factor; and
n = total stories of the building.

Alternate Procedure: Along-Wind Response. Through a simple static analysis, the peak building response in the along-wind direction can be obtained by

$$\hat{r} = \overline{r} + \sqrt{\hat{r}_B^2 + \hat{r}_R^2} \qquad \text{(C26.11-22)}$$

where $\overline{r}$, $\hat{r}_B$, and $\hat{r}_R$ = mean, peak background, and resonant response components of interest, for example, shear forces, moment, or displacement. Once the equivalent static wind load distribution is obtained, any response component, including acceleration, can be obtained using a simple static analysis. It is suggested that caution must be exercised when combining the loads instead of response according to the preceding expression, for example:

$$\hat{P}_j = \overline{P}_j + \sqrt{\hat{P}_{Bj}^2 + \hat{P}_{Rj}^2} \qquad \text{(C26.11-23)}$$

because the background and the resonant load components have normally different distributions along the building height. Additional background can be found in Zhou and Kareem (2001b); Zhou et al. (2002), and Chen and Kareem (2004).

Aerodynamic Loads on Tall Buildings: An Interactive Database. Under the action of wind, tall buildings oscillate simultaneously in the along-wind, across-wind, and torsional directions. While the along-wind loads have been successfully treated in terms of gust loading factors based on quasi-steady and strip theories, the across-wind and torsional loads cannot be treated in this manner because these loads cannot be related in a straightforward manner to fluctuations in the approach flow. As a result, most current codes and standards provide little guidance for the across-wind and torsional response (Zhou et al. 2002; Chen and Kareem 2004; Kwon and Kareem 2013; Bernardini et al. 2013).

To provide some guidance at the preliminary design stages of buildings, an interactive aerodynamic loads database for

Table C26.11-3. Along-Wind, Across-Wind, Torsional Moments, and Acceleration Response

	Survivability Design				Serviceability Design				
	Aerodynamic load coefficient			Base Moments (10^6 kips-ft)	Aerodynamic load coefficient		Accelerations (milli-g or rad/s²)		
								Corner	
Load Components	σ_{CM}	f_1	$C_M(f_1)$	M	f_1	$C_M(f_1)$	σ_a	X	Y
ASCE 7-10	–	–	–	1.73	–	–	1.95	2.77	3.24
Along-wind	0.109	0.193	0.046	1.72	0.292	0.022	2.03		
Across-wind	0.133	0.193	0.093	1.82	0.292	0.024	2.64		
Torsional	0.044	0.337	0.040	0.086	0.512	0.043	0.0001		

Note: As this database is experimental in nature thus has limitation in scope, it can be conveniently expanded using additional data as it becomes available.

assessing dynamic wind-induced loads on a suite of generic isolated buildings is introduced (Zhou et al. 2002; Kwon et al. 2008). Although the analysis based on this experimental database is not intended to replace wind tunnel testing in the final design stages, it provides users a methodology to approximate the previously untreated across-wind and torsional responses in the early design stages. The database consists of high-frequency base balance measurements involving seven rectangular building models, with side ratio (D/B, where D is the depth of the building section along the oncoming wind direction) from 1/3 to 3, and three aspect ratios for each building model in two approach flows, namely, BL_1 ($\bar{\alpha} = 0.16$) and BL_2 ($\bar{\alpha} = 0.35$) corresponding to an open and an urban environment, respectively. The data are accessible with a user-friendly, Java-based Internet applet, the NatHaz Aerodynamic Loads Database, version 2.0 (NALD 2012). Through the use of this interactive portal, users can select the geometry and dimensions of a model building from the available choices and specify an urban or suburban condition. Upon doing so, the aerodynamic load spectra for the along-wind, across-wind, or torsional directions are displayed with a Java interface permitting users to specify a reduced frequency (building frequency × building dimension/wind velocity) of interest and automatically obtain the corresponding spectral value. When coupled with the supporting Web documentation, examples, and concise analysis procedure, the database provides a comprehensive tool for computation of wind-induced response of tall buildings, suitable as a design guide in the preliminary stages.

Example: An example tall building is used to demonstrate the analysis using the database. The building is a square steel tall building with size $H \times W_1 \times W_2 = 656 \times 131 \times 131$ ft (200× 40 × 40 m) and an average radius of gyration of 59 ft (18 m).

The three fundamental mode frequencies, f_1, are 0.2, 0.2, and 0.35 Hz in X, Y, and Z directions, respectively; the mode shapes are all linear, or β is equal to 1.0, and there is no modal coupling. The building density is equal to 0.485 slugs/ft³ (250 kg/m³). This building is located in Exposure A or close to the BL_2 test condition of the Internet-based database (Zhou et al. 2002; Kwon et al. 2008). In this location, the 3-s design gust speed at a 700-year recurrence interval is 115 mi/h (51 m/s) in ASCE 7-16. For serviceability requirements, 3-s design gust speed with 10-year MRI is equal to 76 mi/h (34 m/s) in ASCE 7-16. For the sake of illustration only, the first-mode critical structural damping ratio, ζ_1, is to be 0.01 for both survivability and serviceability design.

Using these aerodynamic data and the procedures provided in Zhou et al. (2002) and Kwon et al. (2008), the wind-load effects are evaluated and the results are presented in Table C26.11-3. This table includes base moments and acceleration response in the along-wind direction obtained by the procedure in ASCE 7-16. It should be pointed out that the building experiences higher across-wind load effects when compared to the along-wind response for this example, which reiterates the significance of wind loads and their effects in the across-wind direction and the need for such database-enabled design approaches.

C26.12 ENCLOSURE CLASSIFICATION

The magnitude and sense of internal pressure are dependent upon the magnitude and location of openings around the building envelope with respect to a given wind direction. Accordingly, the standard requires that a determination be made of the amount of openings in the envelope to assess enclosure classification (enclosed, partially enclosed, or open). *Openings* are specifically defined in this version of the standard as apertures or holes in the building envelope which allow air to flow through the building envelope and which are designed as "open" during design winds. Examples include doors, operable windows, air intake exhausts for air conditioning and/or ventilation systems, gaps around doors, deliberate gaps in cladding, and flexible and operable louvers. The porosity for an "enclosed building" is such that there are not sufficient openings in the exterior building envelope to allow significant air flow into the building. The porosity of a "partially enclosed" building is such that there are sufficient openings in the building envelope windward wall to allow for wind to enter the building; however, there are not sufficient openings in the remaining portions of the building envelope to allow air flow out of the building without a buildup of internal pressure. The porosity for a "partially open" building is such that there exist sufficient openings in the building envelope windward wall to allow for air flow into the building and sufficient openings exist in the remaining portions of the building envelope to allow for some air flow out of the building but with some buildup of internal pressure. The porosity for the "open building" is such that air flow can enter and exit the building without a significant buildup of internal pressure. The classification of a "partially open" building has been added to the standard to help the user in the understanding that a building with openings and significant porosity (such as an open parking garage, for example) that does not meet the requirements of the "partially enclosed" classification does not automatically classify the building as "open" or "enclosed." Once the enclosure classification is known, the designer enters Table 26.13-1 to select the appropriate internal pressure coefficient.

This version of the standard has four terms applicable to enclosure: wind-borne debris regions, glazing, impact-resistant glazing, and impact-protective system. *Wind-borne debris regions* are specified to alert the designer to areas requiring consideration of missile-impact design and potential openings in the building envelope. *Glazing* is defined as any glass or transparent or translucent plastic sheet used in windows, doors, skylights, or curtain walls. *Impact-resistant glazing* is specifically defined as glazing that has been shown by testing to withstand the impact of test missiles. *Impact-protective systems* over glazing can be shutters or screens designed to withstand wind-borne debris impact. Impact resistance of glazing and protective systems can be tested using the test method specified in ASTM E1886, with missiles, impact speeds, and pass/fail criteria specified in ASTM E1996. Glazing in sectional garage doors and rolling doors can be tested for impact resistance with missiles, impact speeds, and pass-fail criteria specified in ANSI/DASMA 115. Other methods involving opening protection of building envelope systems are acceptable when approved by the Authority Having Jurisdiction. Origins of missile-impact provisions contained in these standards are summarized in Minor (1994) and Twisdale et al. (1996).

Section 26.12.3 requires glazing in Category II, III, and IV buildings in wind-borne debris regions to be protected with an impact-protective system or to be made of impact-resistant glazing to reduce the amount of wind and water damage to buildings during design windstorm events.

The standard requires all glazing in the lower 60 ft (18.3 m) of Category II, III, or IV buildings sited in wind-borne debris regions to be protected with an impact-protective system or to be made of impact-resistant glazing. Glazing higher than 60 ft (18.3 m) above grade may be broken by wind-borne debris when a debris source is present, such as aggregate-surfaced roofs on buildings within 1,500 ft (457 m) of the new building. This includes gravel or stone used as ballast that is not protected by a sufficiently high parapet. Accordingly, the glazing in the new building, from 30 ft (9.1 m) above the source building to grade needs to be protected with an impact-protective system or be made of impact-resistant glazing. If loose roof aggregate is proposed for the new building, it too should be considered as a debris source because aggregate can be blown off the roof and be propelled into glazing on the leeward side of the building. Although other types of wind-borne debris can impact glazing higher than 60 ft (18.3 m) above grade, at these higher elevations, loose roof aggregate has been the predominant debris source in previous wind events. The requirement for protection 30 ft (9.1 m) above the debris source is to account for debris that can be lifted during flight. The following references provide further information regarding debris damage to glazing: Beason et al. (1984), Minor (1985, 1994), Kareem (1986), and Behr and Minor (1994).

Although wind-borne debris can occur in just about any condition, the level of risk in comparison to the postulated debris regions and impact criteria may also be lower than that determined for the purpose of standardization. For example, individual buildings may be sited away from likely debris sources that would generate significant risk of impacts similar in magnitude to pea gravel (i.e., as simulated by 2-g steel balls in impact tests) or butt-on 2×4 impacts as required in impact-testing criteria. This situation describes a condition of low vulnerability only as a result of limited debris sources within the vicinity of the building. In other cases, potential sources of debris may be present, but extenuating conditions can lower the risk. These extenuating conditions include the type of materials and surrounding construction, the level of protection offered by surrounding exposure conditions, and the design wind speed. Therefore, the risk of impact may differ from those postulated as a result of the conditions specifically enumerated in the standard and the referenced impact standards. There are vastly differing opinions regarding the significance of these parameters that are not fully considered in developing standardized debris regions or referenced impact criteria.

The definition of the wind-borne debris regions for Risk Category II buildings and structures was chosen such that the coastal areas included in the wind-borne debris regions are approximately consistent with those given in ASCE 7-05 and prior editions. Thus, the new wind speed contours that define the wind-borne debris regions in Section 26.12.3.1 are not direct conversions of the wind speed contours that are defined in ASCE 7-05, as shown in Table C26.5-7.

While the coastal areas included in the wind-borne debris regions for Risk Category II are approximately consistent with those given in ASCE 7-05, significant reductions in the area of wind-borne debris regions for this Risk Category occur around Jacksonville, Florida, in the Florida Panhandle, and inland from the coast of North Carolina.

The introduction in ASCE 7-10 of separate maps for different Risk Categories provides a means for achieving a more risk-consistent approach for defining wind-borne debris regions. The approach selected was to link the geographical definition of the wind-borne debris regions to the wind speed contours in the maps that correspond to the particular Risk Category, resulting in expansion of the wind-borne debris region for some Risk Category III and all Risk Category IV buildings and structures. A review of the types of buildings and structures currently included in Risk Category III suggests that in the expanded wind-borne debris region life-safety issues would be most important for health-care facilities. Consequently, the committee chose to apply the expanded wind-borne debris protection requirement to this type of Risk Category III facilities and not to all Risk Category III buildings and structures.

C26.13 INTERNAL PRESSURE COEFFICIENTS

The internal pressure coefficient values in Table 26.13-1 were obtained from wind tunnel tests (Stathopoulos et al. 1979) and full-scale data (Yeatts and Mehta 1993). Even though the wind tunnel tests were conducted primarily for low-rise buildings, the internal pressure coefficient values are assumed to be valid for buildings of any height. The values $(GC_{pi}) = +0.18$ and -0.18 are for enclosed buildings. It is assumed that the building has no dominant opening or openings and that the small leakage paths that do exist are essentially uniformly distributed over the building's envelope. The internal pressure coefficient values for partially enclosed buildings assume that the building has a dominant opening or openings. For such a building, the internal pressure is dictated by the exterior pressure at the opening and is typically increased substantially as a result. Net loads (i.e., the combination of the internal and exterior pressures) are therefore also significantly increased on the building surfaces that do not contain the opening. Therefore, higher (GC_{pi}) values of $+0.55$ and -0.55 are applicable to this case. These values include a reduction factor to account for the lack of perfect correlation between the internal pressure and the external pressures on the building surfaces not containing the opening (Irwin 1987; Beste and Cermak 1996). Taken in isolation, the internal pressure coefficients can reach values of ± 0.8 (or possibly even higher on the negative side).

For partially enclosed buildings containing a large unpartitioned space, the response time of the internal pressure is increased, and this increase reduces the ability of the internal pressure to respond to rapid changes in pressure at an opening. The gust factor applicable to the internal pressure is therefore reduced. Eq. (26.13-1), which is based on Vickery and Bloxham (1992) and Irwin and Dunn (1994), is provided as a means of adjusting the gust factor for this effect on structures with large internal spaces, such as stadiums and arenas.

Because of the nature of hurricane winds and exposure to debris hazards (Minor and Behr 1993), glazing located below

60 ft (18.3 m) above the ground level of buildings sited in wind-borne debris regions has a widely varying and comparatively higher vulnerability to breakage from missiles, unless the glazing can withstand reasonable missile loads and subsequent wind loading, or the glazing is protected by suitable shutters. [See Section C26.12 for discussion of glazing above 60 ft (18.3 m).] When glazing is breached by missiles, development of higher internal pressure may result, which can overload the cladding or structure if the higher pressure was not accounted for in the design. Breaching of glazing can also result in a significant amount of water infiltration, which typically results in considerable damage to the building and its contents (Surry et al. 1977; Reinhold 1982; Stubbs and Perry 1993).

The influence of compartmentalization on the distribution of increased internal pressure has not been researched. If the space behind breached glazing is separated from the remainder of the building by a sufficiently strong and reasonably airtight compartment, the increased internal pressure would likely be confined to that compartment. However, if the compartment is breached (e.g., by an open corridor door or by collapse of the compartment wall), the increased internal pressure will spread beyond the initial compartment quite rapidly. The next compartment may contain the higher pressure, or it too could be breached, thereby allowing the high internal pressure to continue to propagate. Because of the great amount of air leakage that often occurs at large hangar doors, designers of hangars should consider using the internal pressure coefficients in Table 26.13-1 for partially enclosed buildings

C26.14 TORNADO LIMITATION

Tornadoes have not been considered in the wind load provisions because of their very low probability of occurrence. However, some building owners might want to have portions of their buildings designed to provide a greater level of occupant protection from tornadoes or minimize building damage associated with EF0–EF2 rated tornadoes (see Section C26.14.1 for tornado ratings). Maintaining continuity of building operations for facilities such as emergency operations centers and hospitals is also a factor when considering whether or not to design for tornado wind and wind-borne debris loads. This section provides information and design guidance for those designers and building owners who desire to design for reduced property damage or increased occupant protection, in case the building under consideration is impacted by a tornado.

The following topics are addressed in this section: 1) tornado wind speeds and probabilities, 2) wind pressures induced by tornadoes versus other windstorms, 3) designing for occupant protection, 4) designing to minimize building damage, 5) designing to maintain continuity of building operations, and 6) designing trussed communications towers for wind-borne debris. (Mehta et al. (1976), Minor (1982), Minor et al. (1977), and Wen and Chu (1973) provide early background information on tornadoes.)

C26.14.1 Tornado Wind Speeds and Probabilities. The National Weather Service (NWS) rates tornado severity according to the six levels of observed damage in the Enhanced Fujita Scale (EF Scale). The scale ranges from EF0 to EF5. See Table C26.14-1 for the wind speeds associated with the EF ratings. Damage indicators (DI) and degrees of damage (DOD) are used to establish EF ratings. DIs consist of buildings, other structures, and trees. Information on the EF scale and a description of DIs and DODs can be found in McDonald and Mehta (2006).

NWS data indicate that the median number of tornadoes in the United States between 1990 and 2014 was 1,173 per year. During this time, the fewest number of tornadoes (888) occurred in 2014,

Table C26.14-1 Enhanced Fujita (EF) Scale

EF Number	Wind Speed (mph)	(m/s)
EF0	65–85	29–38
EF1	86–110	39–49
EF2	111–135	50–60
EF3	136–165	61–73
EF4	166–200	74–89
EF5	>200	>89

Note: Speeds are for 3-s peak gust, Exposure C, 33 ft (10 m) above grade. Conversion of mph to m/s: mph × 0.44704 = m/s.
Source: NOAA (http://www.spc.noaa.gov/efscale/ef-scale.html).

and the greatest number (1,817) occurred in 2004. Tornado-related winds have a significantly lower probability of occurrence at a specific location than the high winds associated with meteorological events (frontal systems, thunderstorms, and hurricane winds) responsible for the basic wind speeds given in ASCE 7. The probability of occurrence is a function of the area covered by a tornado and of the specific location. The probability of a site-specific EF0 to EF1 rated tornado strike in the central portion of the United States is on the order of a 4,000-year MRI (Ramsdell and Rishel 2007). (Considering that Risk Category IV buildings and structures have an MRI of 3,000 years in the standard, designing for an EF1 tornado would only result in a small increase in design wind pressures. It would be prudent to design for an EF1 or greater tornado, as discussed in Sections C26.14.4 and C26.14.5.) In the areas of the country where the risk of EF4 and EF5 rated tornadoes is greatest, the annual probabilities that a particular building will be affected by an EF4 or EF5 rated tornado are on the order of 10^{-7} (a 10,000,000-year MRI) (Ramsdell and Rishel 2007). Tornadoes in the West are rare, as illustrated by NWS annual tornado maps from 1952 to 2011 (http://www.spc.noaa.gov/wcm/annualtornadomaps/).

Fig. C26.14-1 shows the recorded EF3–EF5 rated tornadoes between 1950 and 2013. Of the 56,221 recorded tornadoes that occurred between 1950 and 2011, 95% were rated as EF0–EF2 by the NWS, 4% were rated as EF3, and 1% were rated as EF4–EF5.

Damage investigations have indicated that tornado winds are more likely to generate more wind-borne debris compared to nontornadic winds of the same speed. EF0 and EF1 rated tornadoes may generate wind-borne debris that can break unprotected glazing and puncture many types of door, wall, and roof assemblies.

As illustrated in Table C26.14-3, depending on a building's location and Risk Category, EF2 and EF3 rated tornadoes produce wind pressures that range from below to above those derived from ASCE 7-10 for hurricane-prone regions. Hence, for buildings designed for wind pressure in accordance with ASCE 7, the performance of structural elements (i.e., MWFRS), doors, and walls in tornadoes will depend on the relationship between the tornado severity and the basic wind speed. For example, a building in Miami is expected to have greater resistance to strong tornadoes than a building in Orlando, where the basic wind speed is lower than Miami's basic wind speed. However, wind-borne debris can break unprotected glazing and puncture many types of door, wall, and roof assemblies. Even if the glazing is protected from hurricane debris, debris from an EF3 rated tornado may penetrate the glazing because the momentum of debris generated by an EF3 rated tornado may significantly exceed the impact test criteria adopted for hurricane opening protection.

EF3 through EF5 rated tornadoes can produce wind pressures and wind-borne debris loads that are in excess of those derived from the highest design wind speeds for hurricane-prone areas and the wind-borne debris test standards for opening protection

FIGURE C26.14-1 Recorded EF3–EF5 Rated Tornadoes, 1950–2013
Source: FEMA (2015), from National Oceanic and Atmospheric Administration, National Weather Service, Storm Prediction Center

in hurricane prone regions. Fig. C26.14-2 shows the design wind speed for tornado safe rooms recommended by FEMA P-361 (FEMA 2015). This wind speed map is the same as the ICC 500 tornado hazard map. These wind speeds likely have a probability of occurrence that is on the order of 1×10^{-6} to 1×10^{-7} per year (Ramsdell and Rishel 2007).

C26.14.2 Wind Pressures Induced by Tornadoes Versus Other Windstorms. Photo analysis of wind-borne debris shed from buildings indicates that tornado debris has a greater vertical trajectory than hurricane debris. Such observations and tornado simulations suggest that updrafts are greater in tornadoes than in other windstorms. Research by Mishra et al. (2008) indicates that atmospheric pressure drops play a significant role in wind pressures experienced in tornadoes. Laboratory research conducted in a tornado simulator (Haan et al. 2010) measured wind pressures on a gable roof of a residential-size building model. Measured pressures included the pressure drop in the simulated vortex and point pressures, which were integrated to determine MWFRS loads. Results have been compared with loads produced by both ASCE 7 directional and envelope MWFRS wind load provisions. The largest lateral loads determined from the tornado simulator test results are slightly lower than the lateral loads produced using the MWFRS directional method of ASCE 7, Chapter 27, but they are up to about 50% larger than those produced using the MWFRS envelope method of ASCE 7, Chapter 28. For a tightly sealed building where the internal pressure in the building does not rapidly equalize with the atmospheric pressure drop in the tornado vortex, the uplift loads on the building obtained from the simulator tests can be substantially larger than those obtained from either the directional or envelope MWFRS methods of ASCE 7. However, at the other extreme, if an open building is assumed and the atmospheric pressure drop in the vortex is removed from the surface pressures on the building roof, the net uplift MWFRS loads from the simulator tests produces lower uplift loads on the building than those produced using the ASCE 7 directional and envelope MWFRS methods. In consideration of the typical leakage of most buildings, the use of internal pressure coefficients for a partially enclosed building and the fact that the building would have already experienced the highest horizontal winds in the tornado vortex before it was exposed to the largest pressure drop, which occurs in the vortex core, the Wind Load Subcommittee decided that the uplift load coefficients should not be further increased for most buildings to account for atmospheric pressure drop effects in the tornado vortex. This would not be an appropriate decision for nuclear power plant containment structures or other structures where extreme measures are taken to ensure that the interior of the building is sealed from atmospheric conditions.

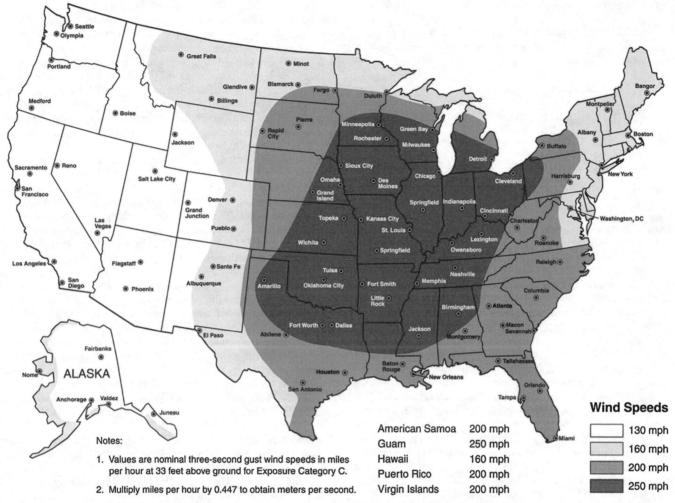

FIGURE C26.14-2 Tornado Safe Room Design Wind Speed Map

Source: FEMA (2008)

C26.14.3 Occupant Protection. A report (CDC 2012) on fatalities during the April 25–28, 2011, tornado outbreak illustrates the importance of providing tornado protection for building occupants. There were 351 recorded tornadoes during the outbreak, where 338 fatalities were caused by 27 of the tornadoes. According to the report, 90% of the people who died were in buildings at the time the tornadoes struck.

ICC 500 provides requirements for design and construction of residential and community storm shelters. FEMA P-320 (2014) provides prescriptive solutions for residential and small business safe rooms that will accommodate up to 16 occupants. FEMA P-361 (2015) provides comprehensive guidance for the design of community and residential safe rooms, as well as for quality assurance and quality control for their design and construction. The criteria in FEMA P-361 and ICC 500 are quite similar. All safe room criteria in FEMA P-361 meet the storm shelter requirements in ICC 500, but a few design and performance criteria in FEMA P-361 are more restrictive than those in ICC 500. The FEMA P-320 and P-361 documents provide additional guidance on design standards that should be followed in order for shelters to potentially qualify for federal grant funding.

Tornado shelters complying with ICC 500 or safe rooms complying with FEMA P-320 (2014) or P-361 (2015) are the recommended methods of protecting occupants from EF3, EF4, or EF5 rated tornadoes. However, if a tornado shelter or safe room is not provided, some level of occupant protection may be achieved if the building is designed to minimize damage (see Section C26.14.4). Depending upon the magnitude of the design enhancements, the quality of construction, maintenance and repair of the building, the potential for roof structure and wall collapse is reduced if the building is struck by an EF0 to EF2 rated tornado. Hence, in addition to minimizing building damage, there will be some level of occupant protection, although the amount of protection is difficult to quantify. The design objective of minimizing damage and providing for occupant protection is discussed in van de Lindt et al. (2013). However, if occupants of such buildings do not have access to a shelter or safe room, they are exposed to a slight risk of death or injury from wind-borne debris generated by EF0 to EF2 rated tornadoes. They are also clearly susceptible to death or injury if the building is struck by a stronger tornado (EF3 to EF5).

For buildings that do not have an ICC 500-compliant storm shelter, or a FEMA P-320 (2014) or P-361 (2015) compliant safe room, FEMA P-431 (2009) provides guidance in identifying best available refuge areas.

C26.14.4 Minimizing Building Damage. Table C26.14-2 shows the comparison between hurricane and tornado wind speeds. By using design strategies consistent with construction in hurricane-prone regions, new buildings can be relatively econo-

Table C26.14-2 Comparison of Hurricane and Tornado Wind Speeds

Hurricane Category	Wind Speed over Land (mi/h)	(m/s)	EF Rating	Wind Speed (mi/h)	(m/s)
1	81–105	36–46	EF0	65–85	29–38
2	106–121	47–54	EF1	86–110	39–49
3	122–143	55–64	EF2	111–135	50–60
4	144–171	65–76	EF3	136–165	61–73
5	>171	>76	EF4	165–200	74–89
			EF5	>200	>89

Note: Speeds are 3-s peak gust, Exposure C, 33 ft (10 m) above grade. Conversion of mph to m/s: mph × 0.44704 = m/s.

mically designed and built to be more resistant to most tornadoes (i.e., EF0–EF2) (Prevatt et al. 2012). Furthermore, published tornado simulator results suggest that the loads on the buildings may be somewhat similar to those from straight-line winds provided the tornado is large enough to engulf the building. With appropriate strengthening and selection of building materials and systems, the cost of tornado repairs can be reduced. However, when designing to minimize tornado damage, it is important to realize that wind-borne debris is likely to cause breaches in the building envelope. If exterior glazing is breached, interior ceilings and/or walls may collapse from wind pressures, and water damage is possible.

If a building is designed to minimize tornado damage, it is recommended that consideration also be given to including a tornado shelter complying with ICC 500 or a safe room complying with FEMA P-320 (2014) or P-361 (2015) for occupant protection. For example, if a medical research lab is designed for EF3 pressures in order to minimize building damage, a portion of the building could also be designed as a shelter to provide occupant protection in the event that the facility was struck by significant tornado-generated debris or winds from an EF4 or EF5 rated tornado.

Tornado Wind Pressure Calculations. There are several differences in the values of variables recommended for use in tornado wind pressure calculations as compared with normal wind pressure calculations. (The pressure calculation recommendations are consistent with those given in FEMA P-361 (2015) and ICC 500.) However, the methodology and the equations are the same as provided in Section 26.10.2, Eq. (26.10-1) (for velocity pressure), Section 27.3.1, Eq. (27.3-1) (for MWFRS), and Section 30.3.2 and Eq. 30.3-1 (for C&C). There are two methods presented in this commentary for the development of tornado wind pressures. Both methods achieve the same answer. The first method (Extended Method) is to use wind pressure calculation parameters (in equations noted above) modified for use in determining tornado pressures. Because this method does require some adjustment to almost every parameter, a second method (Simplified Method) is presented that reduces all of the changed parameters to a single multiplier. The Simplified Method is intended to provide a simple method of accounting for various tornado-related design considerations and for the uncertainties in tornado pressures caused by a lack of field pressure measurements and limited laboratory research.

The following provides a description and discussion of those differences that are then combined to provide a single factor, identified as a Tornado Factor or TF (i.e., the Simplified Method) that can be applied to results of ASCE 7 calculated pressures and wind loads to produce tornado design wind loads for a target tornado intensity. The TF factors shown in Table C26.14-2 reflect a combination of adjustment factors for internal pressure, gust factor, directionality and external pressure coefficient differences, and the uncertainty in these adjustment factors. Because the original design calculations may be based on Exposures B, C, or D and the tornado design pressures and loads assume Exposure C or D, the recommended TF factor is different depending on whether the original calculations were for Exposure Category B or for either Exposure Category C or D. Note that these factors are offered as a way to simplify the assessment of the increases in loads caused by tornado design, and the designer always has the option of using the Extended Method with the revised factors in the wind pressure equation suggested below for designing for tornado wind loads.

Extended Method. A tornado event is considered an ultimate wind speed event, so the tornado design wind speed for the upper end of the target design EF Scale should be used as the design wind speed. Wind speeds for tornado design may also come from ICC 500 or FEMA P-361 (2015). Because the tornado design wind speed is an ultimate wind speed, it is treated the same as the basic wind speeds provided in Fig. 26.5-1 (e.g., the load factor for strength design is 1.0 and the wind loads can be multiplied by 0.6 for use in allowable stress design).

It is recommended that the pressures always be determined based on an Exposure C condition except as noted following, regardless of the actual exposure of the site. It is anticipated that the winds of the tornado are going to create or at least resemble winds from an Exposure C terrain condition. Thus, even in more suburban settings, vegetation will be shredded and some roofs will be torn off, creating a smoother terrain than the original one. If the design exposure is D, the D exposure should also be used for tornado design.

It is recommended that the directionality factor be taken as $K_d = 1.0$ for a tornado. In Section C26.6, the following describes the reason for the directionality factor: "This factor accounts for two effects: (1) The reduced probability of maximum winds coming from any given direction, and (2) the reduced probability of the maximum pressure coefficient occurring for any given wind direction." In the judgment of the Wind Load Subcommittee, it was decided to set $K_d = 1.0$ for wind pressures from tornadoes since neither of the two conditions used as the basis for $K_d = 0.85$ seemed to apply. Considering the rotational winds in a tornado, and the likelihood that at least one building corner or window will experience a coincident worst (GC_p) direction and a maximum or near maximum wind speed, a value of $K_d = 1.0$ seems prudent.

It is recommended that the topographic effect factor be taken as $K_{zt} = 1.0$, since topographic effects on tornadic wind speeds are not well understood.

The gust-effect factor G should be increased from 0.85 to 0.90 for rigid buildings. A $G = 0.90$ value is derived from the gust-effect factor used for rigid buildings of $G = 0.85$ without the 0.925 calibration factor incorporated within that value. Thus, for tornado design, $G = 0.85/0.925 = 0.92$, or it can be rounded down to 0.90. $G = 0.90$ should be the minimum gust-effect factor used.

Component and cladding (C&C) loads in ASCE 7 are based on (GC_p) values that are associated with a wind event that lasts for a period of time that is frequently taken as one hour. Most tornadoes move with a translational speed that limits the strongest winds to a matter of minutes, if not seconds. Consequently, a reduction in the C&C loads appears to be warranted (e.g., Kopp and Morrison 2011). Given the limited amount of research that has been conducted on wind pressures in tornado simulators, the Wind Load Subcommittee judgment is to limit the reduction to 10% until more information is available.

It is recommended that the internal pressure coefficient be taken as $(GC_{pi}) = \pm 0.55$. A breach of the building envelope is likely during a tornado, which will significantly increase the wind-induced internal pressure in the building while helping to promote equalization of atmospheric pressure drop in the vortex if it passes over the building. From the changes in design criteria noted above, the velocity pressure q for tornado pressure is

$$q = 0.00256 K_d K_z K_{zt} K_e V^2 \quad \text{(C26.14-1)}$$

which is reduced to $0.00256 K_z V^2$, where V^2 is selected from the EF rating selected for design and K_z is based on Exposure C or D as appropriate.

Given the nature of the wind profile in a tornado, it is recommended that q be determined at the mean roof height, h, and that q_h be used throughout the pressure calculations as the value of velocity pressure, q.

For the MWFRS pressure, then

$$p = q_h GC_p - (\pm 0.55) \quad \text{(C26.14-2)}$$

where

$G = 0.90$ or higher as determined by the current methods, and
C_p = external pressure coefficient found by the current directional methods of Chapter 27.

For C&C pressures,

$$p = q_h [(0.9 \times (GC_p)) - (\pm 0.55)] \quad \text{(C26.14-3)}$$

where (GC_p) is the external pressure coefficient for C&C found by current methods.

To summarize, the recommended tornado design wind pressures and loads for both MWFRS and C&C elements should be calculated using the following:

- V = upper end of range of wind speed for target EF scale or the speed from ICC 500 or FEMA P-361 (2015) (Note: The EF scale does not provide an upper bound wind speed for EF5-rated tornadoes. When designing for EF5-rated tornadoes, the design wind speeds shown in Fig. C26.14-2 are reasonable upper bound speeds.);
- K_z = velocity pressure exposure coefficient evaluated at mean roof height for Exposure C or D as appropriate;
- $K_d = 1.0$;
- $K_{zt} = 1.0$;
- $K_e = 1.0$ or value from Table 26.9-1;
- $(GC_{pi}) = \pm 0.55$;
- G(MWFRS) = 0.9 or higher if appropriate from current methods;
- C_p(MWFRS) = ASCE 7 values per Chapter 27 and directional analysis; and
- (GC_p)(C&C) = ASCE 7 values for the appropriate zone.

It should be recognized that there is considerable uncertainty when it comes to the development and justification of tornado design wind loads. Nevertheless, it is also important to recognize that the EF scale for tornadoes is essentially a damage scale based on a series of damage indicators (DIs) and degrees of damage (DODs). The expert opinions that were used to establish the wind speeds associated with the DIs and DODs were at least partially shaped by observations of damage in hurricanes where there was a higher degree of confidence in wind speeds than in tornadoes. Consequently, changes in wind loads based on static pressure drop estimates in the middle of tornadoes is probably not justified. Conversely, though, observations of dramatic structural failures in hurricanes are almost always accompanied by observations of breaching of the building envelope that would have produced significant internal pressures on the building or a part of the building. Because breaches in the building envelope are highly likely in tornadoes because of the amount of debris in the circulation and there is likely insufficient time to protect openings unless a passive system is used (which in turn might reduce the opportunity for pressure equalization), it is prudent to design for some measure of additional internal pressurization. This is done by setting the internal pressure coefficient to ± 0.55.

Simplified Method. A series of calculations were carried out to estimate the effect of the various tornado design recommendations outlined above. These calculations are shown in Table C26.14-3, which illustrates quantitative comparisons of MWFRS pressures derived from ASCE 7-10 and estimated pressures induced by EF0–EF4 rated tornadoes. The calculations are based on a 30-ft by 30-ft enclosed building with a mean roof height of 22 ft and a gable roof angle of 35° sited in Exposure C, which is the building size and exposure used in research scaled up to a full-size building (Haan et al. 2010). The tornado calculations assume partially enclosed conditions (caused by broken glazing) and a Tornado Factor (TF) based on tornado design considerations outlined in this section. Based on these calculations of the increase in pressures from tornadoes on both walls and roofs, it was concluded that the use of simple single factors could be applied to the design wind loads calculated using normal ASCE 7 procedures and equations for enclosed buildings as a reasonable way for designers to address tornado design for their clients who wanted to evaluate options and to provide a measure of tornado damage reduction.

The C&C TF factors are conservatively based on the field zones of roofs and walls. These simple, singular factors (TF factors) are summarized in Table C26.14-4 and illustrated in Eq. (C26.14-4).

Note that these adjustments can be applied to either limit state "ultimate" design loads or to allowable stress design pressures or loads calculated for the building to arrive at tornado design pressures or loads.

Applying these TF factors from Table C26.14-4 follows:

$$p = q_i (GC_p - (GC_{pi})) \times TF \quad \text{(C26.14-4)}$$

where the $q_i(GC_p - (GC_{pi}))$ portion of the equation is determined using the current ASCE 7 Chapter 27 directional procedure for

Table C26.14-3 MWFRS Pressure Comparisons – Straight Line Wind versus Tornado

	Wind Speed	Roof Uplift Pressure	Wall Negative Pressure
	mph	psf	psf
ASCE 7-16	115	−30	−22
ASCE 7-16	120	−33	−24
ASCE 7-16	180	−74	−55
ASCE 7-16	190	−82	−61
EF0 (upper end of range)	85	−27	−22
EF1 (upper end of range)	110	−45	−37
EF2 (upper end of range)	135	−68	−55
EF3 (upper end of range)	165	−102	−82
EF4 (upper end of range)	200	−150	−120

Note: Conversion of mph to m/s: mph × 0.44704 = m/s; Conversion of psf to kN/m²: psf × 0.0479 kN/m²

Table C26.14-4 Increases in Design Loads to Address Tornado Risks Using Recommended Tornado Factors

Original Enclosure Classification	Loading	Original Exposure B	Original Exposure C or D
Partially enclosed buildings	MWFRS	1.8	1.2
	C&C	1.6	1.1
Enclosed buildings[a]	MWFRS	2.5	1.6
	C&C	2.1	1.5

[a]The tornado factors to be used to increase the design loads on elements of enclosed buildings are based on the effects of high internal pressures. High internal pressures have a much greater effect on elements that typically receive less wind, so the net effect of these increase factors is typically much higher than would result if the building were designed for the specific tornado loads or if the tornado factors for partially enclosed buildings were used with partially enclosed building designs.

enclosed buildings. The wind speed used in Eq. (C26.14-4) to determine q is the selected design wind speed for the tornado design. Alternatively, the designer can calculate wind pressures for both MWFRS and C&C using the design wind speeds from ASCE 7, Fig. 26.5-1A, B, C, or D, or Fig. 26.5-2A, B, C, or D (or http://windspeed.atcouncil.org/), then scale the pressures for tornado design as follows:

$$p = q_i(GC_p - (GC_{pi})) \times (V_{tornado}/V_{design})^2 \times TF$$

where $V_{tornado}$ is the selected tornado wind speed to be used for this tornado design and TF factors are from Table C16.14-4.

Example.
Description of Building
Building dimensions: 30 ft × 60 ft
Mean roof height = 40 ft

The building is located in Des Moines, Iowa, and is a two-story, light commercial building with a flat roof. The primary building use is computer server storage. The building is located in a commercial office park; the assumed exposure is B.

The owner has decided that he would like the building to be designed for an EF2 tornado. The top end of the EF2 tornado scale is 135 mph. The ASCE 7 wind speed for Risk Category II buildings in this location is 110 mph.

Determine the differences in wall and roof pressures (both MWFRS and C&C) for a tornado design for 135 mph compared to the design pressures using ASCE 7 mapped wind speeds.

Tornado wind pressure determination using the Extended Method
Wind speed $V = 135$ mph;
Mean roof height $h = 40$ ft;
$K_d = 1.0$;
$K_{zt} = 1.0$;
$K_z = 1.04$ (taken at Exposure C and 40 ft);
Velocity pressure $q = 0.00256 K_z K_{zt} K_d V^2 = 0.00256 \times 1.04 \times 1.0 \times 1.0 \times 135^2 = 48.7$ psf; and
$G = 0.9$.
For the MWFRS: $p = q_h(GC_p - (GC_{pi}))$
Table C26.14-5 shows calculations for MWFRS wind pressures from the Extended Method.

Tornado wind pressure determination using the Single TF Factor Method
Wind speed $V = 135$ mi/h;
Mean roof height $h = 40$ ft;
$K_d = 0.85$;
$K_{zt} = 1.0$;
$K_z = 0.76$ (taken at Exposure B and 40 ft);
Velocity pressure $q = 0.00256 K_z K_{zt} K_d V^2 = 0.00256 \times 0.76 \times 1.0 \times 0.85 \times 135^2 = 30.2$ psf; and $G = 0.85$.
For the MWFRS: $p = q_i(GC_p - (GC_{pi}))$
where (GC_{pi}) is for an enclosed condition and thus $= \pm 0.18$.
Table C26.14-6 shows MWFRS wind pressures using ASCE 7 parameters. Table C26.14-7 shows MWFRS wind pressures comparison: extended versus single-factor methods.

Multiplying by the MWFRS TF factor for an original B exposure = 2.5 (from Table C26.14-8).

Alternatively, the wind speeds can be scaled in accordance with Eq. (C26.14-5), where $V_{tornado}$ is the tornado design wind speed (135 mph in this example) and V_{design} is the ASCE 7 mapped wind speed for the location (110 mph in this example).

Wind pressures p are developed using the ASCE 7 design wind speed and the ASCE 7 parameters:

$$(V_{tornado}/V_{design})^2 = (135/110)^2 = 1.51$$

The results using this method are shown in Table C26.14-8.

Load Path. It is necessary to take the pressures through the appropriate load paths so that connections between building elements can be designed. The most important load paths indicated by damage observations are the roof-to-wall connection, the wall-to-floor connection, the floor-to-foundation connection, any connection across floors, and the connection between the exterior walls at the corners. In addition to the exterior

Table C26.14-5 Example Problem MWFRS Wind Pressures from Extended Method

	q (psf)	C_p		$\pm(GC_{pi})$	Pressures with $-(GC_{pi})$	Pressures with $-(GC_{pi})$
MWFRS: Walls						
Windward	48.7	+0.8		±0.55	61.8	8.3
Leeward	48.7	$L/B = 0.5$,	−0.5	±0.55	4.9	−48.7
		$L/B = 2.0$,	−0.3		13.6	−39.9
Sidewalls	48.7	−0.7		±0.55	−3.9	−57.5
MWFRS: Roof						
Wind perpendicular to 30-ft side	48.7	$h/L = 0.67$		±0.55		
		0–h,	−0.9		−12.7	−66.2
		h–2h,	−0.5		4.9	−48.7
Wind perpendicular to 60-ft side	48.7	$h/L = 1.33$		±0.55		
		0–h/2,	−1.3		−30.2	−83.7
		>h/2,	−0.7		−3.9	−57.5

Table C26.14-6 Example Problem MWFRS Wind Pressures Using ASCE 7 Parameters

	q (psf)	C_p	$\pm(GC_{pi})$	Pressures with $-(GC_{pi})$	Pressures with $+(GC_{pi})$
MWFRS: Walls					
Windward (at mean roof height)	30.2	+0.8	±0.18	25.9	15.1
Leeward	30.2	$L/B = 0.5, -0.5$	±0.18	−7.4	−18.2
		$L/B = 2.0, -0.3$		−2.3	−13.1
Sidewalls	30.2	−0.7	±0.18	−12.5	−23.4
MWFRS: Roof					
Wind perpendicular to 30-ft side	30.2	$h/L = 0.67$	±0.18		
		$0–h, -0.9$		−17.6	−28.5
		$h – 2h, -0.5$		−7.4	−18.2
Wind perpendicular to 60-ft side	30.2	$h/L = 1.33$	±0.18		
		$0–h/2, -1.3$		−27.9	−38.7
		$>h/2, -0.7$		−12.5	−23.4

Table C26.14-7 Example Problem MWFRS Wind Pressures Comparison: Extended versus Single-Factor Methods

	Tornado Factor	ASCE 7 Enclosed Condition		Tornado Extended Method		Ratio of the Largest Pressures Single-Factor/Extended
		Pressures with $-(GC_{pi})$	Pressures with $+(GC_{pi})$	Pressures with $-(GC_{pi})$	Pressures with $+(GC_{pi})$	
MWFRS: Walls						
Windward (at mean roof height)	2.5	25.9 × 2.5 = 64.9	15.1 × 2.5 = 37.7	61.8	8.3	1.05
Leeward (largest pressures)	2.5	−7.4 × 2.5 = −18.5	−18.2 × 2.5 = −45.6	4.9	−48.7	0.94
Sidewalls	2.5	−12.5 × 2.5 = −31.3	−23.4 × 2.5 = −58.4	−3.9	−57.5	1.02
MWFRS: Roof						
Wind perpendicular to 30-ft side (largest pressures)	2.5	−17.6 × 2.5 = −44.1	−28.5 × 2.5 = −71.3	−12.7	−66.2	1.08
Wind perpendicular to 60-ft side (largest pressures)	2.5	−27.9 × 2.5 = −69.8	−38.8 × 2.5 = −96.9	−30.2	−83.7	1.16

Table C26.14-8 Example Problem MWFRS Wind Pressures Comparison: Extended versus Single-Factor Methods with Factored Wind Speeds

	ASCE 7 Design Wind Speed		ASCE 7 Design Wind Speed × Correction × TF		Tornado Extended Method		Ratio of the Largest Pressures:Single-Factor/Extended
	Pressures with $-(GC_{pi})$	Pressures with $+(GC_{pi})$	Pressures with $-(GC_{pi})$	Pressures with $+(GC_{pi})$	Pressures with $-GC_{pi}$	Pressures with $+GC_{pi}$	
MWFRS: Walls							
Windward (at mean roof height)	17.2	10.0	17.2 × 2.5 × 1.51 = 65.0	10.0 × 2.5 × 1.51 = 37.8	61.8	8.3	1.05
Leeward (largest pressures)	−4.9	−12.1	−4.9 × 2.5 × 1.51 = −18.5	−12.1 × 2.5 × 1.51 = −45.7	4.9	−48.7	0.94
Sidewalls	−8.3	−15.5	−8.3 × 2.5 × 1.51 = −31.4	−15.5 × 2.5 × 1.51 = −58.6	−3.9	−57.5	1.02
MWFRS: Roof							
Wind perpendicular to 30-ft side (largest pressures)	−11.7	−18.9	−12.8 × 2.5 × 1.51 = −44.2	−18.9 × 2.5 × 1.51 = −71.4	−12.7	−66.2	1.08
Wind perpendicular to 60-ft side (largest pressures)	−18.5	−25.7	−18.5 × 2.5 × 1.51 = −69.9	−25.7 × 2.5 × 1.51 = −97.1	−30.2	−83.7	1.16

walls, additional shear walls in the interior of the building may reduce the tendency for the building to rack and/or overturn. Finally, adequate shear wall design requires proper anchorage at the ends of the shear walls and sometimes at the ends of shear wall segments. This anchorage is inherent in some types of structural systems such as reinforced masonry that follows the latest design requirements but must be explicitly added for other types of structural systems such as wood frame.

If it is desired to avoid collapse of interior walls and/or ceilings in the event that exterior glazing is breached, loads on these interior elements need to be calculated. Resistance to failure may require additional or strengthened connections between the top of interior walls and the roof or ceiling systems and connections between the bottom of interior walls and the floor and foundation systems. Reasonable design loads for use in designing these elements and their connections would be 80% of the exterior wall design loads.

Exterior Glazing. Glazing damage is prevalent during tornadoes (Roueche and Prevatt 2013). To minimize breaching of exterior glazing by EF0–EF2 rated debris, specify glazing assemblies that have been designed and tested in accordance with ASTM E1886, using ASTM E1996 test missile D or preferably E. To avoid breaching of exterior glazing by EF3–EF5 rated debris, specify glazing assemblies that have been tested in accordance with AAMA 512, using test missiles given in ICC 500) or FEMA P-361 (FEMA 2015).

Critical Facilities. For design guidance to minimize building damage to critical facilities, see Tornado Recovery Advisory No. 6, *Critical Facilities Located in Tornado-Prone Regions: Recommendations for Architects and Engineers* (FEMA 2011) and FEMA P-908 (FEMA 2012). This Tornado Recovery Advisory includes detailed recommendations for three levels of enhancement for the MWFRS and C&C to minimize building damage.

C26.14.5 Continuity of Building Operations.
Designing a building to ensure that it will remain operational if struck by an EF4 or EF5 rated tornado is expensive. However, for those facilities such as emergency operations centers and hospitals where it is desired to avoid interrupted operations, see FEMA (2011) in FEMA P-908 (FEMA 2012). This Tornado Recovery Advisory includes detailed recommendations related to the MWFRS, the building envelope, HVAC, water, sewer, and emergency power.

C26.14.6 Trussed Communications Towers.
Damage investigations have shown that wind-borne debris can cling to trussed communications towers (FEMA 2012). Clinging debris such as metal roof panels and chain-link fence with privacy slats can increase A_f, the projected surface area subject to wind loading, thus increasing the net wind load on the structure, with tower collapse as a possible result. To minimize the collapse potential, towers can be designed for the additional wind load caused by clinging debris.

When it is desired to design for clinging debris, consider the potential debris sources in the vicinity of the tower. Based on FEMA (2012) damage observations, as a minimum, design for 40 ft^2 (3.7 m^2) of projected surface area of clinging debris located at midheight of the tower or 50 ft (15 m), whichever is lower.

REFERENCES

American Society of Mechanical Engineers (ASME). (1992). *Steel stacks*. STS-1, New York.

Applied Research Associates, Inc. (2001). *Hazard mitigation study for the Hawaii Hurricane Relief Fund*. ARA Report 0476, Raleigh, NC.

Beason, W. L., Meyers, G. E., and James, R. W. (1984). "Hurricane related window glass damage in Houston." *J. Struct. Eng.* 110(12), 2843–2857.

Behr, R. A., and Minor, J. E. (1994). "A survey of glazing system behavior in multi-story buildings during Hurricane Andrew." *The structural design of tall buildings*. 3, 143–161.

Bernardini, E., Spence, S. M. J., and Kareem, A. (2013a). "A probabilistic approach for the full response estimation of tall buildings with 3D modes using the HFFB." *Struct. Saf.*, 44, 91–101.

Bernardini, E., Spence, S. M. J., and Kareem, A. (2013b). "An efficient performance-based design approach to the high frequency force balance." *Proc., 11th Int. Conf. on Structural Safety and Reliability, ICOSSAR 2013*, G. Deodatis, B. Ellingwood, and D. Frangopol, eds., Taylor & Francis, London, 1777–1784.

Beste, F., and Cermak, J. E. (1996). "Correlation of internal and area-averaged wind pressures on low-rise buildings." Proc., 3rd Int. Colloquium on Bluff Body Aerodynamics and Applications, Virginia Polytechnic Institute, Blacksburg, VA.

Centers for Disease Control and Prevention (CDC). Tornado-related fatalities– Five states, Southeastern United States, April 25-28, 2011. Morbidity and Monthly Weekly Report, July 20, 2012, 61(28).

Chen, X., and Kareem, A. (2004). "Equivalent static wind loads on buildings: New model." *J. Struct. Eng.* 130(10), 1425–1435.

Chock, G., Peterka, J., and Yu, G. (2005). "Topographic wind speed-up and directionality factors for use in the city and county of Honolulu building code." Proc., 10th Americas Conf. on Wind Eng., Louisiana State University, Baton Rouge, LA.

Cleveland, W. S., and Devlin, S. J. (1988). "Locally weighted regression: An approach to regression analysis by local fitting." *J. Am. Stat. Assn.* 83(403): 596–610. Doi:10.2307/2289282.

Comité International des Cheminées Industrielles (CICIND). (1999). *Model code for steel chimneys, Revision 1-1999*, Zurich.

Davenport, A. G., Grimmond, C. S. B., Oke, T. R., and Wieringa, J. (2000). "Estimating the roughness of cities and sheltered country." Preprint, 12th AMS Conf. Applied Climatol., American Meteorological Society, Boston, 96–99.

Donelan, M. A., Haus, B. K., Reul, N., Plant, W. J., Stiassnie, M., Graber, H. C., et al. (2004). "On the limiting aerodynamic roughness of the ocean in very strong winds." *Geophys. Res. Lett.*, 31, 1–5.

Durst, C. S. (1960). "Wind speeds over short periods of time." *Meteorol*, 89, 181–187.

Ellingwood, B. (1981). "Wind and snow load statistics for probabilistic design." *J. Struct. Div.* 107(7), 1345–1350.

Ellingwood, B., MacGregor, J. G., Galambos, T. V., and Cornell, C. A. (1982). "Probability based load criteria: Load factors and load combinations." *J. Struct. Div.* 108(5), 978–997.

Engineering Sciences Data Unit (ESDU). (1982). *Strong winds in the atmospheric boundary layer, Part 1: Mean hourly wind speed*, ESDU 82026, London.

ESDU. (1990). *Characteristics of atmospheric turbulence near the ground. Part II: Single point data for strong winds (neutral atmosphere)*, ESDU 85020, London.

ESDU. (1993). *Strong winds in the atmospheric boundary layer, Part 2: Discrete gust speeds*, ESDU 83045, London.

European Convention for Structural Steelwork (ECCS). (1978). *Recommendations for the calculation of wind effects on buildings and structures*, Technical Committee T12, Brussels.

Federal Emergency Management Agency (FEMA). (2008). *Safe rooms for tornadoes and hurricanes: Guidance for community and residential safe rooms*. P-361, Second Edition. Federal Emergency Management Agency, Washington, DC.

FEMA. (2009). *Tornado protection: Selecting refuge areas in buildings*. P-431. Washington, DC ⟨http://www.fema.gov/media-library/assets/documents/2246⟩.

FEMA. (2011). *Critical facilities located in tornado-prone regions: Recommendations for architects and engineers*, Tornado Recovery Advisory 6 (RA6), Washington, DC ⟨http://www.fema.gov/media-library/assets/documents/25810⟩.

FEMA. (2012). *Mitigation assessment team report–Spring 2011 tornadoes: April 25–28 and May 22; Building performance observations, Recommendations and technical guidance*. P-908. Washington, DC ⟨http://www.fema.gov/media-library/assets/documents/25810⟩.

FEMA. (2014). *Taking shelter from the storm: Building a safe room for your home or small business*. P-320. Washington, DC ⟨http://www.fema.gov/media-library/assets/documents/2009⟩.

FEMA. (2015). *Safe rooms for tornadoes and hurricanes: Guidance for community and residential safe rooms*. P-361, Third Edition. Washington, DC ⟨https://www.fema.gov/media-library/assets/documents/3140⟩.

Goel, R. K., and Chopra, A. K. (1997). "Period formulas for moment-resisting frame buildings." *J. Struct. Eng.* 123(11), 1454–1461.

Goel, R. K., and Chopra, A. K. (1998). "Period formulas for concrete shear wall buildings." *J. Struct. Eng.* 124(4), 426–433.

Gurley, K., and Kareem, A. (1993). "Gust loading factors for tension leg platforms." *Appl. Ocean Res*, 15(3), 137–154.

Haan, F. L., Jr., Balaramudu, V. K., and Sarkar, P. P. (2010) "Tornado-induced wind loads on a low-rise building." *J. Struct. Eng.* 136(1), 106–116.

Harris, R. I., and Deaves, D. M. (1981). "The structure of strong winds." Proc., CIRIA Conf. Wind Eng. in the Eighties, Construction Industry Research and Information Association, London.

Ho, E. (1992). "Variability of low building wind lands." Ph.D. thesis, Univ. of Western Ontario, London, ON.

International Organization for Standardization (ISO). (1997). "Wind actions on structures." *ISO 4354*, Geneva.

Irwin, P. A. (1987). "Pressure model techniques for cladding loads." *J. Wind Eng. Indust. Aerodyn.* 29, 69–78.

Irwin, P. A. (2006). "Exposure categories and transitions for design wind loads." *J. Struct. Eng.* 132(11), 1755–1763.

Irwin, P. A., and Dunn, G. E. (1994). "Review of internal pressures on low-rise buildings." *RWDI Report 93-270 for Canadian Sheet Building Institute*, Feb. 23.

Isyumov, N., Ho, E., and Case, P. (2013). "Influence of wind directionality on wind loads and responses," *Proc., 12th Americas Conf. Wind Eng.*, D. Reed, and A Jain, eds.

Jackson, P. S., and Hunt, J. C. R. (1975). "Turbulent wind flow over a low hill." *Quart. J. Royal Meteorol. Soc.* 101, 929–955.

Jung, S., and Masters, F. J. (2013). "Characterization of open and suburban boundary layer wind turbulence in 2008 Hurricane Ike," *Wind Struct.* 17, 135–162.

Kareem, A. (1985). "Lateral-torsional motion of tall buildings." *J. Struct. Eng.* 111(11), 2479–2496.

Kareem, A. (1986). "Performance of cladding in Hurricane Alicia." *J. Struct. Eng.* 112(12), 2679–2693.

Kareem, A. (1992). "Dynamic response of high-rise buildings to stochastic wind loads." *J. Wind Eng. Indust. Aerodyn.* 41–44.

Kareem, A., Kwon, D. K., and Tamura, Y. (2012). "Cyberbased analysis, modeling and simulation of wind load effects in VORTEX-winds," Proc., 3rd American Assn. Wind Eng. Workshop, Hyannis, MA, August 12–14.

Kijewski-Correa, T., Kareem, A., Guo, Y. L., Bashor, R., and Weigand, T. (2013). "Performance of tall buildings in urban zones: Lessons learned from a decade of full-scale monitoring." *Int. J. High-Rise Bldgs.* 2(3), 179–192.

Kopp, G. A., and Morrison, M. J. (2011). Discussion of "Tornado-induced wind loads on low-rise buildings" by F. L. Haan, V. K. Balaramudu, and P. P. Sarkar, *J. Struct. Eng.* 137, 1620–1622.

Kwon, D.-K., and Kareem, A. (2013). "Comparative study of major international wind codes and standards for wind effects on tall buildings." *Eng. Struct.* 51, 23–25.

Kwon, D.-K., Kijewski-Correa, T., and Kareem, A. (2008). "e-Analysis of high-rise buildings subjected to wind loads." *J. Struct. Eng.* 133(7), 1139–1153.

Lemelin, D. R., Surry, D., and Davenport, A. G. (1988). "Simple approximations for wind speed-up over hills." *J. Wind Eng. Indust. Aerodyn.* 28, 117–127.

Lettau, H. (1969). "Note on aerodynamic roughness element description." *J. Appl. Meteorol.* 8, 828–832.

Lombardo, F., Pintar, A., Vickery, P. J., Simiu, E., and Levitan, M. (2016). Development of new wind speed maps for ASCE 7–16. NIST Special Publication. *In Preparation*.

McDonald, J. R., and Mehta, K. C. (2006). *A recommendation for an enhanced Fujita Scale (EFScale)*, Wind Science and Engineering Center, Texas Tech University, Lubbock, TX. ⟨http://www.depts.ttu.edu/nwi/Pubs/FScale/EFScale.pdf⟩.

Means, B., Reinhold, T. A., and Perry, D. C. (1996). "Wind loads for low-rise buildings on escarpments." In *Building an international community of structural engineers*, S. K. Ghosh, and J. Mohammadi, eds., ASCE, Reston, VA, 1045–1052.

Mehta, K. C., and Marshall, R. D. (1998). *Guide to the use of the wind load provisions of ASCE 7-95*, ASCE, Reston, VA.

Mehta, K. C., McDonald, J. R., and Minor, J. E. (1976). "Windspeeds analyses of April 3–4 1974 tornadoes." *J. Struct. Div.* 102(9), 1709–1724.

Minor, J. E. (1982). "Tornado technology and professional practice." *J. Struct. Div.* 108(11), 2411–2422.

Minor, J. E. (1985). "Window glass performance and hurricane effects." In *Hurricane Alicia: One year later*, A. Kareem, ed., ASCE, Reston, VA, 151–167.

Minor, J. E. (1994). "Windborne debris and the building envelope." *J. Wind Eng. Indust. Aerodyn.*, 53, 207–227.

Minor, J. E., and Behr, R. A. (1993). "Improving the performance of architectural glazing in hurricanes." In *Hurricanes of 1992: Lessons learned and implications for the future*, ASCE, Reston, VA, 476–485.

Minor, J. E., McDonald, J. R., and Mehta, K. C. (1977). "The tornado: An engineering oriented perspective." *TM ERL NSSL-82*, National Oceanic and Atmospheric Administration, Environmental Research Laboratories, Boulder, CO. A 1993 reprint is available through NTIS as PB93148435. ⟨http://www.depts.ttu.edu/nwi/Pubs/Reports/The%20Tornado.pdf⟩.

Mishra, A.R., James, D. J, and Letchford, C. W. (2008). "Physical simulation of a single celled 4 tornado-like vortex. Part A, *JWEIA* 96, 1243–1251; Part B, *JWEIA* 96, 1258–1272.

NatHaz Aerodynamic Loads Database (NALD). (2012). "NatHaz aerodynamic loads database, version 2." ⟨http://aerodata.ce.nd.edu/⟩.

National Hurricane Center (NHC). (2015). Saffir-Simpson Hurricane Wind Scale, National Hurricane Center. ⟨http://www.nhc.noaa.gov/aboutsshws.php⟩.

Peterka, J. A., and Shahid, S. (1998). "Design gust wind speeds in the United States." *J. Struct. Eng.* 124(2), 207–214.

Pickands, James III. (1971). "The two-dimensional poisson process and extremal processes." *J. Appl. Prob.* 8(4) 745–56. Doi: 10.2307/3212238.

Pintar, A., Simiu, E., Lombardo, F., and Levitan, M. (2015). Maps of non-hurricane non-tornadic wind speeds with specified mean recurrence intervals for the contiguous united states using a two-dimensional poisson process extreme value model and local regression. NIST Special Publication 500-301, National Institute of Standards and Technology ⟨http://dx.doi.org/10.6028/NIST.SP.500-301⟩.

Powell, M. D. (1980). "Evaluations of diagnostic marine boundary-layer models applied to hurricanes." *Monthly Weather Rev.*, 108(6), 757–766.

Powell, M. D., Vickery, P. J., and Reinhold, T. A. (2003). "Reduced drag coefficients for high wind speeds in tropical cyclones." *Nature*, 422, 279–283.

Prevatt, D. O., van de Lindt, J. W., Back, E. W., Graettinger, A. J., Shiling, P, and Coulbourne, W, et al. (2012). "Making the case for improved structural design: Tornado outbreaks of 2011." *Leadership Mgmt Eng.* 12(4), 254–270.

Ramsdell, J. V., and Rishel, J.P. (2007). *Tornado Climatology of the Contiguous United States*, U.S. Nuclear Regulatory Commission, NUREG/CR-4461, Rev 2, Washington, DC. http://pbadupws.nrc.gov/docs/ML0708/ML070810400.pdf

Reinhold, T. A., ed. (1982). "Wind tunnel modeling for civil engineering applications." In *Proc., Int. Workshop on Wind Tunnel Modeling Criteria and Techniques in Civil Engineering Applications*, Cambridge University Press, Gaithersburg, MD.

Roueche, D. B., and Prevatt, D. O. (2013). "Residential damage patterns following the 2011 Tuscaloosa, AL and Joplin, MO tornadoes." *Journal of Disaster Research*, 8(6), 1061–1067.

Sataka, N., Suda, K., Arakawa, T., Sasaki, A., and Tamura, Y. (2003). "Damping evaluation using full-scale data of buildings in Japan." *J. Struct. Eng.*, 129(4), 470–477.

Simiu, Emil. (2011). *Design of Buildings for Wind–A guide for ASCE 7-10 Standard Users and Designers of Special Structures*, 2nd ed., John Wiley and Sons, Inc.

Simiu, E., and Scanlan, R. H. (1996). *Wind effects on structures*, 3rd Ed., John Wiley & Sons, New York.

Simiu, E., Vickery, P., and Kareem, A. (2007). "Relation between Saffir-Simpson hurricane scale wind speeds and peak 3-s gust speeds over open terrain," *J. Struct. Eng.*, 133(7), 1043–1045.

Smith, Richard L. (1989). "Extreme Value Analysis of Environmental Time Series: An Application to Trend Detection in Ground-Level Ozone." *Statistical Science*, 4(4): 367–77.

Solari, G. (1993a). "Gust buffeting. I: Peak wind velocity and equivalent pressure." *J. Struct. Eng.*, 119(2), 365–382.

Solari, G. (1993b). "Gust buffeting. II: Dynamic alongwind response." *J. Struct. Eng.*, 119(2), 383–398.

Solari, G., and Kareem, A. (1998). "On the formulation of ASCE 7-95 gust effect factor." *J. Wind Eng. Indust. Aerodyn.*, Vols. 77–78, 673–684.

Spence, S., and Kareem, A. "Tall Buildings and Damping: A Concept-Based Data Driven Model." *J. Struct. Eng.*, ASCE, Vol. 140, No. 5, 2014

Standards Australia. (1994). "Design of steel lattice towers and masts." *AS3995-1994*, Standards Australia, North Sydney, Australia.

Stathopoulos, T., Surry, D., and Davenport, A. G. (1979). "Wind-induced internal pressures in low buildings." In *Proc., 5th Int. Conf. on Wind Engineering*, J. E. Cermak, ed. Colorado State University, Fort Collins, CO.

Stubbs, N., and Perry, D. C. (1993). "Engineering of the building envelope: To do or not to do." In *Hurricanes of 1992: Lessons learned and implications for the future*, R. A. Cook, and M. Sotani, eds., ASCE Press, Reston, VA, 10–30.

Surry, D., Kitchen, R. B., and Davenport, A. G. (1977). "Design effectiveness of wind tunnel studies for buildings of intermediate height." *Can. J. Civ. Eng.*, 4(1), 96–116.

Twisdale, L. A., Vickery, P. J., and Steckley, A. C. (1996). *Analysis of hurricane windborne debris impact risk for residential structures*, State Farm Mutual Automobile Insurance Companies, Bloomington, IL.

U.S. Nuclear Regulatory Commission (NRC). (2011). *Technical basis for regulatory guidance on design-basis hurricane wind speeds for nuclear power plants* (NUREG/CR-7005). US Nuclear Regulatory Commission, Washington, DC.

van de Lindt, J., Pei, S., Prevatt, D., Dao, T., Coulbourne, W., Graettinger, A., and Gupta, R. (2012). "Dual Objective Design Philosophy for Tornado Engineering" In *Structures Congress 2012*, American Society of Civil Engineering, Reston, VA., 965–976

Vickery, B. J., and Bloxham, C. (1992). "Internal pressure dynamics with a dominant opening." *J. Wind Eng. Indust. Aerodyn.*, 41–44, 193–204.

Vickery, P. J., and Skerlj, P. F. (2005). "Hurricane gust factors revisited." *J. Struct. Eng.*, 131(5), 825–832.

Vickery, P. J., and Wadhera, D. (2008a). "Development of design wind speed maps for the Caribbean for application with the wind load provisions of ASCE 7. *ARA Report 18108-1* prepared for Pan American Health Organization, Regional Office for the Americas, World Health Organization, Disaster Management Programme, Washington, DC.

Vickery, P. J., and Wadhera, D. (2008b). "Statistical models of the Holland pressure profile parameter and radius to maximum winds of hurricanes from flight level pressure and H* wind data." *J. Appl. Meteor, 47, 2497-2517.*

Vickery, P. J., Skerlj, P. F., Steckley, A. C., and Twisdale, L. A. (2000). "Hurricane wind field model for use in hurricane simulations." *J. Struct. Eng.*, 126(10), 1203–1221.

Vickery, P. J., Wadhera, D., Galsworthy, J., Peterka, J. A., Irwin, P. A., and Griffis, L. A. (2010). "Ultimate wind load design gust wind speeds in the United States for use in ASCE-7." *J. Struct. Eng.*, 136, 613–625.

Vickery, P. J., Wadhera, D., Powell, M. D., and Chen, Y. (2009a). "A hurricane boundary layer and wind field model for use in engineering applications." *J. Appl. Meteorology*. 48, 381–405.

Vickery, P. J., Wadhera, D., Twisdale, L. A., Jr., and Lavelle, F. M. (2009b). "U.S. hurricane wind speed risk and uncertainty." *J. Struct. Eng.*, 135(3), 301–320.

Walmsley, J. L., Taylor, P. A., and Keith, T. (1986). "A simple model of neutrally stratified boundary-layer flow over complex terrain with surface roughness modulations." *Boundary-Layer Meteorol.*, 36, 157–186.

Wen, Y.-K., and Chu, S.-L. (1973). "Tornado risks and design wind speed." *J. Struct. Div.*, 99(12), 2409–2421.

Wieringa, J., Davenport, A. G., Grimmond, C. S. B., and Oke, T. R. (2001). "New revision of Davenport roughness classification." ⟨www.kcl.ac.uk/ip/suegrimmond/publishedpapers/DavenportRoughness2.pdf⟩ (Jan. 7, 2008).

Wyatt, T. A. (1984). "Sensitivity of lattice towers to fatigue induced by wind gusts." *Eng. Struct.*, 6, 262–267.

Yeatts, B. B., and Mehta, K. C. (1993). "Field study of internal pressures." In *Proc., 7th U.S. Nat. Conf. on Wind Engineering*, 2, 889–897.

Zhou, Y., and Kareem, A. (2001a). "Gust loading factor: New model." *J. Struct. Eng.*, 127(2), 168–175.

Zhou, Y., and Kareem, A. (2001b). "Equivalent static lateral forces on buildings under seismic and wind effects." *J. Wind Eng.*, 89, 605–608.

Zhou, Y., Kijewski, T., and Kareem, A. (2002). "Along-wind load effects on tall buildings: Comparative study of major international codes and standards." *J. Struct. Eng.*, 128(6), 788–796.

OTHER REFERENCES (NOT CITED)

ASCE. (1987). "Wind tunnel model studies of buildings and structures." *ASCE Manuals and Reports of Engineering Practice No. 67*, Reston, VA.

ASTM International. (2006). "Standard specification for rigid poly(vinyl chloride) (PVC) siding." *ASTM D3679-06a*, West Conshohocken, PA.

ASTM. (2007). "Standard test method for wind resistance of sealed asphalt shingles (uplift force/uplift resistance method)." *ASTM D7158-07*, West Conshohocken, PA.

Cook, N. (1985). *The designer's guide to wind loading of building structures, Part I: Background, damage survey, wind data and structural classification*. Building Research Establishment and Butterworths, London.

Defense Civil Preparedness Agency. (1975). "Interim guidelines for building occupants' protection from tornadoes and extreme winds." *TR-83A*, Superintendent of Documents, U.S. Government Printing Office, Washington, DC.

Georgiou, P. N. (1985). "Design wind speeds in tropical cyclone regions." *Doctoral dissertation*, University of Western Ontario, London, Ontario, Canada.

Georgiou, P. N., Davenport, A. G., and Vickery, B. J. (1983). "Design wind speeds in regions dominated by tropical cyclones." *J. Wind Eng. Indust. Aerodyn.*, 13, 139–152.

Holmes, J. D. (2001). *Wind loads on structures*, SPON Press/Taylor & Francis, New York.

Isyumov, N. (1982). "The aeroelastic modeling of tall buildings." In *Proc., Int. Workshop on Wind Tunnel Modeling Criteria and Techniques in Civil Engineering Applications*, T. Reinhold, ed., Cambridge University Press, New York, 373–407.

Jeary, A. P., and Ellis, B. R. (1983). "On predicting the response of tall buildings to wind excitation." *J. Wind Eng. Indust. Aerodyn.*, 13, 173–182.

Kala, S., Stathopoulos, T., and Kumar, K. (2008). "Wind loads on rainscreen walls: Boundary-layer wind tunnel experiments." *J. Wind Eng. Indust. Aerodyn.*, 96(6–7), 1058–1073.

Kareem, A., and Smith, C. E. (1994). "Performance of off-shore platforms in Hurricane Andrew." In *Hurricanes of 1992: Lessons learned and implications for the future*, R. A. Cook, and M. Soltani, eds., ASCE, New York, NY, 577–586.

Kijewski, T., and Kareem, A. (1998). "Dynamic wind effects: A comparative study of provisions in codes and standards with wind tunnel data." *J. Wind and Struct.*, 1(1), 77–109.

Krayer, W. R., and Marshall, R. D. (1992). "Gust factors applied to hurricane winds." *Bull. Am. Meteorol. Soc.*, 73, 613–617.

Kwon, D-K., and Kareem, A. (2013). "A multiple database-enabled design module with embedded features of international codes and standards." *International Journal of High-Rise Buildings*, CTBUH, 2(3), 257–269.

Kwon, K. K., Spence, S. M. J., and Kareem, A. (2014), "A cyberbased data-enabled design framework for high-rise buildings driven by synchronously measured surface pressures." *Adv. Eng. Softw.*, 77, 13–27.

Liu, H. (1999). *Wind engineering: A handbook for structural engineers*, Prentice Hall, New York.

MacDonald, P. A., Kwok, K. C. S., and Holmes, J. H. (1986). "Wind loads on isolated circular storage bins, silos and tanks: Point pressure measurements." *Research Report No. R529*, School of Civil and Mining Engineering, University of Sydney, Sydney, Australia.

Myer, M. F., and White, G. F. (2003). "Communicating damage potentials and minimizing hurricane damage," In *Hurricane! Coping with disaster: Progress and challenges since Galveston, 1900*, R. Simpson, ed., American Geophysical Union, Washington, DC.

Perry, D. C., Stubbs, N., and Graham, C. W. (1993). "Responsibility of architectural and engineering communities in reducing risks to life, property and economic loss from hurricanes." In *Hurricanes of 1992: Lessons learned and implications for the future*, ASCE Press, Reston, VA.

Solari, G. and Kareem, Al. "On the Formulation of ASCE 7-95 Gust Effect Factor." *J. Wind Eng. Indust. Aerodyn.*, Vols. 77-78, 1998, pp 673–684.

Standards Australia. (1989). *Australian standard SAA loading code, Part 2: Wind loads*. Standards House, North Sydney, NSW, Australia.

Stathopoulos, T., Surry, D., and Davenport, A. G. (1980). "A simplified model of wind pressure coefficients for low-rise buildings." In *Proc., 4th Colloquium on Industrial Aerodynamics*, Part 1, Gesellschaft der Freunde der Fachhochschule Aachen, Aachen, pp. 17–31.

Stubbs, N., and Boissonnade, A.C. (1993). "A damage simulation model for building contents in a hurricane environment." In *Proc., 7th U.S. Nat. Conf. on Wind Engineering*, 2, 759–771.

van de Lindt, J. W., Pei, S., Dao, T. N., Graettinger, A., Prevatt, D. O., Gupta, R., and Coulbourne, W. (2013). "Dual Objective-Based Tornado Design Philosophy." *J. Struct. Eng.*, 139(2); 251–263.

Vickery, B. J., Davenport, A. G., and Surry, D. (1984). "Internal pressures on low-rise buildings." In *Proc., 4th Canadian Workshop on Wind Engineering* Canadian Wind Engineering Association, Ottawa, Canada.

Vickery, P. J., and Skerlj, P. F. (1998). "On the elimination of exposure D along the hurricane coastline in ASCE-7." *Report for Andersen Corporation by Applied Research Associates, ARA Project 4667*, Applied Research Associates, Raleigh, NC.

Vickery, P. J., Skerlj, P. F., and Twisdale, L. A. (2000). "Simulation of hurricane risk in the U.S. using empirical track model." *J. Struct. Eng.*, 126(10), 1222–1237.

Vickery, P. J., and Twisdale, L. A. (1995a). "Prediction of hurricane wind speeds in the United States." *J. Struct. Eng.*, 121(11), 1691–1699.

Vickery, P. J., and Twisdale, L. A. (1995b). "Wind-field and filling models for hurricane wind-speed predictions." *J. Struct. Eng.*, 121(11), 1700–1709.

Womble, J. A., Yeatts, B. B., and Mehta, K. C. (1995). "Internal wind pressures in a full and small scale building." In *Proc., 9th Int. Conf. on Wind Engineering*, Wiley Eastern Ltd., New Delhi.

Zhou, Y., and Kareem, A. (2003). "Aeroelastic Balance." *Journal of Engineering Mechanics*, ASCE, 129(3), 283–292.

Zhou, Y., Kareem, A., and Gu, M. (2000). "Equivalent static buffeting loads on structures." *J. Struct. Eng.*, 126(8), 989–992.

Zhou, Y., Kijewski, T., and Kareem, A. (2003). "Aerodynamic loads on tall buildings: Interactive database." *J. Struct. Eng.*, 129(3), 394–404.

CHAPTER C27
WIND LOADS ON BUILDINGS: MAIN WIND FORCE RESISTING SYSTEM (DIRECTIONAL PROCEDURE)

The Directional Procedure is the former "buildings of all heights" provision in Method 2 of ASCE 7-05 for the main wind force resisting system (MWFRS). A simplified method based on this Directional Procedure is provided for buildings up to 160 ft (49 m) in height. The Directional Procedure is considered the traditional approach in that the pressure coefficients reflect the actual loading on each surface of the building as a function of wind direction, namely, winds perpendicular or parallel to the ridge line.

C27.1 SCOPE

C27.1.5 Minimum Design Wind Loads. This section specifies a minimum wind load to be applied horizontally on the entire vertical projection of the building, as shown in Fig. C27.1-1. This load case is to be applied as a separate load case in addition to the normal load cases specified in other portions of this chapter.

PART 1: ENCLOSED, PARTIALLY ENCLOSED, AND OPEN BUILDINGS OF ALL HEIGHTS

C27.3 WIND LOADS: MAIN WIND FORCE RESISTING SYSTEM

C27.3.1 Enclosed and Partially Enclosed Rigid and Flexible Buildings. In Eqs. (27.3-1) and (27.3-2), a velocity pressure term, q_i, appears that is defined as the "velocity pressure for internal pressure determination." The positive internal pressure is dictated by the positive exterior pressure on the windward face at the point where there is an opening. The positive exterior pressure at the opening is governed by the value of q at the level of the opening, not q_h. For positive internal pressure evaluation, q_i may conservatively be evaluated at height $h (q_i = q_h)$. For low buildings, this evaluation does not make much difference, but for the example of a 300-ft- (91.4-m)-tall building in Exposure B with a highest opening at 60 ft (18.2 m), the difference between q_{300} and q_{60} represents a 59% increase in internal pressure. This difference is unrealistic and represents an unnecessary degree of conservatism. Accordingly, $q_i = q_z$ for positive internal pressure evaluation in partially enclosed buildings where height z is defined as the level of the highest opening in the building that could affect the positive internal pressure. For buildings sited in wind-borne debris regions, with glazing that is not impact-resistant or protected with an impact-protective system, q_i should be treated on the assumption that there is an opening.

Fig. 27.3-1. The pressure coefficients for MWFRSs are separated into two categories:

1. Directional Procedure for buildings of all heights (Fig. 27.3-1) as specified in Chapter 27 for buildings that meet the requirements specified therein.
2. Envelope Procedure for low-rise buildings that have a height less than or equal to 60 ft (18.3 m) (Fig. 28.3-1) as specified in Chapter 28 for buildings that meet the requirements specified therein.

In generating these coefficients, two distinctly different approaches were used. For the pressure coefficients given in Fig. 27.3-1, the more traditional approach was followed, and the pressure coefficients reflect the actual loading on each surface of the building as a function of wind direction, namely, winds perpendicular or parallel to the ridge line.

Observations in wind tunnel tests show that areas of very low negative pressure and even slightly positive pressure can occur in all roof structures, particularly as the distance from the windward edge increases and the wind streams reattach to the surface. These pressures can occur even for relatively flat or low-slope roof structures. Experience and judgment from wind tunnel studies have been used to specify either zero or slightly negative pressures (-0.18) depending on the negative pressure coefficient. These values require the designer to consider a zero or slightly positive net wind pressure in the load combinations of Chapter 2.

Fig. 27.3-2. Frame loads on dome roofs are adapted from the Eurocode (1995). The loads are based on data obtained in a modeled atmospheric boundary-layer flow that does not fully comply with requirements for wind tunnel testing specified in this standard (Blessman 1971). Loads for three domes ($h_D/D = 0.5$, $f/D = 0.5$), ($h_D/D = 0$, $f/D = 0.5$), and ($h_D/D = 0$, $f/D = 0.33$) are roughly consistent with data of Taylor (1991), who used an atmospheric boundary layer as required in this standard. Two load cases are defined, one of which has a linear variation of pressure from A to B as in the Eurocode (1995), and one in which the pressure at A is held constant from 0° to 25°; these two cases are based on comparison of the Eurocode provisions with Taylor (1991). Case A (the Eurocode calculation) is necessary in many cases to define maximum uplift. Case B is necessary to properly define positive pressures for some cases, which cannot be isolated with current information and which result in maximum base shear. For domes larger than 200 ft (61 m) in diameter, the designer should consider use of wind tunnel testing. Resonant response is not considered in these provisions. Wind tunnel testing should be used to consider resonant response. Local bending moments in the dome shell may be larger than predicted by this method because of the difference between instantaneous local pressure distributions and those predicted by Fig. 27.3-2.

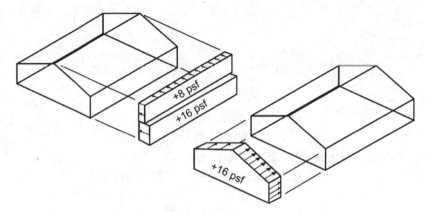

FIGURE C27.1-1. Application of Minimum Wind Load
Note: 1.0 lb/ft² = 0.0479 kN/m²

If the dome is supported on vertical walls directly below, it is appropriate to consider the walls as a "chimney" using Fig. 29.4-1.

Fig. 27.3-3. The pressure and force coefficient values in these tables were taken from ANSI A58.1-1972 (1972). Additional information was added in ANSI A58.1-1982 (1982), which included use of a 1.2 multiplier for component and cladding (C&C) external pressure coefficients, other than for the perimeter areas. That multiplier was changed from 1.2 to 0.87 in ASCE 7-95 (1998), however, no substantiation was provided for the change. The multiplier is changed back to 1.2 in ASCE 7-16. The coefficients specified in these tables are based on wind tunnel tests conducted under conditions of uniform flow and low turbulence, and their validity in turbulent boundary-layer flows has yet to be completely established. Additional pressure coefficients for conditions not specified herein may be found in SIA (1956) and ASCE (1961).

C27.3.2 Open Buildings with Monoslope, Pitched, or Troughed Free Roofs. Figs. 27.3-4 through 27.3-6 and 30.7-1 through 30.7-3 are presented for wind loads on MWFRSs and C&C of open buildings with roofs as shown, respectively. This work is based on the Australian standard AS1170.2-2002, Part 2: Wind Actions, with modifications to the MWFRS pressure coefficients based on recent studies (Altman 2001; Uematsu and Stathopoulos 2003).

Two load cases, A and B, are given in Figs. 27.3-4 through 27.3-6. These pressure distributions provide loads that envelop the results from detailed wind tunnel measurements of simultaneous normal forces and moments. Application of both load cases is required to envelop the combinations of maximum normal forces and moments that are appropriate for the particular roof shape and blockage configuration.

The roof wind loading on open building roofs is highly dependent upon whether goods or materials are stored under the roof and restrict the wind flow. Restricting the flow can introduce substantial upward-acting pressures on the bottom surface of the roof, thus increasing the resultant uplift load on the roof. Figs. 27.3-4 through 27.3-6 and 30.7-1 through 30.7-3 offer the designer two options. Option 1 (clear wind flow) implies that little (less than 50%) or no portion of the cross section below the roof is blocked. Option 2 (obstructed wind flow) implies that a significant portion (more than 75% is typically referenced in the literature) of the cross section is blocked by goods or materials below the roof. Clearly, values would change from one set of coefficients to the other following some sort of smooth, but as yet unknown, relationship. In developing the provisions included in this standard, the 50% blockage value was selected for Option 1, with the expectation that it represents a somewhat conservative transition. If the designer is not clear about usage of the space below the roof or if the usage could change to restrict free air flow, then design loads for both options should be used.

See Section 28.3.5 for explanation of the horizontal wind loads on open buildings with transverse frames and pitched roofs that act in combination with the loads calculated in Section 27.3.3.

C27.3.5 Design Wind Load Cases. Wind tunnel research (Isyumov 1983; Boggs et al. 2000; Isyumov and Case 2000; Xie and Irwin 2000) has shown that torsional load is caused by nonuniform pressure on the different faces of the building from wind flow around the building, interference effects of nearby buildings and terrain, and dynamic effects on more flexible buildings. Load Cases 2 and 4 in Fig. 27.3-8 specifies the torsional loading to 15% eccentricity under 75% of the maximum wind shear for Load Case 2. Although this is more in line with wind tunnel experience on square and rectangular buildings with aspect ratios up to about 2.5, it may not cover all cases, even for symmetric and common building shapes where larger torsions have been observed. For example, wind tunnel studies often show an eccentricity of 5% or more under full (not reduced) base shear. The designer may wish to apply this level of eccentricity at full wind loading for certain more-critical buildings, even though it is not required by the standard. The present more-moderate torsional load requirements can, in part, be justified by the fact that the design wind forces tend to be upper bound for most common building shapes.

In buildings with some structural systems, more severe loading can occur when the resultant wind load acts diagonally to the building. To account for this effect and the fact that many buildings exhibit maximum response in the across-wind direction (the standard currently has no analytical procedure for this case), a structural system should be capable of resisting 75% of the design wind load applied simultaneously along each principal axis, as required by Case 3 in Fig. 27.3-8.

For flexible buildings, dynamic effects can increase torsional loading. Additional torsional loading can occur because of eccentricity between the elastic shear center and the center of mass at each level of the building. Eq. (27.3-4) accounts for this effect.

It is important to note that significant torsion can occur on low-rise buildings also (Isyumov and Case 2000), and therefore, the wind loading requirements of Section 27.3.5 are now applicable to buildings of all heights.

As discussed in Chapter 31, the wind tunnel procedure should always be considered for buildings with unusual shapes, rectangular buildings with larger aspect ratios, and dynamically sensitive buildings. The effects of torsion can more accurately be determined for these cases and for the more normal building shapes using the wind tunnel procedure.

PART 2: ENCLOSED SIMPLE DIAPHRAGM BUILDINGS WITH $h \leq 160$ ft ($h \leq 48.8$ m)

This section was added to ASCE 7-10 to cover the common practical cases of enclosed simple diaphragm buildings up to height $h = 160$ ft ($h = 48.8$ m). Two classes of buildings are covered by this method. Class 1 buildings have $h \leq 60$ ft ($h \leq 18.3$ m) with plan aspect ratios L/B between 0.2 and 5.0. Cases A through F are described in Appendix D to allow the designer to establish the lines of resistance of the MWFRS in each direction so that the torsional load cases of Fig. 27.3-8 need not be considered. Class 2 buildings have 60 ft < $h \leq 160$ ft (18.3 m < $h \leq 48.8$ m) with plan aspect ratios of L/B between 0.5 and 2.0. Cases A through E of Appendix D are described to allow the designer to establish the lines of resistance of the MWFRS so that the torsional load cases of Fig. 27.3-8 need not be considered.

For the type of buildings covered in this method, the internal building pressure cancels out and need not be considered for the design of the MWFRS. Design net wind pressures for roofs and walls are tabulated directly in Tables 27.5-1 and 27.5-2 using the Directional Procedure as described in Part 1. Guidelines for determining the exterior pressures on windward, leeward, and sidewalls are provided in footnotes to Table 27.5-1.

The requirements in Class 2 buildings for natural building frequency ($75/h$) and structural damping ($\beta = 1.5\%$ critical) are necessary to ensure that the gust-effect factor, G_f, which has been calculated and built into the design procedure, is consistent with the tabulated pressures. The frequency of $75/h$ represents a reasonable lower bound to values found in practice. If calculated frequencies are found to be lower, then consideration should be given to stiffening the building. A structural damping value of 1.5%, applicable at the ultimate wind speeds as defined in the new wind speed maps, is conservative for most common building types and is consistent with a damping value of 1% for the ultimate wind speeds divided by $\sqrt{1.6}$, as contained in the ASCE 7-05 wind speed map. Because Class 1 buildings are limited to $h \leq 60$ ft, ($h \leq 18.3$ m) the building can be assumed to be rigid as defined in the glossary, and the gust-effect factor can be assumed to be 0.85. For this class of buildings, frequency and damping need not be considered.

C27.5 WIND LOADS: MAIN WIND FORCE RESISTING SYSTEM

C27.5.1 Wall and Roof Surfaces: Class 1 and 2 Buildings. Wall and roof net pressures are shown in Tables 27.5-1 and 27.5-2 and are calculated using the external pressure coefficients in Fig. 27.3-1. Along-wind net wall pressures are applied to the projected area of the building walls in the direction of the wind, and exterior sidewall pressures are applied to the projected area of the building walls normal to the direction of the wind acting outward, simultaneously with the roof pressures from Table 27.5-2. Distribution of the net wall pressures between windward and leeward wall surfaces is defined in Note 4 of Table 27.5-1. The magnitude of exterior sidewall pressure is determined from Note 2 of Table 27.5-1. It is to be noted that all tabulated pressures are defined without consideration of internal pressures because internal pressures cancel out when considering the net effect on the MWFRS of simple diaphragm buildings. Where the net wind pressure on any individual wall surface is required, internal pressure must be included as defined in Part 1 of Chapter 27.

The distribution of wall pressures between windward and leeward wall surfaces is useful for the design of floor and roof diaphragm elements such as drag strut collector beams, as well as for MWFRS wall elements. The values defined in Note 4 of Table 27.5-1 are obtained as follows: The external pressure coefficient for all windward walls is $C_p = 0.8$ for all L/B values. The leeward wall C_p value is (-0.5) for L/B values from 0.5 to 1.0 and is (-0.3) for $L/B = 2.0$. Noting that the leeward wall pressure is constant for the full height of the building, the leeward wall pressure can be calculated as a percentage of the p_h value in the table. The percentage is $0.5/(0.8 + 0.5) \times 100 = 38\%$ for $L/B = 0.5$ to 1.0. The percentage is $0.3/(0.8 + 0.3) \times 100 = 27\%$ for $L/B = 2.0$. Interpolation between these two percentages can be used for L/B ratios between 1.0 and 2.0. The windward wall pressure is then calculated as the difference between the total net pressure from the table using the p_h and p_0 values and the constant leeward wall pressure.

Sidewall pressures can be calculated in a similar manner to the windward and leeward wall pressures by taking a percentage of the net wall pressures. The C_p value for sidewalls is (-0.7). Thus, for $L/B = 0.5$ to 1.0, the percentage is $0.7/(0.8 + 0.5) \times 100 = 54\%$. For $L/B = 2.0$, the percentage is $0.7/(0.8 + 0.3) \times 100 = 64\%$. Note that the sidewall pressures are constant up the full height of the building.

The pressures tabulated for this method are based on simplifying conservative assumptions made to the different pressure coefficient (GC_p) cases tabulated in Fig. 27.3-1, which is the basis for the traditional all heights building procedure (defined as the Directional Procedure in ASCE 7-10) that has been a part of the standard since 1972. The external pressure coefficients, C_p, for roofs have been multiplied by 0.85, a reasonable gust-effect factor for most common roof framing, and then combined with an internal pressure coefficient for enclosed buildings (± 0.18) to obtain a net pressure coefficient to serve as the basis for pressure calculation. The linear wall pressure diagram has been conceived so that the applied pressures from the table produce the same overturning moment as the more exact pressures from Part 1 of Chapter 27. For determination of the wall pressures tabulated, the actual gust-effect factor has been calculated from Eq. (26.11-10) based on building height, wind speed, exposure, frequency, and the assumed damping value.

C27.5.2 Parapets. The effect of parapet loading on the MWFRS is specified in Section 27.3.5 of Part 1. The net pressure coefficient for the windward parapet is $+1.5$ and for the leeward parapet is -1.0. The combined effect of both produces a net coefficient of $+2.5$ applied to the windward surface to account for the cumulative effect on the MWFRS in a simple diaphragm building. This pressure coefficient compares to a net pressure coefficient of $1.3G_f$ for the tabulated horizontal wall pressure p_h at the top of the building. Assuming that a lower bound gust-effect factor $G_f = 0.85$, the ratio of the parapet pressure to the wall pressure is $2.5/(0.85 \times 1.3) = 2.25$. Thus, a value of 2.25 is assumed as a reasonable constant to apply to the tabulated wall pressure p_h to account for the additional parapet loading on the MWFRS.

C27.5.3 Roof Overhangs. The effect of vertical wind loading on a windward roof overhang is specified in Section 27.3.4 of Part 1. A positive pressure coefficient of +0.8 is specified. This compares to a net pressure coefficient tabulated for the windward edge Zone 3 of -1.06 (derived from $0.85 \times -1.3 \times 0.8 - 0.18$). The 0.85 factor represents the gust-effect factor G, the 0.8 multiplier accounts for the effective wind area reduction to the 1.3 value of C_p specified in Fig. 27.3-1 of Part 1, and the -0.18 is the internal pressure contribution. The ratio of coefficients is $0.8/1.06 = 0.755$. Thus, a multiplier of 0.75 on the tabulated pressure for Zone 3 in Table 27.5-2 is specified.

REFERENCES

Altman, D. R. (2001). "Wind uplift forces on roof canopies." Master's thesis, Dept. of Civil Engineering, Clemson Univ., Clemson, SC.

ASCE. (1961). "Wind forces on structures." *Trans. ASCE*, 126(2), 1124–1198.

ASCE. (1998). *Minimum Design Loads for Buildings and Other Structures*, ASCE 7-95, New York.

American National Standards Institute (ANSI). (1972). *Minimum Design Loads for Buildings and Other Structures*, ANSI A58.1-1972, Washington, DC.

ANSI. (1982). *Minimum Design Loads for Buildings and Other Structures*, ANSI A58.1-1982, Washington, DC.

Blessman, J. (1971). "Pressures on domes with several wind profiles." *Proc., 3rd Int. Conf. on Wind Effects on Buildings and Structures*, Japanese Organizing Committee, Tokyo, 317–326.

Boggs, D. W., Hosoya, N., and Cochran, L. (2000). "Sources of torsional wind loading on tall buildings: Lessons from the wind tunnel." *Proc., Structures Congress 2000: Advanced technology in structural engineering*, P. E. Mohamed Elgaaly, ed., ASCE, Reston, VA.

Eurocode. (1995). "Eurocode 1: Basis of design and actions on structures, Part 2–4: Actions on structures–wind actions." *ENV 1991-2-4*, European Committee for Standardization, Brussels.

Isyumov, N. (1983). "Wind induced torque on square and rectangular building shapes." *J. Wind Eng. Indust. Aerodyn.* 13, 183–186.

Isyumov, N., and Case, P. C. (2000). "Wind-induced torsional loads and responses of buildings." *Proc., Structures Congress 2000: Advanced technology in structural engineering*, P. E. Mohamad Elgaaly, ed., ASCE, Reston, VA.

Standards Australia. (2002). *Structural Design Actions, Part 2: Wind Actions*, AS/NZS 1170.2:2002. Sydney, New South Wales, Australia.

Swiss Society of Engineers and Architects (SIA). (1956). "Normen fur die Belastungsannahmen, die Inbetriebnahme und die Uberwachung der Bauten." *SIA Technische Normen No. 160*, Zurich.

Taylor, T. J. (1991). "Wind pressures on a hemispherical dome." *J. Wind Eng. Indust. Aerodyn.*, 40(2), 199–213.

Uematsu, Y., and Stathopoulos, T. (2003). "Wind loads on freestanding canopy roofs: A review." *J. Wind Eng., Japan Assoc. Wind Eng.*, 95.

Xie, J., and Irwin, P. A. "Key factors for torsional wind response of tall buildings." *Proc., Structures Congress 2000: Advanced technology in structural engineering*, P. E. Mohamed Elgaaly, ed., ASCE, Reston, VA., Chap. 4, Sec. 22.

OTHER REFERENCES (NOT CITED)

Twisdale, L. A., Vickery, P. J., and Steckley, A. C. (1996). *Analysis of hurricane windborne debris impact risk for residential structures*, State Farm Mutual Automobile Insurance Companies, Bloomington, IL.

Vickery, B. J., and Bloxham, C. (1992). "Internal pressure dynamics with a dominant opening." *J. Wind Eng. Indust. Aerodyn.*, 41–44, 193–204.

Yeatts, B. B., and Mehta, K. C. (1993). "Field study of internal pressures." *Proc., 7th U.S. Nat. Conf. on Wind Eng.*, Gary Hart, ed., Vol. 2, 889–897.

CHAPTER C28
WIND LOADS ON BUILDINGS: MAIN WIND FORCE RESISTING SYSTEM (ENVELOPE PROCEDURE)

The Envelope Procedure is the former "Low-Rise Buildings" provision in Method 2 of ASCE 7-05 for Main Wind Force Resisting System (MWFRS). The simplified method in this chapter is derived from the MWFRS provisions of Method 2 and is intended for simple diaphragm buildings up to 60 ft (18.3 m) in height.

PART 1: ENCLOSED AND PARTIALLY ENCLOSED LOW-RISE BUILDINGS

C28.3 WIND LOADS: MAIN WIND FORCE RESISTING SYSTEM

C28.3.1 Design Wind Pressure for Low-Rise Buildings. See commentary to Section C26.10.1.

C28.3.2 Parapets. See commentary to Section C26.10.2.

Loads on Main Wind Force Resisting Systems: The pressure coefficients for MWFRS are basically separated into two categories:

1. Directional Procedure for buildings of all heights (Fig. 27.3-1) as specified in Chapter 27 for buildings meeting the requirements specified therein; and)]
2. Envelope Procedure for low-rise buildings (Fig. 28.3-1) as specified in Chapter 28 for buildings meeting the requirements specified therein.)]

In generating these coefficients, two distinctly different approaches were used. For the pressure coefficients given in Fig. 27.3-1, the more traditional approach was followed and the pressure coefficients reflect the actual loading on each surface of the building as a function of wind direction, namely, winds perpendicular or parallel to the ridge line.

For low-rise buildings, however, the values of (GC_{pf}) represent "pseudo" loading conditions that, when applied to the building, envelop the desired structural actions (bending moment, shear, thrust) independent of wind direction. To capture all appropriate structural actions, the building must be designed for all wind directions by considering in turn each corner of the building as the windward or reference corner shown in the eight sketches of Fig. 28.3-1. At each corner, two load patterns are applied, one for each wind direction range. The end zone creates the required structural actions in the end frame or bracing. Note also that for all roof slopes, all eight load cases must be considered individually to determine the critical loading for a given structural assemblage or component thereof.

To develop the appropriate pseudovalues of (GC_{pf}), investigators at the University of Western Ontario (Davenport et al. 1978) used an approach that consisted essentially of permitting the building model to rotate in the wind tunnel through a full 360° while simultaneously monitoring the loading conditions on each of the surfaces (Fig. C28.3-1). Both Exposures B and C were considered. Using influence coefficients for rigid frames, it was possible to spatially average and time average the surface pressures to ascertain the maximum induced external force components to be resisted. More specifically, the following structural actions were evaluated:

1. Total uplift;)]
2. Total horizontal shear;)]
3. Bending moment at knees (two-hinged frame);)]
4. Bending moment at knees (three-hinged frame); and)]
5. Bending moment at ridge (two-hinged frame).)]

The next step involved developing sets of pseudopressure coefficients to generate loading conditions that would envelop the maximum induced force components to be resisted for all possible wind directions and exposures. Note, for example, that the wind azimuth producing the maximum bending moment at the knee would not necessarily produce the maximum total uplift. The maximum induced external force components determined for each of the preceding five categories were used to develop the coefficients. The end result was a set of coefficients that represent fictitious loading conditions but that conservatively envelop the maximum induced force components (bending moment, shear, and thrust) to be resisted, independent of wind direction.

The original set of coefficients was generated for the framing of conventional pre-engineered buildings, that is, single-story, moment-resisting frames in one of the principal directions and bracing in the other principal direction. The approach was later extended to single-story, moment-resisting frames with interior columns (Kavanagh et al. 1983).

Subsequent wind tunnel studies (Isyumov and Case 1995) have shown that the (GC_{pf}) values of Fig. 28.3-1 are also applicable to low-rise buildings with structural systems other than moment-resisting frames. That work examined the instantaneous wind pressures on a low-rise building with a 4:12 pitched gable roof and the resulting wind-induced forces on its MWFRS. Two different MWFRSs were evaluated. One consisted of shear walls and roof trusses at different spacings. The other had moment-resisting frames in one direction, positioned at the same spacings as the roof trusses, and diagonal wind bracing in the other direction. Wind tunnel tests were conducted for both Exposures B and C. The findings of this study showed that the (GC_{pf}) values of Fig. 28.3-1 provided satisfactory estimates of the wind forces for both types of structural systems. This work confirms the validity of Fig. 28.3-1, which reflects the combined action of wind pressures on different external surfaces of a building and thus takes advantage of spatial averaging.

In the original wind tunnel experiments, both B and C exposure terrains were checked. In these early experiments, Exposure

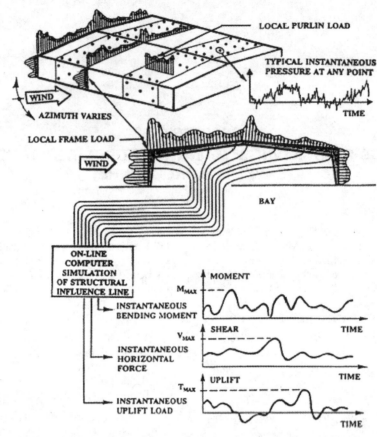

FIGURE C28.3-1 Unsteady Wind Loads on Low Buildings for Given Wind Direction

B did not include nearby buildings. In general, the force components, bending moments, and so forth were found comparable in both exposures, although (GC_{pf}) values associated with Exposure B terrain would be higher than those for Exposure C terrain because of reduced velocity pressure in Exposure B terrain. The (GC_{pf}) values given in Figs. 28.3-1, 30.3-1, 30.3-2A–C, 30.3-3, 30.3-4, 30.3-5A–B, and 30.3-6 are derived from wind tunnel studies modeled with Exposure C terrain. However, they may also be used in other exposures when the velocity pressure representing the appropriate exposure is used.

In comprehensive wind tunnel studies conducted by Ho at the University of Western Ontario (1992), it was determined that when low buildings [$h < 60$ ft ($h < 18.3$ m)] are embedded in suburban terrain (Exposure B, which included nearby buildings), the pressures in most cases are lower than those currently used in existing standards and codes, although the values show a very large scatter because of high turbulence and many variables. The results seem to indicate that some reduction in pressures for buildings located in Exposure B is justified. The Task Committee on Wind Loads believes that it is desirable to design buildings for the exposure conditions consistent with the exposure designations defined in the standard. In the case of low buildings, the effect of the increased intensity of turbulence in rougher terrain (i.e., Exposure B versus C) increases the local pressure coefficients. Beginning in ASCE 7-98, the effect of the increased turbulence intensity on the loads is treated with the truncated profile. Using this approach, the actual building exposure is used and the profile truncation corrects for the underestimate in the loads that would be obtained otherwise.

Fig. 28.3-1 is most appropriate for low buildings with width greater than twice their height and a mean roof height that does not exceed 33 ft (10 m). The original database included low buildings with widths no greater than 5 times their eave heights, and eave height did not exceed 33 ft (10 m). In the absence of more appropriate data, Fig. 28.3-1 may also be used for buildings with mean roof height that does not exceed the least horizontal dimension and is less than or equal to 60 ft (18.3 m). Beyond these extended limits, Fig. 27.3-1 should be used.

All the research used to develop and refine the low-rise building method for MWFRS loads was done on gable-roofed buildings. In the absence of research on hip-roofed buildings, the committee has developed a rational method of applying Fig. 28.3-1 to hip roofs based on its collective experience, intuition, and judgment. This suggested method is presented in Fig. C28.3-2.

Research (Isyumov 1983; Isyumov and Case 2000) indicated that the low-rise method alone underestimates the amount of torsion caused by wind loads. In ASCE 7-02, Note 5 was added to Fig. 28.3-1 to account for this torsional effect and has been carried forward through subsequent editions. The reduction in loading on only 50% of the building results in a torsional load case without an increase in the predicted base shear for the building. This reduction in loading results in equivalent torsion that agrees well with the wind tunnel measurements carried out by Elsharawy et al. (2012, 2015) and Stathopoulos et al. (2013). In general, the provision will have little or no effect on the designs of MWFRS that have well-distributed resistance. However, it will affect the design of systems with centralized resistance, such as a single core in the center of the building. An illustration of the intent of the note on two of the eight load patterns is shown in Fig. 28.3-1. All eight patterns should be modified in this way as a separate set of load conditions in addition to the eight basic patterns.

Internal pressure coefficients (GC_{pi}) to be used for loads on MWFRS are given in Table 26.13-1. The internal pressure load can be critical in one-story, moment-resisting frames and in

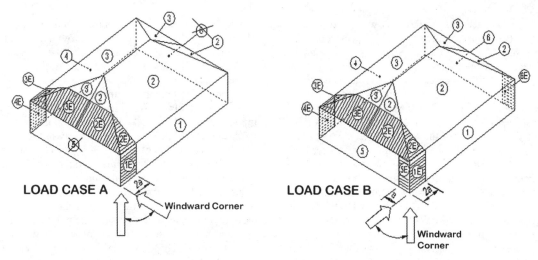

Notes
1. Adapt the loadings shown in Fig. 28.3-1 for hip-roofed buildings as shown above. For a given hip roof pitch, use the roof coefficients from the Case A table for both Load Case A and Load Case B.
2. The total horizontal shear shall not be less than that determined by neglecting the wind forces on roof surfaces.

FIGURE C28.3-2. Hip-Roofed Low-Rise Buildings

the top story of a building where the MWFRS consists of moment-resisting frames. Loading cases with positive and negative internal pressures should be considered. The internal pressure load cancels out in the determination of total lateral load and base shear. The designer can use judgment in the use of internal pressure loading for the MWFRS of high-rise buildings.

The edge strip definition was modified following research (Elsharawy et al. 2014) showing that the definition of dimension "a" in ASCE 7-10 led to unduly large edge strips and end zones for very large buildings.

C28.3.4 Minimum Design Wind Loads. This section specifies a minimum wind load to be applied horizontally on the entire vertical projection of the building, as shown in Fig. C27.1-1. This load case is to be applied as a separate load case in addition to the normal load cases specified in other portions of this chapter.

C28.3.5 Horizontal Wind Loads on Open or Partially Enclosed Buildings with Transverse Frames and Pitched Roofs. In 2016, new provisions have been added for wind loads on the longitudinal MWFRS of open or partially enclosed buildings with pitched roofs as shown in Fig. 28.4-3. based on research at the University of Western Ontario (UWO) (Kopp, Galsworthy, and Oh 2010). This study examined the shielding effect of these multiple transverse frames for an open-sided building that was covered by a roof. The shielding effect adopted in ASCE 7 was conservatively simplified; therefore, the exponential form provided in Kopp, Galsworthy, and Oh is a more accurate and acceptable alternative. Building models consisting of three, six, and nine frames were evaluated. A building with two frames was not tested. Therefore, an extrapolation using $n = 2$ is not necessarily conservative. However, this method can be conservatively used for a building with two frames by using $n = 3$. Examples of evaluating these additional wind forces, are presented in Shoemaker, Kopp, and Galsworthy (2011).

The wind loads calculated using Section 28.4.5 are applicable to buildings with open end walls. end walls with the gable filled with cladding, and with additional end wall cladding; however, the area used is always the total end wall area, A_f. The effective solid area of a frame, A_s, is the projected area of any portion of the end wall that would be exposed to the wind.

The measured peak base shear coefficients were used as the basis for the design drag loads in the direction parallel to the ridge (i.e., wind directions in the range 0° to 45°). These loads include both the effects of friction drag and pressure drag. However, to put this in a format consistent with ASCE 7-10, it was reasonable to use with the enclosed pseudo load coefficients (GC_W) and then apply factors to account for the parameters that affect the load coefficients on open buildings (i.e., building size, solidity ratio, and number of frames. This method yielded conservative results for all experimental wind tunnel data points. The force from Eq. 28.3-3, calibrated to the measured base shear, does not reflect a direct load path from the calculated end wall pressure but is to be used to calculate the longitudinal bracing requirement. For the building configurations evaluated, the UWO study showed that the force measured in the bracing was equal to 70% of the total base shear. The remaining base shear was transferred directly at the column bases.

The wind tunnel studies used to develop the provisions of Section 28.3.5 did not evaluate the effect of obstructed flow due to materials or objects sheltered by the building. Barring an unusual arrangement of materials that could produce a venturi effect, it is judged that obstructed flow would decrease the wind loads on the longitudinal MWFRS. However, as noted in previous studies (Altman 2001, Uematsu and Stathopolous 2003), the roof wind loads are more sensitive to the effect of obstructed flow.

The wind load in the transverse direction (perpendicular to the ridge) for this type of open building is a separate loading case and is due to the horizontal pressure from the roof load calculated using Section 27.3.2, with C_N from Fig. 27.3-5, and additional pressures acting on the projected areas of any surfaces exposed to the transverse wind.

PART 2: ENCLOSED SIMPLE DIAPHRAGM LOW-RISE BUILDINGS

This simplified approach of the Envelope Procedure is for the relatively common low-rise ($h \leq 60$ ft [$h \leq 18.3$ m]), regular-shaped, simple diaphragm building case (see definitions for "simple diaphragm building" and "regular-shaped building") where pressures for the roof and walls can be selected directly from a table. Fig. 28.5-1 provides the design pressures for MWFRS for the specified conditions. Values are provided for enclosed buildings only [$(GC_{pi}) = \pm 0.18$].

Horizontal wall pressures are the net sum of the windward and leeward pressures on vertical projection of the wall. *Horizontal roof pressures* are the net sum of the windward and leeward pressures on vertical projection of the roof. *Vertical roof pressures* are the net sum of the external and internal pressures on the horizontal projection of the roof.

Note that for the MWFRS in a diaphragm building, the internal pressure cancels for loads on the walls and for the horizontal component of loads on the roof. This is true because when wind forces are transferred by horizontal diaphragms (e.g., floors and roofs) to the vertical elements of the MWFRS (e.g., shear walls, X-bracing, or moment frames), the collection of wind forces from windward and leeward sides of the building occurs in the horizontal diaphragms. Once transferred into the horizontal diaphragms by the vertically spanning wall systems, the wind forces become a net horizontal wind force that is delivered to the lateral-force-resisting elements of the MWFRS. There should be no structural separations in the diaphragms. Additionally, there should be no girts or other horizontal members that transmit significant wind loads directly to vertical frame members of the MWFRS in the direction under consideration. The equal and opposite internal pressures on the walls cancel each other in the horizontal diaphragm. This simplified approach of the Envelope Procedure combines the windward and leeward pressures into a net horizontal wind pressure, with the internal pressures canceled.

The user is cautioned to consider the precise application of windward and leeward wall loads to members of the roof diaphragm where openings may exist and where particular members, such as drag struts, are designed. The design of the roof members of the MWFRS for vertical loads is influenced by internal pressures. The maximum uplift, which is controlled by Load Case B, is produced by a positive internal pressure. At a roof slope of approximately 28° and above, the windward roof pressure becomes positive, and a negative internal pressure used in Load Case 2 in the table may produce a controlling case. From 25° to 45°, both positive and negative internal pressure cases (Load Cases 1 and 2, respectively) must be checked for the roof.

For the designer to use this method for the design of the MWFRS, the building must conform to all of the requirements listed in Section 26.8.2; otherwise, the Directional Procedure, Part 1 of the Envelope Procedure, or the Wind Tunnel Procedure must be used. This method is based on Part 1 of the Envelope Procedure, as shown in Fig. 28.3-1, for a specific group of buildings (simple diaphragm buildings). However, the torsional loading from Fig. 28.3-1 is deemed to be too complicated for a simplified method. The last requirement in Section 28.5.2 prevents the use of this method for buildings with lateral systems that are sensitive to torsional wind loading.

Note 5 of Fig. 28.3-1 identifies several building types that are known to be insensitive to torsion and may therefore be designed using the provisions of Section 28.5. Additionally, buildings whose lateral resistance in each principal direction is provided by two shear walls, braced frames, or moment frames that are spaced apart a distance not less than 75% of the width of the building measured normal to the orthogonal wind direction, and other building types and element arrangements described in Section 27.5.1 or 27.5.2 are also insensitive to torsion. This property could be demonstrated by designing the building using Part 1 of Chapter 28, Fig. 28.3-1, and showing that the torsion load cases defined in Note 5 do not govern the design of any of the lateral resisting elements. Alternatively, it can be demonstrated within the context of Part 2 of Chapter 28 by defining torsion load cases based on the loads in Fig. 28.5-1 and reducing the pressures on one-half of the building by 75%, as described in Fig. 28.3-1, Note 5. If none of the lateral elements are governed by these torsion cases, then the building can be designed using Part 2 of Chapter 28; otherwise, the building must be designed using Part 1 of Chapter 27 or Part 1 of Chapter 28.

Values are tabulated for Exposure B at $h = 30$ ft (9.1 m), and $K_{zt} = 1.0$. Multiplying factors are provided for other exposures and heights. The following values have been used in preparation of Fig. 28.5-1: $h = 30$ ft (9.1 m), Exposure B, $K_z = 0.70$, $K_d = 0.85$, $K_{zt} = 1.0$, $(GC_{pi}) = \pm 0.18$ (enclosed building).

Pressure coefficients are from Fig. 28.3-1.

Wall elements resisting two or more simultaneous wind-induced structural actions (e.g., bending, uplift, or shear) should be designed for the interaction of the wind loads as part of the MWFRS. The horizontal loads in Fig. 28.5-1 are the sum of the windward and leeward pressures and are therefore not applicable as individual wall pressures for the interaction load cases. Design wind pressures, p_s for zones A and C, should be multiplied by $+0.85$ for use on windward walls and by -0.70 for use on leeward walls (the plus sign signifies pressures acting toward the wall surface). For sidewalls, p_s for zone C multiplied by -0.65 should be used. These wall elements must also be checked for the various separately acting (not simultaneous) components and cladding (C&C) load cases.

Main wind force resisting roof members spanning at least from the eave to the ridge or supporting members spanning at least from the eave to the ridge are not required to be designed for the higher end zone loads. The interior zone loads should be applied because of the enveloped nature of the loads for roof members.

REFERENCES

Altman, D. R. (2001). "Wind uplift forces on roof canopies." Master's thesis. Department of Civil Engineering, Clemson University, Clemson. SC.

Davenport, A. G., Surry, D., and Stathopoulos, T. (1978). "Wind loads on low-rise buildings." *Final report on Phase III, BLWT-SS4*, Univ. of Western Ontario, London, ON.

Elsharawy, M., Alrawashdeh, H., and Stathopoulos, T. (2014) "Wind loading zones for flat roofs," *Proc., 4th Intl. Structural Specialty Conf.*, CSCE, Halifax, NS, May 28–31.

Elsharawy, M., Galal, K., and Stathopoulos, T. (2015), "Torsional and shear wind loads on flat-roofed buildings." *Eng. Struct.*, 84 (2), 313–324.

Elsharawy, M., Stathopoulos, T., and Galal, K. (2012). "Wind-induced torsional loads on low buildings." *J. Wind Eng. Indust. Aerodyn.*, 40–48, 104–106.

Ho, E. (1992). "Variability of low building wind lands." Ph.D. thesis, Univ. of Western Ontario, London, ON.

Isyumov, N. (1983). "Wind induced torque on square and rectangular building shapes." *J. Wind Eng. Indust. Aerodyn.*, 13, 183–186.

Isyumov, N., and Case, P. (1995). "Evaluation of structural wind loads for low-rise buildings contained in ASCE standard 7-95." *BLWT-SS17-1995*, Univ. of Western Ontario, London, ON.

Isyumov, N., and Case, P. C. (2000). "Wind-induced torsional loads and responses of buildings." In *Advanced technology in structural engineering*, P. E. Mohamad Elgaaly, ed., ASCE, Reston, VA.

Kavanagh, K. T., Surry, D., Stathopoulos, T., and Davenport, A. G. (1983). "Wind loads on low-rise buildings." *Phase IV, BLWT-SS14*, Univ. of Western Ontario, London, ON.

Kopp, G. A., Galsworthy., J., and Oh., J. H. (2010). "Horizontal wind loads on open-frame. low-rise buildings." *J. Struct. Div.*, 136(1), 98–105.

Shoemaker, W. L., Kopp, G. A., and Galsworthy, J. (2011). "Design of braced frames in open buildings for wind loading," *AISC Eng. J.* 3rd quarter, 225–233.

Stathopoulos, T., Elsharawy, M., and Galal, K. (2013). "Wind load combinations including torsion for rectangular medium-rise building." *Int. J. High-Rise Bldgs.* 2(3), 1–11.

Uematsu, Y., and Stathopoulos, T. (2003). "Wind loads on free-standing canopy roofs: A review." *J. Wind Eng., Japan Assoc. Wind Eng.* 95.

OTHER REFERENCES (NOT CITED)

Krayer, W. R., and Marshall, R. D. (1992). "Gust factors applied to hurricane winds." *Bull. Am. Meteorol. Soc.*, 73, 613–617.

CHAPTER C29

WIND LOADS ON BUILDING APPURTENANCES AND OTHER STRUCTURES: MAIN WIND FORCE RESISTING SYSTEM (DIRECTIONAL PROCEDURE)

C29.3 DESIGN WIND LOADS: SOLID FREESTANDING WALLS AND SOLID SIGNS

C29.3.1 Solid Freestanding Walls and Solid Freestanding Signs. See commentary, Section C26.10.1.

The Risk Category for rooftop equipment or appurtenances is required to be not less than that for the building on which the equipment is located, nor that for any other facility to which the equipment provides a necessary service. For example, if a solar array is located on the roof of a hospital, design wind load for the solar array is based on the Risk Category for the hospital, even if the solar array is not needed for functioning of the hospital. In another example, if an antenna provides critical communication service for a hospital and the antenna is located on top of a parking structure separate from the hospital building, design wind load for the antenna is based on the Risk Category for the hospital, which is greater than the Risk Category for the parking structure.

Fig. 29.3-1. The provisions in Fig. 29.3-1 are based on the results of boundary-layer wind tunnel studies (Letchford 1985, 2001; Holmes 1986; Letchford and Holmes 1994; Ginger et al. 1998a, b; Letchford and Robertson 1999; Mehta et al. 2012).

A surface curve fit to Letchford's (2001) and Holmes's (1986) area averaged mean net pressure coefficient data (equivalent to mean force coefficients in this case) is given by the following equation (Fox and Levitan 2005):

$$C_f = \{1.563 + 0.008542 \ln(x) - 0.06148y + 0.009011[\ln(x)]^2 - 0.2603y^2 - 0.08393y[\ln(x)]\}/0.85$$

where $x = B/s$ and $y = s/h$.

The 0.85 term in the denominator modifies the wind tunnel-derived force coefficients into a format where the gust-effect factor as defined in Section 26.11 can be used.

Force coefficients for Cases A and B were generated from the preceding equation, then rounded off to the nearest 0.05. That equation is only valid within the range of B/s and s/h ratios given in the figure for Cases A and B.

Of all the pertinent studies on single-faced signs, only Letchford (2001) specifically addressed eccentricity (i.e., Case B). Letchford reported that his data provided a reasonable match to Cook's (1990) recommendation for using an eccentricity of 0.25 times the average width of the sign. However, the data were too limited in scope to justify changing the existing eccentricity value of 0.2 times the average width of the sign, which is also used in the 2011 Australian/New Zealand Standard (Standards Australia 2011).

Mehta et al. (2012) tested a variety of aspect ratios (B/s) and clearance ratios (s/h) for double-faced signs with all sides enclosed to address current industry practice. The study included both wind tunnel testing and a full-scale field test to calibrate the wind tunnel models (Zuo et al. 2014; Smith et al. 2014). These sign configurations exhibited an average reduction of 16% in mean force coefficients with a range of 9% to 22% as compared to single-faced sign force coefficients given by the equation above. These tests also showed that the eccentricity of 0.2 times the width of the structure is overly conservative. Eccentricities reported in the study ranged from 0.039 to 0.105 times the width of the structure, with an average of 0.061. Testing by Giannoulis et al. (2012) supports the findings in Mehta et al. (2012).

Case C was added to account for the higher pressures observed in both wind tunnel studies (Letchford 1985, 2001; Holmes 1986; Letchford and Holmes 1994; Ginger et al. 1998a, b; Letchford and Robertson 1999) and full-scale studies (Robertson et al. 1997) near the windward edge of a freestanding wall or sign for oblique wind directions. Linear regression equations were fit to the local mean net pressure coefficient data (for wind direction 45°) from the referenced wind tunnel studies to generate force coefficients for square regions starting at the windward edge. Pressures near this edge increase significantly as the length of the structure increases. No data were available on the spatial distribution of pressures for structures with low aspect ratios ($B/s < 2$).

The sample illustration for Case C at the top of Fig. 29.3-1 is for a sign with an aspect ratio $B/s = 4$. For signs of differing B/s ratios, the number of regions is equal to the number of force coefficient entries located below each B/s column heading.

For oblique wind directions (Case C), increased force coefficients have been observed on above ground signs compared to the same aspect ratio walls on ground (Letchford 1985, 2001; Ginger et al. 1998a). The ratio of force coefficients between aboveground and on-ground signs (i.e., $s/h = 0.8$ and 1.0, respectively) is 1.25, which is the same ratio used in the Australian/New Zealand Standard (Standards Australia 2002). Note 5 of Fig. 29.3-1 provides for linear interpolation between these two cases.

For walls and signs on the ground ($s/h = 1$), the mean vertical center of pressure ranged from $0.5h$ to $0.6h$ (Holmes 1986; Letchford 1989; Letchford and Holmes 1994; Robertson et al. 1995, 1996; Ginger et al. 1998a); $0.55h$ was the average value. For aboveground walls and signs, the geometric center best represents the expected vertical center of pressure.

The reduction in C_f caused by porosity (Note 2) follows a recommendation (Letchford 2001). Both wind tunnel and full-scale data have shown that return corners significantly reduce the net pressures in the region near the windward edge of the wall or sign (Letchford and Robertson 1999).

C29.3.2 Solid Attached Signs. Signs attached to walls and subject to the geometric limitations of Section 29.3.2 should experience wind pressures approximately equal to the external pressures on the wall to which they are attached. The dimension requirements for signs supported by frameworks, where there is a small gap between the sign and the wall, are based on the collective judgment of the committee.

C29.4 DESIGN WIND LOADS: OTHER STRUCTURES

Guidance for determining G, C_f, and A_f for structures found in petrochemical and other industrial facilities that are not otherwise addressed in ASCE 7 can be found in *Wind Loads for Petrochemical and Other Industrial Facilities* (ASCE Task Committee on Wind-Induced Forces (2011)).

Figs. 29.4-1, 29.4-2, and 29.4-3. With the exception of Fig. 29.4-3, the pressure and force coefficient values in these tables are unchanged from ANSI A58.1-1972 (ANSI 1972). The coefficients specified in these tables are based on wind tunnel tests conducted under conditions of uniform flow and low turbulence, and their validity in turbulent boundary-layer flows has yet to be completely established. Additional pressure coefficients for conditions not specified herein may be found in two references (SIA 1956; ASCE 1961).

With regard to Fig. 29.4-3, the force coefficients are a refinement of the coefficients specified in ANSI A58.1-1982 (1982) and in ASCE 7-93 (1994). The force coefficients specified are offered as a simplified procedure that may be used for trussed towers and are consistent with force coefficients given in TIA (1991) and force coefficients recommended by Working Group No. 4 (IASS 1981).

It is not the intent of this standard to exclude the use of other recognized literature for the design of special structures, such as transmission and telecommunications towers. Recommendations for wind loads on tower guys are not provided as in previous editions of the standard. Recognized literature should be referenced for the design of these special structures as is noted in Section 29.1.3. For the design of flagpoles, see NAAMM (2007). For the design of structural supports for highway signs, luminaires, and traffic signals, see AASHTO LTS-6 (AASHTO 2013).

C29.4.1 Rooftop Structures and Equipment for Buildings. Wind loads on rooftop structures and equipment are revised in ASCE 7-16 to use Eqs. (29.4-2) and (29.4-3) for buildings of all heights. The change provides an improved representation of the limited amount of research that is available (Hosoya et al. 2001; Kopp and Traczuk 2007). The change also eliminates inconsistencies between equipment on roofs below versus slightly above the 60-ft (18-3 m) height. The research in Hosoya et al. (2001) only treated one value of A_f (0.04Bh). The research in Kopp and Traczuk (2008) treated values of $A_f = 0.02Bh$ and 0.03Bh, and values of $A_r = 0.0067BL$. Because GC_r is expected to approach 1.0 as A_f or A_r approaches that of the building (Bh or BL), a linear interpolation is included as a way to avoid a step function in load if the designer wants to treat other sizes. However, the loads provided by these provisions are best suited for units that are much smaller than the building, $A_f < 0.05Bh$ and $A_r < 0.01BL$. The resulting loads are expected to be overly conservative if applied to linelike structures that extend more than $0.1B$ or $0.1L$ across the roof.

Both research studies showed high uplift forces on the rooftop equipment. Hence, uplift loads are addressed in Section 29.4.1.

Mechanical equipment screens commonly are used to conceal plumbing, electrical, or mechanical equipment from view and are defined as rooftop structures not covered by a roof and located away from the edge of the building roof such that they are not considered parapets. Many configurations and types of screens are available ranging from solid walls to porous panels, which allow some air to flow through. Though the use of equipment screens is prevalent, little research is available to provide guidance for determining wind loads on screen walls and equipment behind screens. Accordingly, rooftop screens, equipment behind screens, and their supports and attachments to buildings should be designed for the full wind load determined in accordance with Section 29.4.1. Where substantiating data have been obtained using the Wind Tunnel Procedure (Chapter 31), design professionals may consider wind load reductions in the design of rooftop screens and equipment. For example, studies by Zuo et al. (2011) and Erwin et al. (2011) suggest that wind loads on some types of screen materials and equipment behind screens may be overestimated by the equations defined in Section 29.4.

The design wind forces for ground-mounted tanks or similar structures (smooth surface such as concrete or steel) with the aspect ratios H/D (height to diameter) in the range of 0.25 to 4 inclusive shall be determined in accordance with Section 29.4.2.

C29.4.2 Design Wind Loads: Circular Bins, Silos, and Tanks with $h \leq 120$ ft ($h \leq 36.5$ m), $D \leq 120$ ft ($D \leq 36.5$ m), and $0.25 \leq H/D \leq 4$. Section 29.4 contains the provisions for determining wind loads on silo and tank walls and roofs. The provisions are largely based on Standards Australia (2011) and the wind tunnel tests of low-rise cylindrical structures carried out at high Reynolds numbers (Re $> 1.0 \times 10^5$) by Sabransky and Melbourne (1987) and Macdonald et al. (1988, 1990). Significant increases in drag forces of grouped silos were found in the wind tunnel tests, so the provisions of grouped tanks and silos are specified in this section.

C29.4.2.1 External Walls of Isolated Circular Bins, Silos, and Tanks. This section specifies the drag coefficient, C_f, for the walls of circular bins, silos, and tanks. The drag coefficient is adopted from Standards Australia (2011). Note that the drag force of 0.63 obtained from an integration of the equations in AS/NZS 1170.2 (Standards Australia 2002) is close to the value of smooth surface type from Fig. 29.4-1.

C29.4.2.2 Roofs of Isolated Circular Bins, Silos, and Tanks. This section specifies the external pressure coefficients (C_p) for the roofs of circular bins, silos, and tanks. Two conditions are covered as shown in Fig. 29.4-5: roofs that have the roof angle $\alpha < 10°$, and roofs that have $10° \leq \alpha < 30°$. The results of $\alpha < 10°$ roofs are based on comparisons of domed roofs and flat roofs from Chapter 27 of ASCE 7-10 for maximum uplift conditions. The results of $10° \leq \alpha < 30°$ roofs are consistent with data of Sabransky and Melbourne (1987) and Macdonald et al. (1988). According to the wind tunnel tests, only suctions are observed for the roofs defined.

C29.4.2.3 Undersides of Isolated Elevated Circular Bins, Silos, and Tanks. This section specifies the external pressure coefficients, C_p, for the underneath sides of circular bins, silos, and tanks. The external pressure coefficients, C_p, are adopted from Standards Australia (2011).

For calculating gust-effect factor G, structural period T should be based on the analysis of the whole structure: tank and support structure.

C29.4.2.4 Roofs and Walls of Grouped Circular Bins, Silos, and Tanks. For grouped silos, C_p and C_f values for roofs and walls are largely based on Standards Australia (2011) and wind

Diagram

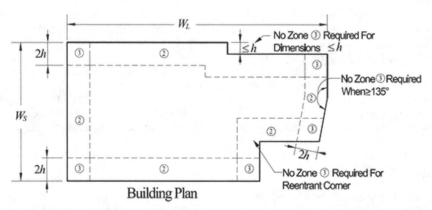

Note: h = height of building, in ft (m); W_L = width of a building on its longest side, in ft (m); W_S = width of a building on its shortest side, in ft (m)
FIGURE C29.4-1 Sketch of Building with Irregular Plan Dimensions for Use with Wind Loads on Roof-Mounted Solar Panels

tunnel tests by Sabransky and Melbourne (1987) and Macdonald et al. (1990). Test results of an in-line group of three silos with a clear spacing of $0.25D$ between nearest adjacent walls ($1.25D$ center-to-center) by Sabransky and Melbourne (1987) indicated that the drag coefficient for the central cylinder in the group increased by approximately 65% relative to that of the isolated cylinder. Pressure coefficients over the entire roof were significantly larger in magnitude than the corresponding pressure coefficients on the isolated model. It was concluded that a clear spacing of $0.25D$ produced the maximum interference between two finite cylinders.

C29.4.3 Rooftop Solar Panels for Buildings of All Heights with Flat Roofs or Gable or Hip Roofs with Slopes Less Than 7°. Section 29.4.3 addresses low-profile solar panels on low-slope roofs. Requirements specific to this type of solar array are provided because such systems (a) are in widespread use and (b) have been subject to wind tunnel testing that provides guidance on appropriate design wind loads. Fig. 29.4-7 intentionally has a limited range of application, with the maximum height above the roof surface (h_2) for the solar panels limited to 4 ft (1.2 m) and the panel chord length (L_p) limited to 6.7 ft (2.0 m). Wind tunnel data (e.g., Kopp 2013) show that increasing the overall height above the roof or panel chord length increases the wind loads. Likewise, the height of the gap between the panels and the roof surface (h_1) is limited to 2 ft (0.61 m); otherwise, the wind flow under the panels can cause uplift exceeding that covered in the figure. The requirements are not applicable to open structures because the applicable test data are from enclosed structures, which have different aerodynamics than open structures. The roof zones shown in Fig. 29.4-7 are larger than those for the roof loads, as explained in Banks (2012) and Kopp (2013). Buildings with nonrectangular plans, such as that sketched in Fig. C29.4-1, adapted from SEAOC (2012), can be used for guidance in applying the requirements. Reentrant corners do not create a Zone 3 because cornering vortices only form at outward or protruded corners. Similarly, outward corners at angles greater than 90° tend to weaken the vortices; as this angle becomes more obtuse, the building corner begins to flatten out and resemble more of an edge condition.

The nominal net pressure coefficient $(GC_{rn})_{nom}$ curves are derived from wind tunnel test data within the range of parameters allowed by Fig. 29.4-7. These curves are created based on a methodology consistent with that used for the ASCE 7 components and cladding (C&C) loads. The net pressure coefficients shown in the design curves of the figure are denoted $(GC_{rn})_{nom}$ since these values need to be adjusted for array edge conditions, parapet size, and solar panel length.

The wind tunnel data indicate that the $(GC_{rn})_{nom}$ values are not linearly related to the panel tilt angle over the full tilt angle range. The data indicate that there is a relatively small change in $(GC_{rn})_{nom}$ values for the lower tilt panels in the 1° to 5° range. Then there is a rapid increase in $(GC_{rn})_{nom}$ values from 5° to 15°. There is again a relatively small change in $(GC_{rn})_{nom}$ values for higher tilt panels in the 15° to 35° range because, for the higher tilt angles, upstream panels create turbulence, which increases the wind loads on all downstream panels (e.g., Kopp et al. 2012; Kopp 2013). Thus, the figure was created with two $(GC_{rn})_{nom}$ curves to address this phenomenon: a $(GC_{rn})_{nom}$ curve for low-tilt panels in the 0° to 5° range and another for high-tilt panels in the 15° to 35° range. For panel tilt angles in the 5° to 15° range, linear interpolation is permitted.

$(GC_{rn})_{nom}$ values are for both positive and negative values. Wind tunnel test data show similar positive and negative pressures for solar panels (which are very different than typical roof member design wind loads).

Parapets typically worsen the wind loads on solar panels, particularly on wider buildings. The parapets lift the vortices higher above the roof surface and push them closer together, inward from the edges. It is not entirely clear why the vortex effects are more severe in this situation, but tests show that this can result in wind loads that are significantly greater than in the absence of a parapet, particularly for unshrouded tilted panels. The parapet height factor, γ_p, accounts for this effect.

Solar panels are typically installed in large arrays with closely spaced rows, and the end rows and panels experience larger wind pressures than interior panels, which are sheltered by adjacent panels. To account for the higher loading at the end rows and panels, an array edge increase factor is applied, taken from SEAOC (2012). However, single rows of solar panels can be determined using this section, taking into account that all solar panels are defined as being exposed.

Rooftop equipment and structures, such as HVAC units, screens, or penthouses, can provide some sheltering benefits to solar arrays located directly downwind of the object; conversely, however, the regions around edges of such structures can have accelerated wind flow under varied wind directions. Accordingly, the edge increase factor ignores such structures and is calculated based on the distance to the building edge or adjacent array, neglecting any intervening rooftop structures. This results in the panels adjacent to rooftop objects being designed for higher wind loads to account for the accelerated wind flow.

The requirements can be used for arrays in any plan orientation relative to building axes or edges; the dimensions d_1 and d_2 are measured parallel to the principal axes of the array being considered. The requirement in Fig. 29.4-7 for array panels to be set back from the roof edge is meant to ensure that the panels are out of the high-speed wind in the separated shear layers at the edge. If the array is made up of a single row of solar panels, or a single panel, then d_2 is undefined and $\gamma_E = 1.5$.

Wind tunnel studies have shown that the wind loads on rooftop solar panels need not be applied simultaneously to the roof C&C wind loads for portions of the roof that are covered by the panel. Where a portion of the span of a roof member is covered by a solar array and the remainder is not covered, then the roof member should be designed with the solar array wind load on the covered portion with simultaneous application of roof C&C load on the uncovered portion. In a separate load case, the member should also be checked for C&C wind loads assuming that the photovoltaic panels are not present. For installations of new panels on existing buildings, this separate load case to check the capacity of the existing roof structure to resist the roof C&C wind loads applied over the entire roof area (i.e., assuming that the solar panels are not present) is not required.

The wind loads here were obtained for solar arrays without aerodynamic treatments such as shrouds or deflectors. Uplifting wind load for arrays with shrouding may be lower, but because of the range of possible results and sensitivity to design details, such arrays would need to be wind tunnel tested in order to use reduced loads from those specified here. It should also be noted that horizontal (drag) loads could increase with the use of shrouds or deflectors.

Procedure for Using Fig. 29.4-7. To simplify the use of the figure, the following is a step-by-step procedure.

Step 1: Confirm applicability of the figure to the solar installation and building.
Step 2: For panels with $\omega \leq 2°$ and $h_2 \leq 10$ in. (254 mm), the procedure using d_2 Section 29.4.4 per Note 4 may be used.
Step 3: Confirm that layout provides minimum distance from roof edge per Note 5.
Step 4: Determine roof zones.
Step 5: Determine effective wind area and normalized wind area for each element being evaluated.
Step 6: Compute $(GC_{rn})_{nom}$ from applicable chart, using linear interpolation for values of ω between 5° and 15°.
Step 7: Apply chord length adjustment factor, γ_c.
Step 8: Apply the Edge Factor d_2, γ_E, if necessary.
Step 9: Apply parapet height factor, γ_p.
Step 10: Calculate (GC_{rn}).
Step 11: Calculate pressure, p, using Eq. (29.4-5).

C29.4.4 Rooftop Solar Panels Parallel to the Roof Surface on Buildings of All Heights and Roof Slopes. Wind loads of roof-mounted, planar solar panels that are close to and parallel to the roof surface tend to be lower than the loads on a bare roof because of pressure equalization (Kopp et al. 2012; Kopp 2013), except on the perimeter of the array. The solar array pressure equalization factor, γ_a, accounts for this reduction, based, in particular, on data from Stenabaugh et al. (2015). For pressure equalization to occur, the panels cannot be too large, there needs to be a minimum gap between the panels, and the height above the roof surface cannot be too large. The current requirements are based on panel sizes up to 6.7 ft (2.0 m) long for heights above the roof surface that are less than 10 in. (254 mm) and a minimum gap around the panels of 0.25 in. (6.35 mm). Larger gaps and lower heights above the roof surface could further decrease the wind loads, but wind tunnel testing would be required to take advantage of this difference. For metal roof panels, the 10-in. (254-mm) maximum distance above the roof surface is measured from the flat portion of the panels, rather than from the top of the panel ribs.

Panels around the edge of the array may experience higher wind loads. The definition of these exposed panels is the same as for tilted panels in Section 29.4.3.

C29.5 PARAPETS

Before the 2002 edition of ASCE 7, no provisions for the design of parapets were included because of the lack of direct research. In the 2002 edition of ASCE 7, a rational method was added based on the committee's collective experience, intuition, and judgment. In the 2005 edition, the parapet provisions were updated as a result of research performed at the University of Western Ontario (Mans et al. 2000, 2001) and at Concordia University (Stathopoulos et al. 2002a, b).

Wind pressures on a parapet are a combination of wall and roof pressures, depending on the location of the parapet and the direction of the wind (Fig. C29.5-1). A windward parapet experiences the positive wall pressure on its front surface (exterior side of the building) and the negative roof edge zone pressure on its back surface (roof side). This behavior is based on the concept that the zone of suction caused by the wind stream separation at the roof eave moves up to the top of the parapet when one is present. Thus, the same suction that acts on the roof edge also acts on the back of the parapet.

The leeward parapet experiences a positive wall pressure on its back surface (roof side) and a negative wall pressure on its front surface (exterior side of the building). There should be no reduction in the positive wall pressure to the leeward parapet caused by shielding by the windward parapet because, typically, they are too far apart to experience this effect. Because all parapets would be designed for all wind directions, each parapet would in turn be the windward and leeward parapet and, therefore, must be designed for both sets of pressures.

For the design of the main wind force resisting system (MWFRS), the pressures used describe the contribution of the parapet to the overall wind loads on that system. For simplicity, the front and back pressures on the parapet have been combined into one coefficient for MWFRS design. The designer should not typically need the separate front and back pressures for MWFRS design. The internal pressures inside the parapet cancel out in the determination of the combined coefficient. The summation of these external and internal, front and back pressure coefficients is a new term (GC_{pn}), the combined net pressure coefficient for a parapet.

For the design of the components and cladding (C&C), a similar approach was used. However, it is not possible to simplify the coefficients because of the increased complexity of the C&C pressure coefficients. In addition, the front and back pressures are not combined because the designer may be designing separate elements on each face of the parapet. The internal pressure is required to determine the net pressures on the windward and leeward surfaces of the parapet. The provisions guide the designer to the correct (GC_p) and velocity pressure to use for each surface, as illustrated in Fig. C29.5-1.

Interior walls that protrude through the roof, such as party walls and fire walls, should be designed as windward parapets for both MWFRS and C&C.

The internal pressure that may be present inside a parapet is highly dependent on the porosity of the parapet envelope.

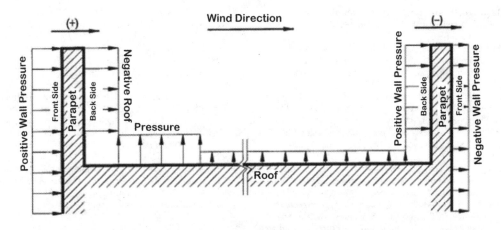

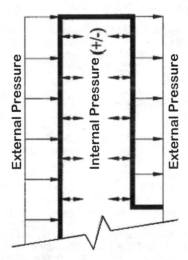

External and Internal Parapet Pressures
(Component and Cladding Only)

FIGURE C29.5-1. Design Wind Pressures on Parapets

In other words, it depends on the likelihood of the wall surface materials to leak air pressure into the internal cavities of the parapet. For solid parapets, such as concrete or masonry, the internal pressure is zero because there is no internal cavity. Certain wall materials may be impervious to air leakage, and as such have little or no internal pressure or suction, so using the value of (GC_{pi}) for an enclosed building may be appropriate. However, certain materials and systems used to construct parapets containing cavities are more porous, thus justifying the use of the (GC_{pi}) values for partially enclosed buildings or higher. Another factor in the internal pressure determination is whether the parapet cavity connects to the internal space of the building, allowing the building's internal pressure to propagate into the parapet. Selection of the appropriate internal pressure coefficient is left to the judgment of the design professional.

C29.7 MINIMUM DESIGN WIND LOADING

This section specifies a minimum wind load to be applied horizontally on the entire vertical projection of the other structures, as shown in Fig. C27.1-1. This load case is to be applied as a separate load case in addition to the normal load cases specified in other portions of this chapter.

REFERENCES

American Association of State Highway and Transportation Officials (AASHTO). (2013). "Standard specifications for structural supports for highway signs, luminaires, and traffic signals," 6th Ed. *AASHTO LTS-6*, Washington, DC.

American National Standards Institute (ANSI). (1972). *Minimum design loads for buildings and other structures*, ANSI A58.1-1972, Washington, DC.

ANSI. (1982). *Minimum design loads for buildings and other structures*, ANSI A58.1-1982, Washington, DC.

ASCE. (1961). "Wind forces on structures." *Trans. ASCE*, 126(2), 1124–1198.

ASCE. (1994). *Minimum design loads for buildings and other structures*, New York.

ASCE Task Committee on Wind-Induced Forces. (2011). *Wind loads for petrochemical and other industrial facilities*, Reston, VA.

Banks, D (2012). "Wind loads on tilted flat panels on commercial roofs: The effects of corner vortices." *Advances in hurricane engineering*, C.P. Jones, and L.G. Griffis, eds. ASCE, Reston, VA.

Cook, N. J. (1990). *The designer's guide to wind loading of building structures, Part II*, Butterworths Publishers, London.

Erwin, J. W., Chowdhury, A. G., and Bitsuamlak, G. (2011). "Wind loads on rooftop equipment mounted on a flat roof." *J. Wind Eng.* 8(1), 23–42.

Fox, T., and Levitan, M. (2005). "A comprehensive look at wind loading on freestanding walls and signs. *Proc., 10th Americas Conf. on Wind Eng.*, Baton Rouge, LA.

Giannoulis, A., Stathopoulos, T., Briassoulis, D., and Mistriotis, A. (2012). "Wind loading on vertical panels with different permeabilities." *J. Wind Eng. Industr. Aerodyn.* 107, 1–16.

Ginger, J. D., Reardon, G. F., and Langtree, B. A. (1998a). "Wind loads on fences and hoardings." Proc., Australasian Struct. Eng. Conf., Engineers Australia Structural College, Barton, Australian Capital Territory, AU, 983–990.

Ginger, J. D., Reardon, G. F., and Langtree, B. L. (1998b). *Wind loads on fences and hoardings.* Cyclone Structural Testing Station, James Cook University, Townsville, Queensland, AU.

Holmes, J. D. (1986). "Wind tunnel tests on free-standing walls at CSIRO." *Internal Report 86/47*, CSIRO Division of Building Research, Clayton, South Victoria, AU.

Hosoya, N., Cermak, J. E., and Steele, C. (2001). "A wind-tunnel study of a cubic rooftop AC unit on a low building." Proc., Americas Conf. on Wind Eng., American Association for Wind Engineering, Fort Collins, CO.

IASS, Working Group No. 4. (1981). *Recommendations for guyed masts.* International Association for Shell and Spatial Structures, Madrid, Spain.

Kopp, G. A. (2013). "Wind loads on low profile, tilted, solar arrays placed on large, flat, low-rise building roofs." *J. Struct. Eng.*, doi: 10.1061/(ASCE) ST.1943-541X.0000821.

Kopp, G. A., Farquhar, S., and Morrison, M. J. (2012). "Aerodynamic mechanisms for wind loads on tilted, roof-mounted, solar arrays." *J. Wind Eng. Indust. Aerodyn.* 111, 40–52.

Kopp, G. A., and Traczuk, G. (2007). "Wind loads on a roof-mounted cube." (BLWT-SS47_2007). The Boundary Layer Wind Tunnel Laboratory, London, ON.

Letchford, C. W. (1985). "Wind loads on free-standing walls." *Report OUEL 1599/85*, Dept. Eng. Sci., Univ. of Oxford, Oxford, UK.

Letchford, C. W. (1989). "Wind loads and overturning moments on free standing walls." Proc., 2nd Asia Pacific Symp. on Wind Eng., International Association for Wind Engineering, Kanagawa, Japan.

Letchford, C. W. (2001). "Wind loads on rectangular signboards and hoardings." *J. Wind Eng. Indust. Aerodyn.* 89, 135–151.

Letchford, C. W., and Holmes, J. D. (1994). "Wind loads on free-standing walls in turbulent boundary layers." *J. Wind Eng. Indust. Aerodyn.* 51(1), 1–27.

Letchford, C. W., and Robertson, A. P. (1999). "Mean wind loading at the leading ends of free-standing walls." *J. Wind Eng. Indust. Aerodyn.* 79(1), 123–134.

Macdonald, P. A., Kwok, K. C. S., and Holmes, J. D. (1988). "Wind loads on circular storage bins, silos and tanks: 1. Point pressure measurements on isolated structures," *J. Wind Eng. Indust. Aerodyn.* 31(2–3), 165–187.

Macdonald, P. A., Holmes, J. D., and Kwok, K. C. S. (1990). Wind loads on circular storage bins, silos and tanks. II. Effect of grouping, *J. Wind Eng. Indust. Aerodyn.* 34(1), 77–95.

Mans, C., Kopp, G., and Surry, D. (2000). "Wind loads on parapets, Part 1." *BLWTL-SS23-2000*, University of Western Ontario, London, ON.

Mans, C., Kopp, G., and Surry, D. (2001). "Wind loads on parapets, Parts 2 and 3." *BLWT-SS37-2001 and BLWT-SS38-2001*, University of Western Ontario, London, ON.

Mehta, K. C., Smith, D. A., and Zuo, D. (2012). "Field and wind tunnel testing of signs, final report, test procedures and outcomes." *Lubbock*, TX, Texas Tech University.

National Association of Architectural Metal Manufacturers (NAAMM). (2007). "Guide specifications for design of metal flagpoles," ANSI/NAAMM FP 1001–13, Glen Ellyn, IL.

Robertson, A. P., Hoxey, R. P., Short, J. L., Ferguson, W. A., and Osmond, S. (1995). "Wind loads on free-standing walls: A full-scale study." *Proc., 9th Intl. Conf. on Wind Eng.*, Wiley Eastern Science, 457–468.

Robertson, A. P., Hoxey, R. P., Short, J. L., Ferguson, W. A., and Osmond, S. (1996). "Full-scale testing to determine the wind loads on free-standing walls." *J. Wind Eng. Indust. Aerodyn.* 60(1), 123–137.

Robertson, A. P., Hoxey, R. P., Short, J. L., and Ferguson, W. A. (1997). "Full scale measurements and computational predictions of wind loads on free standing walls." *J. Wind Eng. Indust. Aerodyn.*, 67-68, 639–646.

Sabransky, I. J., and Melbourne, W. H. (1987). "Design pressure distribution on circular silos with conical roofs." *J. Wind Eng. Indust. Aerodyn.* 26(1), 65–84.

Smith, D. A., Zuo, D., and Mehta, K. C. (2014). "Characteristics of wind induced net force and torque on a rectangular sign measured in the field." *J. Wind Eng. Indust. Aerodyn.* 133(0), 80–91.

Standards Australia. (2002). "Structural design actions, Part 2: Wind actions." *AS/NZS 1170.2:2002*. Standards Australia, Sydney, New South Wales, AU.

Standards Australia. (2011). "Structural design actions—Wind actions." *AS/NZS 1170.2:2011*. Standards Australia, Sydney, New South Wales, AU.

Stathopoulos, T., Saathoff, P., and Bedair, R. (2002a). "Wind pressures on parapets of flat roofs." *J. Arch. Eng.* 8(2), 49–54.

Stathopoulos, T., Saathoff, P., and Du, X. (2002b). "Wind loads on parapets." *J. Wind Eng. Indust. Aerodyn.* 90, 503–514.

Stenabaugh, S. E., Iida, Y., Kopp, G. A., and Karava, P. (2015). "Wind loads on photovoltaic arrays mounted on sloped roofs of low-rise building, parallel to the roof surface." *J. Wind Eng. Indust. Aerodyn.* 139(4), 16–26.

Structural Engineers Association of California (SEAOC). (2012). "Wind loads on low profile solar photovoltaic system on flat roofs." *Report SEAOC-PV2-2012*, Sacramento, CA.

Swiss Society of Engineers and Architects (SIA). (1956). "Normen fur die Belastungsannahmen, die Inbetriebnahme und die Uberwachung der Bauten." *SIA Technische Normen No. 160*, Zurich.

Telecommunications Industry Association (TIA). (1991). "Structural standards for steel antenna towers and antenna supporting structures." *ANSI/EIA/TIA 222-E.*, Arlington, VA.

Zuo, D., Letchford, C. W., and Wayne, S. (2011). "Wind tunnel study of wind loading on rectangular louvered panels," *Wind Struct.* 14(5), 449–463.

Zuo, D., Smith, D. A., and Mehta, K. C. (2014). "Experimental study of wind loading of rectangular sign structures." *J. Wind Eng. Indust. Aerodyn.* 130(0): 62–74.

CHAPTER C30
WIND LOADS: COMPONENTS AND CLADDING

In developing the set of pressure coefficients applicable for the design of components and cladding (C&C) as given in Figs. 30.3-1, 30.3-2A–C, 30.3-3, 30.3-4, 30.3-5A–B, and 30.3-6, an envelope approach was followed but using different methods than for the main wind-force resisting system (MWFRS) of Fig. 28.3-1. Because of the small effective area that may be involved in the design of a particular component (consider, for example, the effective area associated with the design of a fastener), the pointwise pressure fluctuations may be highly correlated over the effective area of interest.

Consider the local purlin loads shown in Fig. C28.3-1. The approach involved spatial averaging and time averaging of the point pressures over the effective area transmitting loads to the purlin while the building model was permitted to rotate in the wind tunnel through 360°. As the induced localized pressures may also vary widely as a function of the specific location on the building, height above ground level, exposure, and, more importantly, local geometric discontinuities and location of the element relative to the boundaries in the building surfaces (e.g., walls, roof lines), these factors were also enveloped in the wind tunnel tests. Thus, for the pressure coefficients given in Figs. 30.3-1, 30.3-2A–C, 30.3-3, 30.3-4, 30.3-5A–B, and 30.3-6, the directionality of the wind and influence of exposure have been removed and the surfaces of the building have been "zoned" to reflect an envelope of the peak pressures possible for a given design application.

For ASCE 7-16, the roof zones and pressure coefficients for Fig. 30.3-2A were modified based on the analysis by Kopp and Morrison (2014), which made use of the extensive wind tunnel database developed by Ho et al. (2005). St. Pierre et al. (2005) provided an evaluation of this database compared to earlier data by Davenport et al. (1977, 1978) and ASCE 7 (2002), while Ho et al. (2005) compared the data to full-scale field data from Texas Tech University (Mehta and Levitan 1998). All source data used in the study are publicly accessible through the National Institute of Standards and Technology's website (see, e.g., Main and Fritz 2006). Compared to previous versions of ASCE 7, the pressure coefficients have been increased and are now more consistent with coefficients for buildings higher than 60 ft (18.3 m). Roof zone sizes are also modified from those of earlier versions in order to minimize the increase of pressure coefficients in Zones 1 and 2. The data indicate that for these low-rise buildings, the size of the roof zones depends primarily on the building height, h. A Zone 1 now occurs for large buildings, which accounts for the lower wind loads in the middle of the roof. Zone 3 (roof corner) is an "L" shape, consistent with the shape of Zone 3 for buildings higher than 60 ft (18.3 m) and consistent with the wind loading data. Four potential zone configurations based on the ratios of the smallest and largest building plan dimensions are illustrated in Fig. C30-1. In addition, when the greatest horizontal dimension is less than $0.4h$ (the building does not correspond to a typical low-rise building shape), there is a single roof zone (Zone 3). Detailed explanations can be found in Kopp and Morrison (2014).

As indicated in the discussion for Fig. 28.3-1, the wind tunnel experiments checked both Exposure B and C terrains. Basically, (GC_p) values associated with Exposure B terrain would be higher than those for Exposure C terrain because of reduced velocity pressure in Exposure B terrain. The (GC_p) values given in Figs. 30.3-1, 30.3-2A–C, 30.3-3, 30.3-4, 30.3-5A–B, and 30.3-6 are associated with Exposure C terrain as obtained in the wind tunnel. However, they may also be used for any exposure when the correct velocity pressure representing the appropriate exposure is used. The (GC_p) values given in Figs. 30.3-2A–C are associated with wind tunnel tests performed in both Exposures B and C.

For Fig. 30.3-2A, the coefficients apply equally to Exposure B and C, based on wind tunnel data that show insignificant differences in (GC_p) for Exposures B and C. Consequently, the truncation for K_z in Table 30.3-1 of ASCE 7-10 is not required for buildings below 30 ft (9.1 m), and the lower K_z values may be used.

The pressure coefficients given in Fig. 30.5-1 for buildings with mean height greater than 60 ft (18.3 m) were developed following a similar approach, but the influence of exposure was not enveloped (Stathopoulos and Dumitrescu-Brulotte 1989). Therefore, exposure categories B, C, or D may be used with the values of (GC_p) in Fig. 30.5-1 as appropriate.

C30.1 SCOPE

C30.1.1 Building Types. Guidance for determining C_f and A_f for C&C of structures found in petrochemical and other industrial facilities that are not otherwise addressed in ASCE 7 can be found in *Wind Loads for Petrochemical and Other Industrial Facilities* (ASCE 2011). The 2011 edition references ASCE 7-05, and the user needs to make appropriate adjustments where compliance with the ASCE 7-10 standard is required.

C30.1.5 Air-Permeable Cladding. Air-permeable roof or wall claddings allow partial air pressure equalization between their exterior and interior surfaces. Examples include siding, pressure-equalized rain screen walls, shingles, tiles (including modular vegetative roof assemblies), concrete roof pavers, and aggregate roof surfacing.

The peak pressure acting across an air-permeable cladding material is dependent on the characteristics of other components or layers of a building envelope assembly. At any given instant, the total net pressure across a building envelope assembly is equal to the sum of the partial pressures across the individual layers, as shown in Fig. C30.1-1. However, the proportion of the total net pressure borne by each layer varies from instant to

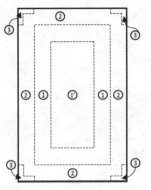 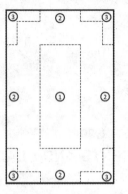 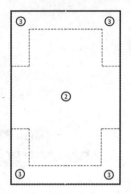 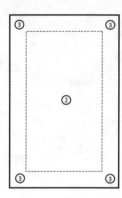

Buildings with least horizontal dimension greater than 2.4h

Buildings with least horizontal dimension greater than 1.2h but less than 2.4h

Buildings with least horizontal dimension less than 1.2h and largest horizontal dimension greater than 1.2h

Buildings with largest horizontal dimension less than 1.2h

FIGURE C30-1 Four Possible Scenarios for Roof Zones, Which Depend on the Ratios of the Least and Largest Horizontal Plan Dimensions to the Mean Roof Height h

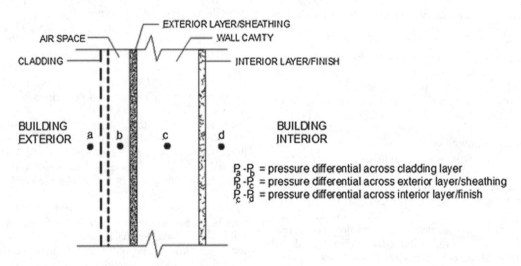

FIGURE C30.1-1 Distribution of Net Components and Cladding Pressure Acting on a Building Surface (Building Envelope) Composed of Three Components (Layers)

instant because of fluctuations in the external and internal pressures and depends on the porosity and stiffness of each layer, as well as the volumes of the air spaces between the layers. As a result, although there is load sharing among the various layers, the sum of the peak pressures across the individual layers typically exceeds the peak pressure across the entire system. In the absence of detailed information on the division of loads, a simple, conservative approach is to assign the entire differential pressure to each layer designed to carry load.

To maximize pressure equalization (reduction) across any cladding system (irrespective of the permeability of the cladding itself), the layer or layers behind the cladding should be

- Relatively stiff in comparison to the cladding material; and
- Relatively air-impermeable in comparison to the cladding material.

Furthermore, the air space between the cladding and the next adjacent building envelope surface behind the cladding (e.g., the exterior sheathing) should be as small as practicable and compartmentalized to avoid communication or venting between different pressure zones of a building's surfaces.

The design wind pressures derived from Chapter 30 represent the pressure differential between the exterior and interior surfaces of the exterior envelope (wall or roof system). Because of partial air-pressure equalization provided by air-permeable claddings, the C&C pressures derived from Chapter 30 can overestimate the load on air-permeable cladding elements. The designer may elect either to use the loads derived from Chapter 30 or to use loads derived by an approved alternative method. If the designer desires to determine the pressure differential across a specific cladding element in combination with other elements comprising a specific building envelope assembly, appropriate pressure measurements should be made on the applicable building envelope assembly, or reference should be made to recognized literature (Cheung and Melbourne 1986; Haig 1990; Baskaran 1992; SBCCI 1994; Peterka et al. 1997; ASTM 2006, 2007; Kala et al. 2008; Baskaran et al. 2012; Kopp and Gavanski 2012; and Cope et al. 2012) for documentation pertaining to wind loads. Such alternative methods may vary according to a given cladding product or class of cladding products or assemblies because each has unique features that affect pressure equalization. It is important to consider the methodology used to determine wind pressure distribution through a multilayered assembly including an air-permeable cladding layer. Recent full-scale wind tunnel tests have shown that an accurate distribution of the wind pressure in a multilayered exterior wall assembly must account for the spatial and temporal (dynamic) fluctuations in wind pressure representative of actual wind flow conditions (Cope et al. 2012). Other factors to consider include the influence of airflow pathways through the assembly (e.g., openings or penetrations through any given layer) and appropriate methods of enveloping peak pressure coefficients for each layer of a multilayered assembly (e.g., Cope et al. 2012) to ensure system reliability and consistency with the characterization of peak pressure coefficients in this standard.

Modular Vegetative Roof Assemblies consist of vegetation and other components integrated as a tray. These trays have vertical air gaps (a minimum of 0.25 in. (6.25 mm)) between the module and roofing system and horizontal air gaps between them. These air gaps allow partial air pressure equalization.

C30.3 BUILDING TYPES

C30.3.1 Conditions. For velocity pressure, see commentary, Section C26.10.1.

C30.3.2 Design Wind Pressures. For velocity pressure, see commentary, Section C26.10.1.

Figs. 30.3-1 and 30.3-2A–C. The pressure coefficient values provided in these figures are to be used for buildings with a mean roof height of 60 ft (18.3 m) or less. The values were obtained from wind tunnel tests conducted at the University of Western Ontario (Davenport et al. 1977, 1978; Ho et al. 2005; St. Pierre et al. 2005; Kopp and Morrison 2014; Vickery et al. 2011; Gavanski et al. 2013). The negative roof (GC_p) values given in these figures are significantly greater (in magnitude) than those given in previous versions (2010 and earlier) but are consistent with those given in Ho et al. (2005). The (GC_p) values given in the figures are given in equation form in Tables C30.3-1 to C30.3-10. Note that the (GC_p) values given in Fig. 30.3-2A–C are a function of the roof slope. Some of the characteristics of the values in the figure are as follows:

1. The values are combined values of (GC_p). The gust-effect factors from these values should not be separated;
2. The velocity pressure, q_h, evaluated at mean roof height should be used with all values of (GC_p);
3. The values provided in the figure represent the upper bounds of the most severe values for any wind direction. The reduced probability that the design wind speed may not occur in the particular direction for which the worst pressure coefficient is recorded has not been included in the values shown in the figure; and
4. The wind tunnel values, as measured, were based on the mean hourly wind speed. The values provided in the figures are the measured values divided by the 3-second dynamic gust pressure at mean roof height to adjust for the reduced pressure coefficient values associated with a 3-s gust speed.

Table C30.3-1. Walls for Buildings with $h \leq 60$ ft ($h \leq 18.3$ m) (Figure 30.3-1)

Positive:	$(GC_p) = 1.0$	for $A = 10$ ft²
Zones 4 and 5	$(GC_p) = 1.1766 - 0.1766 \log A$	for $10 < A \leq 500$ ft²
	$(GC_p) = 0.7$	for $A > 500$ ft²
Negative:	$(GC_p) = -1.1$	for $A = 10$ ft²
Zone 4	$(GC_p) = -1.2766 + 0.1766 \log A$	for $10 < A \leq 500$ ft²
	$(GC_p) = -0.8$	for $A > 500$ ft²
Negative:	$(GC_p) = -1.4$	for $A = 10$ ft²
Zone 5	$(GC_p) = -1.7532 + 0.3532 \log A$	for $10 < A \leq 500$ ft²
	$(GC_p) = -0.8$	for $A > 500$ ft²

Table C30.3-2. Gable Roof, $\theta \leq 7°$ (Figure 30.3-2A)

Positive with and without overhang		
All Zones	$(GC_p) = 0.3$	for $A \leq 10$ ft²
	$(GC_p) = 0.4000 - 0.1000 \log A$	for $10 \leq A \leq 100$ ft²
	$(GC_p) = 0.2$	for $A \geq 100$ ft²
Negative without overhang		
Zone 1'	$(GC_p) = -0.9$	for $A \leq 100$ ft²
	$(GC_p) = -1.9000 + 0.5000 \log A$	for $100 \leq A \leq 1,000$ ft²
	$(GC_p) = -0.4$	for $A \geq 1,000$ ft²
Zone 1	$(GC_p) = -1.7$	for $A \leq 10$ ft²
	$(GC_p) = -2.1120 + 0.4120 \log A$	for $10 \leq A \leq 500$ ft²
	$(GC_p) = -1.0$	for $A \geq 500$ ft²
Zone 2	$(GC_p) = -2.3$	for $A \leq 10$ ft²
	$(GC_p) = -2.8297 + 0.5297 \log A$	for $10 \leq A \leq 500$ ft²
	$(GC_p) = -1.4$	for $A \geq 500$ ft²
Zone 3	$(GC_p) = -3.2$	for $A \leq 10$ ft²
	$(GC_p) = -4.2595 + 1.0595 \log A$	for $10 \leq A \leq 500$ ft²
	$(GC_p) = -1.4$	for $A \geq 500$ ft²
Negative with overhang		
Zone 1 and 1'	$(GC_p) = -1.7$	for $A \leq 10$ ft²
	$(GC_p) = -1.8000 + 0.1000 \log A$	for $10 \leq A \leq 100$ ft²
	$(GC_p) = -3.3168 + 0.8584 \log A$	for $100 \leq A \leq 500$ ft²
	$(GC_p) = -1.0$	for $A \geq 500$ ft²
Zones 2	$(GC_p) = -2.3$	for $A \leq 10$ ft²
	$(GC_p) = -3.0063 + 0.7063 \log A$	for $10 \leq A \leq 500$ ft²
	$(GC_p) = -1.1$	for $A \geq 500$ ft²
Zone 3	$(GC_p) = -3.2$	for $A \leq 10$ ft²
	$(GC_p) = -4.4360 + 1.2360 \log A$	for $10 \leq A \leq 500$ ft²
	$(GC_p) = -1.1$	for $A \geq 500$ ft²

Table C30.3-3. Gable Roof, $7° < \theta \leq 20°$ (Figure 30.3-2B)

Positive with and without overhang		
All Zones	$(GC_p) = 0.7$	for $A \leq 2$ ft²
	$(GC_p) = 0.7709 - 0.2354 \log A$	for $2 \leq A \leq 100$ ft²
	$(GC_p) = 0.3$	for $A \geq 100$ ft²
Negative without overhang		
Zones 1 and 2e	$(GC_p) = -2.0$	for $A \leq 20$ ft²
	$(GC_p) = -4.7920 + 2.1460 \log A$	for $20 \leq A \leq 100$ ft²
	$(GC_p) = -0.5$	for $A \geq 100$ ft²
Zones 2n, 2r, and 3e	$(GC_p) = -3.0$	for $A \leq 10$ ft²
	$(GC_p) = -4.4307 + 1.4307 \log A$	for $10 \leq A \leq 250$ ft²
	$(GC_p) = -1.0$	for $A \geq 250$ ft²
Zone 3r	$(GC_p) = -3.6$	for $A \leq 10$ ft²
	$(GC_p) = -5.4000 + 1.800 \log A$	for $10 \leq A \leq 100$ ft²
	$(GC_p) = -1.8000$	for $A \geq 100$ ft²
Negative with overhang		
Zones 1 and 2e	$(GC_p) = -2.5$	for $A \leq 20$ ft²
	$(GC_p) = -4.3614 + 1.4307 \log A$	for $20 \leq A \leq 100$ ft²
	$(GC_p) = -1.5$	for $A \geq 100$ ft²
Zones 2n and 2r	$(GC_p) = -3.5$	for $A \leq 10$ ft²
	$(GC_p) = -4.5730 + 1.0730 \log A$	for $10 \leq A \leq 250$ ft²
	$(GC_p) = -2.0$	for $A \geq 250$ ft²
Zone 3e	$(GC_p) = -4.1$	for $A \leq 10$ ft²
	$(GC_p) = -5.9599 + 1.8599 \log A$	for $10 \leq A \leq 250$ ft²
	$(GC_p) = -1.5$	for $A \geq 250$ ft²
Zone 3r	$(GC_p) = -4.7$	for $A \leq 10$ ft²
	$(GC_p) = -7.1000 + 2.4000 \log A$	for $10 \leq A \leq 100$ ft²
	$(GC_p) = -2.3$	for $A \geq 100$ ft²

Table C30.3-4. Gable Roofs, $20° < \theta \leq 27°$ (Figure 30.3-2C)

Positive with and without overhang

All Zones	$(GC_p) = 0.7$	for $A \leq 2$ ft²
	$(GC_p) = 0.7709 - 0.2354 \log A$	for $2 \leq A \leq 100$ ft²
	$(GC_p) = 0.3$	for $A \geq 100$ ft²

Negative without overhang

Zones 1 and 2e	$(GC_p) = -1.5$	for $A \leq 20$ ft²
	$(GC_p) = -2.2744 + 0.5952 \log A$	for $20 \leq A \leq 300$ ft²
	$(GC_p) = -0.8$	for $A \geq 300$ ft²
Zones 2n, 2r, and 3r	$(GC_p) = -2.5$	for $A \leq 10$ ft²
	$(GC_p) = -3.6054 + 1.1054 \log A$	for $10 \leq A \leq 150$ ft²
	$(GC_p) = -1.2$	for $A \geq 150$ ft²
Zones 3e	$(GC_p) = -3.6$	for $A \leq 4$ ft²
	$(GC_p) = -4.5880 + 1.6410 \log A$	for $4 \leq A \leq 50$ ft²
	$(GC_p) = -1.8$	for $A \geq 50$ ft²

Negative with overhang

Zones 1 and 2e	$(GC_p) = -2.0$	for $A \leq 20$ ft²
	$(GC_p) = -2.2212 + 0.1701 \log A$	for $20 \leq A \leq 300$ ft²
	$(GC_p) = -1.8$	for $A \geq 300$ ft²
Zones 2n and 2r	$(GC_p) = -3.0$	for $A \leq 10$ ft²
	$(GC_p) = -3.6802 + 0.6802 \log A$	for $10 \leq A \leq 150$ ft²
	$(GC_p) = -2.2$	for $A \geq 150$ ft²
Zone 3r	$(GC_p) = -3.6$	for $A \leq 10$ ft²
	$(GC_p) = -5.2155 + 1.6155 \log A$	for $10 \leq A \leq 150$ ft²
	$(GC_p) = -1.7$	for $A \geq 150$ ft²
Zone 3e	$(GC_p) = -4.7$	for $A \leq 4$ ft²
	$(GC_p) = -6.0173 + 2.1880 \log A$	for $4 \leq A \leq 50$ ft²
	$(GC_p) = -2.3$	for $A \geq 50$ ft²

Table C30.3-5. Gable Roofs, $27° < \theta \leq 45°$ (Figure 30.3-2D)

Positive with and without overhang

All Zones	$(GC_p) = 0.9$	for $A \leq 10$ ft²
	$(GC_p) = 1.3000 - 0.4000 \log A$	for $10 \leq A \leq 100$ ft²
	$(GC_p) = 0.5$	for $A \geq 100$ ft²

Negative without overhang

Zones 1, 2e, and 2r	$(GC_p) = -1.8$	for $A \leq 10$ ft²
	$(GC_p) = -2.8000 + 1.0000 \log A$	for $10 \leq A \leq 100$ ft²
	$(GC_p) = -0.8$	for $A \geq 100$ ft²
Zones 2n and 3r	$(GC_p) = -2.0$	for $A \leq 10$ ft²
	$(GC_p) = -2.7686 + 0.7686 \log A$	for $10 \leq A \leq 200$ ft²
	$(GC_p) = -1.0$	for $A \geq 200$ ft²
Zone 3e	$(GC_p) = -3.2$	for $A \leq 2$ ft²
	$(GC_p) = -3.5043 + 1.0110 \log A$	for $2 \leq A \leq 300$ ft²
	$(GC_p) = -1.0$	for $A \geq 300$ ft²

Negative with overhang

Zones 1, 2e, and 2r	$(GC_p) = -2.6$	for $A \leq 10$ ft²
	$(GC_p) = -3.6000 + 1.0000 \log A$	for $10 \leq A \leq 100$ ft²
	$(GC_p) = -1.6$	for $A \geq 100$ ft²
Zones 2n and 3r	$(GC_p) = -2.8$	for $A \leq 10$ ft²
	$(GC_p) = -3.5686 + 0.7686 \log A$	for $10 \leq A \leq 200$ ft²
	$(GC_p) = -1.8$	for $A \geq 200$ ft²
Zone 3e	$(GC_p) = -4.0$	for $A \leq 2$ ft²
	$(GC_p) = -4.3043 + 1.0110 \log A$	for $2 \leq A \leq 300$ ft²
	$(GC_p) = -1.8$	for $A \geq 300$ ft²

Table C30.3-6. Hip Roofs, No Overhang, $7° < \theta \leq 20°$ (Figure 30.3-2E)

Positive $h/B \geq 0.8$

All Zones	$(GC_p) = 0.7$	for $A \leq 10$ ft²
	$(GC_p) = 1.1000 - 0.4000 \log A$	for $10 \leq A \leq 100$ ft²
	$(GC_p) = 0.3$	for $A \geq 100$ ft²

Negative $h/B \geq 0.8$

Zone 1	$(GC_p) = -1.8$	for $A \leq 20$ ft²
	$(GC_p) = -3.2891 + 1.1445 \log A$	for $20 \leq A \leq 100$ ft²
	$(GC_p) = -1.0$	for $A \geq 100$ ft²
Zone 2r	$(GC_p) = -2.4$	for $A \leq 10$ ft²
	$(GC_p) = -3.2455 + 0.8455 \log A$	for $10 \leq A \leq 200$ ft²
	$(GC_p) = -1.3$	for $A \geq 200$ ft²
Zones 2e and 3	$(GC_p) = -2.6$	for $A \leq 10$ ft²
	$(GC_p) = -3.5223 + 0.9223 \log A$	for $10 \leq A \leq 200$ ft²
	$(GC_p) = -1.4$	for $A \geq 200$ ft²

Positive $h/B \leq 0.5$

All Zones	$(GC_p) = 0.7$	for $A \leq 10$ ft²
	$(GC_p) = 1.1000 - 0.4000 \log A$	for $10 \leq A \leq 100$ ft²
	$(GC_p) = 0.3$	for $A \geq 100$ ft²

Negative $h/B \leq 0.5$

Zone 1	$(GC_p) = -1.3$	for $A \leq 20$ ft²
	$(GC_p) = -1.8584 + 0.4292 \log A$	for $20 \leq A \leq 100$ ft²
	$(GC_p) = -1.0$	for $A \geq 100$ ft²
Zone 2r	$(GC_p) = -2.4$	for $A \leq 10$ ft²
	$(GC_p) = -3.2455 + 0.8455 \log A$	for $10 \leq A \leq 200$ ft²
	$(GC_p) = -1.3$	for $A \geq 200$ ft²
Zones 2e and 3	$(GC_p) = -1.8$	for $A \leq 10$ ft²
	$(GC_p) = -2.3380 + 0.5380 \log A$	for $10 \leq A \leq 200$ ft²
	$(GC_p) = -1.1$	for $A \geq 200$ ft²

Table C30.3-7. Hip Roofs, Overhang, $7° < \theta \leq 20°$ (Figure 30.3-2F)

Negative $h/B \geq 0.8$

Zone 1	$(GC_p) = -2.3$	for $A \leq 20$ ft²
	$(GC_p) = -2.8584 + 0.4292 \log A$	for $20 \leq A \leq 100$ ft²
	$(GC_p) = -2.0$	for $A \geq 100$ ft²
Zone 2r	$(GC_p) = -2.9$	for $A \leq 10$ ft²
	$(GC_p) = -3.3612 + 0.4612 \log A$	for $10 \leq A \leq 200$ ft²
	$(GC_p) = -2.3$	for $A \geq 200$ ft²
Zones 2e	$(GC_p) = -3.1$	for $A \leq 10$ ft²
	$(GC_p) = -3.6380 + 0.5380 \log A$	for $10 \leq A \leq 200$ ft²
	$(GC_p) = -2.4$	for $A \geq 200$ ft²
Zones 3	$(GC_p) = -3.7$	for $A \leq 10$ ft²
	$(GC_p) = -5.0835 + 1.3835 \log A$	for $10 \leq A \leq 200$ ft²
	$(GC_p) = -1.9$	for $A \geq 200$ ft²

Negative $h/B \leq 0.5$

Zone 1	$(GC_p) = -1.8$	for $A \leq 20$ ft²
	$(GC_p) = -1.4277 - 0.2861 \log A$	for $20 \leq A \leq 100$ ft²
	$(GC_p) = -2.0$	for $A \geq 100$ ft²
Zones 2r	$(GC_p) = -2.9$	for $A \leq 10$ ft²
	$(GC_p) = -3.3612 + 0.4612 \log A$	for $10 \leq A \leq 200$ ft²
	$(GC_p) = -2.3$	for $A \geq 200$ ft²
Zones 2e	$(GC_p) = -2.3$	for $A \leq 10$ ft²
	$(GC_p) = -2.4537 + 0.1537 \log A$	for $10 \leq A \leq 200$ ft²
	$(GC_p) = -2.1$	for $A \geq 200$ ft²
Zone 3	$(GC_p) = -2.9$	for $A \leq 10$ ft²
	$(GC_p) = -3.8992 + 0.9992 \log A$	for $10 \leq A \leq 200$ ft²
	$(GC_p) = -1.6$	for $A \geq 200$ ft²

Each C&C element should be designed for the maximum positive and negative pressures (including applicable internal pressures) acting on it. The pressure coefficient values should be determined for each C&C element on the basis of its location on the building and the effective area for the element. Research (Stathopoulos and Zhu 1988, 1990) indicated that the pressure coefficients provided generally apply to facades with architectural features, such as balconies, ribs, and various facade textures.

Overhang pressures were determined by adding the effective uplift (GC_p)s implied in ASCE 7-10. These effective uplift

Table C30.3-8. Hip Roofs, $20° < \theta \leq 27°$ (Figure 30.3-2G)

		Positive $h/B \geq 0.8$	
All Zones	$(GC_p) = 0.7$		for $A \leq 10$ ft^2
	$(GC_p) = 1.1000 - 0.4000 \log A$		for $10 \leq A \leq 100$ ft^2
	$(GC_p) = 0.3$		for $A \geq 100$ ft^2
		Negative without overhang	
Zone 1	$(GC_p) = -1.4$		for $A \leq 10$ ft^2
	$(GC_p) = -2.0000 + 0.6000 \log A$		for $10 \leq A \leq 100$ ft^2
	$(GC_p) = -0.8$		for $A \geq 100$ ft^2
Zones 2e, 2r, and 3	$(GC_p) = -2.0$		for $A \leq 10$ ft^2
	$(GC_p) = -2.7686 + 0.7686 \log A$		for $10 \leq A \leq 200$ ft^2
	$(GC_p) = -1.0$		for $A \geq 200$ ft^2
		Negative with overhang	
Zone 1	$(GC_p) = -1.9$		for $A \leq 10$ ft^2
	$(GC_p) = -2.0000 + 0.1000 \log A$		for $10 \leq A \leq 100$ ft^2
	$(GC_p) = -1.8$		for $A \geq 100$ ft^2
Zones 2e and 2r	$(GC_p) = -2.5$		for $A \leq 10$ ft^2
	$(GC_p) = -2.8843 + 0.3843 \log A$		for $10 \leq A \leq 200$ ft^2
	$(GC_p) = -2.0$		for $A \geq 200$ ft^2
Zone 3	$(GC_p) = -3.1$		for $A \leq 10$ ft^2
	$(GC_p) = -4.3298 + 1.2298 \log A$		for $10 \leq A \leq 200$ ft^2
	$(GC_p) = -1.5$		for $A \geq 200$ ft^2

(GC_p)s were computed by subtracting the (GC_p) values given for the roof (no overhang) case from the (GC_p)s given for the overhang case. The additional (GC_p)s are given in Fig. C30.3-1.

The following guidance is based on the collective judgment of the wind load committee. For "L-shaped," "T-shaped," and other "irregular" shapes, Fig. C30.3-2 depicts the roof and wall zones for use with Figs. 30.3-1, 30.3-2, 30.3-4, 30.3-5, 30.3-6, 30.4-1, and 30.5-1 for wind loads on components and cladding of buildings, showing the applicability to buildings that are rectangular in plan. To address buildings with nonrectangular plans, Fig. C30.3-2 can be used for guidance in applying the requirements. When an outward corner protrudes less than the distance a from the wall, neither Zone 3 nor 5 are required; however, when the outward protrusion is greater than a, Zones 3 and 5 are required. Reentrant (interior) corners do not require Zones 3 or 5. For corners that have an included interior angle greater than 135°, neither Zone 3 nor 5 is required. To determine the length of a, a rectangle which enclosed the building is drawn over the building plan. The dimensions of this rectangle are used to determine the horizontal dimensions for the calculation of a.

Figs. 30.3-4, 30.3-5A, and 30.3-5B. These figures present values of (GC_p) for the design of roof C&C for buildings with

Table C30.3-9. Hip Roofs, $27° < \theta \leq 45°$, No Overhang (Figure 30.3-2H)

		Positive	
All Zones	$(GC_p) = 0.9$		for $A \leq 2$ ft^2
	$(GC_p) = 1.0063 - 0.3532 \log A$		for $2 \leq A \leq 100$ ft^2
	$(GC_p) = 0.3$		for $A \geq 100$ ft^2
		Negative	
Zone 1	$(GC_p) = -0.6175 - 0.0200\theta$		for $A \leq 10$ ft^2
	$(GC_p) = -1.0191 - 0.0250\theta + [0.4016 + 0.0050\theta] \log A$		for $10 \leq A \leq 200$ ft^2
	$(GC_p) = -0.0950 - 0.0135\theta$		for $A \geq 200$ ft^2
Zone 2e	$(GC_p) = 0.2000 - 0.0670\theta$		for $A \leq 2$ ft^2
	$(GC_p) = -0.8000 + \left[\dfrac{\log(280 - 5\theta)(0.0670\theta - 1)}{0.301 - \log(280 - 5\theta)}\right] + \left[\dfrac{1 - 0.0670\theta}{0.3010 - \log(280 - 5\theta)}\right] \log A$		for $2 \leq A \leq [280 - 5\theta]$ ft^2
	$(GC_p) = -0.8$		for $A \geq [280 - 5\theta]$ ft^2
Zones 2r	$(GC_p) = 1.0000 - 0.0820\theta$		for $A \leq 5$ ft^2
	$(GC_p) = 2.0746 - 0.1261\theta + [0.0630\theta - 1.5373] \log A$		for $5 \leq A \leq 100$ ft^2
	$(GC_p) = -1.0000$		for $A \geq 100$ ft^2
Zones 3	$(GC_p) = 1.2500 - 0.1080\theta$		for $A \leq [9 - 0.1350\theta]$ ft^2
	$(GC_p) = \left[\dfrac{0.1835\theta - 3.8230}{\log(9 - 0.1350\theta) - 1.6990}\right] - 1.0 + \left[\dfrac{2.25 - 0.1080\theta}{\log(9 - 0.1350\theta) - 1.6990}\right] \log A$		for $[9 - 0.1350\theta] \leq A \leq 50$ ft^2
	$(GC_p) = -1.0000$		for $A \geq 50$ ft^2

Table C30.3-10. Hip Roofs, $27° < \theta \leq 45°$, Overhang (Figure 30.3-2I)

		Negative	
Zone 1	$(GC_p) = -1.4175 - 0.0200\theta$		for $A \leq 10$ ft^2
	$(GC_p) = -1.8191 - 0.0250\theta + [0.4016 + 0.0050\theta] \log A$		for $10 \leq A \leq 200$ ft^2
	$(GC_p) = -0.8950 - 0.0135\theta$		for $A \geq 200$ ft^2
Zone 2e	$(GC_p) = -0.6000 - 0.0670\theta$		for $A \leq 2$ ft^2
	$(GC_p) = -1.6000 + \left[\dfrac{\log(280 - 5\theta)(0.0670\theta - 1)}{0.301 - \log(280 - 5\theta)}\right] + \left[\dfrac{1 - 0.0670\theta}{0.301 - \log(280 - 5\theta)}\right] \log A$		for $2 \leq A \leq 280 - 5\theta$ ft^2
	$(GC_p) = -1.6000$		for $A \geq 280 - 5\theta$ ft^2
Zones 2r	$(GC_p) = 0.2000 - 0.0820\theta$		for $A \leq 5$ ft^2
	$(GC_p) = 1.2745 - 0.1261\theta + [0.0630\theta - 1.5373] \log A$		for $5 \leq A \leq 100$ ft^2
	$(GC_p) = -1.8000$		for $A \geq 100$ ft^2
Zones 3	$(GC_p) = 0.4500 - 0.1080\theta$		for $A \leq 9 - 0.1350\theta$ ft^2
	$(GC_p) = \left[\dfrac{0.1835\theta - 3.823}{\log(9 - 0.135\theta) - 1.699}\right] - 1.8 + \left[\dfrac{2.25 - 0.108\theta}{\log(9 - 0.135\theta) - 1.699}\right] \log A$		for $9 - 0.1350\theta \leq A \leq 50$ ft^2
	$(GC_p) = -1.8000$		for $A \geq 50$ ft^2

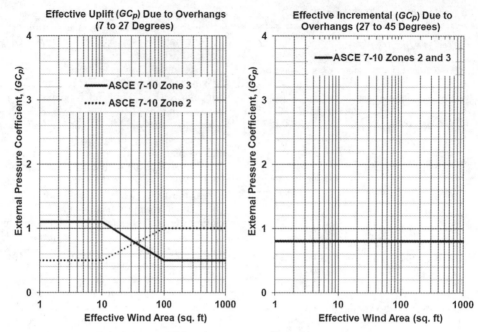

FIGURE C30.3-1 Effective Incremental (GC_p) Caused by Overhangs

multispan gable roofs and buildings with monoslope roofs. The coefficients are based on wind tunnel studies (Stathopoulos and Mohammadian 1986; Surry and Stathopoulos 1988; Stathopoulos and Saathoff 1991).

Fig. 30.3-6. The values of (GC_p) in this figure are for the design of roof C&C for buildings with sawtooth roofs and mean roof height, h, less than or equal to 60 ft (18.3 m). Note that the coefficients for corner zones on segment A differ from those coefficients for corner zones on the segments designated as B, C, and D. Also, when the roof angle is less than or equal to 10°, values of (GC_p) for regular gable roofs (Fig. 30.3-2A) are to be used. The coefficients included in Fig. 30.3-6 are based on wind tunnel studies reported by Saathoff and Stathopoulos (1992).

Fig. 30.3-7. This figure for cladding pressures on dome roofs is based on Taylor (1991). Negative pressures are to be applied to the entire surface because they apply along the full arc that is perpendicular to the wind direction and that passes through the top of the dome. Users are cautioned that only three shapes were available to define values in this figure: $h_D/D = 0.5, f/D = 0.5$; $h_D/D = 0.0, f/D = 0.5$; and $h_D/D = 0.0, f/D = 0.33$.

Fig. 30.5-1. The pressure coefficients shown in this figure reflect the results obtained from comprehensive wind tunnel studies carried out (Stathopoulos and Dumitrescu-Brulotte 1989). The availability of more-comprehensive wind tunnel data has also allowed a simplification of the zoning for pressure coefficients; flat roofs are now divided into three zones; and walls are represented by two zones.

The external pressure coefficients and zones given in Fig. 30.5-1 were established by wind tunnel tests on isolated "boxlike" buildings (Akins and Cermak 1975; Peterka and Cermak 1975). Boundary-layer wind tunnel tests on high-rise buildings (mostly in downtown city centers) have shown that variations in pressure coefficients and the distribution of pressure on the different building facades are obtained (Templin and Cermak 1978). These variations are caused by building geometry, low attached buildings, nonrectangular cross sections, setbacks, and sloping surfaces. In addition, surrounding buildings contribute to the variations in pressure. Wind tunnel tests indicate that pressure coefficients are not distributed symmetrically and can give rise to torsional wind loading on the building.

Boundary-layer wind tunnel tests that include modeling of surrounding buildings permit the establishment of more exact magnitudes and distributions of (GC_p) for buildings that are not isolated or "boxlike" in shape.

PART 1: LOW-RISE BUILDINGS

The C&C tables in Fig. 30.4-1 are a tabulation of the pressures on an enclosed, regular, 30-ft- (9.1-m)-high building with a roof as described. The pressures can be modified to a different exposure and height with the same adjustment factors as the MWFRS pressures. For the designer to use Part 2 for the design of the C&C, the building must conform to all five requirements in Section 30.4; otherwise, one of the other procedures specified in Section 30.1.1 must be used.

PART 3: BUILDINGS WITH $h > 60$ ft ($h > 18.3$ m)

In Eq. (30.5-1) a velocity pressure term, q_i, appears that is defined as the "velocity pressure for internal pressure determination." The positive internal pressure is dictated by the positive exterior pressure on the windward face at the point where there is an opening. The positive exterior pressure at the opening is governed by the value of q at the level of the opening, not q_h. For positive internal pressure evaluation, q_i may conservatively be evaluated at height $h (q_i = q_h)$. For low buildings, this height does not make much difference, but for the example of a 300-ft- (91.4-m)-tall building in Exposure B with the highest opening at 60 ft (18.3 m), the difference between q_{300} and q_{60} represents a 59% increase in internal pressure. This increase is unrealistic and represents an unnecessary degree of conservatism. Accordingly, $q_i = q_z$ for positive internal pressure evaluation in partially enclosed buildings where height z is defined as the level of the highest opening in the building that could affect the positive internal pressure. For buildings sited in wind-borne debris regions, glazing that is not impact-resistant or protected with an impact-protective system, q_i should be treated as an opening.

Diagram

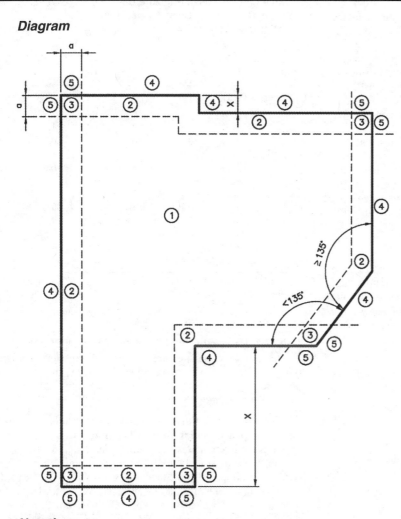

Notation

a = 10% of least horizontal dimension or $0.4h$, whichever is smaller, but not less than either 4% of least horizontal dimension or 3 ft (0.9 m).

X = Offset dimension, in ft (m).

Notes

1. Labels marked on the roof plan indicate roof zones; labels marked outside the roof plan indicate wall zones.
2. If $X \leq a$, then Zone 3 and Zone 5 need not be applied at that corner.
3. If the interior angle is ≥135°, then Zone 3 and Zone 5 need not be applied at that corner.

FIGURE C30.3-2 Plan View of Roof and Wall Zones for Component and Cladding Loads on Buildings with Nonrectangular Plans

PART 4: BUILDINGS WITH 60 ft < h ≤ 160 ft (18.3 m < h ≤ 48.8 m) (SIMPLIFIED)

C30.6 BUILDING TYPES

This section was added to ASCE 7-10 to cover the common practical case of enclosed buildings up to height $h = 160$ ft ($h = 48.8$ m). Table 30.6-2 includes wall and roof pressures for flat roofs ($\theta \leq 7°$), gable roofs, hip roofs, monoslope roofs, and mansard roofs. Pressures are derived from Fig. 30.5-1 (flat roofs) for $h > 60$ ft ($h > 18.3$ m). For flat, gable, hip, monoslope, and mansard roofs with $h \leq 60$ ft ($h \leq 18.3$ m) and all roof slopes, reference is made to the roof and wall pressures tabulated in Fig. 30.4-1. For hip and gable roofs with $h > 60$ ft ($h > 18.3$ m), reference is made to Fig. 30.5-1, Note 6, which permits the use of roof pressure coefficients in Figs. 30.3-2A through 30.3-2I as defined for low-rise buildings for these roof shapes if the appropriate q_h is used. Similarly, the roof pressure coefficients in Fig. 30.3-5A and B for monoslope roofs from Part 3 are permitted. Pressures were selected for each zone that encompasses the largest pressure coefficients for the comparable zones from the different roof shapes. Thus, for some cases, the pressures tabulated are conservative in order to maintain simplicity. The (GC_p) values from these figures were combined with an internal pressure coefficient (±0.18) to obtain a net coefficient from which pressures were calculated. The tabulated pressures are applicable to the entire zone shown in the various figures.

Pressures in Table 30.6-2 are shown for an effective wind area of 10 ft² (0.93 m²). A reduction factor is also shown to obtain pressures for larger effective wind areas. The reduction factors are based on the graph of external pressure coefficients shown in the figures in Part 3 and are based on the most conservative reduction for each zone from the various figures.

Note that the roof pressures tabulated for buildings with $h \leq 60$ ft ($h \leq 18.3$ m) in Fig. 30.4-1 are based on $h = 30$ ft ($h = 9.1$ m) and Exposure B. An adjustment factor λ is applied to the tabulated pressures for other heights and exposures. The tabulated pressures in Table 30.6-2 are based on Exposure C. An adjustment factor (EAF) from Eq. (30.6-1) is applied for other exposures.

C30.6.1 Wind Load: Components and Cladding.

C30.6.1.2 Parapets. Parapet C&C wind pressures can be obtained from the tables as shown in the parapet figures. The pressures obtained are slightly conservative based on the net pressure coefficients for parapets compared to roof zones from Part 3. Two load cases must be considered based on pressures applied to both windward and leeward parapet surfaces, as shown in Fig. 30.6-1.

C30.6.1.3 Roof Overhangs. C&C pressures for roof overhangs for flat and monoslope roofs with $h > 60$ ft ($h > 18.3$ m) can be obtained from the tables as shown in Fig. 30.6-2. These pressures are slightly conservative and are based on the external pressure coefficients from Part 3. Pressures for roof overhangs in flat, gable, hip, monoslope, and mansard roofs with $h \leq 60$ ft ($h \leq 18.3$ m) can be obtained directly from the tabulated pressures in Fig. 30.4-1.

PART 5: OPEN BUILDINGS

C30.7 BUILDING TYPES

In determining loads on C&C elements for open building roofs using Figs. 30.7-1, 30.7-2, and 30.7-3, it is important for the designer to note that the net pressure coefficient, C_N, is based on contributions from the top and bottom surfaces of the roof. This fact implies that the element receives load from both surfaces. Such would not be the case if the surface below the roof were separated structurally from the top roof surface. In this case, the pressure coefficient should be separated for the effect of top and bottom pressures, or conservatively, each surface could be designed using the C_N value from Figs. 30.7-1, 30.7-2, and 30.7-3.

PART 7: NONBUILDING STRUCTURES

C30.12 CIRCULAR BINS, SILOS, AND TANKS WITH $h \leq 120$ ft ($h \leq 36.5$ m)

Section 30.12 contains the provisions for determining wind pressures on silo and tank walls and roofs. The results of isolated and grouped silos are largely based on Australian Standards (Standards Australia 2011), and the wind tunnel tests by Sabransky and Melbourne (1987) and Macdonald et al. (1988, 1990). Significant increases in the mean pressures of grouped silos were found in the wind tunnel tests, so the provisions of grouped tanks and silos are specified in this section.

Table C30.12-1 Mean Pressure Coefficients ($GC_p - GC_{pi}$) for Open-Topped Tanks

Angle α	Aspect Ratio H/D					
	0.25	0.50	1	2	3	4
0°	1.69	1.80	1.9	2	2.07	2.11
15°	1.39	1.50	1.6	1.7	1.77	1.81
30°	0.99	1.10	1.2	1.3	1.37	1.41
45°	0.39	0.50	0.6	0.7	0.77	0.81
60°	−0.01	−0.01	−0.1	−0.1	−0.13	−0.09
75°	−0.11	−0.31	−0.5	−0.7	−0.83	−0.89
90°	−0.11	−0.31	−0.5	−0.7	−0.83	−0.89
105°	−0.01	−0.11	−0.2	−0.3	−0.33	−0.29
120°	0.09	0.10	0.2	0.2	0.27	0.21
135°	0.29	0.30	0.4	0.5	0.47	0.51
150°	0.29	0.40	0.5	0.5	0.57	0.61
165°	0.29	0.40	0.5	0.5	0.57	0.61
180°	0.29	0.40	0.5	0.5	0.57	0.61

Note: D = diameter of circular structure, in ft (m); H = height in ft (m); α = angle from the wind direction to a point on the wall of a circular bin, silo, or tank, in degrees.

C30.12.2 External Walls of Isolated Circular Bins, Silos, and Tanks. This section specifies the external pressure coefficients ($GC_{p(\alpha)}$) for the walls of circular bins, silos, and tanks. The pressure coefficients for isolated silos are adopted from Australian Standards (Standards Australia 2011).

C30.12.3 Internal Surface of Exterior Walls of Isolated Open-Topped Circular Bins, Silos, and Tanks. This section specifies the internal pressure coefficients (GC_{pi}) for the walls of circular bins, silos, and tanks. The internal pressure coefficients (GC_{pi}) are adopted from Standards Australia (2011). Based on the wind tunnel test results, mean pressures on walls for open-topped bins, silos, and tanks are different from the values of circular bins, silos, and tanks with flat or conical roofs. Table C30.12-1 lists the mean pressure coefficients ($GC_p - GC_{pi}$) for open-topped circular bins, silos, and tanks, based on Eqs. (30.12-2) and (30.12-5). The distribution of the external pressure around the perimeter of the wall is shown in Fig. C30.12-1.

C30.12.4 Roofs of Isolated Circular Bins, Silos, and Tanks. This section specifies the external pressure coefficients (GC_p) for the roofs of circular bins, silos, and tanks. Two conditions are covered as shown in Fig. 30.12-2: Class 1 roofs have the roof angle θ < 10°, and Class 2 roofs have $10° \leq \theta < 30°$. Zone 1 pressures are defined differently that either increase with the increment of the silo heights for Class 1 roofs, or with the silo or tank diameters for Class 2 roofs. For cladding design, Zone 3 pressures are specified for the local pressures near the windward edges applicable to all classes, and Zone 4 is specified for the region near the cone apex used for Class 2b roofs only. Fig. C30.12-2 is the graphic presentation of the elevation views for the external pressure coefficients (GC_p). For Class 1 roofs, the external pressure coefficients are based on comparisons of domed roofs and flat roofs from Chapter 27 of ASCE 7-10 for maximum uplift conditions. The results of Class 2 roofs are consistent with data of Sabransky and Melbourne (1987) and Macdonald et al. (1988).

C30.12.6 Roofs and Walls of Grouped Circular Bins, Silos, and Tanks. For grouped silos, (GC_p) values for roofs and walls are largely based on AS/ NZS 1170.2 (Standards Australia 2011) and wind tunnel tests by Sabransky and Melbourne (1987) and

Diagrams

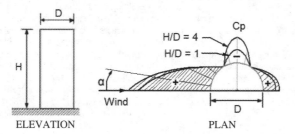

Notation

C_p = External pressure coefficient to be used in determination of wind loads for buildings.
D = Diameter of a circular structure, in ft (m).
H = Height of solid cylinder, in ft (m).
α = Angle from the wind direction to a point on the wall of a circular bin, silo, or tank, in degrees (see Section 30.12.2).

FIGURE C30.12-1 Mean Pressure Coefficients $((GC_p)-(GC_{pi}))$ for Open-Topped Tanks

Diagrams

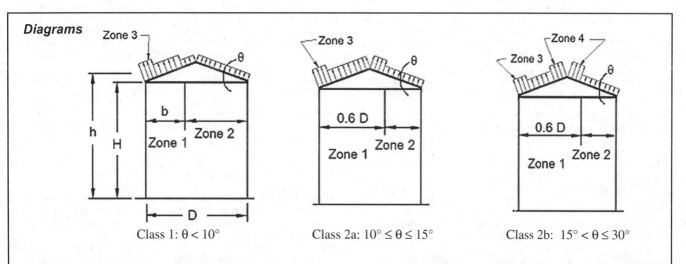

Notation

b = Horizontal dimension specified for Zone 1 of a conical roof, in ft (m). For roof angles less than 10 degrees, b is calculated from the table of external pressure coefficients in Fig. 30.12-2 (e.g., $b = 0.5D$ for $H/D = 0.5$.) For roof angles equal to or larger than 10 degrees, $b = 0.6D$. (So for Class 2a and 2b, $b = 0.6D$.)
D = Diameter of a circular structure, in ft (m).
h = Mean roof height, in ft (m).
H = Height of the solid cylinder, in ft (m).
θ = Angle of plane of roof from horizontal in degrees.

FIGURE C30.12-2 External Pressure Coefficients (GC_p) for Roofs

Macdonald et al. (1990). Test results of an in-line group of three silos with a clear spacing of $0.25D$ between nearest adjacent walls ($1.25D$ center-to-center) by Sabransky and Melbourne (1987) indicated that the mean pressure coefficient between the gaps increased by 70% compared to the one for the isolated silo. A similar result was observed for the roof near the wall of the silo. It was concluded that a clear spacing of $0.25D$ produced the maximum interference between two finite cylinders.

Test results of an in-line group of five silos with various center-to-center spacings by Macdonald et al. (1990) indicated that the region of positive pressure on the windward side spans a larger angular sector of the circumference than that for an isolated silo, and high negative mean pressures occur near the point of shortest distance between the adjacent silos and at the outside corners of the groups.

REFERENCES

Akins, R. E., and Cermak, J. E. (1975). "Wind pressures on buildings." *Technical Report CER 7677REAJEC15*, Fluid Dynamics and Diffusion Lab, Colorado State University, Fort Collins, CO.
ASCE. (2002). "Minimum design loads for buildings and other structures," *ASCE/SEI 7-02*, American Society of Civil Engineers, Reston, VA.
ASCE Task Committee on Wind-Induced Forces. (2011). *Wind Loads for Petrochemical and Other Industrial Facilities*. American Society of Civil Engineers, Reston, VA.

ASTM. (2006). "Standard specification for rigid poly(vinyl chloride) (PVC) siding." *ASTM D3679-06a*, West Conshohocken, PA.

ASTM. (2007). "Standard test method for wind resistance of sealed asphalt shingles (uplift force/uplift resistance method)." *ASTM D7158-07*, West Conshohocken, PA.

Baskaran, A. (1992). "Review of design guidelines for pressure equalized rainscreen walls." *Internal Report No. 629*, National Research Council Canada, Institute for Research in Construction, Ottawa, Ontario, Canada.

Baskaran, A., Molleti, S., Ko, S., and Shoemaker, L. (2012). "Wind uplift performance of composite metal roof assemblies." *J. Archit. Eng.*, 18(1), 2–15.

Cheung, J. C. J., and Melbourne, W. H. (1986). "Wind loadings on porous cladding." *Proc., 9th Australian Conf. on Fluid Mechanics*, Australasian Fluid Mechanics Society, Victoria, Australia, 308.

Cope, A., Crandell, J., Johnston, D., Kochkin, V., Liu, Z., Stevig, L., and Reinhold, T. (2012) "Wind loads on components of multi-layer wall systems with air-permeable exterior cladding." In *Advances in Hurricane Engineering*, American Society of Civil Engineers, Reston, VA, 238–257. doi: 10.1061/9780784412626.022.

Davenport, A. G., Surry, D., and Stathopoulos, T. (1977). "Wind loads on low-rise buildings." *Final Report on Phases I and II*, BLWT-SS8, University of Western Ontario, London, Ontario, Canada.

Davenport, A. G., Surry, D., and Stathopoulos, T. (1978). "Wind loads on low-rise buildings." *Final Report on Phase III*, BLWT-SS4, University of Western Ontario, London, Ontario, Canada.

Gavanski, E., Kordi, B., Kopp, G. A., and Vickery, P. J. (2013). "Wind loads on roof sheathing of houses." *J. Wind Eng. Ind. Aerodyn.* 114, 106–121.

Haig, J. R. (1990). *Wind loads on tiles for USA*, Redland Technology Ltd., Horsham, West Sussex, UK.

Ho, T. C. E., Surry, D., Morrish, D., and Kopp, G. A. (2005). "The UWO contribution to the NIST aerodynamic database for wind loads on low buildings: Part 1. Basic aerodynamic data and archiving." *J. Wind Eng. Ind. Aerodyn.*, vol. 93, pp. 1–30.

Kala, S., Stathopoulos, T., and Kumar, K. (2008). "Wind loads on rainscreen walls: Boundary-layer wind tunnel experiments." *J. Wind Eng. Ind. Aerodyn.*, 96(6–7), 1058–1073.

Kopp, G., and Gavanski, E. (2012). "Effects of pressure equalization on the performance of residential wall systems under extreme wind loads." *J. Struct. Eng.*, 138(4), 526–538.

Kopp, G. A., and Morrison, M. J. (2014). "Component and cladding pressures and zones for the roofs of low-rise buildings." *Boundary Layer Wind Tunnel Report*, University of Western Ontario, London, ON.

Macdonald, P. A., Holmes, J. D., and Kwok, K. C. S. (1990). "Wind loads on circular storage bins, silos and tanks. II. Effect of grouping." *J. Wind Eng. Ind. Aerodyn.*, 34(1), 77–95.

Macdonald, P. A., Kwok, K. C. S., and Holmes, J. D. (1988). "Wind loads on circular storage bins, silos and tanks: 1. Point pressure measurements on isolated structures." *J. Wind Eng. Ind. Aerodyn.*, 31(2-3), 165–187.

Main, J. A., and Fritz, W. P. (2006). Database-assisted design for wind: Concepts, software, and examples for rigid and flexible buildings, National Institute of Standards and Technology: NIST Building Science Series 180.

Mehta, K. C., and Levitan, M. L. (1998). Field experiments for wind pressures, Department of Civil Engineering Progress Report, Texas Tech University.

Peterka, J. A., and Cermak, J. E. (1975). "Wind pressures on buildings: Probability densities." *J. Struct. Div.*, 101(6), 1255–1267.

Peterka, J. A., Cermak, J. E., Cochran, L. S., Cochran, B. C., Hosoya, N., Derickson, R. G., Harper, C., Jones, J., and Metz, B. (1997). "Wind uplift model for asphalt shingles." *J. Arch. Eng.*, 3(4), 147–155.

Saathoff, P. J., and Stathopoulos, T. (1992). "Wind loads on buildings with sawtooth roofs." *J. Struct. Eng.*, 118(2), 429–446.

Sabransky, I. J., and Melbourne, W. H. (1987) "Design pressure distribution on circular silos with conical roofs." *J. Wind Eng. Ind. Aerodyn.*, 26(1), 65–84.

Southern Building Code Congress International (SBCCI). (1994). *Standard building code*, Janesville, WI.

St. Pierre, L. M., Kopp, G. A., Surry, D., Ho, T. C. E. (2005). "The UWO contribution to the NIST aerodynamic database for wind loads on low buildings: Part 2. Comparison of data with wind load provisions." *J. Wind Eng. Ind. Aerodyn.*, 93, 31–59.

Standards Australia. (2011). *Structural design actions—Wind actions*. Standards Australia, North Sydney, Australia, AS/NZS 1170.2:2011.

Stathopoulos, T., and Dumitrescu-Brulotte, M. (1989). "Design recommendations for wind loading on buildings of intermediate height." *Can. J. Civil Eng.*, 16(6), 910–916.

Stathopoulos, T., and Mohammadian, A. R. (1986). "Wind loads on low buildings with mono-sloped roofs." *J. Wind Eng. Ind. Aerodyn.*, 23, 81–97.

Stathopoulos, T., and Saathoff, P. (1991). "Wind pressures on roofs of various geometries." *J. Wind Eng. Ind. Aerodyn.*, 38, 273–284.

Stathopoulos, T., and Zhu, X. (1988). "Wind pressures on buildings with appurtenances." *J. Wind Eng. Ind. Aerodyn.*, 31, 265–281.

Stathopoulos, T., and Zhu, X. (1990). "Wind pressures on buildings with mullions." *J. Struct. Eng.*, 116(8), 2272–2291.

Surry, D., and Stathopoulos, T. (1988). "The wind loading of buildings with monosloped roofs." *Final report*, BLWT-SS38, University of Western Ontario, London, Ontario, Canada.

Taylor, T. J. (1991). "Wind pressures on a hemispherical dome." *J. Wind Eng. Ind. Aerodyn.*, 40(2), 199–213.

Templin, J. T., and Cermak, J. E. (1978). "Wind pressures on buildings: Effect of mullions." *Technical Report CER76-77JTT-JEC24*, Fluid Dynamics and Diffusion Lab, Colorado State University, Fort Collins, CO.

Vickery, P. J., Kopp, G. A., and Twisdale, L. A., Jr. (2011). "Component and cladding wind pressures on hip and gable roofs: Comparisons to the U.S. wind loading provisions." *13th International Conference on Wind Engineering*, Amsterdam, Netherlands, July.

OTHER REFERENCES (NOT CITED)

Batts, M. E., Cordes, M. R., Russell, L. R., Shaver, J. R., and Simiu, E. (1980). "Hurricane wind speeds in the United States." *NBS Building Science Series 124*, National Bureau of Standards, Washington, DC.

Best, R. J., and Holmes, J. D. (1978). "Model study of wind pressures on an isolated single-story house." *Wind Engineering Report 3/78*, James Cook University of North Queensland, Australia.

Beste, F., and Cermak, J. E. (1996). "Correlation of internal and area-averaged wind pressures on low-rise buildings." *Proc., 3rd Int. Colloq. on Bluff Body Aerodynamics and Applications*, Virginia Polytechnic Institute, Blacksburg, VA.

Chock, G., Peterka, J., and Yu, G. (2005). "Topographic wind speed-up and directionality factors for use in the city and county of Honolulu building code." *Proc., 10th Americas Conf. on Wind Engineering*, Baton Rouge, LA.

CSA Group. (2015). "Standard test method for the dynamic wind uplift resistance of vegetated roof assemblies," *CSA A123.24-15*, CSA Group, Toronto, ON.

Davenport, A. G., Grimmond, C. S. B., Oke, T. R., and Wieringa, J. (2000). "Estimating the roughness of cities and sheltered country." *Preprint of the 12th AMS Conf. on Applied Climatology*, American Meteorological Society, Boston, MA, 96–99.

Eaton, K. J., and Mayne, J. R. (1975). "The measurement of wind pressures on two-story houses at Aylesbury." *J. Indust. Aerodyn.*, 1(1), 67–109.

Ellingwood, B. (1981). "Wind and snow load statistics for probabilistic design." *J. Struct. Div.*, 107(7), 1345–1350.

Engineering Sciences Data Unit (ESDU). (1990). "Strong winds in the atmospheric boundary layer. Part 1: Mean hourly wind speeds." *Item Number 82026, with Amendments A to C*, ESDU, London.

Ho, E. (1992). "Variability of low building wind lands." Doctoral dissertation, University of Western Ontario, London, Ontario, Canada.

Marshall, R. D. (1977). "The measurement of wind loads on a full-scale mobile home." *NBSIR 77-1289*, National Bureau of Standards, U.S. Dept. of Commerce, Washington, DC.

McDonald, J. R. (1983). "A methodology for tornado hazard probability assessment." *NUREG/CR3058*, U.S. Nuclear Regulatory Commission, Washington, DC.

Peterka, J. A., and Shahid, S. (1993). "Extreme gust wind speeds in the U.S." In *Proc., 7th U.S. Nat. Conf. on Wind Engineering*, Gary Hart, ed., International Association for Wind Engineering, Kanagawa, Japan, 2, 503–512.

Powell, M. D. (1980). "Evaluations of diagnostic marine boundary-layer models applied to hurricanes." *Monthly Weather Rev.*, 108(6), 757–766.

Sataka, N., Suda, K., Arakawa, T., Sasaki, A., and Tamura, Y. (2003). "Damping evaluation using full-scale data of buildings in Japan." *J. Struct. Eng.*, 129(4), 470–477.

Single Ply Roof Industry (SPRI). (2013). "Wind design standard for vegetative roofing systems." *ANSI/SPRI RP-14*, Single Ply Roofing Industry, Waltham, MA.

Stathopoulos, T. (1981). "Wind loads on eaves of low buildings." *J. Struct. Div.*, 107(10), 1921–1934.

Stathopoulos, T., and Luchian, H. (1992). "Wind-induced forces on eaves of low buildings." *Proc., Wind Engineering Society Inaugural Conf.*, Cambridge, UK.

Stathopoulos, T., and Luchian, H. D. (1990). "Wind pressures on building configurations with stepped roofs." *Can. J. Civil Eng.*, 17(4), 569–577.

Stathopoulos, T., Surry, D., and Davenport, A. G. (1979). "Wind-induced internal pressures in low buildings." In *Proc., 5th Int. Conf. on Wind Engineering*, J. E. Cermak, ed., Colorado State University, Fort Collins, CO.

Stathopoulos, T., Wang, K., and Wu, H. (1999). "Wind standard provisions for low building gable roofs revisited." In *Proc., 10th Int. Conf. on Wind Engineering*, J. E. Cermak, ed., Balkema, Netherlands.

Stathopoulos, T., Wang, K., and Wu, H. (2000). "Proposed new Canadian wind provisions for the design of gable roofs." *Can. J. Civil Eng.*, 27(5), 1059–1072.

Stathopoulos, T., Wang, K., and Wu, H. (2001). "Wind pressure provisions for gable roofs of intermediate roof slope." *Wind and Structures*, 4(2).

Stubbs, N., and Perry, D. C. (1993). "Engineering of the building envelope: To do or not to do." In *Hurricanes of 1992: Lessons learned and implications for the future*, R. A. Cook and M. Sotani, eds., ASCE Press, Reston, VA, 10–30.

Surry, D., Kitchen, R. B., and Davenport, A. G. (1977). "Design effectiveness of wind tunnel studies for buildings of intermediate height." *Can. J. Civil Eng.*, 4(1), 96–116.

Twisdale, L. A., Vickery, P. J., and Steckley, A. C. (1996). *Analysis of hurricane windborne debris impact risk for residential structures*, State Farm Mutual Automobile Insurance Companies, Bloomington, IL.

CHAPTER C31
WIND TUNNEL PROCEDURE

Wind tunnel testing is specified when a building or other structure contains any of the characteristics defined in Sections 27.1.3, 28.1.3, 29.1.3, or 30.1.3 or when the designer wishes to more accurately determine the wind loads. For some building or structure shapes, wind tunnel testing can reduce the conservatism caused by enveloping of wind loads inherent in the Directional Procedure, Envelope Procedure, or Analytical Procedure for Components and Cladding (C&C). Also, wind tunnel testing accounts for shielding or channeling and can more accurately determine wind loads for a complex building or structure shape than can the Directional Procedure, Envelope Procedure, or Analytical Procedure for C&C. It is the intent of the standard that any building or other structure can be allowed to use the wind tunnel testing method to determine wind loads. Requirements for proper testing are given in ASCE 49 (2012).

It is common practice to resort to wind tunnel tests when design data are required for the following wind-induced loads:

1. Curtain wall pressures resulting from irregular geometry;
2. Across-wind and/or torsional loads;
3. Periodic loads caused by vortex shedding; and
4. Loads resulting from instabilities, such as flutter or galloping.

Boundary-layer wind tunnels capable of developing flows that meet the conditions stipulated in Section 31.2 typically have test-section dimensions in the following ranges: width of 6 to 12 ft (2 to 4 m), height of 6 to 10 ft (2 to 3 m), and length of 50 to 100 ft (15 to 30 m). Maximum wind speeds are ordinarily in the range of 25 to 100 mi/h (10 to 45 m/s). The wind tunnel may be either an open-circuit or closed-circuit type.

Three basic types of wind tunnel test models are commonly used. These are designated as follows: (1) rigid pressure model (PM), (2) rigid high-frequency base balance model (H-FBBM), and (3) aeroelastic model (AM). One or more of the models may be used to obtain design loads for a particular building or structure. The PM provides local peak pressures for design of elements, such as cladding and mean pressures, for the determination of overall mean loads. The H-FBBM measures overall fluctuating loads (aerodynamic admittance) for the determination of dynamic responses. When motion of a building or structure influences the wind loading, the AM is used for direct measurement of overall loads, deflections, and accelerations. Each of these models, together with a model of the surroundings (proximity model), can provide information other than wind loads, such as snow loads on complex roofs, wind data to evaluate environmental impact on pedestrians, and concentrations of air pollutant emissions for environmental impact determinations. Several references provide detailed information and guidance for the determination of wind loads and other types of design data by wind tunnel tests (Cermak 1977; Reinhold 1982; ASCE 1999; Boggs and Peterka 1989).

Wind tunnel tests frequently measure wind loads that are significantly lower than required by Chapters 26, 27, 28, 29, and 30 because of the shape of the building or other structure; the likelihood that the highest wind speeds occur at directions where the building or structure's shape or pressure coefficients are less than their maximum values; specific buildings or structures included in a detailed proximity model that may provide shielding in excess of that implied by exposure categories; and necessary conservatism in enveloping load coefficients in Chapters 28 and 30. In some cases, adjacent buildings or structures may shield the subject building or structure sufficiently that removal of one or two of the adjacent buildings or structures could significantly increase wind loads. Additional wind tunnel testing without specific nearby buildings or structures (or with additional buildings or structures if they might cause increased loads through channeling or buffeting) is an effective method for determining the influence of adjacent buildings or structures.

For this reason, the standard limits the reduction that can be accepted from wind tunnel tests to 80% of the result obtained from Part 1 of Chapter 27 or Part 1 of Chapter 28, or Chapter 30, if the wind tunnel proximity model included any specific influential buildings or other objects that, in the judgment of an experienced wind engineer, are likely to have substantially influenced the results beyond those characteristic of the general surroundings. If there are any such buildings or objects, supplemental testing can be performed to quantify their effect on the original results and possibly justify a limit lower than 80%, by removing them from the detailed proximity model and replacing them with characteristic ground roughness consistent with the adjacent roughness. A specific influential building or object is one within the detailed proximity model that protrudes well above its surroundings, or is unusually close to the subject building, or may otherwise cause substantial sheltering effect or magnification of the wind loads. When these supplemental test results are included with the original results, the acceptable results are then considered to be the higher of both conditions.

However, the absolute minimum reduction permitted is 65% of the baseline result for C&C and 50% for the main wind force resisting system (MWFRS). A higher reduction is permitted for MWFRS because C&C loads are more subject to changes caused by local channeling effects when surroundings change, and they can easily be dramatically increased when a new adjacent building is constructed. It is also recognized that cladding failures are much more common than failures of the MWFRS. In addition, for the case of MWFRS, it is easily demonstrated that the overall drag coefficient for certain common building shapes, such as circular cylinders (especially with rounded or domed tops), is one-half or less of the drag coefficient for the rectangular prisms that form the basis of Chapters 27, 28, and 30.

For C&C, the 80% limit is defined by the interior Zones 1 and 4 in Figs. 30.3-1, 30.3-2A–C, 30.3-3, 30.3-4, 30.3-5A–B, 30.3-6, 30.3-7, and 30.4-1. This limitation recognizes that pressures in the edge zones are the ones most likely to be reduced by the specific geometry of real buildings compared with the rectangular prismatic buildings assumed in Chapter 30. Therefore, pressures in edge and corner zones are permitted to be as low as 80% of the interior pressures from Chapter 30 without the supplemental tests. The 80% limit based on Zone 1 is directly applicable to all roof areas, and the 80% limit based on Zone 4 is directly applicable to all wall areas.

The limitation on MWFRS loads is more complex because the load effects (e.g., member stresses or forces, deflections) at any point are the combined effect of a vector of applied loads instead of a simple scalar value. In general, the ratio of forces or moments or torques (force eccentricity) at various floors throughout the building using a wind tunnel study will not be the same as those ratios determined from Chapter 27 and 28, and therefore comparison between the two methods is not well defined. Requiring each load effect from a wind tunnel test to be no less than 80% of the same effect resulting from Chapters 27 and 28 is impractical and unnecessarily complex and detailed, given the approximate nature of the 80% value. Instead, the intent of the limitation is effectively implemented by applying it only to a simple index that characterizes the overall loading. For flexible (tall) buildings, the most descriptive index of overall loading is the base overturning moment. For other buildings, the overturning moment can be a poor characterization of the overall loading, and the base shear is recommended instead.

C31.4 LOAD EFFECTS

C31.4.1 Mean Recurrence Intervals of Load Effects. Examples of analysis methods for combining directional wind tunnel data with the directional meteorological data or probabilistic models based thereon are described in Lepage and Irwin (1985), Rigato et al. (2001), Isyumov et al. (2013), Irwin et al. (2005), Simiu and Filliben (2005), and Simiu and Miyata (2006).

C31.4.2 Limitations on Wind Speeds. Section 31.4.2 specifies that the statistical methods used to analyze historical wind speed and direction data for wind tunnel studies shall be subject to the same limitations specified in Section 31.4.2 that apply to the Analytical Method.

Database-Assisted Design. Wind tunnel aerodynamics databases that contain records of pressures measured synchronously at large numbers of locations on the exterior surface of building models have been developed by wind researchers, such as Simiu et al. (2003) and Main and Fritz (2006). Such databases include data that permit a designer to determine, without specific wind tunnel tests, wind-induced forces and moments in MWFRSs and C&C of selected shapes and sizes of buildings. A public domain set of such databases, recorded in tests conducted at the University of Western Ontario (Ho et al. 2005; St. Pierre et al. 2005) for buildings with gable roofs is available on the National Institute of Standards and Technology website, www.nist.gov/wind (NIST 2012). Interpolation software for buildings with similar shape and with dimensions close to and intermediate between those included in the set of databases is also available on that site. Because the database results are for generic surroundings as permitted in ASCE 49 interpolation or extrapolation from these databases should be used only if Condition 2 of Section 27.1.2 is true. Extrapolations from available building shapes and sizes are not permitted, and interpolations in some instances may not be advisable. For these reasons, the guidance of an engineer experienced in wind loads on buildings and familiar with the usage of these databases is recommended.

All databases must have been obtained using testing methodology that meets the requirements for wind tunnel testing specified in Chapter 31.

C31.4.3 Wind Directionality. The variability of wind speed determined for particular azimuth intervals is greater than that of the wind speed determined regardless of wind direction (Isyumov et al. 2013). Consequently, wind loads and wind-induced effects determined by allowing for wind directionality are inherently less certain. Several methods for combining data from wind tunnel model studies with information on wind speed and direction at the project site are currently in use (Isyumov et al. 2013; Yeo and Simiu 2011; and Simiu 2011). Whichever method is used shall be clearly described to allow scrutiny by the designer and the Authority Having Jurisdiction. A common approach for allowing for uncertainties in the wind direction is to rotate the project wind climate relative to the orientation of the building or structure. This rotation of the wind climate at the building location is intended to ensure that the wind loads determined for design are not unconservative and shall be considered regardless of the method used for arriving at the design wind speeds. The appropriate magnitude of wind climate rotation varies depending on the quality and resolution of the directional wind climate data at the project site.

C31.6 ROOF-MOUNTED SOLAR COLLECTORS FOR ROOF SLOPES LESS THAN 7 DEGREES

C31.6.1 Wind Tunnel Test Requirements. For solar collector installations, it is necessary to model a generic building with the solar collectors on the roof of a scaled building, then generate (GC_{rn}) pressure coefficients that are applicable to any site, a wide range of building sizes, and varied collector layouts. The approach needs to be similar to that used to develop the $(G\hat{C}_p)$ figures in ASCE 7 by modeling the generic buildings with various features to capture a wide range of effects. The objective of such testing is to evaluate aerodynamic effects accounted for by (GC_{rn}) pressure coefficients (in contrast to site-specific wind tunnel testing, which also evaluates the effect of surrounding structures and terrain). Nearby buildings should not be included unless they are to be a part of every design application for this collector.

Wind tunnel testing for roof-mounted solar collectors must include a sufficiently large test matrix to address an appropriate range of the relevant variables that affect wind loads as listed in the provisions. Tests are often performed at model scale of 1:50 or larger where the match of wind tunnel turbulence characteristics is not ideal, resulting in some added requirements for testing, including integral scale limits. The wind tunnel study should provide recommendations for setback distances from larger rooftop equipment, penthouses, clerestories, and other building features. Guidance for testing is provided in ASCE 49, Kopp and Banks (2013), and Kopp et al. (2011, 2012). Wind loads are expressed as coefficients usable in Chapters 27, 29, and 30 to produce loads in engineering units. Alternately, a different formulation of nondimensional load coefficients may be used provided that the analysis procedure is clearly defined in the test report.

C31.6.1.1 Limitations on Wind Loads for Rooftop Solar Collectors. The minimum components and cladding wind load pressures indicated in ASCE 7 are primarily applicable to the building envelope and are not entirely applicable to rooftop solar collectors. The limitations contained herein are to establish the lower bound wind pressures for wind tunnel studies of conditions similar to those addressed by Fig. 29. 4-7 The limits on wind tunnel results shown in Fig. 29.4-7 represent an envelope of wind loads measured in the wind tunnel without deflectors or shrouds that are commonly used to lower wind loads. Specific installations or collector geometries may give significantly lower loads than Fig. 29.4-7; limits are imposed to prevent too much deviation from the enveloped results.

C31.6.1.2 Peer Review Requirements for Wind Tunnel Tests of Roof-Mounted Solar Collectors. Solar collector systems that have aerodynamic devices or more efficient profiles can have wind tunnel based wind loads less than the lower bound thresholds indicated in Sections 31.6.1 and 31.6.2. In order to use these lower values, a peer review of the test and report is required. The peer reviewer qualifications and requirements are included to promote consistencies among the various jurisdictions so that a peer review could be accepted by multiple enforcement agencies. The peer review qualifications are intended to be those of a wind tunnel expert familiar with wind tunnel testing of buildings and the applicability of the ASCE 7 provisions to determine generalized wind design coefficients for roof-mounted solar collectors. One source for peer reviewers is the American Association for Wind Engineering's (AAWE) boundary layer wind tunnels list (http://www.aawe.org/info/wind_tunnels.php).

REFERENCES

ASCE. (1999). "Wind tunnel model studies of buildings and structures." *ASCE Manuals and Reports of Engineering Practice No. 67*, Reston, VA.

ASCE. (2012). "Wind tunnel testing for buildings and other structures." *ASCE/SEI 49-12*, Reston, VA.

Boggs, D. W., and Peterka, J. A. (1989). "Aerodynamic model tests of tall buildings." *J. Eng. Mech.* 115(3), 618–635.

Cermak, J. E. (1977). "Wind-tunnel testing of structures." *J. Eng. Mech. Div.* 103(6), 1125–1140.

Ho, T. C. E., Surry, D., Morrish, D., and Kopp, G. A. (2005). "The UWO contribution to the NIST aerodynamic database for wind loads on low buildings: Part 1. Archiving format and basic aerodynamic data." *J. Wind Eng. Indust. Aerodyn.*, 93, 1–30.

Irwin, P., Garber, J., and Ho, E. (2005). "Integration of wind tunnel data with full scale wind climate." Proc., 10th Americas Conf. on Wind Eng., Baton Rouge, LA. doi: 10.1061/541X.0000654.

Isyumov, N., Ho, E., and Case, P. (2013). "Influence of Wind Directionality on Wind Loads and Responses." Proc., 12th Americas Conf. on Wind Eng. 141(8) doi: 10.1061/541X.0001180, 04014208.

Isyumov, N., Mikitiuk, M., Case, P., Lythe, G., and Welburn, A. (2013). "Predictions of wind loads and responses from simulated tropical storm passages." Proc., 11th Int. Conf. on Wind Eng, D. A. Smith and C. W. Letchford, eds. 19(3), 295–320. doi: 10.12989/was.2014.19.3.295.

Kopp, G., and Banks, D. (2013). "Use of the wind tunnel test method for obtaining design wind loads on roof-mounted solar arrays." *J. Struct. Eng.* 139(2), 284–287.

Kopp, G. A., Farquhar, S., and Morrison, M. J. (2012). "Aerodynamic mechanisms for wind loads on tilted, roof-mounted, solar arrays." *J. Wind Eng. Ind. Aerodyn.* 111, 40–52.

Kopp, G., Maffei, J., and Tilley, C. (2011). "Rooftop solar arrays and wind loading: A primer on using wind tunnel testing as a basis for code compliant design per ASCE 7," Boundary Layer Wind Tunnel Laboratory, Univ. of Western Ontario, Faculty of Engineering.

Lepage, M. F., and Irwin, P. A. (1985). "A technique for combining historical wind data with wind tunnel tests to predict extreme wind loads." *Proc., 5th U.S. Nat. Conf. on Wind Eng.*, M. Mehta, ed. doi: 10.1061/541X.0001625, 04016148.

Main, J. A., and Fritz, W. P. (2006). "Database-assisted design for wind: Concepts, software, and examples for rigid and flexible buildings." *NIST Building Science Series 180*, National Institute of Standards and Technology, Washington, DC.

National Institute of Standards and Technology (NIST). (2012). "Extreme winds and wind effects on structures." ⟨www.nist.gov/wind⟩ (March 5, 2012).

Reinhold, T. A., ed. (1982). "Wind tunnel modeling for civil engineering applications." *Proc., Int. Workshop on Wind Tunnel Modeling Criteria and Techniques in Civil Eng. Applications*, Cambridge University Press, Gaithersburg, MD.

Rigato, A., Chang, P., and Simiu, E. (2001). "Database-assisted design, standardization, and wind direction effects." *J. Struct. Eng.*, 127(8), 855–860.

Simiu, E. (2011). *Design of building for wind*, John Wiley and Sons, Hoboken, NJ.

Simiu, E., and Filliben, J. J. (2005). "Wind tunnel testing and the sector-by-sector approach to wind directionality effects." *J. Struct. Eng.* 131(7), 1143–1145.

Simiu, E., and Miyata, T. (2006). *Design of buildings and bridges for wind: A practical guide for ASCE Standard 7 users and designers of special structures*, John Wiley and Sons, Hoboken, NJ.

Simiu, E., Sadek, F., Whalen, T. A., Jang, S., Lu, L.-W., Diniz, S. M. C., et al. (2003). "Achieving safer and more economical buildings through database-assisted, reliability-based design for wind." *J. Wind Eng. Indust. Aerodyn.*, 91, 1587–1611.

St. Pierre, L. M., Kopp, G. A., Surry, D., and Ho, T. C. E. (2005). "The UWO contribution to the NIST aerodynamic database for wind loads on low buildings: Part 2. Comparison of data with wind load provisions." *J. Wind Eng. Indust. Aerodyn.*, 93, 31–59.

Yeo, D., and Simiu, E. (2011) "High-rise reinforced concrete structures: Database-assisted design for wind," *J. Struct. Eng.* 127, 1340–1349.

APPENDIX C11A
QUALITY ASSURANCE PROVISIONS

[THIS APPENDIX HAS BEEN DELETED IN ITS ENTIRETY FROM THE 2016 EDITION]

APPENDIX C11B
EXISTING BUILDING PROVISIONS

There is no Commentary for Appendix 11B.

APPENDIX CC
SERVICEABILITY CONSIDERATIONS

CC.1 SERVICEABILITY CONSIDERATIONS

Serviceability limit states are conditions in which the functions of a building or other structure are impaired because of local damage, deterioration, or deformation of building components, or because of occupant discomfort. Although safety generally is not an issue with serviceability limit states (one exception would be for cladding that falls off a building caused by excessive story drift under wind load), they nonetheless may have severe economic consequences. The increasing use of the computer as a design tool, the use of stronger (but not stiffer) construction materials, the use of lighter architectural elements, and the uncoupling of the nonstructural elements from the structural frame may result in building systems that are relatively flexible and lightly damped. Limit state design emphasizes the fact that serviceability criteria (as they always have been) are essential to ensure functional performance and economy of design for such building structural systems (Ad Hoc Committee on Serviceability Research 1986, National Building Code of Canada 1990, and West and Fisher 2003).

In general, serviceability is diminished by

1. Excessive deflections or rotation that may affect the appearance, functional use, or drainage of the structure or may cause damaging transfer of load to nonload supporting elements and attachments;
2. Excessive vibrations produced by the activities of building occupants, mechanical equipment, or the wind, which may cause occupant discomfort or malfunction of building service equipment; and
3. Deterioration, including weathering, corrosion, rotting, and discoloration.

In checking serviceability, the designer is advised to consider appropriate service loads, the response of the structure, and the reaction of the building occupants.

Service loads that may require consideration include static loads from the occupants and their possessions, snow or rain on roofs, temperature fluctuations, and dynamic loads from human activities, wind-induced effects, or the operation of building service equipment. The service loads are those loads that act on the structure at an arbitrary point in time. (In contrast, the nominal loads have a small probability of being exceeded in any year; factored loads have a small probability of being exceeded in 50 years.) Appropriate service loads for checking serviceability limit states may be only a fraction of the nominal loads.

The response of the structure to service loads normally can be analyzed assuming linear elastic behavior. However, members that accumulate residual deformations under service loads may require examination with respect to this long-term behavior. Service loads used in analyzing creep or other long-term effects may not be the same as those used to analyze elastic deflections or other short-term or reversible structural behavior.

Serviceability limits depend on the function of the building and on the perceptions of its occupants. In contrast to the ultimate limit states, it is difficult to specify general serviceability limits that are applicable to all building structures. The serviceability limits presented in Sections CC.2.1, CC.2.2, and CC.2.3 provide general guidance and have usually led to acceptable performance in the past. However, serviceability limits for a specific building should be determined only after a careful analysis by the engineer and architect of all functional and economic requirements and constraints in conjunction with the building owner. It should be recognized that building occupants are able to perceive structural deflections, motion, cracking, and other signs of possible distress at levels that are much lower than those that would indicate that structural failure was impending. Such signs of distress may be taken incorrectly as an indication that the building is unsafe and may diminish its commercial value.

CC.2 DEFLECTION, VIBRATION, AND DRIFT

CC.2.1 Vertical Deflections. Excessive vertical deflections and misalignment arise primarily from three sources: (1) gravity loads, such as dead, live, and snow loads; (2) effects of temperature, creep, and differential settlement; and (3) construction tolerances and errors. Such deformations may be visually objectionable; may cause separation, cracking, or leakage of exterior cladding, doors, windows, and seals; and may cause damage to interior components and finishes. Appropriate limiting values of deformations depend on the type of structure, detailing, and intended use (Galambos and Ellingwood 1986). Historically, common deflection limits for horizontal members have been 1/360 of the span for floors subjected to full nominal live load and 1/240 of the span for roof members. Deflections of about 1/300 of the span (for cantilevers, 1/150 of the length) are visible and may lead to general architectural damage or cladding leakage. Deflections greater than 1/200 of the span may impair operation of movable components such as doors, windows, and sliding partitions.

In certain long-span floor systems, it may be necessary to place a limit (independent of span) on the maximum deflection to minimize the possibility of damage of adjacent nonstructural elements (ISO 1977). For example, damage to non-load-bearing partitions may occur if vertical deflections exceed more than about 10 mm (3/8 in.) unless special provision is made for differential movement (Cooney and King 1988); however, many components can and do accept larger deformations.

Load combinations for checking static deflections can be developed using first-order reliability analysis (Galambos and Ellingwood 1986). Current static deflection guidelines for floor

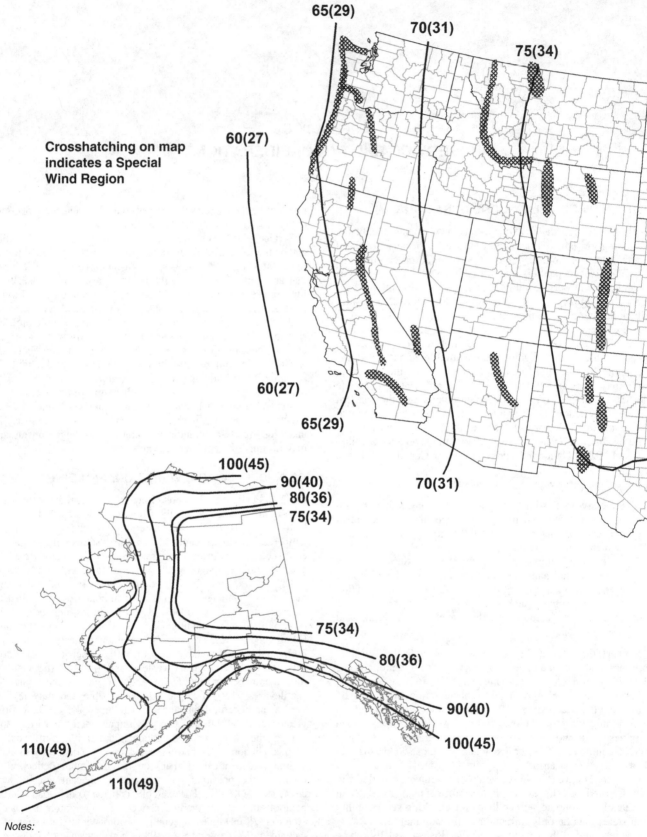

Notes:
1. Values are nominal design 3-s gust wind speeds in mi/h (m/s) at 33 ft (10 m) above ground for Exposure Category C.
2. Linear interpolation between contours is permitted.
3. Islands and coastal areas outside the last contour shall use the last wind speed contour of the coastal area.
4. Mountainous terrain, gorges, ocean promontories, and special wind regions shall be examined for unusual wind conditions.

FIGURE CC.2-1 10-Year MRI 3-s Gust Wind Speed in mi/h (m/s) at 33 ft (10 m) above Ground in Exposure C

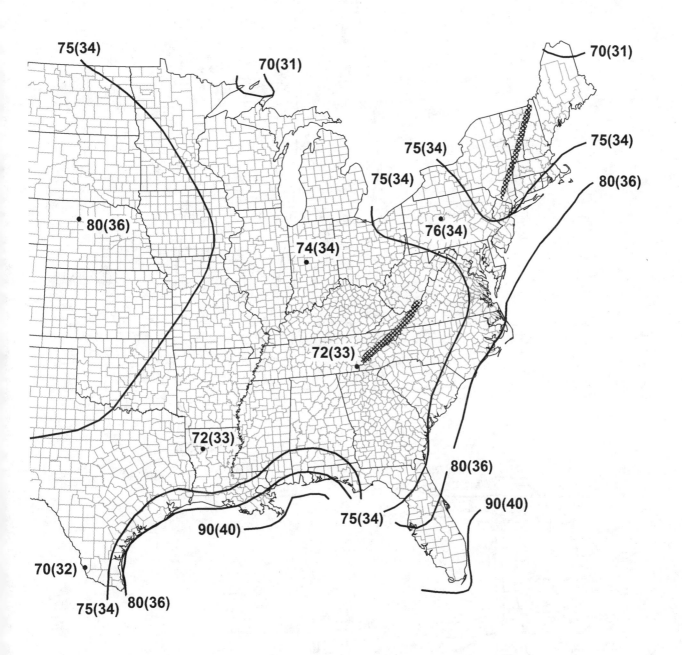
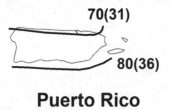

FIGURE CC.2-1 (*Continued*)

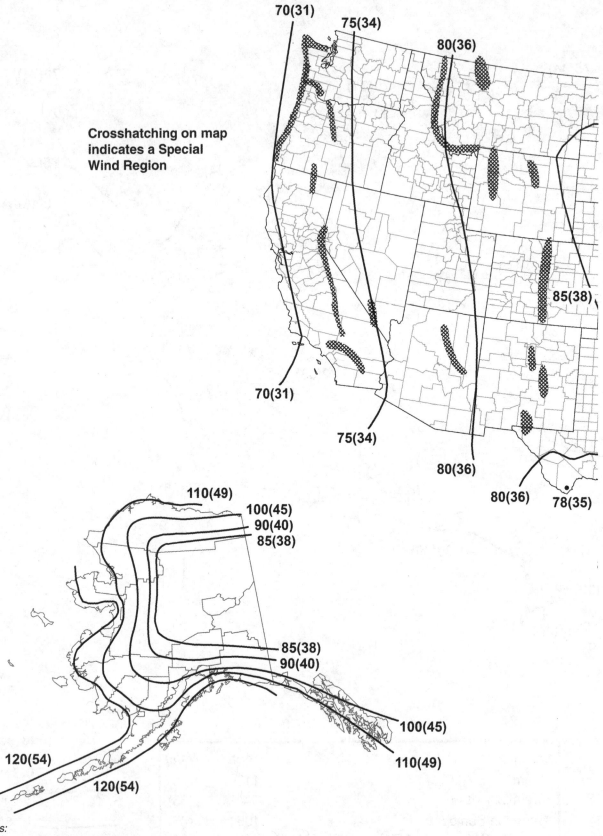

Notes:
1. Values are nominal design 3-s gust wind speeds in mi/h (m/s) at 33 ft (10 m) above ground for Exposure Category C.
2. Linear interpolation between contours is permitted.
3. Islands and coastal areas outside the last contour shall use the last wind speed contour of the coastal area.
4. Mountainous terrain, gorges, ocean promontories, and special wind regions shall be examined for unusual wind conditions.

FIGURE CC.2-2 25-Year MRI 3-s Gust Wind Speed in mi/h (m/s) at 33 ft (10 m) above Ground in Exposure C

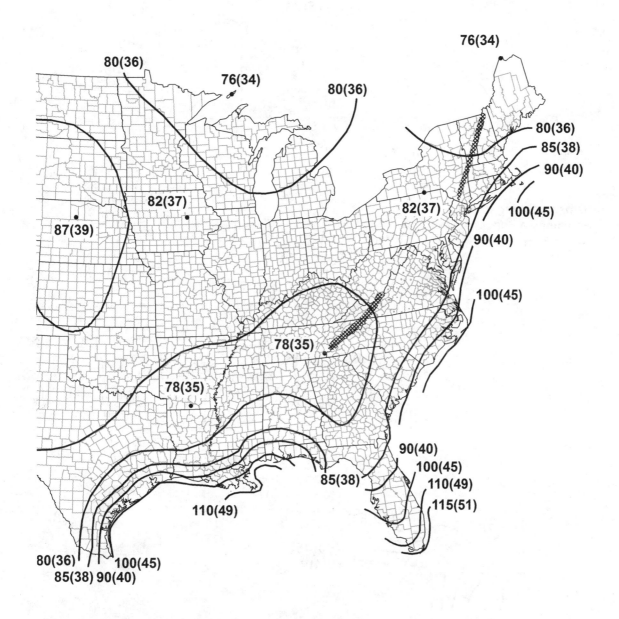

FIGURE CC.2-2 (*Continued*)

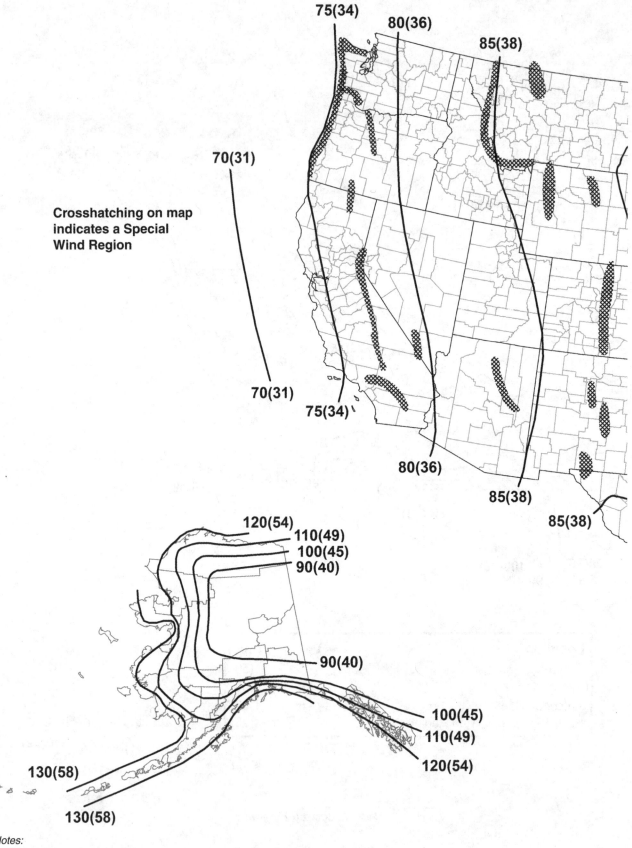

Notes:
1. Values are nominal design 3-s gust wind speeds in mi/h (m/s) at 33 ft (10 m) above ground for Exposure Category C.
2. Linear interpolation between contours is permitted.
3. Islands and coastal areas outside the last contour shall use the last wind speed contour of the coastal area.
4. Mountainous terrain, gorges, ocean promontories, and special wind regions shall be examined for unusual wind conditions.

FIGURE CC.2-3 50-Year MRI 3-s Gust Wind Speed in mi/h (m/s) at 33 ft (10 m) above Ground in Exposure C

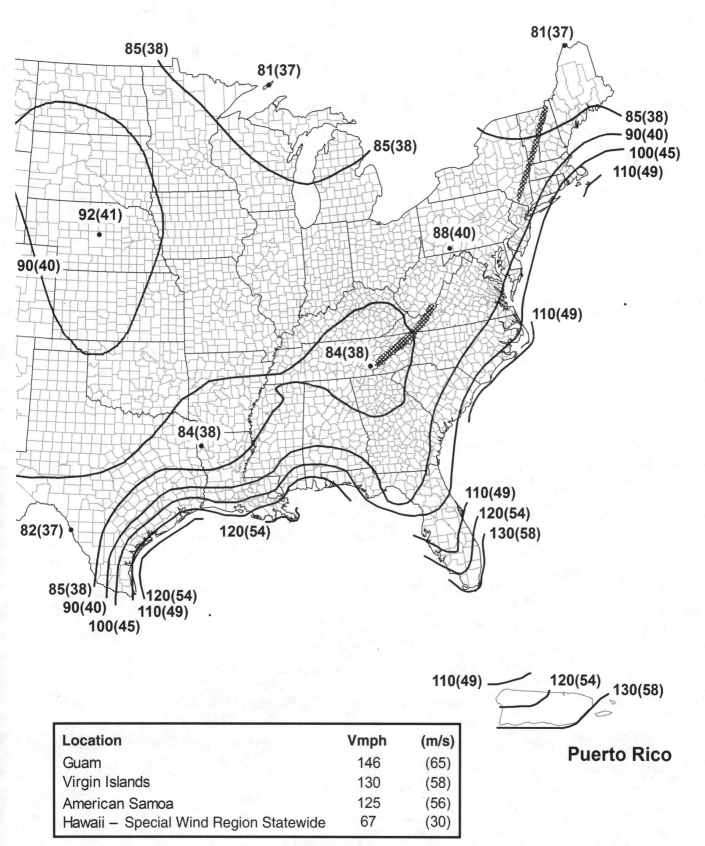

Location	Vmph	(m/s)
Guam	146	(65)
Virgin Islands	130	(58)
American Samoa	125	(56)
Hawaii – Special Wind Region Statewide	67	(30)

FIGURE CC.2-3 (Continued)

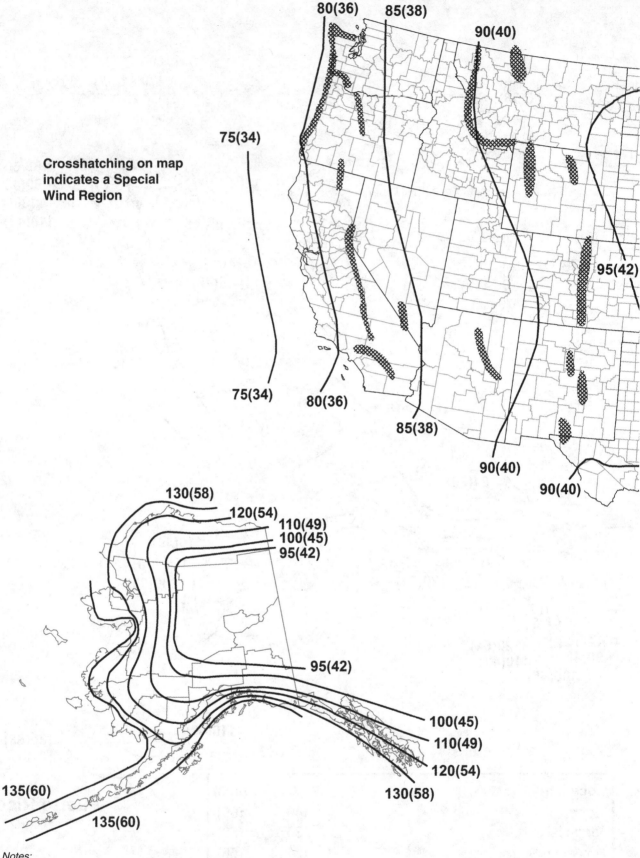

Notes:
1. Values are nominal design 3-s gust wind speeds in mi/h (m/s) at 33 ft (10 m) above ground for Exposure Category C.
2. Linear interpolation between contours is permitted.
3. Islands and coastal areas outside the last contour shall use the last wind speed contour of the coastal area.
4. Mountainous terrain, gorges, ocean promontories, and special wind regions shall be examined for unusual wind conditions.

FIGURE CC.2-4 100-Year MRI 3-s Gust Wind Speed in mi/h (m/s) at 33 ft (10 m) above Ground in Exposure C

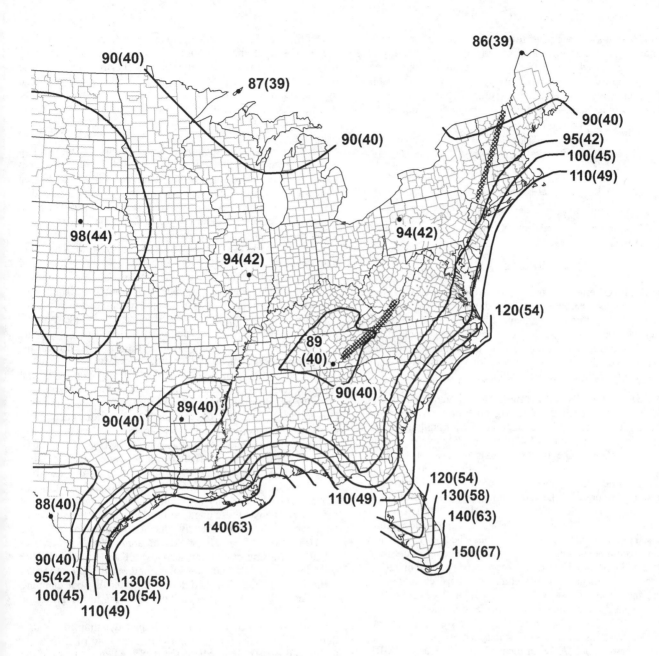

FIGURE CC.2-4 (Continued)

and roof systems are adequate for limiting surficial damage in most buildings. A combined load with an annual probability of 0.05 of being exceeded would be appropriate in most instances. For serviceability limit states involving visually objectionable deformations, reparable cracking or other damage to interior finishes, and other short-term effects, the suggested load combinations are

$$D + L \quad (CC.2\text{-}1a)$$

$$D + 0.5S \quad (CC.2\text{-}1b)$$

For serviceability limit states involving creep, settlement, or similar long-term or permanent effects, the suggested load combination is

$$D + 0.5L \quad (CC.2\text{-}2)$$

The dead load effect, D, used in applying Eqs. (CC.2-1) and (CC.2-2) may be that portion of dead load that occurs after attachment of nonstructural elements. Live load, L, is defined in Chapter 4. For example, in composite construction, the dead load effects frequently are taken as those imposed after the concrete has cured; in ceilings, the dead load effects may include only those loads placed after the ceiling structure is in place.

CC.2.2 Drift of Walls and Frames. Drifts (lateral deflections) of concern in serviceability checking arise primarily from the effects of wind. Drift limits in common usage for building design are on the order of 1/600 to 1/400 of the building or story height (ASCE Task Committee on Drift Control of Steel Building Structures 1988; Griffis 1993). These limits generally are sufficient to minimize damage to cladding and nonstructural walls and partitions. Smaller drift limits may be appropriate if the cladding is brittle. West and Fisher (2003) contains recommendations for higher drift limits that have successfully been used in low-rise buildings with various cladding types. It also contains recommendations for buildings containing cranes. An absolute limit on story drift may also need to be imposed in light of evidence that damage to nonstructural partitions, cladding, and glazing may occur if the story drift exceeds about 10 mm (3/8 in.) unless special detailing practices are made to tolerate movement (Freeman 1977; Cooney and King 1988). Many components can accept deformations that are significantly larger.

Use of the nominal (700-year mean recurrence interval (MRI) or 1,700-year MRI) wind load in checking serviceability is excessively conservative. The following load combination, derived similarly to Eqs. (CC.2-1a) and (CC.2-1b), can be used to check short-term effects:

$$D + 0.5L + W_a \quad (CC.2\text{-}3)$$

in which W_a is wind load based on serviceability wind speeds in Figs. CC.2-1 through CC.2-4. Some designers have used a 10-year MRI (annual probability of 0.1) for checking drift under wind loads for typical buildings (Griffis 1993), whereas others have used a 50-year MRI (annual probability of 0.02) or a 100-year MRI (annual probability of 0.01) for more drift-sensitive buildings. The selection of the MRI for serviceability evaluation is a matter of engineering judgment that should be exercised in consultation with the building client.

The maps included in this appendix are appropriate for use with serviceability limit states and should not be used for strength limit states. Because of its transient nature, wind load need not be considered in analyzing the effects of creep or other long-term actions.

Deformation limits should apply to the structural assembly as a whole. The stiffening effect of nonstructural walls and partitions may be taken into account in the analysis of drift if substantiating information regarding their effect is available. Where load cycling occurs, consideration should be given to the possibility that increases in residual deformations may lead to incremental structural collapse.

CC.2.3 Vibrations. Structural motions of floors or of the building as a whole can cause the building occupants discomfort. In recent years, the number of complaints about building vibrations has been increasing. This increasing number of complaints is associated in part with the more flexible structures that result from modern construction practice. Traditional static deflection checks are not sufficient to ensure that annoying vibrations of building floor systems or buildings as a whole do not occur (Ad Hoc Committee on Serviceability Research 1986). Whereas control of stiffness is one aspect of serviceability, mass distribution and damping are also important in controlling vibrations. The use of new materials and building systems may require that the dynamic response of the system be considered explicitly. Simple dynamic models often are sufficient to determine whether there is a potential problem and to suggest possible remedial measurements (Bachmann and Ammann 1987; Ellingwood 1989).

Excessive structural motion is mitigated by measures that limit building or floor accelerations to levels that are not disturbing to the occupants or do not damage service equipment. Perception and tolerance of individuals to vibration is dependent on their expectation of building performance (related to building occupancy) and to their level of activity at the time the vibration occurs (ANSI 1983). Individuals find continuous vibrations more objectionable than transient vibrations. Continuous vibrations (over a period of minutes) with acceleration on the order of 0.005 g to 0.01 g are annoying to most people engaged in quiet activities, whereas those engaged in physical activities or spectator events may tolerate steady-state accelerations on the order of 0.02 g to 0.05 g. Thresholds of annoyance for transient vibrations (lasting only a few seconds) are considerably higher and depend on the amount of structural damping present (Murray 1991). For a finished floor with (typically) 5% damping or more, peak transient accelerations of 0.05 g to 0.1 g may be tolerated.

Many common human activities impart dynamic forces to a floor at frequencies (or harmonics) in the range of 2 to 6 Hz (Allen and Rainer 1976; Allen et al. 1985; Allen 1990a,b). If the fundamental frequency of vibration of the floor system is in this range and if the activity is rhythmic in nature (e.g., dancing, aerobic exercise, or cheering at spectator events), resonant amplification may occur. To prevent resonance from rhythmic activities, the floor system should be tuned so that its natural frequency is well removed from the harmonics of the excitation frequency. As a general rule, the natural frequency of structural elements and assemblies should be greater than 2.0 times the frequency of any steady-state excitation to which they are exposed unless vibration isolation is provided. Damping is also an effective way of controlling annoying vibration from transient events because studies have shown that individuals are more tolerant of vibrations that damp out quickly than those that persist (Murray 1991).

Several studies have shown that a simple and relatively effective way to minimize objectionable vibrations to walking and other common human activities is to control the floor stiffness, as measured by the maximum deflection independent of span. Justification for limiting the deflection to an absolute value rather than to some fraction of span can be obtained by considering the dynamic characteristics of a floor system modeled as a uniformly loaded simple span. The fundamental frequency of vibration, f_o, of this system is given by

$$f_o = \frac{\pi}{2l^2}\sqrt{\frac{EI}{\rho}} \quad \text{(CC.2-4)}$$

in which EI = flexural rigidity of the floor, l = span, and $\rho = w/g$ = mass per unit length; g = acceleration due to gravity 32.17 ft/s² (9.81 m/s²), and w = dead load plus participating live load. The maximum deflection caused by w is

$$\delta = (5/384)(wl^4/EI) \quad \text{(CC.2-5)}$$

Substituting EI from this equation into Eq. (CC.2-3), we obtain

$$f_o \approx 18/\sqrt{\delta} \quad (\delta \text{ in mm}) \quad \text{(CC.2-6)}$$

This frequency can be compared to minimum natural frequencies for mitigating walking vibrations in various occupancies (Allen and Murray 1993). For example, Eq. (CC.2-6) indicates that the static deflection caused by uniform load, w, must be limited to about 0.2 in. (5 mm), independent of span, if the fundamental frequency of vibration of the floor system is to be kept above about 8 Hz. Many floors that do not meet this guideline are perfectly serviceable; however, this guideline provides a simple means for identifying potentially troublesome situations where additional consideration in design may be warranted.

CC.3 DESIGN FOR LONG-TERM DEFLECTION

Under sustained loading, structural members may exhibit additional time-dependent deformations caused by creep, which usually occur at a slow but persistent rate over long periods of time. In certain applications, it may be necessary to limit deflection under long-term loading to specified levels. This limitation can be done by multiplying the immediate deflection by a creep factor, as provided in material standards, that ranges from about 1.5 to 2.0. This limit state should be checked using load combination in Eq. (CC.2-2).

CC.4 CAMBER

Where required, camber should be built into horizontal structural members to give proper appearance and drainage and to counteract anticipated deflection from loading and potential ponding.

CC.5 EXPANSION AND CONTRACTION

Provisions should be made in design so that if significant dimensional changes occur, the structure will move as a whole and differential movement of similar parts and members that meet at joints will be at a minimum. Design of expansion joints to allow for dimensional changes in portions of a structure separated by such joints should take both reversible and irreversible movements into account. Structural distress in the form of wide cracks has been caused by restraint of thermal, shrinkage, and prestressing deformations. Designers are advised to provide for such effects through relief joints or by controlling crack widths.

CC.6 DURABILITY

Buildings and other structures may deteriorate in certain service environments. This deterioration may be visible upon inspection (e.g., weathering, corrosion, and staining) or may result in undetected changes in the material. The designer should either provide a specific amount of damage tolerance in the design or should specify adequate protection systems and/or planned maintenance to minimize the likelihood that such problems will occur. Water infiltration through poorly constructed or maintained wall or roof cladding is considered beyond the realm of designing for damage tolerance. Waterproofing design is beyond the scope of this standard. For portions of buildings and other structures exposed to weather, the design should eliminate pockets in which moisture can accumulate.

REFERENCES

Ad Hoc Committee on Serviceability Research. (1986). "Structural serviceability: A critical appraisal and research needs." *J. Struct. Engrg.*, 112(12), 2646–2664.

Allen, D. E. (1990a). "Floor vibrations from aerobics." *Can. J. Civ. Engrg.*, 19(4), 771–779.

Allen, D. E. (1990b). "Building vibrations from human activities." *Concrete Int.*, 12(6), 66–73.

Allen, D. E., and Murray, T. M. (1993). "Design criterion for vibrations due to walking." *Eng. J.*, 30(4), 117–129.

Allen, D. E., and Rainer, J. H. (1976). "Vibration criteria for long-span floors." *Can. J. Civ. Engrg.*, 3(2), 165–173.

Allen, D. E., Rainer, J. H., and Pernica, G. (1985). "Vibration criteria for assembly occupancies." *Can. J. Civ. Engrg.*, 12(3), 617–623.

American National Standards Institute (ANSI). (1983). *Guide to the evaluation of human exposure to vibration in buildings, ANSI S3.29-1983*, ANSI, New York.

ASCE Task Committee on Drift Control of Steel Building Structures. (1988). "Wind drift design of steel-framed buildings: State-of-the-art report." *J. Struct. Engrg.*, 114(9), 2085–2108.

Bachmann, H., and Ammann, W. (1987). "Vibrations in structures." 3rd Ed. *Struct. Eng. Doc.*, International Association for Bridge and Structural Engineering, Zurich, Switzerland.

Cooney, R. C., and King, A. B. (1988). "Serviceability criteria for buildings." *BRANZ Report SR14*, Building Research Association of New Zealand, Porirua, New Zealand.

Ellingwood, B. (1989). "Serviceability guidelines for steel structures." *Eng. J.*, 26(1), 1–8.

Freeman, S. A. (1977). "Racking tests of high-rise building partitions." *J. Struct. Div.*, 103(8), 1673–1685.

Galambos, T. U., and Ellingwood, B. (1986). "Serviceability limit states: Deflection." *J. Struct. Engrg.* 112(1), 67–84.

Griffis, L. G. (1993). "Serviceability limit states under wind load." *Eng. J.*, 30(1), 1–16.

International Organization for Standardization (ISO). (1977). "Bases for the design of structures—Deformations of buildings at the serviceability limit states." *ISO 4356*. International Organization for Standardization.

Murray, T. (1991). "Building floor vibrations." *Eng. J.*, 28(3), 102–109.

National Building Code of Canada. (1990). *Commentary A, serviceability criteria for deflections and vibrations*, National Research Council, Ottawa.

West, M., and Fisher, J. (2003). *Serviceability design considerations for steel buildings*, 2nd Ed., Steel Design Guide No. 3, American Institute of Steel Construction, Chicago.

OTHER REFERENCES (NOT CITED)

Ellingwood, B., and Tallin, A. (1984). "Structural serviceability: Floor vibrations." *J. Struct. Engrg.*, 110(2), 401–418.

Ohlsson, S. (1988). "Ten years of floor vibration research—A review of aspects and some results." *Proc., Symposium on Serviceability of Buildings*, National Research Council of Canada, Ottawa, 435–450.

Tallin, A. G., and Ellingwood, B. (1984). "Serviceability limit states: Wind induced vibrations." *J. Struct. Engrg.*, 110(10), 2424–2437.

APPENDIX CD
BUILDINGS EXEMPTED FROM TORSIONAL WIND LOAD CASES

As discussed in Section C27.3.6, a building will experience torsional loads caused by nonuniform pressures on different faces of the building. Because of these torsional loads, the four load cases as defined in Fig. 27.3-8 must be investigated except for buildings with flexible diaphragms and for buildings with diaphragms that are not flexible meeting the requirements for spatial distribution and stiffness of the main wind force resisting system (MWFRS).

The requirements for spatial distribution and stiffness of the MWFRS for the simple cases shown are necessary to ensure that wind torsion does not control the design. Presented in Appendix D are different requirements which, if met by a building's MWFRS, then torsional wind load cases need not be investigated. Many other configurations are also possible, but it becomes too complex to describe their limitations in a simple way.

In general, the designer should place and proportion the vertical elements of the MWFRS in each direction so that the center of pressure from wind forces at each story is located near the center of rigidity of the MWFRS, thereby minimizing the inherent torsion from wind on the building. In buildings with rigid diaphragms, a torsional eccentricity larger than about 5% of the building width should be avoided to prevent large shear forces from wind torsion effects and to avoid torsional story drift that can damage interior walls and cladding.

The following information is provided to aid designers in determining whether the torsional wind loading cases (Fig. 27.3-8, load cases 2 and 4) control the design. Reference is made to Fig. CD-1. The equations shown in the figure for the general case of a square or rectangular building having inherent eccentricity e_1 or e_2 about principal axis 1 and 2, respectively, can be used to determine the required stiffness and location of the MWFRS in each principal axis direction.

Using the equations contained in Fig. CD-1, it can be shown that regular buildings (as defined in Chapter 12, Section 12.3.2), which at each story meet the requirements specified for the eccentricity between the center of mass (or alternatively, center of rigidity) and the geometric center with the specified ratio of seismic to wind design story shears can safely be exempted from the wind torsion load cases of Fig. 27.3-6. It is conservative to measure the eccentricity from the center of mass to the geometric center rather than from the center of rigidity to the geometric center. Buildings that have an inherent eccentricity between the center of mass and center of rigidity and that are designed for code seismic forces have a higher torsional resistance than if the center of mass and rigidity are coincident.

Using the equations contained in Fig. CD-1 and a building drift analysis to determine the maximum displacement at any story, it can be shown that buildings with diaphragms that are not flexible and that are defined as torsionally regular under wind load need not be designed for the torsional load cases of Fig. 27.3-6. Furthermore, it is permissible to increase the basic wind load case proportionally so that the maximum displacement at any story is not less than the maximum displacement under the torsional load case. The building can then be designed for the increased basic loading case without the need for considering the torsional load cases.

Diagram

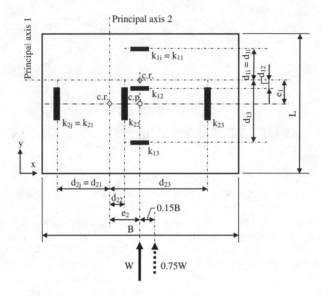

Notation

B	=	Horizontal plan dimension of the building normal to the wind.
L	=	Horizontal plan dimension of the building parallel to the wind.
c.r.	=	Center of rigidity.
c.p.	=	Center of wind pressure.
k_{1i}	=	Stiffness of frame i parallel to major axis 1.
k_{2j}	=	Stiffness of frame j parallel to major axis 2.
d_{1i}	=	Distance of frame i to c.r. perpendicular to major axis 1.
d_{2j}	=	Distance of frame j to c.r. perpendicular to major axis 2.
e_1	=	Distance from c.p. to c.r. perpendicular to major axis 1.
e_2	=	Distance from c.p. to c.r. perpendicular to major axis 2.
J	=	Polar moment of inertia of all MWFRS wind frames in the building.
W	=	Wind load as required by standard.
V_{1i}	=	Wind force in frame i parallel to major axis 1.
V_{2j}	=	Wind force in frame j parallel to major axis 2.
x_0, y_0	=	Coordinates for center of rigidity from the origin of any convenient x, y axes.

Equations

$$x_0 = \frac{\sum_{i=1}^{n} x_{1i} k_{1i}}{\sum_{i=1}^{n} k_{1i}} \qquad y_0 = \frac{\sum_{i=1}^{n} y_{1i} k_{1i}}{\sum_{i=1}^{n} k_{1i}}$$

$$J = \sum_{i=1}^{n} k_{1i} d_{1i}^{\,2} + \sum_{j=1}^{m} k_{2j} d_{2j}^{\,2}$$

$$V_{1i} = \frac{(0.75W) k_{1i}}{\sum_{i=1}^{n} k_{1i}} + \frac{(0.75W)(e_1 + 0.15B) k_{1i} d_{1i}}{J}$$

$$V_{2j} = \frac{(0.75W) k_{2j}}{\sum_{j=1}^{m} k_{2j}} + \frac{(0.75W)(e_2 + 0.15B) k_{2j} d_{2j}}{J}$$

Figure CD-1. Exemption from Torsional Load Cases

APPENDIX CE
PERFORMANCE-BASED DESIGN PROCEDURES FOR FIRE EFFECTS ON STRUCTURES

CE.1 SCOPE

Design approaches that consider fire effects on structures are generally categorized as either (1) standard fire resistance design (also referred to as the prescriptive approach) or (2) performance-based design (PBD). Although this appendix does not pertain to and should not be used for standard fire resistance design, it is discussed relative to PBD in this commentary.

Designers may elect to use PBD procedures for fire effects on a structure to meet stakeholder design objectives as permitted by Section 1.3.6 and alternative materials, design, and methods of construction and equipment provision in the building codes. When PBD procedures are used, the structure is typically designed for primary gravity and environmental loads and is then evaluated for fire exposure. An alternative material, design, or method of construction typically requires approval where the authority having jurisdiction finds that the proposed design is satisfactory and complies with the intent of the building code.

Structural fire resistance is the structure's ability to carry loads during exposure to fire conditions as well as provide a barrier to fire spread. Structural performance during fire exposure is often simply expressed as Fire Resistance > Fire Effects (Buchanan 2002). Three design philosophies are used for comparing fire resistance and fire effects, which are based on measures of time, temperature, and strength.

Time is used for standard fire resistance ratings in the building codes, where a particular configuration is shown, by testing or equivalent analysis, to provide adequate resistance to a standard fire exposure under test conditions for a period of time. *Temperature* is used in situations where it is postulated that a particular temperature will cause failure in a component or subsystem. A maximum allowable temperature is specified, and thermal protection is provided for defined fire exposures to ensure that the limiting temperature is not reached within the specified fire rating or period of time. *Strength* compares applied gravity loads and fire effects (e.g., thermally induced forces and deformations in structural members) to the fire resistance (e.g., temperature-dependent stiffness and strength) of the heated structural members and connections.

CE.2 DEFINITIONS

STANDARD FIRE RESISTANCE DESIGN: Standard fire resistance design methods are based on either the *time* or *temperature*. Fire resistance is most often defined as an hourly rating (e.g., 2-hour fire rated assembly) based on either results of the standard fire test or equivalent analysis methods.

Standard fire resistance testing provides a method of rank ordering through comparative testing of different structural assemblies under controlled laboratory conditions. Each test uses the same standard time–temperature curve, which continually rises in temperature, to heat the structural members and assemblies with an established set of failure criteria. Standard time–temperature curves include ASTM E119 (2012) and ANSI/UL 263 (UL 2011b); international standards include ISO 834 (1999), CAN/ULC S101 (ULC 2007), and BS 476-20 (BSI 1987). There are also standard fire curves for hydrocarbon pool fires, such as ASTM E1529 (2010b) and UL 1709 (UL 2011a).

The standard furnace test has a standard time–temperature curve that provides severe heating conditions for test assemblies representative of field construction. Because of size limitations of furnaces, members and assembly sizes are limited. For instance, floor assemblies are typically tested at spans no greater than 17 ft (5.2 m), whereas an actual floor span may be much greater.

The fire rating of a member or assembly is based on the first failure criterion reached. For thermal response, there are temperature failure criteria measured by thermocouples applied to structural members. For flame and heat passage, the tested assembly cannot allow for ignition of cotton waste on the unexposed surface. For structural response, member deflections cannot become excessive (though excessive deflection is usually not explicitly defined in standard fire test methods). Fire resistance directories (e.g., *UL Fire Resistance Directory* 2013) provide a list of rated assemblies based on standard testing but do not provide information about the failure criterion upon which the listing is based.

When members or assemblies and their passive fire protection are similar to those already tested, methods to calculate equivalent performance for a standard fire test can be used to determine the fire resistance rating (e.g., ASCE/SEI/SFPE 29-05 2007). Analysis methods are available for structural members and fire barrier assemblies made from structural steel, plain concrete, reinforced concrete, timber and wood, concrete masonry, and clay masonry.

Standard fire resistance testing and equivalence computations do not include member connections, structural system response, or natural fire exposure. Standard fire resistance testing and equivalent analyses of hourly ratings do not provide the information needed to predict the actual performance of a structural system during structural design fires.

PERFORMANCE-BASED STRUCTURAL FIRE DESIGN: Performance-based structural fire design is useful for cases where standard fire resistance design would not address the design objectives of stakeholders. For example, a performance-based approach is appropriate for cases where performance of a structure during structural design fires needs to be quantified to properly assess risks to life safety and/or property protection. Building code variances for structural fire protection may require a performance-based approach to demonstrate the adequacy of an alternative design. Performance-based analyses also provide opportunities to develop alternative designs that are optimized for aesthetics, functionality, and/or costs.

Acceptance of performance-based designs is subject to approval by the authority having jurisdiction. The designer must demonstrate that the design provides a level of safety that is equivalent or superior to that which would be attained by a design that conforms to the code's prescriptive criteria. Performance-based structural fire design provides a level of safety that is based on evaluation of structural system demand and capacity under fire conditions. Since the prescriptive approach is based solely on standard testing and does not consider structural system performance, the level of safety provided cannot be quantified. Consequently, there exists no practical method to quantitatively compare the level of safety provided by a PBD to that provided by the prescriptive approach. Therefore, it is the responsibility of the designer to properly demonstrate to the authority having jurisdiction that the PBD satisfies the required performance objectives and complies with the general intent of the building code.

The authority having jurisdiction should be contacted before initiating a PBD process to determine if their office is capable of and willing to support such alternative means, or if they have any concerns or requirements that need to be addressed. A peer review by an independent qualified party may be required by the authority having jurisdiction as part of the PBD process. Section 1.3.1.3 addresses PBD requirements for all types of load and performance requirements.

CE.3 GENERAL REQUIREMENTS

The frequency of major building fires is relatively low because of the small probability of ignition reaching flashover conditions (see Section CE.5.2 for a discussion of flashover). Occupant or fire department intervention and/or fire suppression system extinguishment of fire typically limit fire development before it becomes uncontrolled (Hall 2013). However, certain events and/or circumstances may result in uncontrolled fires that affect the structural system. In such cases, structural integrity should be maintained to ensure occupant life safety (see Section E.4.1). This assumes that the structural system is not significantly damaged by another hazard event, such as an earthquake or an explosion.

The term "fire effects" includes thermal response and corresponding structural deformations and loads induced by heating and cooling of structural systems during fire exposure, as well as temperature-dependent changes in structural stiffness and strength, nonlinear geometric and material responses, and restrained thermal expansion or contraction. All structural materials need to be evaluated for heating effects, but some materials, such as wood, may not experience additional strains or stresses from cooling effects. However, steel connections and fasteners used in wood construction may need to be evaluated for cooling. Fire effects may result in significant forces, rotations, deflections, and deformations of members and connections. Section loss resulting from fire exposure (e.g., because of spalling or charring) may also contribute to these effects.

PBD includes development of quantifiable performance objectives that are evaluated with appropriate analysis methods. The portion of the structure affected by fire, which includes members and connections as well as cooler surrounding sections that may provide restraint against thermal expansion, should be considered to determine the structural system performance and failure modes.

Analysis techniques used to evaluate fire effects on structures range in complexity from single element analyses to finite element models that represent structural systems. Single members (columns or beams) and their connections can be analyzed in isolation for a structural design fire if reasonable assumptions can be made about restraint beyond the member and its connections, such as whether the rest of the structure provides rigid restraint to a heated member or whether the restraint of shear studs in a composite floor should be included or ignored.

Temperature-dependent strength and stiffness properties of materials may be based on peak temperatures for a structural design fire only if thermal expansion, restraint, and cooling effects can be conservatively neglected. Such considerations should include whether inelastic deformations, such as local buckling that is induced by restraint of thermal expansion, affect the member and connection behavior during cooling. For instance, bolt tear-out may occur during cooling of steel framing, which may result in member or subsystem failure. For composite floors, the thermal expansion of steel beams will be restrained as long as their shear stud connections remain intact.

Unlike single element analyses, finite element models of structural systems are able to capture the effects of thermal expansion, alternative load paths, secondary load-carrying mechanisms (e.g., catenary action), nonlinear material response, large displacement response, and connection performance. It is essential that the scope and complexity of the analysis techniques used to analyze structural response to structural design fires address the performance objectives in Section E.4 and are acceptable to the authority having jurisdiction.

The achievement of adequate performance and sufficient continuity and ductility for alternate load paths following member failure due to fire effects caused by a structural design fire requires consideration of connection capacity between structural members and application of the load combination in Eq. (2.5-1) with the appropriate resistance factors and member capacities for the given construction materials, as discussed in Section C2.5.

Since design and evaluation of structures for fire conditions is inherently multidisciplinary, multiple design professionals may be required. Design professionals may include structural engineers, fire protection engineers, architects, and others. Provided that multiple design professionals may be involved, the role and responsibility of each design professional should be clearly stated in contract documents.

For a standard fire resistance design, the architect usually serves as the responsible party for satisfying code requirements for structural fire protection. As such, the architect typically selects qualified fire resistance assemblies from available listings, perhaps with the consultation of fire protection engineers. For performance-based structural fire design, a team consisting of architects, fire protection engineers, and structural engineers is typically required. The fire protection engineer, or a design professional with similar qualifications, quantifies the fuel load, evaluates structural design fires, and estimates the temperature histories of structural systems. The structural engineer's primary responsibility is to evaluate the response of the structural system to fire effects based upon the provided temperature histories. The structural engineer may also assist the fire protection engineer in determining which structural systems should be evaluated for the structural design fires and in computing deformations of structural elements that may adversely affect the integrity of fire resistance rated assemblies, such as fire barriers.

CE.4 PERFORMANCE OBJECTIVES

Performance objectives primarily address structural stability and load path continuity, and requirements related to occupant egress.

Project-specific performance objectives may also need to be considered.

CE.4.1 Structural Integrity. Structural integrity supports life safety during fire in buildings and other structures. Accordingly, structural systems that support evacuation routes (e.g., corridors and exit stairs) and refuge areas should be evaluated for stability and load path continuity during structural design fires. Evaluation of stability and load path continuity should consider all supporting structural members and connections. For instance, a column under fire exposure may become unstable if lateral support elements lose their stiffness or load path continuity.

Stairways, horizontal exits, or even entire building floors may be designated as areas of refuge so that occupants can remain safely within the building during a fire. For instance, mobility-impaired occupants may need to remain within an area of refuge during a fire while awaiting rescue or evacuation assistance from emergency responders. Since areas of refuge are meant to serve as a place of safety according to the building codes, it is essential that structural systems that support these areas maintain stability, provide a continuous load path, and limit deformations throughout the heating and cooling of the structure under fire exposure.

Building codes limit egress travel distances to exits (e.g., stairways) but generally do not limit the total evacuation time. As the vertical remoteness of occupants from the point of discharge to a public way (e.g., a public street) is increased, the time required to evacuate the building increases. Hence, special consideration should be given to cases in which phased evacuation procedures are expected and longer occupant egress times are anticipated. For instance, in very tall buildings, occupants may be expected to remain on upper floors for hours, and even if those occupants are directed to use the stairways, the total evacuation may exceed one hour (SFPE 2013). In these cases, the structural performance of vertical exit stairways may represent a paramount concern to designers.

Building codes typically do not mandate how a building is to be evacuated, but there is often a requirement for certain buildings, such as tall buildings, to develop evacuation plans (SFPE 2013). Determination of the time frame necessary for occupant egress and the intended function of refuge areas generally requires the expertise of a fire protection engineer or a design professional with similar qualifications.

CE.4.2 Project-Specific Performance Objectives. In addition to the minimum requirements for structural integrity, project-specific performance objectives may be required. Project-specific performance objectives may address issues such as resilience aspects that consider recovery, property protection, business continuity, environmental protection, adequate structural support of fire resistance rated assemblies to limit fire and smoke spread, and/or structural support of ingress routes for first responders.

A greater level of structural performance may be required than that specified by Section E.4.1. For instance, the following example performance objectives may be applied to limit structural damage based on Risk Category:

- For buildings and other structures that meet Risk Category I criteria for low risk to human life in the event of failure, it may be necessary to avoid structural collapse from fire effects if the collapse is likely to damage valuable property within the building or surrounding properties, including other buildings and infrastructure systems.
- For buildings and other structures that meet Risk Category II or III criteria, or for any Risk Category that would likely damage surrounding properties if structural collapse occurred, it may be necessary for the structure to endure structural design fires such that the primary structural system (e.g., columns, structural members having direct connections to columns, and lateral bracing members) remains stable with a continuous load path to supporting members during the heating and cooling of the structure. Damage to structural members or assemblies that do not compromise the stability of the primary structural system or continuity of the load path could be allowed.
- For buildings and other structures that meet Risk Category IV criteria, it may be necessary for the structure to endure structural design fires such that the entire structural system remains stable with a continuous load path to supporting members during the heating and cooling of the structure.

When designing to limit structural damage, buildings and other structures that meet Risk Category IV criteria may require that structural integrity be maintained for the entire structural system for structural design fires. By maintaining load path continuity and structural stability, enhanced property protection of adjacent areas of the building may be achieved, allowing for rapid reoccupation of areas not directly affected by fire exposure. For instance, if a critical facility experiences a severe fire in a given area, if structural integrity is maintained during and after the fire, such that there is no localized collapse and smoke damage and flame spread are contained, repair and recovery efforts will be limited primarily to the fire-affected areas.

Environmental protection objectives may include limiting the release and spread of hazardous or toxic chemicals to the air, ground and surfaces, or waterways because of loss of structural integrity.

It is desirable for fire-rated resistance assemblies to remain functional (resist fire spread and maintain adequate strength and stiffness for structural integrity) during structural design fires. Three limit states would need to be considered to evaluate fire resistance rated assemblies (e.g., fire barriers): (1) heat transmission leading to unacceptable rise of temperature on the unexposed surface, (2) breach of the barrier caused by loss of support, cracking, or loss of integrity, and (3) loss of load-bearing capacity. All three contribute to performance of fire resistance rated assemblies.

It is desirable for load-carrying elements (e.g., fire-rated floors and walls) that also serve as fire barriers to not have their fire resistance impaired because of deformations or other fire effects. When a fire resistance rated assembly is not load-bearing, the deformation of structural members supporting the assembly should not compromise its performance. However, criteria for limiting deformation of structural members supporting nonstructural fire resistance rated assemblies is not readily available. Fire resistance rated systems are qualified based on results of standard testing in which the supporting boundaries of the furnace (e.g., concrete floor) do not undergo deformation during heating. Addressing such performance objectives may require the designer to develop specific performance criteria per the discretion of the authority having jurisdiction.

CE.5 THERMAL ANALYSIS OF FIRE EFFECTS

Section E.5 provides requirements for determining structural design fires and the thermal response of structural members and connections. Structural design fires have the potential to affect the integrity and stability of a structure. Development of structural design fires involves consideration of compartment layout, boundary materials, ventilation, and fuel load that combine to create conditions that potentially threaten the structural system.

The designer should consider a sufficient number of structural design fires to properly address the risks with consideration of uncertainty associated with heating parameters. Based upon the time-dependent thermal boundary conditions from fire exposure, the thermal response of structural members and connections can be determined based upon principles of heat transfer.

CE.5.1 Fuel Load. Structural design fires depend on the fuel load and its distribution. Fuel load is commonly expressed as a fuel load density, or fuel load per unit floor area. NFPA 557 (2012b) establishes a basis for selecting fuel load density and distribution. Other methods may be used if acceptable to the authority having jurisdiction. Although the fuel load density concept implies a uniform distribution of combustibles in compartments, the actual distribution of combustibles may need to be addressed for analyses of structural design fires where localized heating effects may be significant.

The fuel load based on the contents of a building, space, or area typically varies because combustible materials release different levels of thermal energy when burning. For instance, plastics generally release more energy per unit mass than wood products. The conversion of building contents into an equivalent mass based upon their potential energy provides a consistent basis for determining the total energy of the fuel load. The fuel load in equivalent mass measured in pound mass (lbm) (kilograms (kg)) can be readily converted into total energy measured in British thermal units (Btu) (kilojoules (kJ)) for use in characterization of structural design fires.

CE.5.2 Structural Design Fires. Structural design fires are structurally significant fires based on the physical parameters, such as building layout, compartment boundary materials (e.g., walls), ventilation openings (e.g., doors and windows), and fuel load that are specific to a particular building space or spaces. Structurally significant fires include those that are not controlled by active fire protection systems, such as automatic fire sprinklers or firefighting efforts. Other design fires considered for fire detection, evacuation, or other fire-related issues may not be structurally significant fires.

Provided that there is sufficient oxygen available to support combustion, the duration of a structural design fire is dependent upon the heat release rate history of the fire and the total energy of the fuel load. A structural design fire reaches burnout when either the available fuel load is fully consumed or there is insufficient oxygen to support combustion. The materials involved in a fire significantly affect the heat release rate history. Hence, the heat release rate history of a fire is not necessarily correlated with the total energy of the fuel load.

Structural design fires should be evaluated using methods acceptable to the authority having jurisdiction. In certain cases, it may be necessary to perform fire modeling to evaluate structural design fires. Fire modeling generally requires the expertise of a fire protection engineer or a design professional with similar qualifications. Most fire models simulate the effects of fire (e.g., hot air and smoke flows) and not the phenomena of combustion and flame spread. Where fire modeling is required to determine time-dependent thermal boundary conditions on the structural system, the designer should substantiate the model according to SFPE G.06 (2011a).

Based upon the fuel load, ignition(s), and arrangement of compartments and ventilation openings, a structural design fire may be broadly defined as one of the following types: enclosure fire, localized fire, exterior fire, or traveling fire. SFPE S.01 (2011b) provides methods to determine time-dependent thermal boundary conditions on a structural system caused by a structural design fire for either enclosure or localized fires. NFPA 80A (2012a) provides similar methods for exterior fires.

Events such as an earthquake or flood may result in enclosure, localized, and/or traveling fires within a building. Postevent fires may have compounding factors such as dispersed flammable contents, electrical malfunctions, interrupted power and/or water supplies, damage to fire sprinkler systems, or overextended emergency responders. Also, structural damage caused by a severe event may be exacerbated by a fire (e.g., reinforcing steel exposed by concrete spalling at a connection or loss of member insulating materials).

Enclosure Fire. An enclosure fire is affected by the compartment(s) in which it is contained. As a hot upper gas layer forms with the progression of fire, it reradiates heat back to the fire and fuel packages. The compartment boundaries may radiate inward as well. These conditions may eventually lead to flashover, at which point the fire is considered to be a fully developed fire. Flashover occurs when there is a rapid transition from localized burning to simultaneous burning of all combustible materials within the enclosure. Flashover can only occur in an enclosed compartment with sufficient fuel and ventilation, where the ceiling can trap hot gases that lead to radiant heating of all fuels to the point of combustion.

In most fires, there is variability in the fire size, depending on which items are ignited first and how the fire grows and spreads. Focusing on fully developed fires eliminates much of this variability since a fully developed (postflashover) fire is less sensitive to which items are ignited first and how the fire grows. For PBD analyses, neglecting the heating of the structure during the growth stage of a fire is usually a reasonable assumption since heating of the structure during the fully developed stage is much greater than heating during the growth stage.

Localized Fire. A localized fire burns combustibles at a given location and does not reach flashover. Localized burning occurs in open exposures, large spaces, areas with high ceilings, or other locations that are not conducive to flashover. This typically occurs in relatively large compartments or spaces, when the fuel is concentrated within a region. Fires that do not reach flashover may produce localized heating on the structure.

Exterior Fire. Exterior fires may lead to ignition and subsequent fire exposure within a building, possibly on multiple floors. For example, buildings in close proximity may mutually increase the risk of a large fire exposure through heat exchange between buildings. Flame impingement and convective heat transfer from exterior fires, sometimes from fires extending out windows from lower floors, may also create a fire hazard.

Traveling Fire. A traveling fire is characterized by the spread of fire from combustible to combustible across an open plan that does not burn simultaneously throughout the entire compartment. These fires move across areas as flames spread, burning over a limited area at any one time. Traveling fires are characterized by areas with combustibles that are not yet burning, a fire front with generally intense heating, and a trailing region where fuels have been largely consumed.

CE.5.3 Heat Transfer Analysis. The thermal response of the structural system depends on the structural design fire and the three modes of heat transfer: conduction, convection, and radiation. All three modes of heat transfer typically occur when a structure is heated by fire.

Heat transfer analysis methods are specific to the material's physical and chemical response to heat and are used for homogeneous and nonhomogeneous materials. Material responses may include charring, intumescing, dehydration, phase changes, and chemical reactions. These material responses and associated properties may significantly affect how heat transfer analyses are conducted.

Relevant material thermal properties for heat transfer analyses include density, thermal conductivity, emissivity for exposed surfaces, and specific heat (which may include heat effects caused by phase change, if any). Many of these material properties have strong temperature dependence. Sources with temperature-dependent thermal properties for steel, concrete, masonry, and timber are listed here:

- ACI/TMS. (2007). *Code Requirements for Determining Fire Resistance of Concrete and Masonry Construction Assemblies*, ACI 216.1-07/TMS-216-07.
- American Wood Council (AWC). (2015). *National Design Specification (NDS) for Wood Construction*, AWC NDS-2015.
- European Committee for Standardisation. (2004a). *Eurocode 2: Design of Concrete Structures. Part 1–2: General Rules—Structural Fire Design*, EN 1992-1-2.
- European Committee for Standardisation. (2005). *Eurocode 3: Design of Steel Structures. Part 1–2: General Rules—Structural Fire Design*, EN 1993-1-2.
- European Committee for Standardisation. (2004b). *Eurocode 5: Design of Timber Structures. Part 1–2: General—Structural Fire Design*, EN 1995-1-1.
- European Committee for Standardisation. (2003). *Eurocode 6: Design of Masonry Structures. Part 1-2: General Rules—Structural Fire Design*, EN 1996-1-2.
- Forest Products Laboratory (FPL). (2010). *Wood Handbook: Wood as an Engineering Material*, Forest Products Laboratory General Technical Report FPL-GTR-190.
- Kodur, V., and Harmathy, T. (2008) *Properties of Building Materials*, SFPE Handbook of Fire Protection Engineering.

For design purposes, constant values of thermal properties can be used if they yield conservative results. Depending on the heating or cooling conditions at exposed surfaces, applicable values for convection heat transfer coefficient should be used. Thermal insulation should be analyzed using the specified minimum thickness.

Heat transfer analyses inherently assume that the materials remain in place during the fire exposure. If insulating materials are expected to fail during a structural design fire, either the heat transfer analyses should account for the resulting increased heating of the structure or the insulation design should be modified. The deformations of structural members during structural design fires may need to be considered as part of the evaluation of mechanical integrity of fire resistance rated assemblies.

The temperature histories of structural members and connections comprising the structural system should be determined using heat transfer analyses as permitted in SFPE S.02 (2014). Other approved methods may be used if acceptable to the authority having jurisdiction.

CE.6 STRUCTURAL ANALYSIS OF FIRE EFFECTS

Structural analysis of fire effects requires consideration of the heated members and connections with consideration of the entire structural system. A single member analysis may be justified where only a single member is affected by a fire without consequential effects from surrounding members. A systems approach requires consideration of thermal expansion of heated sections and restraint by cooler adjacent framing, thermally induced forces and displacements on connections, the response of floor systems, and thermally induced failure modes across the structure (McAllister et al 2013). Structural elements may have large deflections that are an order of magnitude greater than deflection limits normally anticipated for structures (McAllister et al. 2012). Large deflections may induce forces in adjacent structural assemblies (e.g., members and connections).

Thermal Expansion and Restraint. Floor systems may experience thermal restraint during heating from columns and cooler adjacent floor members; interior bays typically experience more thermal restraint than exterior bays. Columns, on the other hand, typically do not experience significant thermal restraint from floor systems. However, if there is lateral bracing in place, it may impose some thermal restraint on the braced column section, depending on the framing geometry and member temperatures, if the bracing members and adjacent column are at a significantly lower temperature.

A temperature gradient through the depth or thickness of a structural element causes differential thermal expansion between the hotter and cooler external surfaces. Differential expansion results in curvature for simply supported members. For members with partial or full rotational end restraint, a temperature gradient results in a strain gradient through the depth of the member because of thermal restraint.

The effect of thermal expansion and contraction needs to be carefully considered. Thermal expansion and contraction of construction materials may generate forces sufficient to cause yielding or fracture, depending on the temperature reached and the degree of restraint provided by the surrounding structural system to the thermally induced actions (Gillie et al. 2002). In fact, thermal restraint may dominate the behavior of framing systems, particularly floor systems, with degradation of stiffness and strength a secondary factor (Bailey et al. 1999). Fire-exposed elements that have experienced plastic deformations caused by weakening and thermal restraint may experience tensile strain as the structure cools and may induce forces in adjacent structural assemblies (e.g., connections), depending on the level of thermal restraint.

Columns. Fire effects on steel columns include loss of strength and stiffness, thermal expansion, and P-delta effects under thermal gradients, which may affect global and local column buckling strength. Design procedures for fire effects on steel compression members are provided in Appendix 4 of AISC (2010). The equations are based on analyses conducted by Takagi and Deierlein (2007). The effects of thermal gradients on the axial-moment capacity of steel wide-flange sections are discussed by Garlock and Quiel (2008). Analyses by Seif and McAllister (2013) indicate when elevated temperatures may result in local and global buckling modes of steel wide-flange sections.

Fire effects on concrete columns include loss of strength and stiffness in both the concrete and reinforcement. The primary causes of fire damage to concrete are deterioration in mechanical properties of the cement paste and aggregate, cracking, and spalling (Khoury 2001). Spalling occurs in normal-weight concrete as well as high-strength concrete (Hertz 2003). Concrete cover serves as insulation for reinforcement, so concrete cracking or spalling allows direct heating of reinforcement. Lie and Irwin (1993) provide models based on experimental data for predicting the performance of reinforced concrete columns with rectangular sections, based on axial deformations and temperature through the concrete section. Kodur and McGrath (2003) present experimental results for high-strength concrete columns that include the effects of concrete materials, loading, and spacing of ties. Kodur and Phan (2007) describe the factors that influence spalling in high-strength concrete members.

Floor Systems. A study of steel beams and composite floor systems exposed to a range of heating scenarios (Moss et al. 2004) found that the system behavior at elevated temperatures caused interrelated changes in the deflected shape, axial force, bending moments, and internal stresses that varied with the type of support condition and thermal restraint.

Composite floor systems tested at Cardington (British Steel 1999) and for the FRACOF (Zanon et al. 2011) and COSSFIRE (Zhao and Roosefid 2011) programs found that composite floors with beam lengths less than 30 ft (9 m) did not experience failures during the heating phase, but connection failures did occur during the cooling phase if significant deformation occurred in the floor beams during heating. Wang et al. (2011) found that connection types and axial restraint for floor beams influenced the response of floor systems, primarily during the cooling phase. In contrast, the WTC 7 numerical analysis of composite floors predicted that shear stud and beam connections failed during the heating phase of long-span composite floor systems with one-sided beam-to-girder floor beams of 50- to 56-ft (15- to 17-m) length (McAllister et al. 2012). This numerical prediction was not verified through observation or physical fire testing because of the large size of the floor bays and inability to scale structural responses to fire. Shear stud failure has not been observed in any fire event or in any structural fire tests. The lack of observations may be caused by the limitations of full-scale structural fire tests to date. In structural fire tests, floor beam length is typically less than 20–30 ft (7–10 m), columns and connections are often protected from heating, and/or substantial restraint of the steel beam and concrete slab prevents thermal expansion of the composite section. Failure modes for many structural fire tests with shorter, restrained floor beams tend to occur during the cooling phase at the beam-to-column connection (Bisby et al. 2013). Bailey et al. (1999) and Elghazouli and Izzuddin (2001) assessed the performance of the Cardington floor systems and identified response mechanisms of the floor system to heating and cooling effects. Bailey (2004) presents a performance-based design approach that considers membrane action in the composite floor system.

Reinforced concrete floors, including cast-in-place and precast/prestressed construction, are typically designed by providing a specified cover thickness over reinforcement. In general, there is little guidance for the PBD of concrete floor systems for structural design fires. However, PCI (2011) provides design guidance for precast and prestressed concrete floor systems. Whereas heating of reinforcement is to be avoided, the heating of prestressed strands in floor systems should be of particular concern, as loss of prestress may significantly degrade the floor system performance.

Floor Connections. The performance of connections needs to be considered in the structural analysis of fire effects, particularly connections in floor systems. For instance, steel shear connections may experience bolt shear, local buckling or tear-out of connection plates, or local buckling of the beam flange near the connection. Parametric studies of single plate shear (fin) connections (Yu et al. 2009; Selamet and Garlock 2010; Hu and Engelhardt 2011) and double angle connections (Pakala et al. 2012) identified critical dimensions and component interactions that control connection behavior at elevated temperatures. Huang et al. (1999) evaluated the role of shear stud connections by comparing models with varying levels of composite action against test data from the Cardington fire tests of composite beams.

Heating of concrete and reinforcement in concrete framing (cast-in-place or precast) may result in concrete spalling that accelerates reinforcement heating and loss of strength in the reinforcement.

Failure Modes. Fire-induced failure modes include large deflections, member buckling (local, global, or lateral torsional), connection failures (bearing, bolt tear-out, bolt shear, or weld failure), reinforcement and anchorage failures, and section loss (concrete spalling, cracking, or crushing or charred timber sections).

As temperatures increase sufficiently to reduce the strength and stiffness of the cross section, yielding or buckling modes may occur at arbitrary point-in-time (service) load levels. If such weakening occurs for members with temperature gradients, the resulting gradient in member stiffness and strength can alter the combined axial load and moment resistance of the member (Garlock and Quiel 2008). The strength and stiffness gradient may also cause the member centroid (i.e., center of strength and stiffness) to shift toward the cooler (i.e., stronger) side of the cross section. This centroidal shift induces moments in axially loaded elements.

CE.6.1 Temperature History for Structural Members and Connections. Temperature histories of structural members and connections depend on the thermal response for structural design fires. Temperature histories may include thermal gradients across a section or along a member length.

Thermal finite element analyses typically use 2D or 3D models with a fine mesh of solid elements (element size on the order of inches (centimeters)), whereas structural analyses typically use a coarser mesh of shell and/or beam elements (element sizes on the order of feet (meters)). Careful consideration should be given to the tradeoff between optimal model features for each analysis versus mapping of temperatures between two sets of nodal data. For example, a simplistic heat transfer model, such as the lumped mass method, provides uniform temperatures; uniform temperatures may be inappropriate for a structural system that is likely to experience significant temperature gradients. There may be situations where it is appropriate to use the same mesh discretization in the thermal and structural models, so that transfer of nodal temperature data is seamless.

In most cases, results from heat transfer models are mapped to significantly fewer nodes of beam and shell elements in structural models. The average rate of temperature changes in structural members and connections typically occurs on the order of minutes, and temperature data sets for the structural system can be input at set intervals to reflect the progress of heating and cooling. Temperatures may be interpolated linearly between the data sets during the structural analysis.

CE.6.2 Temperature-Dependent Properties. At elevated temperatures, the strength and stiffness of the material(s) comprising a structural assembly change. Sources for strength and stiffness degradation for steel (including prestressing steel), concrete, masonry, and timber are listed here:

- American Concrete Institute. (ACI). (2007). *Code Requirements for Determining Fire Resistance of Concrete and Masonry Construction Assemblies*, ACI/TMS.
- ACI. (2001). *Guide for Determining the Fire Endurance of Concrete Elements* (Reapproved 2001), 216R-89.
- American Institute of Steel Construction. (AISC). (2010). *Specification for Structural Steel Buildings*, 14th Ed., Appendix A4, "Structural Design for Fire Conditions."
- American Wood Council. (AWC). (2015). *National Design Specification (NDS) for Wood Construction*, ANSI/AWC.

- AWC. (2003). *Calculating the fire resistance of exposed wood members*, TR10.
- ASTM International. (2006). *Significance of Tests and Properties of Concrete and Concrete-Making Materials*, ASTM STP 169D.
- European Committee for Standardisation. (2005). *Eurocode 3: Design of Steel Structures. Part 1–2: General Rules—Structural Fire Design*, EN 1993-1-2.
- European Committee for Standardisation. (2004a). *Eurocode 2: Design of Concrete Structures. Part 1–2: General Rules—Structural Fire Design*, EN 1992-1-2.
- National Institute of Standards and Technology. (NIST). (2010). *Best Practices Guidelines for Structural Fire Resistant Design of Steel and Concrete Buildings*.
- Precast/Prestressed Concrete Institute. (PCI). (2011). *Design for Fire Resistance of Precast/Prestressed Concrete*, Third Ed., MNL-124-11.

The Eurocode defines separate curves for material properties at elevated temperatures for the proportional limit and the yield stress. However, the yield stress at elevated temperatures is often defined at a 0.02 strain in the Eurocode. For ambient temperatures, the yield stress is defined at a 0.002 offset strain (other methods are also permitted as defined in ASTM A6/A6M (2010a), and the proportional limit is not defined by an ASTM test method. However, values for the proportional limit are given in AISC (2010) Appendix 4 in terms of the ratio to the yield strength. The analyst should be careful to note these distinctions in material models.

Where possible, temperature-dependent material properties should be obtained from consensus standards. Alternative sources of data, such as research studies and independent tests, may provide useful data. However, reliance upon a single test or source of data should be avoided. At a minimum, test data from several sources should be collected and used to develop a representative set of temperature-dependent material properties.

CE.6.3 Load Combinations. For extraordinary events, such as structurally significant fires, load combinations 2.5-1 and 2.5-2 are used to evaluate the structural system performance. This load combination was developed for extraordinary events that may lead to ultimate limit states such as gross inelastic deformation or partial collapse.

Load combination 2.5-1 is used to perform a safety check on a structure designed for the basic load combinations at room temperature conditions and to evaluate the effect of elevated temperatures. The force in structural members caused by fire effects, A_k, has a load factor of 1.0 (Ellingwood and Corotis 1991; Ellingwood 2005). The live load factor of 0.5 is intended for typical occupancies and arbitrary point-in-time live loads that likely exist during a significant fire. The 0.5 live load factor is also used in other load combinations in Section 2.3 when live load is a companion load and not the principal load. Note that the live load in this load combination differs from the approach used in standard fire test methods, where the assembly is loaded to its design limit for member stress during the standard fire exposure, which represents the application of the full dead and live load.

Whereas gravity loads for the structure remain constant during most fires (assuming that most of the building contents are not burning), time-dependent temperature histories may result in time-varying member strength and thermally induced forces, depending on the temperatures reached by structural members.

REFERENCES

American Concrete Institute. (ACI). (2001). "Guide for determining the fire endurance of concrete elements." (Reapproved 2001), *ACI 216R-89*, ACI, Farmington Hills, MI.

ACI. (2007). "Code requirements for determining fire resistance of concrete and masonry construction assemblies." *ACI 216.1-07/TMS-216-07*, ACI, Farmington Hills, MI.

American Institute of Steel Construction. (AISC). (2010). *Specification for structural steel buildings*, 14th Ed., Appendix A4, "Structural design for fire conditions," Chicago, IL.

ASCE. (2007). "Standard calculation methods for structural fire protection." *ASCE/SFPE 29*, ASCE, Reston, VA.

American Wood Council. (AWC). (2015). "National design specification for wood construction." *ASD/LRFD Manual*, AWC, Leesburg, VA.

AWC. (2003). "Calculating the fire resistance of exposed wood members." *TR10-2003*, AWC, Leesburg, VA.

ASTM International. (2006). "Significance of tests and properties of concrete and concrete-making materials." *ASTM STP 169D*, West Conshohocken, PA (see Chapter 27, "Resistance to fire and high temperature," by Stephen S. Szoke).

ASTM. (2010a). "Standard specification for general requirements for rolled structural steel bars, plates, shapes, and sheet piling." *ASTM A6/A6M*, West Conshohocken, PA.

ASTM. (2010b). "Standard test methods for determining effects of large hydrocarbon pool fires on structural members and assemblies." *ASTM E1529-10*, West Conshohocken, PA, doi: 10.1520/E1529-10, www.astm.org.

ASTM. (2012). "Standard test methods for fire tests of building construction and materials." *ASTM E119-12a*, West Conshohocken, PA, doi: 10.1520/E0119-12A, www.astm.org.

Bailey, C. G. (2004). "Membrane action of slab/beam composite floor systems in fire." *Eng. Struct.* 26(12), 1691–1703.

Bailey, C. G., Lennon, T., Moore, D. B. (1999). "The behaviour of full-scale steel-framed buildings subjected to compartment fires." *Struct. Eng.* 77(8), 1182–1192.

Bisby, L., Gales, J., and Maluk, C. (2013). "A contemporary review of large-scale non-standard structural fire testing." *Fire Sci. Rev.* 2(1), doi:10.1186/2193-0414-2-1.

British Standards Institution. (BSI). (1987). "Fire tests on building materials and structures. Method for determination of the fire resistance of elements of construction (general principles)." *BS 476-20*, London.

British Steel. (1999). *The behaviour of multi-storey steel framed buildings in fire*. British Steel PLC, South Yorkshire, UK.

Buchanan, A. H. (2002). *Structural design for fire safety*, John Wiley & Sons, New York.

Elghazouli, A. Y., and Izzuddin, B. A. (2001). "Analytical assessment of the structural performance of composite floors subject to compartment fires." *Fire Safety J.*, 36(8), 769–793.

Ellingwood, B. R. (2005). "Load combination requirements for fire resistant structural design." *J. Fire Protect. Eng.* 15(2), 43–61.

Ellingwood, B. R., and Corotis, R. B. (1991). "Load combinations for buildings exposed to fires." *Eng. J., ASIC*, 28(1), 37–44.

European Committee for Standardisation (CEN). (2003). "Eurocode 6: Design of masonry structures. Part 1–2: General rules—Structural fire design." *EN 1996-1-2*, Brussels, Belgium.

CEN. (2004a). "Eurocode 2: Design of concrete structures. Part 1–2: General rules—Structural fire design." *EN 1992-1-2*, Brussels, Belgium.

CEN. (2004b). "Eurocode 5: Design of timber structures. Part 1–2: General—Structural fire design." *EN 1995-1-1*, Brussels, Belgium.

CEN. (2005). "Eurocode 3: Design of steel structures. Part 1–2: General rules—Structural fire design." *EN 1993-1-2*, Brussels, Belgium.

Forest Products Laboratory. (FPL). (2010). Wood handbook: Wood as an engineering material, *FPLG Technical Report FPL-GTR-190*, USDA Forest Service, Madison, WI.

Garlock, M. E. M., and Quiel, S. E. (2008). "Plastic axial load and moment interaction curves for fire-exposed steel sections with thermal gradients." *J. Struct. Eng.*, 134(6) 874.

Gillie, M., Usmani, A. S., and Rotter, J. M. (2002). "A structural analysis of the Cardington British steel corner test." *J. Constr. Steel Res.* 58(4), 427–442.

Hall, J. (2013). *The total cost of fire in the United States*, National Fire Protection Association, Quincy, MA.

Hertz, K. D. (2003). "Limits of spalling of fire-exposed concrete." *Fire Safety J.*, 38(2), 103–116.

Hu, G., and Engelhardt, M. D. (2011). "Investigations on the behavior of steel single plate beam end framing connections in fire." *J. Struct. Fire Eng.* 2(3), 195–204.

Huang, Z., Burgess, I. W., and Plank, R. J. (1999). "The influence of shear connections on the behavior of composite steel-framed buildings in fire." *J. Construct. Steel Res.*, 51(3), 219–237.

International Standards Organization (ISO). (1999). "Fire resistance tests–Elements of building construction. Part 1: General requirements." *ISO 834-1:1999*, ISO, Geneva, Switzerland.

Khoury, G. A. (2001). "Effect of fire on concrete and concrete structures." *Prog. Struct. Eng. Mater.* 2(4), 429–447.

Kodur, V., and Harmathy, T. (2008). "Properties of building materials." *SFPE handbook of fire protection engineering*. P. J. DiNenno, ed., National Fire Protection Association, Quincy, MA.

Kodur, V., and McGrath, R. (2003). "Fire endurance of high strength concrete columns." *Fire Tech.*, 39, 73–87.

Kodur, V. K. R., and Phan, L. (2007). "Critical factors governing the fire performance of high strength concrete systems." *Fire Safety J.*, 42, 482–488.

Lie, T. T., and Irwin, R. J. (1993). "Method to calculate the fire resistance of reinforced concrete columns with rectangular cross section." *ACI Struct. J.*, 90(1), 52–60.

McAllister, T. P., Gross, J. L., Sadek, F., Kirkpatrick, S., MacNeill, R. A., Zarghamee, M., et al. (2013). "Structural response of World Trade Center buildings 1, 2, and 7 to impact and fire damage." *Fire Tech.*, 49(3), 709–739.

McAllister, T. P., MacNeill, R., Erbay, O. O., Sarawit, A. T., Zarghamee, M. S., Kirkpatrick, S., et al. (2012). "Analysis of structural response of WTC 7 to fire and sequential failures leading to collapse." *J. Struct. Eng.*, 138(1), 109–117.

Moss, J. M., Buchanan, A. H., Septro, J., Wastney, C., and Welsh, R. (2004). "Effect of support conditions on the fire behaviour of steel and composite beams." *Fire Mater.* 28, 159–175.

National Fire Protection Association. (NFPA). (2012a). "Recommended practice for protection of buildings from exterior fire exposures." *NFPA 80A*, Quincy, MA.

NFPA. (2012b) NFPA 557: *Standard for determination of fire loads for use in structural fire protection design*, NFPA, Quincy, MA.

National Institute of Standards and Technology. (NIST). (2010). "Best practices guidelines for structural fire resistant design of steel and concrete buildings." *Tech. Note 1681*, Gaithersburg, MD.

Pakala, P., Kodur, V., and Dwaikat, M. (2012). "Critical factors influencing the fire performance of bolted double angle connections." *Eng. Struct.* 42, 106–114.

Precast/Prestressed Concrete Institute. (PCI). (2011). "Design for fire resistance of precast/prestressed concrete," 3rd Ed., *MNL-124-11*, Chicago.

Seif, M. S., and McAllister, T. P. (2013). "Stability of wide flange structural steel columns at elevated temperatures." *J. Construct. Steel Res.* 84(5), 17–26.

Selamet, S., and Garlock, M. E. (2010). "Robust fire design of single plate shear connections." *Eng. Struct.* 32, 2367–2378.

Society of Fire Protection Engineers. (SFPE). (2011a). "Engineering guidelines for substantiating a fire model for a given application." *SFPE G.06*. SFPE, Gaithersburg, MD.

SFPE. (2011b). "Engineering standard on calculating fire exposures to structures." *SFPE S.01*, SFPE, Gaithersburg, MD.

SFPE. (2013). *Engineering guide: Fire safety for very tall buildings*, SFPE, Gaithersburg, MD.

SFPE. (2014). SFPE S.02: *Engineering standard on the development and use of methodologies to predict the thermal performance of structural and fire resistive assemblies*, SFPE, Gaithersburg, MD.

Takagi, J., and Deierlein, G. G. (2007). "Strength design criteria for steel members at elevated temperatures." *J. Construct. Steel Res.*, 63, 1036–1050.

Underwriters Laboratories. (UL). (2011a). "Standard for rapid rise fire tests of protection materials for structural steel." *ANSI/UL 1709*, Northbrook, IL.

UL. (2011b). "UL standard for safety for fire tests of building construction and materials." *ANSI/UL 263*, Northbrook, IL.

UL. (2013). *UL fire resistance directory*, Northbrook, IL.

Underwriters Laboratories of Canada. (ULC). (2007). "Standard methods of fire endurance tests of building construction and materials." *CAN/ULC-S101*, Ottawa.

Wang, Y. C., Dai, X. H., and Bailey, C. G. (2011). "An experimental study of relative structural fire behaviour and robustness of different types of steel joint in restrained steel frames." *J. Construct. Steel Res.* 67, 1149–1163.

Yu, H., Burgess, I. W., Davison, J. B., and Plank, R. J. (2009). "Experimental investigation of the behaviour of fin plate connections in fire." *J. Construct. Steel Res.* 65(3), 723–736.

Zanon, R., Sommavilla, M., Vassart, O., Zhao, B., and Franssen, J. M. (2011). "FRACOF: Fire resistance assessment of partially protected steel-concrete composite floors." XXIII Giornate italiane della costruzione in acciaio. *Lacco Ameno*, Ischia, Italy, 527–536.

Zhao, B., and Roosefid, M. (2011). "Experimental and numerical investigations of steel and concrete floors subjected to ISO fire condition." *J. Struct. Fire Eng.*, 2(4), 301–310.

INDEX

Provisions appear on pages 1–402.
Commentary appears on pages 405–822.

Page numbers followed by *e*, *f*, or *t* indicate equations, figures or tables.

access floors, seismic design requirements, 127–128, 610
accidental mass eccentricity, 175, 186, 570, 697–698
active fault, defined, 77
additions, 5, 77. *See also* alterations
adequate structural strength, use of term, 410
aerodynamic shade, snow drifts and, 59–61, 59*e*, 62*f*, 63*f*, 497–498, 498*f*, 499*f*
aerodynamic wind loads, on tall buildings, 754
aeroelastic model (AM), wind tunnel test, 793
air-permeable cladding, wind loads, 333, 781–783, 782*f*
air-supported structures, snow loads, 54, 494
aleatory uncertainty, 35
 defined, 466
Alfred P. Murrah Federal Building, 411
"all heights" method. *See* enclosed, partially enclosed, and open buildings of all heights
allowable stress design (ASD)
 defined, 1
 load combinations, 8–9, 420–421, 421*t*
 procedures, 2, 405
 requirement to use, 417
 seismic design criteria, 525
along-wind response. *See* gust effects, wind loads
alterations, 393
 defined, 77
amusement structures, 152
anchorages
 access floors, 603–605
 in concrete, 637
 fall arrest and lifeline, loads and, 16, 436
 in masonry, 143, 144, 150
 nonbuilding structures, 150, 155, 155*t*, 637
 nonstructural components, 125, 603–605, 605*f*
 structural walls, 4, 108–109, 108*e*
 tanks and vessels, 644, 644*f*
appendage, defined, 77
approval, defined, 77
approved, defined, 21, 245
approximate fundamental frequency, 751–753, 752*e*, 752*f*, 753*e*
approximate fundamental period, 102, 102*e*, 561–562, 561*f*
arched roofs, wind loads MWFRS directional procedure, 278*f*
architectural components, seismic design requirements, 125–128, 126*e*, 126*t*, 127*t*, 128*e*, 146, 605–611, 608–609*t*, 611*f*
ASCE Tsunami Design Geodatabase, 25, 26*f*, 32, 33*f*, 456, 465, 465*f*
atmospheric icing, 69–76, 517–523
 definitions, 69, 518–519
 design procedure, 71
 design temperatures for freezing rain, 70*f*, 71, 72–73*f*, 522
 ice loads caused by freezing rain, 70–71, 70*e*, 70*f*, 71*e*, 72–73*f*, 74*f*, 75*f*, 76*f*, 519–521, 520*f*, 521*t*
 partial loading, 76, 522
 site-specific studies, 69
 wind on ice-covered structures, 71, 72–73*f*, 74*f*, 75*f*, 76, 521–522, 522*t*

attached canopy, defined, 245
 on buildings, 380
attachments, defined, 78, 526–527, 527*f*
attics (uninhabitable), live loads, 18

balconies and decks, live loads, 433–434
barrel vault roofs, snow loads, 54, 59, 495, 497
base
 defined, 78
 seismic design criteria, 527–529, 527*f*, 528*f*, 529*f*
base flood, defined, 21
base flood elevation (BFE), defined, 21
base level, defined, 167
base shear, defined, 78
base slab averaging
 defined, 197
 soil-structure interaction, 201, 201*e*, 707, 707*f*
basic wind speed, defined, 245
Bathymetric profile, defined, 25
bearing walls
 alternative structural design criteria, 114–119, 114*f*, 115*t*, 116*e*, 118*e*, 119*e*
 defined, 81
boilers and pressure vessels, seismic design requirements, 133, 159–160, 616
boundary elements, defined, 78
braced frame, defined, 79
breakaway wall
 defined, 21
 flood loads, 21, 440
building, defined, 1, 78
building appurtenances and rooftop structures and equipment
 wind loads, components and cladding, 375–382, 375*e*, 379*f*, 379*t*, 380*e*, 380*f*, 380*t*, 381*f*, 382*t*
 see also wind loads on building appurtenances and other structure, MWFRS (directional procedure)
building envelope products, C&C ratings for, 740
building frame system, defined, 79
building structure. *See* seismic design requirements, building structure
building types, defined
 enclosed, 245, 732
 envelope, 246
 flexible, 246, 732
 low-rise, 245
 open, 245, 732
 partially enclosed, 245, 732
 partially open, 246, 732
 regular-shaped, 246, 732
 rigid, 246, 732
 simple diaphragm, 246
 torsionally regular under wind load, 246

cables
 material-specific seismic design and detailing requirements, 136, 620
 wind on ice-covered, 521–522, 522*t*

camber, 395, 811
cantilevered column system, defined, 78
cantilevered storage racks, seismic design requirements, 638, 638f
ceilings, accoustical tile standard, 610
ceilings, seismic design requirements, 127, 607, 608–609t, 610
change of use provisions, 393
channelized scour, defined, 86
characteristic earthquake, defined, 78
chimneys and stacks, seismic design requirements, 152, 640
circular bins, wind loads
   components and cladding, 382, 382e, 382t, 383f, 384–385f, 385, 386f, 387f, 788–789, 789f
   MWFRS directional procedure, 322, 326f, 327, 327e, 328f, 329f, 776, 776–777
circular sector, defined, 743
classification, of buildings and other structures
   risk categorization, 4–5, 4t, 5t, 412–414, 413f
   toxic, highly toxic, and explosive substances, 5, 414–415
climate data, estimation of basic wind speeds from regional, 249, 740–741
closure ratio (inundated protected area), defined, 25
coastal A-zone, defined, 21, 439–440
coastal high hazard area (V-zone), defined, 21, 439–440
cold-formed steel light frame construction, seismic design and detailing requirements, 135–136, 619–620
collapse, of buildings, 410–411
collapse prevention structural performance level, defined, 25
collector (drag strut, tie, diaphragm strut), defined, 78
component importance factor, 121, 595
component period, 124–125, 124e, 602–603
components and appurtenances
   defined, 69, 78, 526–527, 527f
   see also building appurtenances and rooftop structures and equipment
components and cladding (C&C), 731, 740
   defined, 246, 732
   see also wind loads, components and cladding
composite steel and concrete structures, seismic design and detailing requirements, 142, 627–628
concentrated live loads, 13, 435
concentrically braced frame (CBF), defined, 79
concrete
   anchorages in, 150
   chimneys and stacks, 152, 640
   definitions, 78
concrete, seismic design and detailing requirements
   composite steel and concrete structures, 142, 627–628
   diaphragms, 139–142, 141f, 142f, 622–627, 622f, 623f, 624f, 625f, 626f, 627f
   foundations, 136, 621
   piles, 137–139, 137e, 138–139e, 621–622
   precast structural walls, 136, 621
   shear walls, 136–137, 621
conduits and cable distributions systems, seismic design requirements, 131
connection, defined, 136
connector, defined, 136, 621
construction documents, defined, 78
containment systems (secondary), seismic design requirements, 153, 640–641
continuous beam systems, snow loads, 54, 57
controlled drainage, defined, 65
cool roof slope factor, snow loads and, 54, 59f
counteracting structural actions, basic requirements, 3

coupling beams, 143–144
   defined, 78
crane loads, 17–18, 437
critical action, defined, 78
critical equipment/critical systems, defined, 25
critical facility, defined, 25
curved roofs, snow loads and, 54, 58–59, 60f, 495, 497

damping device, defined, 181
damping system, defined, 181
damping systems, seismic design requirements
   alternative procedures and acceptance criteria, 189–196, 189e, 190e, 191e, 191t, 192e, 193e, 194e, 195e, 195t, 196e, 699–701, 699f, 700e, 700f, 700t
   damping systems distinguished from seismic force-resisting systems, 693, 694f
   design review, 187, 698
   general design requirements, 183–185, 183e, 185e, 693–697, 696f
   nonlinear response history, 186–187, 697–698, 697t
   testing, 187–189, 698–699
data-base assisted design, wind tunnel procedure, 794
dead loads, 11–12, 425, 426–427t
deadweight tonnage (DWT), defined, 25
debris impact loads. *See* impact loads
deformability, elements defined, 78
deformation, types defined, 78
deformation-controlled actions
   defined, 78
   global acceptance criteria, 166, 166t, 670–671, 670t
design earthquake, types defined, 78
design earthquake ground motion, defined, 78
design flood, defined, 21
design flood elevation (DFE), defined, 21
design force, defined, 246
design pressure, defined, 246
design professional, defined, 80
design strength, defined, 1, 27, 80
design tsunami parameters, defined, 27
design wind speeds, ASCE 7-93 to 7-10, 740t
designated seismic systems
   certification required, 597–598
   defined, 78
   design requirements, 122
detailed plain concrete structural wall, defined, 136
deterministic seismic hazard analysis (DSHA), 711, 713, 723
diaphragm boundary, defined, 78
diaphragm chord, defined, 78
diaphragm flexibility, seismic design requirements, 95t, 96–98, 96e, 96f, 97t, 98f, 98t
diaphragms
   buildings exempted from torsional wind load cases, 397–399, 398f, 399f
   concrete, seismic design and detailing requirements, 139–142, 141f, 142f, 622–627, 622f, 623f, 624f, 625f, 626f, 627f
   defined, 246, 732
   steel, seismic design and detailing requirements, 136, 620
   types, defined, 78
diaphragms, chords, and collectors, seismic design, building structures, 90–92t, 95t, 106–108, 106e, 106f, 107e, 107f, 108t
directional procedure
   defined, 246
   see also wind loads on buildings, MWFRS directional procedure; wind loads on building appurtenances and other structures, MWFRS directional procedure

displacement restraint system, defined, 167
displacement-dependent damping device, defined, 181
displacements, seismic demands on nonstructural components, 124, 124e
distribution systems
  defined, 78
  seismic design criteria, 529–530
  seismic design requirements, 131–133, 614
domed roofs
  snow loads, 59, 497
  wind loads, components and cladding, 350f
  wind loads, MWFRS directional procedure, 277f, 767–768
drainage systems. *See* rain loads
drift, defined, 51
drift and deformation, seismic design requirements, 109–110, 109t
drifts on lower roofs (aerodynamic shade), snow and, 59–61, 59e, 62f, 63f, 497–498, 498f, 499f
dual structural system, defined, 79
duct systems, seismic design requirements, 131–132, 614
ductility-governed action, defined, 27
durability, serviceability considerations, 395, 803
dynamic analysis
  isolated structures, 175–176, 176e, 687–689, 688t
  nonstructural components, 123–124, 123e, 601–602, 602f
  see also linear dynamic analysis

earthquakes
  characteristic earthquake, defined, 78
  design, defined, 78
  design earthquake ground motion, defined, 78
  load combinations for, 420–421
  nature of "loads," 525
  performance-based procedures, 407–408
  risk categorization of buildings, 5t
earth-retaining structures, seismic design requirements, 152, 640
eave height, defined, 246
eccentrically braced frame (EBF), defined, 79
effective damping, defined, 167
effective stiffness, defined, 167
effective wind area, defined, 246–247, 732–733.
  See also wind loads, components and cladding
electric power-generating facilities, seismic design requirements, 151, 639
electrical systems, seismic design requirements, 121, 130
element actions, defined, 78–79
elevators and escalators
  impact loads, 16
  seismic design requirements, 133, 616
embedment, soil-structure interaction, 201–202, 201e, 707–708
enclosed, partially enclosed, and open buildings of all heights, MWFRS directional procedure, 273–274, 273t, 274e, 275–276f, 277f, 278f, 279f, 280f, 281f, 282f, 283f, 767–769, 768f
enclosed and partially enclosed low-rise buildings, MWFRS envelope procedure, 311, 311e, 311t, 312–314f, 771–773, 772f, 773f
enclosed building, defined, 245
enclosed simple diaphragm buildings with height <160 ft, wind loads MWFRS directional procedure, 273, 284–285, 284t, 285f, 286f, 288–290t, 291–310t, 769–770
enclosed simple diaphragm low-rise buildings, MWFRS envelope procedure, 315, 315e, 315t, 316–320f, 773–774
enclosure, defined, 79

enclosure classification, wind loads, 270–271, 754–756
enclosure fire, 818
Energy Grade Line Analysis, tsunamis, 458–459, 459t, 461–462, 463f, 464f, 467, 469, 470–471
envelope procedure
  defined, 247
  see also wind loads on buildings, MWFRS envelope procedure
epistemic uncertainty, defined, 466
equipment and equipment platforms, snow loads on, 64
equipment support, defined, 79
equivalent lateral force (ELF) procedure
  building structures, 101–104, 101e, 102e, 102t, 103e, 103f, 104e, 560–567, 560f, 561f, 562f, 564f, 565e, 565f, 566e, 567e
  damping systems, 184
  isolated structures, 172–175, 173e, 174e, 682–687, 683f, 684f, 685t, 686t, 687f
  soil-structure interaction, 198, 198e
erosion and scour, flood loads, 21, 440
escalators. *See* elevators and escalators
escarpments
  defined, 247
  topographic effects on wind speed, 266, 267f, 268, 268e
essential facilities, defined, 1
evacuation, tsunamis and, 48–50, 50f, 447–449, 458, 468–469, 472, 483, 484
expansion and contraction, serviceability considerations, 395, 803
exposure, wind load general requirements, 266, 741–744, 741f, 742t, 743e, 744f, 745f, 746f, 747f, 748f
exposure factor, snow loads, 52, 58t, 493–494
extraordinary events
  basic requirements, 4
  load combinations for, 9, 421–423, 422e

fall arrest and lifeline anchorage, live loads, 16, 436
fasteners, power-actuated, 125, 605
file cabinets, 433
fire, defined, 401
fire effects, performance-based design procedures for
  definitions, 401, 815–816
  general requirements, 401, 816
  performance objectives, 401–402, 816–817
  performance-based structural fire design, 815–816
  structural analysis of fire effects, 402, 819–821
  thermal analysis of fire effects, 402, 817–819
fire exposure, defined, 401
fire resistance
  basic requirements, 3, 410
  defined, 401
fixed ladders
  defined, 13
  live loads and, 16, 435
fixed service equipment, weight of, 11, 425
flat roof snow loads, 51, 51e, 53e, 54, 58f, 493–495
  defined, 51
  examples of establishing, 501–502, 501e, 502e
flexible building or structure, defined, 246, 732
flexible component, defined, 78
flexible connections, defined, 79
flexible diaphragm, defined, 246
flexure-controlled diaphragm, 530
  defined, 78
flood hazard area, defined, 21
flood hazard map, defined, 21

flood insurance rate map (FIRM), defined, 21
flood loads, 21–23, 22*e*, 22*t*, 23*e*, 23*f*, 439–445, 441*e*, 442*e*, 442*f*, 442*t*, 443*f*, 443*t*, 444*e*
   definitions, 21, 439–440
floods
   load combinations for, 7, 8, 418, 421
   performance-based procedures, 406
   risk categorization of buildings, 5*t*
floor responses, seismic design requirements, 123–124, 124*e*, 601–602, 602*f*
floors and foundations, weight of, 12, 431
force-controlled actions
   defined, 78
   global acceptance criteria, 165, 165*e*, 165*t*, 667–670, 667*e*, 668*f*, 668*t*, 669*f*
force-controlled elements, defined, 181
force-sustained action, defined, 27
foundation geotechnical capacity, defined, 79
foundation input motion, defined, 197
foundation structural capacity, defined, 79
foundations
   building structures, 110–114, 111*t*, 112*t*, 113*t*, 587*f*, 588*f*
   concrete, 136, 621
   nonbuilding structures, 150
   soil-structure interaction and damping effects, 199–201, 199*e*, 200*e*, 200*t*, 201*e*, 705–706, 706*f*
   tsunami loads and effects, 46–49, 47*e*, 47*f*, 47*t*, 48*f*, 480–484, 481*e*, 481*f*, 482*e*, 482*f*, 483*f*
frames
   alternative structural design criteria, 114–119, 114*f*, 115*t*, 116*e*, 118*e*, 119*e*
   drift of, serviceability considerations, 395, 802, 802*e*, 804–805*f*, 806–807*f*, 808–809*f*, 810–811*f*
   types, defined, 79
free roof, defined, 247
freeboard, seismic design requirements, 153, 153*e*, 641
free-field, use of term, 703
free-field motion, defined, 197
freezer building, defined, 51
freezing rain
   defined, 69, 518
   design temperatures for, 70*f*, 71, 72–73*f*, 522
   ice loads caused by, 70–71, 70*e*, 70*f*, 71*e*, 72–73*f*, 74*f*, 75*f*, 76*f*, 519–521, 520*f*, 521*t*
friction clips, 125, 605, 605*f*
   defined, 79
Froude number, defined, 27
fuel load
   defined, 401
   performance-based procedures for fire effects, 402, 818
functionality requirements, performance-based procedures, 409–410

gable/hip roofs
   defined, 336*f*
   snow loads, 57–58, 61*f*, 496–497
   wind loads, components and cladding, 337*f*, 338*f*, 339*f*, 340*f*, 341*f*, 342*f*, 343*f*, 344*f*, 351–362*f*, 783*t*, 784*t*, 785*t*
   wind loads, MWFRS directional procedure, 275–276*f*
garage loads
   passenger vehicles, 17, 18, 437
   trucks and buses, 18
gas spheres, seismic design, 160–161, 648
geological hazards and geotechnical investigation, seismic design criteria, 86, 86*e*, 86*t*, 537–539, 537*e*

glass, seismic design requirements, 126, 127*t*, 128, 128*e*, 607, 610–611
glaze, defined, 69
glazed curtain wall, defined, 79
glazed openings, wind loads, 270–271
glazed storefront, defined, 79
glazing
   definitions, 247, 755
   tornado damage and, 763
global acceptance criteria, nonlinear response history analysis
   deformation-controlled actions, 166, 166*t*, 670–671, 670*t*
   element-level acceptance criteria, 165, 667
   elements of gravity force-resisting system, 166, 671
   force-controlled actions, 165, 165*e*, 165*t*, 667–670, 667*e*, 668*f*, 668*t*, 669*f*
   story drift, 165, 666–667
   unacceptable response, 165, 664–666, 665*f*, 665*t*, 666*t*
grab bar systems
   defined, 13
   live loads and, 16, 435
grade plane, defined, 27, 79
granular material storage tanks, seismic design requirements, 158–159, 647–648, 647*e*, 647*f*, 648*f*
gross frontal area, defined, 743
ground motions, nonlinear response history analysis
   amplitude scaling, 164, 661–662, 661*f*
   application to structural model, 164, 662
   ground motion modification, 164, 660–661
   ground motion selection, 164, 659–660
   period range for scaling/matching, 164, 661
   spectral matching, 164, 662
   target response spectrum, 163–164, 658–659, 658*f*
ground snow loads, 51, 52–53*f*, 55*t*, 56*t*, 57*t*, 58*t*, 489, 490–491*t*, 491–493, 492*t*, 493*t*
ground-supported cantilever walls or fences, seismic design requirements, 641–642, 641*f*, 642*f*
gust effects, wind loads, 269–270, 269*e*, 269*t*, 270*e*, 750–754, 751*e*, 751*t*, 752*e*, 752*f*, 752*t*, 753*e*, 754*e*, 755*t*
guys and cables. *See* cables

handrail and guardrail systems
   defined, 13
   live loads and, 13, 16, 435
hazard-consistent tsunami scenario, defined, 27
heat transfer
   defined, 401
   thermal analysis of fire effects, 402, 818–819
heating, ventilating, air-conditioning, and refrigeration (HVACR)
   defined, 79
   seismic design requirements, 130
helipad loads
   helipad defined, 13
   live loads, 18, 437–438
high-deformability element, defined, 78
hills
   defined, 247
   topographic effects on wind speed, 266, 267*f*, 268, 268*e*
hip and gable roofs. *See* gable/hip roofs
hoarfrost
   defined, 69
   formation of, 518
hoists, 16, 435–436
hurricane wind speeds, 735, 735*e*
hurricane-prone areas

basic wind speeds at selected, 736t, 737t, 738t, 739t
  defined, 247
hydraulic structures, seismic design requirements, 153, 640, 640f
hydrodynamic loads
  defined, 27
  during flooding, 22, 22e, 440–441, 441e
  tsunami loads and effects, 42–44, 42e, 42t, 43e, 473–477, 473e, 474e, 475f, 476f, 477f
hydrostatic loads
  defined, 27
  during flooding, 21–22, 440
  tsunami loads and effects, 41–42, 41e, 42e, 473
hydrostatic pressure, soil loads, 12, 425, 430–431t, 431

ice dams and icicles, snow loads, 54, 494, 495, 496f
ice loads
  load combinations for, 7, 8–9, 418–419, 421
  risk categorization of buildings, 5t
  see also atmospheric icing
ice-sensitive structures, defined, 69, 518
immediate occupancy structural performance level, defined, 27
impact loads
  defined, 27
  flooding, 23, 441–444, 442e, 442f, 442t, 443f, 443t, 444e
  live loads, 16, 435–436
  tsunamis, 44–46, 44e, 44t, 45f, 46e, 46t, 477–480, 477t, 478t, 479e
impact-protective system, 247, 755
impact-resistant glazing, 247, 755
importance factor
  defined, 1
  seismic design criteria, 85, 535, 535f
  snow loads and, 52, 493, 494–495
in-cloud icing, defined, 69, 518–519, 519f
inspection, special defined, 79
inspector, special defined, 79
internal pressure coefficients, wind loads, 271, 271e, 271t, 756
inundation depth, defined, 27
inundation depth and flow velocity, tsunamis
  analysis of design, 30–31, 30e, 458–461, 459t, 460e, 460f
  based on runup, 31–32, 31e, 31f, 32t, 461–462, 462e, 463f, 464f
  based on site-specific probabilistic hazard analysis, 32–38, 32e, 33f, 34f, 35f, 35t, 37f, 38f, 462, 465–467, 465f, 467f
inundation elevation, defined, 27
inundation limit, defined, 27
inverted pendulum-type structures, defined, 79
isolated structures. See seismically isolated structures, seismic design requirements
isolation interface, defined, 167
isolation system, defined, 167
isolator unit, defined, 167

joint, defined, 79

kinematic soil-structure interactions, 201–202, 201e, 703, 704, 707–708, 707f
  defined, 197

ladders, fixed, live loads, 435
lateral forces, basic requirements, 4
lateral pressures
  crane loads, 18

soil loads, 12, 431
leeward snowdrifts, 61, 497–498, 498f
library stack rooms, live loads, 18, 433, 438
life safety structural performance level, defined, 27
lifeline anchorages. See fall arrest and lifeline anchorage
light frame wall, defined, 81
light frame wood shear wall, defined, 81
light-frame construction, defined, 79
limit deformation, defined, 78
limit state, defined, 1
limited-deformability element, defined, 78
linear analysis
  damping systems, seismic design requirements, 699–700, 700f
  as precondition to nonlinear response history analysis, 163, 657
linear dynamic analysis
  seismic design, building structures, 104–106, 105e, 567–571
  soil-structure interaction, 198
liquefaction evaluation requirements, seismic design criteria, 538–539
liquefaction scour, defined, 27
liquefiable sites, seismic design requirements, 150, 637
liquids, tanks and vessels for, 155–161, 155–156e, 156t, 157e, 157t, 160, 631f, 644–648, 647f, 648f, 649–652f, 653–654f
live loads, 13–19, 14–16t, 433–439
  concentrated, 13, 16, 435
  crane loads, 17–18, 437
  definitions, 14–16t
  garage loads, 17, 18, 437
  handrail and guardrail systems, 13, 16, 435
  helipad loads, 18, 437–438
  impact loads, 16, 435–436
  library stack rooms, 18, 433, 438
  live load defined, 13
  roof load reduction, 17, 17e, 437
  seating for assembly use, 18–19, 438
  sidewalks, driveways, yards subject to trucking, 19
  solar panels, 19, 438
  stair treads, 19
  uniform load reduction, 16–17, 16e, 436–437, 436f
  uniformly distributed loads, 433–435, 434t
  uninhabitable attics, 18
  unspecified loads, 13
"lives at risk," 413
load and resistance factor (LRFD), 417, 521, 587
load combinations, 7–9, 417–423
  for allowable stress design, 8–9, 420–421, 421t
  for extraordinary events, 9, 421–423, 422e
  for strength design, 7–8, 417–420, 418t, 419e
  for structural integrity loads, 9
load effects, defined, 1
load factor, defined, 1
load path connections, basic requirements, 4
load tests, generally, 5, 415
loads, defined, 1
local coseismic tsunami, defined, 27
local scour, defined, 27
longitudinal forces, crane loads, 18
longitudinal reinforcement ratio, defined, 79
long-period transition. See seismic ground motion, long period transition, and risk coefficient maps
long-term deflection, serviceability considerations, 395, 803
low-deformability element, defined, 78

low-rise buildings
  defined, 245
  wind loads, components and cladding, 333–334, 334e, 335f, 337f, 338f, 339f, 340f, 341f, 342f, 343f, 344f, 345f, 346f, 347f, 348f, 349f, 350f, 786
  wind loads, components and cladding, simplified, 334, 334e, 350t, 351–362f
  see also enclosed and partially enclosed low-rise buildings
low-sloped roofs, snow loads and, 52–53, 495

machinery, impact loads, 16
main wind force resisting system (MWFRS)
  buildings exempted from torsional wind load cases, 397–399, 398f, 399f
  defined, 245, 247, 733
  procedure to determine, 245
  wind tunnel procedure, 793–794
main wind force resisting system (MWFRS) directional procedure, wind loads on building appurtenances and other structures, 321–333, 775–780
  design wind loads, other structures, 322, 322e, 325f, 326f, 327, 328f, 329f, 330–331f, 331, 776–778, 777f
  design wind loads, solid freestanding walls and solid signs, 322, 322e, 323–324f, 775–776
  general requirements, 322
  scope, 321, 321t
main wind force resisting system (MWFRS), directional procedure, wind loads on buildings, 273–310, 767–774
  enclosed, partially enclosed, and open buildings of all heights, 273–274, 273t, 274e, 275–276f, 277f, 278f, 279f, 280f, 281f, 282f, 283f, 288–290t, 767–769, 768f, 773
  enclosed simple diaphragm buildings with height <160 ft, 273, 284–285, 284t, 285f, 286f, 291–310t, 769–770
main wind force resisting system (MWFRS), envelope procedure, wind loads on buildings, 311–320, 767, 771–774
main wind force resisting system (MWFRS), with pitched roofs, 773
  enclosed and partially enclosed low-rise buildings, 311, 311e, 311t, 312–314f, 771–773, 772f, 773f
  enclosed simple diaphragm low-rise buildings, 315, 315e, 315t, 316–320f, 773–774
  scope, 311, 771
mapped risk coefficient spectral response periods. See seismic ground motion, long period transition, and risk coefficient maps
marquee, 433
masonry
  anchorages in, 143, 144, 150
  seismic design and detailing requirements, 142–144, 628
materials and constructions, weight of, 11, 425, 426–427t, 428–430t
material-specific seismic design and detailing requirements, 135–144, 619–629
  composite steel and concrete, 142, 627–628
  concrete, 136–142, 137e, 138–139e, 139f, 141f, 142f, 620–627, 622f, 623f, 624f, 625f, 626f, 627f
  masonry, 142–144, 628
  steel, 135–136, 619–620
  wood, 144, 628
maximum considered earthquake, geometric mean peak ground acceleration, 41, 79, 206–207, 220–221f, 222–223f, 224–225f, 537, 714, 723
  defined, 79
maximum considered tsunami, defined, 27, 456
maximum displacement, defined, 167

mean recurrence intervals of load effects, wind tunnel procedure, 389, 794
mean roof height, defined, 247
means of egress, use of term, 593
mechanical and electrical components, seismic design
  nonbuilding structures, 146
  nonstructural components, 129–134, 129t, 593, 596, 596f, 611–617, 613f, 614f
  premanufactured, 121, 596, 596f
mechanically anchored tanks or vessels, defined, 80
minimum snow load, defined, 51
modeling criteria, seismic design, 100–101
moment frames
  defined, 79
  seismic design, 94–96
momentum flux, defined, 27
monoslope roofs, wind loads
  components and cladding, 347f, 348f
  MWFRS directional procedure, 274, 274e, 279f, 280f, 281f, 282f, 768
Monte Carlo analysis, 407, 460–461, 469, 735
multiple folded plate roofs, snow loads, 54, 495, 497
multispan gable roofs, wind loads, components and cladding, 346f

nearshore profile, defined, 27
nearshore tsunami amplitude, defined, 27
nominal loads, defined, 1
nominal strength, defined, 1, 80
nonbuilding critical facility structure, defined, 27
nonbuilding structures
  defined, 27, 80
  wind loads, components and cladding, 382, 382e, 382t, 383f, 384–385f, 385, 386f, 387t, 788–789, 789f
  see also seismic design requirements, nonbuilding structures
noncritical actions, defined, 79
nonhurricane wind speeds, 734–735, 735e
nonlinear response history analysis, 163–166, 657–671
  damping systems, 186–187, 697–698, 697t
  global acceptance criteria, 165–166, 165e, 165t, 166t, 664–671, 665f, 665t, 666t, 667e, 668f, 668t, 669f, 670t
  ground motions, 163–164, 658–662, 658f, 661f
  linear analysis, 163, 657
  modeling and analysis, 164–165, 662–664, 664f
  soil-structure interaction, 199
  vertical response analysis, 163, 658
nonspecified load combinations, 8, 419–420, 419e, 420e
nonstructural components and systems
  defined, 1, 27, 78
  see also seismic design requirements, nonstructural components

occupancy, defined, 1
occupant protections, tornadoes and, 756–759
offshore tsunami amplitude, 30–32, 35–36, 449, 456, 458–460, 465, 465f, 466
  defined, 28
one- and two-story buildings, exempted from torsional load cases, 397
one-way slabs, live loads, 17, 437
open buildings
  defined, 245
  wind loads, components and cladding, 375, 375e, 376f, 377f, 378f, 788
open structure, defined, 28

open-frame equipment structures, snow loads, 63–64, 63f, 64f, 501
openings, defined, 247
open-top tank, defined, 80
ordinary action, defined, 79
ordinary precast structural wall, defined, 136
orthogonal, defined, 80
other structures, defined, 1
out-of-plane bending, seismic design requirements, 126–127, 607
owner, defined, 80

parapets
   snow load, 61, 61f, 498–500
   wind loads, components and cladding, 364, 374f, 375, 375e, 379f, 379t, 788
   wind loads, MWFRS directional procedure, 274, 274e, 284–285, 286f, 331–332, 769, 778–779, 779f
   wind loads, MWFRS envelope procedure, 314, 314e
partial loading, 13, 435
   atmospheric icing, 76, 522
   snow loads, 54, 57, 61f, 495–496
partially enclosed building, defined, 245
partially open building, defined, 246
partitions
   defined, 80
   live loads and, 13, 435
   seismic design requirements, 128, 610
passenger vehicles
   garage loads, 437
   live loads, 17
P-Delta effect, 104, 104e, 105, 150, 159, 165, 565–566, 569–570, 663
   defined, 1, 80
peaks-over-thresholds (POT) model, 520, 735
performance-based procedures
   defined, 1
   earthquakes, 407–408
   floods, 406
   functionality requirements, 409–410
   strength and stiffness requirements, 2, 405–409, 407f, 407t
   see also fire effects, performance-based design procedures for
petrochemical tanks, seismic design requirements, 158, 646
piers and wharves, seismic design requirements, 152, 639–640
pile cap, defined, 80
pile scour, defined, 28
piles
   composite steel and concrete structures, 142
   concrete, 137–139, 137e, 138–139e, 621–622
   defined, 80
pipe racks, seismic design requirements, 150, 150e, 638, 638t
pipes and cable trays, snow loads, 64, 64f
piping and tubing systems, 132, 614–615, 644
pitched roofs
   wind loads, MWFRS directional procedure, 274, 274e, 279f, 280f, 281f, 282f, 768
   wind loads, MWFRS envelope procedure, 314–315, 314e
plain concrete, defined, 78
plunging scour, defined, 28
ponding, defined, 51, 65
ponding instability and ponding load
   defined, 51, 65
   rain loads, 65, 512, 512e, 512f
   snow loads, 62, 489, 501
pore pressure softening, defined, 28
power failures, risk categorization, 413
power-actuated fasteners, 125, 605

precast concrete diaphragm design options, defined, 136
precast structural walls, seismic design and detailing requirements, 136, 621
premanufactured mechanical and electric components
   defined, 80
   seismic design requirements, 121, 596, 596f
pressure model (PM), wind tunnel test, 793
primary drainage system, defined, 65
primary members, defined, 65
primary structural component, defined, 28
probabilistic seismic hazard analysis (PSHA), tsunami inundation depth and flow velocity based on, 32–38, 32e, 33f, 34f, 35f, 35t, 37f, 38f, 462, 465–467, 465f, 467f
projections and parapets. See parapets

quality assurance, seismic design criteria, 526

raceway, defined, 614
radiation damping
   defined, 197
   soil-structure interaction, 199–201, 199e, 200e, 200t, 201e, 705–706, 706f
rain loads, 65, 65e, 507–513
   controlled drainage, 65, 512–513, 513e, 513f
   design rain loads, 65, 507–511, 508e, 508f, 509e, 509t, 510f, 510t, 511t
   ponding instability and ponding load, 65, 512, 512e, 512f
   roof drainage, 65, 507
rain-on snow surcharge, 62, 500–501
   examples of establishing, 502–504, 502e, 503e, 504e
reasonable probability, use of term, 410
recognized literature, defined, 28, 247
reference sea level, defined, 28
refrigerated gas liquid storage tanks and vessels, 160–161
regional climatic data, estimation of basic wind speeds from, 249, 740–741
registered design professional, defined, 80
reinforced concrete, defined, 78
relative sea level change, defined, 28
reliability analysis, 406–409, 422, 467
required live loads, 13
required strength, defined, 80
resistance factor, defined, 1
ridges
   defined, 247
   topographic effects on wind speed, 266, 267f, 268, 268e
rigid building or structure, defined, 246
rigid component, defined, 78
rigid diaphragm, defined, 246
rigid high-frequency base balance model (H-FBBM), wind tunnel test, 793
rigid pressure model (PM), wind tunnel test, 793
rime, defined, 69
rime ice, 519f
risk categorization, building classification and, 84, 412–414, 413f
risk category
   defined, 1
   see also wind hazard maps
risk-targeted maximum considered earthquake (MCER)
   damping systems, 184–186
   definitions, 79–80
   ground motion hazard analysis, 205–206
   ground motion parameters, 209, 210–211f, 212–213f, 214f, 215f, 216, 217f, 218f, 219f, 717–722, 719t, 720t, 721t, 722–723, 722t

response spectrum, 85, 171–172
seismic ground motion values, 84, 84e, 84t
see also seismic ground motion, long period transition, and risk coefficient maps
Ronan point disaster, 411
roof live loads
  defined, 13
  reduction in, 17, 17e, 437
  solar panels and, 19
roof overhangs
  components and cladding, 364, 374f, 380t, 788
  enclosed simple diaphragm buildings with Height <160 ft, MWFRS, 287f, 287t, 773
  MWFRS directional procedure, 274, 275–276f, 285, 770
  MWFRS envelope procedure, 312–314f, 314
roofs
  open buildings with monoslope, pitched, or troughed free, 274, 274e, 279f, 280f, 281f, 282f
  pressures, 774
  rain loads and drainage, 65, 507
  snow loads, 54, 58–63, 59e, 60f, 62f, 63f, 495, 497–501, 498f, 499f, 501f
  vegetative roofs, 11, 425, 434, 783
  wind loads MWFRS directional procedure, 291–310t, 322, 322e, 776
  see also solar panels, rooftop; wall and roof surfaces
rugged component, defined, 78
runup elevation
  defined, 28
  inundation depth and flow velocity, tsunamis, 31–32, 31e, 31f, 32t, 461–462, 462e, 463f, 464f
R-value
  defined, 51
  snow loads, 54

Saffir-Simpson Hurricane Wind Scale, 735–736, 735t, 736t
sawtooth roofs
  snow loads, 54, 59, 495, 497
  wind loads, 349f
scragging, defined, 167
screen enclosure, defined, 13
scupper, defined, 65
sea level, wind loads and ground elevation above, 268, 268t, 748, 748e
seating for assembly use, live loads, 18–19, 438
secondary drainage system, defined, 65
secondary member, defined, 65
secondary structural component, defined, 28
seismic design category, 85–86, 85t, 535–537
  category A, design requirements, 86, 537
  defined, 80
seismic design criteria, 77–87, 525–541
  allowable stress standards, 525
  alternative materials and methods, 77, 526
  applicability, 77, 526
  definitions, 77–81, 526–530, 527f, 528f, 529f
  earthquake "loads" nature, 525
  federal government construction, 525
  geological hazards and investigation, 86, 86e, 86t, 537–539, 537e
  importance factor and risk category, 85, 535, 535f
  quality assurance, 77, 526
  seismic design category, 80, 85–86, 85t, 535–537
  seismic ground motion values, 83–85, 84e, 84f, 84t, 85e, 530–535, 531f, 532e, 532f, 533f, 534f, 539–540, 540f

  vertical ground motion for, 87, 87e, 87t, 539–540, 540f
seismic design requirements, building structures, 89–119, 543–591
  analysis procedure selection, 100, 100t, 557–558, 558t
  diaphragm flexibility, configuration, irregularities, redundancy, 95t, 96–98, 96e, 96f, 97t, 98f, 98t, 550–552, 552f, 553f, 554, 554f, 555f
  diaphragms, chords, and collectors, 90–92t, 95t, 106–108, 106e, 106f, 107e, 107f, 108t, 571–579, 572f, 573f, 574f, 575f, 576f, 578f, 578t, 579e, 579f, 579t
  direction of loading, 99–100, 557
  drift and deformation, 109–110, 109t, 581–583
  equivalent lateral force (ELF) procedure, 101–104, 101e, 102e, 102t, 103e, 103f, 104e, 560–567, 560f, 561f, 562f, 564f, 565e, 565f, 566e, 567e
  foundation design, 110–114, 111t, 112t, 113t, 587f, 588f
  linear dynamic analysis, 104–106, 105e, 567–571, 571f
  modeling criteria, 100–101, 558–560, 559f
  seismic load effects and combinations, 98–99, 98e, 99e, 555–556
  simplified alternative criteria for simple bearing wall or frame system, 114–119, 114f, 115t, 116e, 118e, 119e, 588–589, 589f
  structural design basis, 89, 543–546, 544f, 545f
  structural system selection, 89–96, 90–92t, 95t, 546–550, 550t
  structural walls and anchorages, 108–109, 108e, 580–581, 580f, 581f
seismic design requirements, nonbuilding structures, 145–161, 631–656
  analysis procedure selection, 145, 147t, 631–634, 632f, 633f, 634f, 635f
  design, 145
  design basis, 631
  structural design requirements, 146–147, 147e, 148t, 149–150, 149e, 636–637
  structures not similar to buildings, 152–153, 640–642, 640f, 641f, 642f
  structures similar to buildings, 150–152, 638–640
  supported by other structures, 146, 634–635
  tanks and vessels, 153–161, 153–154e, 154t, 155e, 155t, 156e, 156t, 157e, 157t, 642–655, 642e, 643e, 644f, 647f, 648f, 649–652f, 653–654f
seismic design requirements, nonstructural components, 121–134, 593–618, 594f
  anchorages, 125, 603–605, 605f
  applied to nonbuilding structures, 596
  architectural components, 125–128, 126e, 126t, 127t, 128e, 593, 594f, 605–611, 608–609t, 611f
  component importance factor, 121, 595
  exemptions, 595–596
  general design requirements, 122–123, 122t, 597–600
  mechanical and electrical components, 129–134, 129t, 593, 596, 596f, 611–617, 613f, 614f
  seismic demands and, 123–125, 123e, 124e, 124f, 600–603, 600f, 601f, 602f, 603f
seismic force-resisting system, defined, 80
seismic forces, defined, 80
seismic ground motion, long period transition, and risk coefficient maps, 209–236, 717–724, 719t, 720t, 721t, 722–723, 722t
  Alaska, 214f, 215f, 222f, 227f, 232f, 235f
  American Samoa, 209, 219f, 231f, 234f
  conterminous US, 210–211f, 212–213f, 220–221f, 225–226f, 230f, 233f

ground motion web tool, 724
Guam and Northern Mariana Islands, 209, 218f, 224f, 231f, 234f
Hawaii, 216f, 222f, 228f, 232f, 235f
Puerto Rico and US Virgin Islands, 217f, 223f, 229f, 232f, 235f
updates, 723
seismic ground motion values, 83–85, 84e, 84f, 84t, 85e, 530–535, 531f, 532e, 532f, 533f, 534f, 539–540, 540f
seismic load effects and combinations, 8, 9, 98–99, 98e, 99e, 420
seismic loading, buildings controlled by, 397
seismic loading direction, 99–100
seismically isolated structures, seismic design requirements, 167–179, 673–692, 673f
    analysis procedure selection, 172, 682
    design review, 176, 689
    dynamic analysis procedures, 175–176, 176e, 687–689, 688t
    equivalent lateral force (ELF) procedure, 172–175, 173e, 174e, 682–687, 683f, 684f, 685t, 686t, 687f
    general design requirements, 168–171, 170–171e, 674–682, 674t, 677e, 677f, 677t, 678f, 678t, 680t, 681f
    seismic ground motion criteria, 171–172, 682
    testing, 176–179, 178e, 178f, 689–692
self-anchored tanks or vessels, defined, 80
self-straining forces and effects
    basic requirements, 3, 410
    load combinations for, 7–8, 9, 419, 421
service loads, defined, 1–2
serviceability considerations
    camber, 395, 811
    deflection, vibration, and drift, 395, 801–803, 804–805f, 806–807f, 808–809f, 810–811f
    performance-based procedures, 409
serviceability design, wind speeds for, 740, 740e, 740t
shear keys, material-specific seismic design and detailing requirements, 144
shear panel, defined, 80
shear wall-frame interactive system, defined, 79
shear walls (concrete), seismic design and detailing requirements, 136–137, 621
shear-controlled diaphragm, defined, 78, 530
shielding, 273, 311, 321, 333, 732
shipping, tsunami debris impact and, 478–479
shoaling, defined, 28
short hangers, bracing exemptions for, 596
sidewalks, driveways, and yards subject to trucking, live loads, 19
signs, wind loads and
    solid attached, 322, 776
    solid freestanding, 322, 322e, 324–325f, 775
silos
    components and cladding, 382, 382e, 382t, 383f, 384–385f, 385, 788–789, 789f
    MWFRS directional procedure, 322, 326f, 327, 328f, 329f, 776, 776–777
simple diaphragm building, defined, 246
site class, defined, 80
site classification procedure
    definitions, 203, 709–710
    generally, 203, 204t, 709
    parameters, 204, 204e, 710
site-specific ground motion procedures, for seismic design, 205–207, 711–715
    design acceleration parameters, 206, 713–714
    design response spectrum, 206, 206e, 713
    maximum considered earthquake, geometric mean peak ground acceleration, 206–207, 714

    maximum considered earthquake, ground motion hazard analysis, 205–206, 206f, 712–713
    site response analysis, 205, 711–712
site-specific probabilistic hazard analysis, tsunami inundation depth and flow velocity, 32–38, 32e, 33f, 34f, 35f, 35t, 37f, 38f, 462, 465–467, 465f
sliding snow, 61–62, 500, 500f, 502, 502e
slippery surface, defined, 51
sloped roofs, snow loads, 51, 54, 54e, 58f, 59f, 60f, 495, 502, 502e
snow, defined, 69
snow loads, 51–76, 489–505
    caused by freezing rain, 519, 519f
    definitions, 51
    drifts on lower roofs (aerodynamic shade), 59–61, 59e, 62f, 63f, 497–498, 498f, 499f
    on equipment and equipment platforms, 64
    examples of establishing, 501–504, 501e, 502e, 502f, 503e, 503f, 504e
    in excess of design value, 489
    on existing roofs, 62–63, 501, 501f
    on flat roofs, 51, 51e, 53e, 54, 58t, 493–495
    on ground, 51, 52–53f, 55t, 56t, 57t, 58t, 489, 490–491t, 491–493, 492t, 493t
    on open-frame equipment structures, 63–64, 63f, 64f
    partial loading, 54, 57, 61f, 495–496
    ponding instability and, 62, 501
    on projections and parapets, 61, 61f, 498–500
    rain-on snow surcharge, 62, 500–501
    sliding snow, 61–62, 500, 500f, 502, 502e
    on sloped roofs, 51, 54, 54e, 58f, 59f, 60f, 495, 502, 502e
    unbalanced loads, 57–59, 61f, 496–497
soil damping, defined, 197
soil hysteretic damping, 199–201, 199e, 200e, 200t, 201e
soil-structure interaction, 705–706
soil loads, 11t, 12
soil loads and hydrostatic pressure, 425, 430–431t, 431
soil-structure interaction (SSI), 197–202, 703–708
    ELF procedure, 198, 198e
    foundation damping effects, 199–201, 199e, 200e, 200t, 201e, 705–706, 706f
    kinematic SSI effects, 201–202, 201e, 703, 704, 707–708, 707f
    SSI adjusted structural demands, 198–199, 198e, 704–705
solar array, defined, 247
solar panels, rooftop
    defined, 247
    live loads, 19, 438
    MWFRS directional procedure, 327, 327e, 330–331, 330–331f, 331f, 777–778, 777f
    seismic design requirements, 133–134, 133e, 595, 616–617
    for slopes < 7 degrees, 389–390, 794–795
    weight of, 12, 425
    wind loads, components and cladding, 385
soliton fission, 36, 462
    defined, 28
space frame system, defined, 79
special flood hazard area, defined, 21
special inspections, defined, 79
special wind regions, 249, 740
sprinkler systems, seismic design requirements, 132–133, 616
stairs and ramps
    live loads, 19
    seismic design requirements, 128, 611, 611f
standard fire resistance design, defined, 401, 815
steel

cables, 136, 620
chimneys and stacks, 152, 640
composite steel and concrete structures, 142, 627–628
seismic design and detailing requirements, 135–136, 619–620
storage racks, 150–151, 638–639, 638f
tubular support structures, for wind turbines, 153, 641
stepped roofs, wind loads, components and cladding, 345f
stiffness. See strength and stiffness requirements
storage racks
  defined, 80
  seismic design requirements, 150–151, 638–639, 638f
story, defined, 80
story above grade plane, defined, 80, 530, 530f
story drift, defined, 80
story drift ratio, defined, 80
story shear, defined, 80
strength, types defined, 80
strength and stiffness requirements, 2–3, 2t, 3t, 405
  performance-based procedures, 2, 405–409, 407f, 407t
strength design
  defined, 2
  load combinations for, 7–8, 417–420, 418t, 419e
  procedures, 405
structural analysis, of fire effects, 402, 819–821
structural component, defined, 28
structural damping, gust effects, 753
structural design fire, 402, 818
  defined, 401
structural design procedures, tsunami loads and effects, 38–41, 38e, 39f, 39t, 40e, 41t, 467–473, 468t, 470f, 471f, 472f
structural height, defined, 80
structural integrity requirements, 3–4, 410–412
  accidents, misuse, sabotage and, 410
  collapse and, 410–411
  design alternatives, 411
  guidelines for, 411–412
  load combinations for, 9
  performance objectives, fire effects, 401, 817
structural observations, defined, 80
structural systems
  defined, 79
  seismic design requirements, 89–96, 90–92t, 95t
structural wall, defined, 28, 81
structure, defined, 80
subdiaphragm, defined, 80
supports, defined, 80, 526–527, 527f
surface roughness, 58t, 266, 268, 493, 741–744, 741f, 745f
surge, defined, 28
susceptible bay, defined, 65
sustained flow scour, defined, 28

tall buildings, aerodynamic wind loads on, 754
tanks
  with height <120ft, wind loads, components and cladding, 382, 382e, 382t, 383f
  wind loads, MWFRS directional procedure, 322, 326f, 327, 327e, 328f, 329f, 776, 776–777
tanks and vessels, seismic design requirements, 151–152, 153, 639, 639f, 642
  anchorages, 644, 644f
  boilers and pressure vessels, 648
  design basis, 153–154, 153–154e, 642–643, 642e, 643e
  elevated, for liquids and granular materials, 631f, 647–648, 647f, 648f
  flexibility of piping attachment, 154–155, 155e, 155t, 644

ground-supported tanks for granular materials, 158–159, 647, 647e
ground-supported tanks for liquids, 155–156e, 155–157, 156t, 157e, 157t, 644–646
horizontal, saddle-supported vessels for liquid and vapor, 648, 655
petrochemical tanks, 158, 646
strength and ductility, 154, 643–644
structural towers for, 151–152
water storage and treatment tanks, 158, 646
telecommunication towers
  seismic design requirements, 153, 641
  tornadoes and, 763–764
telecommunications failures, risk categorization, 413
temporary facilities, defined, 2
testing agency, defined, 81
thermal analysis, of fire effects, 402, 817–819
thermal boundary condition, defined, 401
thermal factor, snow loads, 52, 54, 58t, 63, 494
thermal insulation, defined, 401
thermal response, defined, 401
thermal restraint, defined, 401
topographic effects, wind loads general requirements, 266, 267f, 268, 268e, 744–748
topographic transect, defined, 28
tornado limitation, wind loads general requirements, 270, 271, 756–764, 757t, 758f, 759f, 759t, 760e, 761t, 762t, 763t, 764t
torsional wind load cases, buildings exempted from, 397–399, 398f, 399f, 804f, 816
torsionally regular building under wind load, defined, 246
total maximum displacement, defined, 167
toxic, highly toxic, and explosive substances, building classification and, 5, 414–415
toxic substances, defined, 2
transfer forces, diaphragm, defined, 78, 530
traveling fire, 818
troughed free roofs, wind loads MWFRS directional procedure, 274, 274e, 279f, 280f, 281f, 282f
truck and bus garages, live loads, 18
trussed communication towers, tornadoes and, 763–764
tsunami, defined, 28
tsunami amplitude, defined, 28
tsunami bore, defined, 28
tsunami bore height, defined, 28
tsunami breakaway wall, defined, 28
tsunami design zone, defined, 28
tsunami design zone map, defined, 28
tsunami evacuation map, defined, 28
tsunami loads and effects, 25–50, 447–488
  debris impact loads, 44–46, 44e, 44t, 45f, 46e, 46t, 477–480, 477t, 478t, 479e
  definitions, 25–29, 456
  foundation design, 46–49, 47e, 47f, 47t, 48f, 480–484, 481e, 481f, 482e, 482f, 483f
  hydrodynamic loads, 42–44, 42e, 42t, 43e, 473–477, 473e, 474e, 475f, 476f, 477f
  hydrostatic loads, 41–42, 41e, 42e, 473
  inundation depth and flow velocity, analysis of design, 30–31, 30e, 458–461, 459t, 460e, 460f
  inundation depth and flow velocity, based on runup, 31–32, 31e, 31f, 32t, 461–462, 462e, 463f, 464f
  inundation depth and flow velocity, based on site-specific probabilistic hazard analysis, 31e, 32–38, 32e, 33f, 34f, 35f, 35t, 37f, 38f, 462, 465–467, 465f

nonstructural components and systems, 50, 485
risk categories, 30, 457–458
scope and exceptions, 25, 26f, 27f, 28f, 447–449, 456, 457f
structural countermeasures for, 49, 484
structural design procedures for, 28, 38–41, 38e, 39f, 39t, 40e, 41t, 467–473, 468t, 470f, 471f, 472f
vertical evacuation refuge strategies and structures, 49, 49f, 449, 458, 468, 469, 472, 484–485
Western state hazard exposure, 448t, 449f, 450f, 451f, 452f, 453f, 454f, 455f, 484–485
tsunami risk category, defined, 28
tsunami-prone region, defined, 28
tubular support structures, wind turbines, 641

ultimate deformation, defined, 78
unbalanced roof loads, 57–59, 61f, 496–497, 502, 502e
uniform live loads, reduction in, 16–17, 16e, 436–437, 436f
uniformly distributed loads, 13, 433–435, 434t
utility lines, seismic design requirements, 133, 616

vegetative and landscaped roofs, 11, 425, 434, 738
vehicle barrier systems
  defined, 13
  live loads and, 16, 435
velocity pressure, wind loads general requirements, 268–269, 268e, 268t, 748–750, 748e, 749e, 749f, 750e, 750f, 750t
velocity-dependent damping device, defined, 181
veneers, defined, 81
ventilated roof, defined, 51
vertical deflections, serviceability considerations, 208f, 395, 801–802
vertical evacuation refuge strategies, tsunamis and, 49, 49f
vertical front area, defined, 743
vertical ground motions, 87, 87e, 87t, 145–146, 539–540, 540f, 634
vertical response analysis, 163, 658
vertical response spectrum, seismic design criteria, 86
vibrations, serviceability considerations, 395, 802–803, 803e

wall and roof surfaces, wind loads MWFRS directional procedure, 284, 769
walls
  cantilevered ground-supported, 153
  components and cladding, wind loads, 335f
  drift of, serviceability considerations, 395, 802, 802e, 804–805f, 806–807f, 808–809f, 810–811f
  solid freestanding, wind loads, 322, 322e, 324–325f, 775
  types defined, 81
  wave loads on vertical, 441, 442t
  wind loads MWFRS directional procedure, 287f, 291–292t
  see also bearing walls; wall and roof surfaces
warm roof slope factor, snow loads and, 54
water storage and treatment tanks, seismic design requirements, 158, 646
water-and sewage-treatment facilities, failure of, 413
wave loads, during flooding, 22–23, 22e, 22t, 23e, 23f, 441, 442t
wheel loads, cranes and, 17–18
wind, generally
  on ice-covered structures, 71, 72–73f, 74f, 75f, 521–522, 522t
  load combinations for, 420–421
  risk categorization of buildings, 5t
wind directionality
  wind loads, general requirements, 266, 266t, 741

wind tunnel procedure, 389, 794
wind hazard maps, 740
  basic wind speed, 249, 734
  component and cladding ratings for building envelope products, 740
  design wind speeds ASCE 7-93 to 7-10, 740t
  estimation of basic wind speeds from regional climatic data, 249, 740–741
  hurricane wind speeds, 735, 735e
  nonhurricane wind speeds, 734–735, 735e
  return periods, 734
  Risk Category I, basic wind speeds, 250–251f, 258–259f
  Risk Category II, basic wind speeds, 252–253f, 260–261f, 737t
  Risk Category III, basic wind speeds, 254–255f, 262–263f, 738t
  Risk Category IV, basic wind speeds, 256–257f, 264–265f, 739t
  Saffir-Simpson Hurricane Wind Scale, 735–736, 735t, 736t, 740
  special wind regions, 249, 740
  wind speeds, for serviceability design, 740, 740e, 740t
  wind speeds, selected coastal locations, 736t
wind loads, components and cladding, 245, 333–387, 781–791
  building appurtenances and rooftop structures and equipment, 375–382, 375e, 379f, 379t, 380e, 380f, 380t, 381f, 382t
  building types, 783–786, 783t, 784t, 785t, 786f, 787f
  buildings with height <60ft, 364, 364e, 364t, 365–373t, 375, 783t, 787–788
  buildings with height >60ft, 350, 350e, 362–363t, 363, 786–787
  low-rise buildings, 333–334, 334e, 335f, 337f, 338f, 339f, 340f, 341f, 342f, 343f, 344f, 345f, 346f, 347f, 348f, 349f, 350f, 786
  low-rise buildings, simplified, 334, 334e, 350t, 351–362f, 787–788
  nonbuilding structures, 382, 382e, 382t, 383f, 384–385f, 385, 386f, 387f, 788–789, 789f
  open buildings, 375, 375e, 376f, 377f, 378f, 788
  rooftop solar panels, 385
  rooftop structures and, 380
wind loads, general requirements, 245–271, 731–766
  alternate procedure to calculate, 753
  critical load condition, 249
  definitions, 245–247, 732–733
  enclosure classification, 270–271, 754–756
  exposure, 266, 741–744, 741f, 742t, 743e, 744f, 745f, 746f, 747f, 748f
  ground elevation above sea level, 268, 268t, 748, 748e
  gust effects, 269–270, 269e, 269t, 270e, 750–754, 751e, 751t, 752e, 752f, 752t, 753e, 754e, 755t
  internal pressure coefficients, 271, 271e, 271t, 756
  limitations, 731–732
  procedures, 245, 246f, 731
  sign convention, 249
  topographic effects, 266, 267f, 268, 268e, 744–748
  tornado limitation, 271, 756–764, 757t, 758f, 759f, 759t, 760e, 761t, 762t, 763t, 764t
  velocity pressure, 268–269, 268e, 268t, 748–750, 748e, 749e, 749f, 750e, 750f, 750t
  wind directionality, 266, 266t, 741
  wind pressures acting on opposite faces of each building surface, 249, 734
  see also wind hazard maps

wind loads on building appurtenances and other structures, MWFRS directional procedure, 321–333, 775–780
    design wind loads, other structures, 322, 322e, 325f, 326f, 327, 328f, 329f, 330–331f, 331, 776–778, 777f
    design wind loads, solid freestanding walls and solid signs, 322, 322e, 323–324f, 775–776
wind loads on buildings, MWFRS directional procedure, 273–310, 767–774
    enclosed, partially enclosed, and open buildings of all heights, 273–274, 273t, 274e, 275–276f, 277f, 278f, 279f, 280f, 281f, 282f, 283f, 767–769, 768f
    enclosed simple diaphragm buildings with height <160 ft, 273, 284–285, 284t, 285f, 286f, 291–310t, 769–770
wind loads on buildings, MWFRS envelope procedure, 311–320, 767, 771–774
    enclosed and partially enclosed low-rise buildings, 311, 311e, 311t, 312–314f, 771–773, 772f, 773f
    enclosed simple diaphragm low-rise buildings, 315, 315e, 315t, 316–320f, 773–774
wind pressure calculations, tornadoes, 760–763, 760e, 761e, 762t, 763t
wind tunnel procedure, 389–390, 781, 793–795
    defined, 247
    design data and, 793–794
    dynamic response, 389
    load effects, 389, 794
    minimum reductions permitted, 793–794
    roof-mounted solar collectors for slopes <7 degrees, 389–390, 794–795
    test conditions, 389
    test types, 793
    wind-borne debris, 389
wind turbine generators, seismic design requirements, 153, 641
wind-borne debris
    enclosure classification, 755–756
    wind tunnel procedure, 389
wind-borne debris regions, 270, 733, 755
    defined, 247, 733
wind-restraint system, defined, 167
windstorms, tornadoes versus, 757–758, 759f
windward snow drifts, 61, 497–498, 498f
wood, seismic design and detailing requirements, 144, 628
wood structure panel, defined, 81

Standard 7-16
Minimum Design Loads and Associated Criteria for Buildings and Other Structures

SUPPLEMENT 1
PROVISIONS

Effective: February 1, 2019

This document contains CHANGES to the above title, which are also posted on the ASCE Library at https://doi.org/10.1061/9780784414248

THIS TYPE AND SIZE FONT INDICATES DIRECTIVE TEXT THAT IS NOT PART OF THE TITLE. CHANGES ARE INDICATED USING HIGHLIGHTED, STRIKE-OUT, AND UNDERLINED TEXT. A HORIZONTAL RULE INDICATES A BREAK BETWEEN CHAPTERS.

Chapter 6

SECTION 6.6.1 AS FOLLOWS:

6.6.1 Maximum Inundation Depth and Flow Velocities Based on Runup.

The maximum inundation depths and flow velocities associated with the stages of tsunami flooding shall be determined in accordance with Section 6.6.2. Calculated flow velocity shall not be taken as less than 10 ft/s (3.0 m/s) and need not be taken as greater than the lesser of $1.5(gh_{max})^{1/2}$ and 50 ft/s (15.2 m/s).

Where the maximum topographic elevation along the topographic transect between the shoreline and the inundation limit is greater than the runup elevation, one of the following methods shall be used:

1. The site-specific procedure of Section 6.7.6 shall be used to determine inundation depth and flow velocities at the site, subject to the above range of calculated velocities.
2. For determination of the inundation depth and flow velocity at the site, the procedure of Section 6.6.2, Energy Grade Line Analysis, shall be used, assuming a runup elevation and horizontal inundation limit that has at least 100% of the maximum topographic elevation along the topographic transect.
3. Where the site lies within a completely overwashed area for which Inundation Depth Points are provided in the ASCE Tsunami Design Geodatabase, the inundation elevation profiles shall be determined using the Energy Grade Line Analysis with the following modifications:
   a. The Energy Grade Line Analysis shall be initiated from the inland edge of the overwashed land with an inundation elevation equal to the maximum topographic elevation of the overwashed portion of the transect.
   b. The Froude number shall be 1 at the inland edge of the overwashed land and shall vary linearly with distance to match the value of the Froude number determined at the shoreline per the coefficient α.
   c. The Energy Grade Line Analysis flow elevation profile shall be uniformly adjusted with a vertical offset such that the computed inundation depth at the Inundation Depth Point is at least the depth specified by the ASCE Tsunami Design Geodatabase, but the flow elevation profile shall not be adjusted lower than the topographic elevations of the overwashed land transect.

TABLE 6.10-1 AS FOLLOWS:

Table 6.10-1 Drag Coefficients for Rectilinear Structures

Width to Inundation Depth[a] Ratio B/h_{sx}	Drag Coefficient C_d
<12	1.25
16	1.3
26	1.4
36	1.5
60	1.75
100	1.8
≥ 120	2.0

a Inundation depth for each of the three Load Cases of inundation specified in Section 6.8.3.1. Interpolation shall be used for intermediate values of width to inundation depth ratio B/h_{sx}.

Where building setbacks occur, drag coefficients shall be determined for each portion of a constant width. For each portion along the inundated height of the building, its equivalent inundated depth is taken as its submerged vertical dimension.

SECTION 6.12.4.1 AS FOLLOWS:

6.12.4.1 Fill.

Fill used for structural support and protection shall be placed in accordance with ASCE 24(2005), Sections 1.5.4 and 2.4.1. Structural fill shall be designed to be stable during inundation and to resist the loads and effects specified in Section 6.12.2.

SECTION 6.17 AS FOLLOWS:

6.17 Consensus Standards and Other Referenced Documents

ASCE/SEI 24-1405, *Flood Resistant Design and Construction*, American Society of Civil Engineers, 20152005.

Cited in: Section 6.12.4.1

Chapter 11

TABLES 11.4-1 and 11.4-2 AS FOLLOWS:

Table 11.4-1 Short-Period Site Coefficient, F_a

Site Class	Mapped Risk-Targeted Maximum Considered Earthquake (MCE$_R$) Spectral Response Acceleration Parameter at Short Period					
	$S_S \leq 0.25$	$S_S = 0.5$	$S_S = 0.75$	$S_S = 1.0$	$S_S = 1.25$	$S_S \geq 1.5$
A	0.8	0.8	0.8	0.8	0.8	0.8
B	0.9	0.9	0.9	0.9	0.9	0.9
C	1.3	1.3	1.2	1.2	1.2	1.2
D	1.6	1.4	1.2	1.1	1.0	1.0
E	2.4	1.7	1.3	See Section 11.4.8	See Section 11.4.8	See Section 11.4.8
F	See Section 11.4.8	See Section 11.4.8	See Section 11.4.8	See Section 11.4.8	See Section 11.4.8	See Section 11.4.8

Note: Use straight-line linear interpolation for intermediate values of S_s.

Table 11.4-2 Long-Period Site Coefficient, F_v

Site Class	Mapped Risk-Targeted Maximum Considered Earthquake (MCE$_R$) Spectral Response Acceleration Parameter at 1-s Period					
	$S_1 \leq 0.1$	$S_1 = 0.2$	$S_1 = 0.3$	$S_1 = 0.5$	$S_1 = 0.5$	$S_1 \geq 0.6$
A	0.8	0.8	0.8	0.8	0.8	0.8
B	0.8	0.8	0.8	0.8	0.8	0.8

C	1.5	1.5	1.5	1.5	1.5	1.4
D	2.4	2.2^a	2.0^a	1.9^a	1.8^a	1.7^a
E	4.2	3.3^a ~~See Section 11.4.8~~	2.8^a ~~See Section 11.4.8~~	2.4^a ~~See Section 11.4.8~~	2.2^a ~~See Section 11.4.8~~	2.0^a ~~See Section 11.4.8~~
F	See Section 11.4.8	See Section 11.4.8	See Section 11.4.8	See Section 11.4.8	See Section 11.4.8	See Section 11.4.8

Note: Use ~~straight-line~~ linear interpolation for intermediate values of S_1.

^{a}~~Also s~~See requirements for site-specific ground motions in Section 11.4.8. These values of F_v shall be used only for calculation of T_S.

Chapter 12

SECTION 12.11.2.1 AS FOLLOWS:

12.11.2.1 Wall Anchorage Forces

Where the anchorage is not located at the roof and all diaphragms are not flexible, the value from Eq. (12.11-1) is permitted to be multiplied by the factor $(1+2z/h)/3$, where z is the height of the anchor above the base of the structure and h is the height of the roof above the base; however, F_p shall not be less than required by Section ~~12.11.2~~ 12.11.1 with a minimum anchorage force of $F_p = 0.2W_p$.

SECTION 12.13.9.2 AS FOLLOWS:

12.13.9.2 Shallow Foundations

12.13.9.2.1.1 Foundation Ties. Individual footings shall be interconnected by ties in accordance with Section 12.13.8.2 and the additional requirements of this section. ~~The ties shall be designed to accommodate the differential settlements between adjacent footings per Section 12.13.9.2, item b.~~ Reinforced concrete sections shall be detailed in accordance with Sections 18.6.2.1 and 18.6.4 of ACI 318.

SECTION 12.13.9.3 AS FOLLOWS:

12.13.9.3 Deep Foundations

12.13.9.3.1 Downdrag Design of piles shall incorporate the effects of downdrag caused by liquefaction. For geotechnical design, the liquefaction-induced downdrag shall be determined as the downward skin friction on the pile within and above the liquefied zone(s). The net geotechnical ultimate capacity of the pile shall be the ultimate geotechnical capacity of the pile below the liquefiable layer(s) reduced by the downdrag load. For structural design, downdrag load induced by liquefaction shall be treated <u>and factored as a seismic load, although it need not be considered concurrently with axial loads resulting from inertial response of the structure, determined according to Section 12.4.</u><s>and factored accordingly.</s>

Chapter 15

SECTION 15.5.3.1 AS FOLLOWS:

15.5.3.1 Steel Storage Racks.

Steel storage racks supported at or below grade shall be designed in accordance with ANSI/RMI MH 16.1, <s>and</s> its force and displacement requirements, <u>and the seismic design ground motion values determined according to Section 11.4</u>, except as follows:

Chapter 21

SECTION 21.2.2 AS FOLLOWS:

21.2.2 Deterministic (MCE_R) Ground Motions.

The deterministic spectral response acceleration at each period shall be calculated as an 84th-percentile 5% damped spectral response acceleration in the direction of maximum horizontal response computed at that period. The largest such acceleration calculated for the characteristic earthquakes on all known active faults within the region shall be used. <u>If the largest spectral response acceleration of the resulting deterministic ground motion response spectrum is less than $1.5F_a$, then this response spectrum shall be scaled by a single factor such that the maximum response spectral acceleration equals $1.5F_a$. For Site Classes A, B, C and D, F_a shall be determined using Table 11.4.1, with the value of S_s taken as 1.5; for Site Class E, F_a shall be taken as 1.0.</u> <s>The ordinates of the deterministic ground motion response spectrum shall not be taken as lower than the corresponding ordinates of the response spectrum determined in accordance with Fig. 21.2-1. For the purposes of calculating the ordinates,</s>

<u>**EXCEPTION:** The deterministic ground motion response spectrum need not be calculated when the largest spectral response acceleration of the probabilistic ground motion response spectrum of 21.2.1 is less than $1.2F_a$.</u>

<s>

(i) for Site Classes A, B or C: F_a and F_v shall be determined using Tables 11.4-1 and 11.4-2, with the value of S_S taken as 1.5 and the value of S_1 taken as 0.6;

(ii) for Site Class D: F_a shall be taken as 1.0, and F_v shall be taken as 2.5; and

(iii) for Site Classes E and F: F_a shall be taken as 1.0, and F_v shall be taken as 4.0.

</s>

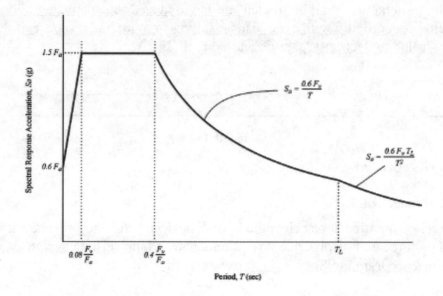

FIGURE 21.2-1 Deterministic Lower Limit on MCE$_R$ Response Spectrum

SECTION 21.2.3 AS FOLLOWS:

21.2.3 Site-Specific MCE$_R$.

The site-specific MCE$_R$ spectral response acceleration at any period, S_{aM}, shall be taken as the lesser of the spectral response accelerations from the probabilistic ground motions of Section 21.2.1 and the deterministic ground motions of Section 21.2.2.

EXCEPTION: The site-specific MCE$_R$ ground motion response spectrum shall be taken as the probabilistic ground motion response spectrum of 21.2.1 when the largest spectral response acceleration of the probabilistic ground motion response spectrum of 21.2.1 is less than $1.2F_a$. For Site Classes A, B, C and D, F_a shall be determined using Table 11.4.1, with the value of S_s taken as 1.5; for Site Class E, F_a shall be taken as 1.0.

The site-specific MCE$_R$ spectral response acceleration at any period shall not be taken less than 150% of the site-specific design response spectrum determined in accordance with 21.3.

SECTION 21.3 AS FOLLOWS:

21.3 DESIGN RESPONSE SPECTRUM

The design spectral response acceleration at any period shall be determined from Eq. (21.3-1):

$$S_a = \frac{2}{3} S_{aM} \qquad (21.3\text{-}1)$$

where S_{aM} is the MCE spectral response acceleration obtained from Section 21.1 or 21.2.

The design spectral response acceleration at any period shall not be taken as less than 80% of S_a determined in accordance with Section 11.4.6, where F_a and F_v are determined as follows:

(i) for Site Class A, B, and C: F_a and F_v are determined using Tables 11.4-1 and 11.4-2, respectively;

(ii) for Site Class D: F_a is determined using Table 11.4-1, and F_v is taken as 2.4 for $S_1 < 0.2$ or 2.5 for $S_1 \geq 0.2$; and

(iii) for Site Class E: F_a is determined using Table 11.4-1 for $S_S < 1.0$ or taken as 1.0 for $S_S \geq 1.0$, and F_v is taken as 4.2 for $S_1 \leq 0.1$ or 4.0 for $S_1 > 0.1$.

For sites classified as Site Class F requiring site-specific analysis in accordance with Section 11.4.~~7~~8, the design spectral response acceleration at any period shall not be less than 80% of S_a determined for Site Class E. ~~in accordance with Section 11.4.5~~.

EXCEPTION: Where a different site class can be justified using the site-specific classification procedures in accordance with Section 20.3.3, a lower limit of 80% of S_a for the justified site class shall be permitted to be used.

Standard 7-16

Minimum Design Loads and Associated Criteria for Buildings and Other Structures

SUPPLEMENT 1
COMMENTARY

Effective: February 1, 2019

This document contains CHANGES to the above title, which are also posted on the ASCE Library at https://doi.org/10.1061/9780784414248

THIS TYPE AND SIZE FONT INDICATES DIRECTIVE TEXT THAT IS NOT PART OF THE TITLE. CHANGES ARE INDICATED USING HIGHLIGHTED, STRIKE-OUT, AND UNDERLINED TEXT. A HORIZONTAL RULE INDICATES A BREAK BETWEEN CHAPTERS.

Chapter C2

SECTION C2.3.2 AS FOLLOWS:

C2.3.2 Load Combinations Including Flood Load.

The nominal flood load, F_a, is based on the 100-year flood (Section 5.1). The recommended flood load factor of 2.0 in V-Zones and Coastal A-Zones is based on a statistical analysis of flood loads associated with hydrostatic pressures, pressures caused by steady overland flow, and hydrodynamic pressures caused by waves, as specified in Section 5.4.

The flood load criteria were derived from an analysis of hurricane-generated storm tides produced along the United States East and Gulf coasts (Mehta et al. 1998), where storm tide is defined as the water level above mean sea level resulting from wind-generated storm surge added to randomly phased astronomical tides. Hurricane wind speeds and storm tides were simulated at 11 coastal sites based on historical storm climatology and on accepted wind speed and storm surge models. The resulting wind speed and storm tide data were then used to define probability distributions of wind loads and flood loads using wind and flood load equations specified in Sections 5.3 and 5.4. Load factors for these loads were then obtained using established reliability methods (Ellingwood et al. 1982; Galambos et al. 1982) and achieve approximately the same level of reliability as do combinations involving wind loads acting without floods. The relatively high flood load factor stems from the high variability in floods relative to other environmental loads. The presence of $2.0F_a$ in both combinations (4) and (6) in V-Zones and Coastal A-Zones is the result of high stochastic dependence between extreme wind and flood in hurricane-prone coastal zones. The $2.0F_a$ also applies in coastal areas subject to northeasters, extratropical storms, or coastal storms other than hurricanes, where a high correlation exists between extreme wind and flood.

Flood loads are unique in that they are initiated only after the water level exceeds the local ground elevation. As a result, the statistical characteristics of flood loads vary with ground elevation. The load factor 2.0 is based on calculations (including hydrostatic, steady flow, and wave forces) with stillwater flood depths ranging from approximately 4 to 9 ft (1.2–2.7 m) (average stillwater flood depth of approximately 6 ft (1.8 m)) and applies to a wide variety of flood conditions. For lesser flood depths, load factors exceed 2.0 because of the wide dispersion in flood loads relative to the nominal flood load. As an example, load factors appropriate to water depths slightly less than 4 ft (1.2 m) equal 2.8 (Mehta et al. 1998). However, in such circumstances, the flood load generally is small. Thus, the load factor 2.0 is based on the recognition that flood loads of most importance to structural design occur in situations where the depth of flooding is greatest.

The variability in hydrostatic loads under flood conditions is small when compared with the variability in wave loads and hydrodynamic loads from overland flooding. For coastal flood situations where overland waves are small (in the Coastal A zone and A zone), application of the load factor of 2.0 to below-grade flood-induced (hydrostatic) loads is too conservative, and the 1.6 load factor specified for H loads in Section 2.3.1 is more appropriate. Fig. C2.3-1 illustrates the flood zones and load factors for Fa and H for flood water above grade and below grade.

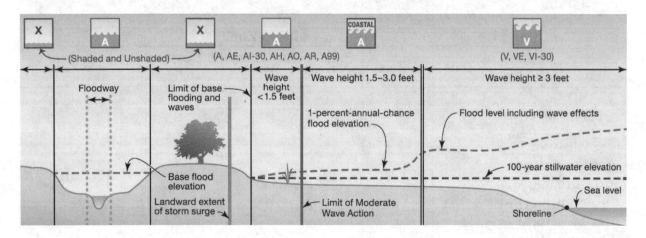

	Flood Zone					
	X Zone	A Zone	X Zone	A Zone	Coastal A Zone	V Zone
LRFD Load Factor on F_a for Structural Elements Above Grade	N/A	1.0	N/A	1.0	2.0	2.0

	Flood Zone					
	X Zone	A Zone	X Zone	A Zone	Coastal A Zone	V Zone
LRFD Load Factor on H (hydrostatic uplift and lateral pressure due to groundwater) for Structural Elements Below Grade	1.6	1.6	1.6	1.6	1.6	1.6

	Flood Zone					
	X Zone	A Zone	X Zone	A Zone	Coastal A Zone	V Zone
ASD Load Factor on F_a for Structural Elements Above Grade	N/A	0.75	N/A	0.75	1.5	1.5

	Flood Zone					
	X Zone	A Zone	X Zone	A Zone	Coastal A Zone	V Zone
ASD Load Factor on H (hydrostatic uplift and lateral pressure due to groundwater) for Structural Elements Below Grade	1.0	1.0	1.0	1.0	1.0	1.0

FIGURE C2.3-1 Illustration of flood zones and ASCE 7-16 load factors for F_a and H for (reading from right to left in figure): coastal flood zones (V Zone, Coastal A Zone, and A Zone), areas outside the 100-yr floodplain (X Zone Shaded are areas inside the 500-yr flood plain and X Zone Unshaded are areas outside the 500-yr flood plain), and riverine flood zone (A Zone).

Chapter C6

SECTION C6.6.1 AS FOLLOWS:

C6.6.1 Maximum Inundation Depth and Flow Velocities Based on Runup.

The Energy Grade Line Analysis stepwise procedure consists of the following steps:

1. Obtain the runup and inundation limit values from the Tsunami Design Zone Map generated by the ASCE Tsunami Design Geodatabase.

2. Approximate the principal topographic transect by a series of $x-z$ grid coordinates defining a series of segmented slopes, in which x is the distance inland from the shoreline to the point and z is the ground elevation of the point. The horizontal spacing of transect points should be less than 100 ft (30.5 m), and the transect elevations should be obtained from a topographic Digital Elevation Model (DEM) of at least 33-ft (10-m) resolution.

3. Compute the topographic slope, ϕ_i, of each segment as the ratio of the increments of elevation and distance from point to point in the direction of the incoming flow.

4. Obtain the Manning's coefficient, n, from Table 6.6-1 for each segment based on terrain analysis.

5. Compute the Froude number at each point on the transect using Eq. (6.6-3).

6. Start at the point of runup with a boundary condition of $E_R = 0$ at the point of runup.

    i. Per Section 6.6.1, where the maximum topographic elevation along the topographic transect between the shoreline and the inundation limit is greater than the runup elevation, use a runup elevation that has at least 100% of the maximum topographic elevation along the topographic transect.*

7. Select a nominally small value of inundation depth [~ 0.1 ft (0.03 m)] h_r at the point of runup.

8. Calculate the hydraulic friction slope, s_i, using Eq. (6.6-2).

9. Compute the hydraulic head, E_i, from Eq. (6.6-1) at successive points toward the shoreline.

10. Calculate the inundation depth, h_i, from the hydraulic head, E_i.

11. Using the definition of Froude number, determine the velocity u. Check against the minimum flow velocity required by Section 6.6.1.

12. Repeat through the transect until the h and u are calculated at the site. These are used as the maximum inundation depth, h_{max}, and the maximum velocity, u_{max}, at the site.

*Where Inundation Depth Points are provided for completely overwashed areas in the ASCE Tsunami Design Geodatabase, where the tsunami flows over an island or peninsula into a second water body, the horizontal distance of the inundation limit shall be taken to be the length of the overwashed land. There are two modifications of Section 6.6.1 given for this case relating to the initial conditions of nonzero depth and velocity at the inland edge of the overwashed area and the Froude number profile that follows a linear interpolation with distance across the overwashed land. To complete the analysis, the inundation elevation profiles are adjusted so that the computed inundation depths are at least the depths specified at the Inundation Depth Points. The adjusted inundation elevation profile should not be lower than the topographic elevation transect. An example is given in Fig. C6.6-3.

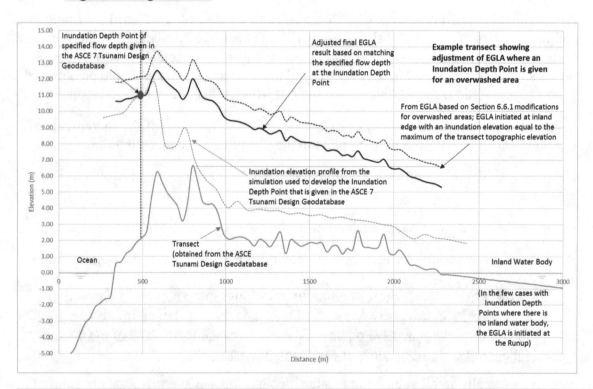

FIGURE C6.6-3 Example EGLA with adjustment where an Inundation Depth Point is specified in the ASCE Tsunami Design Geodatabase

Chapter C12

SECTION C12.13.9.3 AS FOLLOWS:

C12.13.9.3 Deep Foundations.

Pile foundations are intended to remain elastic under axial loadings, including those from gravity, seismic, and downdrag loads. Since geotechnical design is most frequently performed using allowable stress design (ASD) methods, and liquefaction-induced downdrag is assessed at an ultimate level, the requirements state that the downdrag is considered as a reduction in the ultimate capacity. Since structural design is most frequently performed using load and resistance factor design (LRFD) methods, and the downdrag is considered as a load for the pile structure to resist, the requirements clarify that the downdrag is considered as a seismic axial load, to which a factor of 1.0 would be applied for design.

Although downdrag load is to be factored as a seismic load, it is not intended to be considered concurrently with seismic loads because of inertial response of the structure. Significant excess pore pressure dissipation and settlement occurs after the cessation of shaking. This effect has been borne out in the laboratory, as documented by Wilson et al. (1997).

REFERENCES AS FOLLOWS:

REFERENCES

Wilson, D.W., R.W. Boulanger, B.L. Kutter, A. Abghari, (1997) "Aspects of dynamic centrifuge testing of soil-pile-superstructure interaction", ASCE Geotechnical Special Publication No. 64, pp.47-63.

Chapter C21

SECTION C21.2.2 AS FOLLOWS:

C21.2.2 Deterministic (MCE_R) Ground Motions.

Deterministic ground motions are to be based on characteristic earthquakes on all known active faults in a region. The magnitude of a characteristic earthquake on a given fault should be a best estimate of the maximum magnitude capable for that fault but not less than the largest magnitude that has occurred historically on the fault. The maximum magnitude should be estimated considering all seismic-geologic evidence for the fault, including fault length and paleoseismic observations. For faults characterized as having more than a single segment, the potential for rupture of multiple segments in a single earthquake should be considered in assessing the characteristic maximum magnitude for the fault.

For consistency, the same attenuation equations and ground motion variability used in the PSHA should be used in the deterministic seismic hazard analysis (DSHA). Adjustments for directivity and/or directional effects should also be made, when appropriate. In some cases, ground motion simulation methods may be appropriate for the estimation of long-period motions at sites in deep sedimentary basins or from great ($M \geq 8$) or giant ($M \geq 9$) earthquakes, for which recorded ground motion data are lacking.

When the maximum ordinate of the deterministic (MCE$_R$) ground motion response spectrum is less than 1.5F_a, it is scaled up to 1.5F_a to put a lower limit or floor on the deterministic ground motions. A single factor is used to maintain the shape of the response spectrum. The intent of the exception defining site-specific MCE$_R$ ground motions solely in terms of probabilistic MCE$_R$ ground motions (i.e., when peak MCE$_R$ response spectral accelerations are less than 1.2F_a) is to preclude unnecessary calculation of deterministic MCE$_R$ ground motions.

Values of the site coefficients (F_a and F_v) for setting the deterministic (MCE$_R$) ground motion floor are introduced to incorporate both site amplification and spectrum shape adjustment as described in the research study "Investigation of an Identified Short-Coming in the Seismic Design Procedures of ASCE 7-16 and Development of Recommended Improvements for ASCE 7-16" (Kircher 2015). This study found that the shapes of the response spectra of ground motions were not accurately represented by the shape of the design response spectrum of Figure 11.4-1 for the following site conditions and ground motion intensities: (1) Site Class D where values of $S_1 \geq 0.2$; and (2) Site Class E where values of $S_S \geq 1.0$ and/or $S_1 \geq 0.2$. An adjustment of the corresponding values of F_a and F_v was required to account for this difference in spectrum shape, which was causing the design response spectrum to underestimate long-period motions. Two options were considered to address this shortcoming. For the first option, the subject study developed values of new "spectrum shape adjustment" factors (C_a and C_v) that could be used with site factors (F_a and F_v) to develop appropriate values of design ground motions (S_{DS} and S_{D1}). The second option, ultimately adopted by ASCE 7-16, circumvents the need for these new factors by requiring site-specific analysis for Site Class D site conditions where values of $S_1 \geq 0.2$, and for Site Class E site conditions where values of $S_S \geq 1.0$ and/or $S_1 \geq 0.2$ (i.e., new requirements of Section 11.4.8 of ASCE 7-16). The spectrum shape adjustment factors developed by the subject study for Option 1 provide the basis for the values of site coefficients (F_a and F_v) proposed for Section 21.2.2 and Section 21.3 that incorporate both site amplification and adjustment for spectrum shape. Specifically, the proposed value of $F_v = 2.5$ for Site Class D is based on the product of 1.7 (Site Class D amplification at $S_1 = 0.6$, without spectrum shape adjustment) and 1.5 (spectrum shape adjustment factor); the proposed value of $F_v = 4.0$ is based on the product of 2.0 (Site Class E amplification at $S_1 = 0.6$ without spectrum shape adjustment) and 2.0 (spectrum shape adjustment factor), where values of spectrum shape adjustment are taken from Section 6.2.2 (Table 11.4-4) of the subject study. The proposed value of $F_a = 1.0$ is based on the product of 0.8 (Site Class E amplification at $S_S = 1.5$ without spectrum shape

SECTION C21.2.3 AS FOLLOWS:

C21.2.3 Site-Specific MCE_R.

Because of the deterministic lower limit on the MCE_R spectrum (Fig. 21.2-1), the site-specific MCE_R ground motion is equal to the corresponding risk-targeted probabilistic ground motion wherever it is less than the deterministic limit (e.g., $1.5g$ and $0.6g$ for 0.2 and 1.0 s, respectively, and Site Class B). Where the probabilistic ground motions are greater than the lower limits, the deterministic ground motions sometimes govern, but only if they are less than their probabilistic counterparts. On the MCE_R ground motion maps in ASCE/SEI 7-10, the deterministic ground motions govern mainly near major faults in California (like the San Andreas) and Nevada. The deterministic ground motions that govern are as small as 40% of their probabilistic counterparts.

The exception defining site-specific MCE_R ground motions solely in terms of probabilistic MCE_R ground motions (i.e., when peak MCE_R probabilistic ground motions are less than $1.2F_a$) precludes unnecessary calculation of deterministic MCE_R ground motions. Probabilistic MCE_R ground motions are presumed to govern at all periods where the peak probabilistic MCE_R response spectral acceleration (i.e., $< 1.2F_a$) is less than 80% of peak deterministic (MCE_R) response spectral acceleration (i.e., $\geq 1.5F_a$).

The requirement that the site-specific MCE_R response spectrum not be less than 150% of the site-specific design response spectrum of Section 21.3 effectively applies the 80% limits of Section 21.3 to the site-specific MCE_R response spectrum (as well as the site-specific design response spectrum).

SECTION C21.3 AS FOLLOWS:

C21.3 DESIGN RESPONSE SPECTRUM

Eighty percent of the design response spectrum determined in accordance with Section 11.4.6 was established as the lower limit to prevent the possibility of site-specific studies generating unreasonably low ground motions from potential misapplication of site-specific procedures or misinterpretation or mistakes in the quantification of the basic inputs to these procedures. Even if site-specific studies were correctly performed and resulted in ground motion response spectra less than the 80% lower limit, the uncertainty in the seismic potential and ground motion attenuation across the United States was recognized in setting this limit. Under these circumstances, the allowance of up to a 20% reduction in the design response spectrum based on site-specific studies was considered reasonable.

~~As described in Section 21.2.2, values of the site coefficients (F_a and F_v) for setting the deterministic (MCE_R) ground motion floor are introduced to incorporate both site amplification and spectrum shape adjustment.~~

Values of the site coefficients (F_a and F_v) for setting the 80% lower limit are introduced to incorporate both site amplification and spectrum shape adjustment, as described in the research study "Investigation of an Identified Short-Coming in the Seismic Design Procedures of ASCE 7-16 and Development of Recommended Improvements for ASCE 7-16" (Kircher 2015). This study found that the shapes of the response spectra of ground motions were not accurately represented by the shape of the design response spectrum of Fig. 11.4-1 for the following site conditions and ground-motion intensities: (1) Site Class D, where values of $S_1 \geq 0.2$, and (2) Site Class E, where values of $S_S \geq 1.0$ and/or $S_1 \geq 0.2$. An adjustment of the corresponding values of F_a and F_v was required to account for this difference in spectrum shape, which was causing the design response spectrum to underestimate long period motions. Two options were considered to address this shortcoming. For the first option, the subject study developed values of new "spectrum shape adjustment" factors (C_a and C_v) that could be used with site factors (F_a and F_v) to develop appropriate values of design ground motions (S_{DS} and S_{D1}). The second option, ultimately adopted by ASCE 7-16, circumvents the need for these new factors by requiring site-specific analysis for Site Class D site conditions, where values of $S_1 \geq 0.2$, and for Site Class E site conditions, where values of $S_S \geq 1.0$ and/or $S_1 \geq 0.2$ (i.e., new requirements of Section 11.4.8 of ASCE 7-16). The spectrum shape adjustment factors developed by the subject study for Option 1 provide the basis for the values of site coefficients (F_a and F_v) of Section 21.3 that incorporate both site amplification and adjustment for spectrum shape. Specifically, the value of $F_v = 2.5$ for Site Class D is based on the product of 1.7 (Site Class D amplification at $S_1 = 0.6$, without spectrum shape adjustment) and 1.5 (spectrum shape adjustment factor); the value of $F_v = 4.0$ for Site Class E is based on the product of 2.0 (Site Class E amplification at $S_1 = 0.6$ without spectrum shape adjustment) and 2.0 (spectrum shape adjustment factor), where values of spectrum shape adjustment are taken from Section 6.2.2 (Table 11.4-4) of the subject study. The value of $F_a = 1.0$ for Site Class E is based on the product of 0.8 (Site Class E amplification at $S_S = 1.5$ without spectrum shape adjustment) and 1.25 (spectrum shape adjustment factor), where the value of the spectrum shape adjustment is taken from Section 6.2.2 (Table 11.4-3) of the subject study. Site amplification adjusted for spectrum shape effects is approximately independent of ground motion intensity and, for simplicity, the proposed values of site factors adjusted for spectrum shape are assumed to be valid for all ground motion intensities.

Although the 80% lower limit is reasonable for sites not classified as Site Class F, an exception has been introduced at the end of this section to permit a site class other than E to be used in establishing this limit when a site is classified as F. This revision eliminates the possibility of an overly conservative design spectrum on sites that would normally be classified as Site Class C or D.

Minimum Design Loads and Associated Criteria for Buildings and Other Structures
ASCE/SEI 7-16

Errata 1

Effective: July 9, 2018

This document contains errata to the above title, which is posted on the ASCE Library at
https://ascelibrary.org/doi/book/10.1061/9780784414248

THIS TYPE AND SIZE FONT INDICATES DIRECTIVE TEXT THAT IS NOT PART OF THE TITLE. CHANGES ARE INDICATED USING STRIKE-OUT AND UNDERLINE TEXT. A HORIZONTAL RULE INDICATES A BREAK BETWEEN SECTIONS.

Acknowledgments

"SEISMIC TASK SUBCOMMITTEE ON NONLINEAR GENERAL PROVISIONS" AND "SEISMIC TASK COMMITTEE ON SEISMIC ISOLATION" SHOULD ACKNOWLEDGE:

"Reid F. B. Zimmerman, P.E., S.E., M.ASCE"

Chapter 4

IN TABLE 4.3-1, CORRECT THE L_o VALUES AS FOLLOWS:

Roofs

    Roof areas used for assembly purposes   100 (4.70 4.79)

    Vegetative and landscaped roofs

        Roof areas used for assembly purposes   100 (4.70 4.79)

E1

Chapter 6

THE PUBLISHED VERSION OF FIG. C6.6-2 IS ILLEGIBLE. REPLACE WITH:

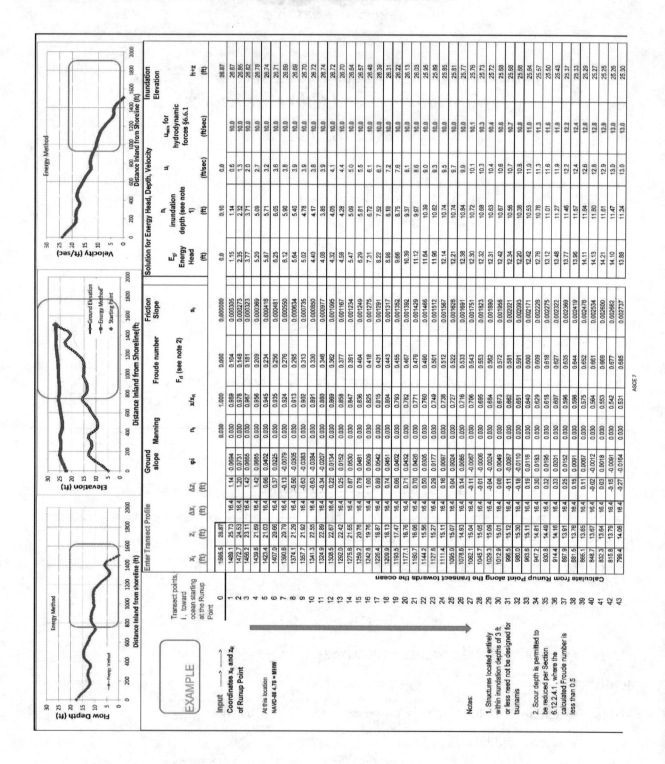

Chapter 7

IN COMMENTARY TABLE C7.2-1, CORRECT THE NEW HAMPSHIRE LOCATION AS FOLLOWS:

New Hampshire

    <u>A</u>ll cities

Chapter 12

REVISE SECTION 12.11.2.1 TO REFERENCE SECTION 12.11.1, AS FOLLOWS:

Where the anchorage is not located at the roof and all diaphragms are not flexible, the value from Eq. (12.11-1) is permitted to be multiplied by the factor $(1 + 2z/h)/3$, where z is the height of the anchor above the base of the structure and h is the height of the roof above the base; however, F_p shall not be less than required by Section ~~12.11.2~~ 12.11.1 with a minimum anchorage force of $F_p = 0.2W_p$.

Chapter 13

REVISE SECTION 13.1.4 ITEMS 5 AND 6 AS FOLLOWS:

5. Mechanical and electrical components in Seismic Design Category C provided that either

    a. The component Importance Factor, I_p, is equal to 1.0 and the component is positively attached to the structure; or

    b. The component weighs 20 lb (89 N) or less ~~or, in the case of a distributed system, 5 lb/ft (73 N/m) or less~~.

6. Discrete mechanical and electrical components in Seismic Design Categories D, E, or F that are positively attached to the structure, provided that either

    a. The component weighs 400 lb (1,779 N) or less, the center of mass is located 4 ft (1.22 m) or less above the adjacent floor level, flexible connections are provided between the component and associated ductwork, piping, and conduit, and the component Importance Factor, I_p, is equal to 1.0; or

    b. The component weighs 20 lb (89 N) or less ~~or, in the case of a distributed system, 5 lb/ft (73 N/m) or less; and~~.

MODIFY SECTION 13.3.2.1 TO DEFINE THE TERMS USED IN EQS. (13.3-7) AND (13.3-8), AS FOLLOWS:

13.3.2.1 Displacements within Structures. For two connection points on the same structure A or the same structural system, one at a height h_x and the other at a height h_y, D_p shall be determined as

$$D_p = \Delta\delta_{xA} - \Delta\delta_{yA} \tag{13.3-7}$$

<u>where</u>

> D_p = relative seismic displacement that the component must be designed to accommodate;
>
> δ_{xA} = deflection at building level x of structure A, determined in accordance with Eq. (12.8-15);
>
> δ_{yA} = deflection at building level y of structure A, determined in accordance with Eq. (12.8-15).

Alternatively, D_p is permitted to be determined using modal procedures described in Section 12.9, using the difference in story deflections calculated for each mode and then combined using appropriate modal combination procedures. D_p is not required to be taken as greater than

(13.3-8)

<u>where</u>

> Δ_{aA} = allowable story drift for structure A as defined in Table 12.12-1;
>
> h_{sx} = story height used in the definition of the allowable drift Δ_a in Table 12.12-1.

MODIFY SECTION 13.5.9.1 AS FOLLOWS:

13.5.9.1 General. Glass in glazed curtain walls, glazed storefronts, and glazed partitions shall meet the relative displacement requirement of Eq. (13.5-2):

$\Delta_{\text{fallout}} \geq 1.25 D_{pI}$ (13.5-2)

or 0.5 in. (13 mm), whichever is greater, where:

Δ_{fallout} = the relative seismic displacement (drift) at which glass fallout from the curtain wall, storefront wall, or partition occurs (Section 13.5.9.2);

D_{pI} = the relative seismic displacement that the component must be designed to accommodate (Section 13.3.2) (D_{pI} shall be applied over the height of the glass component under consideration)~~; and~~

~~I_e = the Importance Factor determined in accordance with Section 11.5.1.~~

MODIFY TABLE 13.6-1 AS FOLLOWS:

- Under "Mechanical and Electrical Components," insert a row for "Manufacturing or process conveyors (nonpersonnel)" with $a_p{}^a = 2\,½$, $R_p{}^b = 3$, and $\Omega_0{}^c = 2$.
- Under "Distribution Systems," change all three instances of "ductwork" to "duct systems".

- Under "Distribution Systems," change "Electrical conduit and cable trays" to "Electrical conduit, cable trays, and raceways"

MODIFY SECTION 13.6.6 AS FOLLOWS:

13.6.6 Distribution Systems: Duct Systems. HVACR and other duct systems shall be designed for seismic forces and seismic relative displacements as required in Section 13.3.

EXCEPTIONS: The following exceptions pertain to ~~ductwork~~ not designed to carry toxic, highly toxic, or flammable gases or not used for smoke control:

1. Design for the seismic forces and relative displacements of Section 13.3 shall not be required for duct systems with $I_p = 1.0$ where flexible connections or other assemblies are provided to accommodate the relative displacement be-tween the duct system and associated components, the duct system is positively attached to the structure, and where one of the following apply:
    a. Trapeze assemblies with 3/8-in. (10-mm) diameter rod hangers not exceeding 12 in. (305 mm) in length from the duct support point to the connection at the supporting structure are used to support duct, and the total weight supported by any single trapeze is ~~less than~~ 100 lb (445 N) ~~10 lb/ft (146 N/m)~~ or less, or
    b. Trapeze assemblies with 1/2-in. (13-mm) diameter rod hangers not exceeding 12 in. (305 mm) in length from the duct support point to the connection at the supporting structure are used to support the duct, and the total weight supported by any single trapeze is 200 lb (890 N) or less, or
    c. Trapeze assemblies with 1/2-in. (13-mm) diameter rod hangers not exceeding 24 in. (610 mm) in length from the duct support point to the connection at the supporting structure are used to support the duct, and the total weight supported by any single trapeze is 100 lb (445 N) or less, or
    d. The duct is supported by individual rod hangers 3/8in. (10 mm) or 1/2in. (13 mm) in diameter, and each hanger in the duct run is 12 in. (305 mm) or less in length from the duct support point to the connection at the supporting structure, and the total weight supported by any single rod is 50 lb (220 N) or less.
2. Design for the seismic forces and relative displacements of Section 13.3 shall not be required where provisions are made to avoid impact with other ducts or mechanical components or to protect the ducts in the event of such impact, the distribution system is positively attached to the structure;, and HVACR ducts have a cross-sectional area of less than 6ft^2 (0.557 m^2) and weigh 20 lb/ft (292 N/m) or less.

Components that are installed in line with the duct system and have an operating weight greater than 75 lb (334 N), such as fans, terminal units, heat exchangers, and humidifiers, shall be supported and laterally braced independent of the duct system, and such braces shall meet the force requirements of Section 13.3.1. Components that are installed in line with the duct system, have an operating weight of 75 lb (334 N) or less, such as small terminal units, dampers, louvers, and diffusers, and are otherwise not independently braced shall be positively attached with mechanical fasteners to the rigid duct on both sides. Piping and conduit attached to in-line equipment shall be provided with adequate flexibility to accommodate the seismic relative displacements of Section 13.3.2.

MODIFY SECTION 13.6.7.3 AS FOLLOWS:

13.6.7.3 Exceptions. Design for the seismic forces of Section 13.3 shall not be required for piping systems where flexible connections, expansion loops, or other assemblies are provided to accommodate the relative displacement between component and piping, where the piping system is positively attached to the structure, and where one of the following apply:

1. ~~Trapeze assemblies are used to support piping whereby no single pipe exceeds the limits set forth in 5a, 5b, or 5c below and the total weight of the piping supported by the trapeze assemblies is less than 10 lb/ft (146 N/m).~~

2́.1. Trapeze assemblies are supported by 3/8-in. (10-mm) diameter rod hangers not exceeding 12 in. (305 mm) in length from the pipe support point to the connection at the supporting structure, do not support piping with I_p greater than 1.0, ~~and~~ no single pipe exceeds the limits set forth in items 4̶5a, 4̶5b, or 4̶5c below, and the total weight supported by any single trapeze is 100 lb (445 N) or less, or

3̶.2. Trapeze assemblies are supported by 1/2-in. (13-mm) diameter rod hangers not exceeding 12 in. (305 mm) in length from the pipe support point to the connection at the supporting structure, do not support piping with I_p greater than 1.0, ~~and~~ no single pipe exceeds the diameter limits set forth in items 4̶5a, 4̶5b, or 4̶5c below, and the total weight supported by any single trapeze is 200 lb (890 N) or less, or

4̶.3. Trapeze assemblies are supported by 1/2-in. (13-mm) diameter rod hangers not exceeding 24 in. (610 mm) in length from the pipe support point to the connection at the supporting structure, do not support piping with I_p greater than 1.0, ~~and~~ no single pipe exceeds the diameter limits set forth in items 4̶5a, 4̶5b, or 4̶5c below, and the total weight supported by any single trapeze is 100 lb (445 N) or less, or

5̶.4. Piping that has an R_p in Table 13.6-1 of 4.5 or greater is either supported by rod hangers and provisions are made to avoid impact with other structural or nonstructural components or to protect the piping in the event of such impact, or pipes with $I_p = 1.0$ are supported by individual rod hangers 3/8 in. (10 mm) or 1/2 in. (13 mm) in diameter; where each hanger in the pipe run is 12 in. (305 mm) or less in length from the pipe support point to the connection at the supporting structure; and the total weight supported by any single hanger is 50 lb (220 N) or less. In addition, the following limitations on the size of piping shall be observed:

   a. In structures assigned to Seismic Design Category C where I_p is greater than 1.0, the nominal pipe size shall be 2 in. (50 mm) or less.

   b. In structures assigned to Seismic Design Categories D, E, or F where I_p is greater than 1.0, the nominal pipe size shall be 1 in. (25 mm) or less.

   c. In structures assigned to Seismic Design Categories D, E, or F where $I_p = 1.0$, the nominal pipe size shall be 3 in. (80 mm) or less.

6̶.5. Pneumatic tube systems supported with trapeze assemblies using 3/8-in. (10-mm) diameter rod hangers not exceeding 12 in. (305 mm) in length from the tube support point to the connection at the supporting structure and the total weight supported by any single trapeze is 100 lb (445 N) or less.

~~7.~~ 6. Pneumatic tube systems supported by individual rod hangers 3/8 in. (10 mm) or ½ in. (13 mm) in diameter, and each hanger in the run is 12 in. (305 mm) or less in length from the tube support point to the connection at the supporting structure, and the total weight supported by any single rod is 50 lb (220 N) or less.

MODIFY SECTION 13.6.8 TO REFERENCE SECTIONS 13.6.5 THROUGH 13.6.8, AS FOLLOWS:

13.6.8 Distribution Systems: Trapezes with a Combination of Systems. Trapezes that support a combination of distribution systems (electrical conduit, raceway, duct, piping, etc.) shall be designed using the most restrictive requirements for the supported distribution systems from Sections 13.6.5 through 13.6.~~8~~7 for the aggregate weight of the supported system. If any distribution system on the trapeze is not exempted, the trapeze shall be braced.

MODIFY THE FIFTH PARAGRAPH OF COMMENTARY SECTION C13.3.1 TO REFERENCE SECTIONS 13.3.1.4.1 AND 13.3.1.4.2, AS FOLLOWS:

Dynamic amplification occurs where the period of a nonstructural component closely matches that of any mode of the supporting structure, although this effect may not be significant depending on the ground motion. For most buildings, the primary mode of vibration in each direction has the most influence on the dynamic amplification for nonstructural components. For long-period structures (such as tall buildings), where the period of vibration of the fundamental mode is greater than 3.5 times T_s, higher modes of vibration may have periods that more closely match the period of nonstructural components. For this case, it is recommended that amplification be considered using such higher mode periods in lieu of the higher fundamental period. This approach may be generalized by computing floor response spectra for various levels that reflect the dynamic characteristics of the supporting structure to determine how amplification varies as a function of component period. Calculation of floor response spectra is described in Section 13.3.1.4.1. This procedure can be complex, but a simplified procedure is presented in Section 13.3.1.4.2. Consideration of nonlinear behavior greatly complicates the analysis.

Chapter 26

MODIFY THE INSET BOX TO FIG. 26.5-1D AS FOLLOWS:

Location	V (mph)	V (m/s)
Guam	~~180~~ 220	~~(80)~~ (98)
Virgin Islands	~~150~~ 180	~~(67)~~ (80)
American Samoa	~~150~~ 180	~~(67)~~ (80)
Hawaii	See Figure 26.5-2D	

EQ. (26.11-6) SHOULD READ AS FOLLOWS:

$$G = 0.925 \left(\frac{1 + 1.7 g_Q I_{\bar{z}} Q}{1 + 1.7 g_v I_{\bar{z}}} \right)$$

IN TABLE 26.11-1, MODIFY THE VALUE FOR FOR EXPOSURE B AS FOLLOWS:

~~1/70~~ 1/7.0

MODIFY THE CAPTION TO COMMENTARY FIG. C26.7-7 AS FOLLOWS:

FIGURE C26.7-7 Exposure D: A Building at the Shoreline ~~(Excluding Shorelines in Hurricane-Prone Regions)~~ with wind Flowing over Open Water for a Distance of at Least One Mile. Shorelines in Exposure D Include Inland Waterways, the Great Lakes, and Coastal Areas of California, Oregon, Washington, and Alaska

Minimum Design Loads and Associated Criteria for Buildings and Other Structures
ASCE/SEI 7-16

Errata 2
Effective: February 13, 2019

This document contains errata to the above title, which is posted on the ASCE Library at https://doi.org/10.1061/9780784414248

THIS TYPE AND SIZE FONT INDICATES DIRECTIVE TEXT THAT IS NOT PART OF THE TITLE. CHANGES ARE INDICATED USING STRIKE-OUT AND HIGHLIGHTED TEXT. A HORIZONTAL RULE INDICATES A BREAK BETWEEN SECTIONS.

Acknowledgments

"SEISMIC TASK SUBCOMMITTEE ON SEISMIC ISOLATION" AND "SEISMIC TASK COMMITTEE ON STEEL" SHOULD ACKNOWLEDGE:

"Su ~~F.~~ Hoa, Ph.D., C.Eng., Aff.M.ASCE"

Chapter 12

REVISE SECTION 12.13.5 AS FOLLOWS:

12.13.5 Strength Design for Foundation Geotechnical Capacity

E9

Where basic combinations for strength design listed in ~~Chapter 2~~ <u>Section 2.3</u> are used, combinations that include earthquake loads, E, are permitted to include reduction of foundation overturning effects defined in Section 12.13.4. The following sections shall apply for determination of the applicable nominal strengths and resistance factors at the soil–foundation interface.

REVISE SECTION 12.13.6 AS FOLLOWS:

12.13.6 Allowable Stress Design for Foundation Geotechnical Capacity

Where basic combinations for allowable stress design listed in Section ~~12.4~~ <u>2.4</u> are used for design, combinations that include earthquake loads, E, are permitted to include reduction of foundation overturning effects defined in Section 12.13.4. Allowable foundation load capacities, Q_{as}, shall be determined using allowable stresses in geotechnical materials that have been determined by geotechnical investigations required by the Authority Having Jurisdiction (AHJ).

Chapter C12

REVISE SECTION C12.12 IN THE TENTH PARAGRAPH AS FOLLOWS:

The allowable story drifts, $Ä_a$, for structures a maximum of four stories above the base are relaxed somewhat, provided that the interior walls, partitions, ceilings, and exterior wall systems have been designed to accommodate story drifts. The type of structure envisioned by footnote ~~d~~ <u>c</u> in Table 12.12-1 would be similar to a prefabricated steel structure with metal skin.

Chapter 13

REVISE SECTION 13.1.4 ITEMS 6b AS FOLLOWS:

6. Discrete mechanical and electrical components in Seismic Design Categories D, E, or F that are positively attached to the structure, provided that either

    a. The component weighs 400 lb (1,779 N) or less, the center of mass is located 4 ft (1.22 m) or less above the adjacent floor level, flexible connections are provided between the component and associated ductwork, piping, and conduit, and the component Importance Factor, I_p, is equal to 1.0; or

    <u>b.</u> The component weighs 20 lb (89 N) or less.

Chapter 26

"REVISE FIGURE 26.8-1 TOPOGRAPHIC FACTOR, K_{ZT}, CORRECT LABLE FOR K_3 MULTIPLIER IN RIGHT COLUMN AS FOLLOWS:

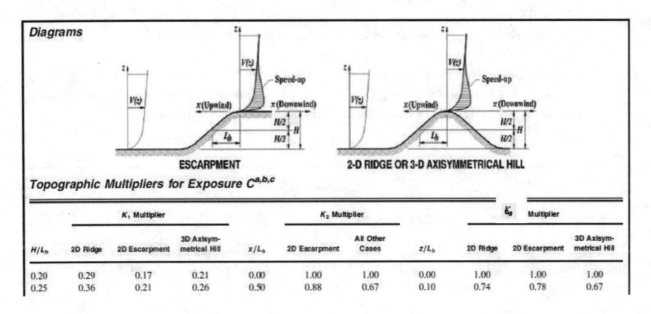

REVISE FIGURE 26.8-1 TOPOGRAPHIC FACTOR, K_{ZT}, CORRECT EQUATION FOR K3 AS FOLLOWS:

$$K_3 = e^{-\gamma z/L_h}$$

Chapter 27

REVISE FIGURE 27.3-3 MAIN WIND FORCE RESISTING SYSTEM AND COMPONENTS AND CLADDING, PART 1 (ALL HEIGHTS): EXTERNAL PRESSURE COEFFICIENTS, Cp, FOR ENCLOSED AND PARTIALLY ENCLOSED BUILDINGS AND STRUCTURES – ARCHED ROOF, CORRECT MISSING NEGATIVE SIGN ("-") FOR ALL LEWARD QUARTER COEFFICIENTS AND ADD MISSING "D" IN NOTE 4, AS FOLLOWS:

External Pressure Coefficient, C_p

Conditions	Rise-to-Span Ratio, r	C_p Windward Quarter	C_p Center Half	C_p Leeward Quarter
Roof on elevated structure	$0 < r < 0.2$	-0.9	$-0.7 - r$	-0.5
	$0.2 \leq r < 0.3^a$	$1.5r - 0.3$	$-0.7 - r$	0.5
	$0.3 \leq r \leq 0.6$	$2.75r - 0.7$	$-0.7 - r$	0.5
Roof springing from ground level	$0 < r \leq 0.6$	$1.4r$	$-0.7 - r$	0.5

^{a}When the rise-to-span ratio is $0.2 \leq r \leq 0.3$, alternate coefficients given by $6r - 2.1$ shall also be used for the windward quarter.

Notes

1. Values listed are for the determination of average loads on main wind-force resisting systems.
2. Plus and minus signs signify pressures acting toward and away from the surfaces, respectively.
3. For wind directed parallel to the axis of the arch, use pressure coefficients from Fig. 27.3-1 with wind directed parallel to ridge.
4. For components and cladding (1) at roof perimeter, use the external pressure coefficients in Fig. 30.3-2A, B, ~~and C~~ C, and D with θ based on springline slope and (2) for remaining roof areas, use external pressure coefficients of this table multiplied by 1.2.

Chapter C28

REVISE SECTION C28.3.2 LAST PARAGRAPH AS FOLLOWS:

The edge strip definition was modified following research (~~Elsharawy et al. 2014~~) (Alrawashdeh and Stathopoulos, 2015) showing that the definition of dimension "a" in ASCE 7-10 led to unduly large edge strips and end zones for very large buildings

REVISE REFERENCES AS FOLLOWS:

~~Elsharawy, M., Alrawashdeh, H., and Stathopoulos, T. (2014) "Wind loading zones for flat roofs," Proc., 4th Intl. Structural Specialty Conf., CSCE, Halifax, NS, May 28–31.~~

Alrawashdeh, H., and Stathopoulos, T. (2015). "Wind pressures on large roofs of low buildings and wind codes and standards" *J. Wind Eng. Indust. Aerodyn.* 147, December, 212–225.

Chapter 29

REVISE FIGURE 29.4-5 DESIGN WIND LOADS (ALL HEIGHTS): ROOFTOP SOLAR PANELS FOR ENCLOSED AND PARTIALLY ENCLOSED BULDINGS, ROOF ANGLE ≤ 7 DEGREES, CORRECT MISSING SUBSCRIPT "n" FOR NORMALIZED WIND AREAS A_n, AS FOLLOWS:

For Array Edge Factors graphs, the Normalized Wind Areas, A_n

---

Chapter 30

REVISE SECTION 30.4.1 CONDITIONS, CORRECT ITEM 5 TO INCLUDE HIP ROOF WITH 45 DEGREES AS FOLLOWS:

5. The building has either a flat roof, a gable roof with θ ≤ 45°, or a hip roof with θ ≤ ~~27~~ 45°.

REVISE FIGURE 30.4-1 COMPONENTS AND CLADDING, CORRECT "a" AND "0.6h" FOR FLAT/HIP/GABLE AS FOLLOWS:

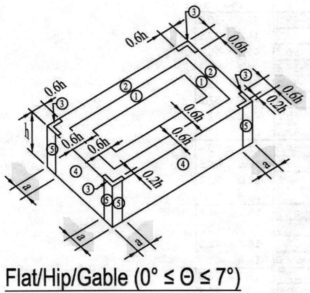

Flat/Hip/Gable (0° ≤ Θ ≤ 7°)

REVISE FIGURE 30.4-1 COMPONENTS AND CLADDING, CORRECT COLUMN FOR NEGATIVE PRESSURES FOR BASIC WIND SPEED OF 150 MPH.

	Zone	Effective wind Area (ft²)	140		150		160	
Walls	4	10	35.3	-38.2	40.5	-43.9	46.1	-50.0
	4	20	33.7	-36.7	38.7	-42.1	44.0	-47.9
	4	50	31.6	-34.6	36.2	-39.7	41.2	-45.1
	4	100	30.0	-33.0	34.4	-37.8	39.2	-43.1
	5	10	35.3	-47.2	40.5	-54.2	46.1	-61.7
	5	20	33.7	-44.0	38.7	-50.5	44.0	-57.5
	5	50	31.6	-39.8	36.2	-45.7	41.2	-52.0
	5	100	30.0	-36.7	34.4	-42.1	39.2	-47.9
Flat/Hip/Gable Roof 0 to 7 Degrees	1	10	14.3	-56.2	16.5	-64.5	18.7	-73.4
	1	20	13.4	-52.5	15.4	-60.2	17.6	-68.5
	1	50	12.3	-47.6	14.1	-54.6	16.0	-62.1
	1	100	11.4	-43.9	13.0	-50.4	14.8	-57.3
	1'	10	14.3	-32.3	16.5	-37.0	18.7	-42.1
	1'	20	13.4	-32.3	15.4	-37.0	17.6	-42.1
	1'	50	12.3	-32.3	14.1	-37.0	16.0	-42.1
	1'	100	11.4	-32.3	13.0	-37.0	14.8	-42.1
	2	10	14.3	-74.1	16.5	-85.1	18.7	-96.8
	2	20	13.4	-69.3	15.4	-79.6	17.6	-90.6
	2	50	12.3	-63.0	14.1	-72.4	16.0	-82.3
	2	100	11.4	-58.3	13.0	-66.9	14.8	-76.1
	3	10	14.3	-101.0	16.5	-115.9	18.7	-131.9
	3	20	13.4	-91.5	15.4	-105.0	17.6	-119.5
	3	50	12.3	-78.9	14.1	-90.5	16.0	-103.0
	3	100	11.4	-69.3	13.0	-79.6	14.8	-90.6
Gable Roof > 7 to 20 Degrees	1	10	21.4	-65.1	24.5	-74.8	27.9	-85.1
	1	20	19.3	-65.1	22.1	-74.8	25.2	-85.1
	1	50	16.5	-39.6	18.9	-45.5	21.5	-51.8
	1	100	14.3	-20.3	16.5	-23.3	18.7	-26.5
	2e	10	21.4	-65.1	24.5	-74.8	27.9	-85.1
	2e	20	19.3	-65.1	22.1	-74.8	25.2	-85.1
	2e	50	16.5	-39.6	18.9	-45.5	21.5	-51.8
	2e	100	14.3	-20.3	16.5	-23.3	18.7	-26.5
	2n	10	21.4	-95.0	24.5	-109.1	27.9	-124.1
	2n	20	19.3	-82.1	22.1	-94.3	25.2	-107.3
	2n	50	16.5	-65.1	18.9	-74.8	21.5	-85.1
	2n	100	14.3	-52.3	16.5	-60.0	18.7	-68.3
	2r	10	21.4	-95.0	24.5	-109.1	27.9	-124.1
	2r	20	19.3	-82.1	22.1	-94.3	25.2	-107.3
	2r	50	16.5	-65.1	18.9	-74.8	21.5	-85.1
	2r	100	14.3	-52.3	16.5	-60.0	18.7	-68.3
	3e	10	21.4	-95.0	24.5	-109.1	27.9	-124.1
	3e	20	19.3	-82.1	22.1	-94.3	25.2	-107.3
	3e	50	16.5	-65.1	18.9	-74.8	21.5	-85.1
	3e	100	14.3	-52.3	16.5	-60.0	18.7	-68.3
	3r	10	21.4	-112.9	24.5	-129.7	27.9	-147.5
	3r	20	19.3	-96.8	22.1	-111.1	25.2	-126.4
	3r	50	16.5	-75.4	18.9	-86.5	21.5	-98.4
	3r	100	14.3	-59.2	16.5	-67.9	18.7	-77.3

Chapter C30

REVISE TABLE C30.3-4 GABLE ROOF 20 DEGREES<Θ≤ 27 DEGREES, CORRECT "3e" AND "3r" AS FOLLOWS:

Table C30.3-4. Gable Roofs, $20° < \theta \leq 27°$ (Figure 30.3-2C)

Negative without overhang		
Zones 2n, 2r, and ~~3r~~ <u>3e</u>	$(GC_p) = -2.5$	for $A \leq 10$ ft^2
	$(GC_p) = -3.6054 + 1.1054 \log A$	for $10 \leq A \leq 150$ ft^2
	$(GC_p) = -1.2$	for $A \geq 150$ ft^2
Zones ~~3e~~ <u>3r</u>	$(GC_p) = -3.6$ for $A \leq 4$ ft^2	
	$(GC_p) = -4.5880 + 1.6410 \log A$	for $4 \leq A \leq 50$ ft^2
	$(GC_p) = -1.8$	for $A \geq 50$ ft^2
Negative with overhang		
Zone ~~3r~~ <u>3e</u>	$(GC_p) = -3.6$	for $A \leq 10$ ft^2
	$(GC_p) = -5.2155 + 1.6155 \log A$	for $10 \leq A \leq 150$ ft^2
	$(GC_p) = -1.7$	for $A \geq 150$ ft^2
Zone ~~3e~~ <u>3r</u>	$(GC_p) = -4.7$	for $A \leq 4$ ft^2
	$(GC_p) = -6.0173 + 2.1880 \log A$	for $4 \leq A \leq 50$ ft^2
	$(GC_p) = -2.3$	for $A \geq 50$ ft^2

REVISE SECTION C30.11 TO INCLUDE MISSING COMMENTARY, AS FOLLOWS:

C30.11 ATTACHED CANOPIES ON BUILDINGS WITH h < 60 ft (18.3 m)

The proposed provisions result from wind tunnel test results on pressures applied on horizontal canopies described in Zisis and Stathopoulos (2010), Zisis et al. (2011) and Candelario et al. (2014). Restrictions to buildings under 60 ft high and to canopies that are essentially flat (maximum slope: 2%) are based upon a lack of test data. Canopies are different from roof overhangs, which are simply extensions of the roof surfaces at the same slope with the roof.

In a canopy with two physical surfaces both Figures 30.11-1A and 30.11-1B would be needed.

Figure 30.11-1A, which provides the coefficients on separate surfaces, would be used to design the fasteners of the top and soffit elements. Figure 30.11-1B would be used to design the structure of the canopy (e.g., joists, posts, building fasteners). In a canopy with one physical surface, only Figure 30.11-1B is needed.

References (to be added on page 790)

Zisis, I., and Stathopoulos, T. (2010). "Wind-induced pressures on patio covers." *Struct. Eng.* 136(9) 1172–1181.

Zisis, I., Stathopoulos, T., and Candelario, J. D. (2011). "Codification of wind loads on a patio cover based on a parametric wind tunnel study." *Proc., 13th Int. Conf. on Wind Engineering*, July 10–15, Amsterdam, The Netherlands. Amsterdam: Multi-Science Publishing Co.

Candelario, J. D., Stathopoulos, T., and Zisis, I. (2014). "Wind loading on attached canopies: Codification study." *Struct. Eng.* 140(5) May, CID: 04014007.